RF Power Amplifiers

RF Power Amplifiers

Second Edition

MARIAN K. KAZIMIERCZUK
Wright State University,
Dayton, Ohio, USA

WILEY

This edition first published 2015

Registered office:
John Wiley & Sons Ltd, The Atrium, Southern Gate, Chichester, West Sussex, PO19 8SQ, United Kingdom

For details of our global editorial offices, for customer services and for information about how to apply for permission to reuse the copyright material in this book please see our website at www.wiley.com.

Library of Congress Cataloging-in-Publication Data

Kazimierczuk, Marian.
RF power amplifiers / Marian K. Kazimierczuk. – Second edition.
pages cm
Includes bibliographical references and index.
ISBN 978-1-118-84430-4
1. Power amplifiers. 2. Amplifiers, Radio frequency. 3. Electric resonators. I. Title.
TK7871.58.P6K39 2015
621.3841′3–dc23

2014022274

A catalogue record for this book is available from the British Library.

ISBN 9781118844304

Set in 9.75/11.75pt TimesLTStd by Laserwords Private Limited, Chennai, India

1 2015

To My Mother

Contents

Preface

The second edition of *RF Power Amplifiers* is designed to be an improvement, updation, and enlargement of the first edition. This book is about RF power amplifiers used in wireless communications and many other RF applications. It is intended as a concept-oriented textbook at the senior and graduate levels for students majoring in electrical engineering, as well as a reference for practicing engineers in the area of RF power electronics. The edition of this book is a thoroughly revised and expanded version of the first edition. The purpose of the book is to provide foundations for RF power amplifiers, efficiency improvement, and linearization techniques. Class A, B, C, D, E, DE, and F RF power amplifiers are analyzed, and design procedures are given. Impedance transformation is covered. Various linearization techniques are explored, such as predistortion, feedforward, and negative feedback techniques. Efficiency improvement methods are also studied, such as dynamic power supply method, envelope elimination and restoration (EER), envelope tracking (ET), Doherty amplifier, and outphasing techniques. Integrated inductors are discussed as well. RF *LC* oscillators are also covered. RF power amplifiers are used as power stages of radio transmitters. Radio transmitters are used in broadcasting systems, mobile wireless communication systems, radars, and satellite communications.

It is assumed that the student is familiar with general circuit analysis techniques, semiconductor devices, linear systems, and electronic circuits. A communications course is also very helpful.

I wish to express my sincere thanks to Laura Bell, Assistant Editor; Richard Davies, Senior Project Editor; and Peter Mitchell, Publisher. It has been a real pleasure working with them. Last but not least, I wish to thank my family for the support.

I am pleased to express my gratitude to Dr. Nisha Kondrath and Dr. Rafal Wojda for MATLAB® figures. The author would welcome and greatly appreciate suggestions and corrections from the readers, for the improvements in the technical content as well as the presentation style.

Prof. Marian K. Kazimierczuk

About the Author

Marian K. Kazimierczuk is Robert J. Kegerreis Distinguished Professor of Electrical Engineering at Wright State University, Dayton, Ohio, USA. He received the M.S., Ph.D., and D.Sc. degrees from the Department of Electronics, Warsaw University of Technology, Warsaw, Poland. He is the author of six books, over 180 archival refereed journal papers, over 210 conference papers, and seven patents.

His research interests are in power electronics, including RF high-efficiency power amplifiers and oscillators, PWM dc–dc power converters, resonant dc–dc power converters, modeling and controls of power converters, high-frequency magnetic devices, electronic ballasts, active power factor correctors, semiconductor power devices, wireless charging systems, renewable energy sources, energy harvesting, green energy, and evanescent microwave microscopy.

He is a Fellow of the IEEE. He served as Chair of the Technical Committee of Power Systems and Power Electronics Circuits, IEEE Circuits and Systems Society. He served on the Technical Program Committees of the IEEE International Symposium on Circuits and Systems (ISCAS) and the IEEE Midwest Symposium on Circuits and Systems. He also served as Associate Editor of the *IEEE Transactions on Circuits and Systems – Part I: Regular Papers*; *IEEE Transactions on Industrial Electronics*; *International Journal of Circuit Theory and Applications*; and *Journal of Circuits, Systems, and Computers*; and he was Guest Editor of the *IEEE Transactions on Power Electronics*. He was an IEEE Distinguished Lecturer.

He received Presidential Award for Outstanding Faculty Member at Wright State University in 1995. He was Brage Golding Distinguished Professor of Research at Wright State University from 1996 to 2000. He received the Trustees' Award from Wright State University for Faculty Excellence in 2004. He received the Outstanding Teaching Award from the American Society for Engineering Education (ASEE) in 2008. He was also honored with the Excellence in Research Award, Excellence in Teaching Awards, and Excellence in Professional Service Award in the College of Engineering and Computer Science, Wright State University. He is listed in *Top Authors in Engineering* and *Top Authors in Electrical & Electronic Engineering*.

He is the author or coauthor of six books: *Resonant Power Converters*, 2nd Ed., Wiley; *Pulse-Width Modulated DC–DC Power Converters*, IEEE Press/Wiley; *High-Frequency Magnetic Components*, 2nd Ed. (translated in Chinese), Wiley; *RF Power Amplifiers*, 2nd Ed., Wiley; *Electronic Devices: A Design Approach*, Pearson/Prentice Hall; and *Laboratory Manual to Accompany Electronic Devices: A Design Approach*, 2nd Ed., Pearson/Prentice Hall.

List of Symbols

A_e	Effective area of antenna
A_v	Voltage gain
a	Coil mean radius
B	Magnetic flux density
B_n	Noise-equivalent bandwidth
BW	Bandwidth
C	Resonant capacitance
C_B	Blocking capacitance
C_c	Coupling capacitance
C_{ds}	Drain-source capacitance of MOSFET
$C_{ds(25V)}$	Drain-source capacitance of MOSFET at $V_{DS} = 25$ V
C_{gd}	Gate-drain capacitance of MOSFET
C_{gs}	Gate-source capacitance of MOSFET
C_{iss}	MOSFET input capacitance at $V_{DS} = 0$, $C_{iss} = C_{gs} + C_{gd}$
C_{oss}	MOSFET output capacitance at $V_{GD} = 0$, $C_{oss} = C_{gs} + C_{ds}$
C_{ox}	Oxide capacitance per unit area
C_{out}	Transistor output capacitance
C_{rss}	MOSFET transfer capacitance, $C_{rss} = C_{gd}$
c_p	Output-power capability of amplifier
D	Coil outer diameter
d	Coil inner diameter
E	Electric field intensity
f	Operating frequency, switching frequency
f_c	Carrier frequency
f_{IF}	Intermediate frequency
f_{LO}	Local oscillator frequency
f_m	Modulating frequency
f_0	Resonant frequency
f_p	Frequency of pole of transfer function
f_r	Resonant frequency of *L–C–R* circuit
f_s	Switching frequency
g_m	Transconductance of transistor
H	Magnetic flux intensity

h	Trace thickness
I_D	dc drain current
I_{DM}	Peak drain current
I_{SM}	Peak current of switch
I_m	Amplitude of current i
I_n	Noise rms current
I_{rms}	RMS value of i
i	Current
i_C	Capacitor current
i_D	Large-signal drain current
i_d	Small-signal drain current
i_L	Inductor current
i_o	ac output current
i_S	Switch current
K	MOSFET parameter
K_n	MOSFET process parameter
K_s	MOSFET parameter at drift velocity saturation of current carriers
k	Boltzmann's constant
k_p	Power gain
L	Resonant inductance, channel length
L_f	Filter inductance
L_{fmin}	Minimum value of filter inductance L_f
L_{RFC}	RF choke inductance
l	Inductor trace length, winding length
m	AM index
m_f	Index of frequency modulation
m_p	Index of phase modulation
N	Number of inductor turns
n	Transformer turns ratio
P_{AM}	Power of AM signal
P_C	Power of carrier
P_D	Power dissipation
PEP	Peak envelope power
P_G	Gate drive power
P_I	dc supply (input) power
P_{Loss}	Power loss
P_{LS}	Power of lower sideband
P_n	Noise power
P_O	ac output power
P_r	Power loss in resonant circuit
P_{rDS}	Conduction power loss in MOSFET on-resistance
P_{rC}	Power loss in resonant capacitor
P_{rL}	Power loss in resonant inductor
P_{sw}	Switching power
P_{US}	Power of upper sideband
P_{tf}	Average power loss due to current fall time t_f
p	Perimeter enclosed by coil
$p_D(\omega t)$	Instantaneous drain power loss
Q_{Co}	Quality factor of capacitor
Q_L	Loaded quality factor at f_o
Q_{Lo}	Quality factor of inductor

Q_o	Unloaded quality factor at f_o
q_A	Reactance factor
q_B	Reactance factor
R	Overall resistance of amplifier without impedance matching
R_{DC}	dc input resistance of amplifier
R_L	Load resistance
R_{Lmin}	Minimum value of R_L
r	Total parasitic resistance
r_C	ESR of resonant capacitor
r_{DS}	On-resistance of MOSFET
r_G	Gate resistance
r_L	ESR of resonant inductor
r_o	Output resistance of transistor
S_i	Current waveform slope
S_v	Voltage waveform slope
s	Trace-to-trace spacing of planar inductor
T	Operating temperature, period of waveform
THD	Total harmonic distortion
t_f	Fall time of MOSFET or BJT
V_A	Channel-modulation voltage
V_{Cm}	Amplitude of the voltage across capacitor
V_c	Amplitude of carrier voltage
V_{DS}	dc drain–source voltage
V_{DSM}	Peak drain–source voltage
V_{GS}	dc gate–source voltage
V_I	dc supply (input) voltage
V_{Lm}	Amplitude of the voltage across inductance
V_{dsm}	Amplitude of small-signal drain–source voltage
V_{gsm}	Amplitude of small-signal gate–source voltage
V_m	Amplitude of modulating voltage
V_n	nth harmonic of the output voltage, noise RMS voltage
V_{rms}	RMS value of v
V_{SM}	Peak voltage of switch
V_t	Threshold voltage of MOSFET
v	Voltage
v_c	Carrier voltage
v_{DS}	Large-signal drain–source voltage
v_{DSsat}	Drain-to-source voltage at the edge of saturation
v_{ds}	Small-signal drain–source voltage
v_{GS}	Large-signal gate–source voltage
v_{gs}	Small-signal gate–source voltage
v_m	Modulating voltage
v_o	ac output voltage
v_{sat}	Saturated carrier drift velocity
W	Energy, channel width
w	Trace width
X	Imaginary part of impedance Z
Y	Input admittance of resonant circuit
Z	Input impedance of resonant circuit
$\|Z\|$	Magnitude of impedance Z
Z_i	Input impedance

Z_o	Characteristic impedance of resonant circuit
α_n	Fourier coefficients of drain current
β	Gain of feedback network
γ_n	Ratio of Fourier coefficients of drain current
Δf	Frequency deviation
δ	Dirac delta impulse function
ε_{ox}	Oxide permittivity
η	Efficiency of amplifier
η_{AV}	Average efficiency of amplifier
η_D	Drain efficiency of amplifier
η_{PAE}	Power-aided efficiency of amplifier
η_r	Efficiency of resonant circuit
θ	Half of drain current conduction angle, mobility degradation coefficient
λ	Wavelength, channel length-modulation parameter
μ	Mobility of current carriers
mu_0	Permeability of free space
μ_n	Mobility of electrons
μ_r	Relative permeability
ξ_n	Ratio of Fourier coefficients of drain-source voltage
ρ	Resistivity
σ	Conductivity
ϕ	Phase, angle, magnetic flux
ψ	Phase of impedance Z
ω	Operating angular frequency
ω_c	Carrier angular frequency
ω_m	Modulating angular frequency
ω_o	Resonant angular frequency

List of Acronyms

AM	Amplitude Modulation
ACLR	Adjacent Channel Leakage Ratio
ASK	Amplitude Shift Keying
BW	Bandwidth
CW	Continuous Wave
CDMA	Code Division Multiple Access
EER	Envelope Elimination and Restoration
ET	Envelope Tracking
FDMA	Frequency Division Multiple Access
FDD	Frequency Division Duplexing
FM	Frequency Modulation
FSK	Frequency Shift Keying
GSM	Global System for Mobile Communications
HEMT	High Electron Mobility Transistor
IMD	Intermodulation Distortion
IP	Intercept Point
LINC	Linear Amplification with Nonlinear Components
LNA	Low Noise Amplifier
LTE	Long-Term Solution
OFDM	Orthogonal Frequency Division Multiplexing
PAR	Peak-to-Average Ratio
PA	Power Amplifier
PAPR	Peak-to-Average Power Ratio
PDF	Probability Density Function
PEP	Peak Envelope Power
PM	Phase Modulation
PWM	Pulse Width Modulation
PSK	Phase Shift Keying
QAM	Quadrature Amplitude Modulation
QPSK	Quaternary Phase Shift Keying
RF	Radio Frequency

RFID	Radio Frequency Identification
SNR	Signal-to-Noise Ratio
TDMA	Time-Division Multiple Access
TDD	Time-Division Duplexing
WCDMA	Wideband Code Division Multiple Access
WLAN	Wireless Local Area Network

1

Introduction

1.1 Radio Transmitters

Radio communication utilizes radio waves as a transmission and receiving medium. A *radio transmitter* consists of information source producing modulating signal, modulator, radio-frequency (RF) power amplifier, and antenna. A *power amplifier* is a circuit that increases the power level of a signal by using energy taken from a power supply [1–28]. Both the efficiency and distortion are critical parameters of power amplifiers. A *radio receiver* consists of an antenna, front end, demodulator, and audio amplifier. A transmitter and receiver combined into one electronic device is called a *transceiver*. A radio transmitter produces a strong RF current, which flows through an antenna. In turn, a transmitter antenna radiates electromagnetic waves (EMWs), called radio waves. Transmitters are used for communication of information over a distance, such as radio and television broadcasting, mobile phones, wireless computer networks, radio navigation, radio location, air traffic control, radars, ship communication, radio-frequency identifications (RFIDs), collision avoidance, speed measurement, weather forecasting, and so on. The information signal is the modulating signal, and it is usually in the form of audio signal from a microphone, video signal from a camera, or digital signal. Modern wireless communication systems include both amplitude-modulated (AM) and phase-modulated (PM) signals with a large peak-to-average ratio (PAR). Typically, the PAR is 6-9 dB for Wideband Code Division Multiple Access (WCDMA) and Orthogonal Frequency Division Multiplexing (OFDM). The main difficulty in transmitters' design is achieving a good linearity and a high efficiency.

An ideal radio transmitter should satisfy the following requirements:

- high efficiency,
- high linearity (i.e., low signal distortion),
- high power gain,
- large dynamic range,

RF Power Amplifiers, Second Edition. Marian K. Kazimierczuk.

- large slew rate,
- low noise level,
- high spectral efficiency,
- wide modulation bandwidth,
- capability of reproducing complex modulated waveforms,
- capability of transmitting high data rate communication,
- capability of transmitting a large diversity of waveforms, and
- portability.

A design of a transmitter with high efficiency and high linearity is a challenging problem. Linear amplification is required when the signal contains both amplitude and phase modulations. Nonlinearities cause imperfect reproduction of the amplified signal.

1.2 Batteries for Portable Electronics

In portable communications, batteries are used as power supplies. The most popular battery technologies are lithium (Li-ion) batteries and nickelcadmium (Ni-Cd) batteries. The nominal output voltage of the Li-ion batteries is 3.6 V. The discharge curve of Li-ion batteries is typically from 4 to 2 V during the period of 5 h of active operation. The nominal output voltage of the Ni-based batteries is 1.25 V. The discharge curve of these batteries is typically from 1.4 to 1 V during the period of 5 h of active operation. These are rechargeable batteries. The energy density of Li-ion batteries is nearly twice that of Ni-based batteries, yielding a smaller battery that stores the same amount of energy. However, the discharge curve of Li-ion batteries is much steeper than that of Ni-based batteries. The slope of the discharge curve of Li-ion batteries is approximately −0.25 V/h, whereas the slope of the discharge curve of Ni-based batteries is approximately −0.04 V/h. Therefore, Li-ion batteries may require a voltage regulator.

1.3 Block Diagram of RF Power Amplifiers

A power amplifier [1–27] is a key element to build a wireless communication system successfully. Its main purpose is to increase the power level of the signal. To minimize interferences and spectral regrowth, transmitters should be linear. A block diagram of an RF power amplifier is shown in Fig. 1.1. It consists of transistor (MOSFET, MESFET, HFET, or BJT), output network, input network, and RF choke. The trend is to replace silicon(Si)-based semiconductor devices with wide band gap (BG) semiconductor devices, such as silicon carbide (SiC) and gallium nitride (GaN) devices. Silicon carbide is also used as a substrate because it has high thermal conductivity, for example, for GaN devices. Gallium nitride semiconductor is used to make high electron mobility transistors (HEMTs). The energy BG of GaN is three times greater than that of silicon, yielding lower performance degradation at high temperatures. The breakdown electric field intensity is six times greater than that of silicon. Also, the carrier saturation velocity is 2.5 greater than that of silicon, resulting in a higher power density.

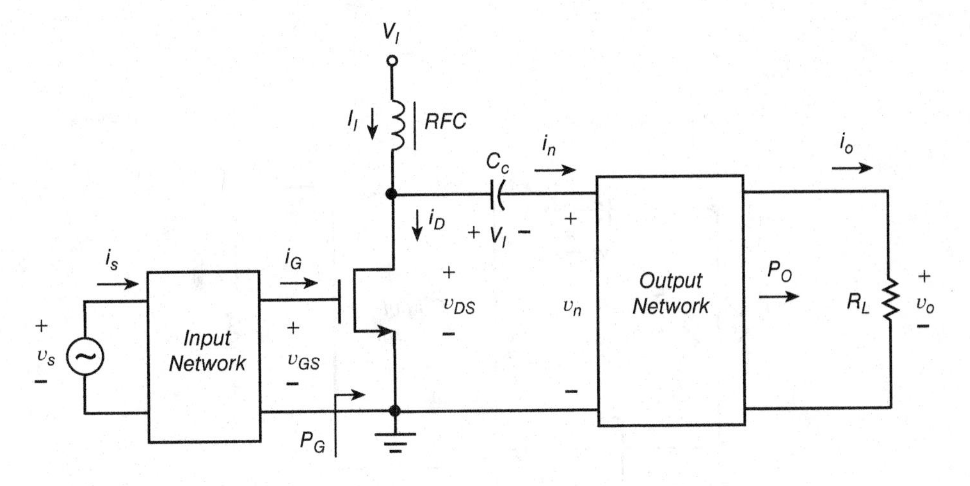

Figure 1.1 Block diagram of RF power amplifier.

In RF power amplifiers, a transistor can be operated

- as a dependent current source,
- as a switch, and
- in overdriven mode (partially as a dependent source and partially as a switch).

Figure 1.2(a) shows a model of an RF power amplifier in which the transistor is operated as a voltage- or current-dependent current source. When a MOSFET is operated as a dependent current source, the drain current waveform is determined by the gate-to-source voltage waveform and the transistor operating point. The drain voltage waveform is determined by the dependent current source and the load network impedance. When a MOSFET is operated as a switch, the switch voltage is nearly zero when the switch is ON and the drain current is determined by the external circuit due to the switching action of the transistor. When the switch is OFF, the switch current is zero and the switch voltage is determined by the external circuit response.

In order to operate the MOSFET as a dependent current source, the transistor cannot enter the ohmic region. It must be operated in the active region, also called the pinch-off region or the saturation region. Therefore, the drain-to-source voltage v_{DS} must be kept higher than the minimum value V_{DSmin}, that is, $v_{DS} > V_{DSmin} = V_{GS} - V_t$, where V_t is the transistor threshold voltage. When the transistor is operated as a dependent current source, the magnitudes of the drain current i_D and the drain-to-source voltage v_{DS} are nearly proportional to the magnitude of the gate-to-source voltage v_{GS}. Therefore, this type of operation is suitable for linear power amplifiers. Amplitude linearity is important for amplification of AM signals.

Figure 1.2(b) shows a model of an RF power amplifier in which the transistor is operated as a switch. To operate the MOSFET as a switch, the transistor cannot enter the active region. It must remain in the ohmic region when it is ON and in the cutoff region when it is OFF. To maintain the MOSFET in the ohmic region, it is required that $v_{DS} < V_{GS} - V_t$. If the gate-to-source voltage v_{GS} is increased at a given load impedance, the amplitude of the drain-to-source voltage v_{DS} will increase, causing the transistor to operate initially in the active region and then in the ohmic region. When the transistor is operated as a switch, the magnitudes of the drain current i_D and the

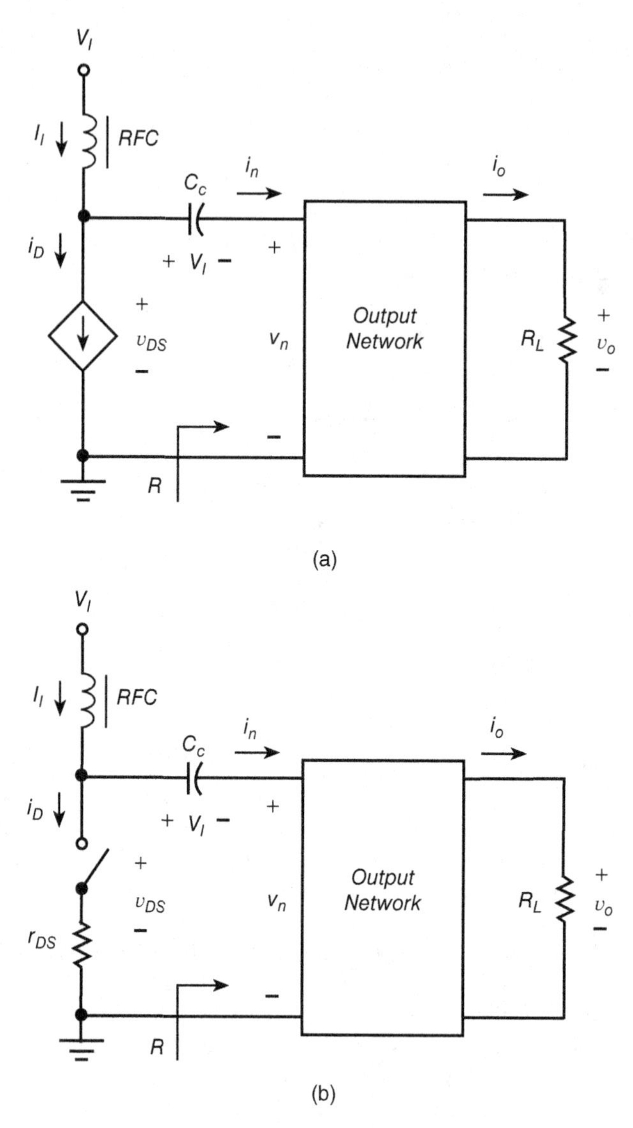

Figure 1.2 Models of operation of transistor in RF power amplifiers: (a) transistor as a dependent current source and (b) transistor as a switch.

drain-to-source voltage v_{DS} are independent of the magnitude of the gate-to-source voltage v_{GS}. In most applications, the transistor operated as a switch is driven by a rectangular gate-to-source voltage v_{GS}. A sinusoidal gate-to-source voltage V_{GS} is used to drive a transistor as a switch at very high frequencies, where it is difficult to generate rectangular voltages. The reason to use the transistors as switches is to achieve high amplifier efficiency. When the transistor conducts a high drain current i_D, the drain-to-source voltage v_{DS} is low, resulting in low power loss.

If the transistor is driven by a sinusoidal voltage v_{GS} of high amplitude, the transistor is overdriven. In this case, it operates in the active region when the instantaneous values of v_{GS} are low and as a switch when the instantaneous values of v_{GS} are high.

The main functions of the output network are as follows:

- Impedance transformation.
- Harmonic suppression.
- Filtering the spectrum of a signal with bandwidth BW to avoid interference with communication signals in adjacent channels.

Modulated signals can be divided into two categories:

- Variable-envelope signals, such as AM and SSB.
- Constant-envelope signals, such as FM, FSK, and CW.

Modern mobile communication systems usually contain both amplitude and phase modulations. Amplification of variable-envelope signals requires linear amplifiers. A linear amplifier is an electronic circuit whose output voltage is directly proportional to its input voltage. The Class A RF power amplifier is a nearly linear amplifier.

1.4 Classes of Operation of RF Power Amplifiers

The classification of RF power amplifiers with a transistor operated as a dependent current source is based on the conduction angle 2θ of the drain current i_D. Waveforms of the drain current i_D of a transistor operated as a dependent source in various classes of operation for sinusoidal gate-to-source voltage v_{GS} are shown in Fig. 1.3. The operating points for various classes of operation are shown in Fig. 1.4.

In Class A, the conduction angle 2θ of the drain current i_D is 360°. The gate-to-source voltage v_{GS} must be greater than the transistor threshold voltage V_t, that is, $v_{GS} > V_t$. This is accomplished by choosing the dc component of the gate-to-source voltage V_{GS} sufficiently greater than the threshold voltage of the transistor V_t such that $V_{GS} - V_{gsm} > V_t$, where V_{gsm} is the amplitude of the ac component of the gate-to-source voltage v_{GS}. The dc component of the drain current I_D must be greater than the amplitude of the ac component I_m of the drain current i_D. As a result, the transistor conducts during the entire cycle. Class A amplifiers are linear, but have low efficiency (lower than 50%).

In Class B, the conduction angle 2θ of the drain current i_D is 180°. The dc component V_{GS} of the gate-to-source voltage v_{GS} is equal to V_t, and the drain bias current I_D is zero. Therefore, the transistor conducts for only half of the cycle.

In Class AB, the conduction angle 2θ is between 180° and 360°. The dc component of the gate-to-source voltage V_{GS} is slightly above V_t, and the transistor is biased at a small drain current I_D. As the name suggests, Class AB is the intermediate class between Class A and Class B. Class AB amplifiers are linear, but have low efficiency (less than 50%).

In Class C, the conduction angle 2θ of the drain current i_D is less than 180°. The operating point is located in the cutoff region because $V_{GS} < V_t$. The drain bias current I_D is zero. The transistor conducts for an interval less than half of the cycle. Class C amplifiers are nonlinear and are only suitable for the amplification of constant-envelope signals, but have a higher efficiency than that Class A and AB amplifiers.

Class A, AB, and B operations are used in audio and RF power amplifiers, whereas Class C is used only in RF power amplifiers and industrial applications.

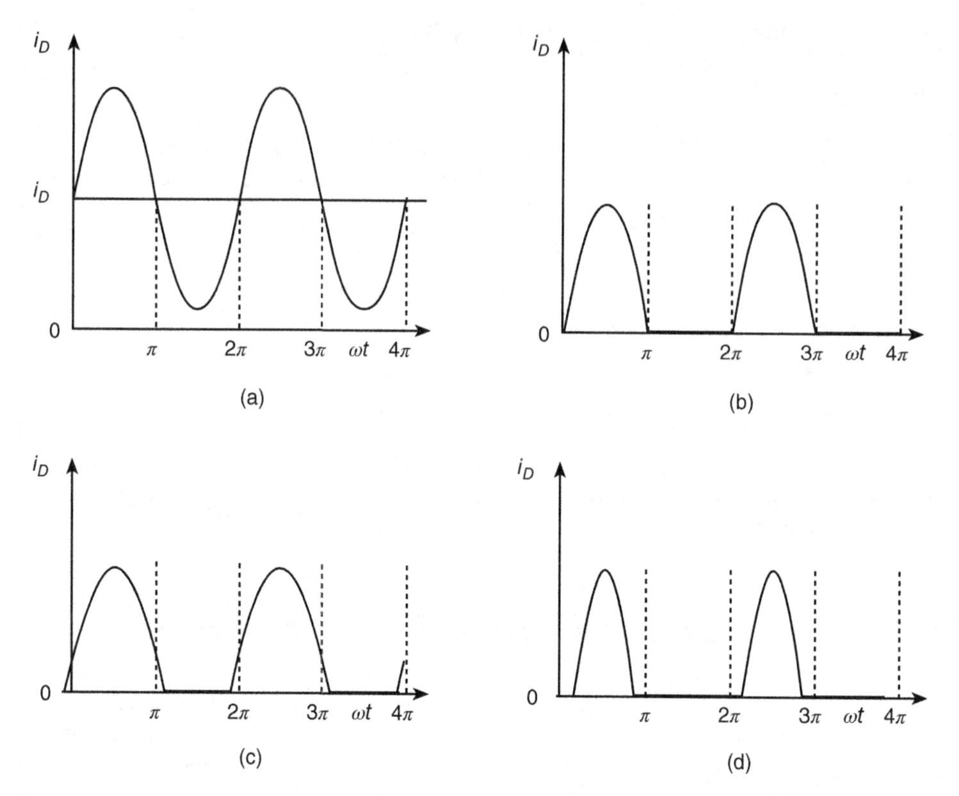

Figure 1.3 Waveforms of the drain current i_D in various classes of operation: (a) Class A, (b) Class B, (c) Class AB, and (d) Class C.

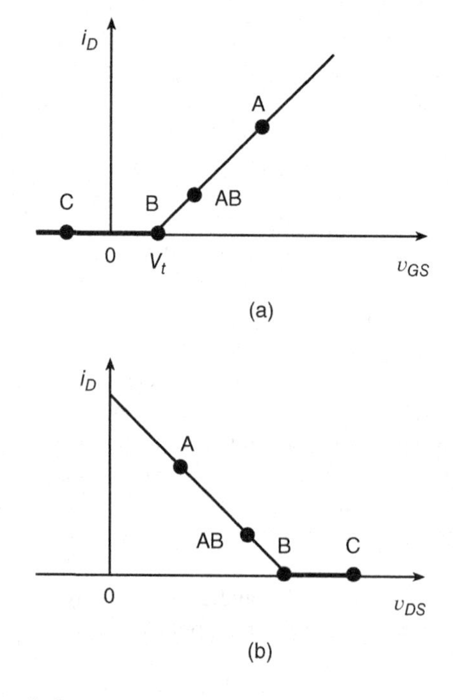

Figure 1.4 Operating points for Classes A, B, AB, and C.

The transistor is operated as a switch in Class D, E, and DE RF power amplifiers. In Class F, the transistor can be operated as either a dependent current source or a switch. RF power amplifiers are used in communications, power generation, and plasma generation.

1.5 Waveforms of RF Power Amplifiers

For steady state, the waveforms of an unmodulated power amplifier are periodic of frequency $f = \omega/(2\pi)$. The drain current waveform can be represented by Fourier series

$$i_D = I_I + \sum_{n=1}^{\infty} I_{mn} \cos(n\omega t + \phi_{in}) = I_{DM} \left[\alpha_0 + \sum_{n=1}^{\infty} \alpha_n \cos(n\omega t + \phi_{in})\right]$$
$$= I_I \left[1 + \sum_{n=1}^{\infty} \frac{I_{mn}}{I_I} \cos(n\omega t + \phi_{in})\right] = I_I \left[1 + \sum_{n=1}^{\infty} \gamma_n \cos(n\omega t + \phi_{in})\right] \quad (1.1)$$

where

$$\alpha_0 = \frac{I_I}{I_{DM}} \quad (1.2)$$

$$\alpha_n = \frac{I_{mn}}{I_{DM}} \quad (1.3)$$

and

$$\gamma_n = \frac{I_{mn}}{I_I} \quad (1.4)$$

The drain-to-source voltage waveform can also be expanded into Fourier series

$$v_{DS} = V_I - \sum_{n=1}^{\infty} V_{mn} \cos(n\omega t + \phi_{vn}) = V_{DSM} \left[\beta_0 - \sum_{n=1}^{\infty} \beta_n \cos(n\omega t + \phi_{vn})\right]$$
$$= V_I \left[1 - \sum_{n=1}^{\infty} \frac{V_{mn}}{V_I} \cos(n\omega t + \phi_{vn})\right] = V_I \left[1 - \sum_{n=1}^{\infty} \xi_n \cos(n\omega t + \phi_{vn})\right] \quad (1.5)$$

where

$$\beta_0 = \frac{V_I}{V_{DSM}} \quad (1.6)$$

$$\beta_1 = \frac{V_{mn}}{V_{DSM}} \quad (1.7)$$

and

$$\xi_n = \frac{V_{mn}}{V_I} \quad (1.8)$$

1.6 Parameters of RF Power Amplifiers

1.6.1 Drain Efficiency of RF Power Amplifiers

When the resonant frequency of the output network f_o is equal to the operating frequency f, the *drain power* (the power delivered by the drain to the output network) is given by

$$P_{DS} = \frac{1}{2} I_m V_m = \frac{1}{2} I_m^2 R = \frac{V_m^2}{2R} \quad \text{for} \quad f = f_o \quad (1.9)$$

where I_m is the amplitude of the fundamental component of the drain current i_D, V_m is the amplitude of the fundamental component of the drain-to-source voltage v_{DS}, and R is the input resistance of the output network at the fundamental frequency. If the resonant frequency f_o is not equal to the operating frequency f, the drain power of the fundamental component is given by

$$P_{DS} = \frac{1}{2} I_m V_m \cos\phi = \frac{1}{2} I_m^2 R \cos\phi = \frac{V_m^2 \cos\phi}{2R} \tag{1.10}$$

where ϕ is the phase shift between the fundamental components of the drain current and the drain-to-source voltage reduced by π.

The *instantaneous drain power dissipation* is

$$p_D(\omega t) = i_D v_{DS} \tag{1.11}$$

The *time-average drain power dissipation* for periodic waveforms is

$$P_D = \frac{1}{2\pi} \int_0^{2\pi} p_D \, d(\omega t) = \frac{1}{2\pi} \int_0^{2\pi} i_D v_{DS} \, d(\omega t) = P_I - P_{DS} \tag{1.12}$$

The dc supply current is

$$I_I = \frac{1}{2\pi} \int_0^{2\pi} i_D \, d(\omega t) \tag{1.13}$$

The dc supply power is

$$P_I = V_I I_I = \frac{V_I}{2\pi} \int_0^{2\pi} i_D \, d(\omega t) \tag{1.14}$$

The *drain efficiency* at a given drain power P_{DS} is

$$\eta_D = \frac{P_{DS}}{P_I} = \frac{P_I - P_D}{P_I} = 1 - \frac{P_D}{P_I} = 1 - \frac{\int_0^{2\pi} i_D v_{DS} \, d(\omega t)}{V_I \int_0^{2\pi} i_D \, d(\omega t)}$$
$$= \frac{1}{2}\left(\frac{I_m}{I_I}\right)\left(\frac{V_m}{I_I}\right) \cos\phi = \frac{1}{2}\gamma_1 \xi_1 \cos\phi \tag{1.15}$$

where $\phi = \phi_{i1} - \phi_{v1}$. When the operating frequency is equal to the resonant frequency $f = f_0$, the drain efficiency is

$$\eta_D = \frac{P_{DS}}{P_I} = \frac{1}{2}\left(\frac{I_m}{I_I}\right)\left(\frac{V_m}{V_I}\right) = \frac{1}{2}\gamma_1 \xi_1 \quad \text{for} \quad f = f_0 \tag{1.16}$$

Efficiency of power amplifiers is maximized by minimizing power dissipation at a desired output power.

For amplifiers in which the transistor is operated as a dependent current source, the highest drain efficiency usually occurs at the *peak envelope power* (PEP). Power amplifiers with time-varying amplitude have a time-varying drain efficiency $\eta_D(t)$. These amplifiers are usually operated below the maximum output power. This situation is called *power backoff*. The peak-to-average power ratio (PAPR) is the ratio of the PEP of the AM waveform *PEP* to the average envelope power for a long time interval $P_{O(AV)}$

$$PAPR = \frac{\text{Peak Power}}{\text{Average Power}} = \frac{PEP}{P_{O(AV)}} = 10\log\left(\frac{PEP}{P_{O(AV)}}\right) \text{ (dB)} \tag{1.17}$$

The *power dynamic range* is the ratio of the largest output power P_{Omax} to the lowest output power P_{Omin} defined as

$$DNR = \frac{\text{Maximum Output Power}}{\text{Minimum Output Power}} = \frac{P_{Omax}}{P_{Omin}} = 10\log\left(\frac{P_{Omax}}{P_{Omin}}\right) \text{ (dB).} \tag{1.18}$$

The *output power* delivered to a resistive load is

$$P_O = \frac{1}{2} I_{om} V_{om} = \frac{1}{2} I_{om}^2 R_L = \frac{V_{om}^2}{2R_L} \tag{1.19}$$

where I_{om} is the amplitude of the output current and V_{om} is the amplitude of the output voltage. The *power loss in the resonant output network* is

$$P_r = P_{DS} - P_O \tag{1.20}$$

The *efficiency of the resonant output network* is

$$\eta_r = \frac{P_O}{P_{DS}} \tag{1.21}$$

The *overall power loss* on the output side of the amplifier (in the transistor(s) and the output network) is

$$P_{Loss} = P_I - P_O = P_D + P_r \tag{1.22}$$

The *efficiency* of the amplifier at a specific output power is

$$\eta = \frac{P_O}{P_I} = \frac{P_O}{P_{DS}} \frac{P_{DS}}{P_I} = \eta_D \eta_r \tag{1.23}$$

The output power level of an amplifier is often referenced to the power level of 1 mW and is expressed as

$$P = 10 \log \frac{P(\mathrm{W})}{0.001} \ (\mathrm{dBm}) = 10 \log P(\mathrm{W}) - 10 \log 0.001 = [10 \log P(\mathrm{W}) + 30] \ (\mathrm{dBm}) \tag{1.24}$$

A dBm or dBW value represents an actual power, whereas a dB value represents a ratio of power, such as the power gain.

1.6.2 Statistical Characterization of Transmitter Average Efficiency

The output power of radio transmitters is a random variable. The average efficiency of a transmitter depends on the statistics of transmitter output power. The statistics is determined by the probability density function (PDF) (or the probability distribution function) of the output power and the dc supply power of a power amplifier. The average output power over a long-time interval $\Delta t = t_2 - t_1$ is

$$P_{O(AV)} = \frac{1}{\Delta t} \int_0^{\Delta t} PDF_{P_O}(t) P_O(t) dt \tag{1.25}$$

and the average supply power over the same time interval $\Delta T = t_2 - t_1$ is

$$P_{I(AV)} = \frac{1}{\Delta t} \int_0^{\Delta t} PDF_{P_I}(t) P_I(t) dt \tag{1.26}$$

Hence, the *long-term average efficiency* is the ratio of the energy delivered to the load (or antenna) E_O to the energy drawn from the power supply E_I over a long period of time Δt

$$\eta_{AV} = \frac{E_O}{E_I} = \frac{E_O/\Delta t}{E_I/\Delta t} = \frac{P_{O(AV)}}{P_{I(AV)}} = \frac{\frac{1}{\Delta t} \int_0^{\Delta t} PDF_{P_O}(t) P_O(t) dt}{\frac{1}{\Delta t} \int_0^{\Delta t} PDF_{P_I}(t) P_I(t) dt} \tag{1.27}$$

This efficiency determines the battery lifetime.

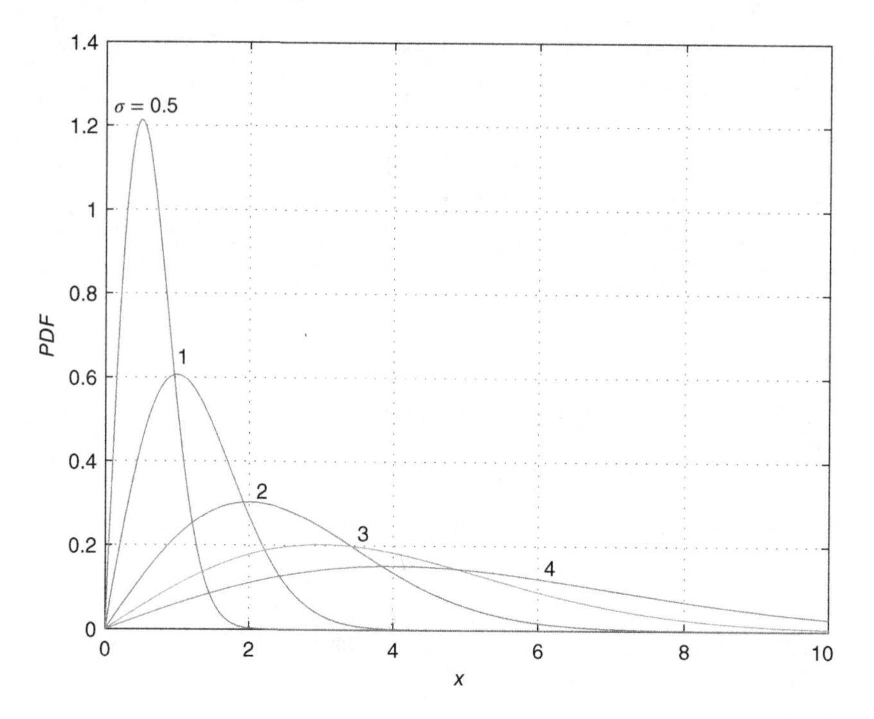

Figure 1.5 Rayleigh's probability density function of the transmitter output power.

The efficiency of power amplifiers in which transistors are operated as dependent current sources increases with the amplitude of the output voltage V_m. It reaches the maximum value at the maximum amplitude of the output voltage, which corresponds to the maximum output power. In practice, power amplifiers with a variable-envelope voltage are usually operated below the maximum output power. For example, the drain efficiency of the Class B power amplifier is $\eta_D = \pi/4 = 78.5\%$ at $V_m = V_I$, but it decreases to $\eta_D = \pi/8 = 39.27\%$ at $V_m = V_I/2$ and to $\eta_D = \pi/16 = 19.63\%$ at $V_m = V_I/4$. The average efficiency is useful for describing the efficiency of radio transmitters with variable-envelope signals, such as AM signals. The probability density function (PDF) of the envelope determines the amount of time an envelope remains at various amplitudes. For multiple carrier transmitters, the PDF may be characterized by Rayleigh's probability distribution.

Rayleigh's PDF of the output power is given by

$$g(P_O) = \frac{P_O(t)}{\sigma^2} e^{-\frac{P_O(t)}{2\sigma^2}} \tag{1.28}$$

where σ is the scale parameter of the distribution. Figure 1.5 shows plots of Rayleigh's PDF for $\sigma = 0.5$, 1, 2, 3, and 4.

1.6.3 Gate-Drive Power

The input impedance of the MOSFET consists of the series combination of the gate resistance r_G and the input capacitance C_i. The input capacitance is given by

$$C_i = C_{gs} + C_{gd}(1 - A_v) \tag{1.29}$$

where C_{gs} is the gate-to-source capacitance, C_{gd} is the gate-to-drain capacitance, and A_v is the voltage gain during the time interval when the drain-to-source voltage v_{DS} decreases. The Miller's capacitance is $C_m = C_{gd}(1 - A_v)$.

The *gate-drive power* is

$$P_G = \frac{1}{2\pi}\int_0^{2\pi} i_G v_{GS}\, d(\omega t) \tag{1.30}$$

For sinusoidal gate current and voltage waveforms, the gate-drive power is

$$P_G = \frac{1}{2} I_{gm} V_{gsm} \cos\phi_G = \frac{r_G I_{gm}^2}{2} \tag{1.31}$$

where I_{gm} is the amplitude of the gate current, V_{gsm} is the amplitude of the gate-to-source voltage, r_G is the gate resistance, and ϕ_G is the phase shift between the fundamental components of the gate current and the gate-to-source voltage. The *total power loss* including the gate-drive power is

$$P_{LS} = P_D + P_r + P_G \tag{1.32}$$

1.6.4 Power-Added Efficiency

The *power gain* of a power amplifier is given by

$$k_p = \frac{P_O}{P_G} = 10\log\left(\frac{P_O}{P_G}\right)\ \text{(dB)} \tag{1.33}$$

The *power-added efficiency* is the ratio of the difference between the output power and the gate-drive power to the dc supply power

$$\eta_{PAE} = \frac{\text{Output Power} - \text{Drive Power}}{\text{DC Supply Power}} = \frac{P_O - P_G}{P_I} = \frac{P_O}{P_I}\left(1 - \frac{P_G}{P_O}\right)$$
$$= \frac{P_O}{P_I}\left(1 - \frac{1}{P_O/P_G}\right) = \frac{P_O}{P_I}\left(1 - \frac{1}{k_p}\right) = \eta\left(1 - \frac{1}{k_p}\right) \tag{1.34}$$

If $k_p = 1$, $\eta_{PAE} = 0$. If $k_p \gg 1$, $\eta_{PAE} \approx \eta$.

For many communication systems, various modulation techniques use variable-envelope voltage and have a very high PAR of the RF output power. Typically, an RF power amplifier achieves a maximum power efficiency at a single operating voltage corresponding to the peak output power. The PAR is usually from 3 to 6 dB for a single-carrier transmitters. For multi-carrier transmitters, the PAR is typically from 6 to 13 dB. The efficiency decreases rapidly as the power is reduced from its maximum value. The *average composite power-added efficiency* is defined as

$$\eta_{AV(PAE)} = \frac{P_{ORF(AV)} - P_{IRF(AV)}}{P_{DC(AV)} + P_{DC(mod)}} = \frac{\int_{V_{min}}^{V_{max}} PDF(V)^*[P_{ORF}(V) - P_{IRF}(V)]dV}{\int_{V_{min}}^{V_{max}} PDF(V)^*[P_{DC(PA)}(V) + P_{DC(mod)}(V)]dV} \tag{1.35}$$

where $PDF(V)^*$ is the PDF of the complex modulated signal, V_{min} and V_{max} are the minimum and maximum voltages of the RF envelope, and $P_{ORF}(V)$, $P_{IRF}(V)$, $P_{DC(PA)}(V)$, and $P_{DC(mod)}(V)$ are all instantaneous power values at a given envelope voltage V.

The overall efficiency of a power amplifier is defined as

$$\eta_{tot} = \frac{P_O}{P_{I(DC)} + P_G + P_{mod}} = \frac{P_O}{P_O + P_O + P_G + P_{mod}} \tag{1.36}$$

where P_{mod} is the power consumption of a modulator.

1.6.5 Output-Power Capability

The *output-power capability* of the RF power amplifier with N transistors is defined as

$$c_p = \frac{P_{Omax}}{NI_{DM}V_{DSM}} = \frac{\eta_{Dmax}P_I}{NI_{DM}V_{DSM}} = \frac{\eta_{Dmax}}{N}\left(\frac{I_I}{I_{DM}}\right)\left(\frac{V_I}{V_{DSM}}\right) = \frac{1}{N}\eta_{Dmax}\alpha_0\beta_0$$
$$= \frac{1}{2N}\left(\frac{I_m}{I_{DM}}\right)\left(\frac{V_m}{V_{DSM}}\right) = \frac{1}{2N}\alpha_{1max}\beta_{1max} \quad (1.37)$$

where I_{DM} is the maximum value of the instantaneous drain current i_D, V_{DSM} is the maximum value of the instantaneous drain-to-source voltage v_{DS}, η_{Dmax} is the amplifier drain efficiency at the maximum output power $P_{O(max)}$, and N is the number of transistors in the amplifier, which are not connected in parallel or in series. For example, a push–pull amplifier has two transistors. The maximum output power of an amplifier with a transistor having the maximum ratings I_{DM} and V_{DSM} is

$$P_{Omax} = c_p N I_{DM} V_{DSM} \quad (1.38)$$

As the output power capability c_p increases, the maximum output power P_{Omax} also increases. The output power capability is useful for comparing different types or families of amplifiers. The larger the c_p, the larger is the maximum output power.

For a single-transistor amplifier, the output-power capability is given by

$$c_p = \frac{P_{Omax}}{I_{DM}V_{DSM}} = \frac{\eta_{Dmax}P_I}{I_{DM}V_{DSM}} = \eta_{Dmax}\left(\frac{I_I}{I_{DM}}\right)\left(\frac{V_I}{V_{DSM}}\right) = \eta_{Dmax}\alpha_0\beta_0$$
$$= \frac{1}{2}\left(\frac{I_{m(max)}}{I_{DM}}\right)\left(\frac{V_{m(max)}}{V_{DSM}}\right) = \frac{1}{2}\alpha_{1max}\beta_{1max} \quad (1.39)$$

1.7 Transmitter Noise

A transmitter contains an oscillator of a carrier frequency. An oscillator is a nonlinear device. It does not generate an ideal single-frequency and constant-amplitude signal. Therefore, the oscillator output power is not only concentrated at a single frequency but also distributed around it. The noise spectra on both sides of the carrier are called noise sidebands. Hence, the voltage and current waveforms contain noise. These waveforms are modulated by noise. There are three categories of noise: AM noise, frequency-modulated (FM) noise, and phase noise. The AM noise results in the amplitude variations of the oscillator output voltage. The FM or PM noise causes the spreading of the frequency spectrum around the carrier frequency. The ratio of a single-sideband noise power contained in 1-Hz bandwidth at an offset from carrier to the carrier power is defined as noise-to-carrier power ratio

$$NCP = \frac{N}{C} = 10\log\left(\frac{N}{C}\right)\left(\frac{\text{dBc}}{\text{Hz}}\right) \quad (1.40)$$

The unit dBc/Hz indicates the number of decibels below the carrier over a bandwidth of 1 Hz. Most of oscillator noise around the carrier is the phase noise. This noise represents the phase jitter. For example, the phase noise is 80 dBc/Hz at 2 kHz offset from the carrier and 110 dBc/Hz at 50 kHz offset from the carrier.

Typically, the output thermal noise of power amplifiers should be below −130 dBm. The purpose of this requirement is to introduce negligible level of noise to the input of the low-noise amplifier (LNA) of the receiver.

Example 1.1

An RF power amplifier has $P_O = 10\,\mathrm{W}$, $P_I = 20\,\mathrm{W}$, and $P_G = 1\,\mathrm{W}$. Find the efficiency, power-added efficiency, and power gain.

Solution. The efficiency of the power amplifier is

$$\eta = \frac{P_O}{P_I} = \frac{10}{20} = 50\% \tag{1.41}$$

The power-added efficiency is

$$\eta_{PAE} = \frac{P_O - P_G}{P_I} = \frac{10-1}{20} = 45\% \tag{1.42}$$

The power gain is

$$k_p = \frac{P_O}{P_G} = \frac{10}{1} = 10 = 10\log(10) = 10\ \mathrm{dB} \tag{1.43}$$

1.8 Conditions for 100% Efficiency of Power Amplifiers

The drain efficiency of any power amplifier is given by

$$\eta_D = \frac{P_{DS}}{P_I} = 1 - \frac{P_D}{P_I} \tag{1.44}$$

The condition for achieving a drain efficiency of 100% is

$$P_D = \frac{1}{T}\int_0^T i_D v_{DS}\,dt = 0 \tag{1.45}$$

For an NMOS transistor, $i_D \geq 0$ and $v_{DS} \geq 0$; for a PMOS transistor, $i_D \leq 0$ and $v_{DS} \leq 0$. In this case, the sufficient condition for achieving a drain efficiency of 100% becomes

$$i_D v_{DS} = 0 \tag{1.46}$$

Thus, the waveforms i_D and v_{DS} should be nonoverlapping for an efficiency of 100%. Nonoverlapping waveforms i_D and v_{DS} are shown in Fig. 1.6.

The drain efficiency of power amplifiers is less than 100% for the following cases:

- The waveforms of $i_D > 0$ and $v_{DS} > 0$ are overlapping (e.g., as in a Class C power amplifier).
- The waveforms of i_D and v_{DS} are adjacent, and the waveform v_{DS} has a jump at $t = t_o$ and the waveform i_D contains an impulse Dirac function, as shown in Fig. 1.7(a).
- The waveforms of i_D and v_{DS} are adjacent, and the waveform i_D has a jump at $t = t_o$ and the waveform v_{DS} contains an impulse Dirac function, as shown in Fig. 1.7(b).

For the case of Fig. 1.7(a), an ideal switch is connected in parallel with a capacitor C. The switch is turned on at $t = t_o$, when the voltage v_{DS} across the switch is nonzero. At $t = t_o$, this voltage can be described by

$$v_{DS}(t_o) = \frac{1}{2}\left[\lim_{t\to t_o^-} v_{DS}(t) + \lim_{t\to t_o^+} v_{DS}(t)\right] = \frac{1}{2}(\Delta V + 0) = \frac{\Delta V}{2} \tag{1.47}$$

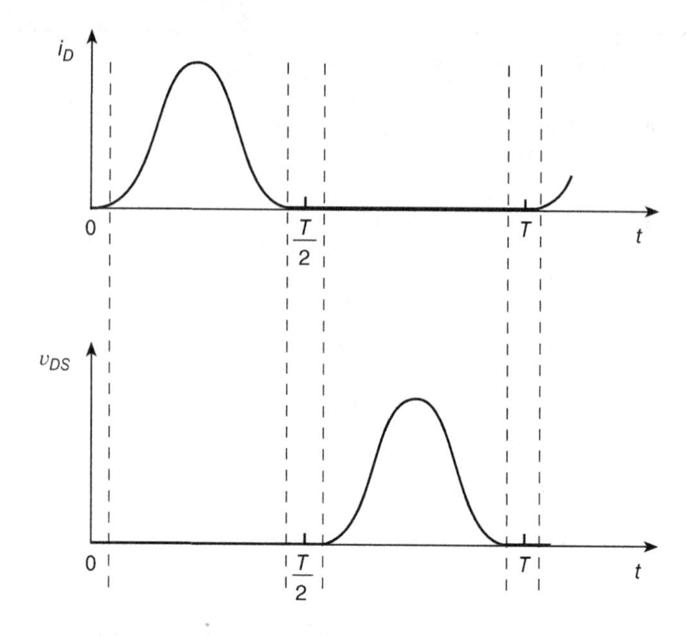

Figure 1.6 Nonoverlapping waveforms of drain current i_D and drain-to-source voltage v_{DS}.

At $t = t_o$, the drain current is given by

$$i_D(t_o) = C\Delta V\delta(t - t_o) \tag{1.48}$$

Hence, the instantaneous power dissipation is

$$p_D(t) = i_D v_{DS} = \begin{cases} \frac{1}{2}C\Delta V^2\delta(t - t_o) & \text{for } t = t_o \\ 0 & \text{for } t \neq t_o \end{cases} \tag{1.49}$$

resulting in the time average power dissipation

$$P_D = \frac{1}{T}\int_0^T i_D v_{DS} dt = \frac{C\Delta V^2}{2T}\int_0^T \delta(t - t_o)dt = \frac{1}{2}fC\Delta V^2 \tag{1.50}$$

and the drain efficiency

$$\eta_D = 1 - \frac{P_D}{P_I} = 1 - \frac{fC\Delta V^2}{2P_I} \tag{1.51}$$

In a real circuit, the switch has a small series resistance, and the current through the switch is an exponential function of time with a finite peak value. Thus, to achieve the efficiency of 100%, either $\Delta V = 0$ or $C = 0$. In a realistic amplifier, the transistor should be turned on at zero drain-to-source voltage v_{DS} so that $\Delta V = 0$. This observation leads to the concept of a zero-voltage switching (ZVS) Class E amplifier.

Example 1.2

An RF power amplifier has a step change in the drain-to-source voltage at MOSFET turn-on $V_{DS} = 5$ V, transistor capacitance $C = 100$ pF, operating frequency $f = 2.4$ GHz, dc supply voltage $V_I = 5$ V, and dc supply current $I_I = 1$ A. Assume that all parasitic resistances are zero. Find the efficiency of the power amplifier.

Solution. The switching power loss is

$$P_D = \frac{1}{2} fC\Delta V_{DS}^2 = \frac{1}{2} \times 2.4 \times 10^9 \times 100 \times 10^{-12} \times 5^2 = 3 \text{ W} \tag{1.52}$$

The dc power loss is

$$P_I = I_I V_I = 1 \times 5 = 5 \text{ W} \tag{1.53}$$

Hence, the drain efficiency of the amplifier is

$$\eta_D = \frac{P_O}{P_I} = \frac{P_I - P_D}{P_I} = 1 - \frac{P_D}{P_I} = 1 - \frac{3}{5} = 40\% \tag{1.54}$$

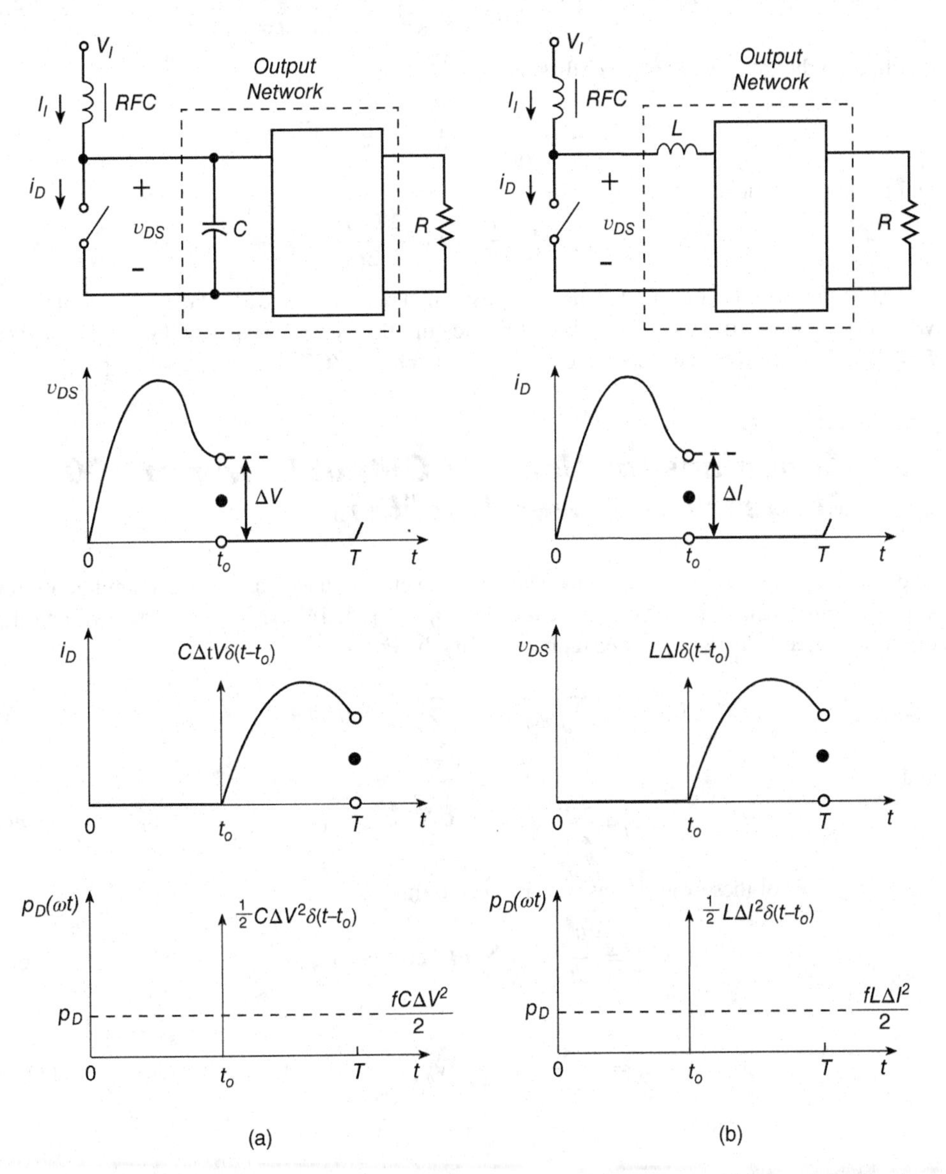

Figure 1.7 Waveforms of drain current i_D and drain-to-source voltage v_{DS} with Dirac delta functions: (a) circuit with the switch in parallel with a capacitor; (b) circuit with the switch in series with an inductor.

For the amplifier of Fig. 1.7(b), an ideal switch is connected in series with an inductor L. The switch is turned on at $t = t_o$, when the current i_D through the switch is nonzero. At $t = t_o$, the switch current can be described by

$$i_D(t_o) = \frac{1}{2}\left[\lim_{t\to t_o^-} i_D(t) + \lim_{t\to t_o^+} i_D(t)\right] = \frac{1}{2}(\Delta I + 0) = \frac{\Delta I}{2} \tag{1.55}$$

At $t = t_o$, the drain current is given by

$$i_D(t_o) = L\Delta I\delta(t - t_o) \tag{1.56}$$

Hence, the instantaneous power dissipation is

$$p_D(t) = i_D v_{DS} = \begin{cases} \frac{1}{2}L\Delta I^2\delta(t - t_o) & \text{for } t = t_o \\ 0 & \text{for } t \neq t_o \end{cases} \tag{1.57}$$

resulting in the time average power dissipation

$$P_D = \frac{1}{T}\int_0^T i_D v_{DS}\, dt = \frac{1}{2}fL\Delta I^2 \tag{1.58}$$

and the drain efficiency

$$\eta_D = 1 - \frac{P_D}{P_I} = 1 - \frac{fL\Delta I^2}{2P_I} \tag{1.59}$$

In reality, the switch in the off-state has a large parallel resistance and a voltage with a finite peak value developed across the switch. The efficiency of 100% can be achieved if either $\Delta I = 0$ or $L = 0$. This leads to the concept of zero-current switching (ZCS) Class E amplifier [3].

1.9 Conditions for Nonzero Output Power at 100% Efficiency of Power Amplifiers

The drain current and drain-to-source voltage waveforms have fundamental limitations for simultaneously achieving 100% efficiency and $P_O > 0$ [13, 14]. The drain current i_D and the drain-to-source voltage v_{DS} can be represented by the Fourier series as

$$i_D = I_I + \sum_{n=1}^{\infty} i_{dsn} = I_I + \sum_{n=1}^{\infty} I_{dn}\sin(n\omega t + \psi_n) \tag{1.60}$$

and

$$v_{DS} = V_I + \sum_{n=1}^{\infty} v_{dsn} = V_I + \sum_{n=1}^{\infty} V_{dsn}\sin(n\omega t + \vartheta_n) \tag{1.61}$$

The derivatives of these waveforms with respect to time are

$$i'_D = \frac{di_D}{dt} = \omega\sum_{n=1}^{\infty} nI_{dn}\cos(n\omega t + \psi_n) \tag{1.62}$$

and

$$v'_{DS} = \frac{dv_{DS}}{dt} = \omega\sum_{n=1}^{\infty} nV_{dsn}\cos(n\omega t + \vartheta_n) \tag{1.63}$$

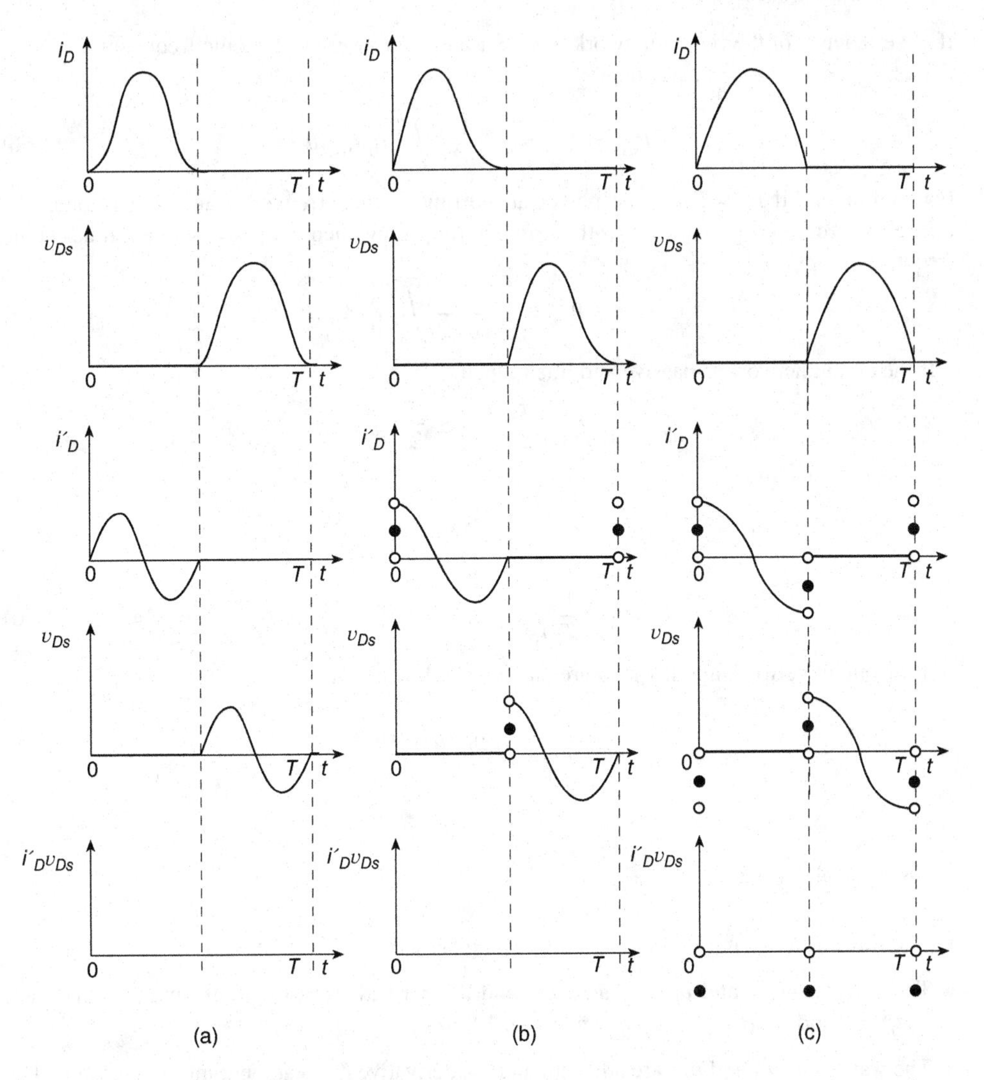

Figure 1.8 Waveforms of power amplifiers with $P_O = 0$.

Hence, the time-average value of the product of the derivatives is

$$\frac{1}{T}\int_0^T i'_D v'_{DS}\,dt = \omega^2 \int_0^T \sum_{n=1}^{\infty} nI_{dn}\cos(n\omega t + \psi_n) \sum_{n=1}^{\infty} nV_{dsn}\cos(n\omega t + \vartheta_n dt)$$

$$= -\frac{\omega^2}{2}\sum_{n=1}^{\infty} n^2 I_{dn} V_{dsn}\cos\phi_n = -\omega^2 \sum_{n=1}^{\infty} n^2 P_{dsn} \tag{1.64}$$

where $\phi_n = \vartheta_n - \psi_n - \pi$ Next,

$$\sum_{n=1}^{\infty} n^2 P_{dsn} = -\frac{1}{4\pi^2 f}\int_0^T i'_D v'_{DS}\,dt \tag{1.65}$$

If the efficiency of the output network is $\eta_n = 1$ and the power at harmonic frequencies is zero, that is, $P_{ds2} = 0$, $P_{ds3} = 0$, …, then

$$P_{ds1} = P_{O1} = -\frac{1}{4\pi^2 f}\int_0^T i'_D v'_{DS}\, dt \tag{1.66}$$

For multipliers, if $\eta_n = 1$ and the power at the fundamental frequency and at harmonic frequencies is zero except that of the nth harmonic frequency, then the power at the nth harmonic frequency is

$$P_{dsn} = P_{On} = -\frac{1}{4\pi^2 n^2}\int_0^T i'_D v'_{DS}\, dt \tag{1.67}$$

If the output network is passive and linear, then

$$-\frac{\pi}{2} \leq \phi_n \leq \frac{\pi}{2} \tag{1.68}$$

In this case, the output power is nonzero

$$P_O > 0 \tag{1.69}$$

if

$$\frac{1}{T}\int_0^T i'_D v'_{DS}\, dt < 0 \tag{1.70}$$

If the output network and the load are passive and linear and

$$\frac{1}{T}\int_0^T i'_D v'_{DS}\, dt = 0 \tag{1.71}$$

then

$$P_O = 0 \tag{1.72}$$

for the following cases:

- The waveforms i_D and v_{DS} are nonoverlapping, as shown in Fig. 1.6.
- The waveforms i_D and v_{DS} are adjacent and the derivatives at the joint time instants t_j are $i'_D(t_j) = 0$ and $v'_{DS}(t_j) = 0$, as shown in Fig. 1.8(a).
- The waveforms i_D and v_{DS} are adjacent and the derivative $i'_D(t_j)$ at the joint time instant t_j has a jump and $v'_{DS}(t_j) = 0$, or *vice versa*, as shown in Fig. 1.8(b).
- The waveforms i_D and v_{DS} are adjacent, and the derivatives of both waveforms $i'_D(t_j)$ and $v'_{DS}(t_j)$ have jumps at the joint time instant t_j, as shown in Fig. 1.8(c).

In summary, ZVS, zero-voltage derivative switching (ZVDS), and ZCS conditions cannot be simultaneously satisfied with a passive load network at a nonzero output power.

1.10 Output Power of Class E ZVS Amplifiers

The Class E ZVS RF power amplifier is shown in Fig. 1.9. Waveforms for the Class E power amplifier under ZVS and zero-derivative switching (ZDS) conditions are shown in Fig. 1.10. Ideally, the efficiency of this amplifier is 100%. Waveforms for the Class E amplifier are shown in Fig. 1.10. The drain current i_D has a jump at $t = t_o$. Hence, the derivative of the drain current at $t = t_o$ is given by

$$i'_D(t_o) = \Delta I \delta(t - t_o) \tag{1.73}$$

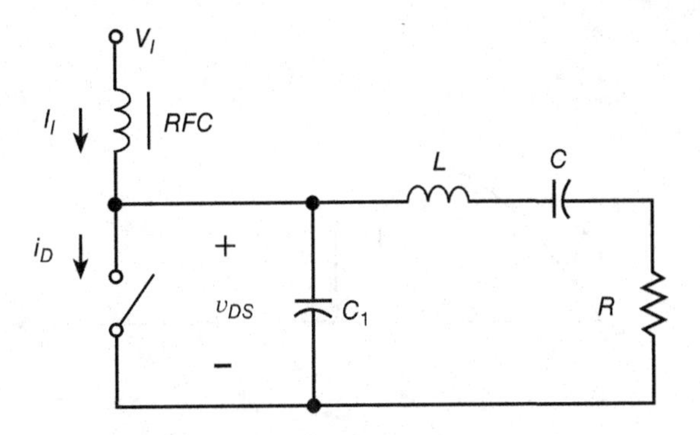

Figure 1.9 Class E ZVS power amplifier.

and the derivative of the drain-to-source voltage at $t = t_o$ is given by

$$v'_{DS}(t_o) = \frac{1}{2}\left[\lim_{t\to t_o^-} v'_{DS}(t) + \lim_{t\to t_o^+} v'_{DS}(t)\right] = \frac{S_v}{2} \tag{1.74}$$

Assuming that $P_r = 0$ and $\eta_r = 1$, the output power of the Class E ZVS amplifier is

$$P_{ds} = P_O = -\frac{1}{4\pi^2 f}\int_0^T i'_D v'_{DS}\,dt = -\frac{1}{4\pi^2 f}\int_0^T v'_{DS}(t_o)\Delta I\delta(t - t_o)dt$$
$$= -\frac{\Delta I S_v}{8\pi^2 f}\int_0^T \delta(t - t_o)dt = -\frac{\Delta I S_v}{8\pi^2 f} \tag{1.75}$$

where $\Delta I < 0$. Since

$$\Delta I = -0.6988 I_{DM} \tag{1.76}$$

and

$$S_v = 11.08 f V_{DSM} \tag{1.77}$$

the output power is

$$P_O = -\frac{\Delta I S_v}{8\pi^2 f} = 0.0981 I_{DM} V_{DSM} \tag{1.78}$$

Hence, the output power capability is

$$c_p = \frac{P_O}{I_{DM} V_{DSM}} = 0.0981 \tag{1.79}$$

Example 1.3

A Class E ZVS RF power amplifier has a step change in the drain current at the MOSFET turn-off $\Delta I_D = -1$ A, a slope of the drain-to-source voltage at the MOSFET turn-off $S_v = 11.08 \times 10^8$ V/s, and the operating frequency is $f = 1$ MHz. Find the output power of the Class E amplifier.

Solution. The output power of the Class E power amplifier is

$$P_O = -\frac{\Delta I S_v}{8\pi^2 f} = -\frac{-1 \times 11.08 \times 10^8}{8\pi^2 \times 10^6} = 14.03 \text{ W} \tag{1.80}$$

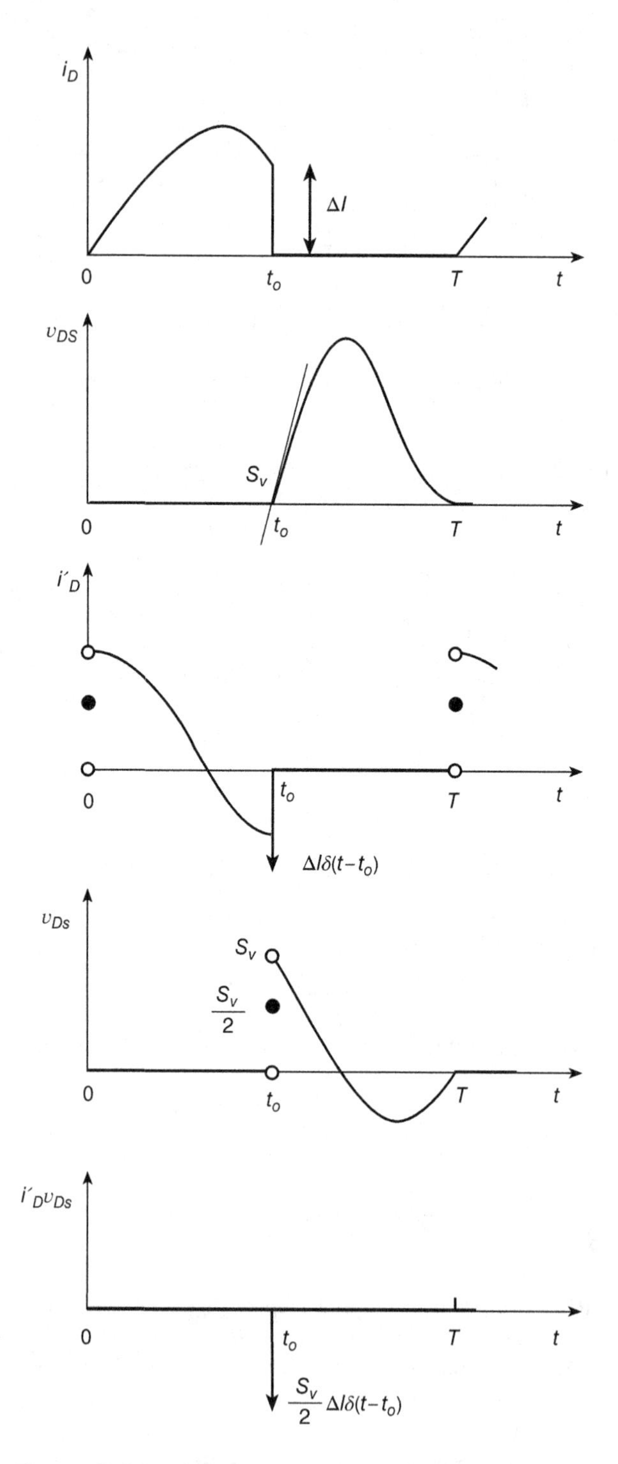

Figure 1.10 Waveforms of Class E ZVS power amplifier.

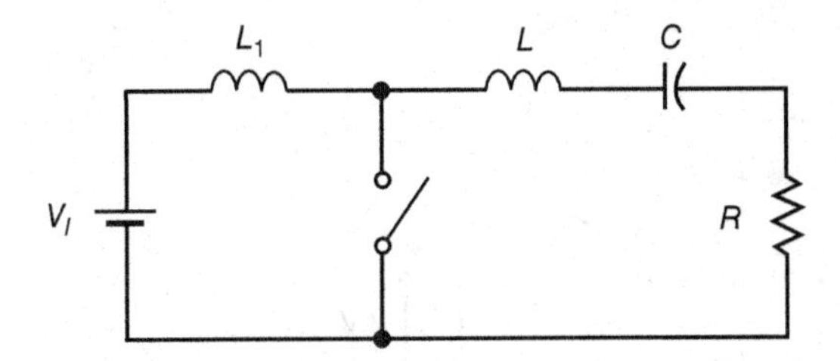

Figure 1.11 Class E ZCS power amplifier.

1.11 Class E ZCS Amplifiers

The Class E ZCS RF power amplifier is depicted in Fig. 1.11. Current and voltage waveforms under ZCS and ZDS conditions are shown in Fig. 1.12. The efficiency of this amplifier with perfect components and under ZCS condition is 100%. The drain-to-source voltage v_{DS} has a jump at $t = t_o$. The derivative of the drain-to-source voltage at $t = t_o$ is given by

$$v'_{DS}(t_o) = \Delta V\delta(t - t_o) \tag{1.81}$$

and the derivative of the drain current at $t = t_o$ is given by

$$i'_D(t_o) = \frac{1}{2}\left[\lim_{t\to t_o^-} i'_D(t) + \lim_{t\to t_o^+} i'_D(t)\right] = \frac{S_i}{2} \tag{1.82}$$

The output power of the Class E ZCS amplifier is

$$\begin{aligned} P_{ds} = P_O &= -\frac{1}{4\pi^2 f}\int_0^T i'_D v'_{DS}\,dt = -\frac{1}{4\pi^2 f}\int_0^T i'_D \Delta V\delta(t - t_o)dt \\ &= -\frac{\Delta V S_i}{8\pi^2 f}\int_0^T \delta(t - t_o)dt = -\frac{\Delta V S_i}{8\pi^2 f} \end{aligned} \tag{1.83}$$

Since

$$\Delta V = -0.6988 V_{DSM} \tag{1.84}$$

and

$$S_i = 11.08 f I_{DM} \tag{1.85}$$

the output power is

$$P_O = -\frac{\Delta V S_i}{8\pi^2 f} = 0.0981 I_{DM} V_{DSM} \tag{1.86}$$

Hence, the output power capability is

$$c_p = \frac{P_O}{I_{DM} V_{DSM}} = 0.0981 \tag{1.87}$$

Example 1.4

A Class E ZCS RF power amplifier has a step change in the drain-to-source voltage waveform at the MOSFET turn-on $\Delta V_{DS} = -100$ V, a slope of the drain current at the MOSFET turn-on $S_i = 11.08 \times 10^7$ V/s, and the operating frequency is $f = 1$ MHz. Find the output power of the Class E amplifier.

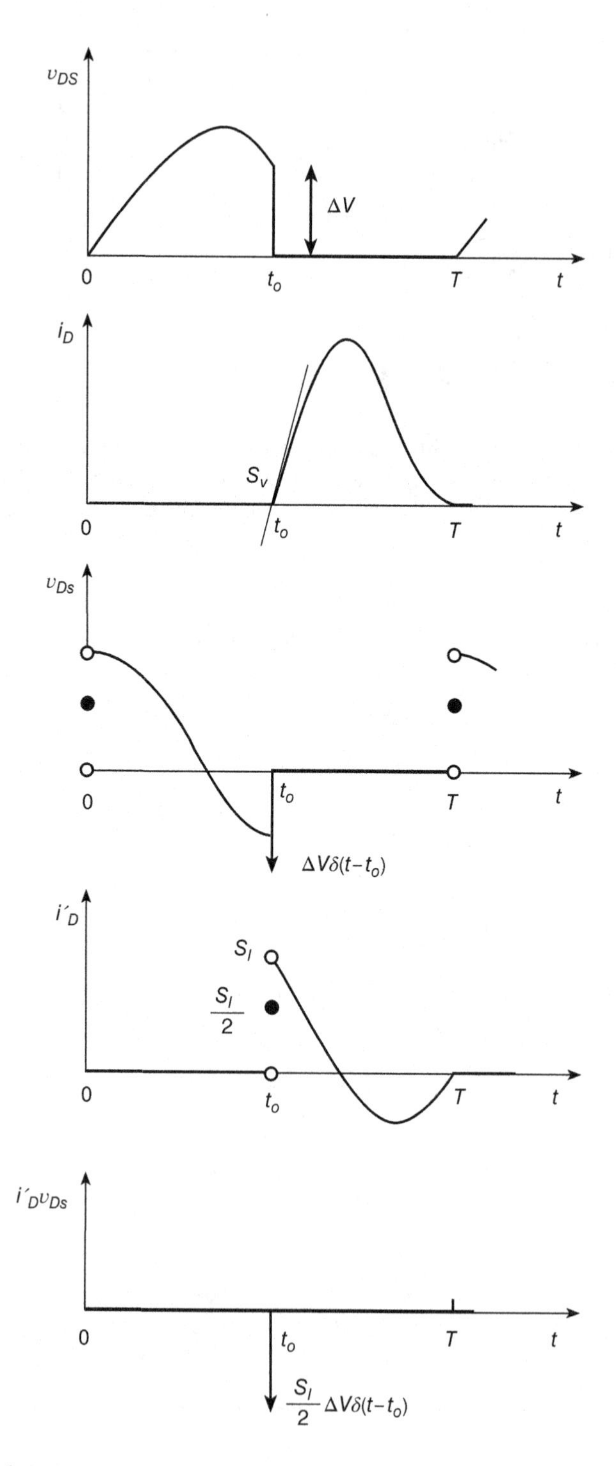

Figure 1.12 Waveforms of the Class E ZCS power amplifier.

Solution. The output power of the Class E power amplifier is

$$P_O = -\frac{\Delta V_{DS} S_i}{8\pi^2 f} = -\frac{-100 \times 11.08 \times 10^7}{8\pi^2 \times 10^6} = 140.320 \text{ W} \tag{1.88}$$

1.12 Antennas

The fundamental principle of wireless communication is based on Ampère–Maxwell's law

$$\Delta \times \mathbf{H} = \mathbf{J} + \frac{\partial \mathbf{D}}{\partial t} \tag{1.89}$$

Radiation. A transmitting antenna is used to radiate EMWs. The displacement current in the conductor of a transmitting antenna is

$$\frac{\partial \mathbf{D}}{\partial t} = 0 \tag{1.90}$$

Hence, Ampère–Maxwell's law becomes

$$\Delta \times \mathbf{H} = \mathbf{J} \tag{1.91}$$

This equation states that a time-varying magnetic field $\mathbf{H}(t)$ around the transmitting antenna is produced by a time-varying current density $\mathbf{J}(t)$ flowing in the transmitting antenna conductor.

Propagation. The conduction current in the air between the transmitting and receiving antennas is zero

$$\mathbf{J} = 0 \tag{1.92}$$

Hence, Ampère–Maxwell's law becomes

$$\Delta \times \mathbf{H} = \epsilon \frac{\partial \mathbf{E}}{\partial t} \tag{1.93}$$

This law states that magnetic and electric fields form an electromagnetic (EM) wave in the propagation process in the air (or other media).

Receiving of EM Wave. A receiving antenna is used to receive an EM wave and convert it into a current. The displacement current in the conductor of a receiving antenna is

$$\frac{\partial \mathbf{D}}{\partial t} = 0 \tag{1.94}$$

Hence, Ampère–Maxwell's law becomes

$$\mathbf{J} = \Delta \times \mathbf{H} \tag{1.95}$$

This equation states that a current density $\mathbf{J}$ is produced in the receiving antenna by a magnetic field $\mathbf{H}$ present around the receiving antenna.

An antenna is a device for radiating or receiving EM radio waves. A transmitting antenna converts an electrical signal into an EM wave. It is a transition structure between a guiding device (such as a transmission line) and free space. A receiving antenna converts an EM wave into an electrical signal. In free space, the EM wave travels at the speed of light $c = 3 \times 10^8$ m/s. The wavelength of an EM wave in free space is given by

$$\lambda = \frac{c}{f} \tag{1.96}$$

Transmitting antennas are used to *radiate* EMWs. The radiation efficiency of antennas is good only if their dimensions are of the same order of magnitude as the wavelength of the carrier frequency f_c. The length of antennas is usually $\lambda/2$ (a half-dipole antenna) or $\lambda/4$ (quarter-wave antenna) and should be higher than $\lambda/10$. The length of antenna depends on the wavelength of the EM wave. The length of quarter-wave antennas is

$$h_a = \frac{\lambda}{4} = \frac{c}{4f} \tag{1.97}$$

For example, the height of a quarter-wave antenna $h_a = 750\,\text{m}$ at $f_c = 100\,\text{kHz}$, $h_a = 75\,\text{m}$ at $f_c = 1\,\text{MHz}$, $h_a = 7.5\,\text{cm}$ at $f_c = 1\,\text{GHz}$, and $h_a = 7.5\,\text{mm}$ at $f_c = 10\,\text{GHz}$. Thus, mobile transmitters and receivers (transceivers) are only possible at high carrier frequencies.

An isotropic antenna is a theoretical point antenna that radiates energy equally in all directions with its power spread uniformly on the surface of a sphere. This results in a spherical wavefront. The uniform radiated power density at a distance r from an isotropic antenna with the output power P_T is given by

$$p(r) = \frac{P_T}{4\pi r^2} \tag{1.98}$$

The power density is inversely proportional to the square of the distance r. The hypothetical isotropic antenna is not practical, but is commonly used as a reference to compare with other antennas. If the transmitting antenna has directivity in a particular direction and efficiency, the power density in that direction is increased by a factor called the *antenna gain* G_T. The power density received by a receiving directive antenna is

$$p_r(r) = G_T \frac{P_T}{4\pi r^2} \tag{1.99}$$

The *antenna efficiency* is the ratio of the radiated power to the total power fed to the antenna $\eta_A = P_{RAD}/P_{FED}$.

A receiving antenna pointed in the direction of the radiated power gathers a portion of the power that is proportional to its cross-sectional area. The *antenna effective area* is given by

$$A_e = G_R \frac{\lambda^2}{4\pi} \tag{1.100}$$

where G_R is the gain of the receiving antenna and λ is the free-space wavelength. Thus, the power received by a receiving antenna is given by the Herald Friis formula for free-space transmission [12]

$$P_{REC} = A_e p(r) = \frac{1}{16\pi^2} G_T G_R P_T \left(\frac{\lambda^2}{r}\right)^2 = \frac{1}{16\pi^2} G_T G_R P_T \left(\frac{c}{rf_c}\right)^2 \tag{1.101}$$

The received power is proportional to $(\lambda/r)^2$ and to the gain of either antenna. As the carrier frequency f_c doubles, the received power decreases by a factor of 4 at a given distance r from the transmitting antenna. The gain of the dish (parabolic) antenna is given by

$$G_T = G_R = 6\left(\frac{D}{\lambda}\right)^2 = 6\left(\frac{Df_c}{c}\right)^2 \tag{1.102}$$

where D is the mouth diameter of the primary reflector. For $D = 3\,\text{m}$ and $f = 10\,\text{GHz}$, $G_T = G_R = 60,000 = 47.8\,\text{dB}$.

The *space loss* is the loss due to spreading the RF energy as it propagates through free space and is defined as

$$S_L = \frac{P_T}{P_{REC}} = \left(\frac{4\pi r}{\lambda}\right)^2 = 10\log\left(\frac{P_T}{P_{REC}}\right) = 20\log\left(\frac{4\pi r}{\lambda}\right) \tag{1.103}$$

There are also other losses such as atmospheric loss, polarization mismatch loss, impedance mismatch loss, and pointing error denoted by L_{syst}. Hence, the link equation is

$$P_{REC} = \frac{L_{syst} G_T G_R P_T}{(4\pi)^2} \left(\frac{\lambda}{r}\right)^2 \tag{1.104}$$

The maximum distance between transmitting and receiving antennas is

$$r_{max} = \frac{\lambda}{4\pi}\sqrt{L_{syst}G_T G_R\left(\frac{P_T}{P_{REC(min)}}\right)}. \tag{1.105}$$

For example, the Global System for Mobile Communications (GSM) cell radius is 35 or 60 km in the 900 MHz band and 20 km in the 1.8 GHz band.

1.13 Propagation of Electromagnetic Waves

EM wave propagation is illustrated in Fig. 1.13. There are three groups of EMWs based on their propagation properties:

- Ground waves (below 2 MHz).
- Sky waves (2–30 MHz).
- Line-of-sight waves, also called space waves or horizontal waves (above 30 MHz).

Ground waves travel parallel to the Earth's surface and suffer little attenuation by smog, moisture, and other particles in the lower part of the atmosphere. Very high antennas are required for transmission of these low-frequency (LF) waves. The approximate transmission distance of ground waves is about 1600 km (1000 miles). Ground wave propagation is much better over water, especially salt water, than over a dry desert terrain. Ground wave propagation is the only way to communicate into the ocean with submarines. Extremely low-frequency (ELF) waves (30–300 Hz) are used to minimize the attenuation of the waves by sea water. A typical frequency is 100 Hz.

The sky waves leave the curved surface of the Earth and are refracted by the ionosphere back to the surface of the Earth; therefore, they are capable of following the Earth's curvature. The altitude of refraction of the sky waves varies from 50 to 400 km. The transmission distance between two transmitters is 4000 km. The ionosphere is a region above the atmosphere, where free ions

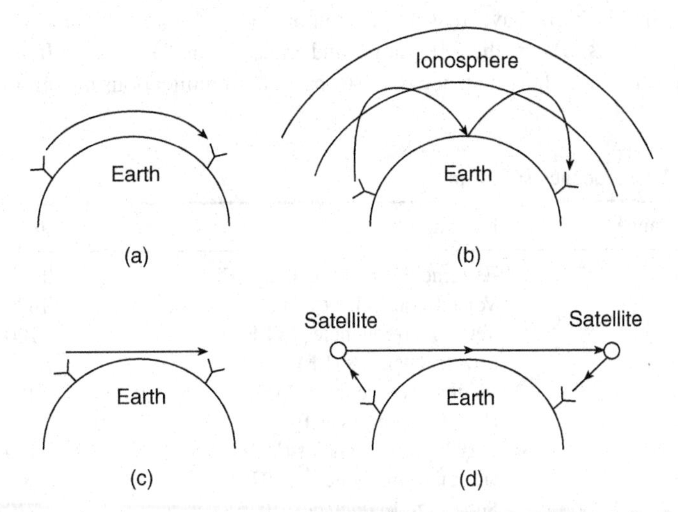

Figure 1.13 Electromagnetic wave propagation: (a) ground wave propagation; (b) sky wave propagation; (c) horizontal wave propagation; and (d) wave propagation in satellite communications.

and electrons exist in sufficient quantity to affect the wave propagation. Ionization is caused by radiation from the Sun. It changes as the position of a point on the Earth with respect to the Sun changes daily, monthly, and yearly. After sunset, the lowest layer of the ionosphere disappears because of rapid recombination of its ions. The higher the frequency, the more difficult the refracting (bending) process. Between the point where the ground wave is completely attenuated and the point where the first wave returns, no signal is received, resulting in a skip zone.

Line-of-sight waves follow straight lines. There are two types of line-of-sight waves: direct waves and ground reflected waves. The direct wave is by far the most widely used for propagation between antennas. Signals of frequencies above the high-frequency (HF) band cannot be propagated for long distances along the surface of the Earth. However, it is easy to propagate these signals through free space. The propagated power depends on distance, orientation of antennas, attenuation by buildings, and multipath [10].

1.14 Frequency Spectrum

Table 1.1 gives the frequency spectrum. In the United States, the allocation of carrier frequencies, bandwidths, and power levels of transmitted EMWs is regulated by the Federal Communications Commission (FCC) for all nonmilitary applications. The communication must occur in a certain part of the frequency spectrum. The carrier frequency f_c determines the channel frequency.

LF EMWs are propagated by ground waves. They are used for long-range navigation, telegraphy, and submarine communication. The medium-frequency (MF) band contains the commercial radio band from 535 to 1705 kHz. This band is used for radio transmission of AM signals to general audiences. The carrier frequencies are from 540 to 1700 kHz. For example, one carrier frequency is at $f_c = 550$ kHz, and the next carrier frequency is at $f_c = 560$ kHz. The modulation bandwidth is 5 kHz. The average power of local stations is from 0.1 to 1 kW. The average power of regional stations is from 0.5 to 5 kW. The average power of clear stations is from 0.25 to 50 kW. A radio receiver may receive power as low as of 10 pW, 1 μV/m, or 50 μV across a $300 - \Omega$ antenna. Thus, the ratio of the output power of the transmitter to the input power of the receiver is on the order of $P_T/P_{REC} = 10^{15}$.

The range from 1705 to 2850 kHz is used for short-distance point-to-point communications for services such as fire, police, ambulance, highway, forestry, and emergency services. The antennas in this band have reasonable height and radiation efficiency. The aeronautical frequency range starts in the MF range and ends in the HF range. It is from 2850 to 4063 kHz and is used for short-distance point-to-point communications and ground–air–ground

Table 1.1 Frequency spectrum.

Frequency range	Band name	Wavelength range
30–300 Hz	Extremely low frequencies (ELF)	10,000–1000 km
300–3000 Hz	Voice frequencies (VF)	1000–100 km
3–30 kHz	Very low frequencies (VLF)	100–10 km
30–300 kHz	Low frequencies (LF)	10–1 km
0.3–3 MHz	Medium frequencies (MF)	1000–100 m
3–30 MHz	High frequencies (HF)	100–10 m
30–300 MHz	Very high frequencies (VHF)	100–10 cm
0.3–3 GHz	Ultrahigh frequencies (UHF)	100–10 cm
3–30 GHz	Superhigh frequencies (SHF)	10–1 cm
30–300 GHz	Extra high frequencies (EHF)	10–1 mm

communications. Aircraft flying scheduled routes are allocated specific channels. The HF band contains the radio amateur band from 3.5 to 4 MHz in the United States. Other countries use this band for mobile and fixed communication services. HFs are also used for long-distance point-to-point transoceanic ground–air–ground communications. HFs are propagated by sky waves. The frequencies from 1.6 to 30 MHz are called short waves.

The very high-frequency (VHF) band contains commercial FM radio and most TV channels. The commercial FM radio transmission is from 88 to 108 MHz. The modulation bandwidth is 15 kHz. The average power is from 0.25 to 100 kW. The transmission distance of VHF TV signals is 160 km (100 miles).

TV channels for analog transmission range from 54 to 88 MHz and from 174 to 216 MHz in the VHF band and from 470 to 890 MHz in the ultrahigh-frequency (UHF) band. The bandwidth of analog TV is 6.7 MHz and digital TV is 10 MHz. The average power is 100 kW for the frequency range from 54 to 88 MHz and 316 kW for the frequency range from 174 to 216 MHz. Broadcast frequency allocations are given in Table 1.2. The UHF and SHF frequency bands are given in Table 1.3. The cellular phone frequency allocation is given in Table 1.4.

The superhigh frequency (SHF) band contains satellite communications channels. Satellites are placed in orbits. Typically, these orbits are 37,786 km in altitude above the equator. Each satellite illuminates about one-third of the Earth. Since the satellites maintain the same position relative to the Earth, they are placed in geostationary orbits. These geosynchronous satellites are called GEO satellites. Each satellite contains a communication system that can receive signals from the Earth or from another satellite and transmit the received signal back to the Earth or to another satellite. This system uses two carrier frequencies. The frequency for transmission from the Earth to the satellite (uplink) is 6 GHz and the transmission from the satellite to the Earth (downlink) is at 4 GHz. The bandwidth of each channel is 500 MHz. Directional antennas are used for radio transmission through free space. Satellite communications are used for TV and

Table 1.2 Broadcast frequency allocation.

Radio or TV	Frequency range	Channel spacing
AM Radio	535–1605 kHz	10 kHz
TV (channels 2–6)	54–72 MHz	6 MHz
TV	76–88 MHz	6 MHz
FM Radio	88–108 MHz	200 kHz
TV (channels 7–13)	174–216 MHz	6 MHz
TV (channels 14–83)	470–806 MHz	6 MHz

Table 1.3 UHF and SHF frequency bands.

Band name	Frequency range (GHz)
L	1–2
S	2–4
C	4–8
X	8–12.4
Ku	12.4–18
K	18–26.5
Ku	26.5–40
V	40–75
W	75–110

Table 1.4 Cellular phone frequency allocation.

System		Frequency range	Channel spacing (MHz)	Multiple access
AMPS	M-B	824–849 MHz	30	FDMA
	B-M	869–894 MHz	30	
GSM-900	M-B	880–915 MHz	0.2	TDMA/FDMA
	B-M	915–990 MHz	0.2	
GSM-1800	M-B	1710–1785 MHz	0.2	TDMA/FDMA
	B-M	1805–1880 MHz	0.2	
PCS-1900	M-B	1850–1910 kHz	30	TDMA
	B-M	1930–1990 kHz	30	
IS-54	M-B	824–849 MHz	30	TDMA
	B-M	869–894 MHz	30	
IS-136	M-B	1850–1910 MHz	30	TDMA
	B-M	1930–1990 MHz	30	
IS-96	M-B	824–849 MHz	30	CDMA
	B-M	869–894 MHz	30	
IS-96	M-B	1850–1910 MHz	30	CDMA
	B-M	1930–1990 MHz	30	

telephone transmission. The electronic circuits in the satellite are powered by solar energy using solar cells that deliver the supply power of about 1 kW. The combination of a transmitter and a receiver is called a transponder. A typical satellite has 12–24 transponders. Each transponder has a bandwidth of 36 MHz. The total time delay for GEO satellites is about 400 ms and the power of received signals is very low. Therefore, a low Earth orbit (LEO) satellite system was deployed for a mobile phone system. The orbits of LEO satellites are from 500 to 1500 km above the Earth. These satellites are not synchronized with the Earth's rotation. The total delay time for LEO satellite is about 250 ms.

1.15 Duplexing

In a two-way communication, a transmitter and a receiver are used. The combination of a transmitter and a receiver is called a *transceiver*. The block diagram of a transceiver is shown in Fig. 1.14. Duplexing techniques are used to allow for both users to transmit and receive signals. The most commonly used duplexing is called time-division duplexing (TDD). The same frequency channel is used for both transmitting and receiving signals, but the system transmits the signal for half of the time and receives for the other half.

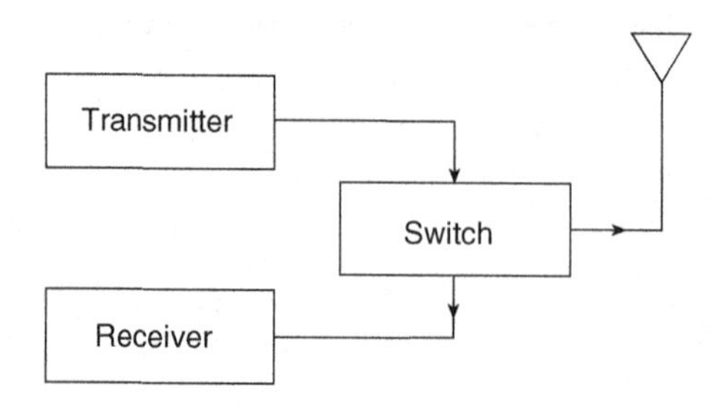

Figure 1.14 Block diagram of a transceiver.

1.16 Multiple-Access Techniques

In multiple-access communications systems, information signals are sent simultaneously over the same channel. Cellular wireless mobile communications use the following multiple-access techniques to allow simultaneous communication among multiple transceivers:

- Time-division multiple access (TDMA).
- Frequency-division multiple access (FDMA).
- Code-division multiple access (CDMA).

In the TDMA, the same frequency band is used by all the users but at different time intervals. Each digitally coded signal is transmitted only during preselected time intervals, called the time slots T_{SL}. During every time frame T_F, each user has access to the channel for a time slot T_{SL}. The signals transmitted from different users do not interfere with each other in the time domain.

In the FDMA, a frequency band is divided into many channels. The carrier frequency f_c determines the channel frequency. Each baseband signal is transmitted at a different carrier frequency. One channel is assigned to each user for a connection period. After the connection is completed, the channel becomes available to other users. In the FDMA, proper filtering must be carried out to provide channel selection. The signals transmitted from different users do not interfere with each other in the frequency domain. The GSM uses both TDMA and FDMA. The uplink is used for mobile transmission and the downlink is used for base station transmission. Each band is divided in 200 kHz slots. Each slot is shared between eight mobiles, each using it in turn. The bandwidth of square pulses is determined by their rise and fall times. Therefore, Gaussian pulse envelope waveforms are generated instead of rectangular pulses. The binary signal is processed by a Gaussian low-pass LP filter to reduce the rise and fall times of pulses, reducing the required bandwidth.

In the CDMA, each user uses a different code (similar to different language). CDMA allows one carrier frequency. Each station uses a different binary sequence to modulate the carrier. The signals transmitted from different users overlap in both the frequency and time domains, but the messages are orthogonal.

There are several standards of cellular wireless communications:

- Advanced Mobile Phone Service (AMPS).
- Global System for Mobile Communications (GSM).
- CDMA wireless standard proposed by Qualcomm.

1.17 Nonlinear Distortion in Transmitters

Linear amplification is required when the signal contains AM. Nonlinearity of amplifiers causes degradation of spectral purity and spectral regrowth. AM signals include multiple carriers. Power amplifiers contain a transistor (MOSFET, MESFET, or BJT), which is a nonlinear device operated under large-signal conditions. The drain current i_D is a nonlinear function of the gate-to-source voltage v_{GS}. Therefore, power amplifiers produce components, which are not present in the amplifier input signal. Another nonlinear mechanism is caused by the saturation of characteristics of the amplifier at a large amplitude of the output voltage. The relationship between the output voltage v_o and the input voltage v_s of a "weakly nonlinear" or a "nearly linear" power amplifier, such as the Class A amplifier, is nonlinear $v_o = f(v_s)$. This relationship can be expanded into Taylor's power series around the operating point Q as

$$v_o = f(v_s) = V_{O(DC)} + a_1 v_s + a_2 v_s^2 + a_3 v_s^3 + a_4 v_s^4 + a_5 v_s^5 + \cdots \tag{1.106}$$

Thus, the output voltage consists of an infinite number of nonlinear terms. Taylor's power series takes into account only the amplitude relationships. Volterra's power series includes both the amplitude and phase relationships.

Nonlinearity of amplifiers produces two types of unwanted signals:

- Harmonics of the carrier frequency $f_h = nf_c$.
- Intermodulation (IDM) products $f_{IDM} = nf_1 \pm mf_2$.

Nonlinear distortion components may corrupt the desired signal, causing errors in signal detection (reproduction) and splitter into adjacent channels. Harmonic distortion (HD) occurs when a single-frequency sinusoidal signal is applied to the power amplifier input. Intermodulation distortion (IMD) occurs when a signal consisting of two or more frequencies is applied at the power amplifier input. To evaluate the linearity of power amplifiers, we can use (1) a single-tone test and (2) and a two-tone test. In a single-tone test, a sinusoidal voltage source is used to drive a power amplifier. In a two-tone test, two sinusoidal sources of different frequencies and connected in series are used as a driver of a power amplifier. The first test will produce harmonics and the second test will produce both harmonics and intermodulation products (*IMP*s). Amplitude-to-phase conversion is caused by nonlinear capacitances. One measure of nonlinearity is the carrier-to-intermodulation (C/I) ratio, which should be 30 dBc or more for communication applications. The traditional measure of nonlinearity is the noise-to-power ratio (NPR).

1.18 Harmonics of Carrier Frequency

To investigate the process of generation of harmonics, let us assume that a power amplifier is driven by a single-tone excitation in the form of a sinusoidal voltage

$$v_s(t) = V_m \cos \omega t \tag{1.107}$$

To gain an insight into generation of harmonics by a nonlinear transmitter, consider an example of a memoryless time-invariant power amplifier modeled by a Taylor series taking a third-order polynomial

$$v_o(t) = a_0 + a_1 v_s(t) + a_2 v_s^2(t) + a_3 v_s^3(t) \tag{1.108}$$

The output voltage of the transmitter is given by

$$\begin{aligned} v_o(t) &= a_0 + a_1 V_m \cos \omega t + a_2 V_m^2 \cos^2 \omega t + a_3 V_m^3 \cos^3 \omega t \\ &= a_0 + a_1 V_m \cos \omega t + \frac{1}{2} a_2 V_m^2 (1 + \cos 2\omega t) + \frac{1}{4} a_3 V_m^3 (3 \cos \omega t + \cos 3\omega t) \\ &= a_0 + \frac{1}{2} a_2 V_m^2 + \left(a_1 V_m + \frac{3}{4} a_3 V_m^3 \right) \cos \omega t + \frac{1}{2} a_2 V_m^2 \cos 2\omega t + \frac{1}{4} a_3 V_m^3 \cos 3\omega t \end{aligned} \tag{1.109}$$

Thus, the output voltage of the power amplifier contains the fundamental components of the carrier frequency $f_1 = f_c$ as well as harmonics $2f_1 = 2f_c$ and $3f_1 = 3f_c, \ldots$, as shown in Fig. 1.15. The amplitude of the nth harmonic is proportional to V_m^n. This harmonics may interfere with other communication channels and must be filtered out to an acceptable low level.

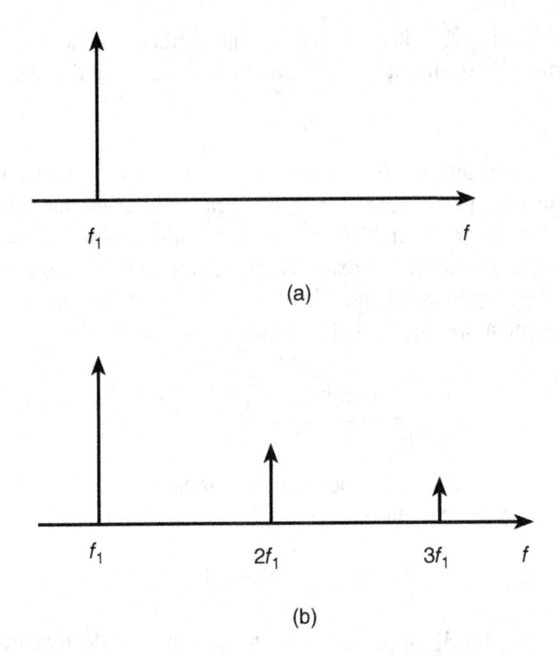

Figure 1.15 Spectrum of the input and output voltages of power amplifiers due to harmonics: (a) spectrum of the input voltage and (b) spectrum of the output voltage because of harmonics.

Even and odd terms produce the following components:

1. The even-order terms produce dc terms and even-order harmonics at $f_h = 2nf$. For example, the second harmonic is produced from the square term. In addition, extra dc terms are produced and the total dc bias increases with the ac signal amplitude V_m. It also varies when the amplitude of the input voltage V_m varies.
2. The odd-order terms produce signals at the fundamental frequency f and at the odd harmonic frequences $f_h = (n+1)f$. For example, the third-order term produces the fundamental component at f and the third harmonic at $3f$. Therefore, there is a nonlinear distortion of the fundamental component by the third-order term.

Let us consider a harmonic distortion in a MOSFET described by the square law. Assume that the total gate-to-source voltage waveform is

$$v_{GS} = V_{GS} + V_{gsm} \sin \omega t. \tag{1.110}$$

The drain current waveform is

$$\begin{aligned} i_D &= K(v_{GS} - V_t)^2 = K[(V_{GS} - V_t)^2 + 2(V_{GS} - V_t)V_{gsm} \sin \omega t + V_{gsm}^2 \sin^2 \omega t] \\ &= K[(V_{GS} - V_t)^2 + 2(V_{GS} - V_t)V_{gsm} \sin \omega t + \frac{1}{2}V_{gsm}^2(1 - \cos 2\omega t)] \\ &= K[(V_{GS} - V_t)^2 + \frac{1}{2}V_{gsm}^2 + 2K(V_{GS} - V_t)V_{gsm} \sin \omega t - \frac{1}{2}V_{gsm}^2 \cos 2\omega t \\ &= I_D + i_{d1} + i_{d2} \quad = I_D + I_{d1m} \sin \omega t - I_{d2} \cos 2\omega t. \end{aligned} \tag{1.111}$$

Hence, the distortion of the drain current waveform by the second harmonic is

$$HD_2 = \frac{I_{dm2}}{I_{dm1}} = \frac{V_{gsm}}{4(V_{GS} - V_t)} = 20 \log \left[\frac{V_{gsm}}{4(V_{GS} - V_t)}\right] = THD. \tag{1.112}$$

Harmonics are always integer multiples of the fundamental frequency. Therefore, the harmonic frequencies of the transmitter output signal with carrier frequency f_c are given by

$$f_h = nf_c \tag{1.113}$$

where $n = 2, 3, 4, \ldots$ is an integer. If a harmonic signal of a sufficiently large amplitude falls within the bandwidth of a nearby receiver, it may cause interference with the reception and cannot be filtered out in the receiver. The harmonics should be filtered out in the transmitter in its bandpass (BP) output network. For example, the output network of a transmitter may offer 37-dB second-harmonic suppression and 55-dB third-harmonic suppression.

The voltage gain of the amplifier at the fundamental frequency f_1 is

$$A_{v1} = \frac{v_{o1}}{v_s} = \frac{\left(a_1 + \frac{3}{4}a_3 V_m^2\right) V_m}{V_m} = a_1 + \frac{3}{4}a_3 V_m^2 \tag{1.114}$$

It can be seen that the voltage gain has not only the linear term a_1 but also an additional term proportional to the square of the input voltage amplitude V_m. In most amplifiers, $a_3 < 0$, yielding

$$A_{v1} = \frac{v_{o1}}{v_s} = a_1 - \frac{3}{4}|a_3| V_m^2 \tag{1.115}$$

and causing the voltage gain A_{v1} to decrease with V_m from the desired linear plot. Therefore, the upper part of the instantaneous output voltage tends to be reduced for large values of V_m. This phenomenon is known as *gain compression* or *amplifier saturation*. The linear part of the voltage gain is limited by the power supply voltage. The range of the input or the output voltage for which the voltage gain is linear is called the *dynamic range*.

Harmonic distortion is defined as the ratio of the amplitude of the nth harmonic V_n to the amplitude of the fundamental V_1

$$HD_n = \frac{V_n}{V_1} = 20 \log \left(\frac{V_n}{V_1}\right) \text{ (dB)} \tag{1.116}$$

The distortion by the second harmonic is given by

$$HD_2 = \frac{V_2}{V_1} = \frac{\frac{1}{2}a_2 V_m^2}{a_1 V_m + \frac{3}{4}a_3 V_m^3} = \frac{a_2 V_m}{2\left(a_1 + \frac{3}{4}V_m^2\right)} \tag{1.117}$$

For $a_1 \gg 3a_3 V_m^2/4$,

$$HD_2 \approx \frac{a_2 V_m}{2a_1} \tag{1.118}$$

The second-harmonic distortion HD_2 is proportional to the input voltage amplitude V_m.

The distortion by the third harmonic is given by

$$HD_3 = \frac{V_3}{V_1} = \frac{\frac{1}{4}a_3 V_m^3}{a_1 V_m + \frac{3}{4}a_3 V_m^3} = \frac{a_3 V_m^2}{4a_1 + 3a_3 V_m^2} \tag{1.119}$$

For $a_1 \gg 3a_3 V_m^2/4$,

$$HD_3 \approx \frac{a_3 V_m^2}{4a_1} \tag{1.120}$$

The third-harmonic distortion HD_3 is proportional to V_m^2. Usually, the amplitudes of harmonics should be -50 to -70 dB below the amplitude of the carrier.

The ratio of the power of an nth harmonic P_n to the carrier power P_c is

$$HD_n = \frac{P_n}{P_c} = 10\log\left(\frac{P_n}{P_c}\right) \text{ (dBc)} \tag{1.121}$$

The term "dBc" refers to the ratio of the power of a spectral distortion component P_n to the power of the carrier P_c.

The harmonic content in a waveform is described by the total harmonic distortion (THD), defined as

$$THD = \sqrt{\frac{P_2 + P_3 + P_4 + \cdots}{P_1}} = \sqrt{\frac{\frac{V_2^2}{2R} + \frac{V_3^2}{2R} + \frac{V_4^2}{2R} + \cdots}{\frac{V_1^2}{2R}}} = \sqrt{\frac{V_2^2 + V_3^2 + V_4^2 + \cdots}{V_1^2}}$$

$$a = \sqrt{\left(\frac{V_2}{V_1}\right)^2 + \left(\frac{V_3}{V_1}\right)^2 + \left(\frac{V_4}{V_1}\right)^2 + \cdots} = \sqrt{HD_2^2 + HD_3^2 + HD_4^2 + \cdots} \tag{1.122}$$

where $HD_2 = V_2/V_1, HD_3 = V_3/V_1, \ldots$ Higher harmonics ($n \geq 2$) are distortion terms.

1.19 Intermodulation Distortion

Intermodulation (IM) occurs when two or more signals of different frequencies are applied to the input of a nonlinear circuit, such as a nonlinear RF transmitter. This results in mixing the components of different frequencies. Therefore, the output signal contains components with additional frequencies, called *IMPs*. The frequencies of the IM products are either the sums or the differences of the carrier frequencies of input signals and their harmonics. For a two-frequency input excitation at frequencies f_1 and f_2, the frequencies of the output signal components are given by

$$f_{IMD} = nf_1 \pm mf_2 \tag{1.123}$$

where $n = 0, 1, 2, 3, \ldots$ and $m = 0, 1, 2, 3, \ldots$ are integers. The order of an IM product for a two-tone signal is the sum of the absolute values of the coefficients n and m given by

$$k_{IMD} = n + m \tag{1.124}$$

If the IM products of sufficiently large amplitudes fall within the bandwidth of a receiver, they will degrade the reception quality. For example, $2f_1 + f_2$, $2f_1 - f_2$, $2f_2 + f_1$, and $2f_2 - f_1$ are the third-order IM products. The third-order IM products usually have components in the system bandwidth. In contrast, the second-order harmonics $2f_1$ and $2f_2$ and the second-order IM products $f_1 + f_2$ and $f_1 - f_2$ are generally out of the system passband and are therefore not a serious problem.

A two-tone (two-frequency) excitation test is used to evaluate IM distortion of power amplifiers. Assume that the input voltage of a power amplifier consists of two sinusoids of equal amplitudes V_m and closely spaced frequencies f_1 and f_2

$$v_s(t) = V_m(\cos\omega_1 t + \cos\omega_2 t) = 2V_m \cos\left(\frac{\omega_2 - \omega_1}{2}\right)\cos\left(\frac{\omega_2 + \omega_1}{2}\right) \tag{1.125}$$

If the power amplifier is a memoryless time-invariant circuit described by a Taylor series using a third-order polynomial at the operating point Q, the output voltage waveform is given by

$$v_o(t) = a_0 + a_1 v_s(t) + a_2 v_s^2(t) + a_3 v_s^3(t)$$
$$= a_0 + a_1 V_m(\cos\omega_1 t + \cos\omega_2 t) + a_2 V_m^2(\cos\omega_1 t + \cos\omega_2 t)^2 + a_3 V_m^3(\cos\omega_1 t + \cos\omega_2 t)^3 + \cdots$$

$$= a_0 + a_1 V_m \cos\omega_1 t + a_1 V_m \cos\omega_2 t + \frac{1}{2}a_2 V_m^2(1+\cos 2\omega_1 t) + \frac{1}{2}a_2 V_m^2(1+\cos 2\omega_2 t)$$
$$+ a_2 V_m^2 \cos(\omega_2-\omega_1)t + a_1 V_m^2 \cos(\omega_1+\omega_2)t + a_3 V_m^3\left(\frac{3}{4}\cos\omega_1 t + \frac{1}{4}\cos 3\omega_1 t\right)$$
$$+ a_3 V_m^3\left(\frac{3}{4}\cos\omega_2 t + \frac{1}{4}\cos 3\omega_2 t\right)$$
$$+ a_3 V_m^3\left[\frac{3}{2}\cos\omega_1 t + \frac{3}{4}\cos(2\omega_2-\omega_1)t + \frac{3}{4}\cos(2\omega_2+\omega_1)t\right]$$
$$+ a_3 V_m^3\left[\frac{3}{2}\cos\omega_2 t + \frac{3}{4}\cos(2\omega_1-\omega_2)t + \frac{3}{4}\cos(2\omega_1+\omega_2)t\right]$$
$$+ \frac{1}{4}a_3 V_m^3(\cos 3\omega_1 t + \cos 3\omega_3 t) + \cdots$$
$$= a_0 + a_2 V_m^2 + \left(a_1 V_m + \frac{9}{4}a_3 V_m^3\right)\cos\omega_1 t + \left(a_1 V_m + \frac{9}{4}a_3 V_m^3\right)\cos\omega_2 t$$
$$+ \frac{1}{2}a_2 V_m^2(\cos 2\omega_1 t + \cos 2\omega_2 t) + a_2 V_m^2[\cos(\omega_2-\omega_1) + \cos(\omega_1+\omega_2)t]$$
$$+ \frac{3}{4}V_m^3\cos(2\omega_2-\omega_1) + \frac{3}{4}V_m^3\cos(2\omega_2+\omega_1) + \frac{1}{4}a_3 V_m^3(\cos 3\omega_1 t + \cos 3\omega_2 t) + \cdots \quad (1.126)$$

Now assume that the input signal of the power amplifier consists of two sinusoids of different amplitudes V_{m1} and V_{m2} and closely spaced frequencies f_1 and f_2

$$v_s(t) = V_{m1}\cos\omega_1 t + V_{m2}\cos\omega_2 t \quad (1.127)$$

If the power amplifier is a memoryless time-invariant circuit described by a Taylor series using a third-order polynomial, the output voltage waveform is given by

$$v_o(t) = a_0 + a_1 v_s(t) + a_2 v_s^2(t) + a_3 v_s^3(t)$$
$$= a_0 + a_1(V_{m1}\cos\omega_1 t + V_{m2}\cos\omega_2 t) + a_2(V_{m2}\cos\omega_1 t + V_{m2}\cos\omega_2 t)^2$$
$$+ a_3(V_{m1}\cos\omega_1 t + V_{m2}\cos\omega_2 t)^3$$
$$= a_0 + a_1 V_{m1}\cos\omega_1 t + a_1 V_{m2}\cos\omega_2 t + a_2 V_{m1}^2\cos^2\omega_1 t + 2a_2 V_{m1}V_{m2}\cos\omega_1 t\cos\omega_2 t$$
$$+ a_2 V_{m2}^2\cos\omega_2 t + a_3 V_{m1}^3\cos^3\omega_1 t + 3a_3 V_{m1}^2 V_{m2}\cos^2\omega_1 t\cos\omega_2 t$$
$$+ 3a_3 V_{m1}V_{m2}^2\cos\omega_1 t\cos^2\omega_2 t + a_3 V_{m2}^3\cos^3\omega_2 t \quad (1.128)$$

Thus,

$$v_o = a_0 + \frac{1}{2}a_2(V_{m1}^2 + V_{m2}^2) + \left(a_1 V_{m1} + \frac{3}{2}a_3 V_{m1}V_{m2}^2 + \frac{3}{4}a_3 V_{m1}^3\right)\cos\omega_1 t$$
$$+ \left(a_1 V_{m2} + \frac{3}{2}a_3 V_{m2}V_{m1}^2 + \frac{3}{4}a_3 V_{m2}^3\right)\cos\omega_2 t + \frac{1}{2}V_m^2\cos 2\omega_1 t + \frac{1}{2}V_m^2\cos 2\omega_2 t$$
$$+ a_2 V_{m1}V_{m2}\cos(\omega_2-\omega_1)t + a_2 V_{m1}V_{m2}\cos(\omega_1+\omega_2)t$$
$$+ \frac{3}{4}a_3 V_{m1}^2 V_{m2}\cos(2\omega_1-\omega_2)t + \frac{3}{4}a_3 V_{m1}^2 V_{m2}\cos(2\omega_1+\omega_2)t$$
$$+ \frac{3}{4}a_3 V_{m1}V_{m2}^2\cos(2\omega_2-\omega_1)t + \frac{3}{4}a_3 V_{m1}V_{m2}^2\cos(2\omega_2+\omega_1)t$$
$$+ \frac{1}{4}a_3 V_{m1}^3\cos 3\omega_1 t + \frac{1}{4}a_3 V_{m2}^3\cos 3\omega_2 t + \cdots \quad (1.129)$$

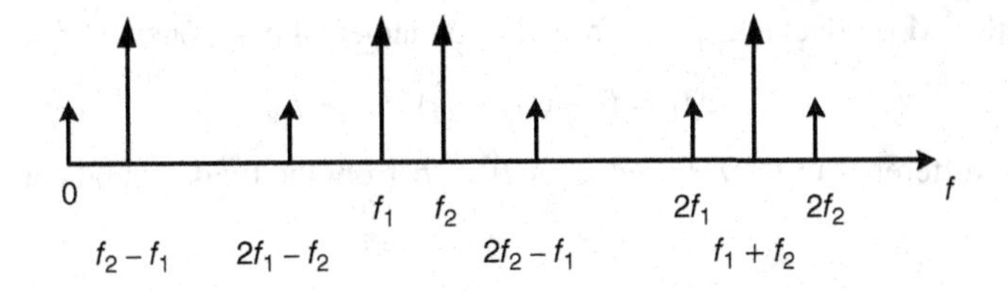

Figure 1.16 Spectrum of the output voltage in power amplifiers due to intermodulation for the two-tone input voltage with $f_2 > f_1$.

Figure 1.16 shows a spectrum of the amplifier output voltage due to intermodulation for two-tone input voltage with $f_2 > f_1$. The output voltage waveform v_o contains the following components:

- A dc component, causing a change in the dc operating current (bias current) of the transistor.
- Fundamental components f_1 and f_2.
- Harmonics of the fundamental components $2f_1, 2f_2, 3f_1, 3f_2, \ldots$
- IM products, which are linear combinations of the input frequencies f_1 and f_2: $nf_1 \pm mf_2$, where $m, n = 0, \pm 1, \pm 2, \pm 3, \ldots$ The IM frequencies are $f_2 - f_1$, $f_1 + f_2$, $2f_1 - f_2$, $2f_2 - f_1$, $2f_1 + f_2$, $2f_2 + f_1$, $3f_1 - 2f_2$, $3f_2 - 2f_1, \ldots$

If the difference between f_2 and f_1 is small, the IM products appear in the close vicinity of f_1 and f_2. The third-order IM products, which are at $2f_1 - f_2$ and $2f_2 - f_1$, are of the most interest because they are the closest to the fundamental components, as illustrated in Fig. 1.17. The frequency

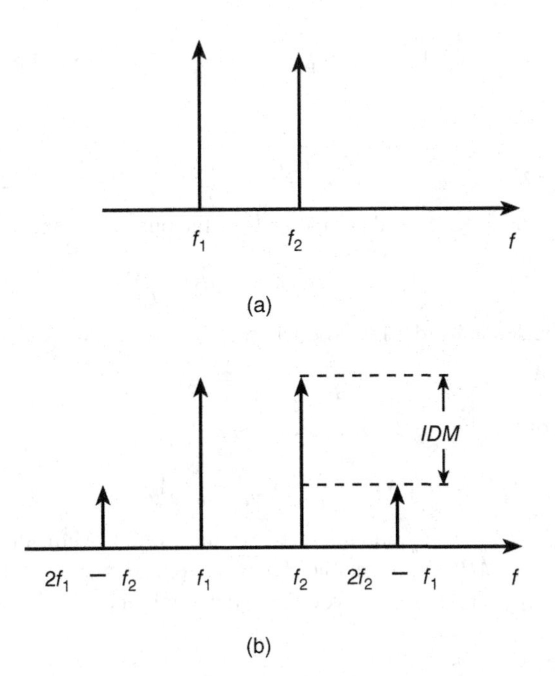

Figure 1.17 Spectrum of the input and output voltages in power amplifiers due to intermodulation: (a) spectrum of the input voltage and (b) several components of the spectrum of the output voltage due to intermodulation.

difference of the IM product at $2f_1 - f_2$ from the fundamental component at f_1 is

$$\Delta f_1 = f_1 - (2f_1 - f_2) = f_2 - f_1 \tag{1.130}$$

The frequency difference of the *IM* product at $2f_2 - f_1$ from the fundamental component at f_2 is

$$\Delta f_2 = (2f_2 - f_1) - f_2 = f_2 - f_1 \tag{1.131}$$

For example, if $f_1 = 800$ kHz and $f_2 = 900$ kHz, then the *IMP*s of particular interest are $2f_1 - f_2 = 2 \times 800 - 900 = 700$ MHz and $2f_2 - f_1 = 2 \times 900 - 800 = 1000$ MHz, resulting in $\Delta f_1 = f_1 - (2f_1 - f_2) = 800 - 700 = 100$ kHz and $\Delta f_2 = (2f_2 - f_1) - f_2 = 1000 - 900 = 100$ kHz. To filter out the unwanted IM components, filters with a very narrow bandwidth are required.

The output voltage at the input signal frequencies f_1 and f_2 is

$$v_{o(f_1,f_2)} = \left(a_1 V_{m1} + \frac{3}{2}a_3 V_{m1} V_{m2}^2 + \frac{3}{4}a_3 V_{m1}^3\right) \cos\omega_1 t + \left(a_1 V_{m2} + \frac{3}{2}a_3 V_{m2} V_{m1}^2 + \frac{3}{4}a_3 V_{m2}^3\right) \cos\omega_2 t \tag{1.132}$$

which for $V_{m1} = V_{m2} = V_m$ becomes

$$v_{o(f_1,f_2)} = \left(a_1 + \frac{9}{2}a_3 V_m^2\right) V_m(\cos\omega_1 t + \cos\omega_2 t) \tag{1.133}$$

The second-order IM products of the output voltage occur at $f_1 - f_2$ and $f_1 + f_2$ and are given by

$$v_{o(f_1-f_2,f_1+f_2)} = a_2 V_{m1} V_{m2}[\cos(\omega_2 - \omega_1)t + \cos(\omega_1 + \omega_2)t] \tag{1.134}$$

which for $V_{m1} = V_{m2} = V_m$ simplifies to the form

$$v_{o(f_1-f_2,f_1+f_2)} = a_2 V_m^2[\cos(\omega_2 - \omega_1)t + \cos(\omega_1 + \omega_2)t] \tag{1.135}$$

The third-order IM products of the output voltage occur at $2f_1 - f_2$ and $2f_2 - f_1$ and are expressed as

$$v_{o(2f_1-f_2,2f_2-f_1)} = \frac{3}{4}a_3 V_{m1}^2 V_{m2} \cos(2\omega_1 - \omega_2)t + \frac{3}{4}a_3 V_{m1} V_{m2}^2 \cos(2\omega_2 - \omega_1)t \tag{1.136}$$

yielding for $V_{m1} = V_{m2} = V_m$

$$v_{o(2f_1-f_2,2f_2-f_1)} = \frac{3}{4}a_3 V_m^3[\cos(2\omega_1 - \omega_2)t + \cos(2\omega_2 - \omega_1)t] \tag{1.137}$$

The amplitudes of the fundamental components of the output voltage are

$$V_{f_1} = V_{f_2} = \left(a_1 + \frac{9}{4}a_3 V_m^2\right) V_m \tag{1.138}$$

the amplitudes of the second-order IM product are

$$V_{f_2-f_1} = V_{f_1+f_2} = a_2 V_m^2 \tag{1.139}$$

and the amplitudes of the third-order IM product are

$$V_{2f_2-f_1} = V_{2f_1-f_2} = \frac{3}{4}a_3 V_m^3 \tag{1.140}$$

Assuming that $V_{m1} = V_{m2} = V_m$ and $a_3 \ll a_1$, the third-order IM distortion by the *IMP* component at $2f_1 \pm f_2$ or the *IMP* component at $2f_2 \pm f_1$ is defined as a ratio of the amplitude of the third-order IM component of the output voltage to the amplitude of the fundamental component of the output voltage

$$IM_3 = \frac{V_{2f_2-f_1}}{V_{f_2}} = \frac{\frac{3}{4}a_3 V_m^3}{\left(a_1 + \frac{9}{4}a_3 V_m^2\right) V_m} = \frac{\frac{3}{4}a_3 V_m^2}{a_1 + \frac{9}{4}a_3 V_m^2} \approx \frac{\frac{3}{4}a_3 V_m^2}{a_1} = \frac{3}{4}\left(\frac{a_3}{a_1}\right) V_m^2 \tag{1.141}$$

for $a_1 \gg \frac{9}{4}a_3V_m^2$. Similarly,

$$IM_2 = \frac{a_2V_m^2}{\left(a_1 + \frac{9}{4}a_3V_m^2\right)V_m} = \frac{a_2V_m}{a_1 + \frac{9}{4}a_3V_m^2} \approx \frac{a_2}{a_1}V_m \tag{1.142}$$

for $a_1 \gg \frac{9}{4}a_3V_m^2$. As the amplitude of the input voltage V_m is increased, the amplitudes of the fundamental components V_{f_1} and V_{f_2} are directly proportional to V_m, whereas the amplitudes of the third-order IM products $V_{2f_1-f_2}$ and $V_{2f_2-f_1}$ are proportional to V_m^3. Therefore, the amplitudes of the IM products increase three times faster on log-log scale than the amplitudes of the fundamentals and have an intersection point.

If the amplitudes are drawn on a log-log scale, they are linear functions of V_m. Usually, the coefficient a_3 is negative, causing the compression of the input voltage versus output voltage characteristic and reducing the amplifier voltage gain. The compression of the voltage gain leads to the compression of the power gain and the characteristic of $P_O = f(P_G)$ saturates. The 1-dB compression point indicates the maximum power value of the linear dynamic power range.

In general, the amplitude of input voltage at which the extrapolated amplitude of the desired output voltage and the nth-order IM product are equal is the nth-order intercept point. Thus, $IM_n = 1$. For the second-order IM component,

$$IM_2 = \frac{a_2}{a_1}V_m = 1 \tag{1.143}$$

which produces

$$V_m = \frac{a_2}{a_1} \tag{1.144}$$

Similarly, for the third-order IM component,

$$IM_3 = \frac{3}{4}\frac{a_2}{a_1}V_m^2 = 1 \tag{1.145}$$

yielding

$$V_m = \sqrt{\frac{4}{3}\frac{a_2}{a_1}} \tag{1.146}$$

Nonlinearities produce power in signal bandwidth. This effect is characterized by the notch power to the total signal power ratio

$$NPR = \frac{\text{Notch power}}{\text{Total signal power}} \tag{1.147}$$

To determine NPR, a power amplifier is driven by Gaussian noise through a notch filter in a portion of its bandwidth.

The effect of the nonlinearities on the adjacent channels is described by the adjacent channel power ratio (ACPR)

$$ACPR = \frac{\text{Power in band outside signal bandwidth}}{\text{Signal power}} = \frac{P_{adjacent-channels}}{P_O} = \frac{\int_{LS} P_O(f)df + \int_{US} P_O(f)df}{\int P_O(f)df} \tag{1.148}$$

The ACPR may be specified for either lower or upper sideband.

1.20 AM/AM Compression and AM/PM Conversion

Assume an AM single-tone (single-frequency) input voltage of an amplifier

$$v_s(t) = V_m(t)\cos\omega t \tag{1.149}$$

In real amplifiers, the voltage gain is a function of the amplitude of the input voltage $k_p(t) = k_p[V_{sm}(t)]$, resulting in an instantaneous voltage gain. In addition, the amplifier transistor contains nonlinear capacitances, dependent on the magnitudes of the transistor voltages. Therefore, the output voltage is

$$v_o(t) = k_p(t)v_s(t) = k_p[V_m(t)]\cos\{\omega t + \phi[V_m(t)]\} \tag{1.150}$$

This results in an AM/AM compression and an AM/PM conversion.

1.21 Dynamic Range of Power Amplifiers

For an ideal amplifier, a plot of the output power as a function of the input power should be a perfect straight line, $P_O = k_p P_i$, where k_p is the power gain and should be constant. Figure 1.18 shows the desired output power $P_O(f_2)$ and the undesired third-order IM product output power $P_O(2f_2 - f_1)$ as functions of the input power P_i on a log-log scale. This characteristic exhibits a linear region and a nonlinear region. As the input power P_i increases, the output power first increases proportionally to the input power and then reaches saturation, causing *power gain compression*. The point at which the power gain of the nonlinear amplifier deviates from that of the ideal linear amplifier by 1 dB is called the *1-dB compression point*. It is used as a measure of the power-handling capability of the power amplifier. The output power at the 1-dB compression point is given by

$$P_{O(1\text{dB})} = k_{p(1\text{dB})} + P_{i(1\text{dB})}\ (\text{dBm}) = k_{po(1\text{dB})} - 1\ \text{dB} + P_{i(1\text{dB})}\ (\text{dBm}) \tag{1.151}$$

where k_{po} is the power gain of an ideal linear amplifier and $k_{p(1\text{dB})}$ is the power gain at the 1-dB compression point.

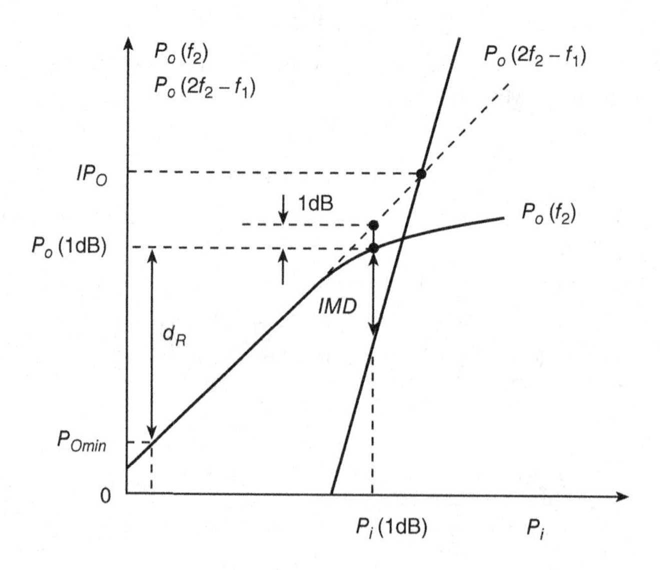

Figure 1.18 Output power $P_O(f_2)$ and $P_O(2f_2 - f_1)$ as functions of input power P_i of power amplifier.

The *dynamic range* of a power amplifier is the region where the amplifier has a linear (i.e., fixed) power gain. It is defined as the difference between the output power $P_{O(1\text{dB})}$ and the minimum detectable power P_{Omin}

$$d_R = P_{O(1\text{dB})} - P_{Omin} \tag{1.152}$$

where the minimum detectable power P_{Omin} is defined as an output power level x dB above the input noise power level P_{On}, usually $x = 3$ dB.

In a linear region of the amplifier characteristic, the desired output power $P_O(f_2)$ is proportional to the input power P_i, for example, $P_O(f_2) = aP_i$. Assume that the third-order IM product output power $P_O(2f_2 - f_1)$ increases proportionally to the third power, for example, $P_O(2f_2 - f_1) = (a/8)^3 P_i$. Projecting the linear region of $P_O(f_2)$ and $P_O(2f_2 - f_1)$ results in an intersection point called the *intercept point* (IP), as shown in Fig. 1.18, where the output power at the IP point is denoted by IP_O.

The IM product is the difference between the desired output power $P_O(f_2)$ and the undesired output power of the IM component $P_O(2f_2 - f_1)$ of a power amplifier

$$IMD = P_O(f_2)\ (\text{dBm}) - P_O(2f_2 - f_1)\ (\text{dBm}) \tag{1.153}$$

Signals with constant envelopes, such as CW, FM, FSK, and GSM, do not require linear amplification.

1.22 Analog Modulation

A *message signal* $v_m(t)$ is usually an LP signal. A signal whose spectrum is in the vicinity $f = 0$ is usually called a *baseband signal*. A baseband signal is usually voice, video, or digital data signal. Modulation converts a message signal $v_m(t)$ from an LP spectrum to a high-pass (HP) spectrum (usually to a BP spectrum) around a carrier frequency f_c. In general, a modulated signal can be described as

$$v(t) = A(t)\cos[2\pi f(t)t + \phi(t)] \tag{1.154}$$

where the amplitude $A(t)$, the frequency $f(t)$, and the phase $\phi(t)$ are modulated. Using the relationship $\cos(\alpha + \beta) = \cos\alpha\cos\beta - \sin\alpha\sin\beta$, any narrow band signal can be presented as simultaneous AM and PM waveform

$$\begin{aligned} v_{RF}(t) &= A(t)\cos[2\pi ft + \phi(t)] = A(t)\cos\phi(t)\cos 2\pi ft - A(t)\sin\phi(t)\sin 2\pi ft \\ &= I(t)\cos 2\pi ft - Q(t)\sin 2\pi ft \end{aligned} \tag{1.155}$$

where

$$I(t) = A(t)\cos\phi(t) \tag{1.156}$$

and

$$Q(t) = A(t)\sin\phi(t) \tag{1.157}$$

$$A(t) = \sqrt{I^2(t) + Q^2(t)} = \sqrt{A^2(t)[\sin^2\phi(t) + \cos^2\phi(t)]} \tag{1.158}$$

and

$$\phi(t) = \arctan\left[\frac{Q(t)}{I(t)}\right] \tag{1.159}$$

The function of a communication system is to transfer information from one point to another through a communication link. Block diagrams of a typical communication system are depicted in Fig. 1.19. The system consists of an upconversion transmitter whose block diagram is shown in Fig. 1.19(a) and a downconversion receiver whose block diagram is shown in Fig. 1.19(b).

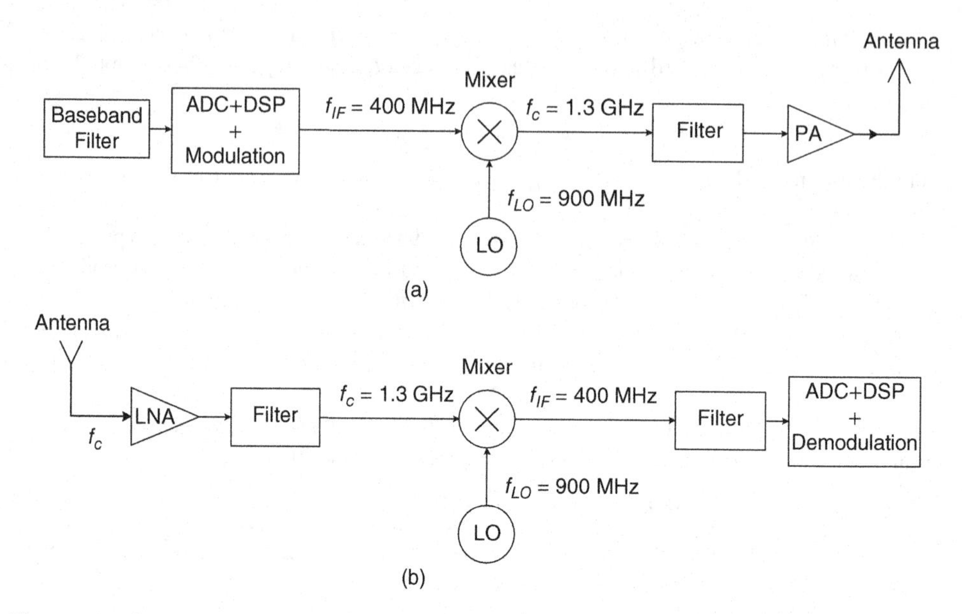

Figure 1.19 Block diagrams of a transmitter and a receiver. (a) Block diagram of an upconversion transmitter. (b) Block diagram of a downconversion receiver.

In a transmitting system, an RF signal is generated, amplified, modulated, and applied to the antenna. A local oscillator generates a signal with a frequency f_{LO}. The signals with an intermediate frequency f_{IF} and the local-oscillation frequency f_{LO} are applied to a mixer. The frequency of the output signal of the mixer and a BP filter is increased (upconversion) from the intermediate frequency f_{IF} to a carrier frequency f_c by adding the local-oscillation frequency f_{LO}

$$f_c = f_{LO} + f_{IF} \tag{1.160}$$

The RF current flows through the antenna and produces EMWs. Antennas produce or collect EM energy. The transmitted signal is received by the antenna, amplified by a LNA, and applied to a mixer. The frequency of the output signal of the mixer and a BP filter is reduced (downconversion) from the carrier frequency f_c to an intermediate frequency f_{IF} by subtracting the local-oscillation frequency f_{LO}

$$f_{IF} = f_c - f_{LO} \tag{1.161}$$

The most important parameters of a transmitter are as follows:

- Spectral efficiency.
- Power efficiency.
- Signal quality in the presence of noise and interference.

A "baseband" signal (modulating signal or information signal) has a nonzero spectrum in the vicinity of $f = 0$ and negligible elsewhere. For example, a voice signal generated by a microphone or a video signal generated by a TV camera are baseband signals. The modulating signal may consist of many, for example, 24 multiplexed telephone channels. In RF systems with analog modulation, the carrier is modulated by an analog baseband signal. The frequency bandwidth occupied by the baseband signal is called the baseband. The modulated signal consists of components with much higher frequencies than the highest baseband frequency. The modulated signal is an RF signal. It consists of components whose frequencies are very close to the frequency of the carrier.

RF modulated signals can be divided into two groups:

- Variable-envelope signals.
- Constant-envelope signals.

Amplification of variable-envelope signals requires linear amplifiers. On the contrary, constant-envelope signals may be amplified by nonlinear amplifiers. However, they require flat frequency response characteristics.

Modulation is the process of placing an information band around a HF carrier for transmission. Modulation conveys information by changing some aspects of a carrier signal in response to a modulating signal. In general, a modulated output voltage is given by

$$v_o(t) = A(t)\cos[2\pi f(t)t + \theta(t)] \tag{1.162}$$

where $A(t)$ is the amplitude of the voltage, and $\theta(t)$ is the phase of the carrier. If the amplitude of the output voltage $A(t)$ is varied with time, it is called amplitude modulation (AM). If the carrier frequency of $f(t)$ is varied with time, it is called frequency modulation (FM). If the phase of $\theta(t)$ is varied, it is called phase modulation (PM). In systems with analog modulation, $A(t), f_c(t)$, and $\theta(t)$ are continuous functions of time. In systems with digital modulation, $A(t), f_c(t)$, and $\theta(t)$ are discrete functions of time.

1.22.1 Amplitude Modulation

In AM, the carrier envelope is varied in proportion to the modulating signal $v_m(t)$. The carrier voltage is usually a sine wave

$$v_c(t) = V_c \cos\omega_c t \tag{1.163}$$

In general, the AM signal is

$$v_{AM}(t) = V(t)\cos\omega_c t = V_c[1 + \alpha(t)]\cos\omega_c t = V_c\cos\omega_c t + \alpha(t)V_c\cos\omega_c t \tag{1.164}$$

where the envelope is

$$V(t) = V_c[1 + \alpha(t)] \tag{1.165}$$

When the modulating voltage is a single-frequency sinusoid

$$v_m(t) = V_m\cos\omega_m t \tag{1.166}$$

the envelope is given by

$$V(t) = V_c + v_m(t) = V_c + V_m\cos\omega_m t = V_c\left(1 + \frac{V_m}{V_c}\cos\omega_m t\right) = V_c(1 + m\cos\omega_m t) \tag{1.167}$$

In this case, the AM voltage is

$$\begin{aligned} v_{AM}(t) = v_o(t) &= V(t)\cos\omega_c t = V_c\cos\omega_c t + v_m(t)\cos\omega_c t = [V_c + v_m(t)]\cos\omega_c t \\ &= V_c\cos\omega_c t + V_m\cos\omega_m t\cos\omega_c t = (V_c + V_m\cos\omega_m t)\cos\omega_c t \\ &= V_c\left(1 + \frac{V_m}{V_c}\cos\omega_m t\right)\cos\omega_c t = V_c(1 + m\cos\omega_m t)\cos\omega_c t \end{aligned} \tag{1.168}$$

where the *modulation index* (or the *modulation depth*) is

$$m = \frac{V_m}{V_c} = \frac{[V_{om(max)} - V_{om(min)}]/2}{[V_{om(max)} + V_{om(min)}]/2} = \frac{V_{om(max)} - V_{om(min)}}{V_{om(max)} + V_{om(min)}} \leq 1 \qquad (1.169)$$

and

$$\alpha(t) = \frac{V_m}{V_c} \cos \omega_m t = m \cos \omega_m t \qquad (1.170)$$

The instantaneous amplitude $V(t)$ is directly proportional to the modulating voltage $v_m(t)$. At $\cos \omega_m t = 1$, $m = m_{max} = 1$, yielding

$$V_{o(MAX)} = V_c(1 + m_{max}) = V_c(1 + 1) = 2V_c \qquad (1.171)$$

Figure 1.20 shows waveforms of modulating signal $v_m(t)$, carrier signal $v_c(t)$, and AM signal $v_{AM}(t)$ at $m = 0.8, f_m = 1$ kHz, and $f_c = 10$ kHz. For $m > 1$, the signal is overmodulated. When $V_m > V_c$, *overmodulation* occurs, causing distortion of the modulated signal. Figures 1.21–1.23 shows the waveforms of AM signals at $m = 0.5, 1,$ and 2.

Applying the trigonometric identity,

$$\cos \omega_c t \cos \omega_m t = \frac{1}{2}[\cos(\omega_c - \omega_m)t + \cos(\omega_c + \omega_m)t] \qquad (1.172)$$

we can express the AM signal as

$$\begin{aligned} v_{AM}(t) = v_o(t) &= V_c \cos \omega_c t + mV_c \cos \omega_m t \cos \omega_c t \\ &= V_c \cos \omega_c t + \frac{V_m}{2} \cos(\omega_c - \omega_m)t + \frac{V_m}{2} \cos(\omega_c + \omega_m)t \\ &= V_c \cos \omega_c t + \frac{mV_c}{2} \cos(\omega_c - \omega_m)t + \frac{mV_c}{2} \cos(\omega_c + \omega_m)t \end{aligned} \qquad (1.173)$$

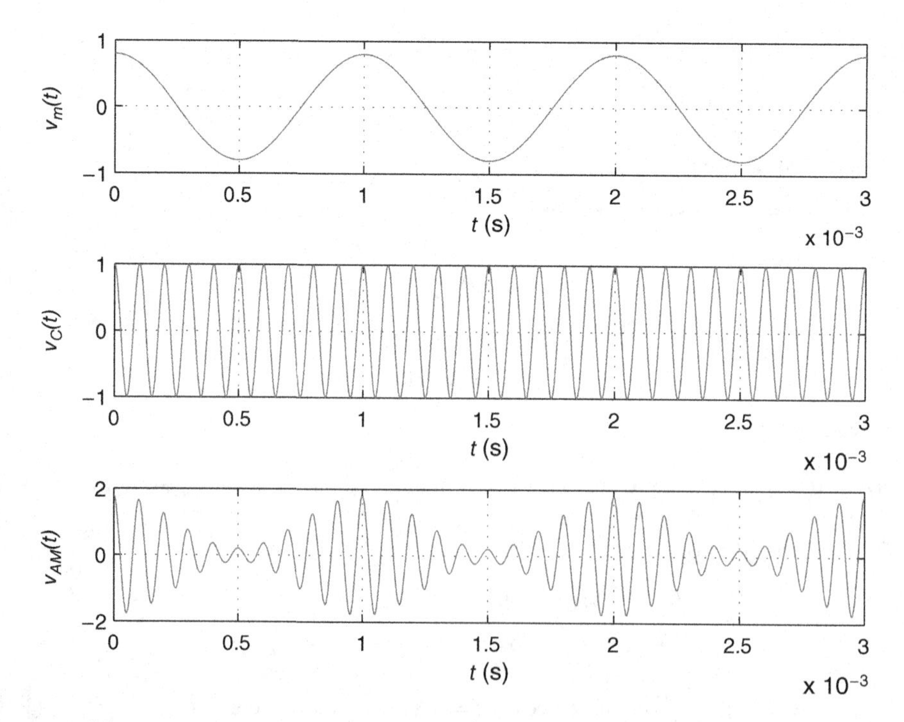

Figure 1.20 Waveforms of modulating signal $v_m(t)$, carrier signal $v_c(t)$, and AM signal $v_{AM}(t)$ at $m = 0.8, f_m = 1$ kHz, and $f_c = 10$ kHz.

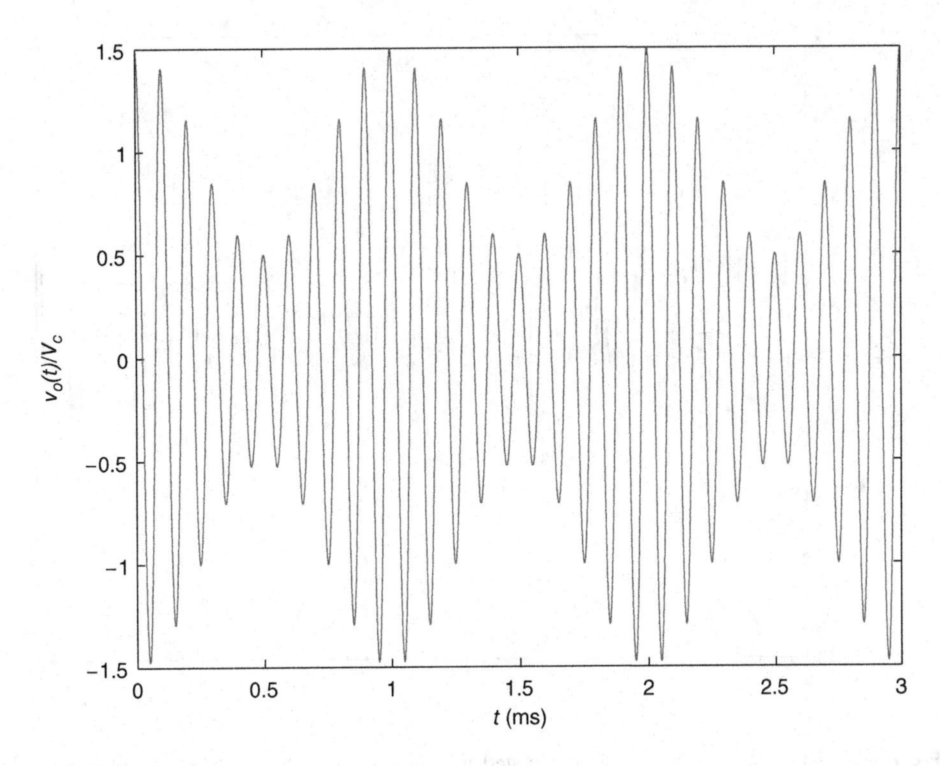

Figure 1.21 AM signal $v_o(t)/V_c$ modulated with a single sinusoid f_m at $m = 0.5$.

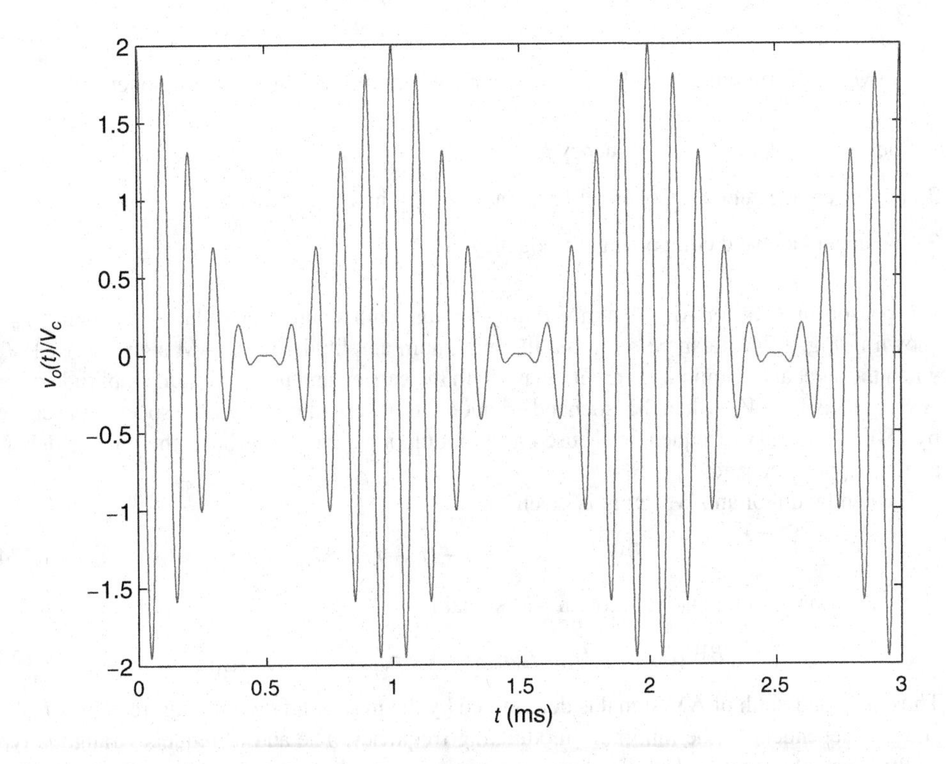

Figure 1.22 AM signal $v_o(t)/V_c$ modulated with a single sinusoid f_m at $m = 1$.

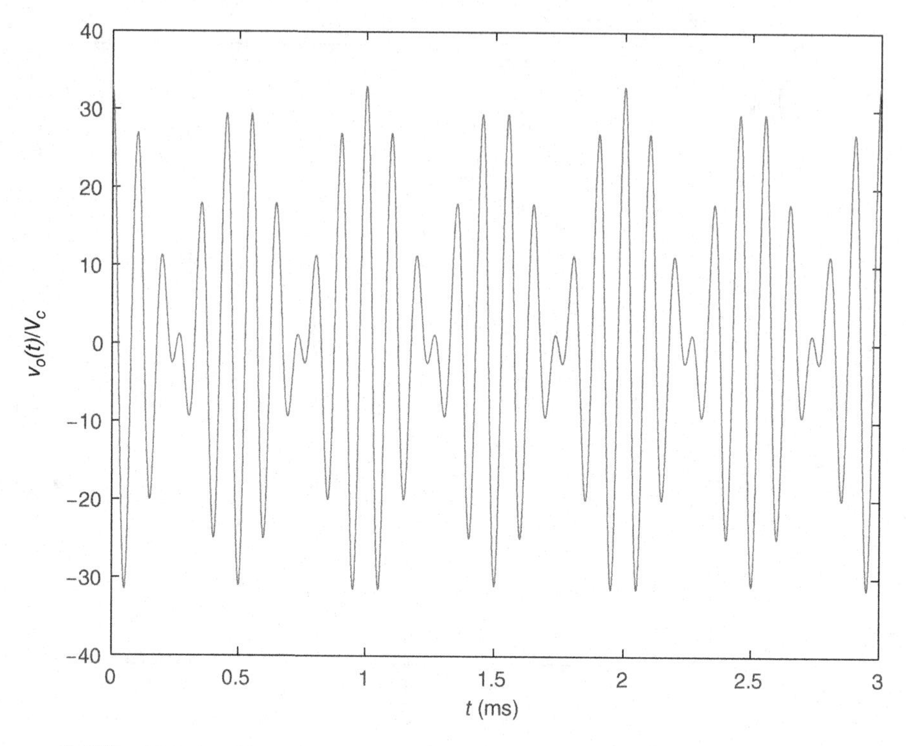

Figure 1.23 AM signal $v_o(t)/V_c$ modulated with a single sinusoid f_m at $m = 2$ (overmodulated signal).

The AM waveform modulated by a single tone consists of the following components:

1. the carrier component at frequency f_c,
2. the lower sideband component at frequency $f_c - f_m$, and
3. the upper sideband component at frequency $f_c + f_m$.

A phasor diagram for AM by a modulating voltage with a single-modulating frequency f_m is shown in Fig. 1.24. It can be seen that the maximum amplitude of the AM signal is $V_c + V_m$, when the sideband components are in phase with the carrier, and the minimum amplitude of the AM signal is $V_c - V_m$, when the sideband components are out of phase with respect to the carrier by 180°. If a band of frequencies is used as a modulating signal, we obtain the lower sideband and the upper sideband.

The bandwidth of an AM signal is given by

$$BW_{AM} = (f_c + f_m) - (f_c - f_m) = 2f_m \tag{1.174}$$

Hence, the maximum bandwidth of an AM signal is

$$BW_{AM(max)} = (f_c + f_{m(max)}) - (f_c - f_{m(max)}) = 2f_{m(max)} \tag{1.175}$$

Thus, the bandwidth of AM signal is determined by the maximum modulating frequency $f_{m(max)}$ and is independent of the minimum modulating frequency. The audio frequency range is typically from 20 Hz to 20 kHz. The frequency range of the human voice from 100 to 3000 Hz contains about 95% of the total energy. The carrier frequency f_c is much higher than $f_{m(max)}$,

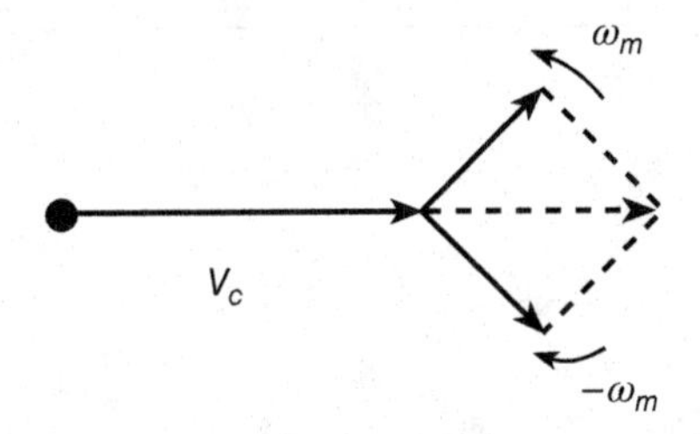

Figure 1.24 Phasor diagram for amplitude modulation by a single modulating frequency.

for example, $f_c/f_{m(max)} = 200$. The AM broadcasting is in the frequency range from 335 to 1605 kHz. It consists of 107 channels. The bandwidth of each channel is 10 kHz.

The time-average power of the carrier is

$$P_C = \frac{V_c^2}{2R} \tag{1.176}$$

The amplitude of modulating voltage is

$$V_m = \frac{mV_c}{2} \tag{1.177}$$

The time-average power of the lower sideband P_{LS} over the cycle of the modulating period $T_m = 1/f_m$ is equal to the average power of the upper sideband P_{US}

$$P_{LS} = P_{US} = \frac{V_m^2}{2R} = \frac{m^2 V_c^2}{8R} = \frac{m^2}{4} P_C \tag{1.178}$$

The total average power of the AM signal is given by

$$P_{AM} = P_C + P_{LS} + P_{US} = P_C + P_m = \left(1 + \frac{m^2}{4} + \frac{m^2}{4}\right) P_C = \left(1 + \frac{m^2}{2}\right) P_C \tag{1.179}$$

where

$$P_m = P_{LS} + P_{US} = \frac{m^2}{2} P_C \tag{1.180}$$

For $m = 1$, the total average power of the AM signal is

$$P_{AMmax} = P_C + P_{LS} + P_{US} = \left(1 + \frac{1}{4} + \frac{1}{4}\right) P_C = \frac{3}{2} P_C \tag{1.181}$$

The typical modulation index is $m = 0.25$, yielding the typical average power of the AM signal

$$P_{AM(typ)} = \left(1 + \frac{m^2}{2}\right) P_C = \left(1 + \frac{0.25^2}{2}\right) P_C = 1.03125 P_C \tag{1.182}$$

The instantaneous output power of the AM voltage modulated by a single sinusoidal voltage is

$$p_{o(AM)}(t) = \frac{v_o^{2(t)}}{2R} = \frac{V_m^2(t)\cos^2\omega_c t}{2R} = \frac{V_c^2(1 + m\cos\omega_m t)^2 \cos^2\omega_c t}{2R}$$
$$= P_C(1 + m\cos\omega_m t)^2 \cos^2\omega_c t \tag{1.183}$$

The PEP is

$$P_{PEP} = P_C(1 + m)^2 \tag{1.184}$$

yielding the maximum PEP

$$P_{PEP(max)} = P_c(1 + m_{max})^2 = P_c(1 + 1)^2 = 4P_C \tag{1.185}$$

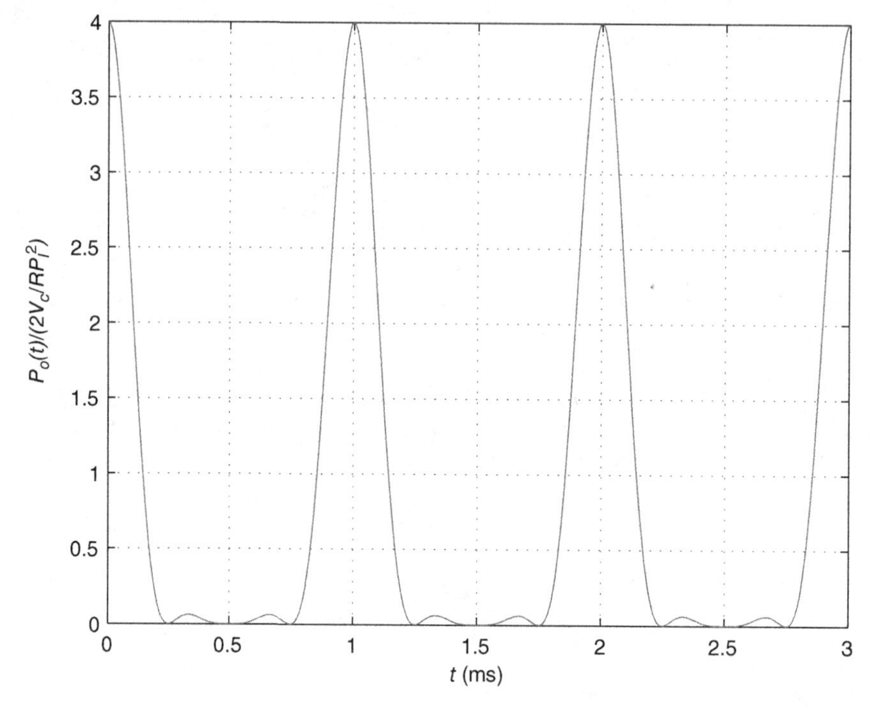

Figure 1.25 Instantaneous power of the AM signal $P_o(t)$ modulated with a single sinusoid f_m at $m = 1$.

Figure 1.25 shows the instantaneous power of the AM signal $P_o(t)$ modulated with a single sinusoid f_m at $m = 1$.

The efficiency of the AM is

$$\eta_{AM} = \frac{m^2}{2 + m^2} \tag{1.186}$$

resulting in

$$\eta_{AM(max)} = \frac{m^2}{2 + m^2} = \frac{m_{max}^2}{2 + m_{max}^2} = \frac{m_{max}^2}{2 + m_{max}^2} = \frac{1}{2 + 1} = \frac{1}{3} \tag{1.187}$$

For $m > 1$, $V_m > V_c$ and *overmodulation* of the AM signal occurs, causing distortion of the envelope. In this case, the shape of the detected signal in the receiver is different from the modulating signal.

Random electrical variations add to the AM signal and alter the original envelope of modulated signal. This is the most important disadvantage of AM systems.

A high-power AM voltage may be generated using the following methods:

- Amplification of the AM signal using a linear RF power amplifier.
- Drain amplitude modulation.
- Gate-to-source bias operating point modulation.

Variable-envelope signals, such as AM signals, usually require linear power amplifiers. Figure 1.26 shows the amplification process of an AM signal by a linear power amplifier. In this case, the carrier and the sidebands are amplified by a linear amplifier. Drain AM signal can be generated by connecting a modulating voltage source in series with the drain dc voltage

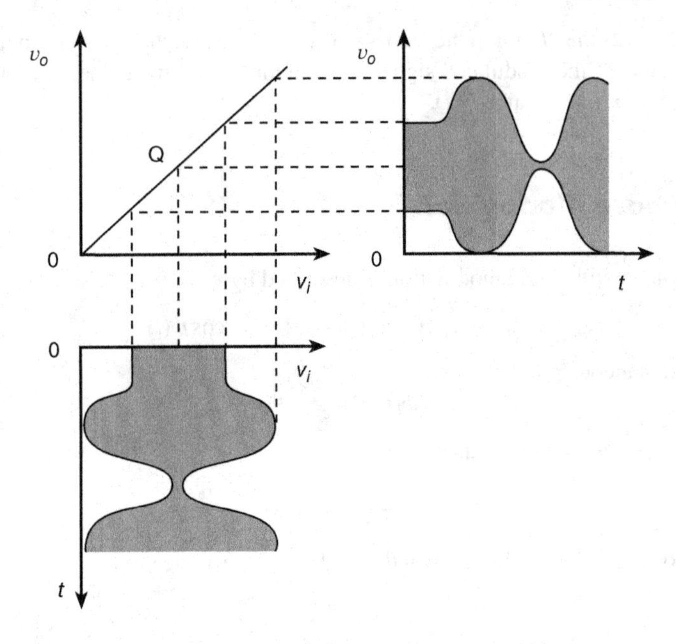

Figure 1.26 Amplification of AM signal by linear power amplifier.

supply V_I. When the MOSFET is operated as a current source such as in Class C power amplifiers, it should be driven into the ohmic region. If the transistor is operated as switch, the amplitude of the output voltage is proportional to the supply voltage and a high-fidelity AM signal is achieved. However, the modulating signal v_m must be amplified to a high power level before it is applied to modulate the supply voltage.

AM is used in commercial radio broadcasting and to transmit the video information in analog TV. Variable-envelope signals are also used in modern wireless communication systems.

Quadrature amplitude modulation (QAM) is a modulation method, which is based on modulating the amplitude of two orthogonal carrier sinusoidal waves. The two waves are out of phase with each other by 90° and, therefore, are called quadrature carriers. The QAM signal (I/Q signal) is expressed as in the Cartesian (or rectangular) form

$$v_{QAM} = I(t) \cos \omega_c t + Q(t) \sin \omega_c t \tag{1.188}$$

where $I(t)$ and $Q(t)$ are modulating signals. The QAM signal can also be expressed in the polar form

$$v_{QAM} = A(t) e^{j\phi(t)} \tag{1.189}$$

where the time-varying amplitude is

$$A(t) = \sqrt{[I(t)]^2 + [Q(t)]^2} \tag{1.190}$$

and the time-varying phase is

$$\phi(t) = \arctan \left[\frac{Q(t)}{I(t)} \right] \tag{1.191}$$

The received signal can be demodulated by multiplying the modulated signal v_{QAM} by a cosine wave of carrier frequency f_c

$$\begin{aligned} v_{DEM} &= v_{QAM} \cos \omega_c t = I(t) \cos \omega_c t \cos \omega_c t + Q(t) \sin \omega_c t \cos \omega_c t \\ &= \frac{1}{2} I(t) + \frac{1}{2} [I(t) \cos(2\omega_c t) + Q(t) \sin(2\omega_c t)] \end{aligned} \tag{1.192}$$

An LP filter removes the $2f_c$ components, leaving only the $I(t)$ term, which is uneffected by $Q(t)$. On the other hand, if the modulated signal v_{QAM} is multiplied by a sine wave and transmitted through an LP filter, one obtains $Q(t)$.

1.22.2 Phase Modulation

The output voltage with angle modulation is described by

$$v_o = V_c \cos[\omega_c t + \theta(t)] = V_c \cos \phi(t) \tag{1.193}$$

where the instantaneous phase is

$$\phi(t) = \omega_c t + \theta(t) \tag{1.194}$$

and the instantaneous angular frequency is

$$\omega(t) = \frac{d\phi(t)}{dt} = \omega_c + \frac{d\theta(t)}{dt} \tag{1.195}$$

Depending on the relationship between $\theta(t)$ and $v_m(t)$, the angle modulation can be categorized as follows:

- Phase modulation (PM).
- Frequency modulation (FM).

In PM systems, the phase of the carrier is changed by the modulating signal $v_m(t)$. In FM systems, the frequency of the carrier is changed by the modulating signal $v_m(t)$. The angle-modulated systems are inherently immune to amplitude fluctuations due to noise. In addition to the high degree of noise immunity, they require less RF power because the transmitter output power contains only the power of the carrier. Therefore, PM and FM systems are suitable for high-fidelity music broadcasting and for mobile radio wireless communications. However, the bandwidth of an angle-modulated signal is much wider than that of an AM signal.

The modulating voltage, also called the message signal, is expressed as

$$v_m(t) = V_m \sin \omega_m t \tag{1.196}$$

The phase of PM signal is given by

$$\theta(t) = k_p v_m(t) = k_p V_m \sin \omega_m t = m_p \sin \omega_m t \tag{1.197}$$

where the PM index is

$$m_p = k_p V_m \tag{1.198}$$

Hence, the PM output voltage is

$$v_{PM}(t) = v_o(t) = V_c \cos[\omega_c t + k_p v_m(t)] = V_c \cos(\omega_c t + k_p V_m \sin \omega_m t)$$
$$= V_c \cos(\omega_c t + m_p \sin \omega_m t) = V_c \cos(\omega_c t + \Delta\phi \sin \omega_m t) = V_c \cos \phi(t) \tag{1.199}$$

where the *index of phase modulation* is the maximum phase shift caused by the modulating voltage V_m.

$$m_p = k_p V_m = \Delta\phi \tag{1.200}$$

Hence, the maximum PM is

$$\Delta\phi_{max} = k_p V_{m(max)} \tag{1.201}$$

Since

$$\phi(t) = \omega_c t + \theta(t) \tag{1.202}$$

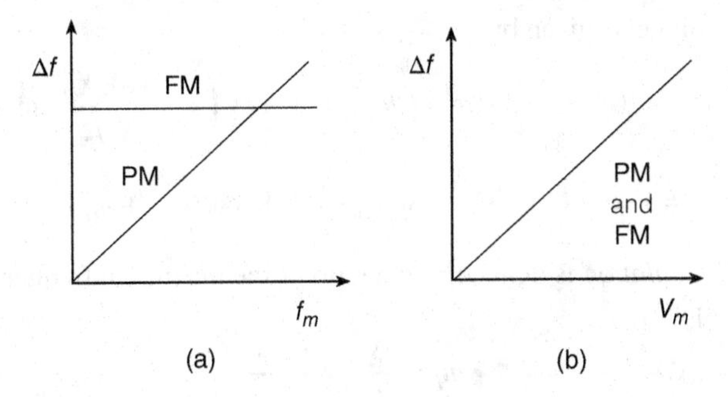

Figure 1.27 Frequency deviation Δf as functions of the modulating frequency f_m and the amplitude of the modulating voltage V_m: (a) frequency deviation Δf as a function of the modulating frequency f_m and (b) frequency deviation Δf as a function of the modulating voltage V_m.

the instantaneous frequency is

$$\omega(t) = \frac{d\phi(t)}{dt} = \omega_c + \frac{d\theta(t)}{dt} = \omega_c + k_p V_m \omega_m \cos \omega_m t = \omega_c + m_p \omega_m \cos \omega_m t \tag{1.203}$$

Thus, the phase change results in a frequency change. The frequency deviation is

$$\Delta f = f_{max} - f_c = f_c + k_p V_m f_m - f_c = k_p V_m f_m = m_p f_m \tag{1.204}$$

The frequency deviation Δf is directly proportional to the modulating frequency f_m. Figure 1.27 shows plots of the FM Δf as functions of the modulating frequency f_m and the amplitude of the modulating voltage V_m.

1.22.3 Frequency Modulation

The modulating voltage is

$$v_m(t) = V_m \sin \omega_m t \tag{1.205}$$

The instantaneous frequency is directly proportional to the amplitude of modulating voltage V_m

$$f(t) = f_c + \Delta f_{max} \frac{v_m(t)}{V_{m(max)}} = f_c + \Delta f_{max} \frac{V_m}{V_{m(max)}} \sin \omega_m t = f_c + \Delta f \sin \omega_m t \tag{1.206}$$

where Δf_{max} is the maximum frequency deviation from the carrier frequency (or rest frequency) f_c.

The derivative of the phase is

$$\frac{d\theta(t)}{dt} = 2\pi k_f v_m(t) = 2\pi k_f V - m \cos \omega_m t \tag{1.207}$$

Thus, the instantaneous phase is

$$\theta(t) = 2\pi \int k_f v_m(t) dt = k_f V_m \int \cos \omega_m t \, dt = \frac{k_f V_m}{f_m} \sin \omega_m t \tag{1.208}$$

In general, the waveform of an FM signal is given by

$$v_o(t) = V_c \cos \left(\omega t + \Delta\omega \int_0^t v_m(t) dt \right) \tag{1.209}$$

The FM output voltage is given by

$$v_{FM}(t) = v_o(t) = V_c \cos[\omega_c t + \theta(t)] = V_c \cos\left(\omega_c t + \frac{k_f V_m}{f_m} \sin \omega_m t\right)$$

$$= V_c \cos\left(\omega_c t + \frac{\Delta f}{f_m} \sin \omega_m t\right) = V_c \cos(\omega_c t + m_f \sin \omega_m t). \quad (1.210)$$

The *index of FM modulation* is defined as the ratio of the maximum frequency deviation Δf to the modulating frequency f_m

$$m_f = \frac{\Delta f}{f_m} = \frac{k_f V_m}{f_m} \quad (1.211)$$

Theoretically, the range of m_f is not limited: m_f can take on any value from 0 to infinity. The index of FM modulation m_f is a function of the modulating voltage V_m and the modulating frequency f_m. If $\Delta f = 75$ kHz, $m_f = 3750$ at $f_m = 20$ Hz and $m_f = 3.75$ at $f_m = 20$ kHz. The index of FM modulation m_f is a function of the modulating voltage amplitude V_m and the modulating frequency f_m.

The amplitude of the FM wave and, therefore, the power remain constant during the modulation process. Figure 1.28 shows that the waveform of the FM signal $v_o(t)/V_c$ modulated a single sinusoid f_m at $m_f = 10$ along with the modulating signal v_m.

The instantaneous frequency of an FM signal is given by

$$f(t) = f_c + k_f V_m \sin \omega_m t \quad (1.212)$$

where the frequency deviation of the FM signal is the maximum change of frequency caused by the modulating voltage V_m

$$\Delta f = f_{max} - f_c = f_c + k_f V_m - f_c = k_f V_m \quad (1.213)$$

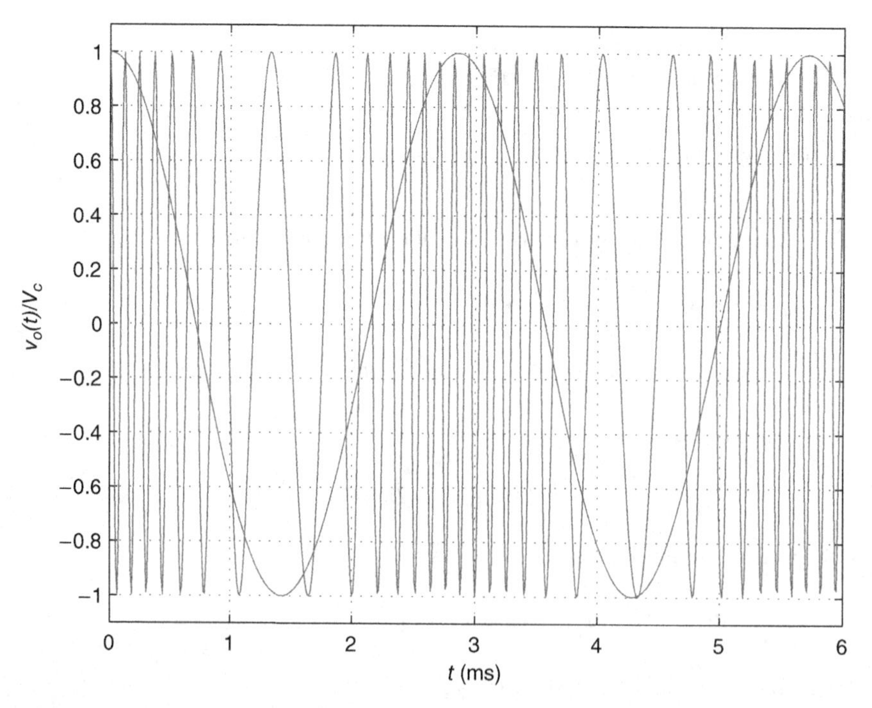

Figure 1.28 Waveforms of the modulating voltage $v_m(t)$ and the FM signal $v_o(t)/V_c$ modulated a single sinusoid f_m at $m_f = 10$.

The maximum frequency deviation is

$$\Delta f_{max} = k_f V_{m(max)} \quad (1.214)$$

The frequency deviation of the FM signal is independent of the modulating frequency f_m.

The angle-modulated signal contains the components at the frequencies

$$f_n = f_c \pm n f_m \qquad n = 0, 1, 2, 3, \ldots \quad (1.215)$$

Therefore, the bandwidth of an angle-modulated signal is infinite. However, the amplitudes of components with large values of n is very small. Hence, the effective bandwidth of the FM signal that contains 98% of the signal power is given by Carson's rule

$$BW_{FM} = 2(\Delta f + f_m) = 2\left(\frac{\Delta f}{f_m} + 1\right) f_m = 2(m_f + 1) f_m \quad (1.216)$$

yielding the maximum bandwidth of the FM signal

$$BW_{FM(max)} = 2(\Delta f + f_{m(max)}) \quad (1.217)$$

If the bandwidth of the modulating signal is BW_m, the FM index is

$$m_f = \frac{\Delta f_{max}}{BW_m} \quad (1.218)$$

and the bandwidth of the FM signal is

$$BW_{FM} = 2(m_f + 1) BW_m \quad (1.219)$$

The FCC's rules and regulations limit the FM broadcast band transmitters to a maximum frequency deviation of 75 kHz to prevent interference of adjacent channels. Commercial FM radio stations with an allowed maximum frequency deviation Δf of 75 kHz and a maximum modulation frequency f_m of 20 kHz require a bandwidth

$$BW_{FM} = 2 \times (75 + 20) = 190 \text{ kHz} \quad (1.220)$$

For $f_c = 100$ MHz and $\Delta f = 75$ kHz,

$$\frac{\Delta f}{f_c} = \frac{75 \times 10^3}{100 \times 10^6} = 0.075\% \quad (1.221)$$

Standard broadcast FM systems use a 200 kHz bandwidth for each station. There are guard bands between the channels.

In general,

$$\phi(t) = \int \omega(t) = 2\pi \int f(t) \quad (1.222)$$

and

$$\omega(t) = 2\pi f(t) = \frac{d\phi(t)}{dt} \quad (1.223)$$

The only difference between PM and FM is that for PM the phase of the carrier varies with the modulating signal and for FM the carrier phase depends on the ratio of the modulating signal amplitude V_m to the modulating frequency f_m. Thus, FM is not sensitive to the modulating frequency f_m, but PM is. If the modulating signal is integrated and used to modulate the carrier, then an FM signal is obtained. This method is used in the Armstrong's indirect FM system. The amount of deviation is proportional to the modulating frequency amplitude V_m.

The relationship between PM and FM is illustrated in Fig. 1.29. If the amplitude of the modulating signal applied to the phase modulator is inversely proportional to the modulating

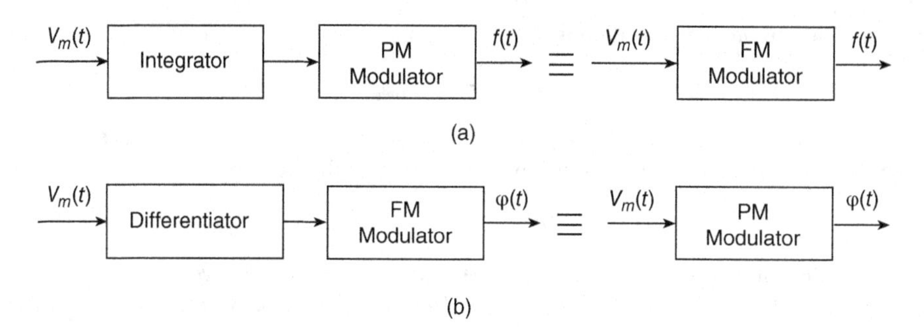

Figure 1.29 Relationship between FM and PM.

frequency f_m, then the phase modulator produces an FM signal. The modulating voltage is given by

$$v_m(t) = \frac{V_m}{\omega_m} \sin \omega_m t \tag{1.224}$$

The modulated voltage is

$$v_o(t) = V_v \cos\left(\omega_c t + \frac{k_p V_m}{\omega_m} \sin \omega_m t\right) = V_c \cos \phi(t) \tag{1.225}$$

Hence, the instantaneous frequency is

$$\omega(t) = \frac{d\phi(t)}{dt} = \omega_c + k_p V_m \cos \omega_m t \tag{1.226}$$

where

$$\Delta f = \frac{k_p V_m}{2\pi} \tag{1.227}$$

This approach is applied in the Armstrong's frequency modulator and is used for commercial FM transmission.

The output power of the transmitter with FM and PM is constant and independent of the modulation index because the envelope is constant and equal to the amplitude of the carrier V_c. Thus, the transmitter output power is

$$P_O = \frac{V_c^2}{2R} \tag{1.228}$$

FM is used for commercial radio and for audio in analog TV.

Both FM and PM signals modulated with a pure sine wave contain a sine as an argument of a cosine, having the modulation index $m = m_f = m_p$, and can be represented as

$$\begin{aligned} v_o(t) &= V_c \cos(\omega_c t + m \sin \omega_m t) = V_c\{J_0(m) \cos \omega_c t \\ &\quad + J_1(m)[\cos(\omega_c + \omega_m)t - \cos(\omega_c - \omega_m)t] + J_2(m)[\cos(\omega_c + 2\omega_m)t - \cos(\omega_c - 2\omega_m)t] \\ &\quad + J_3(m)[\cos(\omega_c + 3\omega_m)t - \cos(\omega_c - 3\omega_m)t] + \cdots\} = \sum_{n=0}^{\infty} A_c J_n(m) \cos[2\pi(f_c \pm n f_m)t] \end{aligned} \tag{1.229}$$

where $n = 0, 1, 2, \ldots$ and $J_n(m)$ are the Bessel functions of the first kind of order n and can be described by

$$J_n(m) = \left(\frac{m}{2}\right) n \left[\frac{1}{n!} - \frac{(m/2)^2}{1!(n+1)!} + \frac{(m/2)^4}{2!(n+2)!} - \frac{(m/2)^6}{3!(n+3)!} + \cdots\right] = \sum_{k=0}^{\infty} \frac{(-1)^k \left(\frac{m}{2}\right)^{n+2k}}{k!(n+k)!} \tag{1.230}$$

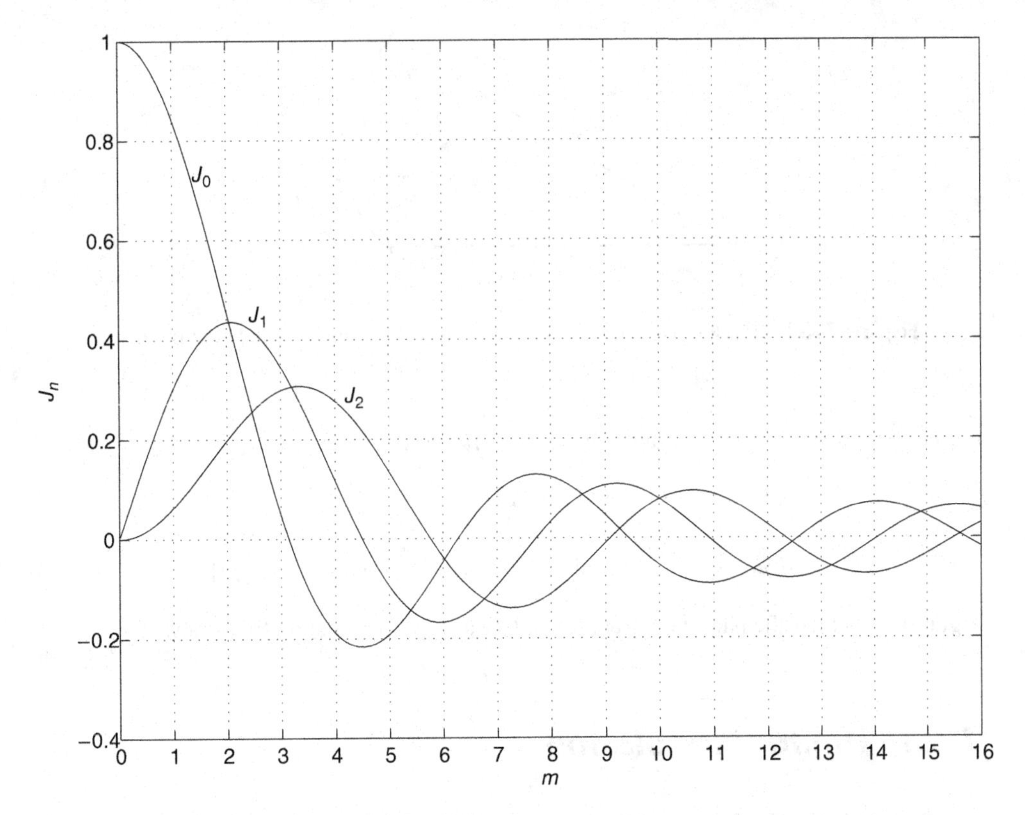

Figure 1.30 Amplitudes of spectrum components J_n for angle modulation as a function of modulation index m.

For $m \ll 1$,

$$J_n(m) \approx \frac{1}{n!}\left(\frac{m}{2}\right)^n \tag{1.231}$$

The angle-modulated voltage consists of a carrier and an infinite number of components (sidebands) placed at multiples of the modulating frequency f_m below and above the carrier frequency f_c, that is, at $f_c \pm nf_m$. The amplitude of the sidebands decreases with n, which allows FM transmission within a finite bandwidth. Figure 1.30 shows the amplitudes of spectrum components for angle modulation as functions of modulation index m. The amplitudes of the carrier (rest) frequency component and the sideband pairs are dependent on the modulation index m and are given by the Bessel function coefficients $J_n(m)$, where n is the order of the sideband pair. J_0 represents the rest-frequency amplitude, J_1 represents the amplitude of the first sideband pairs, and so on. The bandwidth of an FM signal may be written as

$$BW_{FM} = 2(f_{m(max)}) \times (\text{number of significant sideband pairs}) \tag{1.232}$$

The frequency range of the FM broadcasting systems is from 88 to 108 MHz. The FM business band is from 150 to 174 MHz. The average output power of the transmitted FM signal is

$$P_o = (J_0^2 + J_1^2 + J_2)2^2 + J_3^1)P_T \tag{1.233}$$

where P_T is the power of the nonmodulated signal, that is, at $m = 0$.

The angle-modulated signal is

$$v_o(t) = V_c \cos(\omega_c t + m\cos\omega_m t)) = V_c[\cos\omega_c t \cos(m\cos\omega_m t) - \sin\omega_c t \sin(m\cos\omega t)] \tag{1.234}$$

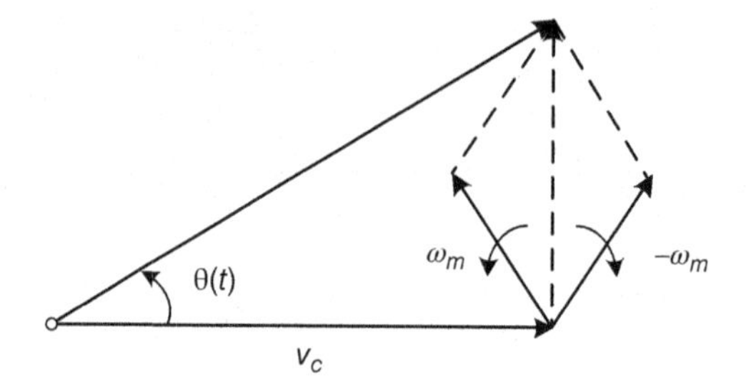

Figure 1.31 Phasor diagram for phase modulation by a single modulating frequency.

For $m \ll 1$, $\cos(m \cos \omega_m t) \approx 1$ and $\sin(m \cos \omega_m t) \approx m \cos \omega t$. Hence,

$$v_o(t) \approx V_c \cos \omega_c t - m \sin \omega_c t \cos \omega_m t$$
$$= V_c \cos \omega_c t - \frac{mV_m}{2} \sin[(\omega_c - \omega_m)t] - \frac{mV_m}{2} \sin[(\omega_c + \omega_m)t] \quad (1.235)$$

Figure 1.31 depicts the phasor diagram for PM by a single modulating frequency.

1.23 Digital Modulation

Digital modulation is actually a special case of analog modulation. In RF systems with digital modulation, the carrier is modulated by a digital baseband signal, which consists of discrete values. Digital modulation offers many advantages over analog modulation and is widely used in wireless communications and digital TV. The main advantage is the quality of the signal, which is measured by the bit error rate (BER). BER is defined as the ratio of the average number of erroneous bits at the output of the demodulator to the total number of bits received in a unit of time in the presence of noise and other interference. It is expressed as a probability of error. Typically, BER $> 10^{-3}$. The waveform of a digital binary baseband signal is given by

$$v_D = \sum_{n=1}^{n=m} b_n v(t - nT_c) \quad (1.236)$$

where b_n is the bit value, equal to 0 or 1. Digital counterparts of analog modulation are amplitude-shift keying (ASK), frequency-shift keying (FSK), and phase-shift keying (PSK).

1.23.1 Amplitude-Shift Keying

Digital AM is called ASK or ON-OFF keying (OOK). It is the oldest type of modulation used in *radio telegraph transmitters*. The binary AM signal is given by

$$v_{BASK} = \begin{cases} V_c \cos \omega_c t & \text{for } v_m = 1 \\ 0 & \text{for } v_m = 0 \end{cases} \quad (1.237)$$

Waveforms illustrating the ASK are shown in Fig. 1.32. The ASK is sensitive to amplitude noise and, therefore, is rarely used in RF applications.

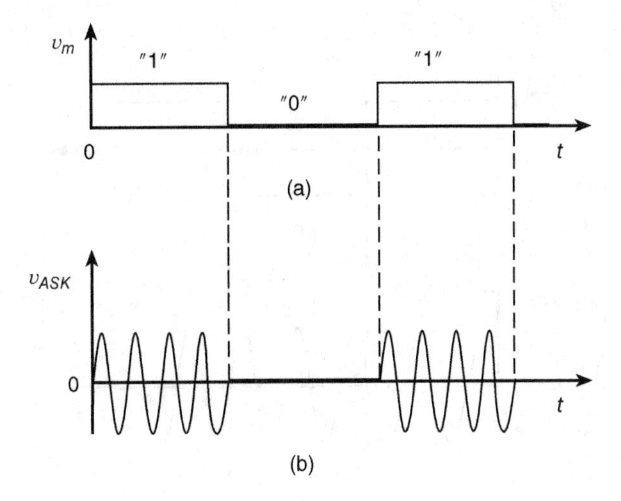

Figure 1.32 Waveform of amplitude-shift keying (ASK).

1.23.2 Phase-Shift Keying

Digital PM is called PSK. One of the most fundamental types of PSK is binary phase-shift keying (BPSK), where the phase shift is $\Delta\phi = 180°$. The BPSK has two phase states. Waveforms for BPSK are shown in Fig. 1.33. When the waveform is in phase with the reference waveform, it represents a logic "1." When the waveform is out of phase with respect to the reference waveform, it represents a logic "0." The modulating binary signal is

$$v_m = \begin{cases} 1 \\ -1 \end{cases} \tag{1.238}$$

The binary PM signal is given by

$$v_{BPSK} = v_m \times v_c = (\pm 1) \times v_c = \begin{cases} v_c = V_c \cos \omega_c t & \text{for } v_m = 1 \\ -v_c = -V_c \cos \omega_c t & \text{for } v_m = -1 \end{cases} \tag{1.239}$$

Another type of PSK is the quaternary phase-shift keying (QPSK), where the phase shift is 90°. The QPSK uses four phase states. The systems 8PSK and 16PSK are also widely used. The PSK is used in transmission of digital signals such as digital TV.

In general, the output voltage with digital PM is given by

$$v_o = V_c \sin\left[\omega_c t + \frac{2\pi(i-1)}{M}\right] \tag{1.240}$$

where $i = 1, 2, \ldots, M$, $M = 2^N$ is the number of phase states, N is the number of data bits needed to determine the phase state. For $M = 2$ and $N = 1$, a BPSK is obtained. For $M = 4$ and $N = 2$, we have a QPSK with the following combinations of digital logic: (00), (01), (11), and (10). For $M = 8$ and $N = 3$, the 8PSK is obtained.

1.23.3 Frequency-Shift Keying

The digital FM is called *FSK*. Waveforms for FSK are shown in Fig. 1.34. FSK uses two carrier frequencies, f_1 and f_2. The lower frequency f_2 may represent a logic "0" and the higher frequency

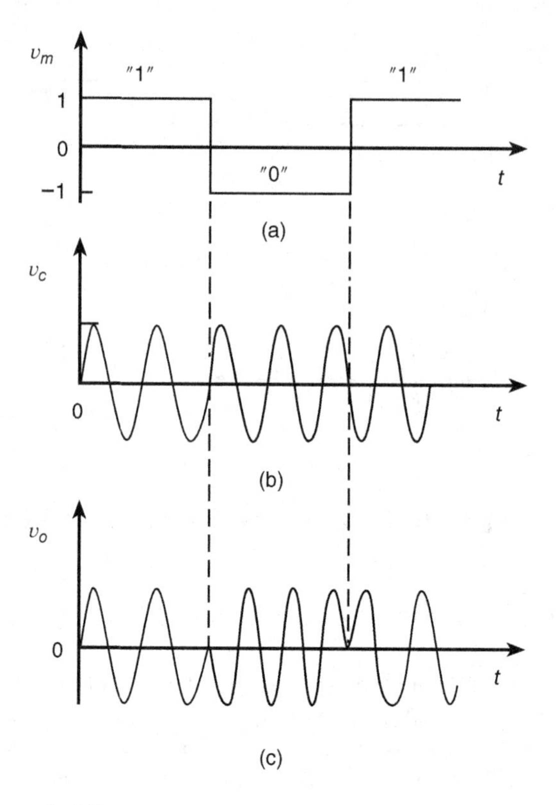

Figure 1.33 Waveforms for binary phase-shift keying (BPSK).

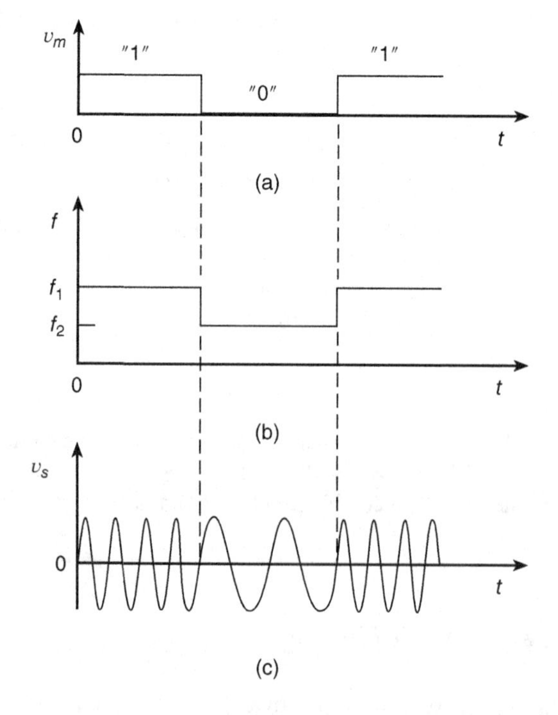

Figure 1.34 Waveforms for frequency-shift keying (FSK).

f_1 may represent a logic "1." The binary FM signal is given by

$$v_{BFSK} = \begin{cases} V_c \cos \omega_{c1} t & \text{for } v_m = 1 \\ V_c \cos \omega_{c2} t & \text{for } v_m = -1 \end{cases} \quad (1.241)$$

For example, GSM uses FSK to transmit a binary signal, where $f_1 = f_c + \Delta f = f_c + 77.708$ kHz and $f_1 = f_c - \Delta f = f_c - 77.708$ kHz, where f_c is the carrier center frequency. The channel spacing is 200 kHz. There are 8 users per channel.

Gaussian frequency-shift keying (GFSK) is used in the Bluetooth. A binary "one" is represented by a positive frequency deviation and a binary "zero" is represented by a negative frequency deviation. The GFSK employs a constant-envelope RF voltage. Thus, a high-efficiency, switching-mode RF power amplifiers can be used in transmitters.

1.24 Radars

Radar was developed during World War II for military applications. The term "radar" is derived from *ra*dio *d*etection *a*nd *r*anging. A radar consists of a transmitter, a receiver, and a transmitting and receiving antenna. A duplexer is used to separate the transmitting and receiving signals. A switch or a circulator can be used as a duplexer. A radar is a system used for detection and location of reflecting targets, such as aircraft, spacecraft, ships, vehicles, people, and other objects. It radiates EMWs (energy) into space and then detects the echo signal reflected from an object. The echo signal contains information about the location, range, direction, and velocity of an reflecting object. A radar can operate in darkness, fog, haze, rain, and snow. Only a very small portion of the transmitted energy is reflected by the object and reradiated back to the radar, where it is received, amplified, and processed.

Today, radars are used in military applications and in a variety of commercial applications, such as navigation, air traffic control, meteorology as weather radars, terrain avoidance, automobile collision avoidance, law enforcement (in speed guns), astronomy, RFID, and automobile automatic toll collection. Military applications of radars include surveillance and tracking of land, air, sea, and space objects from land, air, ship, and space platforms. In monostatic radar systems, the transmitter and the receiver are at the same location. The process of locating an object requires three coordinates: distance, horizontal direction, and angle of elevation. The distance to the object is determined from the time it takes for the transmitted waves to travel to the object and return back. The angular position of the object is found from the arrival angle of the returned signal. The relative motion of the object is determined from the Doppler shift in the carrier frequency of the returned signal. The radar antenna transmits a pulse of EM energy toward a target. The antenna has two basic purposes: radiates RF energy and provides beam focusing. A wide beam pattern is required for search and a narrow beam pattern is needed for tracking. A typical waveform of the radar signal is shown in Fig. 1.35. A radar transmitter typically operates in pulsed mode with an FM (chirped) signal. The amplitude of the pulses is usually constant. Therefore, nonlinear high-efficiency RF amplifiers can be used in radar transmitters. The efficiency of these amplifiers at peak power is usually very high.

The average power of the radar transmitter is

$$P_{AV} = DP_{pk} \quad (1.242)$$

where D is the duty ratio. The duty ratio of the waveform is very low, for example, 0.1%. Therefore, the ratio of the peak power to the average power is very large. For example, the peak RF power is $P_{pk} = 200$ kW and the average power is $P_{AV} = 200$ W. Some portion of this energy is reflected (or scattered) by the target. The echo signal (some of the reflected energy) is received by the radar antenna. The direction of the antenna's main beam determines the location of the

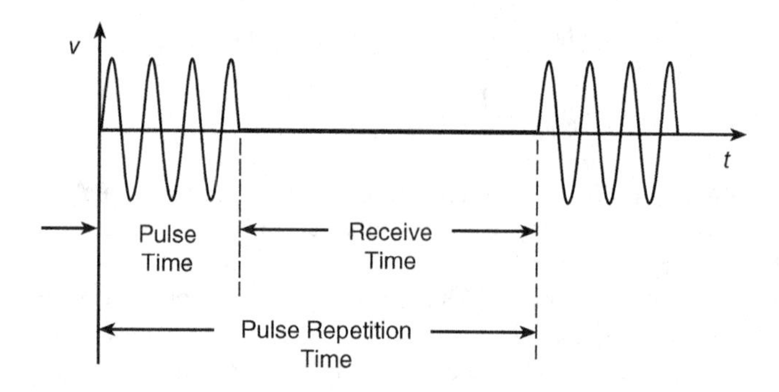

Figure 1.35 Waveform of radar signal.

target. The distance to the target is determined by the time between transmitting and receiving the EM pulse. The speed of the target with respect to the radar antenna is determined by the frequency shift in the EM signal, that is, the Doppler effect. A highly directive antenna is necessary. The radar equation is

$$\frac{P_{rec}}{P_{rad}} = \frac{\sigma_s \lambda^2}{(4\pi)^3 R^4} D^2 \tag{1.243}$$

where P_{rad} is the power transmitted by the radar antenna, P_{rec} is the power received by the radar antenna, σ_s is the radar cross section, λ is the free-space wavelength, and D is the directive gain of the antenna.

1.25 Radio-Frequency Identification

A RFID system consists of a *tag* placed on the item to be identified and a *reader* used to read the tag. The most common carrier frequency of RFID systems is 13.56 MHz, but it can be in the range of 135 kHz–5.875 GHz. The tags are passive or active. The ISO 1443A/B is the international standard for contactless IC cards transmitting at 13.56 MHz in close proximity using a radar antenna. A passive tag contains an RF rectifier, which rectifies a received RF signal from the reader. The rectified voltage is used to supply the tag circuitry. They exhibit long lifetime and durability. An active tag is supplied by an onboard battery. The reading range of active tags is much larger than that of passive tags and are more expensive. A block diagram of an RFID sensor system with a passive tag is shown in Fig. 1.36. The reader transmits a signal to the tag. The memory in the tag contains identification information unique to the particular tag. The microcontroller exports this information to the switch on the antenna. A modulated signal is transmitted to the reader and the reader decodes the identification information.

The tag may be placed in a plastic bag and attached to store merchandise to prevent theft or for checking the store inventory.

Tags may be mounted on windshields inside cars and used for automatic toll collection or checking the parking permit without stopping the car. Passive tags may be used for checking the entire inventory in libraries in seconds or locating a misplaced book. Tags along with pressure sensors may be placed in tires to alert the driver if the tire looses pressure. Small tags may be inserted beneath the skin for animal tracking. IRFD systems are capable of operating in adverse environments.

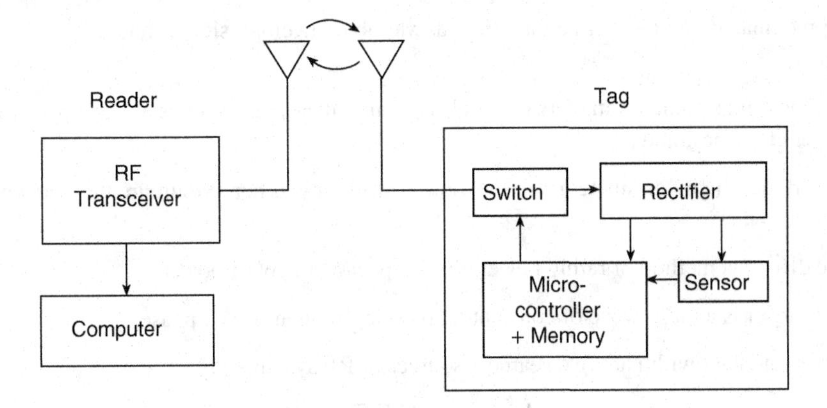

Figure 1.36 Block diagram of an RFID sensor system with a passive tag.

1.26 Summary

- In RF power amplifiers, the transistor can be operated as a dependent current source or as a switch.
- When the transistor is operated as a dependent current source, the drain current depends on the gate-to-source voltage. Therefore, variable-envelope signals can be amplified by the amplifiers in which transistors are operated as dependent current sources.
- When the transistor is operated as a switch, the drain current is independent of the gate-to-source voltage.
- When the transistor is operated as a switch, the drain-to-source voltage in the on-state is low, yielding low conduction loss and high efficiency.
- When the transistor is operated as a switch, switching losses limit the efficiency and the maximum operating frequency of RF power amplifiers.
- Efficiency of a power amplifier is maximized by minimizing power losses at a derided output power.
- Switching power loss can be reduced by using the ZVS technique or ZCS technique.
- In Class A, B, C, and F power amplifiers, the transistor is operated as a dependent current source.
- In Class D, E, and DE power amplifiers, the transistor is operated as a switch.
- When the drain current and the drain-to-source voltage are completely displaced, the output power of an amplifier is zero.
- ZVS, ZVDS, and ZCS conditions cannot be simultaneously satisfied with a passive load network at a nonzero output power.
- The received power P_{REC} is proportional to the square of the ratio $(\lambda/r)^2$. As the carrier's frequency doubles, the received power decreases by a factor of 4 at a given distance r in free space.

- RF modulated signals can be classified as variable-envelope signals and constant-envelope signals.
- The most important parameters of wireless communications systems are spectral efficiency, power efficiency, and signal quality.
- Power amplifiers dissipate more dc power than any other circuit in the transmitter of transceiver.
- The dBm is a method of rating power with respect to 1 mW of power.
- The carrier is a radio wave of constant amplitude, frequency, and phase.
- Power and bandwidth are two scarce resources in RF systems.
- The major analog modulation schemes are AM, FM, and PM.
- AM is the process of superimposing an LF modulating signal on a HF carrier so that the instantaneous changes in the amplitude of the modulating signal produce corresponding changes in the amplitude of the HF carrier.
- In AM transmission, two-thirds of the transmitted power is in the carrier at $m = 1$. No information is transmitted by the carrier.
- The main advantages of AM transmission are narrow bandwidth and simple circuits of transmitters and receivers.
- The main disadvantage of AM transmission is a high level of signal distortion and low efficiency.
- Constant-amplitude signals do not require linear amplification.
- FM is a the process of superimposing the modulating signal on a HF carrier so that the carrier frequency departs from the carrier frequency by an amount proportional to the amplitude of the modulating signal.
- Frequency deviation is the amount of carrier frequency increase or decrease around its center reference value.
- The main advantages of FM transmission are a high signal-to-noise ratio, constant envelope, high efficiency of the transmitters, and low-radiated power.
- Typically, the signal-to-noise ratio is 25 dB lower for FM transmission than that for AM transmission.
- The main disadvantages of FM transmission are a wide bandwidth and the complex circuits of transmitters and receivers.
- The bandwidth of FM transmission is about 10 times wider than that of the AM transmission.
- The major digital modulation schemes are ASK, FSK, and PSK.
- BPSK is a form of PSK in which the binary "1" is represented as no phase shift, and the binary "0" is represented by phase inversion of the carrier signal waveform.
- A transceiver is a combination of a transmitter and a receiver in one electronic device.
- Duplexing techniques allow users to transmit and receive signals simultaneously.
- The major multiple-access techniques are as follows: TDMA technique, FDMA technique, and CDMA technique.

1.27 Review Questions

1.1 What are the modes of operation of transistors in RF power amplifiers?

1.2 What are the classes of operation of RF power amplifiers?

1.3 What are the necessary conditions to achieve 100% efficiency of a power amplifier?

1.4 What are the necessary conditions to achieve nonzero output power in RF power amplifiers?

1.5 Explain the principle of operation of ZVS in power amplifiers.

1.6 Explain the principle of operation of ZCS in power amplifiers.

1.7 What is a ground wave?

1.8 What is a sky wave?

1.9 What is a line-of-sight wave?

1.10 What is an RF transmitter?

1.11 What is a transceiver?

1.12 What is duplexing?

1.13 List multiple-access techniques.

1.14 When are harmonics generated?

1.15 What is the effect of harmonics on communications channels?

1.16 Define THD?

1.17 What are intermodulation products?

1.18 When intermodulation products are generated?

1.19 What is the effect of intermodulation products on communications channels?

1.20 What is intermodulation distortion?

1.21 Define the 1-dB compression point.

1.22 What is the interception point?

1.23 What is the dynamic range of power amplifiers?

1.24 What determines the bandwidth of emission for AM transmission?

1.25 Define the modulation index for FM signals.

1.26 Give an expression for the total power of an AM signal modulated with a single-modulating frequency?

1.27 What is QAM?

1.28 What is phase modulation?

1.29 What is frequency modulation?

1.30 What is the difference between frequency modulation and phase modulation?

1.31 What is FSK?

1.32 What is PSK?

1.33 Explain the principle of operation of a radar.

1.34 Explain the principle of operation of an RFID system.

1.28 Problems

1.1 The rms value of the voltage at the carrier frequency is 100 V. The rms value of the voltage at the second harmonic of the carrier frequency is 1 V. The load resistance is 50 Ω. Neglecting all other harmonics, find *THD*.

1.2 The carrier frequency of an RF transmitter is 4.8 GHz. What is the height of a quarter-wave antenna?

1.3 An AM transmitter has an output power of 10 kW at the carrier frequency. The modulation index at $f_m = 1$ kHz is $m = 0.5$. Find the total output power of the transmitter.

1.4 The amplitude of the carrier of an AM transmitter is 25 V. The antenna input resistance is 50 Ω. The modulation index is $m = 0.5$.

(a) Find the amplitude of the modulating voltage.

(b) Find the output power of the AM transmitter.

1.5 A standard AM broadcast station is allowed to transmit waves with modulating frequencies up to 5 kHz. The carrier frequency is $f_c = 550$ kHz. Determine the following:

(a) the maximum upper sideband frequency,

(b) the minimum sideband frequency, and

(c) the bandwidth of the AM station.

1.6 An intermodulation product occurs at (a) $3f_1 - 2f_2$ and (b) $3f_1 + 2f_2$. What is the order of intermodulation?

1.7 The carrier frequency of an RF transmitter is 2.4 GHz. What is the height of a quarter-wave antenna?

1.8 The peak power of a radar transmitter is 100 kW. The duty ratio is 0.1%. What is the average power of the radar transmitter?

1.9 Find the total output power of the AM voltage when the carrier power is 10 W and $m = 1$.

1.10 Sketch the modulating, carrier, and AM voltage waveforms at $m = 0.5$.

1.11 Find the power of each sideband for a sinusoidal modulating voltage at $m = 0.3$ and the power of the carrier of 1 kW.

1.12 Find the channel bandwidth of an AM transmitter with a maximum modulating frequency of 10 kHz.

1.13 A transmitter applies an FM signal to a 50 Ω antenna. The amplitude of the signal is $V_m = 1000$ V. What is the output power of the transmitter?

1.14 An FM signal is modulated with the modulating frequency $f_m = 10$ kHz and has a maximum frequency deviation $\Delta f = 20$ kHz. Find the modulation index.

1.15 An FM signal has a carrier frequency $f_c = 100.1$ MHz, modulation index $m_f = 2$, and modulating frequency $f_m = 15$ kHz. Find the bandwidth of the FM signal.

1.16 How much bandwidth is necessary to transmit a Chopin's piano sonata over an FM radio broadcasting system with high fidelity?

References

[1] K. K. Clarke and D. T. Hess, *Communications Circuits: Analysis and Design*. Reading, MA: Addison-Wesley Publishing Co., 1971; reprinted Malabar, FL: Krieger, 1994.

[2] H. L. Krauss, C. W. Bostian, and F. H. Raab, *Solid State Radio Engineering*, New York, NY, John Wiley & Sons, 1980.

[3] M. Kazimierczuk, "Class E tuned amplifier with shunt inductor," *IEEE Journal of Solid-State Circuits*, vol. 16, no. 1, pp. 2–7, 1981.

[4] E. D. Ostroff, M. Borakowski, H. Thomas, and J. Curtis, *Solid-State Radar Transmitters*, Boston, MA:Artech House, 1985.

[5] M. K. Kazimierczuk and D. Czarkowski, *Resonant Power Converters*, 2nd Ed. New York, NY: IEEE Press/John Wiley & Sons, 2011.

[6] S. C. Cripps, *RF Power Amplifiers for Wireless Communications*, 2nd Ed. Norwood, MA: Artech House, 2006.

[7] M. Albulet, *RF Power Amplifiers*. Atlanta, GA: Noble Publishing Co., 2001.

[8] T. H. Lee, *The Design of CMOS Radio-Frequency Integrated Circuits*, 2nd Ed. New York, NY: Cambridge University Press, 2004.

[9] P. Reynaret and M. Steyear, *RF Power Amplifiers for Mobile Communications*. Dordrecht, The Netherlands: Springer, 2006.

[10] H. I. Bartoni, *Radio Propagation for Modern Wireless Systems*. Englewood Clifs, NJ: Prenice-Hall, 2000.

[11] N. Levanon and E. Mozeson, *Radar Signals*, 3rd Ed. New York, NY: John Wiley & Sons, 2004.

[12] H. T. Friis, "A note on a simple transmission formula," *IEEE IRA*, vol. 41, pp. 254–256, 1966.

[13] B. Molnar, "Basic limitations of waveforms achievable in single-ended switching-mode tuned (Class E) power amplifiers," *IEEE Journal of Solid-State Circuits*, vol. SC-19, pp. 144–146, 1984.

[14] M. K. Kazimierczuk, "Generalization of conditions for 100-percent efficiency and nonzero output power in power amplifiers and frequency multipliers," *IEEE Transactions on Circuits and Systems*, vol. CAS-33, pp. 805–807, 1986.

[15] F. H. Raab, "Intermodulation disrortion in Khan technique transmitters," *IEEE Transactions on Microwave Theory and Techniques*, vol. 44, pp. 2273–2278, 1996.

[16] M. K. Kazimierczuk and N. O. Sokal, "Cause of instability of power amplifiers," *IEEE Journal of Solid-State Circuits*, vol. 19, pp. 541–542, 1984.

[17] B. Razavi, *RF Microelectronics*. Upper Saddle River, NJ: Prentice Hall, 1989.

[18] P. B. Kenington, *High-Linearity RF Power Amplifier Design*. Norwood, MA: Artech House, 2000.

[19] J. Groe and L. Larson, *CDMA Mobile Radio Design*. Norwood, MA: Artech House, 2000.

[20] A. Grebennikov, *RF and Microwave Power Amplifier Design*. New York, NY: McGraw-Hill, 2005.

[21] A. Grebennikov and N. O. Sokal, *Switchmode RF Power Amplifiers*. Amsterdam: Amsterdam, 2007.

[22] P. Colantonio, F. Giannini, and E. Limiti, *High Efficiency RF and Microwave Solid State Power Amplifiers*. Chichester, UK: John Wiley & Sons, 2009.

[23] J. L. B. Walker, Edited, *Handbook of RF and Microwave Power Amplifiers*. Cambridge, UK: Cambridge University Press, 2012.

[24] A. Raghavan, N. Srirattana, and J. Lasker, *Modeling and Design Techniques for RF Power Amplifiers*. New York, NY: IEEE Press, John Wiley & Sons, 2008.

[25] R. Baxley and G. T. Zhou, "Peak savings analysis of peak-to-average power ratio in OFDM," *IEEE Transactions on Consumer Electronics*, vol. 50, no. 3, pp. 792–798, 1998.

[26] B. S. Yarman, *Design of Ultra Wideband Power Transfer Networks*. Berlin, Springer-Verlag, 2008.

[27] M. K. Kazimierczuk, *Pulse-width Modulated DC-DC Power Converters*. 2nd Ed. Chichester, UK: John Wiley & Sons, 2014.

[28] M. K. Kazimierczuk, *High-Frequency Magnetic Components*, 2nd Ed. Chichester, UK: John Wiley & Sons, 2014.

2

Class A RF Power Amplifier

2.1 Introduction

The Class A RF power amplifier [1–15] is theoretically a linear amplifier. A linear amplifier is supposed to produce an amplified replica of the input voltage or current waveform. It provides an accurate reproduction of both the envelope and the phase of the input signal. The input signal may contain audio, video, and data information. The transistor in the Class A RF power amplifier is operated as a dependent current source. The conduction angle of the drain or collector current is 360°. The efficiency of the Class A RF power amplifier is very low. The maximum drain efficiency with perfect components is 50%. However, the Class A RF power amplifier is a nearly linear circuit with a low degree of nonlinearity. Therefore, it is commonly used as a preamplifier and an output power stage of radio transmitters, especially for amplification of signals with a variable amplitude of the output voltage, for example, in amplitude-modulated (AM) systems. In this chapter, the basic characteristics of the Class A RF power amplifier are analyzed. The amplifier circuits, biasing, current and voltage waveforms, power losses, efficiency, bandwidth, and impedance matching are studied.

2.2 Power MOSFET Characteristics

2.2.1 Square Law for MOSFET Drain Current

A MOSFET is used as a dependent current source in Class A RF power amplifiers. For low drain currents, the drain current of an n-channel enhancement-type MOSFET with a long channel and operating in the pinch-off region (or the saturation region) with a low electric field in the channel and constant carrier mobility in the channel is given by a *square law*

$$i_D = \frac{1}{2}\mu_{n0}C_{ox}\left(\frac{W}{L}\right)(v_{GS} - V_t)^2 = K(v_{GS} - V_t)^2 \quad \text{for} \quad v_{GS} \geq V_t \quad \text{and} \quad v_{DS} \geq v_{GS} - V_t \tag{2.1}$$

RF Power Amplifiers, Second Edition. Marian K. Kazimierczuk.

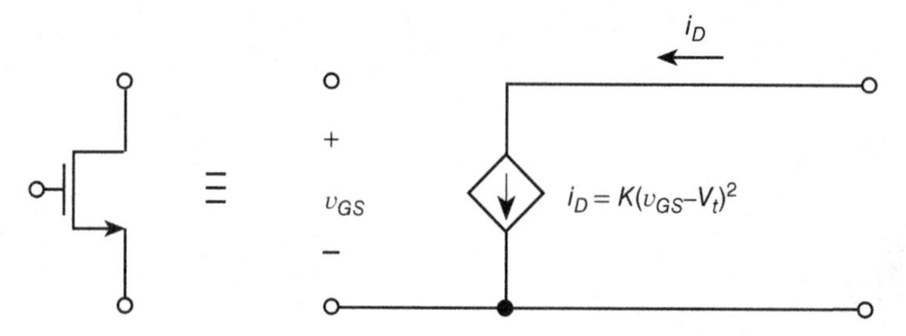

Figure 2.1 Low-frequency large-signal model of power MOSFETs for the pinch-off region at low drain currents described by the square law (i.e., at constant carrier mobility μ_{n0}) with neglected transistor capacitances.

where

$$K = \frac{1}{2}\mu_{n0}C_{ox}\left(\frac{W}{L}\right) = \frac{1}{2}K_n\left(\frac{W}{L}\right) \tag{2.2}$$

μ_{n0} is the low-field surface electron mobility in the channel, $C_{ox} = \epsilon_{ox}/t_{ox}$ is the oxide capacitance per unit area of the gate, t_{ox} is the oxide thickness, $\epsilon_{ox} = 0.345$ pF/cm is the silicon oxide permittivity, L is the channel length, W is the channel width, and V_t is the threshold voltage. Typically, $t_{ox} = 0.1$ μm, $K_n = \mu_{n0}C_{ox} = 20$ μA/V^2, $C_{ox} = \frac{1}{3}$ mF/m^2, and $\mu_{n0} = 600$ cm^2/V · s is the low-field surface electron mobility of silicon at room temperature. The electron mobility in the channel is substantially lower than that in bulk silicon and can be as low as 200 cm^2/V·s. The low-frequency large-signal model of the long-channel MOSFETs for the pinch-off region at low drain current described by the square law and with neglected capacitances is shown in Fig. 2.1. This model represents equation (2.1).

The maximum saturation drain current is

$$i_{Dsat(max)} = \frac{1}{2}\mu_{n0}C_{ox}\left(\frac{W}{L}\right)(v_{GSmax} - V_t)^2 \tag{2.3}$$

Hence, the aspect ratio for a MOSFET described by the square law required to achieve a specified maximum saturation drain current $i_{Dsat(max)}$ at a maximum gate-to-source voltage is given by

$$\frac{W}{L} = \frac{2i_{Dsat(max)}}{\mu_{n0}C_{ox}(v_{GSmax} - V_t)^2} \tag{2.4}$$

The maximum saturation drain current $i_{Dsat(max)}$ must be higher than that of the maximum instantaneous drain current I_{DM} in an amplifier

$$i_{Dsat(max)} > I_{DM} \tag{2.5}$$

2.2.2 Channel-Length Modulation

The drain current i_D is inversely proportional to the channel length L. Once the channel pinches off at saturation drain-to-source voltage $v_{DSsat} = v_{GS} - V_t$, the channel charge nearly vanishes at the drain end and the lateral electric field becomes very high. An increase in drain-to-source voltage v_{DS} beyond saturation voltage v_{DSsat} causes the high-field region at the drain end to widen by a small distance ΔL, as shown in Fig. 2.2. The voltage $v_{DS} - v_{DSsat}$ drops across ΔL and the drain-to-source voltage drops across $L - \Delta L$. As the drain-to-source voltage v_{DS} increases in the pinch-off region, the channel pinch-off point is moved slightly away from the drain toward the source. Thus, the effective channel length L_e is shortened as follows:

$$L_e = L - \Delta L = L\left(1 - \frac{\Delta L}{L}\right) \approx \frac{L}{1 + \frac{\Delta L}{L}} \tag{2.6}$$

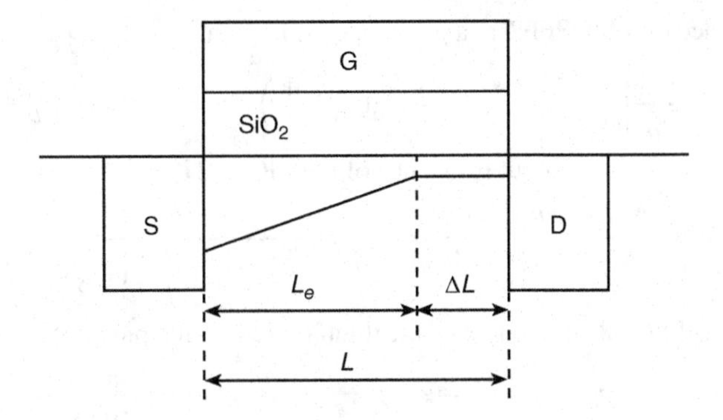

Figure 2.2 Effective channel length in the pinch-off region.

where the fractional change in the channel length is proportional to the drain-to-source voltage v_{DS} and is given by

$$\frac{\Delta L}{L} = \lambda v_{DS} = \frac{v_{DS}}{V_A} \tag{2.7}$$

$\lambda = 1/V_A$ is the channel-length modulation parameter, and V_A is a voltage similar to the early voltage used in bipolar junction transistor (BJT) theory. The phenomenon described earlier is known as *channel-length modulation* and is more significant for short channels. The drain current for MOSFETs with a short channel is given by

$$i_D = \frac{1}{2}\mu_{n0}C_{ox}\left(\frac{W}{L-\Delta L}\right)(v_{GS}-V_t)^2 = \frac{1}{2}\mu_{n0}C_{ox}\left[\frac{W}{L\left(1-\frac{\Delta L}{L}\right)}\right](v_{GS}-V_t)^2$$

$$\approx \frac{1}{2}\mu_{n0}C_{ox}\left(\frac{W}{L}\right)(v_{GS}-V_t)^2\left(1+\frac{\Delta L}{L}\right) = K(v_{GS}-V_t)^2(1+\lambda v_{DS}) \tag{2.8}$$

Another short-channel effect is the reduction of the threshold voltage V_t as the channel length decreases.

2.2.3 Low- and Mid-Frequency Small-Signal Model of MOSFETs

Transistors are often operated under small-signal conditions in preamplifiers in RF applications. The transconductance for low drain currents at a given operating point, determined by the dc components of gate-to-source voltage V_{GS}, drain current I_D, and drain-to-source voltage V_{DS}, is given by

$$g_m = \left.\frac{di_D}{dv_{GS}}\right|_{v_{GS}=V_{GS}} = 2K(v_{GS}-V_t)\Big|_{v_{GS}=V_{GS}} = 2K(V_{GS}-V_t) = 2\sqrt{KI_D}$$

$$= \sqrt{2\mu_{n0}C_{ox}\left(\frac{W}{L}\right)I_D} \tag{2.9}$$

Hence, the ac small-signal component of the drain current of an ideal MOSFET ($\lambda = 0$) with horizontal i_D-v_{DS} characteristics for low dc drain currents I_D is

$$i_d' = g_m v_{gs} = 2K(V_{GS}-V_t)v_{gs} = \mu_{n0}C_{ox}\left(\frac{W}{L}\right)(V_{GS}-V_t)v_{gs} = 2\sqrt{KI_D}v_{gs} \tag{2.10}$$

The horizontal i_D-v_{DS} characteristics of real MOSFETs have a slope in the pinch-off region. As the drain-to-source voltage v_{DS} increases, the drain current i_D slightly increases. The small-signal output conductance of a MOSFET at a given operating point is

$$g_o = \frac{1}{r_o} = \frac{di_D}{dv_{DS}}\bigg|_{v_{GS}=V_{GS}} = \frac{i''_d}{v_{ds}} = \frac{1}{2}\mu_{n0}C_{ox}\left(\frac{W}{L}\right)(V_{GS} - V_t)^2\lambda \approx \lambda I_D = \frac{I_D}{V_A} \tag{2.11}$$

resulting in the small-signal output resistance of the MOSFET

$$r_o = \frac{\Delta v_{DS}}{\Delta i_D} = \frac{v_{ds}}{i''_d} = \frac{1}{\lambda I_D} = \frac{V_A}{I_D} = \frac{V_A}{\frac{1}{2}\mu_{n0}C_{ox}\left(\frac{W}{L}\right)(V_{GS} - V_t)^2} \tag{2.12}$$

Hence, the ac small-signal component of the drain current in the pinch-off region is

$$i_d = i'_d + i''_d = g_m v_{gs} + g_o v_{ds} = g_m v_{gs} + \frac{v_{ds}}{r_o} \tag{2.13}$$

A low-frequency small-signal model of power MOSFETs in the pinch-off region (with neglected transistor capacitances) representing (2.13) is shown in Fig. 2.3.

2.2.4 High-Frequency Small-Signal Model of the MOSFET

A high-frequency small-signal model of power MOSFETs for the pinch-off region is depicted in Fig. 2.4, where $C_{gd} = C_{rss}$ is the gate-to-drain capacitance, $C_{gs} = C_{iss} - C_{rss}$ is the gate-to-source capacitance, and $C_{ds} = C_{oss} - C_{rss}$ is the drain-to-source capacitance. Capacitances C_{rss}, C_{iss}, and C_{oss} are given in manufacturer's data sheets of power MOSFETs.

2.2.5 Unity-Gain Frequency

Consider a high-frequency small-signal model of a MOSFET with a short circuit at the output and with an ideal current source i_g driving the gate, as shown in Fig. 2.5. The drain current is

$$I_d(s) = g_m V_{gs}(s) - sC_{gd}V_{gs}(s) \approx g_m V_{gs}(s) \tag{2.14}$$

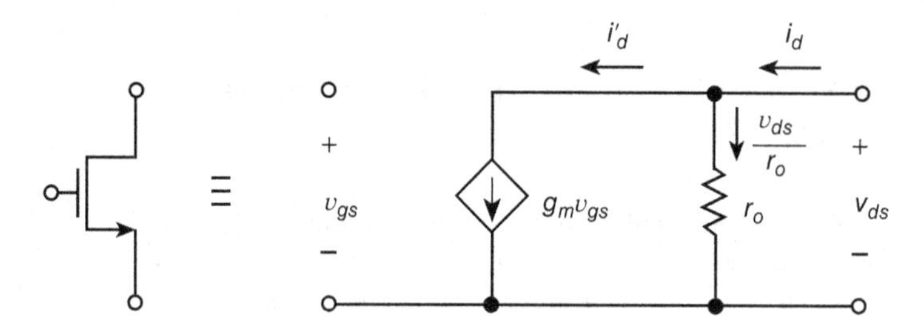

Figure 2.3 Low-frequency small-signal model of power MOSFETs for the pinch-off region.

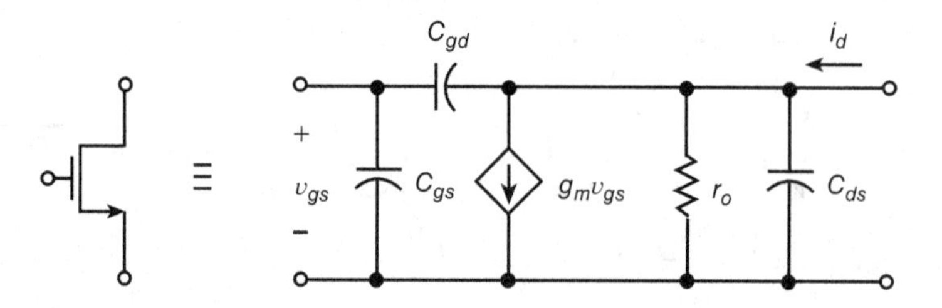

Figure 2.4 High-frequency small-signal model of power MOSFETs for the pinch-off region.

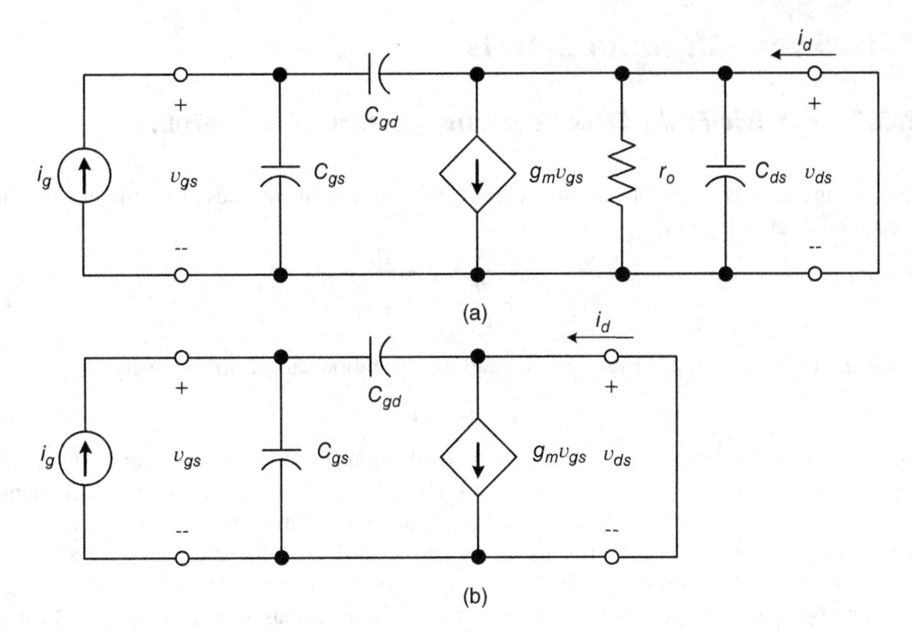

Figure 2.5 High-frequency small-signal model of a power MOSFET to determine the unity-gain frequency f_T. (a) Model. (b) Simplified model.

and the gate current is

$$I_g(s) = s(C_{gs} + C_{gd})V_{gs}(s) \tag{2.15}$$

Hence, the MOSFET current gain is

$$A_i(s) = \frac{I_d(s)}{I_g(s)} = \frac{g_m}{s(C_{gs} + C_{gd})} \tag{2.16}$$

Setting $s = j\omega$, we get

$$A_i(j\omega) = \frac{g_m}{j\omega(C_{gs} + C_{dg})} = |A_i|e^{j\phi_{Ai}} \tag{2.17}$$

The magnitude of the MOSFET current gain is

$$|A_i(\omega)| = \frac{g_m}{\omega(C_{gs} + C_{gd})} \tag{2.18}$$

The unity-gain frequency of a MOSFET is defined to be the frequency at which the magnitude of the short-circuit gain is equal to 1

$$|A_i(\omega_T)| = \frac{g_m}{\omega_T(C_{gs} + C_{gd})} = 1 \tag{2.19}$$

yielding the unity-gain frequency of a MOSFET

$$f_T = \frac{g_m}{2\pi(C_{gs} + C_{gd})} = \frac{K(V_{GS} - V_t)}{\pi(C_{gs} + C_{gd})} = \frac{\sqrt{KI_D}}{\pi(C_{gs} + C_{gd})} = \frac{\sqrt{\frac{1}{2}\mu_{n0}C_{ox}\left(\frac{W}{L}\right)I_D}}{\pi(C_{gs} + C_{gd})} \tag{2.20}$$

The frequency f_T is a figure-of-merit of MOSFETs. A weakness of f_T as an indicator of transistor performance is that it neglects the drain-to-source capacitance C_{ds} and the gate resistance r_G.

2.3 Short-Channel Effects

2.3.1 Electric Field Effect on Charge Carriers Mobility

The average drift velocity of current carriers in a semiconductor caused by the electric field intensity E is given by [14]

$$v = \frac{\mu_{n0}E}{1 + \frac{E}{E_{sat}}} = \frac{\mu_{n0}E}{1 + \frac{\mu_{n0}E}{v_{sat}}} \tag{2.21}$$

where μ_{n0} is the low-field carrier mobility and the saturation carrier drift velocity is

$$v_{sat} = \mu_{n0}E_{sat} \tag{2.22}$$

Figure 2.6 shows a plot of average drift velocity of electrons v as a function of electric field intensity E for silicon at $v_{sat} = 0.8 \times 10^7$ cm/s and $\mu_{n0} = 600$ cm^2/V·s. As the electric field intensity E increases, the average drift velocity v is proportional to E at low values of E and then saturates at E_{sat}. Typically, $v_{sat} = 0.8 \times 10^7$ cm/s for electrons in silicon (Si) and $v_{sat} = 2.7 \times 10^7$ cm/s for electrons in silicon carbide (SiC). The surface mobility of electrons in silicon is $\mu_{n0} =$ 600 cm^2/V·s. The average mobility of current carriers μ decreases when electric field intensity E, the doping concentration, and temperature T increase. The mobility of current carriers (e.g., electrons) as a function of the electric field intensity E is given by

$$\mu_n = \frac{v}{E} = \frac{\mu_{n0}}{1 + \frac{E}{E_{sat}}} = \frac{\mu_{n0}}{1 + \frac{\mu_{n0}E}{v_{sat}}} \tag{2.23}$$

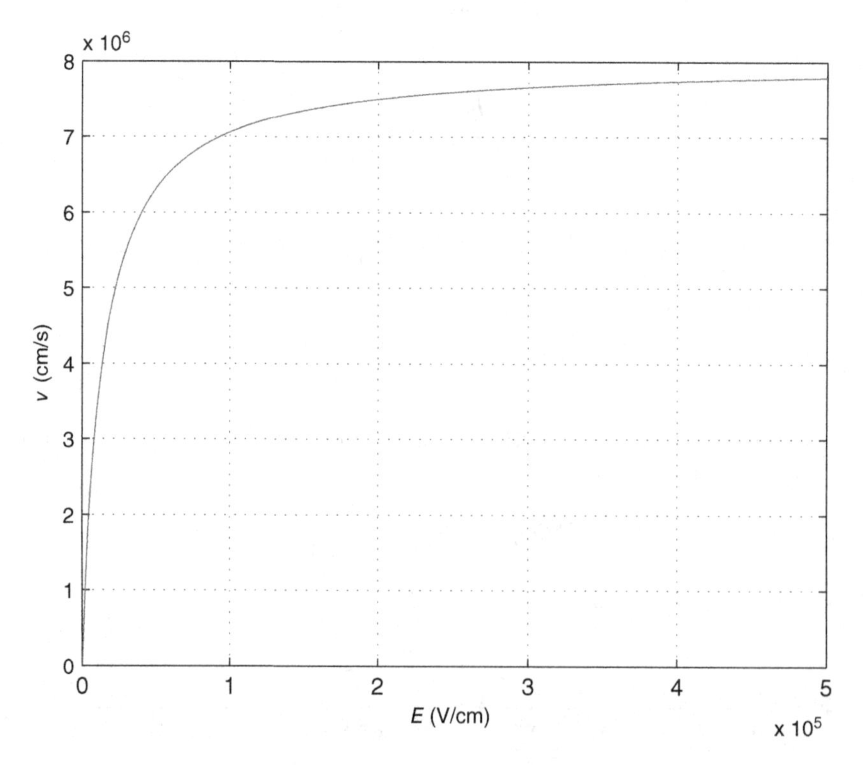

Figure 2.6 Average drift velocity of electrons v as a function of electric field intensity E for silicon at $v_{sat} = 0.8 \times 10^7$ cm/s and $\mu_{n0} = 600$ cm^2/V·s.

The range of the electric field intensity E from point of view of the average mobility μ can be divided into three ranges: low field, intermediate field, and high field [14].

In the low-field range, the average carrier mobility for silicon is approximately constant

$$\mu_n \approx \mu_{n0} \quad \text{for} \quad E < 3 \times 10^3 \text{ V/cm} \tag{2.24}$$

Therefore, the low-field average carrier drift velocity for silicon is directly proportional to the electric field intensity E

$$v \approx \mu_{n0} E \quad \text{for} \quad E < 3 \times 10^3 \text{ V/cm} \tag{2.25}$$

For intermediate-field range, the carrier mobility μ_n decreases when the electric field intensity E increases and is given by (2.23). Consequently, the intermediate-field average carrier drift velocity increases at a lower rate than that of the low-field average drift velocity. This range of the electric field intensity for silicon is

$$3 \times 10^3 \text{ V/cm} \le E \le 6 \times 10^4 \text{ V/cm} \tag{2.26}$$

As the electric field intensity enters the intermediate range, the collisions of electrons with the semiconductor lattice are so frequent and the time between the collisions is so short that the current carriers do not accelerate as much as they do at low electric field intensity, slowing down the movement of the current carriers as a group.

In the high-field range, the average carrier drift velocity reaches a constant value called the saturation drift velocity and is independent of the electric field intensity E

$$v \approx v_{sat} \quad \text{for} \quad E > 6 \times 10^4 \text{ V/cm} \tag{2.27}$$

In this region, the carrier mobility μ_n is inversely proportional to the electric field intensity E. The collision rate is so dramatically increased and the time between the collisions is so much reduced that the velocity of the movement of current carriers is independent of the electric field intensity.

2.3.2 Ohmic Region

In the ohmic region, $v_{DS} \le v_{GS} - V_t$. The electric field intensity in the channel is

$$E = \frac{v_{DS}}{L} \tag{2.28}$$

As the channel length L decrease, E increases at a given voltage v_{DS}. The electron mobility in the channel at any electric field intensity E is

$$\mu_n = \frac{\mu_{n0}}{1 + \dfrac{\mu_{n0} E}{v_{sat}}} = \frac{\mu_{n0}}{1 + \dfrac{\mu_{n0}}{v_{sat}} \dfrac{v_{DS}}{L}} \tag{2.29}$$

The average carrier drift velocity at any electric field intensity E is

$$v_n = \mu_n E = \frac{\mu_{n0} E}{1 + \dfrac{\mu_{n0} E}{v_{sat}}} = \frac{\mu_{n0} E}{1 + \dfrac{\mu_{n0}}{v_{sat}} \dfrac{v_{DS}}{L}} \tag{2.30}$$

The drain current at any average electron drift velocity v_n for the ohmic region is

$$i_D = \mu_n C_{ox} \left(\frac{W}{L}\right) \left[(v_{GS} - V_t) v_{DS} - \frac{v_{DS}^2}{2}\right]$$

$$= \frac{\mu_{n0} C_{ox}}{1 + \dfrac{\mu_{n0}}{v_{sat}} \dfrac{v_{DS}}{L}} \left(\frac{W}{L}\right) \left[(v_{GS} - V_t) v_{DS} - \frac{v_{DS}^2}{2}\right] \quad \text{for} \quad v_{DS} < v_{GS} - V_t \tag{2.31}$$

In can be observed that the drain current i_D decreases as the electron mobility decreases. The drain current is also given by

$$i_D = \mu_n C_{ox} \left(\frac{W}{L}\right) \left(v_{GS} - V_t - \frac{v_{DS}}{2}\right) v_{DS} = \mu_n C_{ox} W \left(v_{GS} - V_t - \frac{v_{DS}}{2}\right) \frac{v_{DS}}{L}$$
$$= C_{ox} W \left(v_{GS} - V_t - \frac{v_{DS}}{2}\right) \mu_n E = C_{ox} W \left(v_{GS} - V_t - \frac{v_{DS}}{2}\right) \frac{\mu_{n0} E}{1 + \frac{\mu_{n0}}{v_{sat}} \frac{v_{DS}}{L}}$$
$$= C_{ox} W \left(v_{GS} - V_t - \frac{v_{DS}}{2}\right) v_n \quad \text{for} \quad v_{DS} < v_{GS} - V_t \tag{2.32}$$

For high electric field in the channel, $v_n = v_{sat}$ and the drain current in the ohmic region is given by

$$i_D = C_{ox} W \left(v_{GS} - V_t - \frac{v_{DS}}{2}\right) v_n \quad \text{for} \quad v_{DS} < v_{GS} - V_t \tag{2.33}$$

Hence, the drain current at $v_n = v_{sat}$ becomes

$$I_{Dsat} = C_{ox} W v_{sat} v_{DSsat} = C_{ox} W v_{sat} (v_{GS} - V_t) \tag{2.34}$$

For instance, the minimum length of a long channel of a silicon MOSFET for low-field operation at $v_{DS} = 0.3 \text{ V} \leq v_{DSsat}$ is

$$L > L_{min} = \frac{v_{DS}}{E_{LFmax}} = \frac{0.3}{3 \times 10^3} = 1 \ \mu\text{m} \tag{2.35}$$

The maximum length of a short channel of a silicon MOSFET for high-field operation at $v_{DS} = 0.3 \text{ V} < v_{DSsat}$ is

$$L < L_{max} = \frac{v_{DSsat}}{E_{HFmin}} = \frac{0.3}{6 \times 10^6} = 50 \text{ nm} \tag{2.36}$$

The intermediate channel length in this example is in the range: $50 \text{ nm} < L < 1 \ \mu\text{m}$.

2.3.3 Pinch-Off Region

At the boundary between the ohmic region and the saturation region,

$$v_{DS} = v_{DSsat} = v_{GS} - V_t \tag{2.37}$$

The electric field intensity in the channel is

$$E = \frac{v_{DSsat}}{L} = \frac{v_{GS} - V_t}{L} \tag{2.38}$$

The carrier mobility in the channel for the n-channel MOSFET can be expressed as

$$\mu_n = \frac{\mu_{n0}}{1 + \frac{\mu_{n0} E}{v_{sat}}} = \frac{\mu_{n0}}{1 + \frac{\mu_{n0} v_{DSsat}}{v_{sat} L}} = \frac{\mu_{n0}}{1 + \frac{\mu_{n0} (v_{GS} - V_t)}{v_{sat} L}} = \frac{\mu_{n0}}{1 + \theta v_{DSsat}} = \frac{\mu_{n0}}{1 + \theta (v_{GS} - V_t)} \tag{2.39}$$

where the *mobility degradation coefficient* is

$$\theta = \frac{\mu_{n0}}{v_{sat} L} \tag{2.40}$$

The average carrier drift velocity in an n-channel is

$$v_n = \mu_n E = \mu_n \frac{v_{DSsat}}{L} = \mu_n \frac{v_{GS} - V_t}{L} \tag{2.41}$$

The drain current at the intermediate electric field intensity for the saturation region is

$$i_D = \frac{1}{2}\mu_n C_{ox}\left(\frac{W}{L}\right)(v_{GS}-V_t)^2 = \frac{1}{2}C_{ox}W(v_{GS}-V_t)\mu_n\left(\frac{v_{DSsat}}{L}\right)$$
$$= \frac{1}{2}C_{ox}W(v_{GS}-V_t)\mu_n E$$
$$= \frac{1}{2}C_{ox}W(v_{GS}-V_t)v_n = \frac{1}{2}C_{ox}W(v_{GS}-V_t)\frac{\mu_{n0}E}{1+\frac{\mu_{n0}E}{v_{sat}}}$$
$$= \frac{1}{2}\mu_{n0}C_{ox}\left(\frac{W}{L}\right)(v_{GS}-V_t)^2\frac{1}{1+\frac{\mu_{n0}}{v_{sat}}\frac{v_{DSsat}}{L}}$$
$$= \frac{1}{2}\frac{\mu_{n0}}{1+\frac{\mu_{n0}}{v_{sat}}\frac{(v_{GS}-V_t)}{L}}C_{ox}\left(\frac{W}{L}\right)(v_{GS}-V_t)^2$$
$$= \frac{1}{2}\frac{\mu_{n0}}{1+\theta(v_{GS}-V_t)}C_{ox}\left(\frac{W}{L}\right)(v_{GS}-V_t)^2 \quad \text{for} \quad v_{GS} \geq V_t$$
$$\text{and} \quad v_{DS} \geq v_{GS}-V_t \tag{2.42}$$

The shorter the channel, the higher the mobility degradation coefficient. The drain current i_D follows the average carrier drift velocity v_n. The decrease in the current carrier mobility μ_n in the channel is called the "short-channel effect." At $v_n = v_{sat}$,

$$i_D = \frac{1}{2}C_{ox}Wv_{sat}(v_{GS}-V_t) \quad \text{for} \quad v_{DS} \geq v_{GS}-V_t \tag{2.43}$$

For example, the minimum length of a long channel of a silicon MOSFET for low-field operation at $v_{DSsat} = 0.6$ V is

$$L > L_{min} = \frac{v_{DS}}{E_{LFmax}} = \frac{0.6}{3\times 10^3} = 2\ \mu\text{m} \tag{2.44}$$

The maximum channel length of a silicon MOSFET for high-field operation at $v_{DSsat} = 0.6$ V is

$$L < L_{max} = \frac{v_{DSsat}}{E_{HFmin}} = \frac{0.6}{6\times 10^3} = 0.1\ \mu\text{m} = 100\ \text{nm} \tag{2.45}$$

The channel length is intermediate for 100 nm $< L <$ 2 μm

Taking into account the channel-length modulation, the drain current for the intermediate electric field intensity in the pinch-off region is

$$i_D = \frac{1}{2}\frac{\mu_{n0}}{1+\theta(v_{GS}-V_t)}C_{ox}\left(\frac{W}{L_e}\right)(v_{GS}-V_t)^2$$
$$= \frac{1}{2}\frac{\mu_{n0}}{1+\theta(v_{GS}-V_t)}C_{ox}\left[\frac{W}{L(1-\Delta L/L)}\right](v_{GS}-V_t)^2$$
$$= \frac{1}{2}\frac{\mu_{n0}}{1+\theta(v_{GS}-V_t)}C_{ox}\left(\frac{W}{L}\right)(v_{GS}-V_t)^2\left(1+\frac{\Delta L}{L}\right)$$
$$= \frac{1}{2}\frac{\mu_{n0}}{1+\theta(v_{GS}-V_t)}C_{ox}\left(\frac{W}{L}\right)(v_{GS}-V_t)^2(1+\lambda v_{DS})$$
$$\text{for} \quad v_{GS} \geq V_t \quad \text{and} \quad v_{DS} \geq v_{GS}-V_t \tag{2.46}$$

where

$$\frac{\Delta L}{L} = \lambda v_{DS} \tag{2.47}$$

For high drain currents, the average carrier drift velocity is $v_n = v_{sat}$; therefore, the i_D–v_{GS} characteristic of an n-channel enhancement-type power MOSFET for the pinch-off region (or the saturation region) is nearly linear and is given by a *linear law* [14]

$$i_D = Qv_{sat} = \frac{1}{2}C_{ox}Wv_{sat}(v_{GS} - V_t) = K_s(v_{GS} - V_t)(1 + \lambda v_{DS})$$

$$\text{for} \quad v_{GS} \geq V_t \quad \text{and} \quad v_{DS} \geq v_{GS} - V_t \tag{2.48}$$

where

$$K_s = \frac{1}{2}C_{ox}Wv_{sat} \tag{2.49}$$

Thus, the large-signal model of the MOSFET for high drain currents in the constant drift velocity range consists of the voltage-dependent current source driven by the voltage $v_{GS} - V_t$ for $v_{GS} > V_t$, and the coefficient of this source is K_s. The i_D–v_{GS} characteristics are displayed in Fig. 2.7 using square law, short-channel law, and linear law. The i_D–v_{DS} characteristics are displayed in Fig. 2.8 using square law, short-channel law, and linear law. The low-frequency large-signal model of power MOSFETs (with neglected capacitances) for constant carrier drift velocity $v \approx v_{sat}$ is shown in Fig. 2.9.

The maximum saturation drain current of the MOSFET described by the linear law is

$$i_{Dsat(max)} = \frac{1}{2}C_{ox}Wv_{sat}(v_{GSmax} - V_t) \tag{2.50}$$

Hence, the MOSFET width to achieve the required maximum saturation drain current is

$$W = \frac{2i_{Dsat(max)}}{C_{ox}v_{sat}(v_{GSmax} - V_t)} \tag{2.51}$$

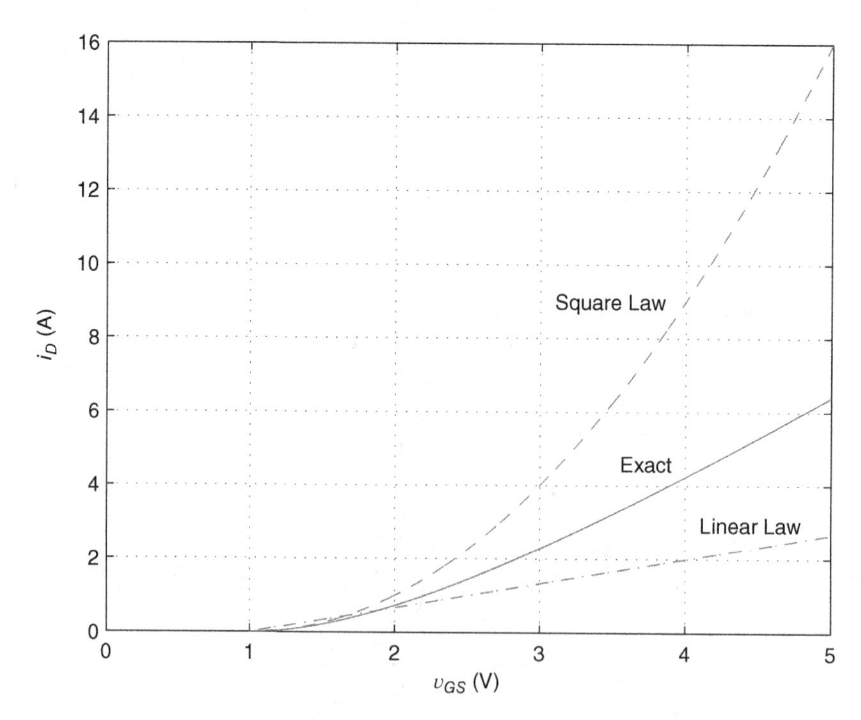

Figure 2.7 MOSFET i_D-v_{GS} characteristics described by a square law, linear law, and exact equation for $V_t = 1$ V, $\mu_{n0} = 600$ cm²/V·s, $C_{ox} = \frac{1}{3}$ mF/m², $W/L = 10^5$, $v_{sat} = 8 \times 10^6$ cm/s, and $\lambda = 0$. Exact plot: $L = 2$ μm (long channel) and $\theta = 0.375$. Linear law: $L = 0.5$ μm (short channel) and $W = 0.5 \times 10^5$ μm.

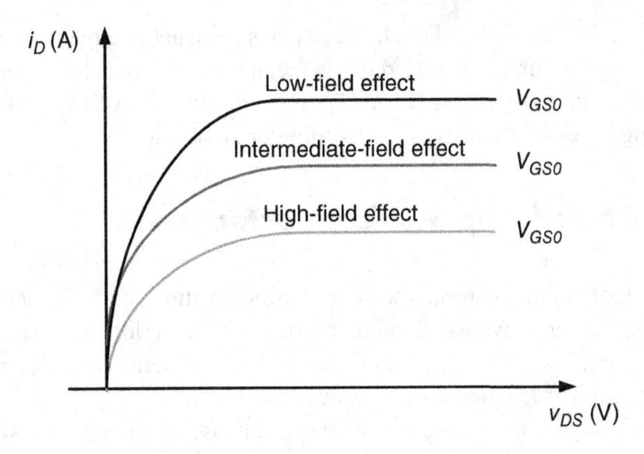

Figure 2.8 MOSFET i_D-v_{DS} characteristics described by a square law, linear law, and exact equation at a fixed voltage V_{GS0} for $V_t = 1$ V, $\mu_{n0} = 600$ cm^2/V·s, $C_{ox} = \frac{1}{3}$ mF/m^2, $W/L = 10^5$, $v_{sat} = 8 \times 10^6$ cm/s, and $\lambda = 0$. Exact plot: $L = 2$ μm (long channel) and $\theta = 0.375$. Linear law: $L = 0.5$ μm (short channel) and $W = 0.5 \times 10^5$ μm.

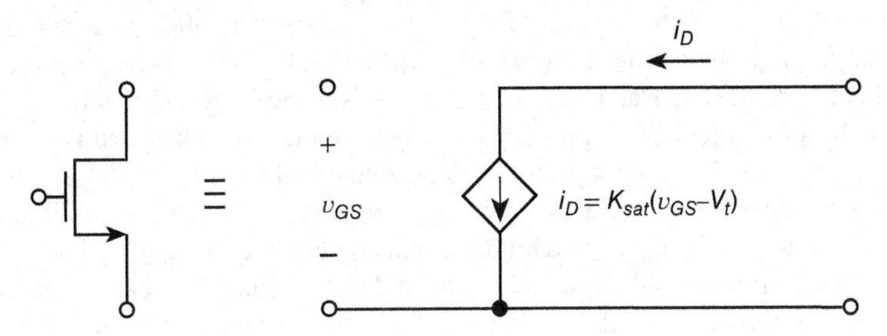

Figure 2.9 Low-frequency large-signal model of power MOSFETs for constant carrier drift velocity $v \approx v_{sat}$.

The transconductance of the MOSFET for high drain currents is

$$g_m = \left.\frac{di_D}{dv_{GS}}\right|_{v_{GS}=V_{GS}} = K_s = \frac{1}{2}C_{ox}Wv_{sat} \tag{2.52}$$

For high drain currents, g_m is constant and independent of the operating point if $v_{GS} > V_t$. The ac model of the MOSFET consists of a voltage-dependent current source i_d driven by the gate-to-source voltage v_{gs}. The ac component of the drain current is

$$i_d' = g_m v_{gs} = K_s v_{gs} = \frac{1}{2}C_{ox}Wv_{sat}v_{gs} \tag{2.53}$$

The small-signal output conductance of a MOSFET at a given operating point is

$$g_o = \frac{1}{r_o} = \left.\frac{di_D}{dv_{DS}}\right|_{v_{GS}=V_{GS}} = K_s(V_{GS} - V_t)\lambda = \frac{1}{2}C_{ox}Wv_{sat}(V_{GS} - V_t)\lambda \approx \lambda I_D = \frac{I_D}{V_A} \tag{2.54}$$

resulting in the small-signal output resistance of the MOSFET

$$r_o = \frac{\Delta v_{DS}}{\Delta i_D} = \frac{1}{\lambda I_D} = \frac{V_A}{I_D} = \frac{V_A}{\frac{1}{2}C_{ox}Wv_{sat}(V_{GS} - V_t)} \tag{2.55}$$

The drain current is given by (2.13). The low-frequency small-signal model for $v \approx v_{sat}$ is displayed in Fig. 2.3, where g_m and r_o are given by (2.52) and (2.54), respectively.

Power MOSFETs operate with either characteristics intermediate between the constant carrier mobility and the constant drift velocity regions (such as in Class A RF power amplifiers) or in all the three regions (such as in Class B and C power amplifiers). A SPICE model of a MOSFET can be used for high-frequency large-signal computer simulation.

2.3.4 Wide Band Gap Semiconductor Devices

Properties of semiconductor materials have a significant impact on the performance of semiconductor devices. Silicon power semiconductor devices are rapidly approaching the theoretical limits of performance. Speed and thermal characteristics of semiconductor materials are especially important in high-frequency applications.

Wide band gap (WBG) semiconductors offer significant advantages over silicon. These semiconductors include gallium nitride (GaN), silicon carbide (SiC), and semiconducting diamond. WBG semiconductors are defined as semiconductors with a significant band gap (BG) energy E_G, usually $E_G > 2E_{G(Si)} = 2.24$ eV. The only WBG semiconductors in group IV are diamond and silicon carbide (SiC). Most WBG semiconductors are compound semiconductors that belong to group III–V, for example, gallium nitride (GaN). The compounds in group II–V are insulators, for example, SiO_2. WBG materials enable the development of smaller semiconductor devices.

WBG devices are very attractive for high-frequency power applications that are subject to high voltages and high temperatures in RF transmitters. Table 2.1 lists major properties of silicon (Si), silicon carbide (SiC), and gallium nitride (GaN). The GaN semiconductor can be doped with silicon (Si) or oxygen (O) to produce the n-type semiconductor or with magnesium (Mn) to obtain the p-type semiconductor. It can be deposited on silicon carbide or sapphire. The GaN semiconductor has very high breakdown electric field E_{BD}, high electron mobility μ_n, high electron drift saturation velocity $v_{sat(n)}$, and high thermal conductivity k_{th}. Higher breakdown electric field E_{BD} allows for thinner and more highly doped devices. Since they can be made thinner and doped higher, faster switching can be achieved. GaN power devices can meet rapidly evolving demands for high energy efficiency, high power, high power density, good linearity, small size, low weight, small board space, high thermal conductivity, and good reliability. Semiconductor GaN HEMTs, MOSFETs, and MESFETs can be used in high-frequency, high-temperature RF power amplifiers to build high-speed data transmission wireless RF transmitters. Other applications include high-speed wide bandwidth dynamic power supplies, amplitude modulators for RF transmitters, radars, transponders, and aerospace circuits. Power circuits have been demonstrated with 700 W pulsed peak power, 21 dB power gain, and 70% energy efficiency at frequencies of 1–3 GHz.

Silicon carbide (SiC) has very high breakdown electric field E_{BD}, high thermal conductivity k_{th}, and high maximum operating temperature T_{Jmax}. SiC conducts heat far better than silicon does. SiC devices are used in high-power, high-voltage, and high-temperature power conversion.

Table 2.1 Properties of semiconductors.

Property	Symbol	Unit	Si	SiC	GaN
Band gap energy	E_G	eV	1.12	3.26	3.42
Electron mobility	μ_n	$cm^2/V \cdot s$	1360	900	2000
Hole mobility	μ_p	$cm^2/V \cdot s$	480	120	300
Breakdown electric field	E_{BD}	V/cm	2×10^5	2.2×10^6	3.5×10^6
Saturation electron drift velocity	v_{sat}	cm/s	10^7	2.7×10^7	2.5×10^7
Dielectric constant	ϵ_r	–	11.7	9.7	9
Intrinsic concentration at $T = 300$ K	n_i	cm^{-3}	10^{10}	10^{-8}	1.9×10^{-7}
Thermal conductivity	k_{th}	W/K·cm	1.5	4.56	1.3

Johnson's figure-of-merit (JFOM) of a semiconductor material is defined as [16]

$$JFOM = \frac{E_{BD} \upsilon_{sat}}{2\pi} \tag{2.56}$$

where E_{BD} is the critical electric field for breakdown in the semiconductor and υ_{sat} is the saturated drift velocity of current carriers (electrons or holes). Figure 2.10 compares JFOM for Si, SiC, and GaN.

Baliga's figure-of-merit (BFOM) of a semiconductor material is [17]

$$BFOM = \epsilon \mu_n E_G^3 = \epsilon_r \epsilon_0 \mu_n E_G^3 \tag{2.57}$$

where E_G is the BG energy of the semiconductor, μ is the electron mobility at low electric field, $\epsilon_0 = 1/(36\pi) = 8.85 \times 10^{-12}$ F/m is the permittivity of free space, and ϵ_r is the relative permittivity of the semiconductor. Figure 2.11 compares BFOM for Si, SiC, and GaN.

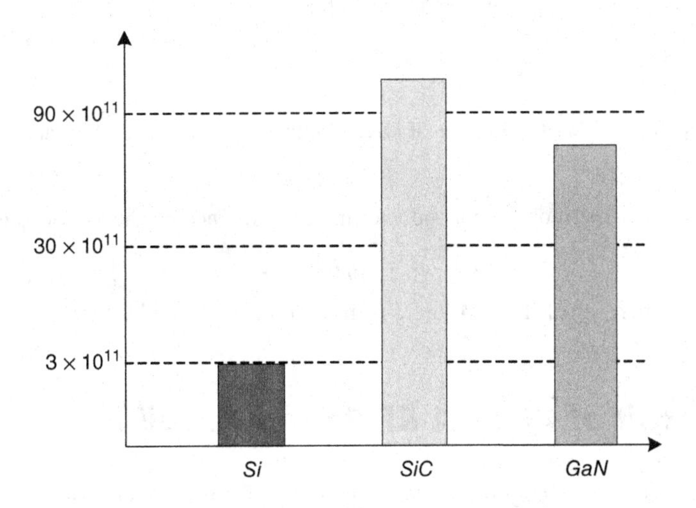

Figure 2.10 Comparison of Johnson's figure-of-merit (JFOM) for Si, SiC, and GaN.

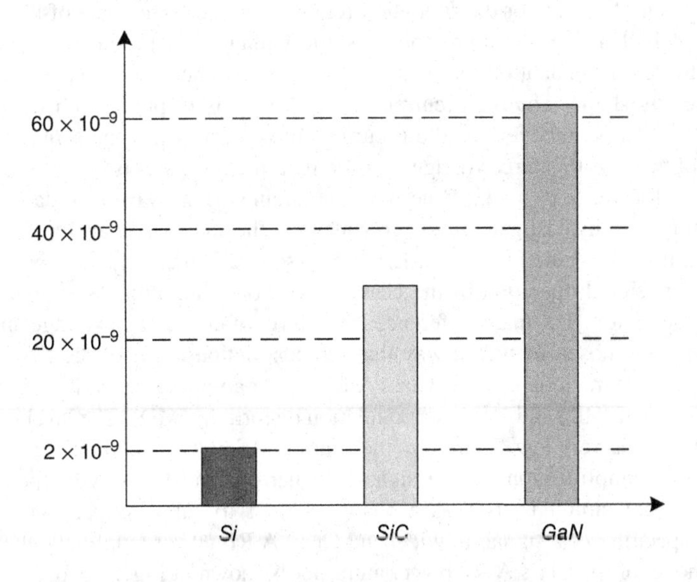

Figure 2.11 Comparison of Baliga's figure-of-merit (BFOM) for Si, SiC, and GaN.

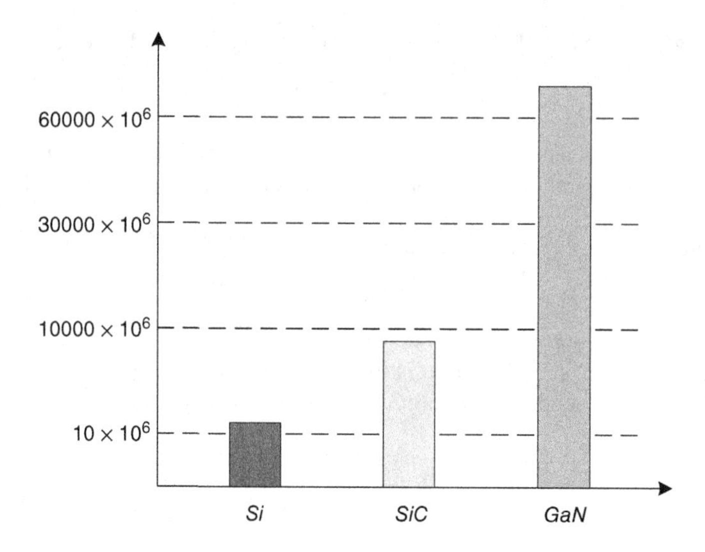

Figure 2.12 Comparison of MKFOM figure-of-merit for Si, SiC, and GaN.

Another figure-of-merit of a semiconductor material, defined by the author, is given by

$$MKFOM = \epsilon \mu_n E_G^3 = \epsilon_r \epsilon_0 \mu_n E_{BD}^3 \tag{2.58}$$

Figure 2.12 compares the MKFOM figure-of-merit for Si, SiC, and GaN.

2.4 Circuit of Class A RF Power Amplifier

The Class A RF power amplifier is shown in Fig. 2.13(a). It consists of a transistor, a parallel-resonant circuit L-C, an RF choke (RFC) L_f, and a coupling capacitor C_c. The amplifier is loaded by a resistance R. The operating point of the transistor is located in the active region (the pinch-off region or the saturation region). The dc component of the gate-to-source voltage V_{GS} is higher than the transistor threshold voltage V_t. The transistor is operated as a voltage-controlled dependent current source. The ac component of the gate-to-source voltage v_{gs} can have any shape. The drain current waveform i_D is in phase with the gate-to-source voltage v_{GS}. The ac component of drain current waveform i_d has the same shape as the ac component of the gate-to-source voltage v_{gs} as long as the transistor is operated in the pinch-off region, that is, for $v_{DS} > v_{GS} - V_t$. Otherwise, the drain current waveform flattens at the crest. We will assume a sinusoidal gate-to-source voltage in the subsequent analysis. At the resonant frequency f_0, the drain current i_D and the drain-to-source voltage v_{DS} are shifted in phase by 180°. The large-signal operation of the Class A RF power amplifier is similar to that of the small-signal operation. The main difference is the level of current and voltage amplitudes. The operating point is chosen in such a way that the conduction angle of the drain current 2θ is 360°. The v_{GS}-v_o characteristic of the Class A RF power amplifier is nearly linear, yielding low harmonic distortion (HD) and low intermodulation distortion (IMD). The level of harmonics in the output voltage is very low. Therefore, the Class A RF power amplifier is a linear amplifier. It is suitable for amplification of AM signals. In narrowband Class A RF power amplifiers, a parallel-resonant circuit is used as a bandpass filter to suppress harmonics and select a narrowband spectrum of a signal. In wideband Class A RF power amplifiers, the filters are not needed. A model of the Class A RF power amplifier is shown in Fig. 2.13(b).

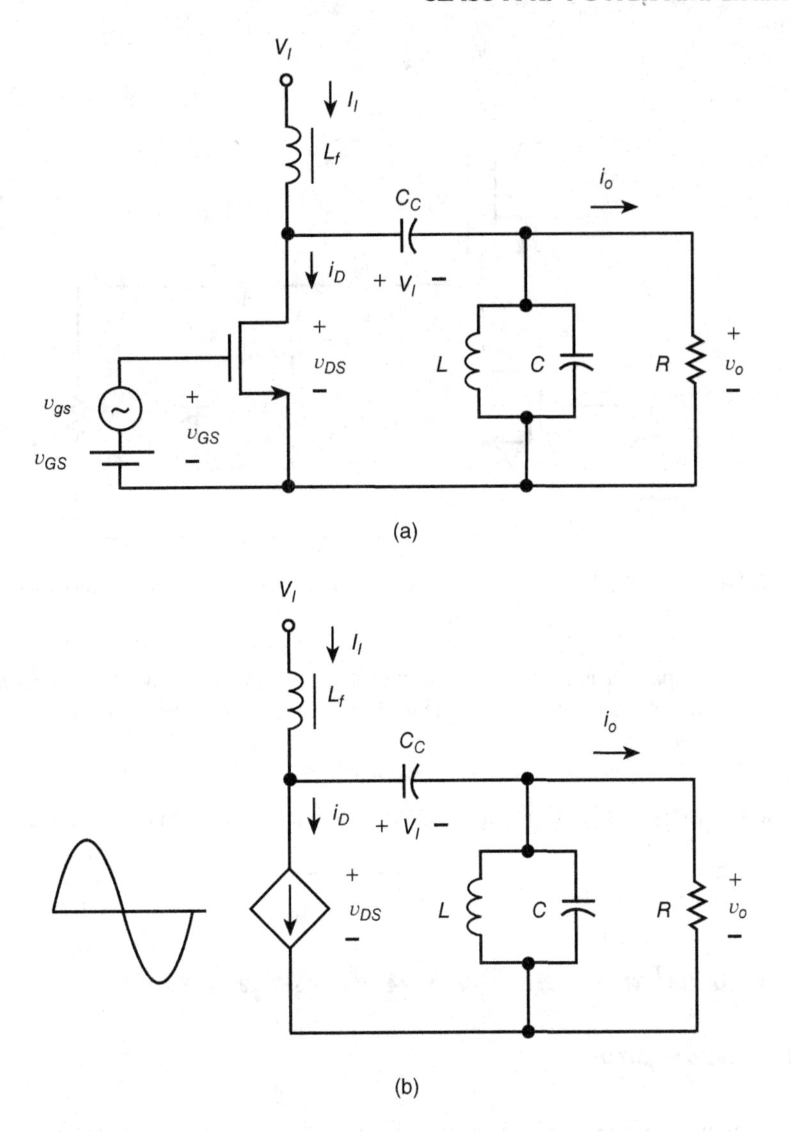

Figure 2.13 Class A RF power amplifier. (a) Circuit. (b) Equivalent circuit.

The RFC can be replaced by a quarter-wavelength transmission line, as shown in Fig 2.14. The input impedance of the transformer is given by

$$Z_i = \frac{Z_o^2}{Z_L} \tag{2.59}$$

The load impedance of the transmission line in the form of filter capacitor C_B of the power supply at the operating frequency is very low, nearly a short circuit. Therefore, the input impedance of the transmission line at the drain for the fundamental component of the drain current is very high, nearly an open circuit.

The current ripple of the choke inductor L_f is low, at least 10 times lower than the dc component I_I, which is equal to the dc component of the drain current I_D. Hence, $X_{Lf} \gg R$, resulting in the design equation

$$X_{Lf} = \omega L_f \geq 10R \tag{2.60}$$

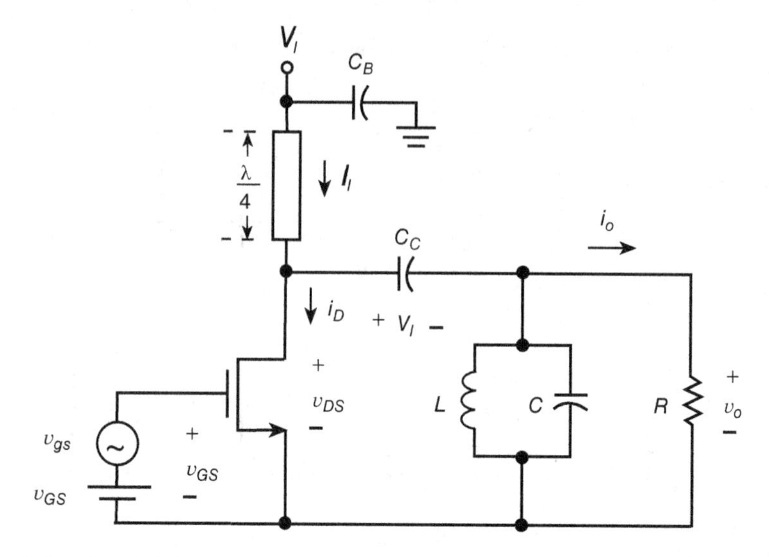

Figure 2.14 Class A RF power amplifier with a quarter-wavelength transformer used in place of RF choke.

The coupling capacitor is high enough so that its ac component is nearly zero. Since the dc component of the voltage across the inductor L in steady state is zero, the voltage across the coupling capacitor is

$$V_{Cc} = V_I \tag{2.61}$$

The ripple voltage across the coupling capacitor is low if $X_{Cc} \ll R$. The design equation is

$$X_{Cc} = \frac{1}{\omega C_c} \leq \frac{R}{10} \tag{2.62}$$

2.5 Waveforms in Class A RF Amplifier

2.5.1 Assumptions

The analysis of the Class A RF power amplifier is initially carried out under the following assumptions that the MOSFET is ideal:

1. $v_{DSsat} = 0$, and therefore $V_{m(max)} = V_I - v_{DSsat} = V_I$.
2. The maximum swing of the drain-to-source voltage v_{DS} is from 0 to $2V_I$ and the maximum swing of the drain current is from 0 to $2I_{DQ}$.
3. $\lambda = 0$, and therefore $g_o = 0$ and $r_o = \infty$, resulting in horizontal characteristics $i_D = f(v_{DS})$.
4. The drain current is given by linear law $i_D = K_s(v_{GS} - V_t)$, that is, it is directly proportional to $v_{GS} - V_t$. Therefore, the drain current waveform does not contain harmonics and intermodulation products.
5. The output capacitance of the MOSFET C_o is linear and is absorbed into the resonant capacitance.
6. The passive components are ideal, that is, resistances of inductors and capacitors are zero, self-capacitances of inductors are zero, and self-inductances of capacitors are zero.

2.5.2 Current and Voltage Waveforms

Figure 2.15 shows current and voltage waveforms in the Class A RF power amplifier. The gate-to-source voltage is given by

$$v_{GS} = V_{GS} + v_{gs} = V_{GS} + V_{gsm} \cos \omega t \tag{2.63}$$

where V_{GS} is the dc gate-to-source bias voltage, V_{gsm} is the amplitude of the ac component of the gate-to-source voltage v_{gs}, and $\omega = 2\pi f$ is the operating angular frequency. The dc component of the voltage V_{GS} is higher than the MOSFET threshold voltage V_t such that the conduction angle of the drain current is $2\theta = 360°$. To keep the transistor in the active region at all times, the following condition must be satisfied:

$$V_{GS} - V_{gsm} > V_t \tag{2.64}$$

Assuming a linear law, the drain current of the MOSFET in the Class A RF power amplifier is

$$i_D = K_s(V_{GS} + V_{gsm} \cos \omega t - V_t) = K_s(V_{GS} - V_t) + K_s V_{gsm} \cos \omega t \tag{2.65}$$

where K_s is the MOSFET parameter given by (2.49) and V_t is the MOSFET threshold voltage. The drain current is expressed as

$$i_D = I_D + i_d = I_D + I_m \cos \omega t = I_I + I_m \cos \omega t > 0 \tag{2.66}$$

where the dc component of the drain current I_D is equal to the dc supply current I_I

$$I_D = I_I = K_s(V_{GS} - V_t) \tag{2.67}$$

The ac component of the drain current is

$$i_d = I_m \cos \omega t \tag{2.68}$$

where the amplitude of the ac component of the drain current is given by

$$I_m = K_s V_{gsm} \tag{2.69}$$

To avoid distortion of the sinusoidal ac component of the drain current, the maximum amplitude of the ac component of the drain current is given by

$$I_{m(max)} = I_I = I_D \tag{2.70}$$

An ideal (lossless) parallel-resonant circuit L-C presents an infinite reactance at the resonant frequency $f_0 = 1/(2\pi\sqrt{LC})$. Therefore, the output current of the parallel-resonant circuit is

$$i_o = i_{Cc} = I_I - i_D = I_I - I_I - i_d = -i_d = -I_m \cos \omega t \tag{2.71}$$

The fundamental component of the drain-to-source voltage is

$$v_{ds} = v_o = -Ri_o = -RI_m \cos \omega t = -V_m \cos \omega t \tag{2.72}$$

where $V_m = RI_m$. The drain-to-source voltage at the resonant frequency f_0 of the parallel-resonant circuit is

$$v_{DS} = V_{DS} + v_{ds} = V_{DS} - V_m \cos \omega t = V_I - V_m \cos \omega t \tag{2.73}$$

where the dc component of the drain-to-source bias voltage V_{DS} is equal to the dc supply voltage V_I

$$V_{DS} = V_I \tag{2.74}$$

Since operation of an ideal transistor as a dependent current source can be maintained only when v_{DS} is positive, it is necessary to limit the amplitude V_m to a value lower than V_I

$$V_{m(max)} = RI_{m(max)} = V_I = V_{DS} \tag{2.75}$$

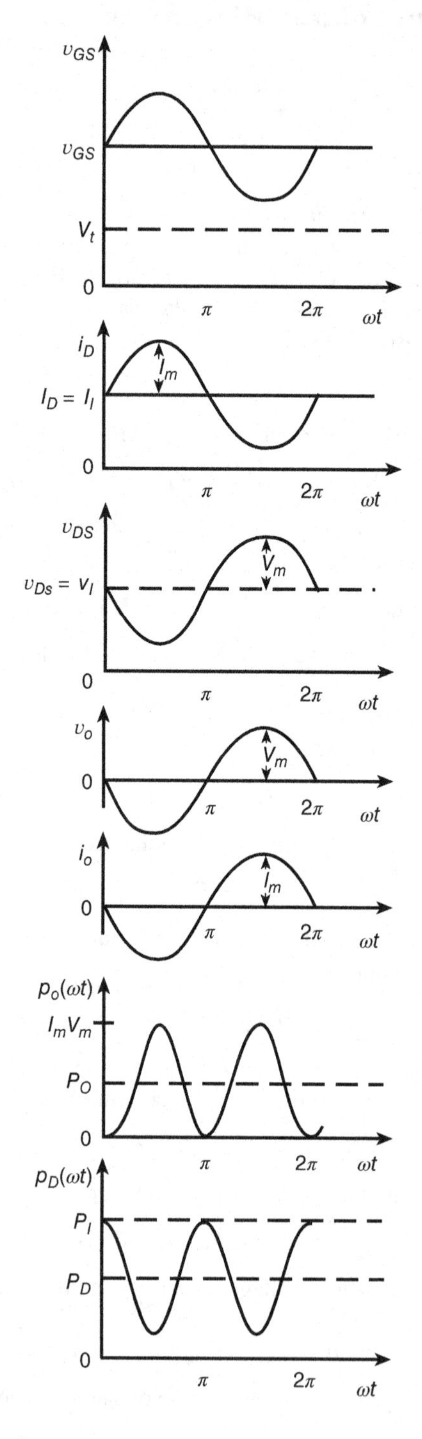

Figure 2.15 Waveforms in Class A RF power amplifier.

The output voltage at the resonant frequency f_0 is

$$v_o = v_{DS} - V_{Cc} = V_I + v_{ds} - V_I = v_{ds} = -V_m \cos \omega t \tag{2.76}$$

where the amplitude of the ac component of the drain-to-source voltage and the output voltage is

$$V_m = RI_m \tag{2.77}$$

The maximum drain-to-source voltage is

$$V_{DS(max)} = V_I + V_{m(max)} \approx 2V_I \tag{2.78}$$

The maximum drain current is

$$I_{DM(max)} = I_I + I_{m(max)} \approx 2I_I \tag{2.79}$$

The maximum MOSFET saturation voltage is

$$v_{DSsat(max)} = v_{GSmax} - V_t = V_{GS} + V_{gsm} - V_t \tag{2.80}$$

Using the square law, the MOSFET aspect ratio is

$$\frac{W}{L} = \frac{I_{DM(max)}}{K_n v^2_{DSsat(max)}} \tag{2.81}$$

2.5.3 Output Power Waveforms

The instantaneous drain power delivered by the drain to the load at the operating frequency $f = f_0 = 1/(2\pi\sqrt{LC})$ is given by

$$p_{ds}(\omega t) = p_o(\omega t) = i_o v_o = (-I_m \cos \omega t)(-V_m \cos \omega t) = I_m V_m \cos^2 \omega t$$
$$= \frac{I_m V_m}{2}(1 + \cos 2\omega t), \tag{2.82}$$

where the double-angle relationship is $\cos^2 \omega t = \frac{1}{2}(1 + \cos 2\omega t)$. The average drain power over a cycle is

$$P_{DS} = P_O = \frac{1}{2\pi}\int_0^{2\pi} p_o d(\omega t) = \frac{1}{2\pi}\int_0^{2\pi} \frac{I_m V_m}{2}(1 + \cos 2\omega t) d(\omega t) = \frac{1}{2} I_m V_m$$
$$= \frac{1}{2} R I_m^2 = \frac{V_m^2}{2R} \tag{2.83}$$

When the operating frequency f is not equal to the resonant frequency f_0, the current through the coupling capacitor is

$$i_{Cc}(\omega t) = -I_m \cos \omega t \tag{2.84}$$

and the ac component of the drain-to-source voltage is

$$v_{ds}(\omega t) = -V_m \cos(\omega t + \phi) \tag{2.85}$$

where ϕ is the phase shift between the current i_{Cc} and the voltage v_{ds}. Using the relationship $\cos \alpha \cos \beta = \frac{1}{2}[\cos(\alpha - \beta) + \cos(\alpha + \beta)]$, the instantaneous power delivered by the drain to the resonant circuit is

$$p_{ds}(\omega t) = i_{C_c}(\omega t) v_{ds}(\omega t) = (-I_m \cos \omega t)[-V_m \cos(\omega t + \phi)] = I_m V_m \cos \omega t \cos(\omega t + \phi)$$
$$= \frac{V_m I_m}{2}[\cos \phi + \cos(2\omega t + \phi)] \tag{2.86}$$

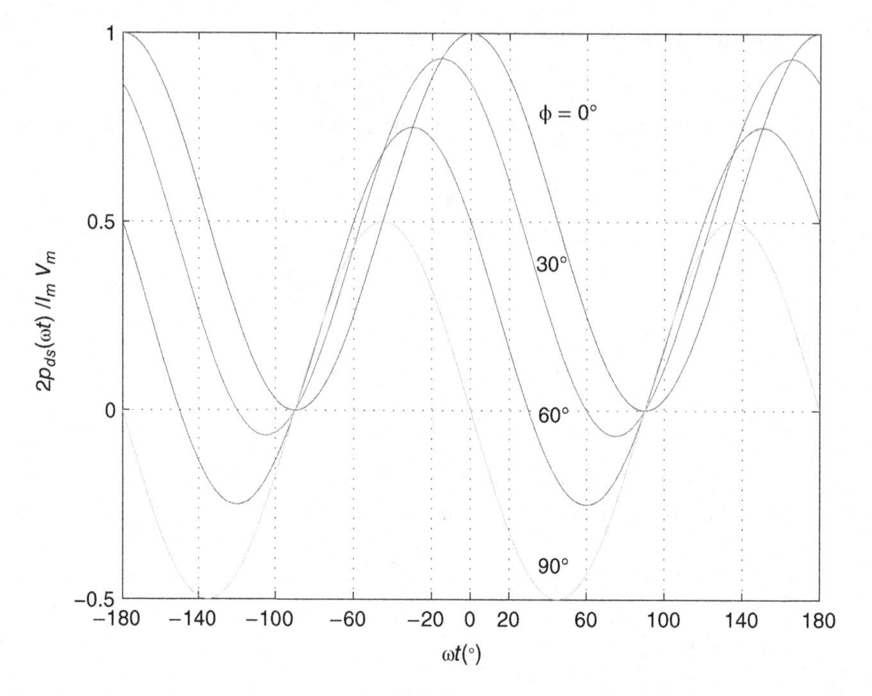

Figure 2.16 Normalized instantaneous drain power $p_{DS}(\omega t)/V_m I_m$ in the Class A RF power amplifier at $f < f_0$ and selected values of the phase shift ϕ.

The time-average drain power is

$$P_{DS} = P_O = \frac{1}{2\pi}\int_0^{2\pi} p_o d(\omega t) = \frac{1}{2\pi}\int_0^{2\pi} \frac{I_m V_m}{2}[\cos\phi + \cos(2\omega t + \phi)]d(\omega t)$$
$$= \frac{1}{2} I_m V_m \cos\phi = \frac{1}{2} R I_m^2 \cos\phi = \frac{V_m^2}{2R}\cos\phi \tag{2.87}$$

Figure 2.16 shows the normalized instantaneous drain power $p_{DS}(\omega t)/V_m I_m$ in the Class A RF power amplifier at $f < f_0$ and selected values of the phase shift ϕ.

2.5.4 Transistor Power Loss Waveforms

The instantaneous power dissipation in the transistor at $f = f_0$ is

$$p_D(\omega t) = i_D v_{DS} = (I_I + I_m \cos\omega t)(V_I - V_m \cos\omega t)$$
$$= I_I V_I - I_m V_m \cos^2\omega t + (V_I I_m - I_I V_m)\cos\omega t$$
$$= I_I V_I - \frac{I_m V_m}{2}(1 + \cos 2\omega t) + (I_m V_I - V_m I_I)\cos\omega t$$
$$= I_I\left(1 + \frac{I_m}{I_I}\cos\omega t\right) V_I\left(1 - \frac{V_m}{V_I}\cos\omega t\right)$$
$$= P_I(1 + \gamma_1 \cos\omega t)(1 - \zeta_1 \cos\omega t) \tag{2.88}$$

where the normalized amplitude of the fundamental component of the drain current is

$$\gamma_1 = \frac{I_m}{I_I} \tag{2.89}$$

and the normalized amplitude of the fundamental component of the drain-to-source voltage is

$$\zeta_1 = \frac{V_m}{V_I} \tag{2.90}$$

Both γ_1 and ζ_1 increase from 0 to ideally 1 as the amplitude of the gate-to-source voltage V_{gsm} increases from 0 to a maximum value. The waveforms of the normalized instantaneous drain power loss $p_D(\omega t)/P_I$ in the Class A RF power amplifier at $\gamma_1 = I_m/I_I = 1, f = f_0$, and selected values of $\zeta_1 = V_m/V_I$ are shown in Fig. 2.17.

The ratio $\gamma_1 = I_m/I_I$ remains constant and $\zeta_1 = V_m/V_I = RI_m/V_I$ varies when the amplitude of the gate-to-source voltage V_{gsm} remains constant and the load resistance R varies. In this case,

$$\begin{aligned} p_D(\omega t) &= i_D v_{DS} = (I_I + I_m \cos \omega t)(V_I - V_m \cos \omega t) = (I_I + I_m \cos \omega t)(V_I - RI_m \cos \omega t) \\ &= I_I \left(1 + \frac{I_m}{I_I} \cos \omega t\right) V_I \left(1 - \frac{V_m}{V_I} \cos \omega t\right) \\ &= I_I \left(1 + \frac{I_m}{I_I} \cos \omega t\right) V_I \left(1 - \frac{RI_m}{V_I} \cos \omega t\right) \\ &= P_I \left(1 + \frac{K_s V_{gsm}}{I_I} \cos \omega t\right) \left(1 - \frac{RK_s V_{gsm}}{V_I} \cos \omega t\right) \end{aligned} \tag{2.91}$$

The average power loss in the transistor over a cycle is

$$\begin{aligned} P_D &= \frac{1}{2\pi} \int_0^{2\pi} p_D \, d(\omega t) = \frac{1}{2\pi} \int_0^{2\pi} [I_I V_I - I_m V_m \cos^2 \omega t + (V_I I_m - I_I V_m) \cos \omega t] d(\omega t) \\ &= I_I V_I - \frac{I_m V_m}{2} = P_I - P_{DS} \end{aligned} \tag{2.92}$$

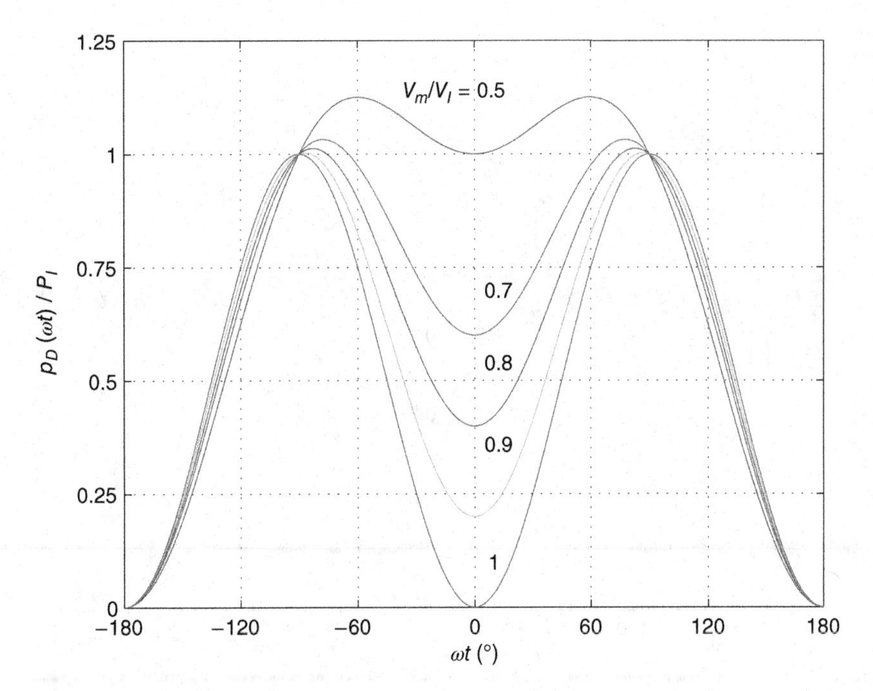

Figure 2.17 Normalized instantaneous drain power loss $p_D(\omega t)/P_I$ in the Class A RF power amplifier for $\gamma_1 = I_m/I_I = 1, f = f_0$, and selected values of $\zeta_1 = V_m/V_I$.

For $I_m = I_I$ and $V_m = V_I$, the instantaneous power dissipation in the transistor at $f = f_0$ becomes

$$p_D(\omega t) = I_I(1 + \cos\ \omega t)V_I(1 - \cos\ \omega t) = P_I(1 - \cos^2\omega t) = \frac{P_I}{2}(1 - \cos\ 2\omega t) \quad (2.93)$$

When the operating frequency f is different from the resonant frequency f_0, the drain-to-source voltage is

$$v_{DS} = V_I - V_m \cos(\omega t + \phi) \quad (2.94)$$

Using the equation $\cos\ \omega t \cos(\omega t + \phi) = \frac{1}{2}[\cos\ \phi + \cos(2\omega t + \phi)]$, the instantaneous power dissipation in the transistor is found as

$$\begin{aligned} p_D(\omega t) &= i_D v_{DS} = (I_I + I_m \cos\ \omega t)[V_I - V_m \cos\ (\omega t + \phi)] \\ &= I_I V_I - I_m V_m \cos\ \omega t \cos(\omega t + \phi) + V_I I_m \cos\ \omega t - I_I V_m \cos(\omega t + \phi) \\ &= I_I V_I - \frac{I_m V_m}{2}[\cos\ \phi + \cos(2\omega t + \phi)] + V_I I_m \cos\ \omega t - I_I V_m \cos(\omega t + \phi) \\ &= I_I \left(1 + \frac{I_m}{I_I} \cos\ \omega t\right) V_I \left[1 - \frac{V_m}{V_I} \cos\ (\omega t + \phi)\right] \\ &= P_I(1 + \gamma_1 \cos\ \omega t)[1 - \xi_1 \cos\ (\omega t + \phi)] \end{aligned} \quad (2.95)$$

where ϕ is the phase shift between the drain current i_D and the drain-to-source voltage v_{DS}. The waveforms of the normalized instantaneous drain power loss $p_D(\omega t)/P_I$ in the Class A RF power amplifier at $\gamma_1 = I_m/I_I = 1$ and selected values for $\phi = 10°$ and $\phi = -10°$ are shown in Figs. 2.18 and 2.19, respectively. The time-average transistor power dissipation is

$$P_D = \frac{1}{2\pi}\int_0^{2\pi} \phi_D(\omega t)d(\omega t) = P_I - \frac{I_m V_m}{2} \cos\ \phi \quad (2.96)$$

As the angle ϕ increases, the time-average transistor power dissipation P_D also increases.

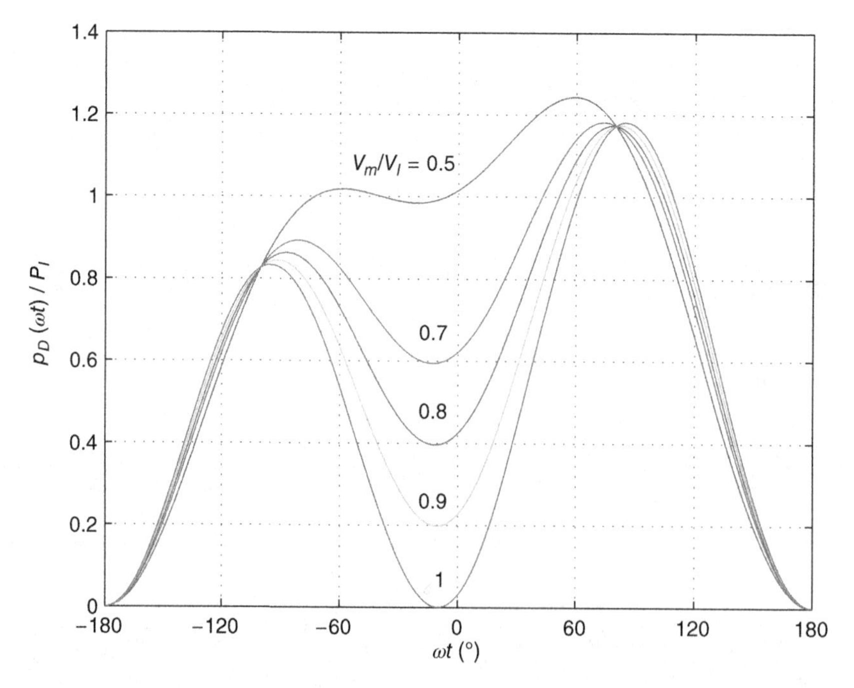

Figure 2.18 Normalized instantaneous drain power loss $p_D(\omega t)/P_I$ in the Class A RF power amplifier for $\gamma_1 = I_m/I_I = 1, f < f_0, \phi = 10°$, and selected values of $\xi_1 = V_m/V_I$.

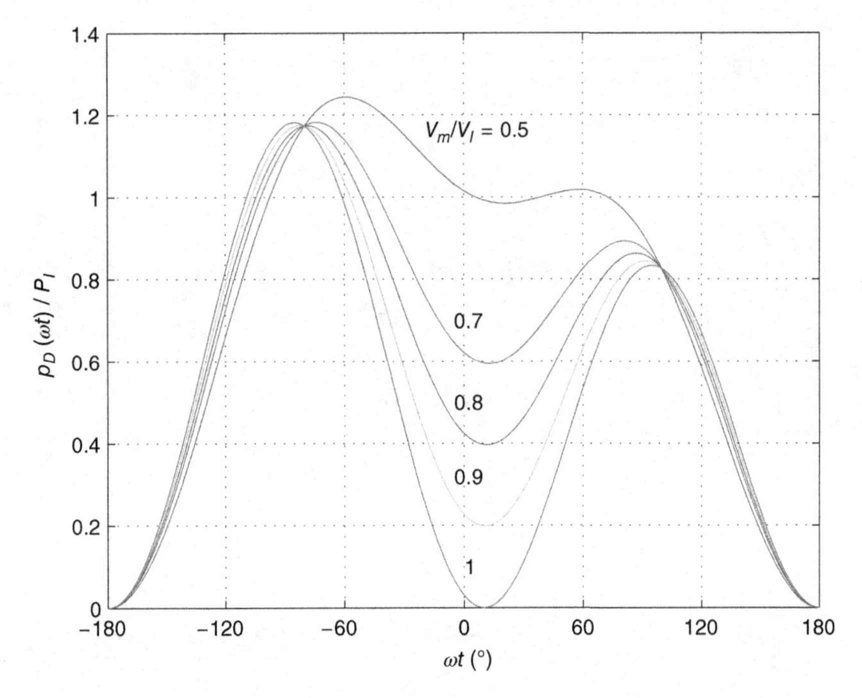

Figure 2.19 Normalized instantaneous drain power loss $p_D(\omega t)/P_I$ in the Class A RF power amplifier for $\gamma_1 = I_m/I_I = 1, f > f_0, \phi = -10°$, and selected values of $\xi_1 = V_m/V_I$.

When the load resistance R remains constant, but the amplitude of the gate-to-source voltage varies, we get

$$p_D(\omega t) = i_D v_{DS} = (I_I + I_m \cos\ \omega t)(V_I - V_m \cos\ \omega t)$$
$$= I_I\left(1 + \frac{RI_m}{RI_I}\cos\ \omega t\right) V_I\left(1 - \frac{V_m}{V_I}\cos\ \omega t\right)$$
$$= I_I\left(1 + \frac{V_m}{V_I}\cos\ \omega t\right) V_I\left(1 - \frac{V_m}{V_I}\cos\ \omega t\right)$$
$$= P_I\left[1 - \left(\frac{V_m}{V_I}\right)\cos^2 \omega t\right] = P_I(1 - \xi_1^2\cos^2 \omega t) = P_I\left[1 - \frac{\xi_1}{2}(1 + \cos\ 2\omega t)\right] \quad (2.97)$$

Figure 2.20 shows normalized instantaneous drain power loss $p_D(\omega t)/P_I$ in the Class A RF power amplifier for $f = f_0$ and selected values of $\gamma_1 = \xi_1 = V_m/V_I$. Figures 2.21 and 2.22 show normalized instantaneous drain power loss $p_D(\omega t)/P_I$ in the Class A RF power amplifier at selected values of $\gamma_1 = \xi_1 = V_m/V_I$ for $\phi = 10°$ and $\phi = -10°$, respectively.

2.6 Energy Parameters of Class A RF Power Amplifier

2.6.1 Drain Efficiency of Class A RF Power Amplifier

The operating point (the bias point) $Q(V_{DSQ}, I_{DQ})$ of the Class A RF power amplifier should be selected at the midpoint of the linear region of the transistor, where

$$V_{DSQ} = \frac{V_I}{2} \quad (2.98)$$

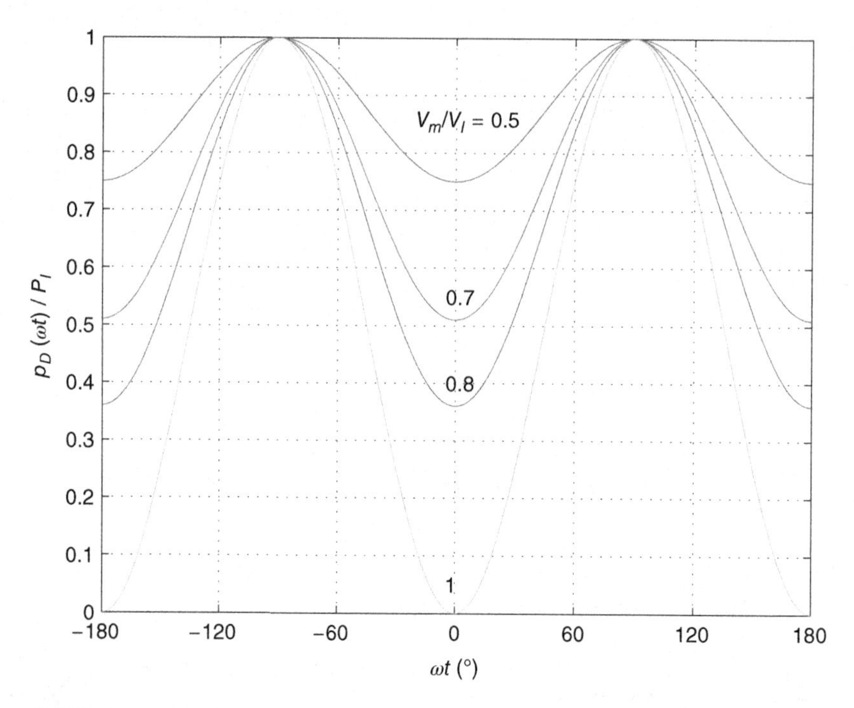

Figure 2.20 Normalized instantaneous drain power loss $p_D(\omega t)/P_I$ in the Class A RF power amplifier for $f = f_0$ and selected values of $\gamma_1 = \xi_1 = V_m/V_I$.

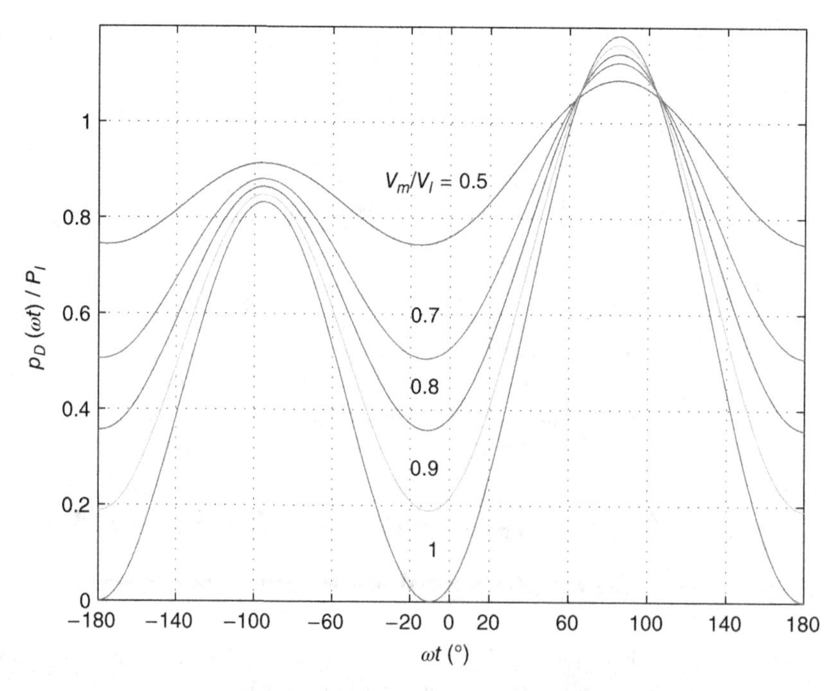

Figure 2.21 Normalized instantaneous drain power loss $p_D(\omega t)/P_I$ in the Class A RF power amplifier for $f < f_0$, $\phi = 10°$, and selected values of $\gamma_1 = \xi_1 = V_m/V_I$.

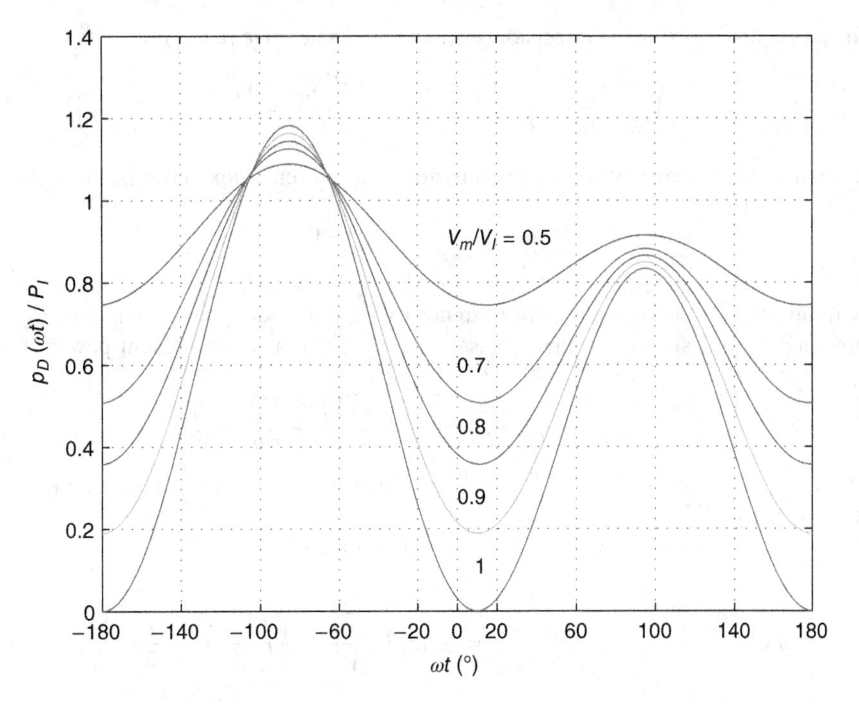

Figure 2.22 Normalized instantaneous drain power loss $p_D(\omega t)/P_I$ in the Class A RF power amplifier for $f > f_0$, $\phi = -10°$, and selected values of $\gamma_1 = \xi_1 = V_m/V_I$.

and

$$I_{DQ} = I_I = \frac{I_{DSmax}}{2} \tag{2.99}$$

The dc supply current of the Class A RF power amplifier is

$$I_I = I_{m(max)} = \frac{V_{m(max)}}{R} = \frac{V_{DS}}{R} = \frac{V_I}{R} \tag{2.100}$$

The dc resistance seen by the power supply V_I is

$$R_{DC} = \frac{V_I}{I_I} = R \tag{2.101}$$

The dc supply power (or the dc input power) in the Class A RF power amplifier is given by

$$P_I = I_D V_{DS} = I_{m(max)} V_I = I_I V_I = R I_I^2 = \frac{V_I^2}{R} \tag{2.102}$$

The dc supply power is constant and independent of the amplitude of the output voltage V_m. The power delivered by the MOSFET to the output RF load at $f = f_0$ is given by

$$P_{DS} = \frac{I_m V_m}{2} = \frac{R I_m^2}{2} = \frac{V_m^2}{2R} = \frac{1}{2}\left(\frac{V_I^2}{R}\right)\left(\frac{V_m}{V_I}\right)^2 = \frac{P_I}{2}\left(\frac{V_m}{V_I}\right) \tag{2.103}$$

For $V_{m(max)} = V_I$, the drain power reaches its maximum value

$$P_{DSmax} = \frac{V_I^2}{2R} = \frac{P_I}{2} \tag{2.104}$$

The output power is proportional to the square of the output voltage amplitude V_m.

The power dissipated in the transistor (excluding the gate drive power) is

$$P_D = P_I - P_{DS} = I_I V_I - \frac{V_m^2}{2R} = \frac{V_I^2}{R} - \frac{V_m^2}{2R} = \frac{V_I^2}{R} - \frac{V_m^2 V_I^2}{2RV_I^2} = \frac{V_I^2}{R}\left[1 - \frac{1}{2}\left(\frac{V_m}{V_I}\right)^2\right] \tag{2.105}$$

The maximum power dissipation in the transistor occurs at zero output power, that is, at $P_{DS} = 0$ (i.e., for $V_m = 0$)

$$P_{Dmax} = P_I = I_I V_I = \frac{V_I^2}{R} \tag{2.106}$$

A heat sink for the transistor should be designed for the maximum power dissipation P_{Dmax}. The minimum power dissipation in the transistor occurs at the maximum output power, that is, at P_{DSmax}

$$P_{Dmin} = P_I - P_{DSmax} = \frac{V_I^2}{R} - \frac{V_I^2}{2R} = \frac{V_I^2}{2R} = \frac{P_I}{2} \tag{2.107}$$

Figure 2.23 shows plots of P_I, P_{DS}, and P_D as functions of the normalized amplitude of the output voltage V_m/V_I.

The drain efficiency of the Class A RF power amplifier at $f = f_0$ is given by

$$\eta_D = \frac{P_{DS}}{P_I} = \frac{1}{2}\left(\frac{I_m}{I_I}\right)\left(\frac{V_m}{V_I}\right) = \frac{1}{2}\gamma_1\xi_1 = \frac{\frac{V_m^2}{2R}}{\frac{V_I^2}{R}} = \frac{1}{2}\left(\frac{V_m}{V_I}\right)^2 = \frac{1}{2}\xi_1^2 \tag{2.108}$$

The drain efficiency η_D as a function of V_m/V_I is shown in Fig. 2.24. The drain efficiency is proportional to $(V_m/V_I)^2$. For example, for $V_m/V_I = \frac{1}{2}$, $\eta_D = 12.5\%$. The maximum theoretical

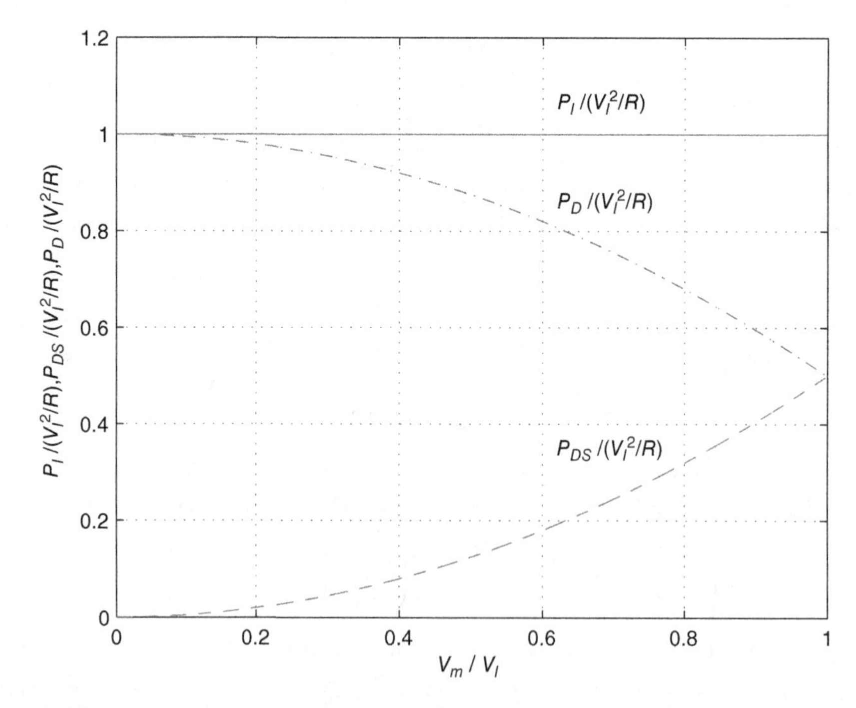

Figure 2.23 Normalized DC supply power $P_I(V_I^2/R)$, drain power $P_{DS}/V_I^2/R)$, and dissipated power $P_D/(V_I^2/R)$ for the Class A RF power amplifier with choke inductor as functions of the normalized output voltage amplitude V_m/V_I.

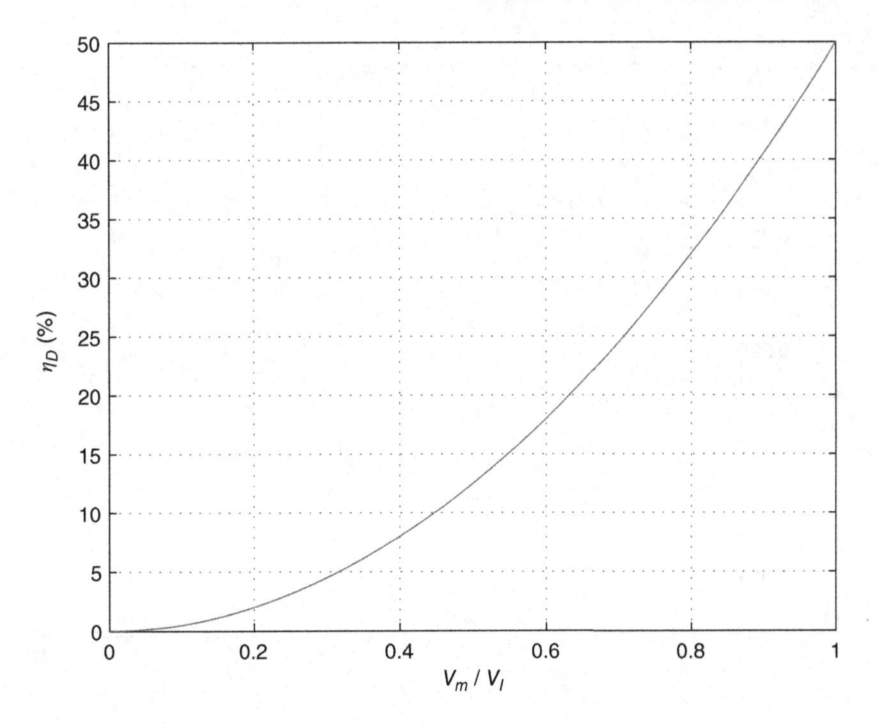

Figure 2.24 Drain efficiency η_D of the Class A RF power amplifier (with choke inductor) as a function of the normalized output voltage amplitude V_m/V_I.

efficiency of the Class A RF amplifier with choke L_f at $v_{DSmin} = v_{DSsat} = 0$ is

$$\eta_{Dmax} = \frac{1}{2}\left(\frac{V_{m(max)}}{V_I}\right)^2 = 0.5 \tag{2.109}$$

2.6.2 Statistical Characterization of Class A RF Power Amplifier

Assuming Rayleigh's PDF of the output voltage amplitude V_m,

$$h(V_m) = \frac{V_m}{\sigma^2} e^{-\frac{V_m^2}{2\sigma^2}} \tag{2.110}$$

one obtains the product

$$\eta_D h(V_m) = \frac{1}{2}\left(\frac{V_m}{V_I}\right)^2 \frac{V_m}{\sigma^2} e^{-\frac{V_m^2}{2\sigma^2}} \tag{2.111}$$

and the long-term average drain efficiency

$$\eta_{D(AV)} = \int_0^{V_I} \eta_D h(V_m) dV_m = \frac{1}{2V_I^2\sigma^2} \int_0^{V_I} V_m^3 e^{-\frac{V_m^2}{2\sigma^2}} dV_m = \frac{\sigma^2}{V_I^2}\left(1 - \frac{\frac{V_I^2}{2\sigma^2} + 1}{e^{\frac{V_I^2}{2\sigma^2}}}\right) \tag{2.112}$$

Figure 2.25 shows plots of drain efficiency η_D, Rayleigh's PDF $h(V_m)$, and their product $\eta_D h(V_m)$ for the Class A RF power amplifier (with choke inductor) as a function of the amplitude of voltage V_m at $V_I = 10$ V and $\sigma = 3$. Figure 2.26 shows the long-term average drain efficiency $\eta_{D(AV)}$ as function of σ for the Class A RF power amplifier (with choke inductor) at $V_I = 10$ V.

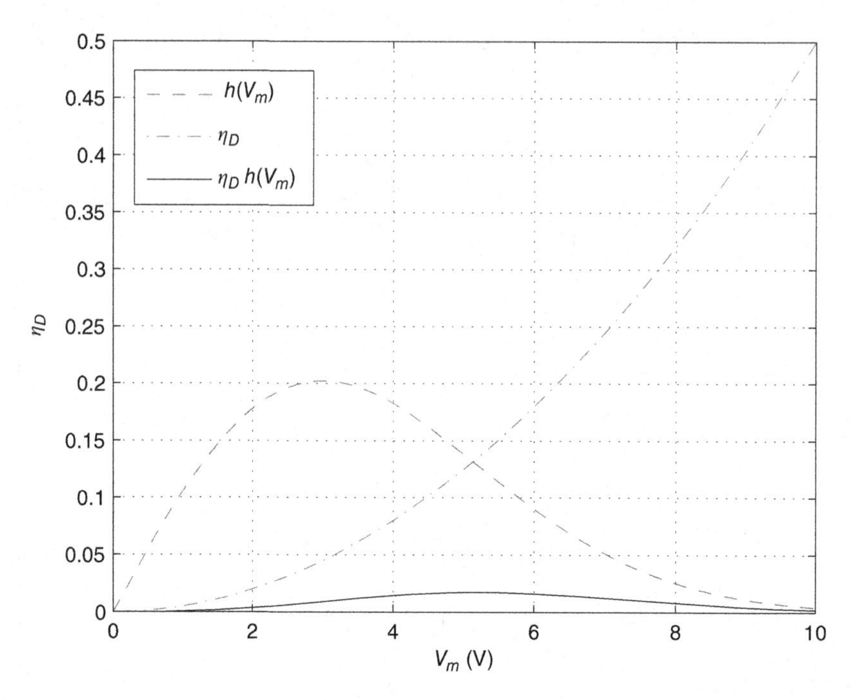

Figure 2.25 Drain efficiency η_D, Rayleigh's probability-density function (PDF) $h(V_m)$, and their product $\eta_D h(V_m)$ for the Class A RF power amplifier (with choke inductor) as a function of the amplitude of voltage V_m at $V_I = 10$ V and $\sigma = 3$.

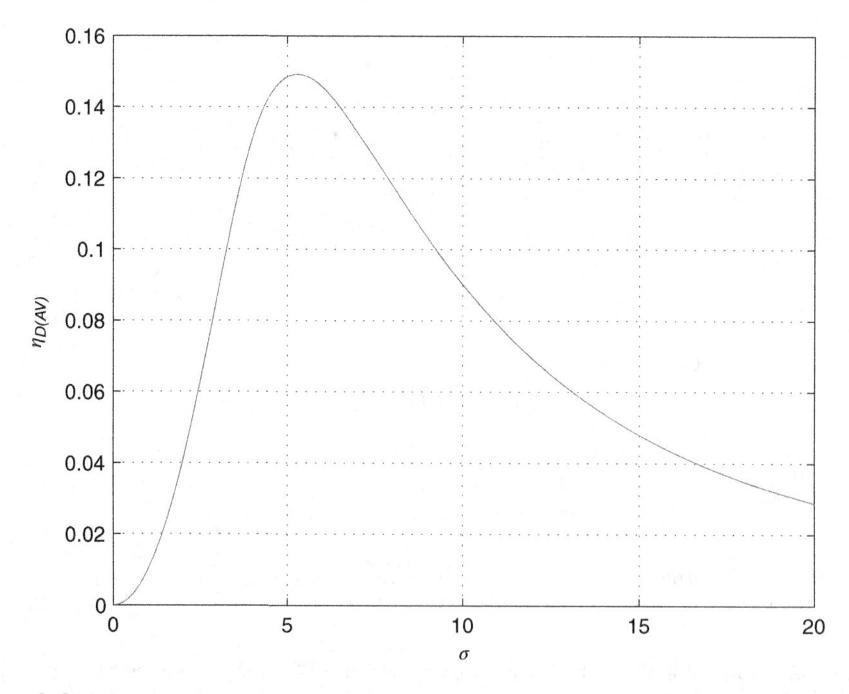

Figure 2.26 Long-term average drain efficiency $\eta_{D(AV)}$ as function of σ for the Class A RF power amplifier (with choke inductor) at $V_I = 10$ V.

In practice, the maximum efficiency of the Class A RF power amplifier is about 35%, but the long-term average efficiency is only about 5%.

The output power is

$$P_O = \frac{V_m^2}{2R} \tag{2.113}$$

The maximum output power occurs at $V_{m(max)} = V_I$ and is given by

$$P_{Omax} = \frac{V_{m(max)}^2}{2R} = \frac{V_I^2}{2R} = \frac{P_I}{2} \tag{2.114}$$

yielding

$$P_I = 2P_{Omax} \tag{2.115}$$

To achieve the specified maximum output power P_{Omax} at a given power supply voltage V_I, the required load resistance is given by

$$R = \frac{V_{m(max)}^2}{2P_{Omax}} \tag{2.116}$$

The drain efficiency is

$$\eta_D = \frac{P_O}{P_I} = \frac{P_O}{2P_{Omax}} = \frac{1}{2}\left(\frac{P_O}{P_{Omax}}\right) \tag{2.117}$$

Figure 2.27 shows the efficiency of the Class A RF power amplifier as a function of the normalized output power P_O/P_{Omax}. In the Class A RF amplifier, the efficiency is directly proportional to the output power. This is because the dc supply power P_I is constant.

Assuming Rayleigh's PDF of the output power P_O

$$g(P_O) = \frac{P_O}{\sigma^2} e^{-\frac{P_O^2}{2\sigma^2}} \tag{2.118}$$

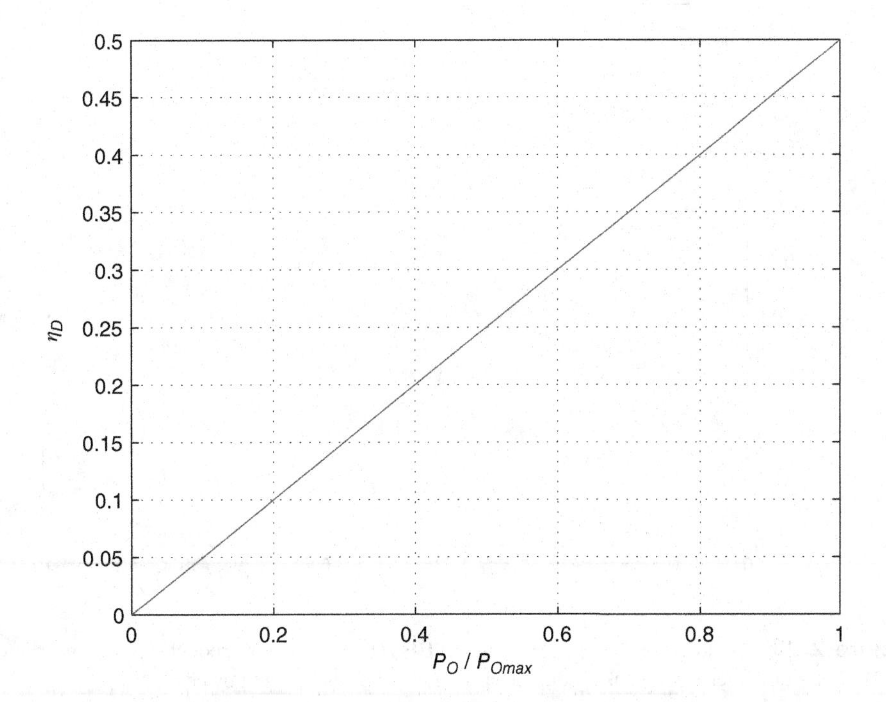

Figure 2.27 Drain efficiency η_D of the Class A RF power amplifier (with choke inductor) as a function of the normalized output power P_O/P_{Omax}.

the product is obtained

$$\eta_D g(P_O) = \frac{1}{2}\left(\frac{P_O}{P_{Omax}}\right)\frac{P_O}{\sigma^2}e^{-\frac{P_O^2}{2\sigma^2}} \tag{2.119}$$

and the long-term average drain efficiency is obtained

$$\eta_{D(AV)} = \int_0^{P_{Omax}} \eta_D g(P_O)dP_O = \int_0^{P_{Omax}} \frac{1}{2}\left(\frac{P_O}{P_{Omax}}\right)\frac{P_O}{\sigma^2}e^{-\frac{P_O^2}{2\sigma^2}}dP_O$$

$$= \frac{1}{2P_{Omax}\sigma^2}\int_0^{P_{Omax}} P_O^2 e^{-\frac{P_O^2}{2\sigma^2}}dP_O = \frac{\sigma}{2P_{Omax}}\sqrt{\frac{\pi}{2}} \tag{2.120}$$

Figure 2.28 shows plots of drain efficiency η_D, Rayleigh's PDF $g(P_O)$, and their product $\eta_D g(P_O)$ for the Class A RF power amplifier (with choke inductor) as a function of the output power P_O at $P_{Omax} = 10$ W and $\sigma = 3$. For $P_{Omax} = 10$ W and $\sigma = 3$, the long-term average efficiency is

$$\eta_{D(AV)} = \frac{\sigma}{2P_{Omax}}\sqrt{\frac{\pi}{2}} = \frac{3}{2\times 10}\sqrt{\frac{\pi}{2}} = 18.8\% \tag{2.121}$$

For $f \neq f_0$, the drain efficiency is

$$\eta_D = \frac{P_{DS}}{P_I} = \frac{1}{2}\left(\frac{V_m}{V_I}^2\right)\cos\phi \tag{2.122}$$

Figure 2.29 shows plots of drain efficiency η_D as a function of V_m/V_I at selected values of ϕ for the Class A RF power amplifier.

As ϕ increases, the drain efficiency η_D also decreases.

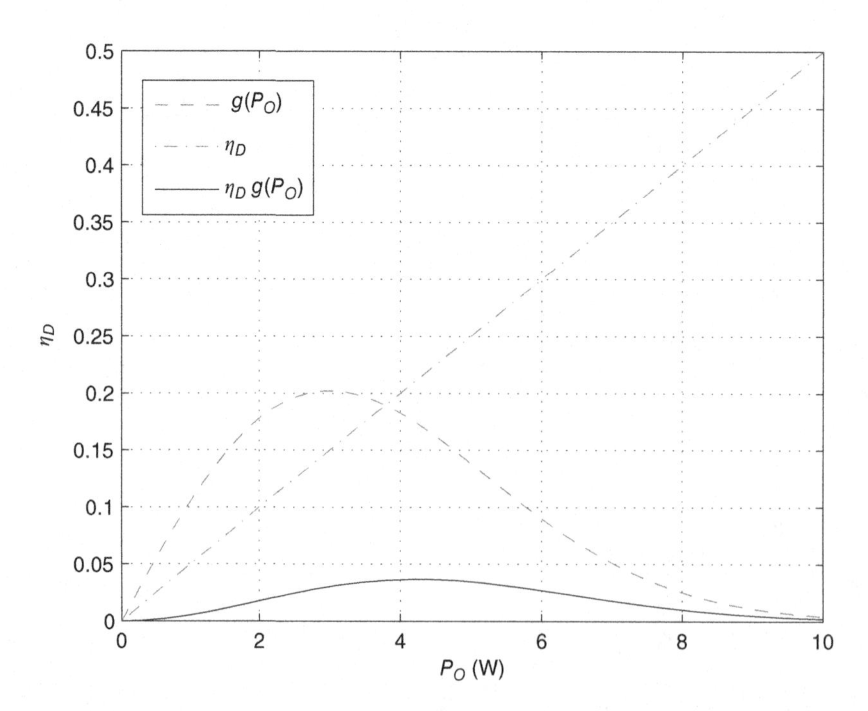

Figure 2.28 Drain efficiency η_D, Rayleigh's PDF $g(P_O)$, and their product $\eta_D g(P_O)$ for the Class A RF power amplifier (with choke inductor) as a function of the output power P_O at $P_{Omax} = 10$ W and $\sigma = 3$.

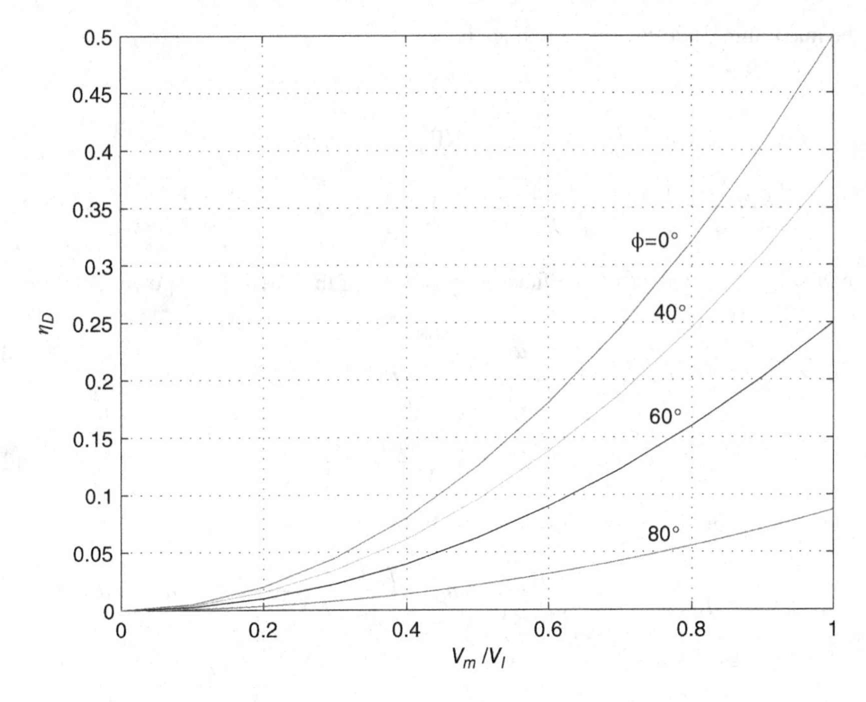

Figure 2.29 Drain efficiency η_D as a function of V_m/V_I at selected values of ϕ for the Class A RF power amplifier.

2.6.3 Drain Efficiency at Nonzero Minimum Drain-to-Source Voltage

In reality, a MOSFET behaves as a dependent current source above a minimum drain-to-source voltage

$$v_{DSmin} = v_{DSsat} = v_{GS} - V_t = V_{GS} + V_{gsm} - V_t \tag{2.123}$$

and therefore the maximum amplitude of the ac component of v_{DS} is given by

$$V_{m(max)} = V_I - v_{DSmin} = V_I - V_{GS} - V_{gsm} + V_t \tag{2.124}$$

resulting in the maximum drain efficiency

$$\eta_{Dmax} = \frac{1}{2}\left(\frac{V_{m(max)}}{V_I}\right)^2 = \frac{1}{2}\left(\frac{V_I - v_{DSmin}}{V_I}\right)^2 = \frac{1}{2}\left(1 - \frac{v_{DSmin}}{V_I}\right)^2 \leq \frac{1}{2} \tag{2.125}$$

The efficiency of the Class A RF power amplifier is low. Therefore, this amplifier is used in low power applications or as an intermediate stage in a cascaded amplifier. The maximum drain efficiency of the Class A RF amplifier is 50%. In contrast, the maximum efficiency of Class A audio power amplifiers is 25%. This is because the RFC is replaced by a resistor at low frequencies, which dissipates power.

2.6.4 Output-Power Capability of Class A RF Power Amplifier

In the Class A RF power amplifier, $I_{m(max)} = I_I$ and $V_{m(max)} = V_I$, the maximum drain current is

$$I_{DM} = I_I + I_{m(max)} = 2I_I \tag{2.126}$$

and the maximum drain-to-source voltage is

$$V_{DSM} = V_I + V_{m(max)} = 2V_I \tag{2.127}$$

The output-power capability of the Class A RF power amplifier is

$$c_p = \frac{P_{Omax}}{V_{DSM}I_{DM}} = \frac{1}{2}\left(\frac{I_{m(max)}}{I_{DM}}\right)\left(\frac{V_{m(max)}}{V_{DSM}}\right) = \frac{1}{2}\alpha_{1max}\beta_{1max} = \frac{1}{2} \times \frac{1}{2} \times \frac{1}{2} = \frac{1}{8} = 0.125 \tag{2.128}$$

where the maximum values of coefficients α_1 and β_1 for the Class A RF power amplifier are

$$\alpha_{1max} = \frac{I_{m(max)}}{I_{DM}} = \frac{1}{2} \tag{2.129}$$

and

$$\beta_{1max} = \frac{V_{m(max)}}{V_{DSM}} = \frac{1}{2} \tag{2.130}$$

Alternatively, the output power capability can be described as

$$c_p = \frac{P_{Omax}}{I_{DM}V_{DSM}} = \frac{\eta_{Dmax}P_I}{I_{DM}V_{DSM}} = \eta_{Dmax}\left(\frac{I_I}{I_{DM}}\right)\left(\frac{V_I}{V_{DSM}}\right) = \eta_{Dmax}\alpha_0\beta_0$$
$$= \frac{1}{2} \times \frac{1}{2} \times \frac{1}{2} = \frac{1}{8} = 0.125 \tag{2.131}$$

where $\eta_{Dmax} = 0.5$ and the coefficients α_0 and β_0 for the Class A RF power amplifier are

$$\alpha_0 = \frac{I_I}{I_{DM}} = \frac{1}{2} \tag{2.132}$$

and

$$\beta_0 = \frac{V_I}{V_{DSM}} = \frac{1}{2} \tag{2.133}$$

2.6.5 Gate Drive Power

The gate drive power is

$$P_G = \frac{1}{2}I_{gm}V_{gsm}\cos\phi_G = \frac{1}{2}I_{gm}^2R_G\cos\phi_G = \frac{V_{gsm}^2}{2R_G}\cos\phi_G \tag{2.134}$$

where R_G is the gate resistance and ϕ_G is the phase shift between the gate voltage and the gate current. The gate impedance consists of a gate resistance R_G and a gate input capacitance $C_g = C_{gs} + (1 - A_m)C_{gd}$, where C_{gs} is the gate-to-source capacitance, C_{gd} is the gate-to-drain capacitance, and A_m is the gate-to-drain voltage gain. The gate input capacitance is increased by the Miller effect.

The power gain of the amplifier is

$$k_p = \frac{P_O}{P_G} = \frac{0.5V_m^2/R}{0.5V_{gsm}^2\cos\phi_G/R_G} = \left(\frac{V_m}{V_{gsm}}\right)^2\left(\frac{R_G}{R}\right)\frac{1}{\cos\phi_G} \tag{2.135}$$

2.7 Parallel-Resonant Circuit

2.7.1 Quality Factor of Parallel-Resonant Circuit

The resonant angular frequency of the resonant circuit is

$$\omega_0 = \frac{1}{\sqrt{LC}} \tag{2.136}$$

The characteristic impedance of the parallel-resonant circuit is

$$Z_o = \sqrt{\frac{L}{C}} = \omega_0 L = \frac{1}{\omega_0 C} \tag{2.137}$$

The loaded quality factor is

$$Q_L = 2\pi \frac{\text{Maximum instantaneous energy stored at } f_0}{\text{Total energy lost per cycle at } T_0 = 1/f_0} = 2\pi \frac{[w_L(\omega_0 t) + w_C(\omega_0 t)]_{max}}{P_O T_0}$$
$$= \omega_0 \frac{[w_L(\omega_0 t) + w_C(\omega_0 t)]_{max}}{P_O} = \omega_0 \frac{w_L(\omega_0 t)_{max}}{P_O} = \omega_0 \frac{w_C(\omega_0 t)_{max}}{P_O} \tag{2.138}$$

where the instantaneous energy stored in the capacitor C is

$$w_C(\omega_0 t) = \frac{1}{2} C v_o^2 = \frac{1}{2} C V_m^2 \cos^2 \omega_0 t \tag{2.139}$$

and the instantaneous energy stored in the inductor L is

$$w_L(\omega_0 t) = \frac{1}{2} L i_L^2 = \frac{1}{2} L I_m^2 \sin^2 \omega_0 t = \frac{1}{2} L \left(\frac{V_m}{\omega_0 L}\right)^2 \sin^2 \omega_0 t = \frac{1}{2} C V_m^2 \sin^2 \omega_0 t \tag{2.140}$$

Hence,

$$Q_L = 2\pi \frac{\frac{1}{2} C V_m^2 (\sin^2 \omega_0 t + \cos^2 \omega_0 t)}{\frac{V_m^2}{2R f_0}} = \omega_0 C R = \frac{R}{Z_o} = \frac{R}{\omega_0 L} = R\sqrt{\frac{C}{L}}$$
$$= \omega_0 \frac{\frac{1}{2} C V_m^2}{P_O} = \omega_0 \frac{W_{Cm}}{P_O} = \omega_0 \frac{\frac{1}{2} L I_m^2}{P_O} = \omega_0 \frac{W_{Lm}}{P_O} \tag{2.141}$$

The amplitude of the current through the inductor L and the capacitor C is

$$I_{Lm} = I_{Cm} = \frac{V_m}{\omega_0 L} = \frac{R I_m}{\omega_0 L} = \frac{V_m}{1/\omega_0 C} = \frac{R I_m}{1/\omega_0 C} = \frac{R I_m}{Z_o} = Q_L I_m \tag{2.142}$$

2.7.2 Impedance of Parallel-Resonant Circuit

For the basic parallel-resonant circuit L-C-R, the output voltage v_o is equal to the fundamental component of the drain-to-source voltage v_{ds}

$$v_o = v_{ds} \tag{2.143}$$

For the ideal parallel-resonant circuit, $i_{C_c} = i_o = -i_d$, yielding the relationship among the fundamental components $I_{C_c1} = I_o = -I_d$. Hence, the admittance of the parallel-resonant circuit in the s-domain is

$$Y(s) = \frac{I_{C_c}}{V_{ds}} = \frac{1}{R} + sC + \frac{1}{sL} = \frac{LCRs^2 + Ls + R}{LRs} = \frac{C\left(s^2 + s\frac{1}{RC} + \frac{1}{LC}\right)}{s} \tag{2.144}$$

and the impedance of the parallel-resonant circuit in the s-domain is

$$Z(s) = \frac{1}{Y(s)} = \frac{s}{C\left(s^2 + s\frac{1}{RC} + \frac{1}{LC}\right)} = \frac{s}{C(s^2 + 2\zeta\omega_0 s + \omega_0^2)} \tag{2.145}$$

where $2\zeta\omega_0 = 1/RC$. Setting $s = j\omega$, one obtains the admittance of the parallel-resonant circuit in the frequency domain

$$Y(j\omega) = \frac{1}{R} + j\left(\omega C - \frac{1}{\omega L}\right) = \frac{1}{R}\left[1 + jQ_L\left(\frac{\omega}{\omega_0} - \frac{\omega_0}{\omega}\right)\right] = |Y|e^{j\phi_Y} \tag{2.146}$$

in which

$$|Y| = \frac{1}{R}\sqrt{1 + Q_L^2\left(\frac{\omega}{\omega_0} - \frac{\omega_0}{\omega}\right)^2} \tag{2.147}$$

and

$$\phi_Y = \arctan\left[Q_L\left(\frac{\omega}{\omega_0} - \frac{\omega_0}{\omega}\right)\right] \tag{2.148}$$

The impedance of the parallel-resonant circuit is

$$Z(j\omega) = \frac{1}{Y(j\omega)} = \frac{1}{\frac{1}{R} + j\left(\omega C - \frac{1}{\omega L}\right)} = \frac{R}{1 + jQ_L\left(\frac{\omega}{\omega_0} - \frac{\omega_0}{\omega}\right)} = |Z|e^{j\phi_Z} \tag{2.149}$$

where

$$|Z| = \frac{R}{\sqrt{1 + Q_L^2\left(\frac{\omega}{\omega_0} - \frac{\omega_0}{\omega}\right)^2}} \tag{2.150}$$

and

$$\phi_Z = -\arctan\left[Q_L\left(\frac{\omega}{\omega_0} - \frac{\omega_0}{\omega}\right)\right] \tag{2.151}$$

Figures 2.30 and 2.31 show the normalized input impedance $|Z|/R$ and phase ϕ_Z of the parallel-resonant circuit at different values of Q_L.

2.7.3 Bandwidth of Parallel-Resonant Circuit

The coupling capacitor current to the output voltage transfer function (which is a transresistance) is given by

$$Z = \frac{V_o}{I_{C_c}} = \frac{V_{ds}}{I_{C_c}} = \frac{R}{1 + jQ_L\left(\frac{\omega}{\omega_0} - \frac{\omega_0}{\omega}\right)} = |Z|e^{j\phi_Z} \tag{2.152}$$

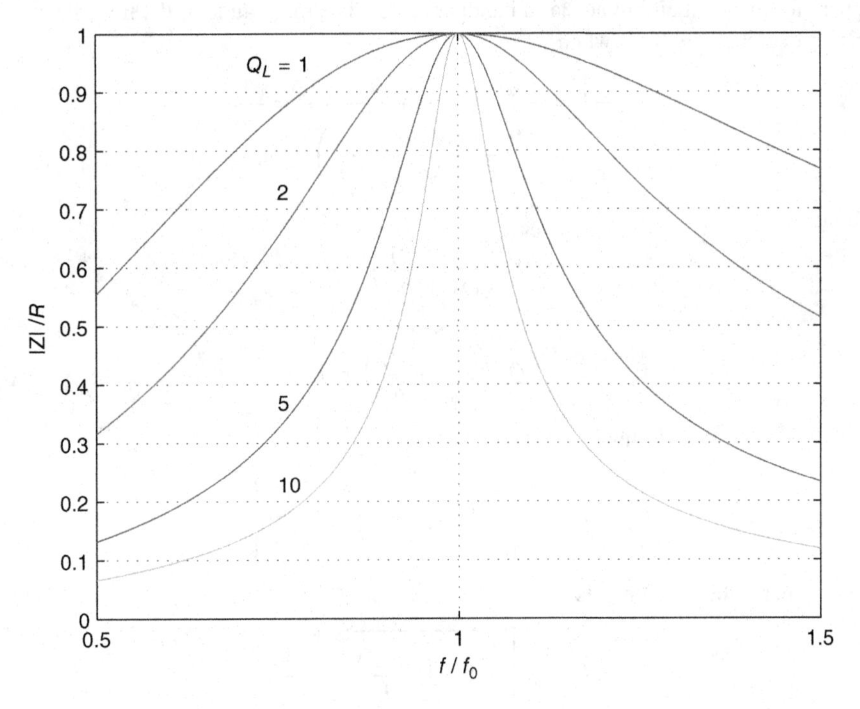

Figure 2.30 Normalized magnitude of the input impedance of the parallel-resonant circuit $|Z|/R$ at different values of Q_L.

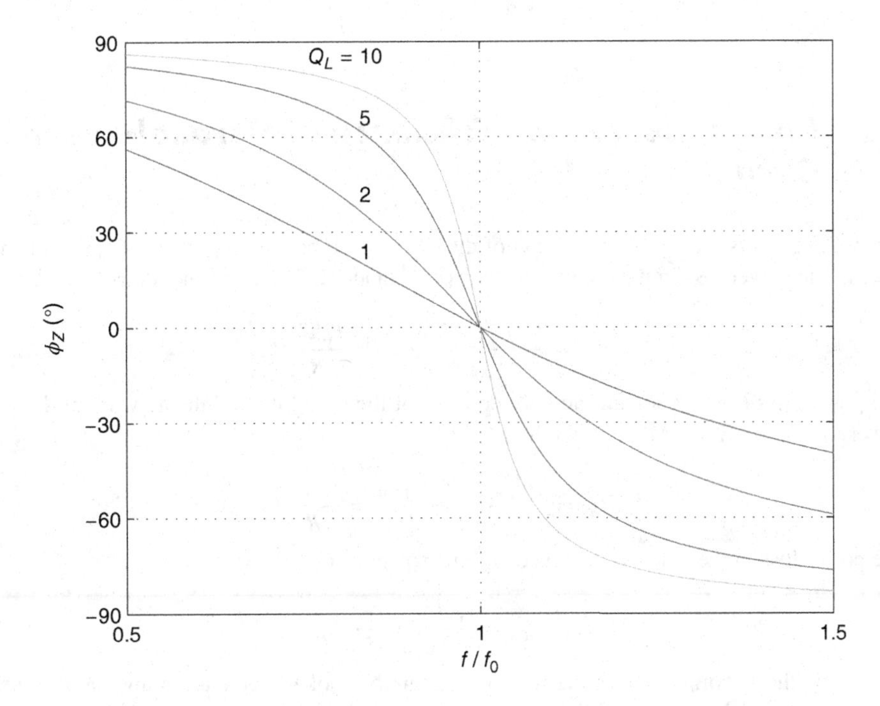

Figure 2.31 Phase of the input impedance of the parallel-resonant circuit ϕ_Z at different values of Q_L.

The parallel-resonant circuit acts as a bandpass filter. The magnitude of the impedance Z of the amplifier decreases by 3 dB when

$$\frac{|Z|}{R} = \frac{1}{\sqrt{1 + Q_L^2\left(\frac{\omega}{\omega_0} - \frac{\omega_0}{\omega}\right)^2}} = \frac{1}{\sqrt{2}} \tag{2.153}$$

yielding

$$Q_L\left(\frac{\omega}{\omega_0} - \frac{\omega_0}{\omega}\right) = -1 \tag{2.154}$$

and

$$Q_L\left(\frac{\omega}{\omega_0} - \frac{\omega_0}{\omega}\right) = 1 \tag{2.155}$$

Hence, the lower 3-dB frequency is

$$f_L = f_0\left[\sqrt{1 + \left(\frac{1}{2Q_L}\right)^2} - \frac{1}{2Q_L}\right] \tag{2.156}$$

and the upper 3-dB frequency is

$$f_H = f_0\left[\sqrt{1 + \left(\frac{1}{2Q_L}\right)^2} + \frac{1}{2Q_L}\right] \tag{2.157}$$

The 3-dB bandwidth of the resonant circuit is

$$BW = f_H - f_L = \frac{f_0}{Q_L} \tag{2.158}$$

2.8 Power Losses and Efficiency of Parallel-Resonant Circuit

Figure 2.32 shows the equivalent circuit of the Class A RF power amplifier with parasitic resistances. The power loss in the equivalent series resistance (ESR) of the inductor r_L is

$$P_{rL} = \frac{r_L I_{Lm}^2}{2} = \frac{r_L Q_L^2 I_m^2}{2} = \frac{r_L Q_L^2}{R} P_O \tag{2.159}$$

The power loss in the resonant inductor consists of the core loss and the ac winding loss. The power loss in the ESR of the capacitor r_C is

$$P_{rC} = \frac{r_C I_{Cm}^2}{2} = \frac{r_C Q_L^2 I_m^2}{2} = \frac{r_C Q_L^2}{R} P_O \tag{2.160}$$

The power loss in the ESR r_{Cc1} of the coupling capacitor C_{c1} is

$$P_{rCc} = \frac{r_{Cc} I_m^2}{2} = \frac{r_{Cc}}{R} P_O \tag{2.161}$$

Neglecting the ac component of the current through the choke L_f and assuming that $I_I \approx I_m$, the power loss in the dc resistance of the choke is

$$P_{rLf} = r_{Lf} I_I^2 = \frac{r_{Lf}}{R} R I_I^2 = \frac{r_{Lf}}{R} P_I \approx \frac{r_{Lf}}{R} P_O \tag{2.162}$$

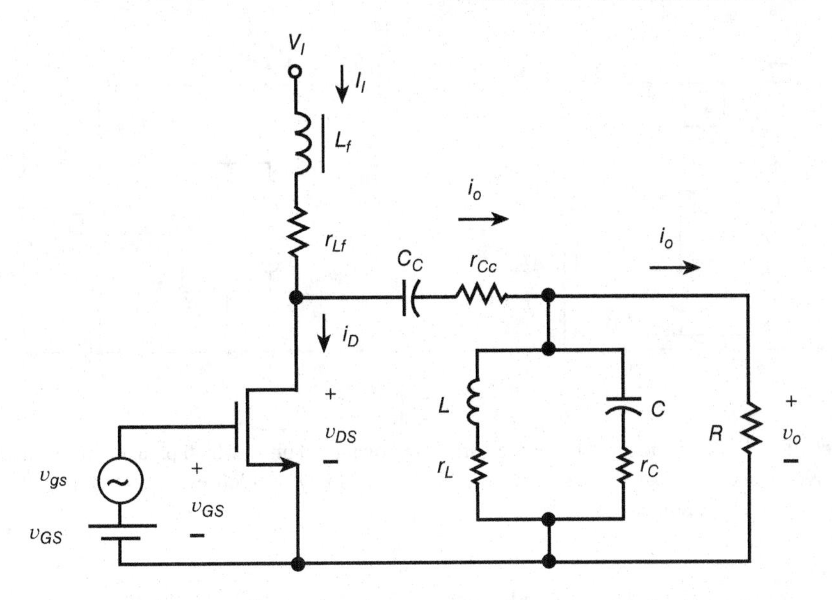

Figure 2.32 Equivalent circuit of the Class A RF power amplifier with parasitic resistances.

The power loss in the choke is approximately the dc winding loss. The overall power loss in the resonant circuit is

$$P_r = P_{rL} + P_{rC} + P_{rCc} + P_{rLf} = \frac{(r_L + r_C)Q_L^2 I_m^2}{2} + \frac{r_{Cc} I_m^2}{2} + r_{Lf} I_I^2$$

$$= \left[\frac{Q_L^2(r_L + r_C) + r_{Cc} + r_{Lf}}{R}\right] P_O \quad (2.163)$$

resulting in the efficiency of the resonant circuit

$$\eta_r = \frac{P_O}{P_O + P_r} = \frac{1}{1 + \frac{P_r}{P_O}} = \frac{1}{1 + \frac{Q_L^2(r_L + r_C) + r_{Cc} + r_{Lf}}{R}} \quad (2.164)$$

The overall power loss of the transistor, the resonant circuit, the coupling capacitor, and the RFC is

$$P_{LS} = P_D + P_r = P_D + P_{rL} + P_{rC} + P_{rCc} + P_{rLf} \quad (2.165)$$

The overall efficiency of the amplifier is

$$\eta = \frac{P_O}{P_I} = \frac{P_O}{P_{DS}}\frac{P_{DS}}{P_I} = \eta_D \eta_r = \frac{P_O}{P_O + P_{LS}} = \frac{1}{1 + \frac{P_{LS}}{P_O}} \quad (2.166)$$

Figure 2.33 shows an equivalent circuit of the parallel-resonant circuit in which parasitic resistances r_L and r_C are represented by a parallel resistance R_r. The quality factor of the resonant inductance is

$$Q_{Lo} = \frac{\omega_o L}{r_L} \quad (2.167)$$

yielding the equivalent parallel resistance of r_L

$$R_{rL} = r_L(1 + Q_{Lo}^2). \quad (2.168)$$

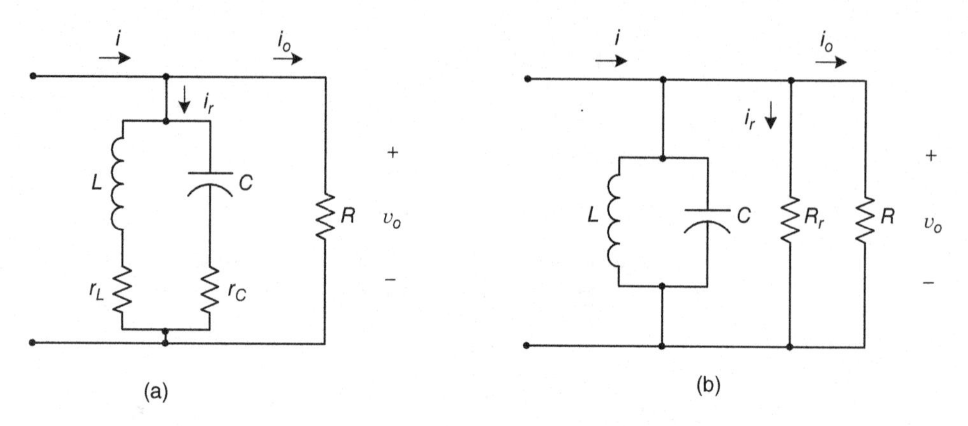

Figure 2.33 Equivalent circuit of the parallel-resonant circuit in which parasitic resistances r_L and r_C are represented by a parallel resistance R_r. (a) Original lossy parallel-resonant circuit. (b) Equivalent parallel-resonant circuit.

The quality factor of the resonant inductance is

$$Q_{Co} = \frac{1}{\omega_o L r_L} \tag{2.169}$$

yielding the equivalent parallel resistance of r_L

$$R_{rC} = r_L(1 + Q_{Co}^2) \tag{2.170}$$

The parallel equivalent resistance of r_L and r_C is

$$R_r = \frac{R_{rL}R_{rC}}{R_{rL} + R_{Co}} = \frac{r_L(1 + Q_{Lo})^2 r_C(1 + Q_{Co})^2}{r_L(1 + Q_{Lo}^2) + r_C(1 + Q_{Co})^2} \tag{2.171}$$

The input power of the resonant circuit is

$$P_{DS} = \frac{1}{2}I_{dm}^2(R_r||R) = \frac{1}{2}I_{dm}^2\left(\frac{R_r R}{R_r + R}\right) = \frac{1}{2}I_{dm}^2 R_t \tag{2.172}$$

where the total parallel resistance is

$$R_t = \frac{R_r R}{R_r + R} \tag{2.173}$$

The output power is

$$P_O = \frac{1}{2}I_m^2 R \tag{2.174}$$

The relationship between the amplitudes of the drain current I_{dm} and the amplitude of the output current I_m is

$$I_m = I_{dm}\frac{R_r}{R + R_r} \tag{2.175}$$

The efficiency of the resonant circuit is

$$\eta_r = \frac{P_O}{P_O + P_r} = \frac{R_r}{R + R_r} = \frac{1}{1 + \frac{R}{R_r}} \tag{2.176}$$

The gate drive power is

$$P_G = \frac{I_{gm}V_{gsm}}{2}\cos\phi_G \tag{2.177}$$

where I_{gm} is the amplitude of the gate current. The power-added efficiency is

$$\eta_{PAE} = \frac{P_O - P_G}{P_I} = \frac{P_O - \frac{P_O}{k_p}}{P_I} = \frac{P_O}{P_I}\left(1 - \frac{1}{k_p}\right) = \eta\left(1 - \frac{1}{k_p}\right) \tag{2.178}$$

where the power gain is

$$k_p = \frac{P_O}{P_G} \tag{2.179}$$

For $k_p = 1$, $\eta_{PAE} = 0$.

2.9 Class A RF Power Amplifier with Current Mirror

Figure 2.34 shows the Class A RF power amplifier with a current mirror biasing. Figure 2.35 depicts an equivalent circuit for the dc component of the Class A RF power amplifier of Fig. 2.34. The biasing circuit of the power MOSFET consists of a MOSFET Q_1 and a dc current source I_{ref}.

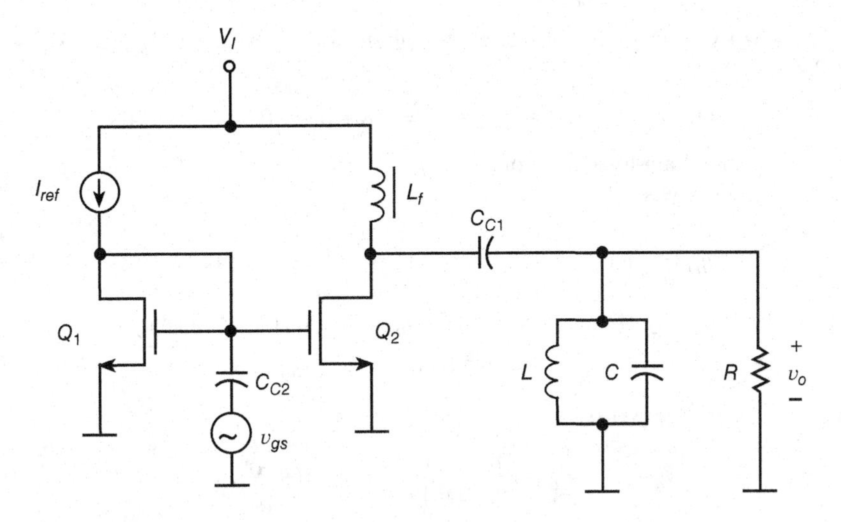

Figure 2.34 Class A RF power amplifier biased with a current mirror.

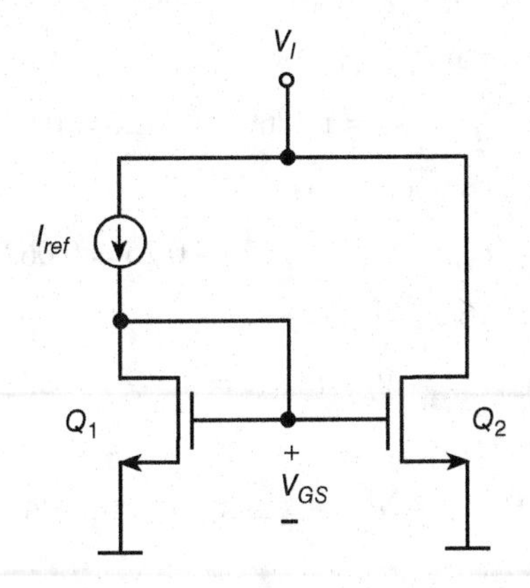

Figure 2.35 Equivalent circuit for the dc component of the Class A RF power amplifier biased with current mirror.

This circuit develops a dc component of the gate-to-source voltage V_{GS} for the power transistor Q_2. The dc reference current source I_{ref} can be replaced by a resistor

$$R = \frac{V_I - V_{GS}}{I_{ref}} \tag{2.180}$$

Example 2.1

Design a basic Class A RF power amplifier to meet the following specifications: $P_O = 0.25$ W, $V_I = 3.3$ V, $f = 1$ GHz, $BW = 100$ MHz, $V_t = 0.356$ V, $K_n = \mu_{n0}C_{ox} = 0.142$ mA/V^2, $r_L = 0.05\ \Omega$, $r_C = 0.01\ \Omega$, $r_{Cc} = 0.07\ \Omega$, and $r_{Lf} = 0.02\ \Omega$.

Solution. The amplifier will be designed for the maximum power $P_O = 0.25$ W. Assume that the MOSFET dc gate-to-source voltage is $V_{GS} = 0.8$ V. Hence, the maximum gate-to-source amplitude is

$$V_{gsm(max)} = V_{GS} - V_t = 0.8 - 0.356 = 0.444 \text{ V} \tag{2.181}$$

Pick $V_{gsm} = 0.4$ V. The maximum saturation drain-to-source voltage at $v_{GSmax} = V_{GS} + V_{gsm} = 0.8 + 0.4 - 1.2$ V is

$$v_{DSsat} = v_{GS} - V_t = V_{GS} + V_{gsm} - V_t = 0.8 + 0.4 - 0.356 = 0.844 \text{ V} \tag{2.182}$$

Pick the minimum drain-to-source voltage $v_{DSmin} = 1$ V.

The drain efficiency is

$$\eta_D = \frac{1}{2}\left(1 - \frac{v_{DSmin}}{V_I}\right)^2 = \frac{1}{2}\left(1 - \frac{1}{3.3}\right)^2 = 0.2429 = 24.29\% \tag{2.183}$$

Assuming the efficiency of the resonant circuit $\eta_r = 0.8$, the total efficiency is

$$\eta = \eta_D\eta_r = 0.2429 \times 0.8 = 0.1943 = 19.43\% \tag{2.184}$$

Hence, the dc supply power is

$$P_I = \frac{P_O}{\eta} = \frac{0.25}{0.1943} = 1.2869 \text{ W} \tag{2.185}$$

The drain ac power is

$$P_{DS} = \eta_D P_I = 0.2429 \times 1.2869 = 0.3126 \text{ W} \tag{2.186}$$

The power loss in the transistor at P_{Omax} is

$$P_D = P_I - P_{DS} = 1.2869 - 0.3126 = 0.9743 \text{ W} \tag{2.187}$$

The power loss in the reactive components is

$$P_r = P_{DS} - P_O = 0.3126 - 0.25 = 0.0626 \text{ W} \tag{2.188}$$

The amplitude of the output voltage is

$$V_{m(max)} = V_I - v_{DSmin} = 3.3 - 1 = 2.3 \text{ V} \tag{2.189}$$

The maximum drain-to-source voltage is

$$V_{DSmax} = V_I + V_m = 3.3 + 2.3 = 5.6 \text{ V} \tag{2.190}$$

The load resistance is

$$R = \frac{V_m^2}{2P_O} = \frac{2.3^2}{2 \times 0.25} = 10.58\ \Omega \tag{2.191}$$

The amplitude of the load current and the ac component of the drain current is

$$I_m = \frac{V_m}{R} = \frac{2.3}{10.58} = 0.2174 \text{ A} \tag{2.192}$$

The dc component of the drain current is

$$I_D = I_I = 1.25 I_m = 1.25 \times 0.2174 = 0.27175 \text{ A} \tag{2.193}$$

The maximum drain current is

$$I_{DM} = I_D + I_m = 0.27175 + 0.2174 = 0.48915 \text{ A} \tag{2.194}$$

The dc resistance that the amplifier presents to the power supply is

$$R_{DC} = \frac{V_I}{I_I} = \frac{3.3}{0.27175} = 12.14\ \Omega. \tag{2.195}$$

The loaded quality factor of the parallel-resonant circuit at $f = f_0$ is

$$Q_L = \frac{f_0}{BW} = \frac{10^9}{10^8} = 10 \tag{2.196}$$

The resonant inductance and capacitance are

$$L = \frac{R}{\omega_0 Q_L} = \frac{10.58}{2\pi \times 10^9 \times 10} = 168.47 \text{ pH} \tag{2.197}$$

and

$$C = \frac{Q_L}{\omega_0 R} = \frac{10}{2\pi \times 10^9 \times 10.58} = 150.43 \text{ pF} \tag{2.198}$$

The reactance of the RFC is

$$X_{Lf} = 100R = 100 \times 10.58 = 1058\ \Omega \tag{2.199}$$

resulting in the choke inductance

$$L_f = \frac{X_{Lf}}{\omega_0} = \frac{1058}{2\pi \times 10^9} = 168.39 \text{ nH} \tag{2.200}$$

The reactance of the coupling capacitor C_{c1} is

$$X_{Cc1} = \frac{R}{100} = \frac{10.58}{100} = 0.1058\ \Omega \tag{2.201}$$

yielding

$$C_{c1} = \frac{1}{\omega_0 X_{Cc1}} = \frac{1}{2\pi \times 10^9 \times 0.1058} = 1.5 \text{ nF} \tag{2.202}$$

The power loss in the inductor ESR is

$$P_{rL} = \frac{r_L I_{Lm}^2}{2} = \frac{r_L Q_L^2 I_m^2}{2} = \frac{0.05 \times 10^2 \times 0.2174^2}{2} = 0.118 \text{ W} \tag{2.203}$$

and the power loss in the capacitor ESR is

$$P_{rC} = \frac{r_C I_{Cm}^2}{2} = \frac{r_C Q_L^2 I_m^2}{2} = \frac{0.01 \times 10^2 \times 0.2174^2}{2} = 0.0236 \text{ W} \tag{2.204}$$

The power loss in the ESR r_{Cc} of the coupling capacitor C_c is

$$P_{rCc} = \frac{r_{Cc} I_m^2}{2} = \frac{0.07 \times 0.2174^2}{2} = 0.00165 \text{ W} \tag{2.205}$$

The power loss in the ESR of the choke is

$$P_{rLf} = r_{Lf} I_I^2 = 0.02 \times 0.27175^2 = 0.001477 \text{ W} \tag{2.206}$$

The total power loss in resonant circuit, RFC, and coupling capacitor is

$$P_r = P_{rL} + P_{rC} + P_{rCc} + P_{rLf} = 0.118 + 0.0236 + 0.00165 + 0.001477 = 0.144727 \text{ W} \tag{2.207}$$

The efficiency of the resonant circuit is

$$\eta_r = \frac{P_O}{P_O + P_r} = \frac{0.25}{0.25 + 0.144727} = 63.33\% \tag{2.208}$$

The efficiency of the amplifier is

$$\eta = \eta_D \eta_r = 0.2429 \times 0.6333 = 15.38\% \tag{2.209}$$

The power loss in the transistor and the resonant circuit is

$$P_{LS} = P_D + P_r = 0.6063 + 0.144727 = 0.751 \text{ W} \tag{2.210}$$

The maximum power is dissipated in the transistor at $P_O = 0$. Therefore, the heat sink should be designed for $P_{LSmax} = P_I \approx 1.3$ W. Assuming the gate drive power $P_G = 0.025$ W, the power gain is

$$k_p = \frac{P_O}{P_G} = \frac{0.25}{0.025} = 10 \tag{2.211}$$

The power-added efficiency is

$$\eta_{PAE} = \frac{P_O - P_G}{P_I} = \frac{0.25 - 0.025}{1.2869} = 0.1748 = 17.48\% \tag{2.212}$$

Let us use a power MOSFET Q_2 with the following parameters: $K_n = \mu_{n0} C_{ox} = 0.142 \text{ mA/V}^2$, $V_t = 0.356$ V, and $L = 0.35$ μm. The gate-to-source voltage at the operating point $V_{GS} = 0.8$ V. Hence, the aspect ratio of the power MOSFET Q_2 is given by

$$\frac{W}{L} = \frac{2I_D}{K_n (V_{GS} - V_t)^2} = \frac{2 \times 0.27175}{0.142 \times 10^{-3} \times (0.8 - 0.356)^2} = 19,415 \tag{2.213}$$

Let us assume $W/L = 20,000$. Hence, the MOSFET channel width is

$$W = \left(\frac{W}{L}\right) L = 20,000L = 15,860 \times 0.35 \times 10^{-6} = 7 \text{ mm} \tag{2.214}$$

Hence,

$$K = \frac{1}{2} \mu_{n0} C_{ox} \left(\frac{W}{L}\right) = \frac{1}{2} K_n \left(\frac{W}{L}\right) = \frac{1}{2} \times 0.142 \times 10^{-3} \times 20,000 = 1.42 \text{ A/V}^2 \tag{2.215}$$

The dc drain current is

$$I_D = K(V_{GS} - V_t)^2 = 1.42 \times (0.8 - 0.356)^2 = 0.28 \text{ A} \tag{2.216}$$

The transconductance of the power MOSFET Q_2 at the operating point is

$$g_m = \sqrt{2K_n \left(\frac{W}{L}\right) I_D} = \sqrt{2 \times 0.142 \times 10^{-3} \times 20,000 \times 0.28} = 1.26 \text{ A/V} \tag{2.217}$$

The amplitude of the ac component of the gate-to-source voltage is

$$V_{gsm} = \frac{I_m}{g_m} = \frac{0.2174}{1.26} = 0.1725 \text{ V} \tag{2.218}$$

A current mirror can be used to bias the power transistor, as shown in Fig. 2.34. The reference current is

$$I_{ref} = \frac{1}{2} \mu_n C_{ox} \left(\frac{W_1}{L_1}\right) (V_{GS} - V_t)^2 \tag{2.219}$$

Assuming the same length of the channel of both transistors (i.e., $L = L_1$), the dc current gain is the dc current gain of the current mirror

$$A_I = \frac{I_D}{I_{ref}} = \frac{W}{W_1} = 100 \tag{2.220}$$

The reference current is

$$I_{ref} = \frac{I_D}{100} = \frac{0.28}{100} = 2.8 \text{ mA} \tag{2.221}$$

Hence, the width of the channel of the current mirror MOSFET Q_1 is given by

$$W_1 = \frac{W}{100} = \frac{7,000 \times 10^{-6}}{100} = 70 \text{ μm} \tag{2.222}$$

Thus, the aspect ratio of the biasing transistor is

$$\frac{W_1}{L} = \frac{70}{0.35} = 200. \tag{2.223}$$

The power consumption by the current mirror is

$$P_{Q1} = I_{ref} V_I = 2.8 \times 10^{-3} \times 3.3 = 9.24 \text{ mW} \tag{2.224}$$

The choke inductance L_f and the coupling capacitance form a high-pass filter whose corner frequency is

$$f_L = \frac{1}{2\pi\sqrt{L_f C_c}} = \frac{1}{2\pi\sqrt{168.47 \times 10^{-9} \times 1.5 \times 10^{-9}}} = 10.016 \text{ MHz} \tag{2.225}$$

To reduce the magnitude of peaking of the transfer function of the high-pass filter, the coupling capacitance C_c should be large and the choke inductance L_f should be low.

2.10 Impedance Matching Circuits

Figure 2.36 shows parallel and series two-terminal networks. These two networks are equivalent at a given frequency f. The reactance factor of these networks is given by Kazimierczuk and Czarkowski [5]

$$q = \frac{x}{r} = \frac{R}{X} \tag{2.226}$$

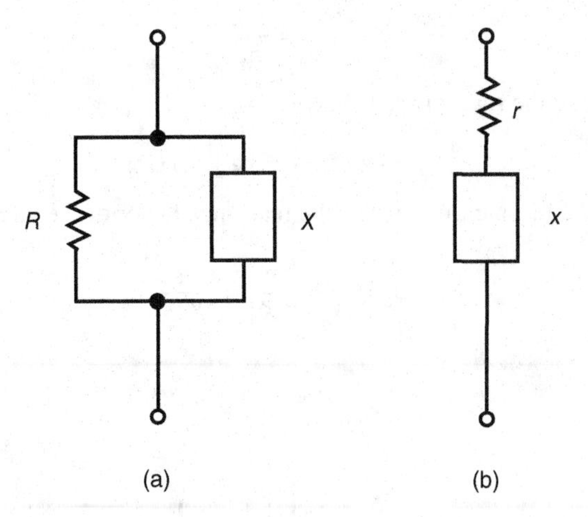

Figure 2.36 Parallel and series two-terminal equivalent networks. (a) Parallel two-terminal network. (b) Series two-terminal network.

The impedance of the parallel two-terminal network is

$$Z = \frac{RjX}{R+jX} = \frac{RjX(R-jX)}{(R+jX)(R-jX)} = \frac{RX^2 + jR^2X}{R^2+X^2} = \frac{R}{1+\left(\frac{R}{X}\right)^2} + j\frac{X}{1+\left(\frac{X}{R}\right)^2}$$

$$= \frac{R}{1+q^2} + j\frac{X}{1+\frac{1}{q^2}} = r + jx \tag{2.227}$$

Hence, the ESR at a given frequency is

$$r = \frac{R}{1+q^2} \tag{2.228}$$

and the equivalent series reactance at a given frequency is

$$x = \frac{X}{1+\frac{1}{q^2}} \tag{2.229}$$

Consider the matching circuit $\pi 1$ shown in Fig. 2.37. The reactance factor of the section to the right of line A is

$$q_A = \frac{R_A}{X_A} = \frac{x_A}{r_A} \tag{2.230}$$

The ESR and reactance of the circuit to the right of line A are

$$r_A = r_B = \frac{R_A}{1+q_A^2} = \frac{R_L}{1+q_A^2} \tag{2.231}$$

$$x_A = \frac{X_A}{1+\frac{1}{q_A^2}} \tag{2.232}$$

and

$$q_A = \sqrt{\frac{R_L}{r_A} - 1} \tag{2.233}$$

The loaded quality factor is

$$Q_L = \frac{x_{AB}}{r_A} \tag{2.234}$$

The series reactance to the right of the B line is

$$x_B = x_{AB} - x_A = (Q_L - q_A)r_A \tag{2.235}$$

Since $X_B = X_{BC}$ at the resonant frequency, the reactance factor of the circuit to the right of line B is

$$q_B = \frac{x_B}{r_B} = \frac{R_B}{X_B} = \frac{R}{X_{BC}} \tag{2.236}$$

resulting in

$$R_B = R = r_B(1 + q_B^2) \tag{2.237}$$

and

$$X_B = x_B\left(1 + \frac{1}{q_B^2}\right) \tag{2.238}$$

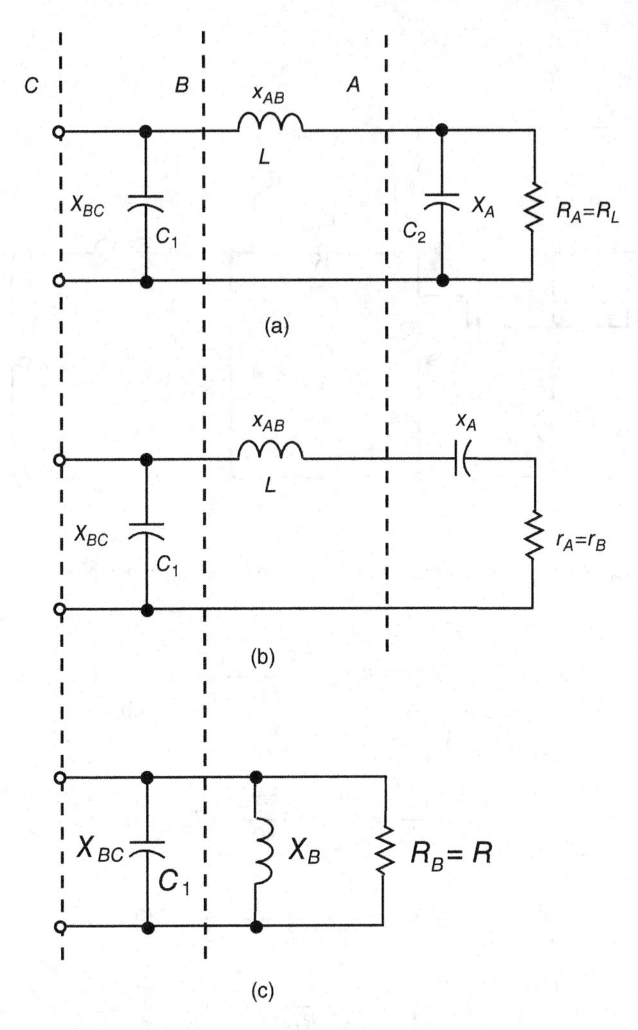

Figure 2.37 Impedance matching resonant circuit π1. (a) Impedance matching circuit. (b) Equivalent circuit after the conversion of the parallel circuit R_A-X_A into the series circuit r_A-x_A. (c) Equivalent circuit after the conversion of the series circuit r_A-($x_A + x_{AB}$) into the parallel circuit R_B-X_B.

Example 2.2

Design an impedance matching circuit to meet the following specifications: $R_L = 50\ \Omega$, $R = 10.58\ \Omega$, and $f = 1$ GHz.

Solution. Assume $q_B = 3$. The reactance X_{BC} is

$$X_{BC} = \frac{1}{\omega_0 C_1} = \frac{R}{q_B} = \frac{10.58}{3} = 3.527\ \Omega \tag{2.239}$$

producing

$$C_1 = \frac{1}{\omega_0 X_{BC}} = \frac{1}{2\pi \times 10^9 \times 3.526} = 45.14\ \text{pF} \tag{2.240}$$

Now

$$r_B = r_A = \frac{R}{1 + q_B^2} = \frac{10.58}{3^2 + 1} = 1.058\ \Omega \tag{2.241}$$

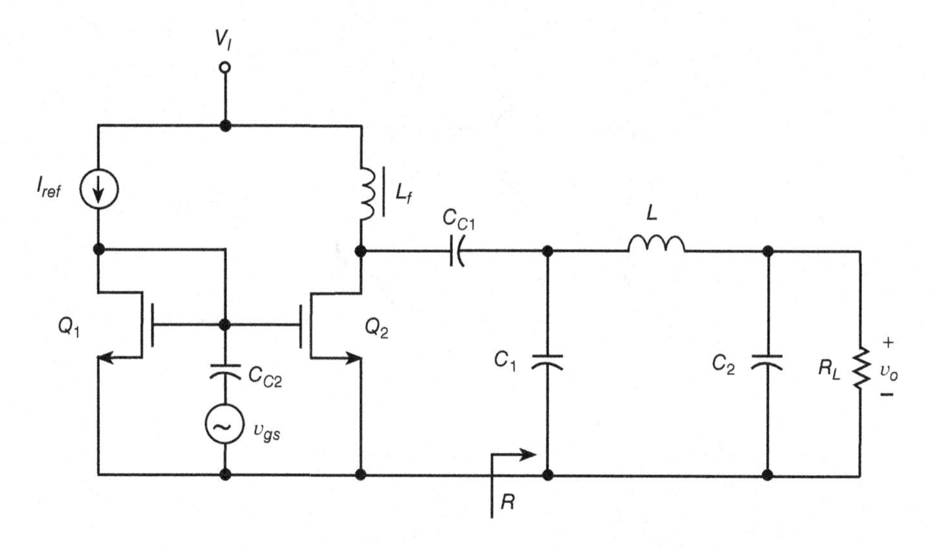

Figure 2.38 Class A RF power amplifier with an impedance matching resonant circuit $\pi 1$.

which gives

$$q_A = \sqrt{\frac{R_L}{r_A} - 1} = \sqrt{\frac{50}{1.058} - 1} = 6.801 \tag{2.242}$$

The reactance of C_2 is

$$X_A = \frac{1}{\omega_0 C_2} = \frac{R_L}{q_A} = \frac{50}{6.801} = 7.3519\ \Omega \tag{2.243}$$

resulting in

$$C_2 = \frac{1}{\omega_0 X_A} = \frac{1}{2\pi \times 10^9 \times 7.3519} = 21.65\ \text{pF} \tag{2.244}$$

The amplitude of the output voltage is

$$V_{om} = \sqrt{2R_L P_O}$$
$$= \sqrt{2 \times 50 \times 0.2}$$
$$= 5\ \text{V}.$$

The loaded quality factor is

$$Q_L = \frac{x_{AB}}{r_A} = \frac{x_B}{r_A} + \frac{x_A}{r_A} = q_B + q_A = 3 + 6.801 = 9.801 \tag{2.245}$$

Thus,

$$x_{AB} = \omega_0 L = Q_L r_A = 9.801 \times 1.058 = 10.3695\ \Omega \tag{2.246}$$

and

$$L = \frac{x_{AB}}{\omega_0} = \frac{10.3695}{2\pi \times 10^9} = 1.65\ \text{nH} \tag{2.247}$$

The bandwidth of the output circuit is

$$BW = \frac{f_0}{Q_L} = \frac{10^9}{9.801} = 102\ \text{MHz} \tag{2.248}$$

Figure 2.38 shows a circuit of a Class A RF power amplifier with a matching network.

2.11 Class A RF Linear Amplifier

2.11.1 Amplifier for Variable-Envelope Signals

Modern wireless communication systems use power amplifiers to amplify variable-envelope signals. Linear amplifiers are required to amplify variable-envelope signals, such as AM signals. The Class A amplifier is a linear amplifier. Figure 2.39 shows a block diagram of an AM transmitter with a linear RF power amplifier that amplifies an AM voltage. A low-frequency audio amplifier amplifies a modulating signal. The carrier-fixed amplitude signal of frequency f_c and the modulating signal of frequency f_m are applied to an AM modulator to obtain an AM signal. The linear RF power amplifies the AM signal. This signal is radiated by an antenna to be propagated to receivers. Figure 2.40 shows waveforms of the gate-to-source voltage v_{GS} and the drain-to-source voltage v_{DS} in an ideal linear amplifier.

The modulating voltage is given by

$$v_{im}(t) = V_{im} \cos \omega_m t \tag{2.249}$$

The carrier voltage is

$$v_{ic} = V_{ic} \cos \omega_c t \tag{2.250}$$

where $\omega_c \gg \omega_m$. The gate-to-source AM voltage at the input of a Class A RF amplifier is

$$\begin{aligned} v_{GS} &= V_{GSQ} + V(t) \cos \omega_c t = V_{GSQ} + [V_{ic} + v_{im}(t)] \cos \omega_c t \\ &= V_{GSQ} + (V_{ic} + V_{im} \cos \omega_m t) \cos \omega_c t \\ &= V_{GSQ} + V_{ic}(1 + m \cos \omega_m t) \cos \omega_c t \\ &= V_{SGQ} + V_{ic} \cos \omega_c t + \frac{mV_{ic}}{2} \cos(\omega_c - \omega_m)t + \frac{mV_{ic}}{2} \cos(\omega_c + \omega_m)t \end{aligned} \tag{2.251}$$

where V_{GSQ} is the dc component of the gate-to-source voltage and the modulation index is

$$m = \frac{V_{im}}{V_{ic}} \tag{2.252}$$

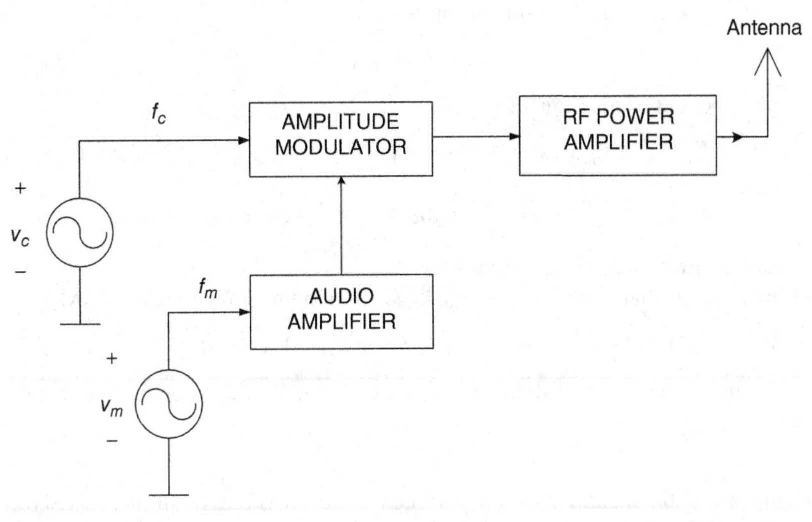

Figure 2.39 Block diagram of an AM transmitter with a linear RF power amplifier that amplifies an AM voltage.

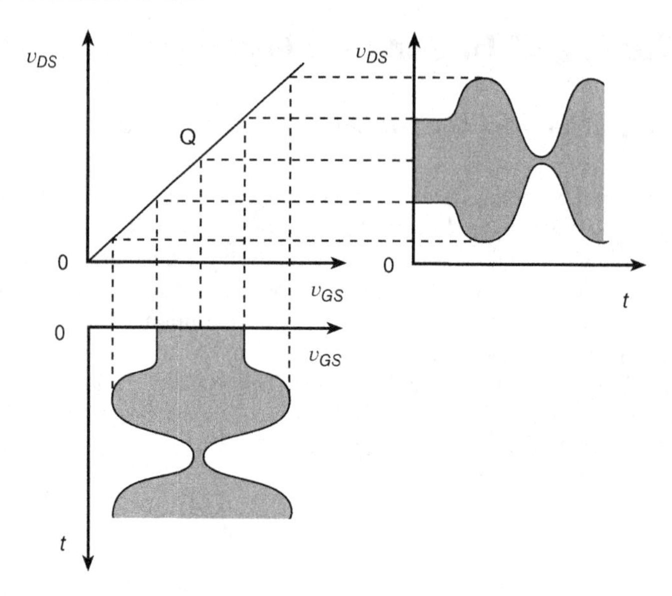

Figure 2.40 Amplification of an AM signal in the Class A linear amplifier.

To avoid distortion of the AM signal, the transistor must be on at all times. The maximum amplitude of the ac component of the gate-to-source voltage occurs at $m = 1$

$$V_{gsm(max)} = V_{ic}(1 + m_{max}) = V_{ic}(1 + 1) = 2V_{ic} \tag{2.253}$$

The dc gate-to-source voltage at the operating point Q must satisfy the following condition:

$$V_{GSQ} > 2V_{ic} \tag{2.254}$$

Assuming a linear law, the drain current is

$$\begin{aligned} i_D &= K_s(v_{GSQ} - V_t) = K_s[V_{GSQ} + V(t)\cos \omega t - V_t] \\ &= K_s\{V_{GSQ} + [V_{ic} + v_m(t)]\cos \omega t - V_t\} = I_{DQ} + i_d \end{aligned} \tag{2.255}$$

where the dc component of the drain current is

$$I_{DQ} = K_s(V_{GSQ} - V_t) \tag{2.256}$$

and the ac component of the drain current is

$$\begin{aligned} i_d &= K_s[(V_{ic} + V_{im}\cos \omega_m t)\cos \omega_c t] = K_s[(V_{ic}(1 + m\cos \omega_m t)\cos \omega_c t] \\ &= K_s[V_{ic}\cos \omega_c t + \frac{mV_{ic}}{2}\cos(\omega_c - \omega_m)t + \frac{mV_{ic}}{2}\cos(\omega_c + \omega_m)t] \end{aligned} \tag{2.257}$$

Thus, the drain current waveform exhibits the AM.

Assuming a linear relationship $v_{DS} = A_v v_{GS}$, we obtain the drain-to-source AM voltage

$$v_{DS} = V_{DSQ} + A_v V_{ic}(1 + m\cos \omega_m t)\cos \omega_c t = V_{DSQ} + V_c(1 + m\cos \omega_m t)\cos \omega_c t \tag{2.258}$$

where A_v is the voltage gain of the amplifier and the amplitude of the carrier at the amplifier output is

$$V_c = A_v V_{ic} \tag{2.259}$$

To avoid distortion, the dc drain-to-source voltage at the operating point must satisfy the condition:

$$V_{DSQ} > 2V_c + V_{GSQ} - V_t \tag{2.260}$$

The output AM voltage of the amplifier modulated with a single sinusoidal voltage is

$$v_o = V_c(1 + m\cos\,\omega_m t)\cos\,\omega_c t \tag{2.261}$$

The instantaneous output AM power of the amplifier with a fixed load resistance and modulated with a single sinusoidal voltage is

$$P_o(t) = \frac{v_o^2(t)}{2R} = \frac{V_c^2(1 + m\cos\,\omega_m t)^2 \cos^2\omega_c t}{2R} \tag{2.262}$$

The dc power P_I drawn from the power supply is constant. Hence, the instantaneous efficiency of the Class A power amplifier is

$$\eta(t) = \frac{P_o(t)}{P_I} = \frac{V_c^2(1 + m\cos\,\omega_m t)^2 \cos^2\omega_c t}{2RP_I} \tag{2.263}$$

The average efficiency of the Class A amplifier with an AM output voltage is

$$\eta_{AV} = \frac{P_{oAV(AM)}}{P_I} = \frac{\int_{V_{min}}^{V_{max}} PDF(V)P_o(V)dV}{2RP_I} \tag{2.264}$$

where PDF is the power density function.

2.11.2 Amplifiers for Constant-Envelope Signals

Class A amplifier can be used to amplify constant-envelope signals, such as angle-modulated signals (FM and PM signals). The gate-to-source FM voltage is

$$v_{gs} = V_{DSQ} + V_{ic}\cos(\omega_c t + m_f \sin\,\omega_m t) \tag{2.265}$$

where

$$V_{DSQ} > V_{ic} \tag{2.266}$$

Hence, the drain-to-source FM voltage is

$$v_{DS} = V_{DSQ} + A_v V_c \cos(\omega_c t + m_f \sin\,\omega_m t) = V_{DSQ} + V_c \cos(\omega_c t + m_f \sin\,\omega_m t) \tag{2.267}$$

where

$$m_f = \frac{\Delta f}{f_m} \tag{2.268}$$

$$V_c = A_v V_{ic} \tag{2.269}$$

and

$$V_{DSQ} > V_{ic} + V_{GSQ} - V_t \tag{2.270}$$

2.12 Summary

- The conduction angle of the drain current is $2\theta = 360°$ in the Class A power amplifier.
- The operating point of the transistor in the Class A amplifier is above the transistor threshold voltage V_t such that the gate-to-source voltage waveform is above V_t for the entire voltage swing ($V_{gsm} < V_{GS} - V_t$ or $V_{GS} - V_{gsm} > V_t$). Therefore, the transistor never enters the cutoff region.
- The RFC in the Class A RF amplifier can be replaced by a quarter-wavelength transformer.

- The quality factor of the resonant circuit is proportional to the total energy stored divided by the energy dissipated per cycle.
- The transistor is biased at a dc current greater than the peak value of the ac component of the drain current ($I_m < I_D$).
- There is no sharp dividing boundary between Class A power amplifiers and small-signal amplifiers.
- In the Class A RF power amplifier, the dc supply voltage and the dc supply current are constant.
- The dc supply power P_I in the Class A RF power amplifier is constant and is independent of the amplitude of the output voltage V_m.
- In the Class A RF power amplifier, the drain efficiency is proportional to the output power.
- Power loss in the transistor of the Class A RF power amplifier is high.
- The Class A power amplifier dissipates the maximum power at zero output power. In this case, $P_D = P_I$.
- The maximum drain efficiency of Class A RF power amplifiers with the RFC is 50%. It occurs at $I_m = I_I$ and $V_m = V_I$.
- In practice, the efficiency of the Class A RF power amplifier at its peak output power is approximately 40%
- Linear amplification is required when the amplified signal contains both amplitude and phase modulation.
- Class A power amplifiers exhibit weak nonlinearity. Therefore, the HD and IMD in Class A power amplifiers are low, compared to the amplifiers. These circuits are nearly linear amplifiers, suitable for the amplification of AM signals.
- The Class A power amplifier suffers from a low efficiency, but provides a higher linearity than other power amplifiers.
- Negative feedback can be used to suppress the nonlinearity.
- The amplitude of the current in the parallel-resonant circuit is high.
- Power losses in the parallel-resonant circuit are high because the amplitude of the current flowing through the resonant inductor and the resonant capacitor is Q_L times higher than that of the output current I_m.
- The maximum drain efficiency of the Class A audio amplifier, in which the RFC is replaced by a resistor, is only 25% at $I_m = I_I$ and $V_m = V_I$.
- The power losses in the parasitic resistances of the parallel-resonant circuit are Q_L^2 times higher than those in series-resonant circuit at the same amplitude of the output current.
- The amplitudes of harmonics in the load current in the Class A amplifier are low.
- Class A power amplifiers are used as low-power drivers of high-power amplifiers and as linear power amplifiers.
- The dc gate-to-source voltage V_{GS} of the power MOSFET can be developed by a current mirror.

2.13 Review Questions

2.1 What is the channel-length modulation in MOSFETs?

2.2 How does the average velocity of current carriers depend on the electric field intensity in semiconductor devices?

2.3 What is the saturation of the average velocity of current carriers in semiconductors?

2.4 What is the short-channel effect?

2.5 Sketch the i_D versus v_{GS} characteristics of MOSFETs with long, intermediate, and short channels.

2.6 Sketch the i_D versus v_{DS} characteristics at the same voltage v_{GS} of MOSFETs with long, intermediate, and short channels.

2.7 What are wide band gap semiconductor materials?

2.8 What is the value of the conduction angle of the drain current in the Class A power amplifier?

2.9 Explain the principle of operation of a quarter-wavelength transformer used in place of RF choke.

2.10 What is the location of the transistor operating point for the Class A RF power amplifier?

2.11 Does the dc supply power of the Class A amplifier depend on the output voltage amplitude V_m?

2.12 Is the power loss in the transistor of the Class A amplifier low?

2.13 Are the power losses in the parallel-resonant circuit high?

2.14 Is the efficiency of the Class A RF power amplifier high?

2.15 Is the linearity of the Class A RF power amplifier good?

2.16 Are the harmonics in the load current of the Class A RF power amplifier high?

2.17 What are the upper and lower limits of the output voltage in the Class A amplifier for linear operation?

2.18 Explain the operation of the Class A power amplifier with a current mirror?

2.14 Problems

2.1 Calculate the maximum channel length of a silicon MOSFET at which the transistor is operated at high field (i.e., with a short channel) at $v_{DS} = 0.2$ V.

2.2 Determine the drain efficiency of a Class A amplifier with an RF choke to meet the following specifications:

(a) $V_I = 20$ V and $V_m = 10$ V.

(b) $V_I = 20$ V and $V_m = 18$ V.

2.3 Determine the maximum power loss in the transistor of the Class A amplifier with an RF choke, which has $V_I = 10$ V and $I_I = 1$ A.

2.4 Design a Class A RF power amplifier to meet the following specifications: $P_O = 0.25$ W, $V_I = 1.5$ V, and $f = 2.4$ GHz.

2.5 Design an impedance matching circuit for the Class A RF power amplifier with $R_L = 50\ \Omega$, $R = 25\ \Omega$, and $f = 2.4$ GHz.

2.6 Determine the drain power loss and the drain efficiency of a Class A RF power amplifier with $I_I = 1$ A and $V_m = 8$ V.

(a) For $V_I = 20$ V.

(b) For $V_I = 10$ V.

2.7 A Class A RF power amplifier has $V_I = 20$ V. Determine the drain efficiency.

(a) For $V_m = 0.9V_I$.

(b) For $V_m = 0.5V_I$.

(c) For $V_m = 0.1V_I$.

References

[1] L. Gray and R. Graham, *Radio Transmitters*, New York, NY: McGraw-Hill, 1961.
[2] E. W. Pappenfus, *Single Sideband Principles and Circuits*, New York, NY: McGraw-Hill, 1964.
[3] K. K. Clarke and D. T. Hess, *Communications Circuits: Analysis and Design*. Reading, MA: Addison-Wesley Publishing Co., 1971; reprinted Malabar, FL: Krieger, 1994.
[4] H. L. Krauss, C. W. Bostian, and F. H. Raab, *Solid State Radio Engineering*, New York, NY: John Wiley & Sons, 1980.
[5] M. K. Kazimierczuk and D. Czarkowski, *Resonant Power Converters*, 2nd Ed. New York, NY: John Wiley & Sons, 2012.
[6] P. B. Kingston, *High-Linearity RF Amplifier Design*, Norwood, MA: Artech House, 2000.
[7] S. C. Cripps, *RF Power Amplifiers for Wireless Communications*, 2nd Ed. Norwood, MA: Artech House, 2006.
[8] M. Albulet, *RF Power Amplifiers*. Atlanta, GA: Noble Publishing Co., 2001.
[9] T. H. Lee, *The Design of CMOS Radio-Frequency Integrated Circuits*, 2nd Ed. New York, NY: Cambridge University Press, 2004.
[10] B. Razavi, *RF Microelectronics*. Upper Saddle River, NJ: Prentice-Hall, 1998.
[11] A. Grebennikov, *RF and Microwave Power Amplifier Design*. New York, NY: McGraw-Hill, 2005.
[12] A. Grebennikov and N. O. Sokal, *Switchmode Power Amplifiers*. Amsterdam: Elsevier, 2007.
[13] J. Aguilera and R. Berenguer, *Design and Test of Integrated Inductors for RF Applications*. Boston, MA: Kluwer Academic Publishers, 2003.
[14] M. K. Kazimierczuk, *Pulse-Width Modulated PWM DC-DC Power Converters*, Chichester, UK: John Wiley & Sons, 2008.
[15] M. K. Kazimierczuk, *High-Frequency Magnetic Components*, 2nd Ed. Chichester, UK: John Wiley & Sons, 2014.
[16] E. O. Johnson, "Physical limitations on frequency and power parameters of transistors," *RCA Review*, vol. 26, pp. 163–177, 1965.
[17] B. J. Baliga, "Power semiconductor devices figure of merit for high-frequency applications," *IEEE Electron Device Letters*, vol. 10, no. 10, pp. 455–457, 1989.

3

Class AB, B, and C RF Power Amplifiers

3.1 Introduction

The Class B radio-frequency (RF) power amplifier [1–10] consists of a transistor and a parallel-resonant circuit. The transistor is operated as a dependent current source. The conduction angle of the drain or collector current in the Class B power amplifier is 180°. The parallel-resonant circuit acts as a bandpass (BP) filter and selects only the fundamental component. The efficiency of the Class B power amplifier is higher than that of the Class A power amplifier. The circuit of the Class C power amplifier is the same as that of the Class B amplifier. However, the operating point is such that the conduction angle of the drain current is less than 180°. The conduction angle of the drain current in the Class AB power amplifier is between 180° and 360°. Class B and C power amplifiers are usually used for RF amplification in radio and TV transmitters as well as in mobile phones. In this chapter, we will present Class AB, B, and C RF power amplifiers with their principle of operation, analysis, and design examples.

3.2 Class B RF Power Amplifier

3.2.1 Circuit of Class B RF Power Amplifier

The circuit of a Class B RF power amplifier is shown in Fig. 3.1. It consists of a transistor (MOSFET, MESFET, or BJT), parallel-resonant circuit, and RF choke. The operating point of the transistor is located exactly at the boundary between the cutoff region and the active region (the saturation region or the pinch-off region). The dc component of the gate-to-source voltage V_{GS} is equal to the transistor threshold voltage V_t. Therefore, the conduction angle of the drain current 2θ is 180°. The transistor is operated as a dependent voltage-controlled current

RF Power Amplifiers, Second Edition. Marian K. Kazimierczuk.

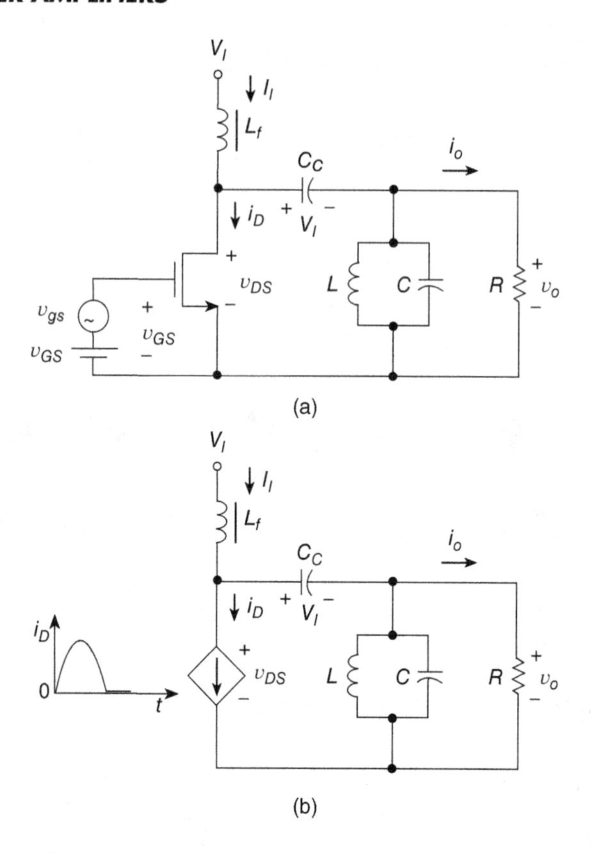

Figure 3.1 Class AB, B, or C RF power amplifiers.

source. Voltage and current waveforms in the Class B power amplifier are shown in Fig. 3.2. The ac component of the gate-to-source voltage v_{gs} is a sine wave. The drain current is a half sine wave and contains the dc component, fundamental component, and even harmonics. The parallel-resonant circuit acts as a BP filter, which attenuates all the harmonics. The "purity" of the output sine wave is a function of the selectivity of the BP filter. The higher the loaded quality factor Q_L, the lower is the harmonic content of the sine wave output current and voltage. The parallel-resonant circuit may be more complex to serve as an impedance matching network.

3.2.2 Waveforms in Class B RF Power Amplifier

The gate-to-source voltage waveform is given by

$$v_{GS} = V_{GS} + v_{gs} = V_{GS} + V_{gsm}\cos\omega t = V_t + V_{gsm}\cos\omega t \tag{3.1}$$

where $V_{GS} = V_t$ in Class B amplifiers. The saturation drain-to-source voltage is

$$v_{DSsat} = v_{GS} - V_t = V_{GS} + V_{gsm} - V_t = V_t + V_{gsm} - V_t = V_{gsm} \tag{3.2}$$

The minimum drain-to-source voltage v_{DSmin} should be equal to or slightly higher than v_{DSsat}. The maximum amplitude of the drain-to-source voltage waveform is

$$V_m = V_I - v_{DSmin} \leq V_I - v_{DSsat} = V_I - V_{gsm} \tag{3.3}$$

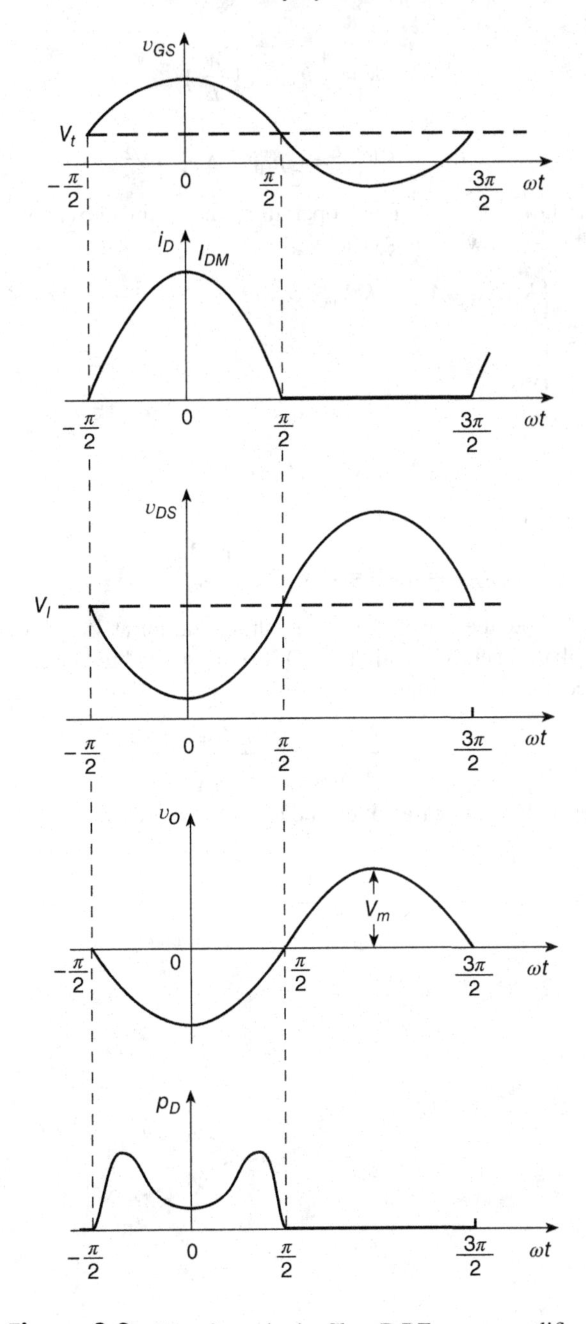

Figure 3.2 Waveforms in the Class B RF power amplifier.

Using the square law for large-signal operation, the drain current is

$$i_D = \begin{cases} K(v_{GS} - V_t)^2 = KV_{gsm}^2 \cos^2 \omega t = I_{DM} \cos^2 \omega t & \text{for} \quad -\frac{\pi}{2} < \omega t \leq \frac{\pi}{2} \\ 0 & \text{for} \quad \frac{\pi}{2} < \omega t \leq \frac{3\pi}{2} \end{cases} \tag{3.4}$$

where

$$K = \frac{1}{2}\mu_{n0}C_{ox}\left(\frac{W}{L}\right) \tag{3.5}$$

and

$$I_{DM} = KV_{gsm}^2 = \frac{1}{2}\mu_{n0}C_{ox}\left(\frac{W}{L}\right)V_{gsm}^2. \tag{3.6}$$

Using the linear law for large-signal operation, the drain current is proportional to the gate-to-source voltage v_{GS}, when v_{GS} is above V_t

$$i_D = \begin{cases} K_s(v_{GS} - V_t) = K_s V_{gsm}\cos\omega t & \text{for} \quad -\frac{\pi}{2} < \omega t \le \frac{\pi}{2} \\ 0 & \text{for} \quad \frac{\pi}{2} < \omega t \le \frac{3\pi}{2} \end{cases} \tag{3.7}$$

where

$$K_s = \frac{1}{2}C_{ox}Wv_{sat} \tag{3.8}$$

and the peak value of the drain current is

$$I_{DM} = i_D(0) = K_s V_{gsm} = \frac{1}{2}C_{ox}Wv_{sat}V_{gsm} \tag{3.9}$$

Figures 3.3 and 3.4 show the waveforms of the drain current at $V_{gsm} = 0.4$ V and $V_{gsm} = 1$ V. The ratio of the peak drain current with the MOSFET described by the square law to that of the MOSFET described by the linear law is

$$\frac{I_{DM(Sq)}}{I_{DM(Lin)}} = \frac{\mu_{n0}V_{gsm}}{Lv_{sat}} \tag{3.10}$$

This ratio increases as V_{gsm} increases. It can be lower or greater than 1.

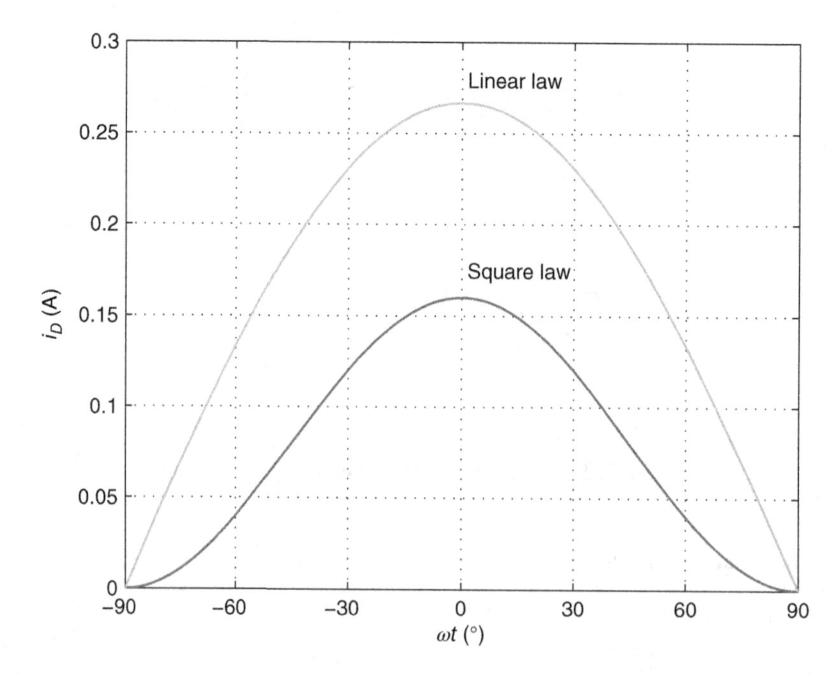

Figure 3.3 Waveforms of the drain currents using the square law and the linear law in the Class B RF power amplifier at $V_{gsm} = 0.4$ V, $\mu_{n0}C_{ox} = 20$ μA/V^2, $C_{ox} = \frac{1}{3}$ mF/m^2, $v_{sat} = 8 \times 10^6$ cm/s, $W/L = 10^5$, and $W = 0.5 \times 10^5$ μm.

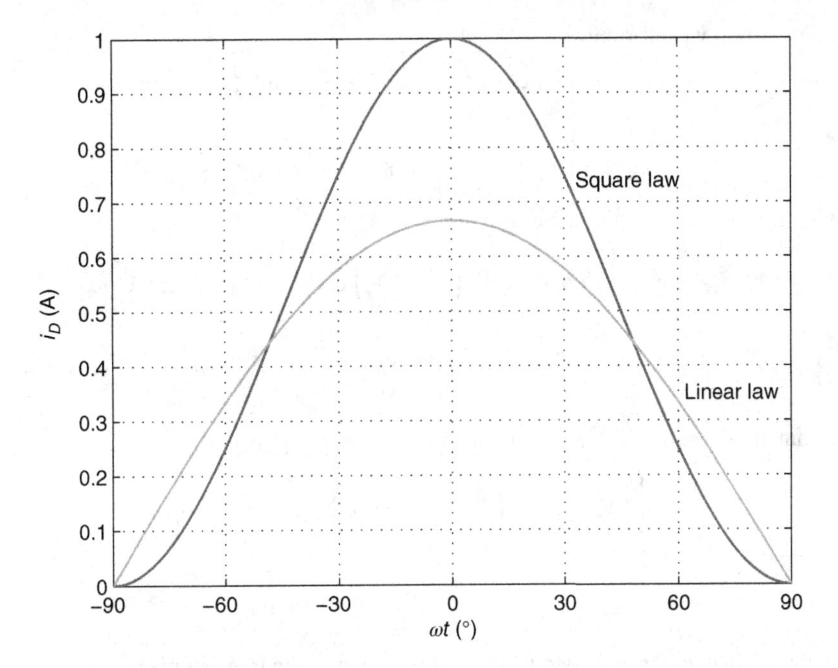

Figure 3.4 Waveforms of the drain currents using the square law and the linear law in the Class B RF power amplifier at $V_{gsm} = 1$ V, $\mu_{n0}C_{ox} = 20$ μA/V^2, $C_{ox} = \frac{1}{3}$ mF/m^2, $\upsilon_{sat} = 8 \times 10^6$ cm/s, $W/L = 10^5$, and $W = 0.5 \times 10^5$ μm.

The drain current in the Class B RF power amplifier is a half sine wave and is given by

$$i_D = \begin{cases} I_{DM} \cos \omega t & \text{for} \quad -\frac{\pi}{2} < \omega t \leq \frac{\pi}{2} \\ 0 & \text{for} \quad \frac{\pi}{2} < \omega t \leq \frac{3\pi}{2} \end{cases} \tag{3.11}$$

The drain-to-source voltage is expressed as

$$\upsilon_{DS} = V_I - V_m \cos \omega t \tag{3.12}$$

The instantaneous power dissipation in the transistor is

$$p_D(\omega t) = i_D \upsilon_{DS} = \begin{cases} I_{DM} \cos \omega t (V_I - V_m \cos \omega t) & \text{for} \quad -\frac{\pi}{2} < \omega t \leq \frac{\pi}{2} \\ 0 & \text{for} \quad \frac{\pi}{2} < \omega t \leq \frac{3\pi}{2} \end{cases} \tag{3.13}$$

The amplitude of the fundamental component of the drain current is

$$I_m = \frac{1}{\pi} \int_{-\frac{\pi}{2}}^{\frac{\pi}{2}} i_D \cos \omega t d(\omega t) = \frac{1}{\pi} \int_{-\frac{\pi}{2}}^{\frac{\pi}{2}} I_{DM} \cos^2 \omega t d(\omega t) = \frac{I_{DM}}{2}. \tag{3.14}$$

The dc supply current is

$$I_I = \frac{1}{2\pi} \int_{-\frac{\pi}{2}}^{\frac{\pi}{2}} i_D d(\omega t) = \frac{1}{2\pi} \int_{-\frac{\pi}{2}}^{\frac{\pi}{2}} I_{DM} \cos \omega t d(\omega t) = \frac{1}{\pi} \int_0^{\frac{\pi}{2}} I_{DM} \cos \omega t d(\omega t)$$

$$= \frac{I_{DM}}{\pi} = \frac{2}{\pi} I_m = \frac{2}{\pi} \frac{V_m}{R} \tag{3.15}$$

The drain current waveform is

$$i_D = \begin{cases} I_I \pi \cos \omega t & \text{for} \quad -\frac{\pi}{2} < \omega t \le \frac{\pi}{2} \\ 0 & \text{for} \quad \frac{\pi}{2} < \omega t \le \frac{3\pi}{2} \end{cases} \tag{3.16}$$

The drain-to-source voltage waveform at $f = f_0$ is

$$v_{DS} = V_I - V_m \cos \omega t = V_I \left(1 - \frac{V_m}{V_I}\right) \cos \omega t = V_I(1 - \xi_1) \cos \omega t \tag{3.17}$$

where

$$\xi_1 = \frac{V_m}{V_I} \tag{3.18}$$

The instantaneous power dissipation in the transistor at $f = f_0$ is

$$p_D(\omega t) = i_D v_{DS} = \begin{cases} I_I V_I \pi \cos \omega t \left(1 - \frac{V_m}{V_I} \cos \omega t\right) & \text{for} \quad -\frac{\pi}{2} < \omega t \le \frac{\pi}{2} \\ 0 & \text{for} \quad \frac{\pi}{2} < \omega t \le \frac{3\pi}{2} \end{cases} \tag{3.19}$$

Hence, the normalized instantaneous power dissipation in the transistor is

$$\frac{p_D(\omega t)}{P_I} = \begin{cases} \pi \cos \omega t \left(1 - \frac{V_m}{V_I} \cos \omega t\right) & \text{for} \quad -\frac{\pi}{2} < \omega t \le \frac{\pi}{2} \\ 0 & \text{for} \quad \frac{\pi}{2} < \omega t \le \frac{3\pi}{2} \end{cases} \tag{3.20}$$

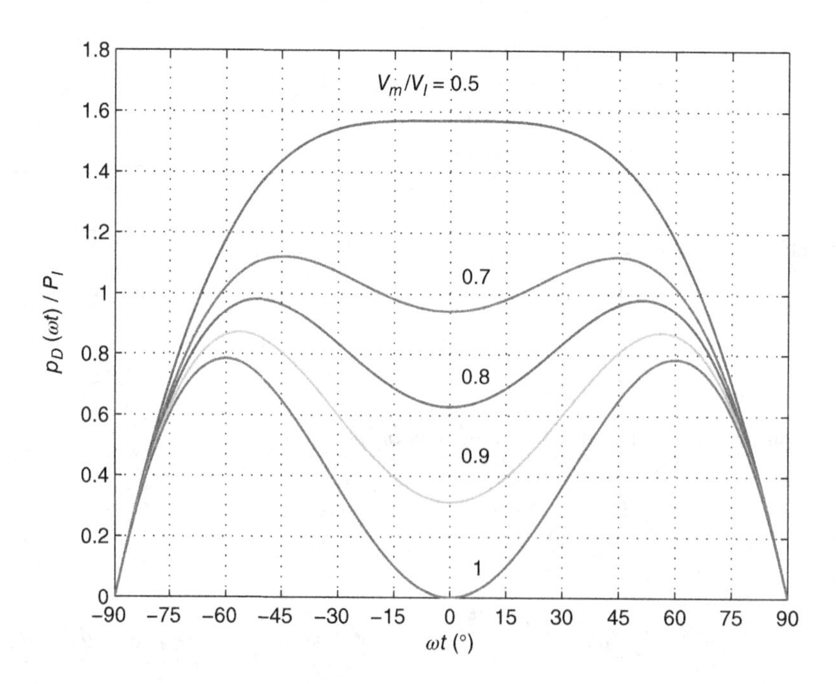

Figure 3.5 Normalized instantaneous power dissipation $p_D(\omega t)/P_I$ at various values of V_m/V_I and $f = f_0$ for the Class B RF power amplifier.

The normalized instantaneous power loss in the transistor at various values of V_m/V_I and at $f = f_0$ is shown in Fig. 3.5 for the Class B power amplifier. As the ratio V_m/V_I increases, the peak values of $p_D(\omega t)/P_I$ decrease, yielding higher drain efficiency.

The normalized instantaneous power dissipation in the transistor at $f \neq f_0$ is

$$\frac{p_D(\omega t)}{P_I} = \begin{cases} \pi \cos \omega t \left[1 - \frac{V_m}{V_I} \cos(\omega t + \phi)\right] & \text{for} \quad -\frac{\pi}{2} < \omega t \leq \frac{\pi}{2} \\ 0 & \text{for} \quad \frac{\pi}{2} < \omega t \leq \frac{3\pi}{2} \end{cases} \tag{3.21}$$

3.2.3 Power Relationships for Class B RF Amplifier

The dc resistance seen by the dc power supply V_I is

$$R_{DC} = \frac{V_I}{I_I} = \frac{\pi}{2} \frac{V_I}{V_m} R. \tag{3.22}$$

For $V_m = V_I$,

$$I_{Imax} = \frac{2}{\pi} \frac{V_I}{R} \tag{3.23}$$

resulting in the dc resistance seen by the power supply V_I

$$R_{DCmin} = \frac{V_I}{I_{Imax}} = \frac{\pi}{2} R \tag{3.24}$$

The maximum value of R_{DC} occurs at $V_m = 0$ and its value is $R_{DCmax} = \infty$. The amplitude of the output voltage is

$$V_m = RI_m = \frac{RI_{DM}}{2} = \frac{1}{4} \mu_{n0} C_{ox} \left(\frac{W}{L}\right) RV_{gsm} \tag{3.25}$$

The dc supply power is

$$P_I = I_I V_I = \frac{I_{DM}}{\pi} V_I = \frac{2}{\pi} V_I I_m = \frac{2}{\pi} \frac{V_I V_m}{R} = \frac{2}{\pi} \left(\frac{V_I^2}{R}\right) \left(\frac{V_m}{V_I}\right) \tag{3.26}$$

Neglecting power losses in passive components, the output power P_O is equal to the drain power P_{DS}. The drain power is

$$P_O = P_{DS} = \frac{I_m V_m}{2} = \frac{V_m^2}{2R} = \frac{1}{2} \left(\frac{V_I^2}{R}\right) \left(\frac{V_m}{V_I}\right)^2 \tag{3.27}$$

The power dissipation at the drain of the MOSFET is

$$P_D = \frac{1}{2\pi} \int_{-\frac{\pi}{2}}^{\frac{\pi}{2}} p_D(\omega t) d(\omega t) = \frac{1}{2\pi} \int_{-\frac{\pi}{2}}^{\frac{\pi}{2}} i_D v_{DS} d(\omega t) = P_I - P_O = \frac{2}{\pi} \frac{V_I V_m}{R} - \frac{V_m^2}{2R}$$
$$= \frac{2}{\pi} \left(\frac{V_I^2}{R}\right) \left(\frac{V_m}{V_I}\right) - \frac{1}{2} \left(\frac{V_I^2}{R}\right) \left(\frac{V_m}{V_I}\right)^2 \tag{3.28}$$

The maximum power dissipation can be determined by taking the derivative of P_D with respect to V_m and setting it to zero. Thus,

$$\frac{dP_D}{dV_m} = \frac{2}{\pi} \frac{V_I}{R} - \frac{V_m}{R} = 0 \tag{3.29}$$

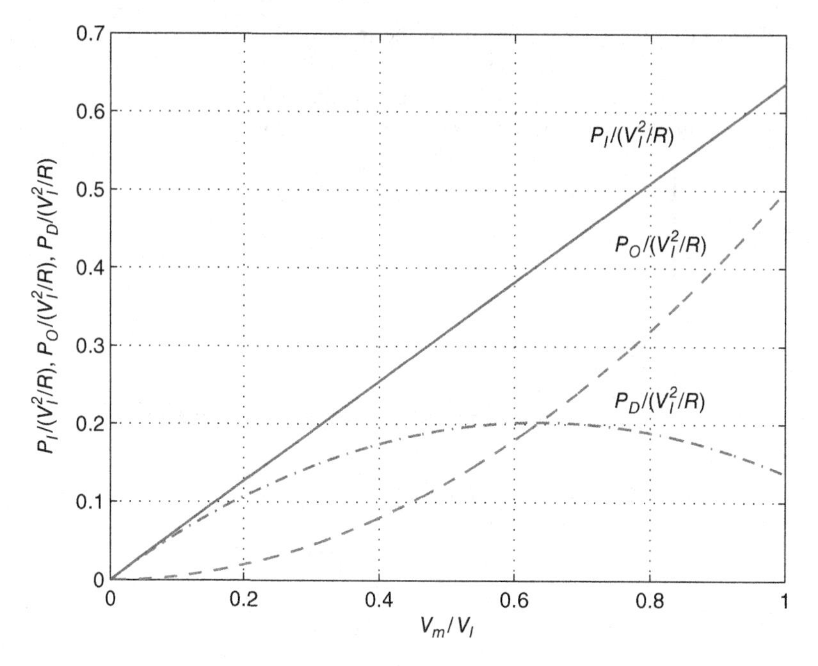

Figure 3.6 Normalized dc supply power $P_I/(V_I^2/R)$, normalized drain power $P_O/(V_I^2/R)$, and normalized drain power dissipation $P_D/(V_I^2/R)$ as functions of V_m/V_I for the Class B RF power amplifier.

The critical value of V_m at which the maximum power dissipation occurs is

$$V_{m(cr)} = \frac{2V_I}{\pi} \tag{3.30}$$

Hence, the maximum power dissipated in the drain of the transistor is

$$P_{Dmax} = \frac{4}{\pi^2}\frac{V_I^2}{R} - \frac{2}{\pi^2}\frac{V_I^2}{R} = \frac{2}{\pi^2}\frac{V_I^2}{R} \tag{3.31}$$

Figure 3.6 shows plots of $P_I/(V_I^2/R)$, $P_O/(V_I^2/R)$, and $P_D/(V_I^2/R)$ as functions of V_m/V_I.

3.2.4 Drain Efficiency of Class B RF Power Amplifier

The drain efficiency of the Class B RF power amplifier is

$$\eta_D = \frac{P_{DS}}{P_I} = \frac{V_m^2/(2R)}{2V_I V_m/(\pi R)} = \frac{\pi}{4}\left(\frac{V_m}{V_I}\right). \tag{3.32}$$

For $V_{m(max)} = V_I$, we obtain the maximum drain efficiency

$$\eta_{Dmax} = \frac{P_{DSmax}}{P_I} = \frac{\pi}{4} \approx 78.54\% \tag{3.33}$$

The drain efficiency η_D as a function of V_m/V_I for the Class B RF power amplifier is depicted in Fig. 3.7. Using a real MOSFET, $V_{m(max)} = V_I - v_{DSmin} \leq V_I - v_{DSsat}$, one obtains the maximum drain efficiency

$$\eta_{Dmax} = \frac{\pi}{4}\left(\frac{V_{m(max)}}{V_I}\right) = \frac{\pi}{4}\frac{(V_I - v_{DSmin})}{V_I} = \frac{\pi}{4}\left(1 - \frac{v_{DSmin}}{V_I}\right) \tag{3.34}$$

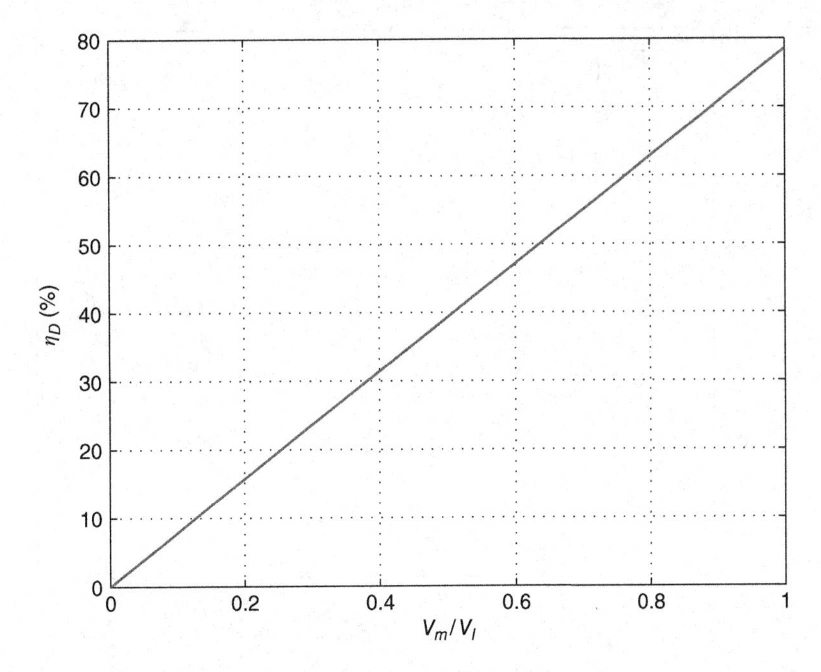

Figure 3.7 Drain efficiency η_D as a function of V_m/V_I for the Class B RF power amplifier.

3.2.5 Statistical Characterization of Drain Efficiency for Class B RF Power Amplifier

Assuming Rayleigh's probability density function (PDF) of the output voltage

$$h(V_m) = \frac{V_m}{\sigma^2} e^{-\frac{V_m^2}{2\sigma^2}} \tag{3.35}$$

the product of instantaneous drain efficiency η_D and the PDF $h(V_m)$ is

$$\eta_D h(V_m) = \frac{\pi}{4}\left(\frac{V_m}{V_I}\right)\frac{V_m}{\sigma^2} e^{-\frac{V_m^2}{2\sigma^2}} \tag{3.36}$$

Thus, the long-term average drain efficiency is determined by

$$\eta_{D(AV)} = \int_0^{V_I} \eta_D h(V_m) dV_m = \int_0^{V_I} \frac{\pi}{4}\left(\frac{V_m}{V_I}\right)\frac{V_m}{\sigma^2} e^{-\frac{V_m^2}{2\sigma^2}} dV_m \tag{3.37}$$

Figure 3.8 shows drain efficiency η_D, Rayleigh's PDF $h(V_m)$, and the product $\eta_D h(V_m)$ as functions of V_m at $V_I = 10$ V and $\sigma = 3$ for the Class B RF power amplifier.

The drain power or output power is

$$P_{DS} = P_O = \frac{V_m^2}{2R} \tag{3.38}$$

The maximum drain or output power is

$$P_{DSmax} = P_{Omax} = \frac{V_I^2}{2R} \tag{3.39}$$

Hence,

$$\frac{P_O}{P_{Omax}} = \left(\frac{V_m}{V_I}\right)^2 \tag{3.40}$$

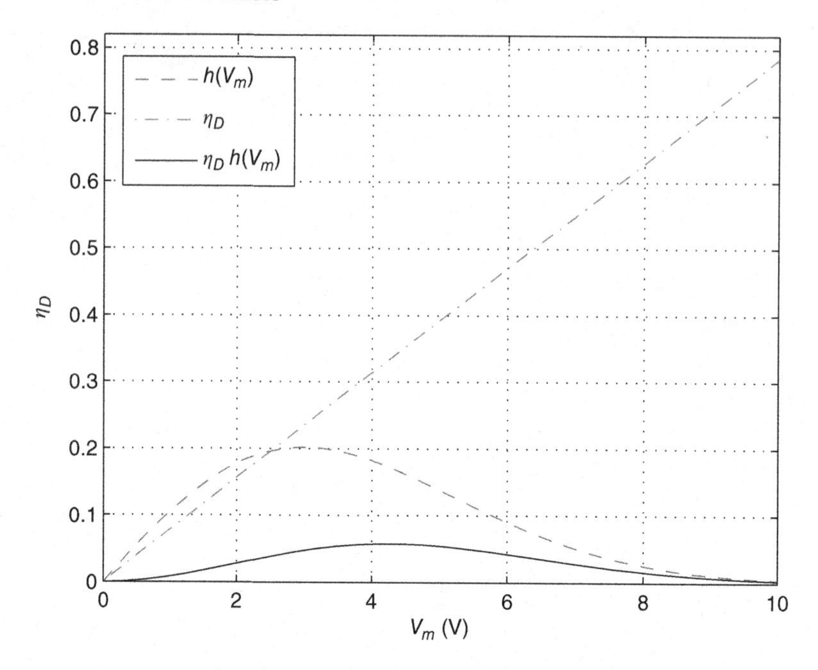

Figure 3.8 Drain efficiency η_D, Rayleigh's probability-density function $h(V_m)$, and the product $\eta_D h(V_m)$ as functions of V_m at $V_I = 10$ V and $\sigma = 3$. for the Class B RF power amplifier.

yielding

$$\frac{V_m}{V_I} = \sqrt{\frac{P_O}{P_{Omax}}} \tag{3.41}$$

Substitution of this equation into (3.32) produces the drain efficiency

$$\eta_D = \frac{\pi}{4}\sqrt{\frac{P_O}{P_{Omax}}} \tag{3.42}$$

Figure 3.9 shows drain efficiency η_D as a function of P_O/P_{Omax} for the Class B RF power amplifier. The drain efficiency η_D of the Class B RF power amplifier is proportional to the square root of the normalized output power P_O/P_{Omax}.

With Rayleigh's PDF of the output power

$$g(P_O) = \frac{P_O}{\sigma^2}e^{-\frac{P_O^2}{2\sigma^2}} \tag{3.43}$$

the product of the drain efficiency η_D and the PDF is

$$\eta_D g(P_O) = \frac{\pi}{4}\sqrt{\frac{P_O}{P_{Omax}}}\frac{P_O}{\sigma^2}e^{-\frac{P_O^2}{2\sigma^2}} \tag{3.44}$$

Hence, the long-term average drain efficiency is determined by

$$\eta_{D(AV)} = \int_0^{P_{Omax}} \eta_D g(P_O)dP_O = \int_0^{P_{Omax}} \frac{\pi}{4}\sqrt{\frac{P_O}{P_{Omax}}}\frac{P_O}{\sigma^2}e^{-\frac{P_O^2}{2\sigma^2}}dP_O \tag{3.45}$$

Figure 3.10 shows drain efficiency η_D, Rayleigh's PDF $g(P_O)$, and the product $\eta_D g(P_O)$ as functions of P_O/P_{Omax} for the Class B RF power amplifier.

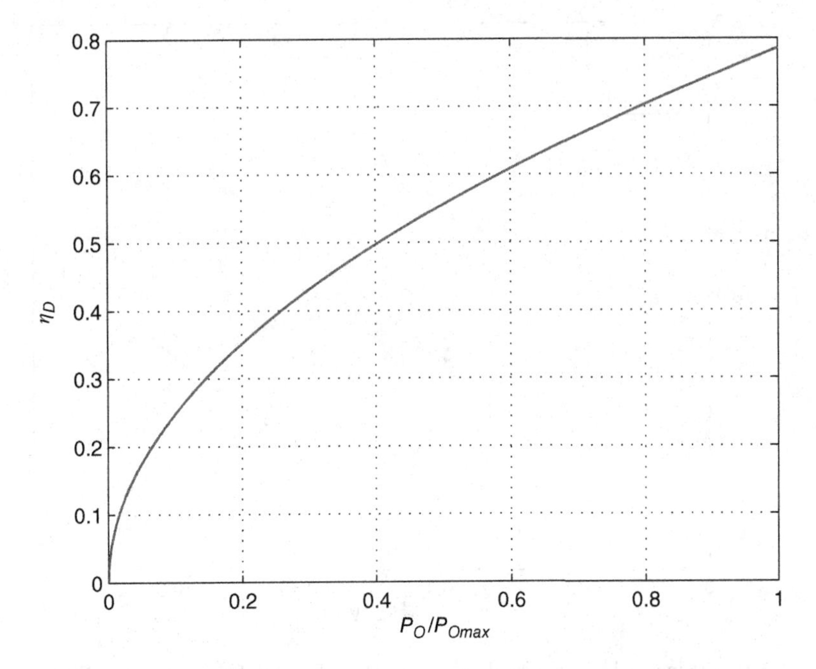

Figure 3.9 Drain efficiency η_D as a function of P_O/P_{Omax} for the Class B RF power amplifier.

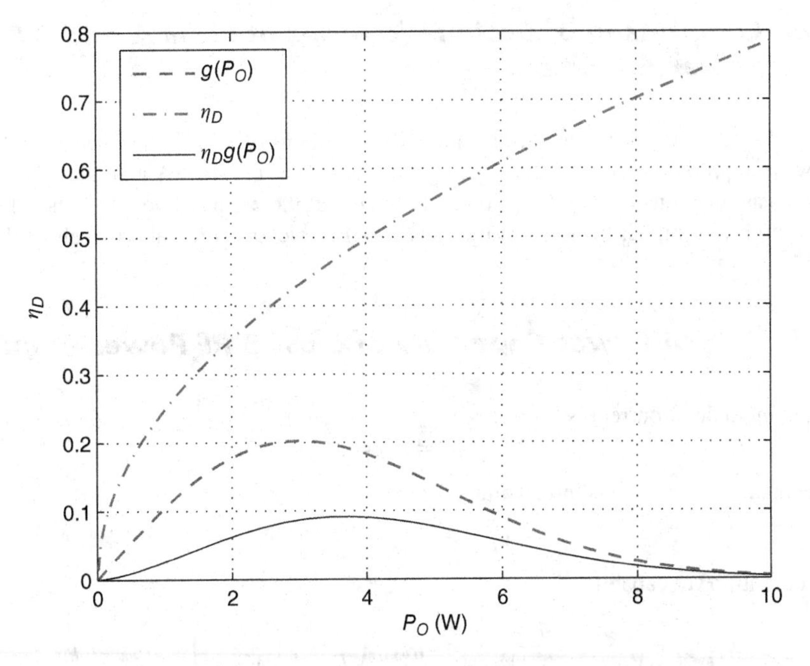

Figure 3.10 Drain efficiency η_D, Rayleigh's PDF $g(P_O)$, and the product $\eta_D g(P_O)$ as functions of P_O at $P_{Omax} = 10$ W and $\sigma = 3$ for the Class B RF power amplifier.

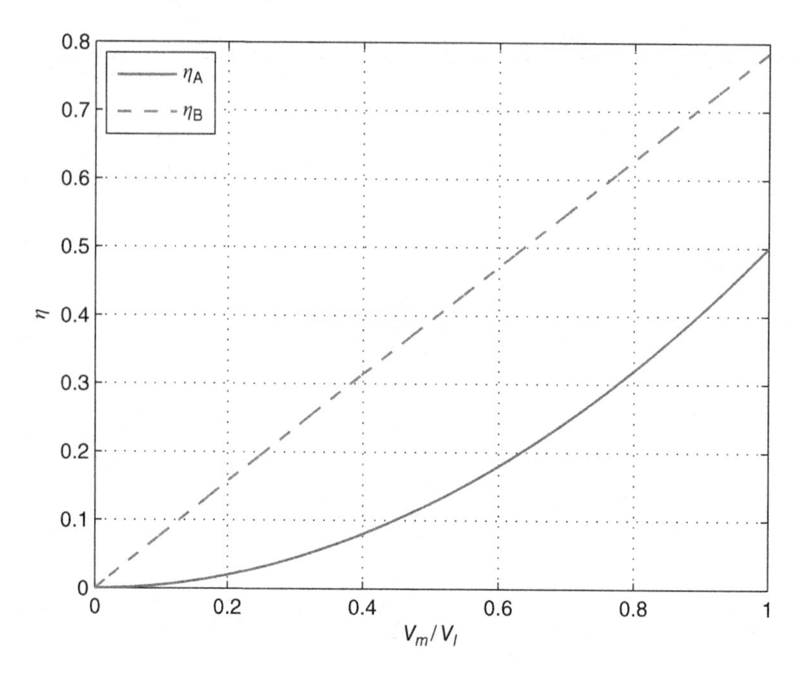

Figure 3.11 Drain efficiencies η_A and η_B as functions of V_m/V_I for Class A and Class B RF power amplifiers.

3.2.6 Comparison of Drain Efficiencies of Class A and B RF Power Amplifiers

Figure 3.11 shows drain efficiencies η_A and η_B as functions of V_m/V_I for Class A and Class B RF power amplifiers. Figure 3.12 compares the efficiencies of Class A and Class B RF power amplifiers as functions of P_O/P_{Omax}. It can be seen that the drain efficiency of the Class B RF power amplifier is higher than that of the Class A RF power amplifier at all levels of the output power P_O.

3.2.7 Output-Power Capability of Class B RF Power Amplifier

The maximum drain current is

$$I_{DM} = \pi I_I = 2I_m \tag{3.46}$$

and the maximum drain-to-source voltage is

$$V_{DSM} = 2V_I = 2V_{m(max)} \tag{3.47}$$

The output-power capability is

$$c_p = \frac{P_{Omax}}{I_{DM}V_{DSM}} = \frac{\eta_{Dmax}P_I}{I_{DM}V_{DSM}} = \eta_{Dmax}\left(\frac{I_I}{I_{DM}}\right)\left(\frac{V_I}{V_{DSM}}\right) = \eta_{Dmax}\alpha_0\beta_0$$
$$= \frac{\pi}{4} \times \frac{1}{\pi} \times \frac{1}{2} = \frac{1}{8} = 0.125 \tag{3.48}$$

where the coefficients α_0 and β_0 for the Class B RF power amplifier are

$$\alpha_0 = \frac{I_I}{I_{DM}} = \frac{1}{\pi} \tag{3.49}$$

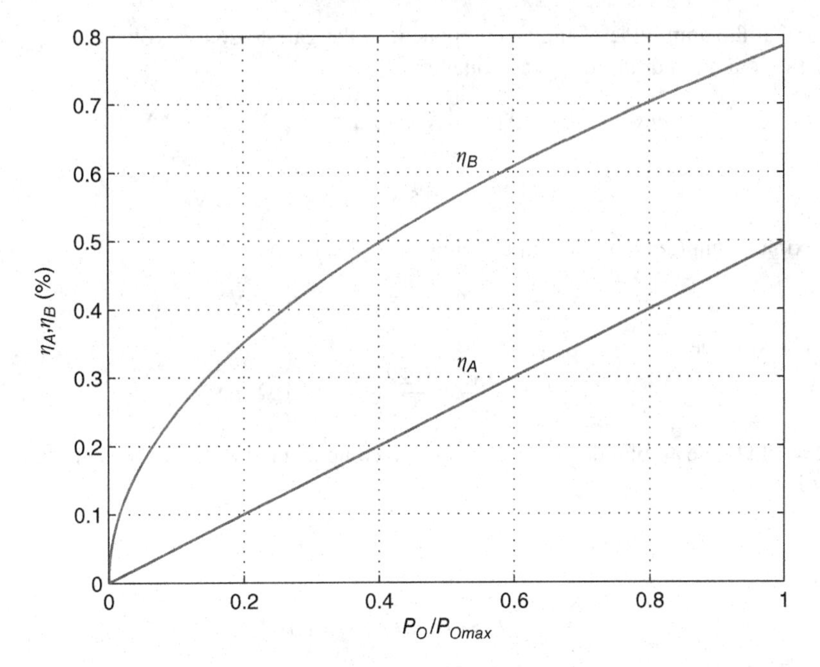

Figure 3.12 Drain efficiencies η_A and η_B as functions of P_O/P_{Omax} for Class A and Class B RF power amplifiers.

and

$$\beta_0 = \frac{V_I}{V_{DSM}} = \frac{1}{2}. \tag{3.50}$$

Another method to determine c_p is

$$c_p = \frac{P_{Omax}}{I_{DM} V_{DSM}} = \frac{1}{2}\left(\frac{I_{m(max)}}{I_{DM}}\right)\left(\frac{V_{m(max)}}{V_{DSM}}\right) = \frac{1}{2}\alpha_{1max}\beta_{1max} = \frac{1}{2} \times \frac{1}{2} \times \frac{1}{2} = 0.125 \tag{3.51}$$

where the maximum values of the coefficients α_1 and β_1 for the Class B RF power amplifier are

$$\alpha_{1max} = \frac{I_{m(max)}}{I_{DM}} = \frac{1}{2} \tag{3.52}$$

and

$$\beta_{1max} = \frac{V_{m(max)}}{V_{DSM}} = \frac{1}{2} \tag{3.53}$$

It is interesting to note that the value of c_p of the Class B amplifier is the same as that of the Class A amplifier.

Example 3.1

Design a Class B RF power amplifier to deliver a power of 20 W at $f = 2.4$ GHz. The bandwidth is $BW = 480$ MHz. The power supply voltage is $V_I = 24$ V.

Solution. Assume that the threshold voltage of the MOSFET is $V_t = 1$ V. For the Class B RF power amplifier, the dc component of the gate-to-source voltage is

$$V_{GS} = V_t = 1 \text{ V}. \tag{3.54}$$

Assume that the amplitude of the ac component of the gate-to-source voltage is $V_{gsm} = 1.5$ V. Hence, the saturation drain-to-source voltage is

$$v_{DSsat} = v_{GS} - V_t = V_{GS} + V_{gsm} - V_t = V_{gsm} = 1.5 \text{ V} \tag{3.55}$$

Pick

$$v_{DSmin} = v_{DSsat} + 0.5 \text{ V} = 1.5 + 0.5 = 2 \text{ V} \tag{3.56}$$

The maximum amplitude of the output voltage is

$$V_m = V_I - v_{DSmin} = 24 - 2 = 22 \text{ V} \tag{3.57}$$

The load resistance is

$$R = \frac{V_m^2}{2P_O} = \frac{22^2}{2 \times 20} = 12.1\ \Omega \tag{3.58}$$

Pick $R = 12\ \Omega$. The amplitude of the fundamental component of the drain current and the output current is given by

$$I_m = \frac{V_m}{R} = \frac{22}{12} = 1.833 \text{ A} \tag{3.59}$$

The dc supply current is

$$I_I = \frac{2}{\pi} I_m = \frac{2}{\pi} \times 1.833 = 1.167 \text{ A} \tag{3.60}$$

The dc resistance presented by the amplifier to the dc power supply is

$$R_{DC} = \frac{V_I}{I_I} = \frac{24}{1.167} = 20.566\ \Omega \tag{3.61}$$

The maximum drain current is

$$I_{DM} = \pi I_I = \pi \times 1.167 = 3.666 \text{ A} \tag{3.62}$$

The maximum drain-to-source voltage is

$$V_{DSM} = 2V_I = 2 \times 24 = 48 \text{ V} \tag{3.63}$$

Assuming that $\mu_{n0} C_{ox} = 0.142 \times 10^{-3}$ A/V^2, $V_t = 1$ V, and $L = 0.35$ μm, the power MOSFET channel width-to-length ratio is

$$\frac{W}{L} = \frac{2I_{DM}}{\mu_{n0} C_{ox} v_{DSsat}^2} = \frac{2 \times 3.666}{0.142 \times 10^{-3} \times 1.5^2} = 22,948 \tag{3.64}$$

Let us assume $W/L = 23,000$. The channel width of the MOSFET is

$$W = \left(\frac{W}{L}\right) L = 23,000 \times 0.35 \times 10^{-6} = 8.05 \text{ mm} \tag{3.65}$$

Hence,

$$K = \frac{1}{2} \mu_{n0} \left(\frac{W}{L}\right) = \frac{1}{2} \times 0.142 \times 10^{-3} \times 23,000 = 1.633 \text{ A/V} \tag{3.66}$$

The dc supply power is

$$P_I = I_I V_I = 1.167 \times 24 = 28 \text{ W} \tag{3.67}$$

The drain power dissipation is

$$P_D = P_I - P_O = 28 - 20 = 8 \text{ W} \tag{3.68}$$

The drain efficiency is

$$\eta_D = \frac{P_O}{P_I} = \frac{20}{28} = 71.43\% \tag{3.69}$$

The loaded quality factor is

$$Q_L = \frac{f}{BW} = \frac{2.4}{0.48} = 5 \tag{3.70}$$

The reactances of the resonant circuit components are

$$X_L = X_C = \frac{R}{Q_L} = \frac{12}{5} = 2.4\ \Omega \tag{3.71}$$

yielding

$$L = \frac{X_L}{\omega_c} = \frac{2.4}{2\pi \times 2.4 \times 10^9} = 159\ \text{pH} \tag{3.72}$$

and

$$C = \frac{1}{\omega_c X_C} = \frac{1}{2\pi \times 2.4 \times 10^9 \times 2.4} = 27.6\ \text{pF} \tag{3.73}$$

The reactance of the RF choke is

$$X_{Lf} = 10R = 10 \times 12 = 120\ \Omega \tag{3.74}$$

which gives

$$L_f = \frac{X_{Lf}}{\omega_c} = \frac{120}{2\pi \times 2.4 \times 10^9} = 8\ \text{nH} \tag{3.75}$$

The reactance of the coupling capacitor is

$$X_{Cc} = \frac{R}{10} = \frac{12}{10} = 1.2\ \Omega \tag{3.76}$$

which produces

$$C_c = \frac{1}{\omega_c X_{Cc}} = \frac{1}{2\pi \times 2.4 \times 10^9 \times 1.2} = 54.8\ \text{pF} \tag{3.77}$$

3.3 Class AB and C RF Power Amplifiers

3.3.1 Waveforms in Class AB and C RF Power Amplifiers

The circuit of the Class C power amplifier is the same as that of the Class B RF power amplifier. The operating point of the transistor is located in the cutoff region. The dc component of the gate-to-source voltage V_{GS} is less than the transistor threshold voltage V_t. Therefore, the conduction angle of the drain current 2θ is less than 180°. Voltage and current waveforms in the Class C power amplifier are shown in Fig. 3.13. The only difference is the conduction angle of the drain current, which is determined by the operating point. The gate-to-source voltage waveform is

$$v_{GS} = V_{GS} + v_{gs} = V_{GS} + V_{gsm} \cos \omega t \tag{3.78}$$

Since

$$v_{GS}(\theta) = V_{GS} + V_{gsm} \cos \theta = V_t \tag{3.79}$$

the cosine of the conduction angle θ is given by

$$\cos \theta = \frac{V_t - V_{GS}}{V_{gsm}} \tag{3.80}$$

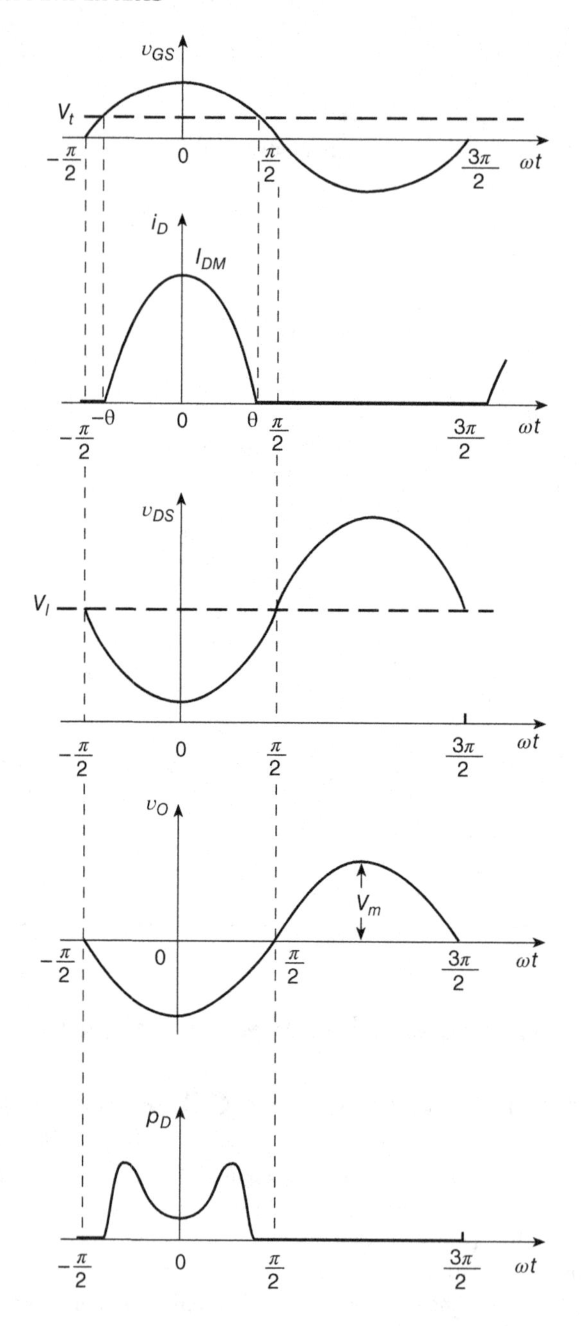

Figure 3.13 Waveforms in Class C RF power amplifier.

and the conduction angle of the drain current waveform is

$$\theta = \begin{cases} \arccos\left(\dfrac{V_t - V_{GS}}{V_{gsm}}\right) & \text{for} \quad \dfrac{V_t - V_{GS}}{V_{gsm}} > 0 \quad \text{(Class C)} \\ 180^\circ - \arccos\left(\dfrac{V_t - V_{GS}}{V_{gsm}}\right) & \text{for} \quad \dfrac{V_t - V_{GS}}{V_{GS}} < 0 \quad \text{(Class AB)} \end{cases} \tag{3.81}$$

Using the square law, the drain current waveform is

$$i_D = K(v_{GS} - V_t)^2 = K(V_{GS} + V_{gsm}\cos\omega t - V_t)^2 \quad \text{for} \quad v_{GS} > V_t \tag{3.82}$$

The maximum value of the drain current is

$$I_{DM} = K(V_{GS} + V_{gsm} - V_t)^2 = Kv_{DSsat}^2 \tag{3.83}$$

resulting in

$$K = \frac{1}{2}\mu_{n0}C_{ox}\left(\frac{W}{L}\right) = \frac{2I_{DM}}{v_{DSsat}^2} = \frac{2I_{DM}}{(V_{GS} + V_{gsm} - V_t)^2} \tag{3.84}$$

Hence,

$$\frac{W}{L} = \frac{2K}{\mu_{n0}C_{ox}} \tag{3.85}$$

Using the linear law, the drain current waveform is expressed by

$$i_D = K_s(v_{GS} - V_t) = K_s(V_{GS} + V_{gsm}\cos\omega t - V_t) \quad \text{for} \quad -\frac{\pi}{2} < \omega t \le \frac{\pi}{2} \tag{3.86}$$

where

$$K_s = \frac{1}{2}C_{ox}Wv_{sat} \tag{3.87}$$

The drain current waveform for any conduction angle θ, that is, for Class AB, Class B, and Class C RF power amplifiers, is given by

$$i_D = \begin{cases} I_{DM}\dfrac{\cos\omega t - \cos\theta}{1 - \cos\theta} & \text{for} \quad -\theta < \omega t \le \theta \\ 0 & \text{for} \quad \theta < \omega t \le 2\pi - \theta. \end{cases} \tag{3.88}$$

The drain current waveform is an even function of ωt and satisfies the condition: $i_D(\omega t) = i_D(-\omega t)$. The drain current waveform can be expanded into Fourier series (Appendix D)

$$i_D(\omega t) = I_I + \sum_{n=1}^{\infty} I_{mn}\cos n\omega t = I_{DM}\left(\alpha_0 + \sum_{n=1}^{\infty}\alpha_n\cos n\omega t\right) \tag{3.89}$$

The dc component of the drain current waveform is

$$I_I = \frac{1}{2\pi}\int_{-\theta}^{\theta} i_D d(\omega t) = \frac{1}{\pi}\int_0^{\theta} i_D d(\omega t) = \frac{I_{DM}}{\pi}\int_0^{\theta}\frac{\cos\omega t - \cos\theta}{1 - \cos\theta}d(\omega t)$$
$$= I_{DM}\frac{\sin\theta - \theta\cos\theta}{\pi(1 - \cos\theta)} = \alpha_0 I_{DM} \tag{3.90}$$

where

$$\alpha_0 = \frac{I_I}{I_{DM}} = \frac{\sin\theta - \theta\cos\theta}{\pi(1 - \cos\theta)} \tag{3.91}$$

The amplitude of the fundamental component of the drain current waveform is given by

$$I_m = \frac{1}{\pi}\int_{-\theta}^{\theta} i_D\cos\omega t d(\omega t) = \frac{2}{\pi}\int_0^{\theta} i_D\cos\omega t d(\omega t)$$
$$= \frac{2I_{DM}}{\pi}\int_0^{\theta}\frac{\cos\omega t - \cos\theta}{1 - \cos\theta}\cos\omega t d(\omega t) = I_{DM}\frac{\theta - \sin\theta\cos\theta}{\pi(1 - \cos\theta)} = \alpha_1 I_{DM} \tag{3.92}$$

where

$$\alpha_1 = \frac{I_m}{I_{DM}} = \frac{\theta - \sin\theta\cos\theta}{\pi(1 - \cos\theta)} \tag{3.93}$$

The amplitude of the nth harmonic of the drain current waveform is

$$I_{mn} = \frac{1}{\pi}\int_{-\theta}^{\theta} i_D \cos n\omega t d(\omega t) = \frac{2}{\pi}\int_{0}^{\theta} i_D \cos n\omega t d(\omega t)$$
$$= \frac{2I_{DM}}{\pi}\int_{0}^{\theta} \frac{\cos\omega t - \cos\theta}{1-\cos\theta} \cos n\omega t d(\omega t)$$
$$= I_{DM}\frac{2}{\pi}\frac{\sin n\theta\cos\theta - n\cos n\theta \sin\theta}{n(n^2-1)(1-\cos\theta)} = \alpha_n I_{DM} \quad \text{for} \quad n \geq 2 \tag{3.94}$$

where

$$\alpha_n = \frac{I_{mn}}{I_{DM}} = \frac{2}{\pi}\frac{\sin n\theta\cos\theta - n\cos n\theta\sin\theta}{n(n^2-1)(1-\cos\theta)} \quad \text{for} \quad n \geq 2. \tag{3.95}$$

Figure 3.14 shows the Fourier coefficients α_n as functions of the conduction angle θ of the drain current waveform i_D.

The ratio of the amplitude of the fundamental component to the dc component of the drain current waveform is given by

$$\gamma_1 = \frac{I_m}{I_I} = \frac{I_m/I_{DM}}{I_I/I_{DM}} = \frac{\alpha_1}{\alpha_0} = \frac{\theta - \sin\theta\cos\theta}{\sin\theta - \theta\cos\theta} \tag{3.96}$$

The ratio $\gamma_1 = I_m/I_I$ as a function of conduction angle θ of the drain current waveform is shown in Fig. 3.15.

The drain current waveform in terms of dc supply current I_I is given by

$$i_D = \begin{cases} I_I \dfrac{\pi(\cos\omega t - \cos\theta)}{\sin\theta - \theta\cos\theta} & \text{for} \quad -\theta < \omega t \leq \theta \\ 0 & \text{for} \quad \theta < \omega t \leq 2\pi - \theta \end{cases} \tag{3.97}$$

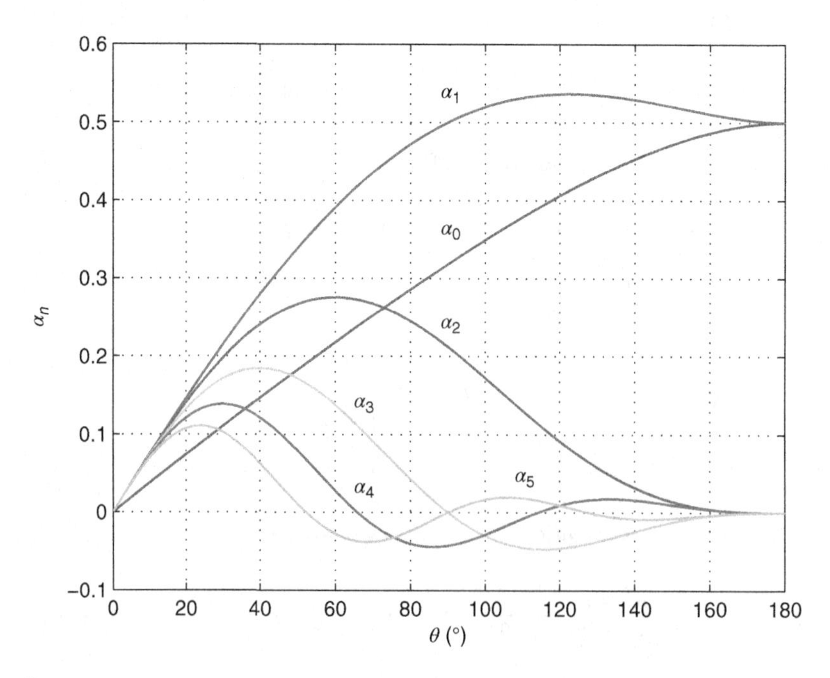

Figure 3.14 Fourier coefficients α_n of the drain current i_D as a function of the conduction angle θ for Class AB, B, and C RF power amplifiers.

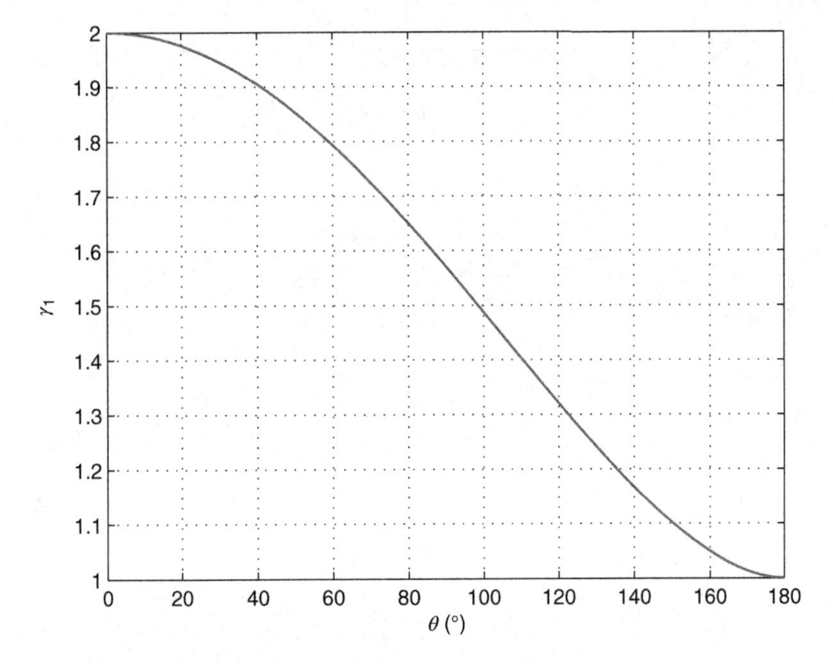

Figure 3.15 Ratio of the fundamental component amplitude to the dc component $\gamma_1 = I_m/I_I$ of the drain current i_D as a function of conduction angle θ for Class AB, B, and C RF power amplifiers.

The drain-to-source voltage waveform at $f = f_0$ is

$$v_{DS} = V_I - V_m \cos \omega t = V_I \left(1 - \frac{V_m}{V_I}\right) \cos \omega t = V_I(1 - \xi_1) \cos \omega t \tag{3.98}$$

where $\xi_1 = V_m/V_I$. The waveform of the normalized drain power loss at the resonant frequency $f = f_0$ is

$$\frac{p_D(\omega t)}{P_I} = \frac{i_D v_{DS}}{P_I} = \begin{cases} \dfrac{\pi(\cos \omega t - \cos \theta)}{\sin \theta - \theta \cos \theta} \left(1 - \dfrac{V_m}{V_I} \cos \omega t\right) & \text{for} \quad -\theta < \omega t \leq \theta \\ 0 & \text{for} \quad \theta < \omega t \leq 2\pi - \theta \end{cases} \tag{3.99}$$

The waveforms of the normalized drain power loss for $\theta = 120°, 60°$, and $45°$ at $f = f_0$ are shown in Figs. 3.16, 3.17, and 3.18, respectively. As the conduction angle θ decreases, the peak values of $p_D(\omega t)/P_I$ increase.

The waveform of the normalized drain power loss at f different from f_0 is given by

$$\frac{p_D(\omega t)}{P_I} = \begin{cases} \dfrac{\pi(\cos \omega t - \cos \theta)}{\sin \theta - \theta \cos \theta} \left[1 - \dfrac{V_m}{V_I} \cos(\omega t + \phi)\right] & \text{for} \quad -\theta < \omega t \leq \theta \\ 0 & \text{for} \quad \theta < \omega t \leq 2\pi - \theta \end{cases} \tag{3.100}$$

where ϕ is the phase shift between the peak value of the drain current and the minimum value of the drain-to-source voltage. The waveforms of the normalized drain power loss for $\theta = 60°$ at $\phi = 15°$ are shown in Fig. 3.19.

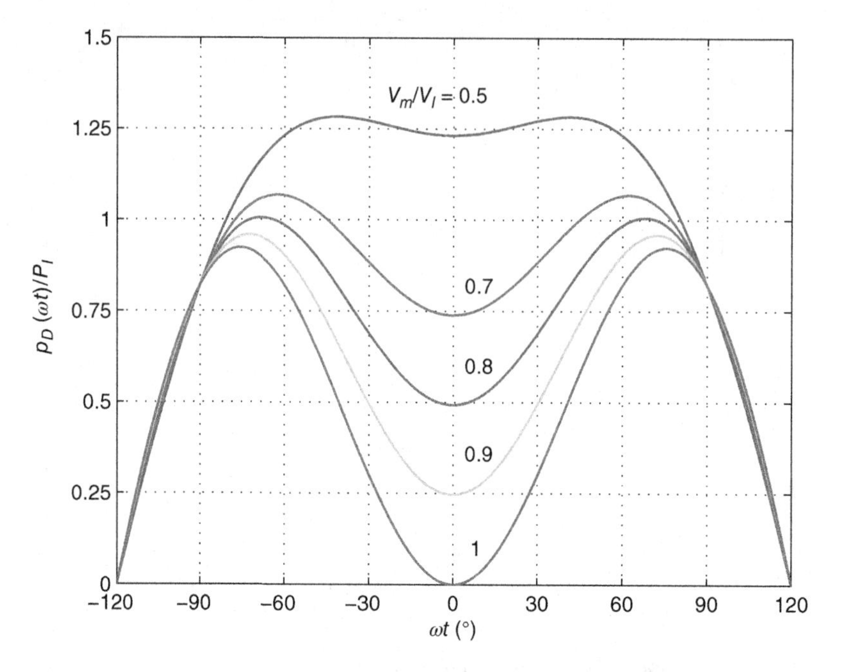

Figure 3.16 Normalized drain power loss waveform $p_D(\omega t)/P_I$ at conduction angle $\theta = 120°$ and $f = f_0$ for the Class AB RF power amplifier.

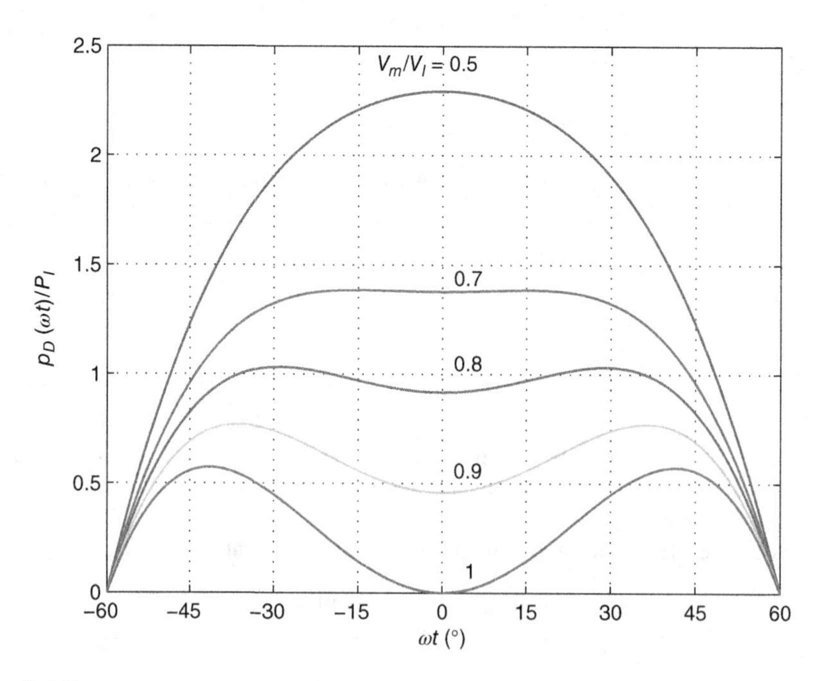

Figure 3.17 Normalized drain power loss waveform $p_D(\omega t)/P_I$ at conduction angle $\theta = 60°$ and $f = f_0$ for the Class C RF power amplifier.

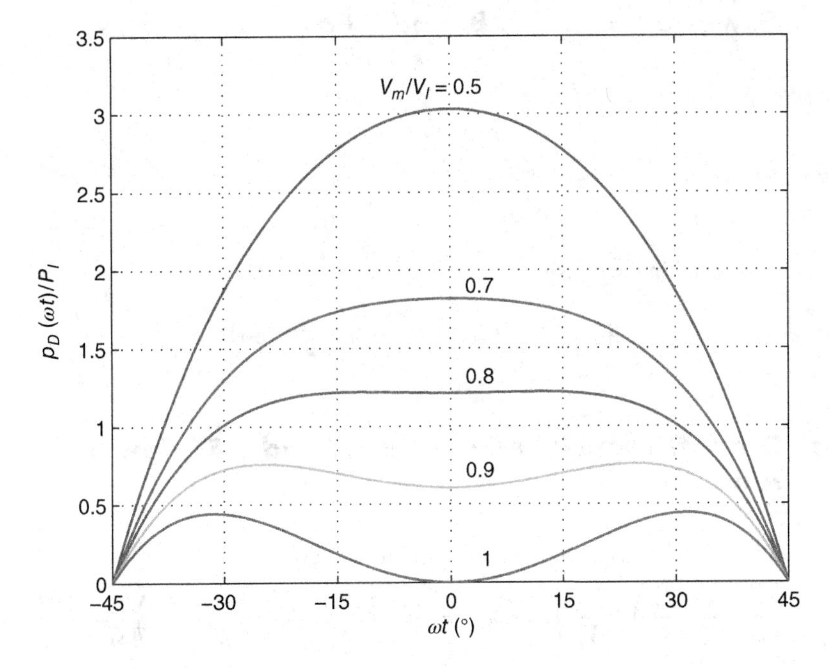

Figure 3.18 Normalized drain power loss waveform $p_D(\omega t)/P_I$ at conduction angle $\theta = 45°$ and $f = f_0$ for the Class C RF power amplifier.

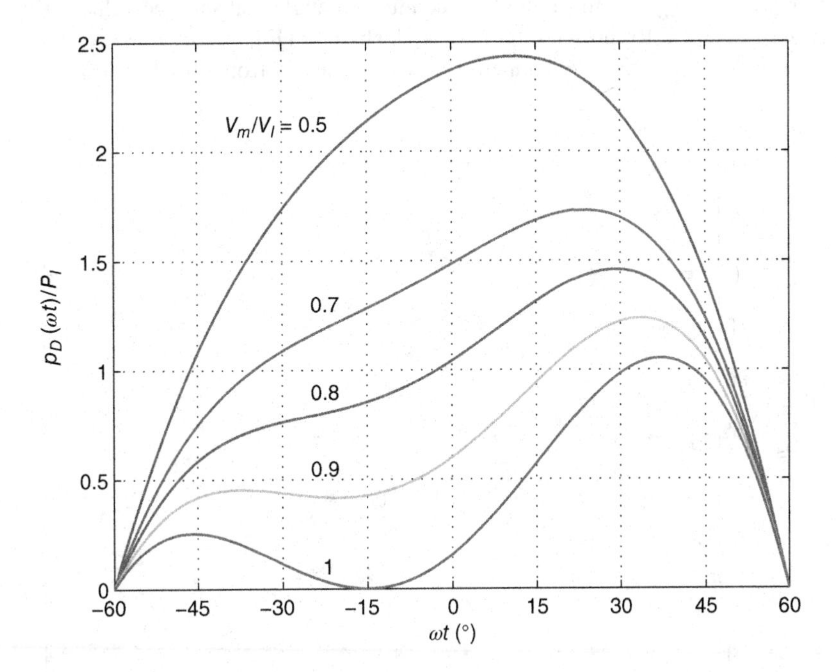

Figure 3.19 Normalized drain power loss waveform $p_D(\omega t)/P_I$ at conduction angle $\theta = 60°$ and $\phi = 15°$ for the Class C RF power amplifier.

3.3.2 Power of Class AB, B, and C Power Amplifiers

The dc supply power is

$$P_I = I_I V_I = \alpha_0 I_{DM} V_I \tag{3.101}$$

The drain power is

$$P_{DS} = \frac{1}{2} I_m V_m = \frac{1}{2} \alpha_1 I_{DM} V_m \tag{3.102}$$

The power dissipated in the transistor is

$$P_D = P_I - P_{DS} = \alpha_0 I_{DM} V_I - \frac{1}{2} \alpha_1 I_{DM} V_m \tag{3.103}$$

3.3.3 Drain Efficiency of Class AB, B, and C RF Power Amplifiers

The drain efficiency of Class AB, B, and C amplifiers is given by

$$\eta_D = \frac{P_{DS}}{P_I} = \frac{1}{2}\left(\frac{I_m}{I_I}\right)\left(\frac{V_m}{V_I}\right) = \frac{1}{2}\gamma_1 \xi_1 = \frac{1}{2}\left(\frac{\alpha_1}{\alpha_0}\right)\left(\frac{V_m}{V_I}\right) = \frac{1}{2}\left(\frac{V_m}{V_I}\right)\frac{\theta - \sin\theta\cos\theta}{\sin\theta - \theta\cos\theta}$$
$$= \frac{\theta - \sin\theta\cos\theta}{2(\sin\theta - \theta\cos\theta)}\left(1 - \frac{v_{DSmin}}{V_I}\right) \tag{3.104}$$

The drain efficiency η_D as a function of the conduction angle θ at selected values of V_m/V_I for Class A, AB, B, and C RF power amplifiers is illustrated in Fig. 3.20. As the conduction angle θ decreases from 180° to 0°, the drain efficiency η_D increases from 50% to 100% at $V_m = V_I$.

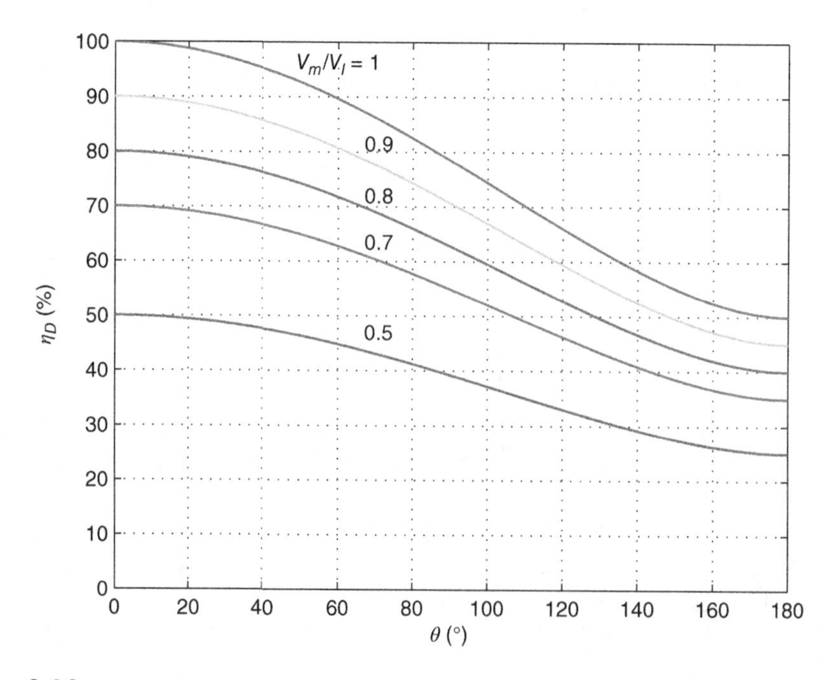

Figure 3.20 Drain efficiency η_D as a function of conduction angle θ at various values of V_m/V_I for Class A, AB, B, and C RF power amplifiers.

3.3.4 Output-Power Capability of Class AB, B, and C RF Power Amplifiers

Since $V_{m(maxc)} \approx V_I$, the maximum drain-to-source voltage is

$$V_{DSM} = V_I + V_{m(max)} \approx 2V_I \approx 2V_{m(max)} \tag{3.105}$$

The maximum drain current is

$$I_{DM} = \frac{I_m}{\alpha_1} \tag{3.106}$$

Thus, the output-power capability is

$$c_p = \frac{P_{Omax}}{V_{DSM}I_{DM}} = \frac{\frac{1}{2}V_{m(max)}I_{m(max)}}{V_{DSM}I_{DM}} = \frac{1}{2}\left(\frac{V_{m(max)}}{2V_I}\right)\left(\frac{I_{m(max)}}{I_{DM}}\right) = \frac{I_m}{4I_{DM}} = \frac{\alpha_{1max}}{4}$$
$$= \frac{\theta - \sin\theta\cos\theta}{4\pi(1-\cos\theta)} \tag{3.107}$$

where $V_{DSM} = 2V_I$. The output-power capability c_p as a function of conduction angle θ for Class A, AB, B, and C RF power amplifiers is shown in Fig. 3.21. As the conduction angle θ approaches zero, c_p also approaches zero.

3.3.5 Parameters of Class AB RF Power Amplifier at $\theta = 120°$

For $90° \leq \theta \leq 180°$, we obtain the Class AB power amplifier. The Fourier coefficients of the drain current waveform at the typical conduction angle $\theta = 120°$ are

$$\alpha_0 = \frac{I_I}{I_{DM}} = \frac{3\sqrt{3}+2\pi}{9\pi} \approx 0.406 \tag{3.108}$$

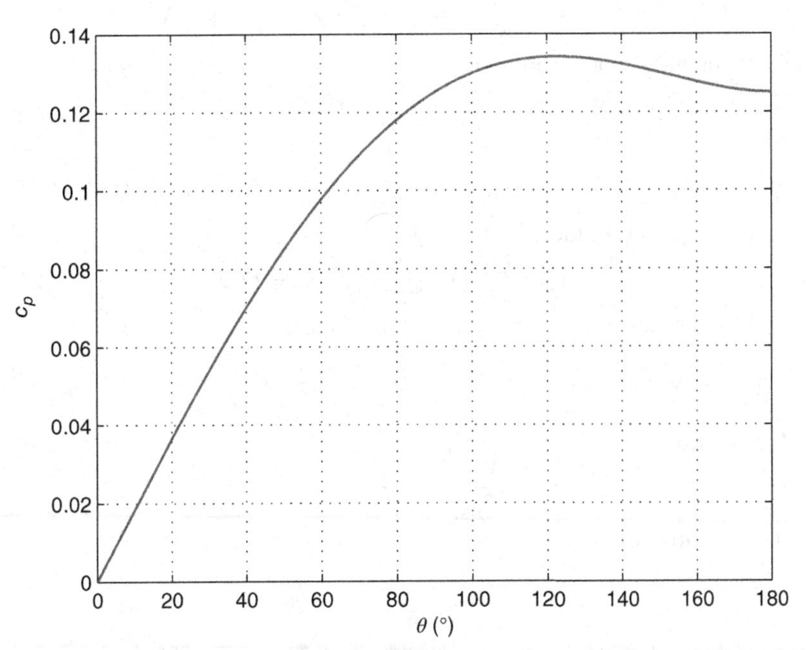

Figure 3.21 Output-power capability c_p as a function of conduction angle θ for Class A, AB, B, and C RF power amplifiers.

and

$$\alpha_1 = \frac{I_m}{I_{DM}} = \frac{3\sqrt{3}+8\pi}{18\pi} \approx 0.5363 \tag{3.109}$$

The ratio of the fundamental component amplitude to the dc component of the drain current waveform is given by

$$\gamma_1 = \frac{I_m}{I_I} = \frac{\alpha_1}{\alpha_0} = \frac{3\sqrt{3}+8\pi}{2(3\sqrt{3}+2\pi)} = 1.321 \tag{3.110}$$

The drain efficiency at $\theta = 120°$ is

$$\eta_D = \frac{P_{DS}}{P_I} = \frac{\eta_{Dmax}P_O}{V_{DSM}I_{DM}} = \frac{3\sqrt{3}+8\pi}{4(3\sqrt{3}+2\pi)}\left(1-\frac{v_{DSmin}}{V_I}\right) \approx 0.6605\left(1-\frac{v_{DSmin}}{V_I}\right) \tag{3.111}$$

The output-power capability at $\theta = 120°$ is

$$c_p = \frac{P_O}{V_{DSM}I_{DM}} = \frac{\eta_{Dmax}P_I}{V_{DSM}I_{DM}} = \eta_{Dmax}\frac{V_I I_I}{V_{DSM}I_{DM}} = \frac{3\sqrt{3}+8\pi}{72\pi} \approx 0.13408 \tag{3.112}$$

Example 3.2

Design a Class AB power amplifier to deliver a power of 12 W at $f = 5$ GHz. The bandwidth is $BW = 500$ MHz and the conduction angle is $\theta = 120°$. The power supply voltage is $V_I = 24$ V.

Solution. Assume that the MOSFET threshold voltage is $V_t = 1$ V, and the dc component of the gate-to-source voltage is $V_{GS} = 1.5$ V. Hence, the amplitude of the ac component of the gate-to-source voltage is

$$V_{gsm} = \frac{V_t - V_{GS}}{\cos\theta} = \frac{1-1.5}{\cos 120°} = \frac{1-1.5}{-0.5} = 1\ \text{V} \tag{3.113}$$

The minimum drain-to-source voltage is

$$v_{DSsat} = v_{GS} - V_t = V_{GS} + V_{gsm} - V_t = 1.5 + 1 - 1 = 1.5\ \text{V} \tag{3.114}$$

Let

$$v_{DSmin} = v_{DSsat} + 0.5\ \text{V} = 1.5 + 0.5 = 2\ \text{V} \tag{3.115}$$

The maximum amplitude of the output voltage is

$$V_m = V_I - v_{DSmin} = 24 - 2 = 22\ \text{V} \tag{3.116}$$

Assuming the efficiency of the resonant circuit $\eta_r = 0.8$, the drain power is

$$P_{DS} = \frac{P_O}{\eta_r} = \frac{12}{0.8} = 15\ \text{W} \tag{3.117}$$

The load resistance is

$$R = \frac{V_m^2}{2P_{DS}} = \frac{22^2}{2\times 15} = 16.133\ \Omega \tag{3.118}$$

The amplitude of the output current is

$$I_m = \frac{V_m}{R} = \frac{22}{16.133} = 1.3637\ \text{A} \tag{3.119}$$

The dc supply current is

$$I_I = \frac{I_m}{\gamma_1} = \frac{1.3637}{1.321} = 1.0323\ \text{A} \tag{3.120}$$

For $\theta = 120°$, the maximum drain current is

$$I_{DM} = \frac{I_I}{\alpha_0} = \frac{1.0323}{0.406} = 2.5426 \text{ A} \tag{3.121}$$

The maximum drain-to-source voltage is

$$V_{DSM} = 2V_I = 2 \times 24 = 48 \text{ V} \tag{3.122}$$

Using the square law and assuming $\mu_{n0}C_{ox} = 0.142\,\text{mA/V}^2$, $V_t = 1$ V, and $L = 0.35$ μm, the MOSFET aspect ratio is

$$\frac{W}{L} = \frac{2I_{DM}}{\mu_{n0}C_{ox}v_{DSsat}^2} = \frac{2 \times 2.5426}{0.142 \times 10^{-3} \times 1.5^2} = 15,916 \tag{3.123}$$

Pick $W/L = 16,000$. The MOSFET channel width is

$$W = \left(\frac{L}{W}\right)L = 16,000 \times 0.35 \times 10^{-6} = 5.6 \text{ mm} \tag{3.124}$$

Hence,

$$K = \frac{1}{2}\mu_{n0}C_{ox}\left(\frac{W}{L}\right) = \frac{1}{2} \times 0.142 \times 10^{-3} \times 16,000 = 1.136 \text{ A/V} \tag{3.125}$$

The dc supply power is

$$P_I = V_I I_I = 24 \times 1.0323 = 24.775 \text{ W} \tag{3.126}$$

The drain power dissipation is

$$P_D = P_I - P_{DS} = 24.775 - 12 = 12.775 \text{ W} \tag{3.127}$$

The drain efficiency is

$$\eta_D = \frac{P_{DS}}{P_I} = \frac{15}{24.775} = 60.055 \tag{3.128}$$

The loaded quality factor of the resonant circuit is

$$Q_L = \frac{f_0}{BW} = \frac{5}{0.5} = 10 \tag{3.129}$$

The reactances of the resonant circuit components are

$$X_L = X_C = \frac{R}{Q_L} = \frac{16.133}{10} = 1.6133\ \Omega \tag{3.130}$$

yielding

$$L = \frac{X_L}{\omega_c} = \frac{1.6133}{2\pi \times 5 \times 10^9} = 51.35 \text{ pH} \tag{3.131}$$

and

$$C = \frac{1}{\omega_c X_C} = \frac{1}{2\pi \times 5 \times 10^9 \times 1.6133} = 19.73 \text{ pF} \tag{3.132}$$

The reactance of the RF choke is

$$X_{Lf} = 10R = 10 \times 16.133 = 161.33\ \Omega \tag{3.133}$$

which gives

$$L_f = \frac{X_{Lf}}{\omega_c} = \frac{161.133}{2\pi \times 5 \times 10^9} = 5.135 \text{ nH} \tag{3.134}$$

The reactance of the coupling capacitor is

$$X_{Cc} = \frac{R}{10} = \frac{161.33}{10} = 16.133\ \Omega \tag{3.135}$$

which produces

$$C_c = \frac{1}{\omega_c X_{Cc}} = \frac{1}{2\pi \times 2.4 \times 10^9 \times 16.133} = 411 \text{ pF} \tag{3.136}$$

3.3.6 Parameters of Class C RF Power Amplifier at $\theta = 60°$

For $\theta < 90°$, we obtain the Class C amplifier. The Fourier coefficients of the drain current at the typical conduction angle $\theta = 60°$ are

$$\alpha_0 = \frac{I_I}{I_{DM}} = \frac{\sqrt{3}}{\pi} - \frac{1}{3} \approx 0.218 \tag{3.137}$$

and

$$\alpha_1 = \frac{I_m}{I_{DM}} = \frac{2}{3} - \frac{\sqrt{3}}{2\pi} \approx 0.391 \tag{3.138}$$

The ratio of the fundamental component amplitude to the dc component of the drain current waveform is given by

$$\gamma_1 = \frac{I_m}{I_I} = \frac{\alpha_1}{\alpha_0} = \frac{4\pi - 3\sqrt{3}}{2(3\sqrt{3} - \pi)} = 1.7936 \tag{3.139}$$

The drain efficiency at $\theta = 60°$ is

$$\eta_D = \frac{P_{DS}}{P_I} = \frac{1}{2}\left(\frac{I_m}{I_I}\right)\left(\frac{V_m}{V_I}\right) = \frac{4\pi - 3\sqrt{3}}{4(3\sqrt{3} - \pi)}\left(1 - \frac{v_{DSmin}}{V_I}\right) \approx 0.8968\left(1 - \frac{v_{DSmin}}{V_I}\right) \tag{3.140}$$

The output power capability at $\theta = 60°$ is

$$c_p = \frac{P_{DS}}{V_{DSM}I_{DM}} = \frac{\eta_{Dmax}P_I}{V_{DSM}I_{DM}} = \eta_{Dmax}\frac{V_I I_I}{V_{DSM}I_{DM}} = \frac{1}{6} - \frac{\sqrt{3}}{8\pi} \approx 0.09777 \tag{3.141}$$

Example 3.3

Design a Class C power amplifier to deliver a power of 6 W at $f = 2.4$ GHz. The bandwidth is $BW = 240$ MHz and the drain current conduction angle is $\theta = 60°$. The power supply voltage is $V_I = 12$ V. The gate-drive power is $P_G = 0.6$ W.

Solution. Assume that the MOSFET threshold voltage is $V_t = 1$ V and the dc component of the gate-to-source voltage is $V_{GS} = 0$. The amplitude of the ac component of the gate-to-source voltage is

$$V_{gsm} = \frac{V_t - V_{GS}}{\cos\theta} = \frac{1 - 0}{\cos 60°} = \frac{1}{0.5} = 2 \text{ V} \tag{3.142}$$

The saturation drain-to-source voltage is

$$v_{DSsat} = v_{GS} - V_t = V_{GS} + V_{gsm} - V_t = 0 + 2 - 1 = 1 \text{ V} \tag{3.143}$$

Assume the minimum drain-to-source voltage to be $v_{DSmin} = 1.2$ V. The maximum amplitude of the output voltage is

$$V_m = V_I - v_{DSmin} = 12 - 1.2 = 10.8 \text{ V} \tag{3.144}$$

Assume that the efficiency of the resonant circuit is $\eta_r = 70\%$. Hence, the drain power is

$$P_{DS} = \frac{P_O}{\eta_r} = \frac{6}{0.7} = 8.571 \text{ W} \tag{3.145}$$

The resistance seen by the drain is

$$R = \frac{V_m^2}{2P_{DS}} = \frac{10.8^2}{2 \times 8.571} = 6.8\ \Omega \tag{3.146}$$

The amplitude of the fundamental component of the drain current is

$$I_m = \frac{V_m}{R} = \frac{10.8}{6.8} = 1.588 \text{ A} \tag{3.147}$$

The dc supply current is

$$I_I = \frac{I_m}{\gamma_1} = \frac{1.588}{1.7936} = 0.8854 \text{ A} \tag{3.148}$$

For $\theta = 60°$, the maximum drain current is

$$I_{DM} = \frac{I_I}{\alpha_0} = \frac{0.8854}{0.218} = 4.061 \text{ A} \tag{3.149}$$

The maximum drain-to-source voltage is

$$V_{DSM} = 2V_I = 2 \times 12 = 24 \text{ V} \tag{3.150}$$

Using the process with $K_n = \mu_{n0} C_{ox} = 0.142 \times 10^{-3}$ A/V^2, $V_t = 1$ V, and $L = 0.35$ μm, we get the MOSFET aspect ratio

$$\frac{W}{L} = \frac{2I_{DM}}{\mu_{n0} C_{ox} v_{DSsat}^2} = \frac{2 \times 4.061}{0.142 \times 10^{-3} \times 1^2} = 57,197 \tag{3.151}$$

Select $W/L = 57,200$. Hence, the MOSFET channel width is

$$W = \left(\frac{W}{L}\right) L = 57,200 \times 0.35 = 20.02 \text{ mm} \tag{3.152}$$

Hence,

$$K = \frac{1}{2}\mu_{n0} C_{ox} \left(\frac{W}{L}\right) = \frac{1}{2} \times 0.142 \times 10^{-3} \times 57,200 = 4.047 \text{ A/V} \tag{3.153}$$

The dc supply power is

$$P_I = I_I V_I = 0.8854 \times 12 = 10.6248 \text{ W} \tag{3.154}$$

The drain power dissipation is

$$P_D = P_I - P_{DS} = 10.6248 - 8.571 = 2.0538 \text{ W} \tag{3.155}$$

The drain efficiency is

$$\eta_D = \frac{P_{DS}}{P_I} = \frac{8.571}{10.6248} = 81.52\% \tag{3.156}$$

The efficiency of the RF power amplifier is

$$\eta = \frac{P_O}{P_I} = \frac{6}{10.6248} = 56.47\% \tag{3.157}$$

The power-added efficiency is

$$\eta_{PAE} = \frac{P_O - P_G}{P_I} = \frac{6 - 0.6}{10.6248} = 50.82\% \tag{3.158}$$

The power gain is

$$k_p = \frac{P_O}{P_G} = \frac{6}{0.6} = 10 \tag{3.159}$$

The loaded quality factor is

$$Q_L = \frac{f_0}{BW} = \frac{2.4}{0.24} = 10 \tag{3.160}$$

The reactances of the resonant circuit components are

$$X_L = X_C = \frac{R}{Q_L} = \frac{6.8}{10} = 0.68\ \Omega \tag{3.161}$$

yielding

$$L = \frac{X_L}{\omega_0} = \frac{0.68}{2\pi \times 2.4 \times 10^9} = 45.09\ \text{pH} \tag{3.162}$$

and

$$C = \frac{1}{\omega_0 X_C} = \frac{1}{2\pi \times 2.4 \times 10^9 \times 0.68} = 97.52\ \text{pF} \tag{3.163}$$

The reactance of the RF choke inductor is

$$X_{Lf} = 10R = 10 \times 6.8 = 68\ \Omega \tag{3.164}$$

yielding

$$L_f = \frac{X_{Lf}}{\omega_0} = \frac{68}{2\pi \times 2.4 \times 10^9} = 4.509\ \text{nH} \tag{3.165}$$

The reactance of the coupling capacitor is

$$X_{Cc} = \frac{R}{10} = \frac{6.8}{10} = 0.68\ \Omega \tag{3.166}$$

producing

$$C_c = \frac{1}{\omega_0 X_{Cc}} = \frac{1}{2\pi \times 2.4 \times 10^9 \times 0.68} = 97.52\ \text{pF} \tag{3.167}$$

3.3.7 Parameters of Class C RF Power Amplifier at $\theta = 45°$

The Fourier coefficients of the drain current at the conduction angle $\theta = 45°$ are

$$\alpha_0 = \frac{I_I}{I_{DM}} = \frac{\sqrt{2}(4-\pi)}{4\pi(2-\sqrt{2})} \approx 0.16491 \tag{3.168}$$

and

$$\alpha_1 = \frac{I_m}{I_{DM}} = \frac{\pi - 2}{2\pi(2-\sqrt{2})} \approx 0.31016 \tag{3.169}$$

The ratio I_m/I_I is

$$\gamma_1 = \frac{I_m}{I_I} = \frac{\alpha_1}{\alpha_0} = \frac{\sqrt{2}(\pi - 2)}{4-\pi} = 1.8808 \tag{3.170}$$

The drain efficiency at $\theta = 45°$ is

$$\eta_D = \frac{P_{DS}}{P_I} = \frac{\pi - 2}{\sqrt{2}(4-\pi)}\left(1 - \frac{v_{DSmin}}{V_I}\right) \approx 0.940378\left(1 - \frac{v_{DSmin}}{V_I}\right) \tag{3.171}$$

The output power capability at $\theta = 45°$ is

$$c_p = \frac{P_O}{V_{DSM} I_{DM}} = \frac{\pi - 2}{8\pi(2-\sqrt{2})} \approx 0.07754 \tag{3.172}$$

Coefficients for Class AB, B, and C power amplifiers are given in Table 3.1. The power loss and the efficiency of the resonant circuit as well as the impedance matching are the same as for the Class A RF power amplifier (Chapter 2).

Table 3.1 Coefficients for Class AB, B, and C power amplifiers.

$\theta(°)$	α_0	α_1	γ_1	η_D	c_p
10	0.0370	0.0738	1.9939	0.9967	0.01845
20	0.0739	0.1461	1.9756	0.9879	0.03651
30	0.1106	0.2152	1.9460	0.9730	0.05381
40	0.1469	0.2799	1.9051	0.9526	0.06998
45	0.1649	0.3102	1.8808	0.9404	0.07750
50	0.1828	0.3388	1.8540	0.9270	0.08471
60	0.2180	0.3910	1.7936	0.8968	0.09775
70	0.2525	0.4356	1.7253	0.8627	0.10889
80	0.2860	0.4720	1.6505	0.8226	0.11800
90	0.3183	0.5000	1.5708	0.7854	0.12500
100	0.3493	0.5197	1.4880	0.7440	0.12993
110	0.3786	0.5316	1.4040	0.7020	0.13290
120	0.4060	0.5363	1.3210	0.6605	0.13409
130	0.4310	0.5350	1.2414	0.6207	0.13376
140	0.4532	0.5292	1.1675	0.5838	0.13289
150	0.4720	0.5204	1.1025	0.5512	0.13010
160	0.4868	0.5110	1.0498	0.5249	0.12775
170	0.4965	0.5033	1.0137	0.5069	0.12582
180	0.5000	0.5000	1.0000	0.5000	0.12500

Example 3.4

Design a Class C power amplifier to deliver a power of 1 W at $f = 2.4$ GHz. The bandwidth is $BW = 240$ MHz and the conduction angle is $\theta = 45°$. The power supply voltage is $V_I = 5$ V.

Solution. Assume that the MOSFET threshold voltage is $V_t = 1$ V and the dc component of the gate-to-source voltage is $V_{GS} = 0$. The amplitude of the ac component of the gate-to-source voltage is

$$V_{gsm} = \frac{V_t - V_{GS}}{\cos\theta} = \frac{1 - 0}{\cos 45°} = 1.414 \text{ V} \tag{3.173}$$

The saturation drain-to-source voltage is

$$v_{DSsat} = v_{GS} - V_t = V_{GS} + V_{gsm} - V_t = V_{gsm} - V_t = 1.414 - 1 = 0.414 \text{ V} \tag{3.174}$$

The maximum amplitude of the output voltage is

$$V_m = V_I - v_{DSmin} = 5 - 0.2 = 4.8 \text{ V} \tag{3.175}$$

The load resistance is

$$R = \frac{V_m^2}{2P_O} = \frac{4.8^2}{2 \times 1} = 11.52\ \Omega \tag{3.176}$$

The amplitude of the drain current is

$$I_m = \frac{V_m}{R} = \frac{4.8}{11.52} = 0.4167 \text{ A} \tag{3.177}$$

The dc supply current is

$$I_I = \frac{I_m}{\gamma_1} = \frac{0.4167}{1.8808} = 0.2216 \text{ A} \tag{3.178}$$

For $\theta = 45°$, the maximum drain current is

$$I_{DM} = \frac{I_I}{\alpha_0} = \frac{0.2216}{0.16491} = 1.3438 \text{ A} \tag{3.179}$$

The maximum drain-to-source voltage is

$$V_{DSM} = 2V_I = 2 \times 5 = 10 \text{ V} \tag{3.180}$$

Let $\mu_{n0}C_{ox} = 0.142 \text{ mA/V}^2$ and $V_t = 1$ V. The MOSFET aspect ratio is

$$\frac{W}{L} = \frac{2I_{DM}}{\mu_{n0}C_{ox}v_{DSsat}^2} = \frac{2 \times 1.3438}{0.142 \times 10^{-3} \times 0.414^2} = 110,427 \tag{3.181}$$

Choose $W/L = 110,500$. The channel length of the MOSFET is

$$L = \left(\frac{W}{L}\right) L = 110,500 \times 0.35 \times 10^{-6} = 28.675 \text{ mm} \tag{3.182}$$

Thus,

$$K = \frac{1}{2}\mu_{n0}\left(\frac{W}{L}\right) = \frac{1}{2} \times 0.142 \times 10^{-3} \times 110,500 = 7.8455 \text{ mm} \tag{3.183}$$

The dc supply power is

$$P_I = I_I V_I = 0.2216 \times 5 = 1.108 \text{ W} \tag{3.184}$$

The drain power dissipation is

$$P_D = P_I - P_O = 1.108 - 1 = 0.108 \text{ W} \tag{3.185}$$

The drain efficiency is

$$\eta_D = \frac{P_O}{P_I} = \frac{1}{1.108} = 90.25\% \tag{3.186}$$

The loaded quality factor is

$$Q_L = \frac{f}{BW} = \frac{2.4}{0.24} = 10 \tag{3.187}$$

The reactances of the resonant circuit components are

$$X_L = X_C = \frac{R}{Q_L} = \frac{11.52}{10} = 1.152 \text{ }\Omega \tag{3.188}$$

yielding

$$L = \frac{X_L}{\omega_c} = \frac{1.152}{2\pi \times 2.4 \times 10^9} = 0.07639 \text{ nH} \tag{3.189}$$

and

$$C = \frac{1}{\omega_c X_C} = \frac{1}{2\pi \times 2.4 \times 10^9 \times 1.152} = 57.565 \text{ pF} \tag{3.190}$$

The reactance of the RF choke is

$$X_{Lf} = 10R = 10 \times 20.1667 = 11.52 \text{ }\Omega \tag{3.191}$$

which gives

$$L_f = \frac{X_{Lf}}{\omega_c} = \frac{11.52}{2\pi \times 2.4 \times 10^9} = 763.9 \text{ pH} \tag{3.192}$$

The reactance of the coupling capacitor is

$$X_{Cc} = \frac{R}{10} = \frac{11.52}{10} = 1.152\ \Omega \tag{3.193}$$

which produces

$$C_c = \frac{1}{\omega_c X_{Cc}} = \frac{1}{2\pi \times 2.4 \times 10^9 \times 1.152} = 57.565\ \text{pF} \tag{3.194}$$

3.4 Push–Pull Complementary Class AB, B, and C RF Power Amplifiers

3.4.1 Circuit of Push–Pull RF Power Amplifier

A circuit of push–pull Class AB, B, or C RF power amplifiers is shown in Fig. 3.22. It consists of complementary pair of transistors (NMOS and PMOS), parallel-resonant circuit, and coupling capacitor C_c. The transistors should have matched characteristics and are operated as voltage-dependent current sources. Since complementary transistors are used, the circuit is called a *complementary push–pull amplifier* or a *complementary-symmetry push–pull amplifier*. If MOSFETs are used, the circuit is called a *push–pull power amplifier*. The circuit may also employ complementary bipolar junction transistors (CBJT): an *npn* transistor and a *pnp* transistor. A Class B push–pull amplifier uses one transistor to amplify the positive portion of the input voltage and another transistor to amplify the negative portion of the input voltage. The coupling capacitor C_c blocks the dc voltage from the load. It also maintains the dc voltage $V_I/2$ and supplies the PMOS transistor when the NMOS transistor is not conducting. Also, two supply

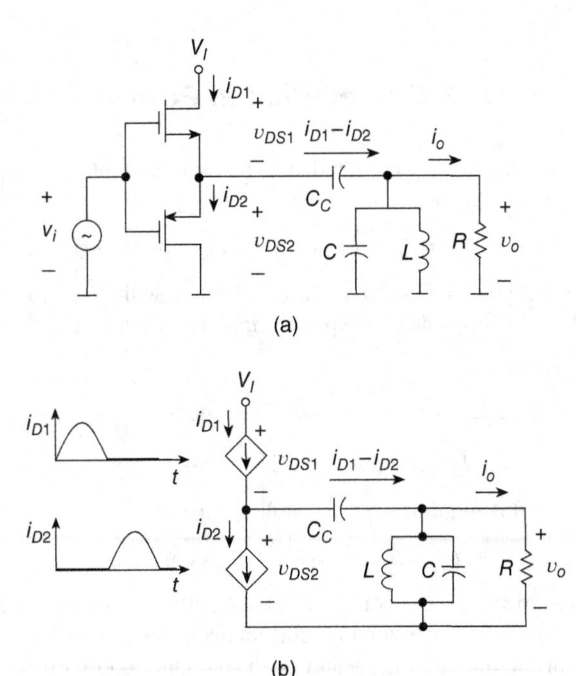

Figure 3.22 Circuit of push–pull Class AB, B, and C RF power amplifiers.

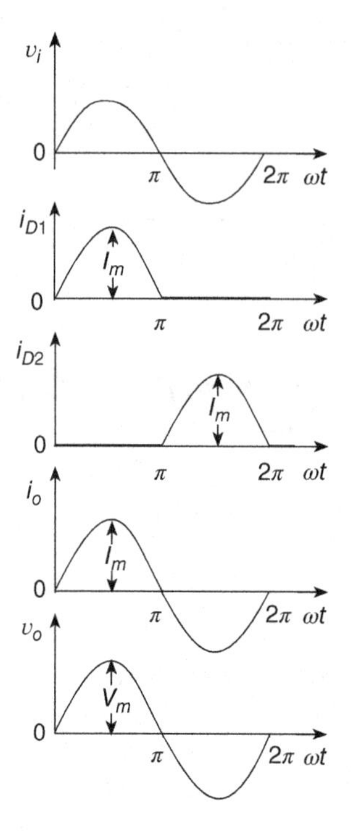

Figure 3.23 Current and voltage waveforms in push–pull Class B RF power amplifier.

voltages V_I can be connected to the drains of both transistors. Current and voltage waveforms for the push–pull Class B RF power amplifier are shown in Fig. 3.23. Similar waveforms can be drawn for Class AB and C amplifiers.

3.4.2 Even Harmonic Cancellation in Push–Pull Amplifiers

Let us assume that both transistors are identical. The drain current of the upper MOSFET can be expanded into a Fourier series

$$i_{D1} = I_D + i_{d1} + i_{d2} + i_{d3} + \cdots = I_D + I_{dm1} \cos \omega t + I_{dm2} \cos 2\omega t + I_{dm3} \cos 3\omega t + \cdots \quad (3.195)$$

The drain current of the lower MOSFET is shifted in phase with respect to the drain current of the upper MOSFET by 180° and can be expanded into a Fourier series

$$\begin{aligned} i_{D2} &= i_{D1}(\omega t - 180^\circ) \\ &= I_D + I_{dm1} \cos(\omega t - 180^\circ) + I_{dm2} \cos 2(\omega t - 180^\circ) + I_{dm3} \cos 3(\omega t - 180^\circ) + \cdots \\ &= I_D - I_{dm1} \cos \omega t + I_{dm2} \cos 2\omega t - I_{dm3} \cos 3\omega t + \cdots \end{aligned} \quad (3.196)$$

Hence, the load current flowing through the coupling capacitor C_B is

$$i_{D1} - i_{D2} = 2I_{dm1} \cos \omega t + 2I_{dm3} \cos 3\omega t + \cdots \quad (3.197)$$

Figure 3.24 shows the spectra of both drain currents and the difference of these currents. Thus, cancellation of all even harmonics of the load current takes place in push–pull power amplifiers, reducing distortion of the output voltage and the THD. Only odd harmonics must be filtered out by the parallel-resonant circuit, which is a BP filter. The same property holds true for all push–pull amplifiers operating in all classes.

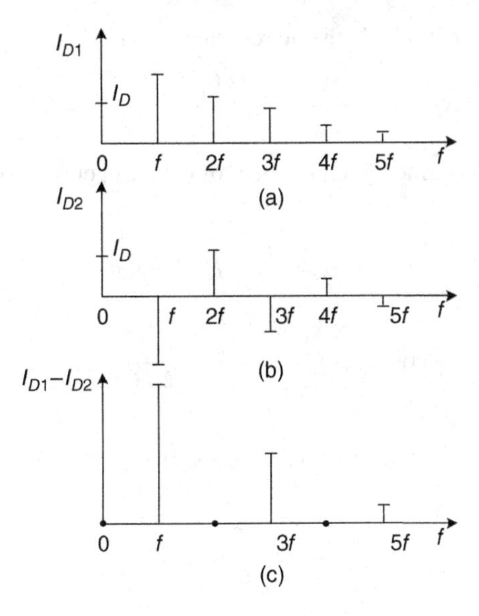

Figure 3.24 Spectra of the drain currents and the output current in push–pull Class B RF power amplifier: (a) spectrum of drain i_{D1}; (b) spectrum of drain i_{D2}; and (c) spectrum of the difference of drain currents $i_{D1} - i_{D2}$.

3.4.3 Power Relationships for Push–Pull RF Power Amplifier

The drain current waveform of the n-channel MOSFET in the Class B CMOS push–pull RF power amplifier is

$$i_{D1} = \begin{cases} I_m \cos \omega t & \text{for} \quad -\dfrac{\pi}{2} < \omega t \leq \dfrac{\pi}{2} \\ 0 & \text{for} \quad \dfrac{\pi}{2} < \omega t \leq \dfrac{3\pi}{2} \end{cases} \tag{3.198}$$

where I_m is the amplitude of the output current equal to the peak value of the drain current of each transistor I_{DM}.

The output voltage is

$$v_o = V_m \cos \omega t \tag{3.199}$$

The dc component of the drain current of the n-channel MOSFET is equal to the dc supply current

$$I_{D1} = I_I = \frac{1}{2\pi} \int_{-\pi/2}^{\pi/2} i_{D1} d(\omega t) = \frac{1}{2\pi} \int_{-\pi/2}^{\pi/2} I_m \cos \omega t d(\omega t) = \frac{1}{\pi} \int_0^{\pi/2} I_m \cos \omega t d(\omega t)$$
$$= \frac{I_m}{\pi} = \frac{V_m}{\pi R} \tag{3.200}$$

The dc supply power of the amplifier is

$$P_I = 2V_I I_I = \frac{2V_I I_m}{\pi} = \frac{2V_I V_m}{\pi R} \tag{3.201}$$

The dc resistance seen by the dc supply voltage source V_I is

$$R_{DC} = \frac{2V_I}{I_I} = \frac{2\pi V_I}{V_m} R = 2\pi R \tag{3.202}$$

For $V_m = 0$, $R_{DC} = \infty$. For $V_m = V_I$, the dc resistance is

$$R_{DCmin} = \frac{2V_I}{I_{Imax}} = \frac{2\pi V_I}{V_{m(max)}} = \frac{2\pi V_m}{I_m} = 2\pi R \tag{3.203}$$

The amplitude of the fundamental component of the drain current of the n-channel MOSFET is

$$I_{dm1} = \frac{1}{\pi}\int_{-\pi/2}^{\pi/2} i_{D1}\cos\omega t d(\omega t) = \frac{1}{\pi}\int_{-\pi/2}^{\pi/2} I_m \cos^2\omega t d(\omega t) = \frac{2}{\pi}\int_{0}^{\pi/2} I_m \cos^2\omega t d(\omega t)$$

$$= \frac{2}{\pi}\int_{0}^{\pi/2} \frac{I_m}{2}(1 + \cos\ 2\omega t) d(\omega t) = \frac{I_m}{2} = \frac{V_m}{2R} \tag{3.204}$$

The output current is

$$i_o = I_m \cos\omega t \tag{3.205}$$

where

$$I_m = \frac{V_m}{R} = 2I_{dm1} \tag{3.206}$$

The ac output power is

$$P_O = \frac{V_m I_m}{2} = \frac{V_m^2}{2R} = \frac{RI_m^2}{2} \tag{3.207}$$

The drain power dissipated in both transistors is

$$P_D = P_I - P_O = \frac{2V_I V_m}{\pi R} - \frac{V_m^2}{2R} \tag{3.208}$$

Setting the derivative of P_D with respect to V_m to zero,

$$\frac{dP_D}{dV_m} = \frac{2V_I}{\pi R} - \frac{V_m}{R} = 0 \tag{3.209}$$

we obtain the critical value of V_m at which maximum value of the drain power loss P_{Dmax} occurs

$$V_{m(cr)} = \frac{2V_I}{\pi}. \tag{3.210}$$

Hence, the maximum power dissipation in both transistors is

$$P_{Dmax} = \frac{2V_I^2}{\pi^2 R}. \tag{3.211}$$

and the maximum power dissipation in each transistor is

$$P_{Dmax(Q_N)} = P_{Dmax(Q_P)} = \frac{P_{Dmax}}{2} = \frac{V_I^2}{\pi^2 R} \tag{3.212}$$

The drain efficiency of the amplifier is

$$\eta_D = \frac{P_{DS}}{P_I} = \frac{P_O}{P_I} = \frac{\pi}{4}\left(\frac{V_m}{V_I}\right) \tag{3.213}$$

For $V_m = V_I$,

$$\eta_{Dmax} = \frac{\pi}{4} = 78.5\% \tag{3.214}$$

For $V_{m(max)} = V_I - v_{DSsat}$,

$$\eta_{Dmax} = \frac{\pi}{4}\left(\frac{V_{m(max)}}{V_I}\right) = \frac{\pi}{4}\left(1 - \frac{v_{DSsat}}{V_I}\right) \tag{3.215}$$

3.4.4 Device Stresses

The transistor current and voltage stresses are

$$I_{DM} = I_m = \pi I_I \tag{3.216}$$

and

$$V_{DSM} = 2V_I \tag{3.217}$$

The amplitudes of the currents through the resonant inductance and capacitance are

$$I_{Lm} = I_{Cm} = Q_L I_m \tag{3.218}$$

The output-power capability is

$$c_p = \frac{P_{Omax}}{2I_{DM}V_{DSM}} = \frac{1}{4}\left(\frac{I_{m(max)}}{I_{DM}}\right)\left(\frac{V_{m(max)}}{V_{DSM}}\right) = \frac{1}{4} \times 1 \times \frac{1}{2} = \frac{1}{8} \tag{3.219}$$

Example 3.5

Design a Class B CMOS push–pull RF power amplifier to deliver a power of 50 W at $f =$ 1.8 GHz. The bandwidth is $BW = 180$ MHz. The power supply voltage is $V_I = 48$ V.

Solution. Assume that the MOSFET threshold voltage is $V_t = 1$ V. Hence, the dc component of the gate-to-source voltage is

$$V_{GS} = V_t = 1 \text{ V} \tag{3.220}$$

Assume the amplitude of the gate-to-source sinusoidal voltage to be $V_{gsm} = 1$ V. Thus,

$$v_{DSsat} = V_{GS} + V_{gsm} - V_t = V_t + V_{gsm} - V_t = V_{gsm} = 1 \text{ V} \tag{3.221}$$

Assuming the efficiency of the resonant circuit $\eta_r = 0.95$, the drain power is

$$P_{DS} = \frac{P_O}{\eta_r} = \frac{50}{0.95} = 52.632 \text{ W} \tag{3.222}$$

The amplitude of the output voltage is equal to the amplitude of the drain-to-source voltage

$$V_m = V_{dsm} = V_I - v_{DSsat} = 48 - 1 = 47 \text{ V} \tag{3.223}$$

The load resistance is

$$R = \frac{V_m^2}{2P_{DS}} = \frac{47^2}{2 \times 52.632} = 20.985 \ \Omega \tag{3.224}$$

The drain efficiency is

$$\eta_D = \frac{\pi}{4}\left(1 - \frac{v_{DSsat}}{V_I}\right) = \frac{\pi}{4}\left(1 - \frac{1}{48}\right) = 76.86\% \tag{3.225}$$

The dc input power is

$$P_I = \frac{P_{DS}}{\eta_D} = \frac{52.632}{0.7686} = 68.4778 \text{ W} \tag{3.226}$$

The dc input current is

$$I_I = \frac{P_I}{V_I} = \frac{68.4778}{48} = 1.4267 \text{ A} \tag{3.227}$$

The amplitude of the fundamental component of the drain current waveform is

$$I_m = I_{DM} = \frac{V_m}{R} = \frac{47}{20.985} = 2.2395 \text{ A} \quad (3.228)$$

The total efficiency is

$$\eta = \frac{P_O}{P_I} = \frac{50}{68.4778} = 73.02\% \quad (3.229)$$

The maximum power loss in both transistors is

$$P_{Dmax} = \frac{2V_I^2}{\pi^2 R} = \frac{2 \times 48^2}{\pi^2 \times 20.09} = 23.24 \text{ W} \quad (3.230)$$

The maximum power loss in each transistor is

$$P_{Dmax(Q_N)} = P_{Dmax(Q_P)} = \frac{V_I^2}{\pi^2 R} = \frac{48^2}{\pi^2 \times 20.09} = 11.62 \text{ W} \quad (3.231)$$

The maximum power loss in each transistor is

$$P_{Dmax} = \frac{V_I^2}{\pi^2 R} = \frac{48^2}{\pi^2 \times 20.985} = 11.124 \text{ W}. \quad (3.232)$$

The maximum drain current is

$$I_{DM} = I_{dm} = 2.2395 \text{ A} \quad (3.233)$$

The maximum drain-to-source voltage is

$$V_{DSM} = 2V_I = 2 \times 48 = 96 \text{ V} \quad (3.234)$$

The loaded quality factor is

$$Q_L = \frac{f_c}{BW} = \frac{1800}{180} = 10 \quad (3.235)$$

The resonant inductance is

$$L = \frac{R_L}{\omega_c Q_L} = \frac{20.985}{2\pi \times 1.8 \times 10^9 \times 10} = 0.1855 \text{ nH} \quad (3.236)$$

The resonant capacitance is

$$C = \frac{Q_L}{\omega_c R_L} = \frac{10}{2\pi \times 1.8 \times 10^9 \times 20.985} = 42.1 \text{ pF} \quad (3.237)$$

The reactance of the blocking capacitor is

$$X_{C_c} = \frac{R}{10} = \frac{20.985}{10} = 2.0985 \ \Omega \quad (3.238)$$

resulting in

$$C_c = \frac{1}{\omega_c X_{Cc}} = \frac{1}{2\pi \times 1.8 \times 10^9 \times 2.0985} = 42.13 \text{ pF} \quad (3.239)$$

Using the process $K_n = \mu_{n0} C_{ox} = 0.142 \times 10^3$ A/V^2 and $L = 0.35$ μm, the aspect ratio of the n-channel MOSFET is

$$\left(\frac{W}{L}\right)_{Q_N} = \frac{2I_{DM}}{K_n v_{DSsat}^2} = \frac{2 \times 2.2396}{0.142 \times 10^{-3} \times 1^2} = 209,517 \quad (3.240)$$

Let $W/L = 210,000$. The channel width of the n-channel MOSFET is

$$W_{Q_N} = \left(\frac{W}{L}\right)_{Q_N} L = 210,000 \times 0.35 \times 10^{-3} = 73.5 \text{ mm} \quad (3.241)$$

The aspect ratio of the p-channel MOSFET is

$$\left(\frac{W}{L}\right)_{Q_P} = \frac{\mu_{n0}}{\mu_{p0}}\left(\frac{W}{L}\right)_{Q_N} = 2.7\left(\frac{W}{L}\right)_{Q_N} = 2.7 \times 210,000 = 567,000 \tag{3.242}$$

The channel width of the p-channel MOSFET is

$$W_{Q_P} = \left(\frac{W}{L}\right)_{Q_P} L = 567,000 \times 0.35 \times 10^{-3} = 198.45 \text{ mm} \tag{3.243}$$

3.5 Transformer-Coupled Class B Push–Pull RF Power Amplifier

3.5.1 Waveforms

A circuit of the transformer-coupled push–pull Class AB, B, and C RF power amplifiers is shown in Fig. 3.25. Current and voltage waveforms in this amplifier are depicted in Fig. 3.26. The output current is

$$i_o = I_m \sin \omega t = nI_{dm} \sin \omega t \tag{3.244}$$

where I_m is the amplitude of the output current, I_{dm} is the peak drain current, and n is the transformer turns ratio, which is equal to the ratio of number of turns of one primary to the number of turns of the secondary

$$n = \frac{I_m}{I_{dm}} = \frac{V_{dm}}{V_m} \tag{3.245}$$

The output voltage is

$$v_o = V_m \sin \omega t = \frac{V_{dm}}{n} \sin \omega t \tag{3.246}$$

where the amplitude of the output current is

$$V_m = I_m R_L \tag{3.247}$$

The resistance seen by each transistor across each primary winding with the other primary winding open is

$$R = n^2 R_L \tag{3.248}$$

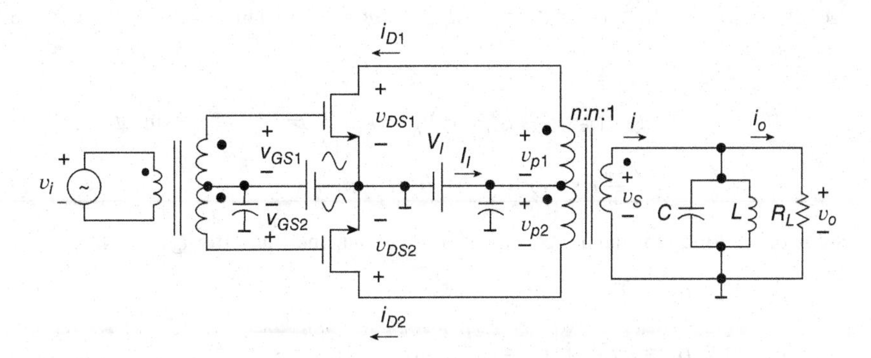

Figure 3.25 Circuit of transformer-coupled push–pull Class AB, B, and C RF power amplifiers.

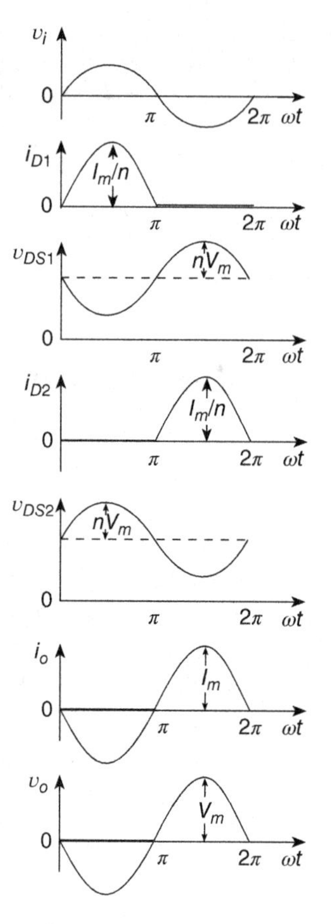

Figure 3.26 Waveforms in transformer-coupled push–pull Class AB, B, and C RF power amplifiers.

When the drive voltage is positive, transistor Q_1 is ON and transistor Q_2 is OFF. The waveforms are

$$i_{D1} = I_{dm} \sin \omega t = \frac{I_m}{n} \sin \omega t \quad \text{for} \quad 0 < \omega t \leq \pi \tag{3.249}$$

$$i_{D2} = 0 \tag{3.250}$$

$$v_{p1} = -i_{D1}R = -i_{D1}n^2R_L = v_{p2} = -RI_{dm} \sin \omega t = -\frac{I_m}{n}R \sin \omega t = -nI_m R_L \sin \omega t \tag{3.251}$$

and

$$v_{DS1} = V_I + v_{p1} = V_I - i_{D1}R = V_I - i_{D1}n^2R_L = V_I - I_{dm}R \sin \omega t$$
$$= V_I - \frac{I_m}{n}R \sin \omega t = V_I - nI_m R_L \sin \omega t \tag{3.252}$$

When the drive voltage is negative, transistor Q_1 is OFF and transistor Q_2 is ON,

$$i_{D1} = 0 \tag{3.253}$$

$$i_{D2} = -I_{dm} \sin \omega t = -\frac{I_m}{n} \sin \omega t \quad \text{for} \quad \pi < \omega t \leq 2\pi \tag{3.254}$$

$$v_{p2} = i_{D2}R = i_{D2}n^2R_L = v_{p2} = -I_{dm}R \sin \omega t = -\frac{I_m}{n}R \sin \omega t$$
$$= -nI_mR_L \sin \omega t \tag{3.255}$$

and

$$v_{DS2} = V_I - v_{p2} = V_I - i_{D2}R = V_I - i_{D2}n^2R_L = V_I + I_{dm}R \sin \omega t = V_I - \frac{I_m}{n}R_L \sin \omega t$$
$$= V_I + nI_mR_L \sin \omega t \tag{3.256}$$

The voltage between the drains of the MOSFETs is

$$v_{D1D2} = v_{p1} + v_{p2} = -n^2R_L(i_{D1} - i_{D2}) = -2nR_LI_m \sin \omega t \tag{3.257}$$

Ideally, all even harmonics cancel out as shown in Section 3.4.1. The voltage across the secondary winding is

$$v_s = \frac{v_{D1D2}}{2n} = -\frac{nR_L}{2}(i_{D1} - i_{D2}) = -R_LI_m \sin \omega t \tag{3.258}$$

The current through the dc voltage source V_I is a full-wave rectified sinusoid given by

$$i_I = i_{D1} + i_{D2} = I_{dm}|\sin \omega t| = \frac{I_m}{n}|\sin \omega t|. \tag{3.259}$$

The dc supply current is

$$I_I = \frac{1}{2\pi}\int_0^{2\pi} I_{dm}|\sin \omega t| d(\omega t) = \frac{1}{2\pi}\int_0^{2\pi} \frac{I_m}{n}|\sin \omega t| d(\omega t) = \frac{2}{\pi}\frac{I_m}{n} = \frac{2}{\pi}\frac{V_m}{nR_L} \tag{3.260}$$

3.5.2 Power Relationships

The dc supply power is given by

$$P_I = V_I I_I = \frac{2}{\pi}\frac{V_IV_m}{nR_L}. \tag{3.261}$$

The output power is

$$P_O = \frac{V_m^2}{2R_L} \tag{3.262}$$

The drain power dissipation is

$$P_D = P_I - P_O = \frac{2}{\pi}\frac{V_IV_m}{nR_L} - \frac{V_m^2}{2R_L} \tag{3.263}$$

The derivative of the power P_D with respect of the output voltage amplitude V_m is

$$\frac{dP_D}{dV_m} = \frac{2}{\pi}\frac{V_I}{nR_L} - \frac{V_m}{R_L} = 0 \tag{3.264}$$

resulting in the critical value of V_m at which the maximum value of the drain power loss P_{Dmax} occurs

$$V_{m(cr)} = \frac{2}{\pi}\frac{V_I}{n} \tag{3.265}$$

The maximum drain power loss in both transistors is

$$P_{Dmax} = \frac{2}{\pi^2}\frac{V_I^2}{n^2R_L} = \frac{2}{\pi^2}\frac{V_I^2}{R} \tag{3.266}$$

The amplitude of the drain-to-source voltage is

$$V_{dm} = \frac{V_m}{n} = \frac{2}{\pi} V_I \tag{3.267}$$

The drain efficiency is

$$\eta_D = \frac{P_O}{P_I} = \frac{\pi}{4} \frac{V_{dm}}{V_I} = \frac{\pi}{4} \frac{nV_m}{V_I} \tag{3.268}$$

For $V_{dm} = nV_m = V_I$, the dc supply power is

$$P_I = V_I I_I = \frac{2}{\pi} \frac{V_I^2}{n^2 R_L} = \frac{2}{\pi} \frac{V_I^2}{R} \tag{3.269}$$

the output power is

$$P_O = \frac{V_m^2}{2R_L} = \frac{n^2 V_I^2}{2R_L} \tag{3.270}$$

and the drain efficiency is

$$\eta_{Dmax} = \frac{\pi}{4} = 78.5\% \tag{3.271}$$

3.5.3 Device Stresses

The MOSFET current and voltage stresses are

$$I_{DM} = I_{dm} = \frac{I_m}{n} \tag{3.272}$$

and

$$V_{DSM} = 2V_I \tag{3.273}$$

The output-power capability is

$$c_p = \frac{P_{Omax}}{2I_{DM} V_{DSM}} = \frac{P_{Omax}}{4I_{DM} V_I} = \frac{1}{8} \left(\frac{I_{dm(max)}}{I_{SM}} \right) \left(\frac{V_{m(max)}}{V_I} \right) = \frac{1}{8} \times 1 \times 1 = \frac{1}{8} \tag{3.274}$$

A circuit of the push–pull Class AB, B, and C RF power amplifiers with tapped capacitor is depicted in Fig. 3.27. This figure shows the distribution of the various components through the circuit and the cancellation of even harmonics in the load network. A circuit of the push–pull

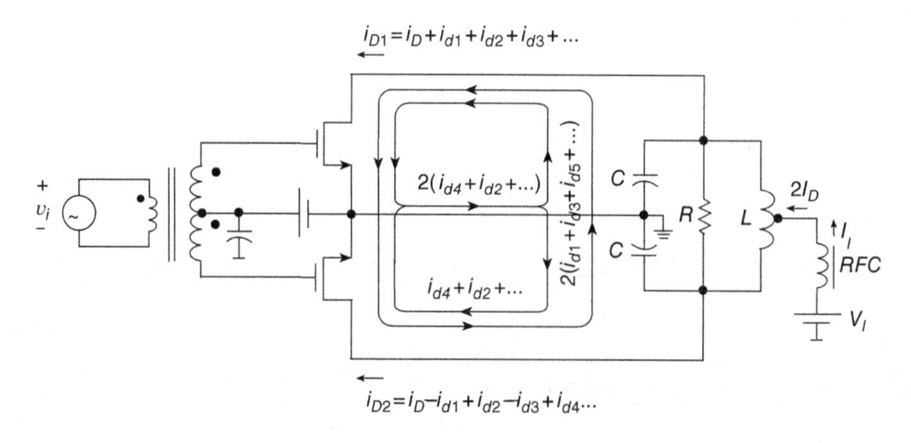

Figure 3.27 A circuit of push–pull Class AB, B, and C RF power amplifiers with tapped capacitor.

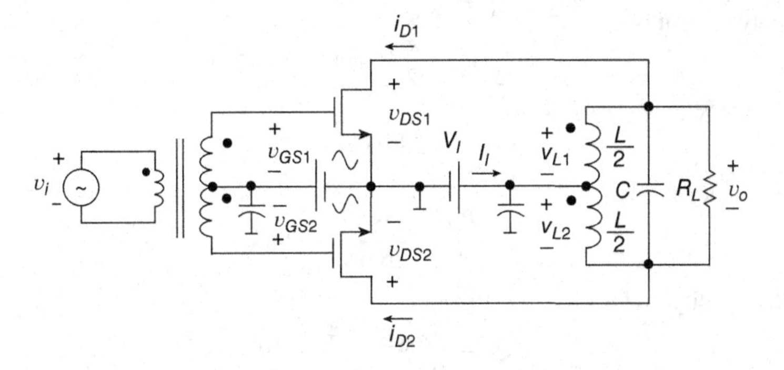

Figure 3.28 A circuit of push–pull Class AB, B, and C RF power amplifiers with tapped inductor.

Class AB, B, and C RF power amplifiers with tapped inductor is shown in Fig. 3.28. All equations for these two amplifiers can be obtained by setting $n = 1$ in the equations for the push–pull amplifier shown in Fig. 3.25.

Example 3.6

Design a transformer-coupled Class B power amplifier to deliver a power of 25 W at $f = 2.4\,\text{GHz}$. The bandwidth is $BW = 240\,\text{MHz}$. The power supply voltage is $V_I = 28\,\text{V}$ and $v_{DSmin} = 1\,\text{V}$.

Solution. Assuming the efficiency of the resonant circuit $\eta_r = 0.94$, the drain power is

$$P_{DS} = \frac{P_O}{\eta_r} = \frac{25}{0.94} = 26.596\ \text{W} \tag{3.275}$$

The resistance seen by each transistor across one part of the primary winding is

$$R = \frac{\pi^2}{2}\frac{V_I^2}{P_{DS}} = \frac{\pi^2}{2} \times \frac{28^2}{26.596} = 145.469\ \Omega \tag{3.276}$$

The transformer turns ratio is

$$n = \sqrt{\frac{R}{R_L}} = \sqrt{\frac{145.469}{50}} = 1.7 \approx \frac{7}{4} \tag{3.277}$$

The maximum value of the drain-to-source voltage is

$$V_{dm} = \pi V_I = \pi \times 28 = 87.965\ \text{V} \tag{3.278}$$

The amplitude of the drain current is

$$I_{dm} = \frac{V_{dm}}{R} = \frac{87.965}{145.469} = 0.605\ \text{A} \tag{3.279}$$

The amplitude of the output voltage is

$$V_m = \frac{V_{dm}}{n} = \frac{87.965}{1.7} = 51.74\ \text{V} \tag{3.280}$$

The amplitude of the output current is

$$I_m = \frac{V_m}{R_L} = \frac{51.74}{50} = 1.0348\ \text{A} \tag{3.281}$$

The dc supply current is

$$I_I = \frac{\pi}{2} I_{dm} = \frac{\pi}{2} \times 0.605 = 0.95 \text{ A} \tag{3.282}$$

The dc supply power is

$$P_I = V_I I_I = 28 \times 0.95 = 26.6 \text{ W} \tag{3.283}$$

The total efficiency is

$$\eta = \frac{P_O}{P_I} = \frac{25}{26.6} = 93.98\% \tag{3.284}$$

The loaded quality factor is

$$Q_L = \frac{f_c}{BW} = \frac{2.4}{0.24} = 10 \tag{3.285}$$

The resonant inductance is

$$L = \frac{R_L}{\omega_c Q_L} = \frac{50}{2\pi \times 2.4 \times 10^9 \times 10} = 0.3316 \text{ nH}. \tag{3.286}$$

The resonant capacitance is

$$C = \frac{Q_L}{\omega_c R_L} = \frac{10}{2\pi \times 2.4 \times 10^9 \times 50} = 13.26 \text{ pF} \tag{3.287}$$

3.6 Class AB, B, and C RF Power Amplifiers with Variable-Envelope Signals

A block diagram of a transmitter with a drain or collector AM is shown in Fig. 3.29. The low-frequency signal modulating is amplified by an audio power amplifier and is used to vary the dc supply voltage of the RF power amplifier. Ideally, the amplitude of the RF power amplifier is directly proportional to the dc supply voltage. Figure 3.30 shows a circuit of Class AB,

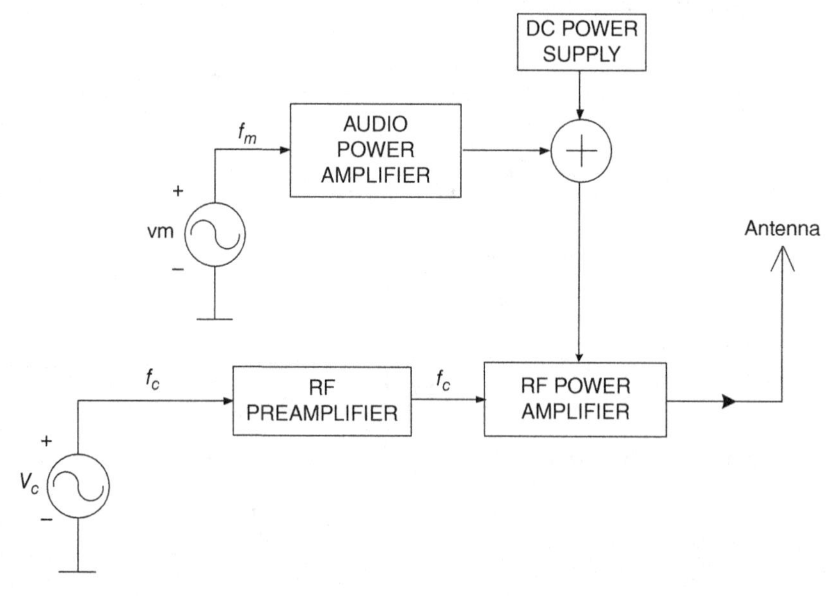

Figure 3.29 Block diagram of an amplitude-modulated (AM) transmitter with a drain or collector AM.

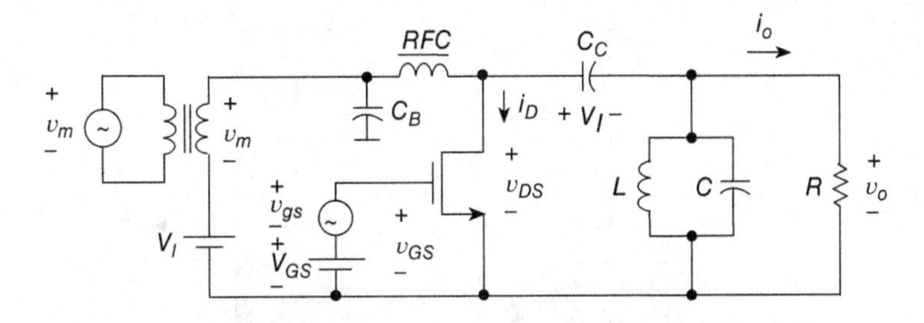

Figure 3.30 RF transmitter with drain AM.

B, and C power amplifiers with AM. In these amplifiers, an AM signal is generated by placing the modulating voltage source v_m in series with the drain dc supply voltage source V_I of a Class C RF power amplifier that is driven into the ohmic (triode) region of the MOSFET. The gate-to-source voltage v_{GS} has a constant amplitude V_{gsm} and its frequency is equal to the carrier frequency f_c. The dc gate-to-source voltage V_{GS} is fixed. Therefore, the conduction angle θ of the drain current i_D is also fixed. These circuits require an audio frequency transformer.

Class AB, B, and C amplifiers can be used to amplify variable-envelope signals, such as AM signals. The ac component of the AM gate-to-source voltage is

$$v_{gs(AM)} = V_{gsm}(1 + m_{in} \cos \omega_m t) \cos \omega_c t \tag{3.288}$$

The overall AM gate-to-source voltage is

$$v_{GS} = V_{GS} + v_{gs(AM)} = V_{GS} + V_{gsm}(1 + m_{in} \cos \omega_m t) \cos \omega_c t \tag{3.289}$$

The AM drain current waveform of a power MOSFET is

$$\begin{aligned} i_D &= \frac{1}{2} C_{ox} W v_{sat} (v_{GS} - V_t) \\ &= \frac{1}{2} C_{ox} W v_{sat} [V_{GS} - V_t + V_{gsm}(1 + m_{in} \cos \omega_m t) \cos \omega_c t] \quad \text{for} \quad v_{GS} > V_t \end{aligned} \tag{3.290}$$

Assuming that the impedance of the parallel-resonant circuit Z is equal to R (i.e., $Z \approx R$), we obtain the AM drain-to-source voltage

$$\begin{aligned} v_{DS} &\approx R i_D \\ &= \frac{1}{2} C_{ox} W v_{sat} R [V_{GS} - V_t + V_{gsm}(1 + m_{in} \cos \omega_m t) \cos \omega_c t] \quad \text{for} \quad v_{GS} > V_t \end{aligned} \tag{3.291}$$

The choice of the class of operation, that is, the operating point Q, has an important effect on nonlinear distortion of variable-envelope signals in RF power amplifiers in which transistors are operated as voltage-dependent current sources. The amplification process of AM signals in Class AB, B, and C amplifiers is illustrated in Fig. 3.31. Figure 3.31(a) illustrates the amplification of an AM signal in the Class AB amplifier, which produces an AM output voltage with $m_{out} > m_{in}$. The output voltage exhibits shallower modulation than the input signal. Figure 3.31(b) illustrates the amplification of an AM signal in the Class B amplifier, where $m_{out} = m_{in}$. The Class B amplifier behaves as a linear RF power amplifier. Its characteristic $v_{DS} = f(v_{GS} - V_t)$ is nearly linear and starts at the origin. Figure 3.31(c) illustrates the amplification of an AM signal in the Class C amplifier, which produces an AM output signal with $m_{out} < m_{in}$. The output voltage exhibits deeper modulation than the input signal. Class AB and C amplifiers can be used for amplifying AM signals with a small modulation index m.

Class AB, B, and C power amplifiers can be used to amplify constant-envelope signals, such as FM and PM signals.

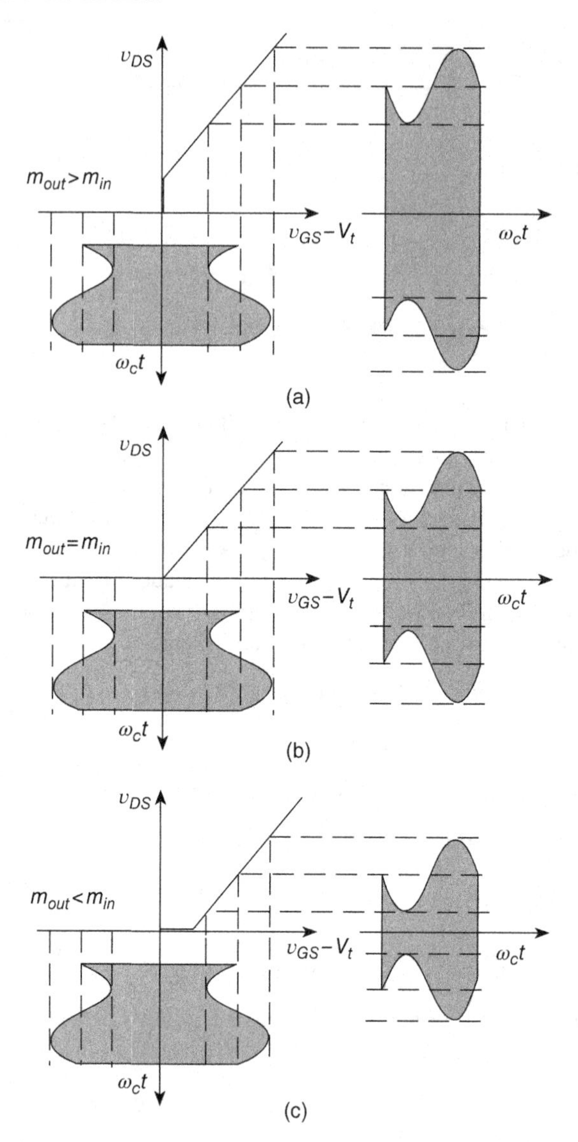

Figure 3.31 Amplification of AM signals in Class AB, B, and C RF power amplifiers: (a) Class AB amplifier ($m_{out} > m_{in}$); (b) Class B linear RF power amplifier ($m_{out} = m_{in}$); and (c) Class C amplifier ($m_{out} < m_{in}$).

3.7 Summary

- The Class B RF power amplifier consists of a transistor, parallel-resonant circuit, RF choke, and blocking capacitor.
- The transistor in the Class B power amplifier is operated as a dependent current source.
- The conduction angle of the drain or collector current 2θ in the Class B power amplifier is 180°.
- The drain efficiency of the Class B power amplifier η_D is high.

- The maximum drain efficiency of the Class B RF power amplifier is $\eta_D = 78.5\%$.
- The drain efficiency of the Class B RF power amplifier is proportional to the square root of the normalized output power P_O/P_{Omax}.
- The drain efficiency of the Class B RF power amplifier is proportional to the normalized amplitude of the ac component of the drain-to-source voltage V_m/V_I.
- The conduction angle of the drain or collector current 2θ in the Class AB RF power amplifier is between 180° and 360°.
- The conduction angle of the drain or collector current 2θ in the Class C RF power amplifier is less than 180°.
- The transistor is operated as a dependent current source in the Class C RF power amplifier.
- The drain efficiency of Class AB RF power amplifier η_D increases from 50% to 78.5% as the conduction angle 2θ decreases from 360° to 180° at $V_m = V_I$.
- The drain efficiency of the Class C RF power amplifier increases from 78.5% to 100% as the conduction angle 2θ decreases from 180° to 0° at $V_m = V_I$.
- The drain efficiency of the Class C RF power amplifier is 89.68% at $\theta = 60°$ and $V_m = V_I$.
- The drain efficiency of the Class C RF power amplifier is 94.04% at $\theta = 45°$ and $V_m = V_I$.
- Class B and C RF power amplifiers are narrow-band circuits because the resonant circuit acts as a BP filter.
- Class B and C RF power amplifiers are linear (or semilinear) because the amplitude of the drain current is proportional to the amplitude of the gate-to-source voltage.
- Class B and C RF power amplifiers are used as medium and high-power narrow-band power amplifiers.
- A push–pull Class B RF power amplifier uses one transistor to amplify the positive portion of the input voltage and another transistor to amplify the negative portion of the input voltage. One transistor is "pushing" the current into the load and the other transistor is "pulling" the current from the load.
- In push–pull topology, even harmonics are cancelled in the load, reducing distortion of the output voltage and the THD.
- The transformer in the transformer-coupled push–pull power amplifier performs the impedance matching function.

3.8 Review Questions

3.1 What is the motivation to operate RF power amplifiers at the drain current conduction angle $2\theta < 360°$?

3.2 List the components of the Class B RF power amplifier.

3.3 What is the transistor mode of operation in the Class B RF power amplifier?

3.4 What is the location of the operating point of the Class B RF power amplifier?

3.5 How high is the drain efficiency of the Class B RF power amplifier?

3.6 What is the mode of operation of the transistor in the Class C RF power amplifier?

3.7 What is the location of the operating point of the Class C RF power amplifier?

3.8 What is the drain efficiency of the Class C RF power amplifier at $\theta = 60°$ and $\theta = 45°$?

3.9 Are the drain current and output voltage dependent on the gate-to-source voltage in Class B and C amplifiers?

3.10 Explain the principle of operation of the push–pull Class B RF power amplifier.

3.11 Explain the principle of operation of the transformer-coupled Push–pull Class B RF power amplifier.

3.12 What types of harmonics are present in the load in push–pull amplifiers?

3.13 How does the transformer contribute to the impedance matching in power amplifiers?

3.9 Problems

3.1 Design a Class B RF power amplifier to meet the following specifications: $V_I = 3.3$ V, $P_O = 1$ W, $BW = 240$ MHz, and $f = 2.4$ GHz.

3.2 Design a Class AB RF power amplifier to meet the following specifications: $V_I = 48$ V, $P_O = 22$ W, $\theta = 120°$, $BW = 90$ MHz, and $f = 0.9$ GHz.

3.3 Design a Class C RF power amplifier to meet the following specifications: $V_I = 3.3$ V, $P_O = 0.25$ W, $\theta = 60°$, $BW = 240$ MHz, and $f = 2.4$ GHz.

3.4 Design a Class C RF power amplifier to meet the following specifications: $V_I = 12$ V, $P_O = 6$ W, $\theta = 45°$, $BW = 240$ MHz, and $f = 2.4$ GHz.

3.5 Design a Class C RF power amplifier to meet the following specifications: $V_I = 28$ V, $P_O = 20$ W, $R_L = 50\ \Omega$, $\theta = 60°$, $f_c = 800$ MHz, and $BW = 80$ MHz.

3.6 Sketch the spectra of the currents $|I_{D1}|$, $|I_{D2}|$, and $|I_{D1} - I_{D2}|$ for the push–pull amplifier.

3.7 Design a Class C RF power amplifier to meet the following specifications: $V_I = 10$ V, $P_O = 5$ W, $R_L = 50\ \Omega$, $\theta = 45°$, $f_c = 10$ GHz, and $BW = 1$ GHz.

3.8 Design a Class C RF power amplifier to meet the following specifications: $P_O = 5$ W, $V_I = 10$ V, $R_L = 50\ \Omega$, $P_G = 0.5$ W, and $f_c = 10$ GHz. Assume $\eta_r = 0.65$ and $\theta = 45°$. Find the component values, device stresses, drain power dissipation, drain efficiency, amplifier efficiency, and power-added efficiency.

References

[1] K. L. Krauss, C. V. Bostian, and F. H. Raab, *Solid State Radio Engineering*. New York, NY: John Wiley & Sons, 1980.

[2] M. Albulet, *RF Power Amplifiers*. Atlanta, GA: Noble Publishing Co., 2001.

[3] S. C. Cripps, *RF Power Amplifiers for Wireless Communications*. Norwood, MA: Artech House, 1999.

[4] M. K. Kazimierczuk and D. Czarkowski, *Resonant Power Converters*. New York, NY: John Wiley & Sons, 2011.

[5] T. H. Lee, *The Design of CMOS Radio-Frequency Integrated Circuits*, 2nd Ed. New York, NY: Cambridge University Press, 2004.

[6] A. Grebennikov, *RF and Microwave Power Amplifier Design*. New York, NY: McGraw-Hill, 2005.

[7] L. B. Hallman, "A Fourier analysis of radio-frequency power amplifier waveforms," *Proceedings of the IRE*, vol. 20, no. 10, pp. 1640–1659, 1932.

[8] F. H. Raab, "Class-E, Class-C, and Class-F power amplifiers based upon a finite number of harmonics," *IEEE Transactions on Microwave Theory and Techniques*, vol. 49, no. 8, pp. 1462–1468, 2001.

[9] H. L. Krauss and J. F. Stanford, "Collector modulation of transistor amplifiers," *IEEE Transactions on Circuit Theory*, vol. 12, pp. 426–428, 1965.

[10] M. K. Kazimierczuk, *High-Frequency Magnetic Components*, 2nd Ed., Chichester, UK: John Wiley & Sons, 2014.

4

Class D RF Power Amplifiers

4.1 Introduction

Class D radio-frequency (RF) resonant power amplifiers [1–20], also called Class D dc–ac resonant power inverters, were invented in 1959 by Baxandall [1] and have been widely used in various applications [2–19] to convert dc energy into ac energy. Examples of applications of resonant amplifiers are radio transmitters, dc–dc resonant converters, solid-state electronic ballasts for fluorescent lamps, LED drivers, induction heating appliances, high-frequency electric heating applied in induction welding, surface hardening, soldering and annealing, induction sealing for tamper-proof packaging, fiber-optics production, and dielectric heating for plastic welding. In Class D amplifiers, transistors are operated as switches. Class D amplifiers can be classified into two groups:

- Class D voltage-switching (or voltage-source) amplifiers.
- Class D current-switching (or current-source) amplifiers.

amplifiers

Class D voltage-switching amplifiers are fed by a dc voltage source. They employ (1) a series-resonant circuit or (2) a resonant circuit that is derived from the series-resonant circuit. If the loaded quality factor is sufficiently high, the current through the resonant circuit is sinusoidal and the current through the switches is a half-sine wave. The voltage waveforms across the switches are square waves.

In contrast, Class D current-switching amplifiers are fed by a dc current source in the form of an RF choke (RFC) and a dc voltage source. These amplifiers contain a parallel-resonant circuit or a resonant circuit that is derived from the parallel-resonant circuit. The voltage across the resonant circuit is sinusoidal for high values of the loaded quality factor. The voltage across the switches is a half-wave sinusoid, and the current through the switches is square wave.

One main advantage of Class D voltage-switching amplifiers is the low voltage across each transistor, which is equal to the supply voltage. This makes these amplifiers suitable for high-voltage applications. For example, a 220 or 277 V rectified line voltage is used to supply

RF Power Amplifiers, Second Edition. Marian K. Kazimierczuk.

the Class D amplifiers. In addition, low-voltage MOSFETs can also be used. Such MOSFETs have a low on-resistance, reducing the conduction losses and operating junction temperature. This yields a high efficiency. The MOSFETs on-resistance r_{DS} increases considerably with increasing junction temperature. This causes the conduction loss $r_{DS}I_{rms}^2$ to increase, where I_{rms} is the rms value of the drain current. Typically, r_{DS} doubles as the temperature rises by 100 °C (e.g., from 25 to 125 °C), doubling the conduction loss. The MOSFET's on-resistance r_{DS} increases with temperature T because both the mobility of electrons $\mu_{n0} \approx K_1/T^{2.5}$ and the mobility of holes $\mu_p \approx K_2/T^{2.7}$ decrease with T for 100 K $\leq T \leq$ 400 K, where K_1 and K_2 are constants. In many applications, the output power or the output voltage can be controlled by varying the operating frequency f (FM control) or phase shift (phase control).

In this chapter, we will study Class D half-bridge and full-bridge series-resonant amplifiers. The design procedure for the Class D amplifier is illustrated with detailed examples.

4.2 MOSFET as a Switch

The i_D-v_{DS} characteristics of a MOSFET in the ohmic region are given by

$$i_D = \mu_{n0} C_{ox} \left(\frac{W}{L}\right) \left[(v_{GS} - V_t)v_{DS} - \frac{v_{DS}^2}{2}\right] \quad \text{for} \quad v_{GS} > V_t \quad \text{and} \quad v_{DS} < v_{GS} - V_t \tag{4.1}$$

where μ_{n0} is the low-field electron mobility in the channel. The large-signal channel resistance of a MOSFET in the ohmic region is given by

$$r_{DS} = \frac{v_{DS}}{i_D} = \frac{1}{\dfrac{i_D}{v_{DS}}} = \frac{1}{\mu_{n0} C_{ox} \left(\dfrac{W}{L}\right)(v_{GS} - V_t - v_{DS}/2)} \quad \text{for} \quad v_{DS} < v_{GS} - V_t \tag{4.2}$$

which simplifies to the form

$$r_{DS} \approx \frac{1}{\mu_{n0} C_{ox} \left(\dfrac{W}{L}\right)(v_{GS} - V_t)} \quad \text{for} \quad v_{DS} \ll 2(v_{GS} - V_t) \tag{4.3}$$

The drain saturation current is

$$I_{Dsat} = \frac{1}{2}\mu_{n0} C_{ox} \left(\frac{W}{L}\right)(V_{GSH} - V_t)^2 = \frac{1}{2}K_n v_{DSsat}^2 \tag{4.4}$$

where V_{GSH} is the high-level gate-to-source voltage when the MOSFET is ON, $v_{DCsat} = V_{OH} - V_t$, and $K_n = \mu_{n0} C_{ox}$. For the MOSFET to operate as a switch, the drain peak current must be sufficiently lower than the drain saturation current

$$I_{Dsat} = aI_{DM} \tag{4.5}$$

where $a > 1$.

Integrated MOSFETs may be used at low power levels and discrete power MOSFETs at high power levels. In integrated MOSFETs, the source and the drain are on the same side of the chip surface, causing horizontal current flow from drain to source when the device is ON. When the device is OFF, the depletion region of the reverse-biased drain-to-body *pn* junction diode spreads into the lightly doped short channel, resulting in a low punch-through breakdown voltage between the drain and the source. The breakdown voltage is proportional to the channel length L, whereas the maximum drain current is inversely proportional to the channel length L. If an integrated MOSFET is designed to have a high breakdown voltage, its channel length L must be increased, which reduces the device aspect ratio W/L and decreases the maximum drain current. In integrated MOSFETs, two contradictory requirements are imposed on the region between the drain and the source: a short channel to achieve a high drain current when the device is ON and a long channel to achieve a high breakdown voltage when the device is OFF.

The drain-to-source resistance of discrete power MOSFETs consists of the channel resistance R_{Ch}, accumulation region resistance R_a, neck region resistance R_n, and drift region resistance R_{DR} [19]. For high-voltage power MOSFETs, the doping level must be low and the width of the drift region must be large. Therefore, the drift resistance is a dominant component of the drain-to-source resistance r_{DS}. For an n-channel silicon power MOSFET, the donor doping level in the drift region is $N_D = 1.293 \times 10^{17}/V_{BD}\ \text{cm}^{-3}$, and the minimum width of the drift region is $W_{Dmin}\ (\mu\text{m}) = V_{BD}/10$, where V_{BD} is the MOSFET breakdown voltage.

4.3 Circuit Description of Class D RF Power Amplifier

Figure 4.1 shows the circuit of a Class D voltage-switching (voltage-source) RF power amplifier with a pulse transformer driver. The circuit consists of two n-channel MOSFETs, a series-resonant circuit, and a driver. It is difficult to drive the upper MOSFET because a high-side gate driver is required. A pulse transformer can be used to drive the MOSFETs. The noninverting output of the transformer drives the upper MOSFET and the inverting output of the transformer drives the lower MOSFET. A pump-charge IC driver can be also used. A circuit of the Class D voltage-switching RF power amplifier with two power supplies V_I and $-V_I$ is depicted in Fig. 4.2 [15].

Figure 4.3 shows a circuit of the Class D CMOS RF power amplifier, in which a PMOS MOSFET Q_P and an NMOS MOSFET Q_N are used as switching devices in a similar way as in digital circuits. This circuit can be integrated for high-frequency applications, such as RF transmitters for wireless communications. The CMOS Class D RF power amplifier requires only one driver. However, cross-conduction of both transistors during the MOSFETs transitions may cause spikes in the drain currents. Nonoverlapping gate-to-source voltages may reduce this problem, but the driver will become more complex [17, 18]. The peak-to-peak value of the gate-to-source drive voltage v_G is equal or close to the dc supply voltage V_I, such as in CMOS digital circuits. Therefore, this circuit is appropriate only for low values of the dc supply voltage V_I, usually below 20 V. At high values of the dc supply voltage V_I, the gate-to-source voltage should be also high, which may cause voltage breakdown of the gate oxide SiO_2.

Figure 4.4 shows the circuit of a Class D RF power amplifier with a voltage mirror driver or voltage level shifter [5]. The dc supply voltage V_I can be much higher than the peak-to-peak voltage of the gate-to-source voltage. Therefore, this circuit can be used for high-voltage applications.

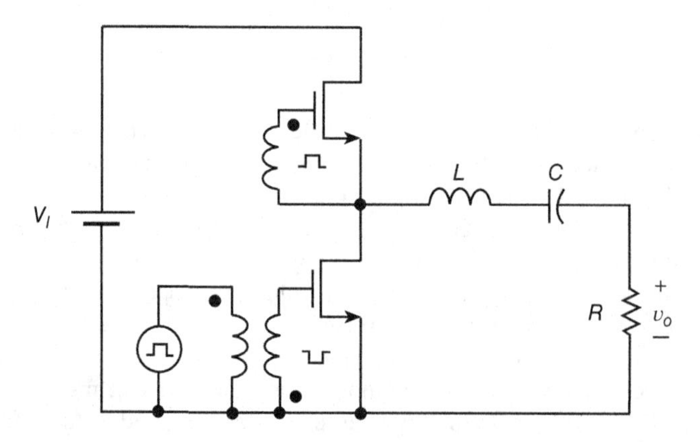

Figure 4.1 Class D half-bridge voltage-switching (voltage-source) RF power amplifier with a series-resonant circuit and a pulse transformer driver.

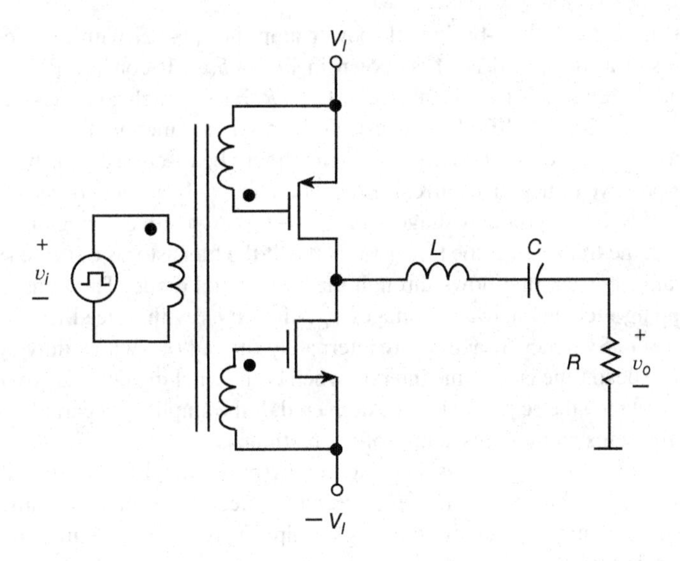

Figure 4.2 Class D half-bridge voltage-switching (voltage-source) RF power amplifier with a series-resonant circuit and a pulse transformer driver.

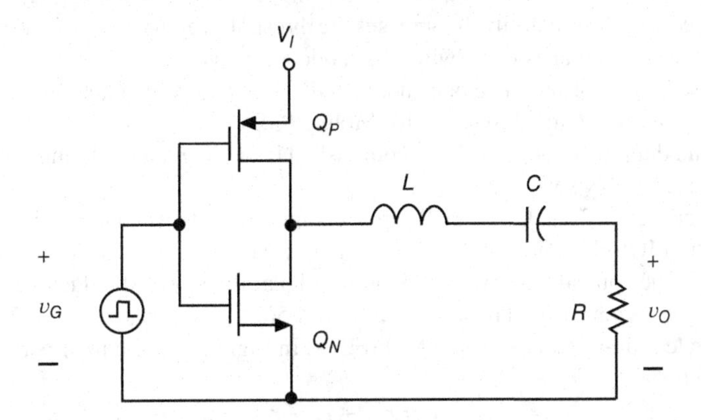

Figure 4.3 Class D CMOS half-bridge voltage-switching (voltage-source) RF power amplifier with a series-resonant circuit.

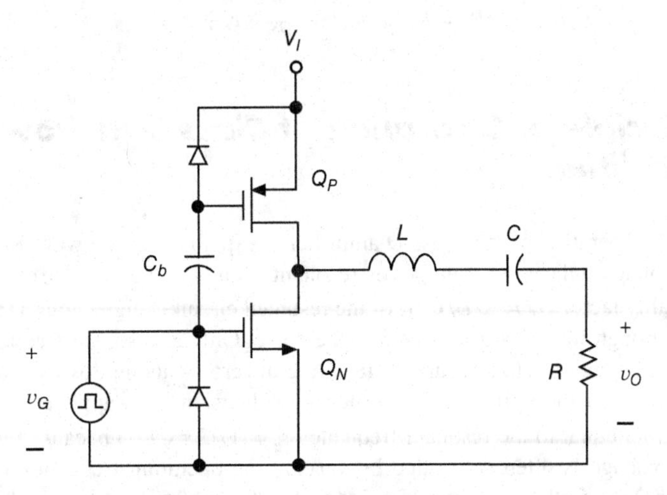

Figure 4.4 Class D half-bridge RF power amplifier with a voltage mirror driver [5].

A circuit of the Class D half-bridge RF power amplifier [1–12] with a series-resonant circuit (and a pulse transformer driver) is shown in Fig. 4.5(a). It consists of two bidirectional switches S_1 and S_2 and a series-resonant circuit L–C–R. Each switch consists of a transistor and an antiparallel diode. The MOSFET's intrinsic body–drain *pn* junction diode may be used as an antiparallel diode in case of an inductive load, which will be discussed shortly. The switch can conduct either positive or negative current. However, it can only accept voltages higher than the negative value of a forward diode voltage $-V_{Don} \approx -1$ V. A positive or negative switch current can flow through the transistor if the transistor is ON. If the transistor is OFF, the switch can conduct only negative current that flows through the antiparallel diode. The transistors are driven by nonoverlapping rectangular-wave voltages v_{GS1} and v_{GS2} with a dead time at an operating frequency $f = 1/T$. Switches S_1 and S_2 are alternately ON and OFF with a duty cycle of 50% or slightly less. The dead time is the time interval when both switching devices are OFF. Resistance R is an ac load to which the ac power is to be delivered. If the amplifier is part of a dc–dc resonant converter, R represents an input resistance of the rectifier.

There has been an increasing interest in designing RF power amplifiers in digital technology in the age of a system-on-chip (SoC). There is a trend to integrate a complete transceiver together with the digital baseband subsystem on a single chip. There are two main issues in designing power amplifiers in submicron CMOS technology: oxide breakdown and hot carrier effects. Both of these problems become worse as the technology is scaled down. The oxide breakdown is a catastrophic effect and sets a limit on the maximum signal swing on the MOSFET drain. The hot carrier effect reduces the reliability. It increases the threshold voltage and consequently degrades the performance of the transistors. Switching-mode RF power amplifiers offer high efficiency and can be used for constant-envelope modulated signals, such as Gaussian minimum- shift keying (GMSK) used in Global Systems for Mobile (GSM) Communications.

Amplitude modulation (AM) may be accomplished by adding a modulating voltage source in series with the dc supply source V_I.

Equivalent circuits of the Class D RF power amplifier are shown in Fig. 4.5(b)–(d). In Fig. 4.5(b), the MOSFETs are modeled by switches whose on-resistances are r_{DS1} and r_{DS2}. Resistance r_L is the equivalent series resistance (ESR) of the physical inductor L and resistance r_C is the ESR of the physical capacitor C. In Fig. 4.5(c), $r_{DS} \approx (r_{DS1} + r_{DS2})/2$ represents the average equivalent on-resistance of the MOSFETs. In Fig. 4.5(d), the total parasitic resistance is represented by

$$r = r_{DS} + r_L + r_C \tag{4.6}$$

which yields an overall resistance in the series-resonant circuit

$$R_t = R + r = R + r_{DS} + r_L + r_C \tag{4.7}$$

4.4 Principle of Operation of Class D RF Power Amplifier

The principle of operation of the Class D amplifier is explained by the waveforms sketched in Fig. 4.6. The voltage at the input of the series-resonant circuit is a square wave of magnitude V_I. If the loaded quality factor $Q_L = \sqrt{L/C}/R$ of the resonant circuit is high enough (e.g., $Q_L \geq 2.5$), the current i through this circuit is nearly a sine wave. Only at resonant frequency $f = f_o$, the MOSFETs turn on and off at zero current, resulting in zero switching losses and an increase in efficiency. In this case, the antiparallel diode never conducts. In many applications, the operating frequency f is not equal to the resonant frequency $f_o = 1/(2\pi\sqrt{LC})$ because the output power or the output voltage is often controlled by varying the operating frequency f (FM control). Figure 4.6(a)–(c) shows the waveforms for $f < f_o$, $f = f_o$, and $f > f_o$, respectively. The tolerance

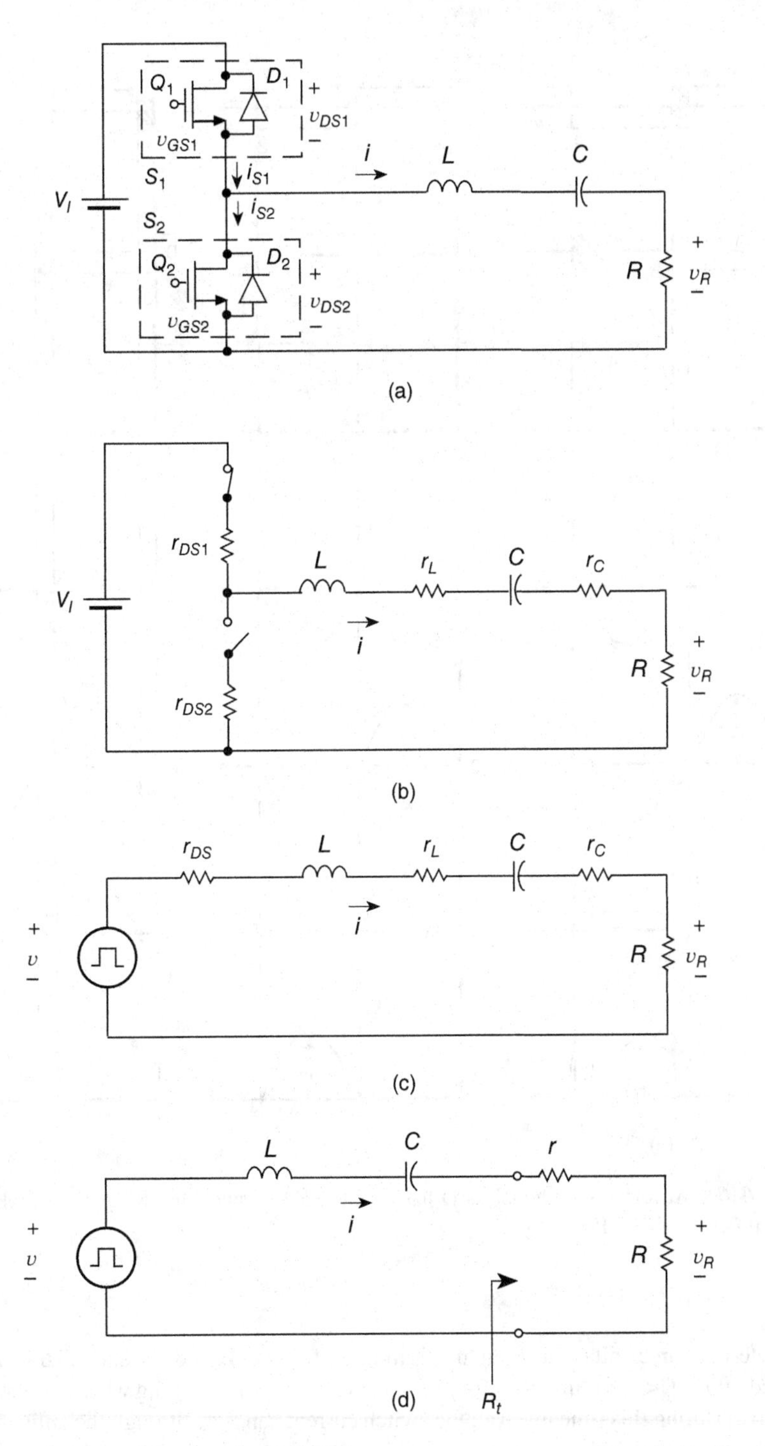

Figure 4.5 Class D half-bridge RF power amplifier with a series-resonant circuit. (a) Circuit. (b)–(d) Equivalent circuits.

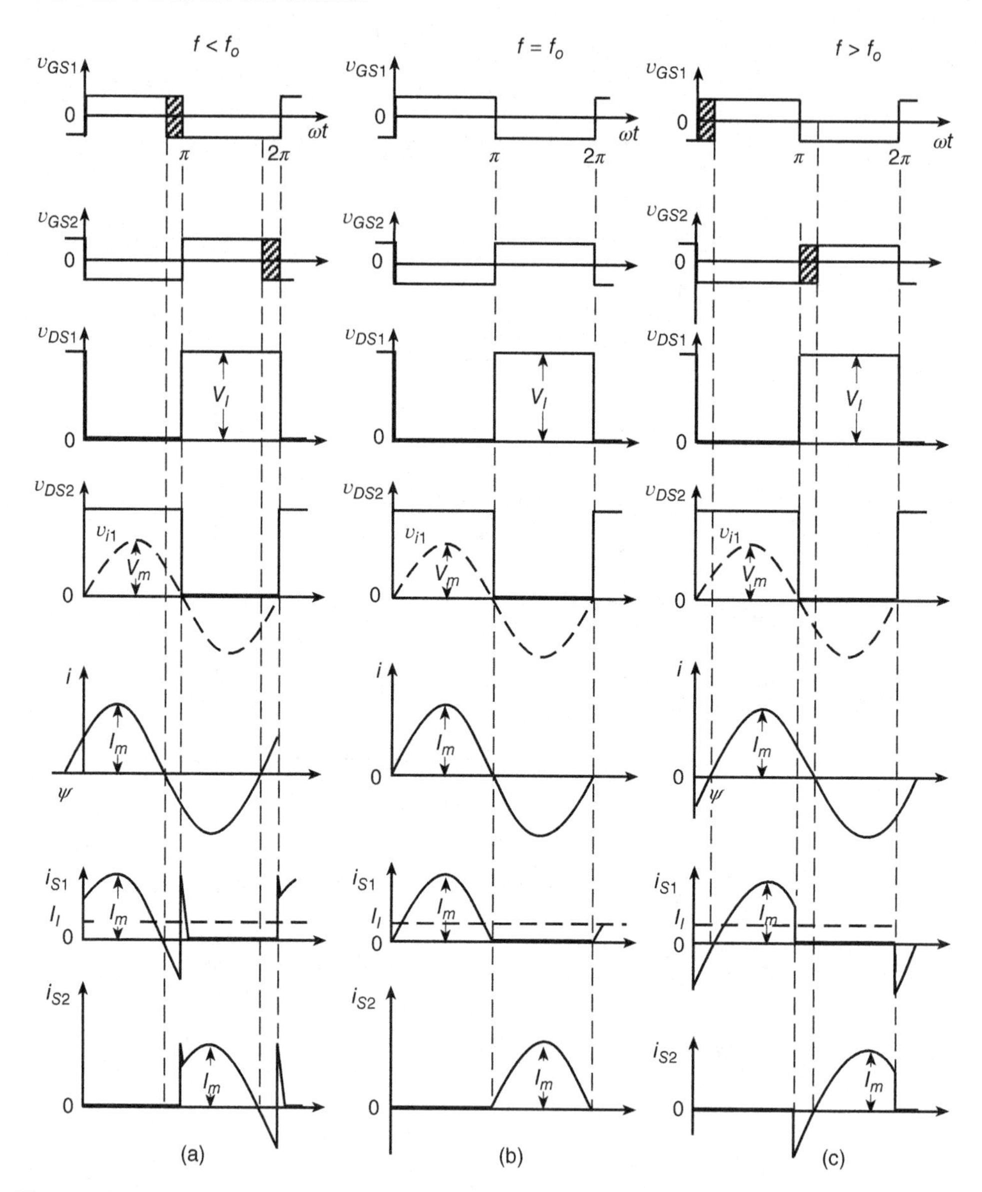

Figure 4.6 Waveforms in the Class D half-bridge voltage-switching RF power amplifier. (a) For $f < f_o$. (b) For $f = f_o$. (c) For $f > f_o$.

of the gate-to-source voltage turn-on time is indicated by the shaded areas. Each transistor should be turned off for $f < f_o$ or turned on for $f > f_o$ in the time interval during which the switch current is negative. During this time interval, the switch current can flow through the antiparallel diode. To prevent cross-conduction (also called shoot-through current), the waveforms of the drive voltages v_{GS1} and v_{GS2} should be non-overlapping and have a sufficient dead time (not shown in Fig. 4.6). At turn-off, MOSFETs have a delay time and bipolar junction transistors (BJTs) have a storage time. If the dead time of the gate-to-source voltages of the two transistors is too short, one transistor still remains ON while the other already turns ON. Consequently, both the transistors are ON at the same time and the voltage power supply V_I is short-circuited by the transistor on-resistances r_{DS1} and r_{DS2}. For this reason, cross-conduction current pulses of magnitude $I_{pk} = V_I/(r_{DS1} + r_{DS2})$ flow through the transistors. For example, if $V_I = 200$ V and $r_{DS1} = r_{DS2} = 0.5\ \Omega$, $I_{pk} = 200$ A. The excessive current stress may cause immediate failure of the devices. The dead time should not be too long, which will be discussed in Sections 4.4.1 and 4.4.2. The maximum dead time increases as f/f_o increases for $f > f_o$ or decreases for $f < f_o$. This is because the time interval during which the switch current is negative becomes longer. The shortest dead time must be at $f = f_o$. There are commercial IC drivers available, which have an adjustable dead time, for example, TI 2525.

4.4.1 Operation Below Resonance

For $f < f_o$, the series-resonant circuit represents a capacitive load. This means that the current through the resonant circuit i leads the fundamental component v_{i1} of the voltage v_{DS2} by the phase angle $|\psi|$, where $\psi < 0$. Therefore, the switch current i_{S1} is positive after the switch turns on at $\omega t = 0$ and i_{S1} is negative before the switch turns off at $\omega t = \pi$. The conduction sequence of the semiconductor devices is Q_1–D_1–Q_2–D_2. Notice that the current in the resonant circuit is diverted from the diode of one switch to the transistor of the other switch (Fig. 4.5). This causes a lot of problems, which is explained shortly. Consider the time when the switch S_2 is turned on, as shown in Fig. 4.6. Prior to this transition, the current i flows through the antiparallel diode D_1 of the switch S_1. When transistor Q_2 is turned on by the drive voltage v_{GS2}, v_{DS2} is decreased, causing v_{DS1} to increase. Therefore, the diode D_1 turns off and the current i is diverted from D_1 to Q_2. There are three undesired effects, when the MOSFET turns on:

1. Reverse recovery of the antiparallel diode of the opposite switch.
2. Discharging the transistor output capacitance.
3. Miller's effect.

The most severe drawback of operation below resonance is the diode reverse-recovery stress when the diode turns off. The MOSFET's intrinsic body–drain diode is a minority carrier device. Each diode turns off at a very large dv/dt, resulting in a large di/dt. It generates a large reverse-recovery current spike (turned upside down). This spike flows through the other transistor because it cannot flow through the resonant circuit. The resonant inductor L does not permit an abrupt current change. Consequently, the spikes occur in the switch current waveform at both turn-on and turn-off transitions of the switch. The magnitude of these spikes can be much (e.g., 10 times) higher than the magnitude of the steady-state switch current. High current spikes may destroy the transistors and always cause a considerable increase in switching losses and noise. During a part of the reverse-recovery interval, the diode voltage increases from -1 V to V_I and both the diode current and voltage are simultaneously high, causing a high reverse-recovery power loss.

The turn-off failure of the power MOSFETs may be caused by the *second breakdown* of the parasitic bipolar transistor. This parasitic bipolar transistor is an integral part of a power MOSFET structure. The body region serves as the base of the parasitic BJT, the source as the BJT emitter, and the drain as the BJT collector. If the body–drain diode is forward biased just prior to the sudden application of drain voltage, the turn-on process of the parasitic bipolar transistor may be initiated by the reverse-recovery current of the antiparallel diode. The second breakdown of the parasitic BJT may destroy the power MOSFET structure. The second breakdown voltage is usually one-half of the voltage at which the device fails if the diode is forward biased. This voltage is denoted as V_{DSS} by manufacturers. For the reasons given earlier, operation at $f < f_o$ should be avoided if power MOSFETs are to be used as switches. For example, the current spikes can be reduced by adding Schottky antiparallel diodes. Silicon Schottky diodes have low breakdown voltages, typically below 100 V. However, silicon carbide Schottky diodes have high breakdown voltages. Since the forward voltage of the Schottky diode is lower than that of the *pn* junction body diode, most of the negative switch current flows through the Schottky diode, reducing the reverse-recovery current of the *pn* junction body diode. Another circuit arrangement is to connect a diode in series with the MOSFET and an ultrafast diode in parallel with the series combination of the MOSFET and diode. This arrangement does not allow the intrinsic diode to conduct and to store the excess minority charge. However, the higher number of components, the additional cost, and the voltage drop across the series diode (which reduces the efficiency) are undesirable. Also, the peak voltages of the transistor and the series diode may become much higher than V_I. Transistors with a higher permissible voltage and a higher on-resistance should be used. High-voltage MOSFETs have high on-resistance. However, high on-resistance increases conduction loss. This method may reduce, but cannot eliminate the spikes. Snubbers should be used to slow down the switching process, and reverse-recovery spikes can be reduced by connecting a small inductance in series with each power MOSFET.

For $f < f_o$, the turn-off switching loss is zero, but the turn-on switching loss is not zero. The transistors are turned on at a high voltage, equal to V_I. When the transistor is turned on, its output capacitance is discharged, causing switching loss. Suppose that the upper MOSFET is initially ON and the output capacitance C_o of the upper transistor is initially discharged. When the upper transistor is turned off, the energy drawn from the dc input voltage source V_I to charge the output capacitance C_o from 0 to V_I is given by

$$W_I = \int_0^T V_I i_{Co} dt = V_I \int_0^T i_{Co} dt = V_I Q \tag{4.8}$$

where i_{Cout} is the charging current of the output capacitance and Q is the charge transferred from the source V_I to the capacitor. This charge equals the integral of the current i_{Cout} over the charging time interval, which is usually much shorter than the time period T of the operating frequency f. Equation (4.8) holds true for both linear and nonlinear capacitances. If the transistor output capacitance is assumed to be linear,

$$Q = C_o V_I \tag{4.9}$$

and (4.8) becomes

$$W_I = C_o V_I^2 \tag{4.10}$$

The energy stored in the linear output capacitance at the voltage V_I is

$$W_C = \frac{1}{2} C_o V_I^2 \tag{4.11}$$

The charging current flows through a resistance that consists of the on-resistance of the bottom MOSFET and lead resistances. The energy dissipated in this resistance is

$$W_R = W_I - W_C = \frac{1}{2} C_o V_I^2 \tag{4.12}$$

which is the same amount of energy as that stored in the capacitance. Note that charging a linear capacitance from a dc voltage source through a resistance requires twice the energy stored in the capacitance. When the upper MOSFET is turned on, its output capacitance is discharged through the on-resistance of the upper MOSFET, dissipating energy in that transistor. Thus, the energy dissipated during charging and discharging of the transistor output capacitance is

$$W_{sw} = C_o V_I^2 \tag{4.13}$$

Accordingly, the turn-on switching loss per transistor is

$$P_{ton} = \frac{W_C}{T} = f W_C = \frac{1}{2} f C_o V_I^2 \tag{4.14}$$

The total power loss associated with the charging and discharging of the transistor output capacitance of each MOSFET is

$$P_{sw} = \frac{W_{sw}}{T} = fW_{sw} = fC_o V_I^2 \tag{4.15}$$

The charging and discharging process of the output capacitance of the bottom transistor is similar. In reality, the drain–source *pn* step junction capacitance is nonlinear. An analysis of turn-on switching loss with a nonlinear transistor output capacitance is given in Section 4.11.2.

Another effect that should be considered at turn-on of the MOSFET is Miller's effect. Since the gate–source voltage increases and the drain–source voltage decreases during the turn-on transition, Miller's effect is significant, by increasing the transistor input capacitance, the gate drive charge, and the drive power requirements, reducing the turn-on switching speed.

The advantage of operating below resonance is that the transistors are turned off at nearly zero voltage, resulting in *zero turn-off switching loss*. For example, the drain–source voltage v_{DS1} is held at nearly -1 V by the antiparallel diode D_1 when i_{S1} is negative. During this time interval, transistor Q_1 is turned off by drive voltage v_{GS1}. The drain–source voltage v_{DS1} is almost zero and the drain current is low during the MOSFET turn-off, yielding zero turn-off switching loss in the MOSFET. Since v_{DS1} is constant, Miller's effect is absent during turn-off, the transistor input capacitance is not increased by Miller's effect, the gate drive requirement is reduced, and the turn-off switching speed is enhanced. In summary, for $f < f_o$, there is a turn-on switching loss in the transistor and a turn-off (reverse-recovery) switching loss in the diode. The transistor turn-off and the diode turn-on are lossless.

As mentioned earlier, the drive voltages v_{GS1} and v_{GS2} are nonoverlapping and have a dead time. However, this dead time should not be made too long. If the transistor Q_1 is turned off too early when the switch current i_{S1} is positive, diode D_1 cannot conduct and diode D_2 turns on, decreasing v_{DS2} to -0.7 V and increasing v_{DS1} to V_I. When the current through diode D_2 reaches zero, diode D_1 turns on, v_{DS1} decreases to -0.7 V, and v_{DS2} increases to V_I. These two additional transitions of each MOSFET voltage would result in switching losses. Note that only the turn-on transition of each switch is *forced* and is directly controlled by the driver, while the turn-off transition is caused by the turn-on of the opposite transistor (i.e., it is automatic).

In very high-power applications, thyristors with antiparallel diodes can be used as switches in Class D amplifiers with a series-resonant circuit topologies. The advantage of such an arrangement is that, for operation below resonance, thyristors are turned off naturally when the switch current crosses zero. Thyristors, however, require more complicated and powerful drive circuitry, and their operating frequency range is limited to 20 kHz. Such low frequencies make the size and weight of the resonant components large, increasing the conduction losses.

4.4.2 Operation Above Resonance

For $f > f_o$, the series-resonant circuit represents an inductive load and the current i lags behind the voltage v_{i1} by the phase angle ψ, where $\psi > 0$. Hence, the switch current is negative after turn-on (for part of the switch "on" interval) and positive before turn-off. The conduction sequence of the semiconductor devices is D_1–Q_1–D_2–Q_2. Consider the turn-off of switch S_1. When transistor Q_1 is turned off by the drive voltage v_{GS1}, v_{DS1} increases, causing v_{DS2} to decrease. As v_{DS2} reaches -0.7 V, D_2 turns on and the current i is diverted from transistor Q_1 to diode D_2. Thus, the turn-off switch transition is *forced* by the driver, while the turn-on transition is caused by the turn-off transition of the opposite transistor, not by the driver. Only the turn-off transition is directly controllable.

The transistors are turned on at *zero voltage*. In fact, there is a small negative voltage of the antiparallel diode, but this voltage is negligible in comparison to the input voltage V_I.

For example, transistor Q_2 is turned on by v_2 when i_{S2} is negative. Voltage v_{DS2} is maintained at nearly −1 V by the antiparallel diode D_2 during the transistor turn-on transition. Therefore, the turn-on switching loss is eliminated, Miller's effect is absent, transistor input capacitance is not increased by Miller's effect, the gate drive power is low, and the turn-on switching speed is high. The diodes turn on at a very low di/dt. The diode reverse-recovery current is a fraction of a sine wave and becomes a part of the switch current when the switch current is positive. Therefore, the antiparallel diodes can be slow, and the MOSFET's body–drain diodes are sufficiently fast as long as the reverse-recovery time is less than one-half of the cycle. The diode voltage is kept at a low voltage of the order of 1 V by the transistor in on-state during the reverse-recovery interval, reducing the diode reverse-recovery power loss. The transistor can be turned on when the switch current is not only negative but also positive and the diode is still conducting because of the reverse-recovery current. Therefore, the range of the on-duty cycle of the gate–source voltages and the dead time can be larger. If the dead time is too long, the current will be diverted from the recovered diode D_2 to diode D_1 of the opposite transistor until transistor Q_2 is turned on, causing extra transitions of both switch voltages, current spikes, and switching losses.

For $f > f_o$, the turn-on switching loss is zero, but there is a turn-off loss in the transistor. Both the switch voltage and current waveforms overlap during turn-off, causing a turn-off switching loss. Also, Miller's effect is considerable, increasing the transistor input capacitance, the gate drive requirements, and reducing the turn-off speed. An approximated analysis of the turn-off switching loss is given in Section 4.11.3. In summary, for $f > f_o$, there is a turn-off switching loss in the transistor, while the turn-on transitions of the transistor and the diode are lossless. The turn-off switching loss can be eliminated by adding a shunt capacitor to one of the transistors and using a dead time in the drive voltages.

4.5 Topologies of Class D Voltage-Source RF Power Amplifiers

Figure 4.7 shows a Class D voltage-switching amplifier with various resonant circuits. These resonant circuits are derived from a series-resonant circuit. In Fig. 4.7(b), C_c is a large coupling capacitor, which can also be connected in series with the resonant inductor. The resonant frequency (i.e., the boundary between the capacitive and inductive load) for the circuits of Fig. 4.7(b)–(g) depends on the load. The resonant circuit shown in Fig. 4.7(b) is employed in a parallel resonant converter. The circuit of Fig. 4.7(d) is used in a series-parallel resonant converter. The circuit of Fig. 4.7(e) is used in a CLL resonant converter. Resonant circuits of Fig. 4.7(a), (f), and (g) supply a sinusoidal output current. Therefore, they are compatible with current-driven rectifiers. The amplifiers of Figs. 4.7(b)–(e) produce a sinusoidal voltage output and are compatible with voltage-driven rectifiers. A high-frequency transformer can be inserted in places indicated in Fig. 4.7. Half-bridge topologies of the Class D voltage-switching amplifier are depicted in Fig. 4.8. They are equivalent for ac components to the basic topology of Fig. 4.5. Figure 4.8(a) shows a half-bridge amplifier with two dc voltage sources. The bottom voltage source $V_I/2$ acts as a short circuit for the current through the resonant circuit, resulting in the circuit of Fig. 4.8(b). A drawback of this circuit is that the load current flows through the internal resistances of the dc voltage sources, reducing efficiency. In Fig. 4.8(c), blocking capacitors $C_B/2$ act as short circuits for the ac component. The dc voltage across each of them is $V_I/2$, but the ac power is dissipated in the ESRs of the capacitors. An equivalent circuit for the amplifier of Fig. 4.8(c) is shown in Fig. 4.8(d). This is a useful circuit if the dc power supply contains a voltage doubler. The voltage stress across the filter capacitors is lower than that in the basic circuit of Fig. 4.5. In Fig. 4.8(e), the resonant capacitor is split into two halves, which are connected in parallel for the ac component. This is possible because the dc input voltage source

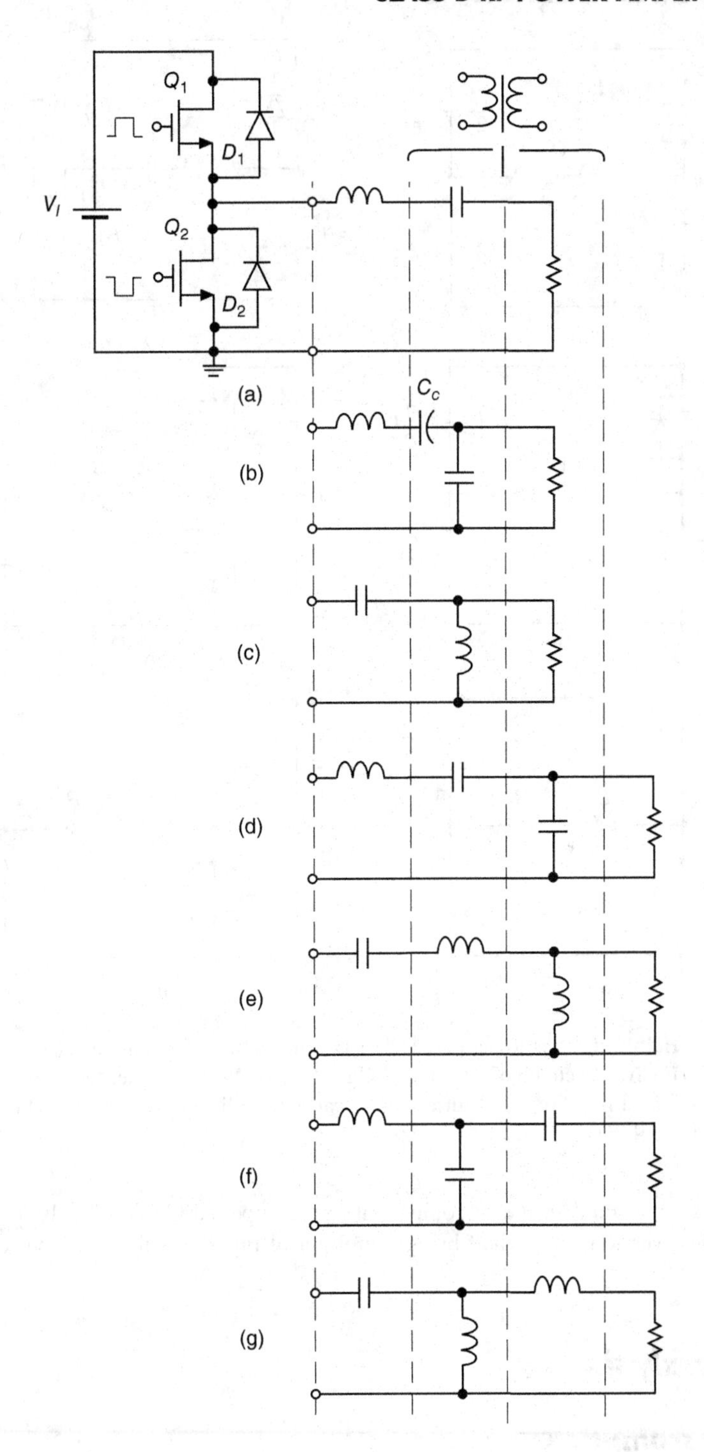

Figure 4.7 Class D voltage-switching amplifier with various resonant circuits.

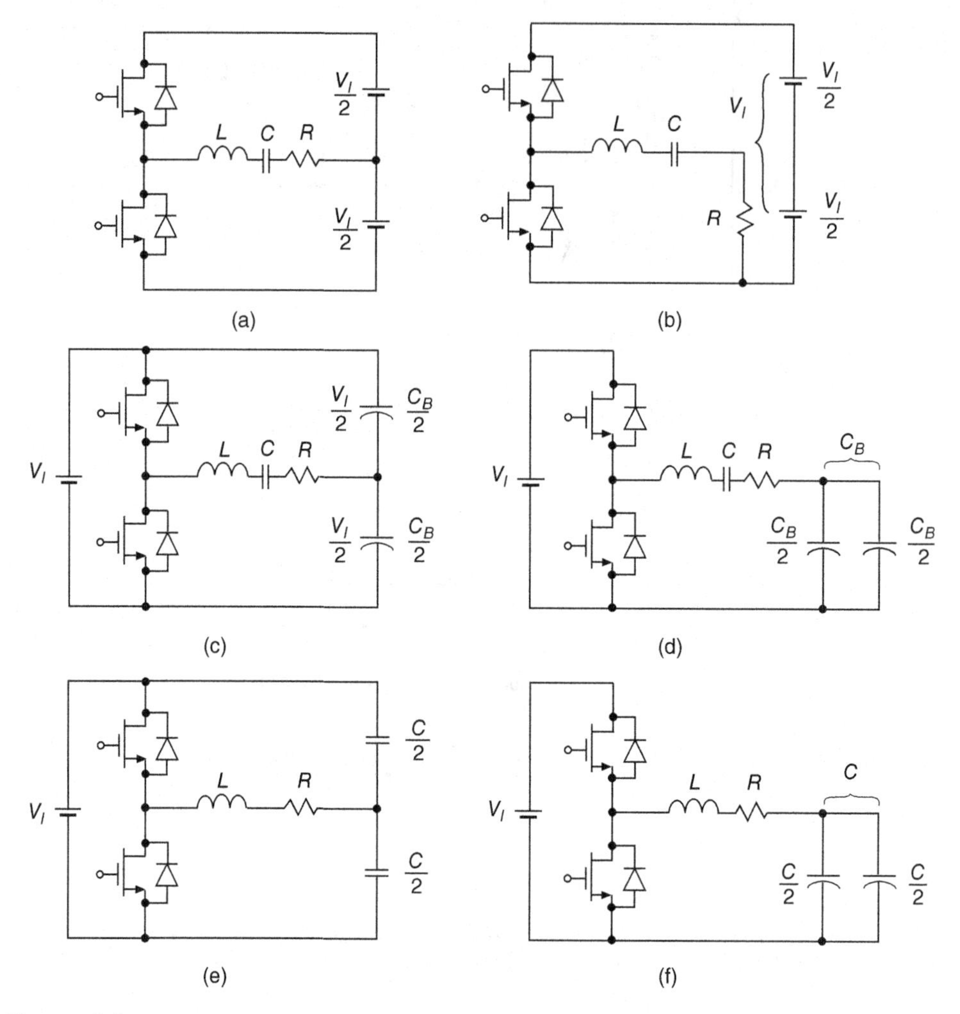

Figure 4.8 Half-bridge topologies of the Class D voltage-switching amplifier. (a) With two dc voltage sources. (b) Equivalent circuit of amplifier of Fig. 4.8(a). (c) With two filter capacitors. (d) Equivalent circuit of amplifier of Fig. 4.8(c). (e) With a resonant capacitor split into two halves. (f) Equivalent circuit of amplifier of Fig. 4.8(e).

V_I acts as a short circuit for the ac component of the upper capacitor. The disadvantage of all transformerless versions of the half-bridge amplifier of Fig. 4.8 is that the load resistance R is not grounded.

4.6 Analysis

4.6.1 Assumptions

The analysis of the Class D amplifier of Fig. 4.5 is based on the equivalent circuit of Fig. 4.5(d) and the following assumptions:

1. The transistor and the diode form a resistive switch whose on-resistance is linear, the parasitic capacitances of the switch are neglected, and the switching times are zero.

2. The elements of the series-resonant circuit are passive, linear, time invariant, and do not have parasitic reactive components.
3. The loaded quality factor Q_L of the series-resonant circuit is high enough so that the current i through the resonant circuit is sinusoidal.

4.6.2 Input Voltage of Resonant Circuit

A dc voltage source V_I and two switches S_1 and S_2 form a square-wave voltage source with offset $V_I/2$. When S_1 is ON and S_2 is OFF, $v = V_I$. When S_1 is OFF and S_2 is ON, $v = 0$. Referring to Fig. 4.5(d), the input voltage of the series-resonant circuit is a square wave

$$v \approx v_{DS2} = \begin{cases} V_I & \text{for} \quad 0 < \omega t \leq \pi \\ 0 & \text{for} \quad \pi < \omega t \leq 2\pi \end{cases} \tag{4.16}$$

This voltage can be expressed by trigonometric Fourier series

$$v \approx v_{DS2} = V_I \left[\frac{1}{2} + \frac{2}{\pi} \sum_{n=1}^{\infty} \frac{1-(-1)^n}{2n} \sin(n\omega t) \right] = V_I \left\{ \frac{1}{2} + \frac{2}{\pi} \sum_{k=1}^{\infty} \frac{\sin[(2k-1)\omega t]}{2k-1} \right\}$$
$$= V_I \left(\frac{1}{2} + \frac{2}{\pi} \sin \omega t + \frac{2}{3\pi} \sin 3\omega t + \frac{2}{5\pi} \sin 5\omega t + \cdots \right) \tag{4.17}$$

Ideally, even harmonics ($n = 2, 4, 6, \ldots$) in a square-wave signal are zero.

The waveform of voltage v can be presented as a symmetrical function of time

$$v \approx v_{DS2} = \begin{cases} V_I & \text{for} \quad -\frac{\pi}{2} < \omega t \leq \frac{\pi}{2} \\ 0 & \text{for} \quad \frac{\pi}{2} < \omega t \leq \frac{3\pi}{2} \end{cases} \tag{4.18}$$

In this case, the trigonometric Fourier series of voltage v is

$$v \approx v_{DS2} = V_I \left[\frac{1}{2} + \frac{2}{\pi} \sum_{n=1}^{\infty} \frac{\sin\left(\frac{n\pi}{2}\right)}{n} \cos(n\omega t) \right]$$
$$= V_I \left(\frac{1}{2} + \frac{2}{\pi} \cos \omega t - \frac{2}{3\pi} \cos 3\omega t + \frac{2}{5\pi} \cos 5\omega t + \cdots \right) \tag{4.19}$$

The fundamental component of voltage v at the input of the resonant circuit is

$$v_{i1} = V_m \sin \omega t \tag{4.20}$$

where its amplitude is given by

$$V_m = \frac{2V_I}{\pi} \approx 0.637 V_I \tag{4.21}$$

This leads to the rms value of voltage v_{i1}

$$V_{rms} = \frac{V_m}{\sqrt{2}} = \frac{\sqrt{2} V_I}{\pi} \approx 0.45 V_I \tag{4.22}$$

and

$$\xi_1 = \frac{V_m}{V_I} = \frac{2}{\pi} \approx 0.637 \tag{4.23}$$

The instantaneous power losses in the ideal transistors (for zero on-resistances and zero switching times) are

$$p_D(\omega t) = i_{S1} v_{DS1} = i_{S2} v_{DS2} = 0 \tag{4.24}$$

Hence, the drain efficiency of an ideal Class D amplifier is 100%.

4.6.3 Series-Resonant Circuit

The parameters of the series-resonant circuit are defined as follows:

- the resonant frequency

$$\omega_o = \frac{1}{\sqrt{LC}} \tag{4.25}$$

- the characteristic impedance

$$Z_o = \sqrt{\frac{L}{C}} = \omega_o L = \frac{1}{\omega_o C} \tag{4.26}$$

- the loaded quality factor

$$Q_L = \frac{\omega_o L}{R + r} = \frac{1}{\omega_o C(R + r)} = \frac{\sqrt{\frac{L}{C}}}{R + r} = \frac{Z_o}{R + r} = \frac{Z_o}{R_t} \tag{4.27}$$

- the unloaded quality factor

$$Q_o = \frac{\omega_o L}{r} = \frac{1}{\omega_o C r} = \frac{Z_o}{r} \tag{4.28}$$

where

$$r = r_{DS} + r_L + r_C \tag{4.29}$$

and

$$R_t = R + r \tag{4.30}$$

The series-resonant circuit acts as a second-order bandpass filter. The bandwidth of this circuit is

$$BW = \frac{f_o}{Q_L}. \tag{4.31}$$

The loaded quality factor is defined as

$$Q_L \equiv 2\pi \frac{\text{Total average magnetic and electric energy stored at resonant frequency } f_o}{\text{Energy dissipated and delivered to load per cycle at resonant frequency } f_o}$$

$$= 2\pi \frac{\text{Peak magnetic energy at } f_o}{\text{Energy lost in one cycle at } f_o} = 2\pi \frac{\text{Peak electric energy at } f_o}{\text{Energy lost in one cycle at } f_o}$$

$$= 2\pi \frac{W_s}{T_o P_{Rt}} = 2\pi \frac{f_o W_s}{P_{Rt}} = \frac{\omega_o W_s}{P_O + P_r} = \frac{Q}{P_O + P_r} \tag{4.32}$$

where W_s is the total energy stored in the resonant circuit at the resonant frequency $f_o = 1/T_o$, and $Q = \omega_o W_s$ is the reactive power of inductor L or capacitor C at the resonant frequency f_o.

The current waveform through the inductor L is given by

$$i_L(\omega t) = I_m \sin(\omega t - \psi) \tag{4.33}$$

resulting in the instantaneous energy stored in the inductor

$$w_L(\omega t) = \frac{1}{2}LI_m^2 \sin^2(\omega t - \psi). \tag{4.34}$$

The voltage waveform across the capacitor C is given by

$$v_C(\omega t) = V_{Cm}\cos(\omega t - \psi) \tag{4.35}$$

resulting in the instantaneous energy stored in the capacitor

$$w_C(\omega t) = \frac{1}{2}CV_{Cm}^2 \cos^2(\omega t - \psi) \tag{4.36}$$

The total energy stored in the resonant circuit at any frequency is given by

$$w_s(\omega t) = w_L(\omega t) + w_C(\omega t) = \frac{1}{2}[LI_m^2 \sin^2(\omega t - \psi) + CV_{Cm}^2 \cos^2(\omega t - \psi)]. \tag{4.37}$$

Since $V_{Cm} = X_C I_m = I_m/(\omega C)$, the total energy stored in the resonant circuit at any frequency can be expressed as

$$w_s(\omega t) = \frac{1}{2}LI_m^2 \left[\sin^2(\omega t - \psi) + \frac{1}{\omega^2 LC}\cos^2(\omega t - \psi)\right]$$

$$= \frac{1}{2}LI_m^2 \left[\sin^2(\omega t - \psi) + \frac{1}{\left(\frac{\omega}{\omega_o}\right)^2}\cos^2(\omega t - \psi)\right] \tag{4.38}$$

For $\omega = \omega_o$, the total energy stored in the resonant circuit W_s is equal to the peak magnetic energy stored in the inductor L

$$w_s(\omega t) = \frac{1}{2}LI_m^2[\sin^2(\omega t - \psi) + \cos^2(\omega t - \psi)] = \frac{1}{2}LI_m^2 = W_s = W_{Lmax}. \tag{4.39}$$

Because $I_m = V_{Cm}/X_C = \omega C V_{Cm}$, the total energy stored in the resonant circuit at any frequency becomes

$$w_s(\omega t) = \frac{1}{2}CV_{Cm}^2[\omega^2 LC \sin^2(\omega t - \psi) + \cos^2(\omega t - \psi)]$$

$$= \frac{1}{2}CV_{Cm}^2 \left[\left(\frac{\omega}{\omega_o}\right)^2 \sin^2(\omega t - \psi) + \cos^2(\omega t - \psi)\right] \tag{4.40}$$

For $\omega = \omega_o$, the total energy stored in the resonant circuit W_s is equal to the peak electric energy stored in the capacitor C

$$w_s(\omega t) = \frac{1}{2}CV_{Cm}^2[\sin^2(\omega t - \psi) + \cos^2(\omega t - \psi)] = W_s = W_{Cmax}. \tag{4.41}$$

Figure 4.9 shows waveforms of magnetic energy, electric energy, and total energy stored in the resonant circuit at the resonant frequency f_o.

For steady-state operation at the resonant frequency f_o, the total instantaneous energy stored in the resonant circuit is constant and equal to the maximum energy stored in the inductor

$$W_s = W_{Lmax} = \frac{1}{2}LI_m^2 \tag{4.42}$$

or, using (4.25), in the capacitor

$$W_s = W_{Cmax} = \frac{1}{2}CV_{Cm}^2 = \frac{1}{2}C\frac{I_m^2}{(\omega_o C)^2} = \frac{1}{2}\frac{I_m^2}{(C\omega_o^2)} = \frac{1}{2}LI_m^2 = W_{Lmax} \tag{4.43}$$

Substitution of (4.42) and (4.43) into (4.32) produces

$$Q_L = \frac{\omega_o LI_m^2}{2P_{Rt}} = \frac{\omega_o CV_{Cm}^2}{2P_{Rt}} \tag{4.44}$$

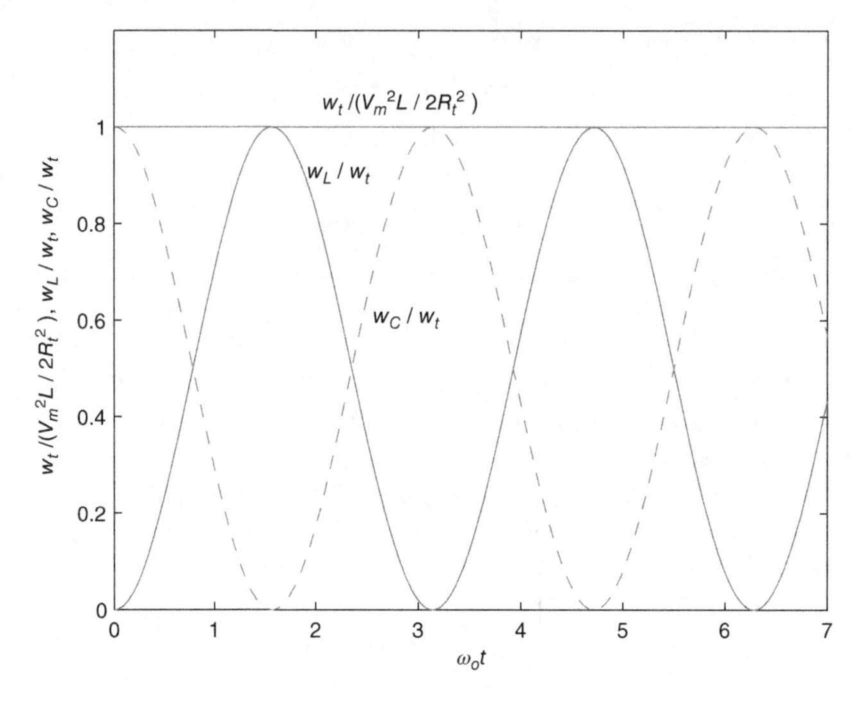

Figure 4.9 Waveforms of magnetic energy, electric energy, and total energy stored in the resonant circuit at the resonant frequency f_o.

The reactive power of the inductor at f_o is $Q = (1/2)V_{Lm}I_m = (1/2)\omega_o LI_m^2$ and of the capacitor is $Q = (1/2)I_m V_{Cm} = (1/2)\omega_o CV_{Cm}^2$. Thus, the quality factor can be defined as a ratio of the reactive power of the inductor or the capacitor to the real power dissipated in the form of heat in all resistances at the resonant frequency f_o. The total power dissipated in $R_t = R + r$ is

$$P_{Rt} = \frac{1}{2}R_t I_m^2 = \frac{1}{2}(R + r)I_m^2 \tag{4.45}$$

Substitution of (4.42) or (4.43), and (4.45) into (4.32) gives

$$Q_L = \frac{\omega_o L}{R + r} = \frac{1}{\omega_o C(R + r)} \tag{4.46}$$

The average magnetic energy stored in the resonant circuit is

$$W_{L(AV)} = \frac{1}{2\pi}\int_0^{2\pi} w_L d(\omega_o t) = \frac{1}{2}LI_m^2 \times \frac{1}{2\pi}\int_0^{2\pi} \sin^2(\omega_o t)d(\omega t) = \frac{1}{4}LI_m^2 \tag{4.47}$$

Similarly, the average electric energy stored in the resonant circuit is

$$W_{C(AV)} = \frac{1}{2\pi}\int_0^{2\pi} w_C d(\omega_o t) = \frac{1}{2}CV_{Cm}^2 \times \frac{1}{2\pi}\int_0^{2\pi} \cos^2(\omega_o t)d(\omega t) = \frac{1}{4}CV_{Cm}^2 \tag{4.48}$$

Hence, the total average energy stored in the resonant circuit is

$$W_s = W_{L(AV)} + W_{C(AV)} = \frac{1}{4}LI_m^2 + \frac{1}{4}CV_{Cm}^2 = \frac{1}{2}LI_m^2 = \frac{1}{2}CV_{Cm}^2 \tag{4.49}$$

This leads to the loaded quality factor.

For $R = 0$,

$$P_r = \frac{1}{2}rI_m^2 \tag{4.50}$$

and the unloaded quality factor is defined as

$$Q_o \equiv \frac{\omega_o W_s}{P_r} = \frac{\omega_o L}{r} = \frac{\omega_o L}{r_{DS} + r_L + r_C} = \frac{1}{\omega_o C r} = \frac{1}{\omega_o C(r_{DS} + r_L + r_C)} \tag{4.51}$$

Similarly, the quality factor of the inductor is

$$Q_{Lo} \equiv \frac{\omega_o W_s}{P_{rL}} = \frac{\omega_o L}{r_L} \tag{4.52}$$

and the capacitor is

$$Q_{Co} \equiv \frac{\omega_o W_s}{P_{rC}} = \frac{1}{\omega_o C r_C} \tag{4.53}$$

4.6.4 Input Impedance of Series-Resonant Circuit

The input impedance of the series-resonant circuit is

$$\mathbf{Z} = R + r + j\left(\omega L - \frac{1}{\omega C}\right) = (R+r)\left[1 + jQ_L\left(\frac{\omega}{\omega_o} - \frac{\omega_o}{\omega}\right)\right]$$
$$= Z_o\left[\frac{R+r}{Z_o} + j\left(\frac{\omega}{\omega_o} - \frac{\omega_o}{\omega}\right)\right] = |Z|e^{j\psi} = R + r + jX \tag{4.54}$$

where

$$|Z| = (R+r)\sqrt{1 + Q_L^2\left(\frac{\omega}{\omega_o} - \frac{\omega_o}{\omega}\right)^2} = Z_o\sqrt{\left(\frac{R+r}{Z_o}\right)^2 + \left(\frac{\omega}{\omega_o} - \frac{\omega_o}{\omega}\right)^2}$$
$$= Z_o\sqrt{\frac{1}{Q_L^2} + \left(\frac{\omega}{\omega_o} - \frac{\omega_o}{\omega}\right)^2} \tag{4.55}$$

$$\psi = \arctan\left[Q_L\left(\frac{\omega}{\omega_o} - \frac{\omega_o}{\omega}\right)\right] \tag{4.56}$$

$$R_t = R + r = |Z|\cos\psi \tag{4.57}$$

$$X = |Z|\sin\psi. \tag{4.58}$$

The reactance of the resonant circuit becomes zero at the resonant frequency f_o. Figures 4.10 and 4.11 show plots of $|Z|/R_t$ as functions of f/f_o and Q_L. From (4.56),

$$\cos\psi = \frac{1}{\sqrt{1 + Q_L^2\left(\frac{\omega}{\omega_o} - \frac{\omega_o}{\omega}\right)^2}} \tag{4.59}$$

Figure 4.12 shows a three-dimensional representation of $|Z|/Z_o$ as a function of the normalized frequency f/f_o and the loaded quality factor Q_L. Plots of $|Z|/Z_o$ and ψ as a function of f/f_o at fixed values of Q_L are shown in Figs 4.13 and 4.14.

For $f < f_o$, ψ is less than zero, which means that the resonant circuit represents a capacitive load to the switching part of the amplifier. For $f > f_o$, ψ is greater than zero, which indicates that the resonant circuit represents an inductive load. The magnitude of the impedance $|Z|$ is usually normalized with respect to R_t, but it is not a good normalization if R changes, and therefore $R_t = R + r$ is variable at fixed resonant components L and C.

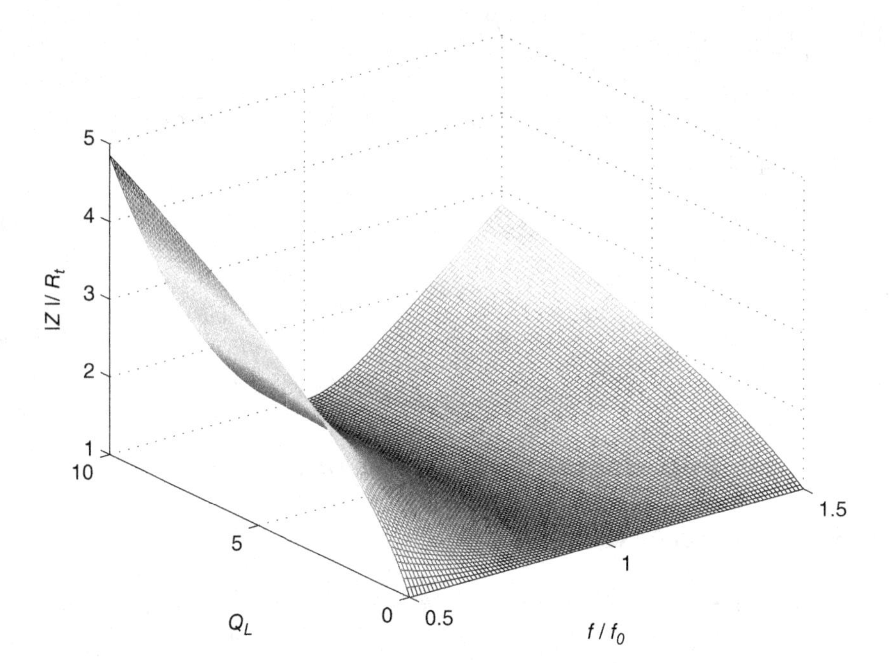

Figure 4.10 Modulus of the normalized input impedance $|Z|/R_t$ as a function of f/f_o and Q_L.

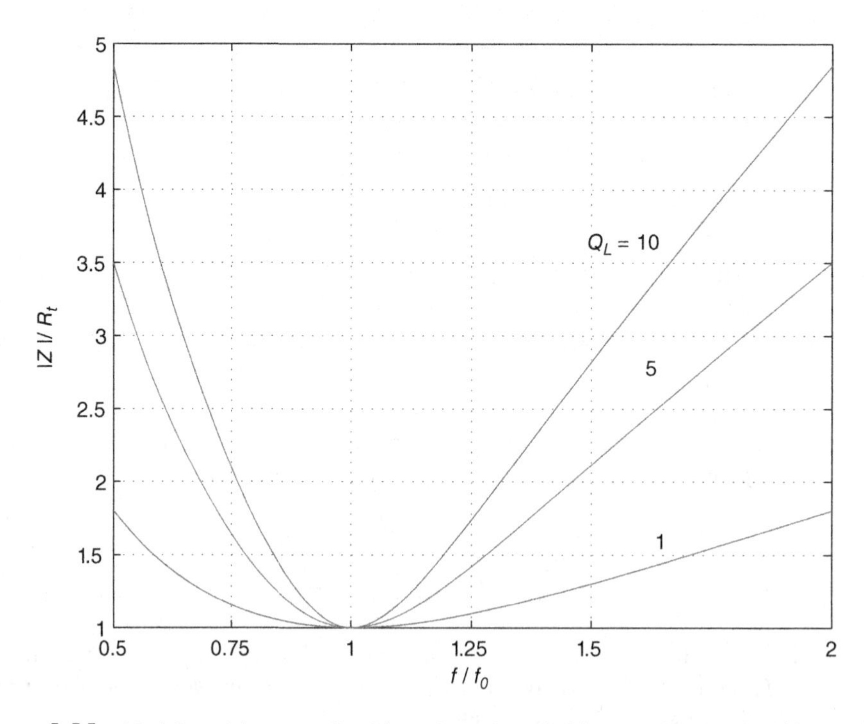

Figure 4.11 Modulus of the normalized input impedance $|Z|/R_t$ as a function of f/f_o at constant values of Q_L.

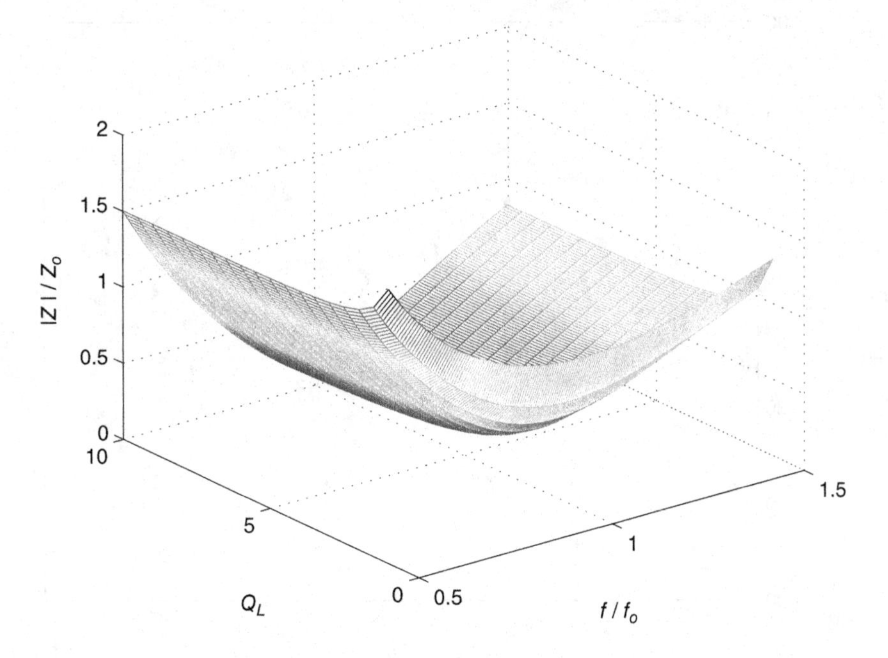

Figure 4.12 Plots of $|Z|/Z_o$ as a function of f/f_o and Q_L.

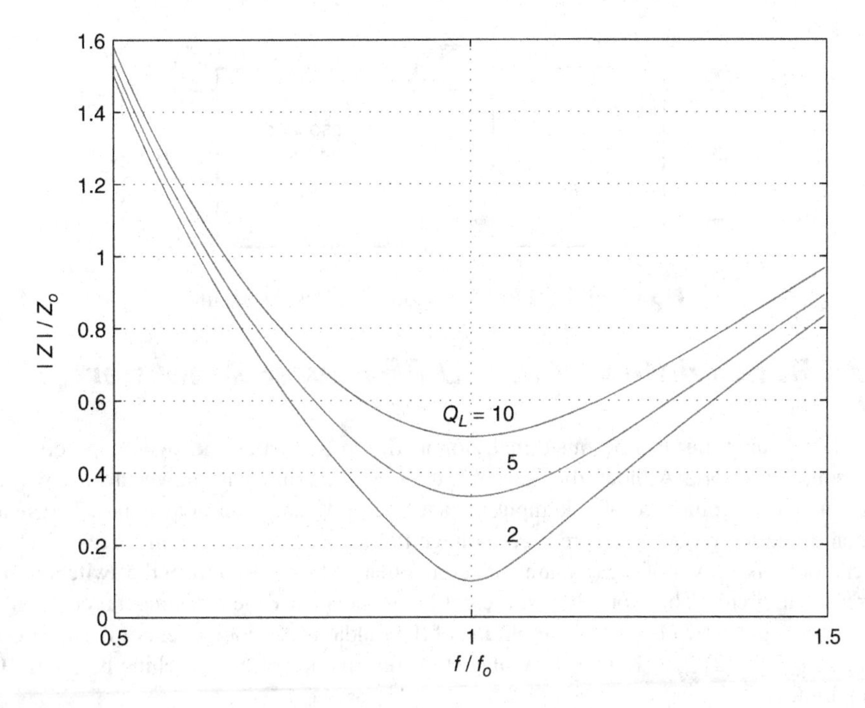

Figure 4.13 Modulus of the normalized input impedance $|Z|/Z_o$ as a function of f/f_o at constant values of Q_L.

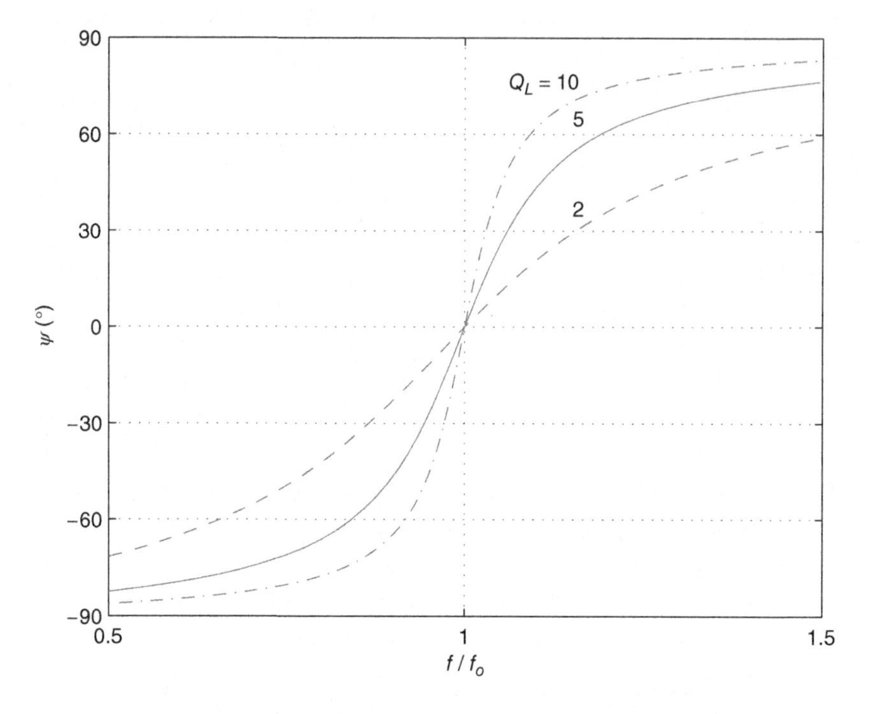

Figure 4.14 Phase of the input impedance ψ as a function of f/f_o at constant values of Q_L.

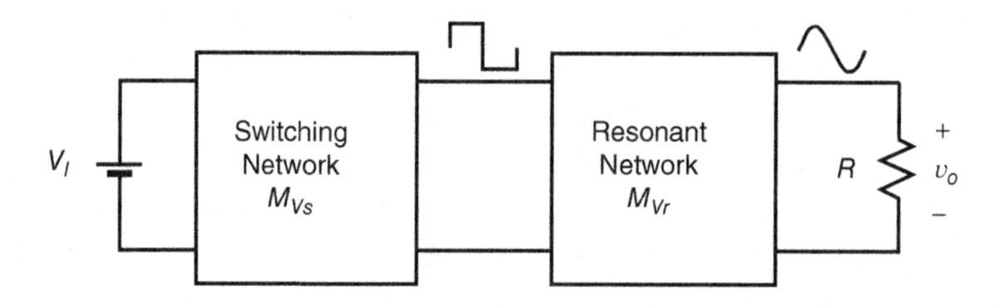

Figure 4.15 Block diagram of a Class D amplifier.

4.7 Bandwidth of Class D RF Power Amplifier

The Class D amplifier can be functionally divided into two parts: the switching network and the resonant network. A block diagram of the Class D amplifier is shown in Fig. 4.15. The switching part is composed of a dc input voltage source V_I and a set of switches. The switches are controlled to produce a square-wave voltage v. Since a resonant circuit forces a sinusoidal current, only the power of the fundamental component is transferred from the switching part to the resonant circuit. Therefore, it is sufficient to consider only the fundamental component of the voltage v given by (4.20). The amplitude of the fundamental component of the square-wave voltage v is $V_m = 2V_m/\pi$. Hence, the voltage transfer function of the switching part of the Class D amplifier is

$$M_{Vs} \equiv \frac{V_m}{V_I} = \frac{2}{\pi} \tag{4.60}$$

where V_m is the amplitude of the fundamental component v_{i1} of the voltage v. The resonant network of the amplifier converts the square-wave voltage v into a sinusoidal current or voltage signal.

The voltage transfer function of the series-resonant circuit shown in Fig. 4.5(d) for the fundamental component is

$$M_{Vr} = \frac{V_{om}}{V_m} = \frac{R}{Z} = \frac{R}{R + r + j\left(\omega L - \frac{1}{\omega C}\right)} = \frac{R}{R+r}\frac{1}{1 + jQ_L\left(\frac{\omega}{\omega_o} - \frac{\omega_o}{\omega}\right)} = |M_{Vr}|e^{\phi_{MVr}} \tag{4.61}$$

where

$$|M_{Vr}| = \frac{R}{|Z|} = \frac{R}{R+r}\frac{1}{\sqrt{1 + Q_L^2\left(\frac{\omega}{\omega_o} - \frac{\omega_o}{\omega}\right)^2}} = \frac{\eta_{Ir}}{\sqrt{1 + Q_L^2\left(\frac{\omega}{\omega_o} - \frac{\omega_o}{\omega}\right)^2}} \tag{4.62}$$

where the efficiency is $\eta_{Ir} = R/(R + r)$. Figure 4.16 illustrates (4.62) in a three-dimensional space. Figure 4.17 shows the voltage transfer function $|M_{Vr}|$ as a function of f/f_o at fixed values of Q_L for $\eta_{Ir} = 0.95$.

The magnitude of the dc-to-ac voltage transfer function for the Class D amplifier with a series-resonant circuit is

$$|M_{VI}| = \frac{V_{om}}{V_I} = \frac{V_{om}}{V_m}\frac{V_{om}}{V_I} = M_{Vs}|M_{Vr}| = \frac{2\eta_{Ir}}{\pi\sqrt{1 + Q_L^2\left(\frac{\omega}{\omega_o} - \frac{\omega_o}{\omega}\right)^2}} \tag{4.63}$$

The maximum value of M_{VI} occurs at $f/f_o = 1$ and $M_{VImax} = 2\eta_{Ir}/\pi \approx 0.637\eta_{Ir}$. Thus, the values of M_{VI} range from 0 to $2\eta_{Ir}/\pi$.

The 3-dB bandwidth satisfies the following condition:

$$\frac{1}{\sqrt{1 + Q_L^2\left(\frac{f}{f_o} - \frac{f_o}{f}\right)^2}} = \frac{1}{\sqrt{2}} \tag{4.64}$$

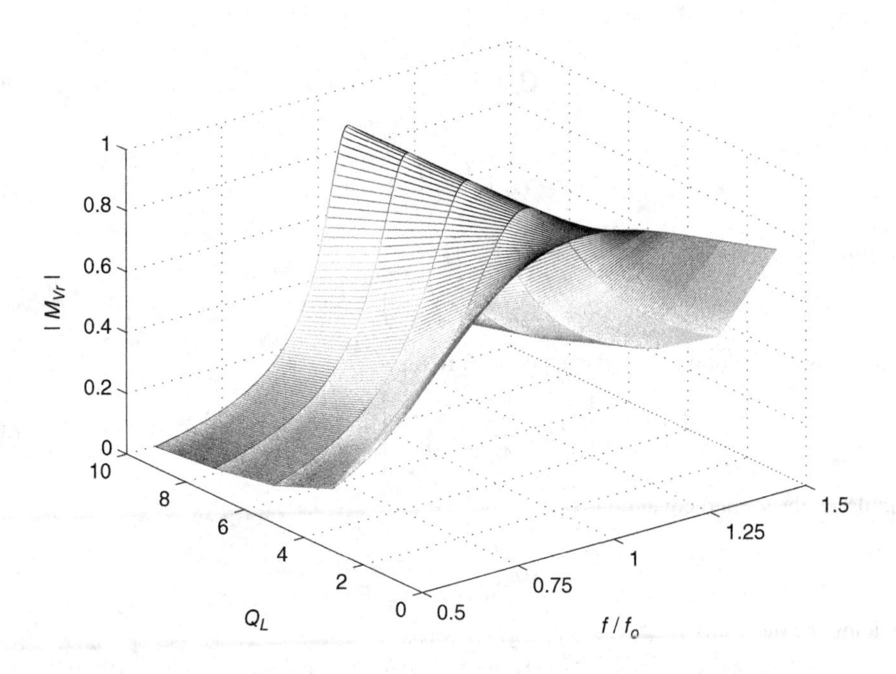

Figure 4.16 Three-dimensional representation of $|M_{Vr}|$ as a function of f/f_o and Q_L for $\eta_{Ir} = 0.95$.

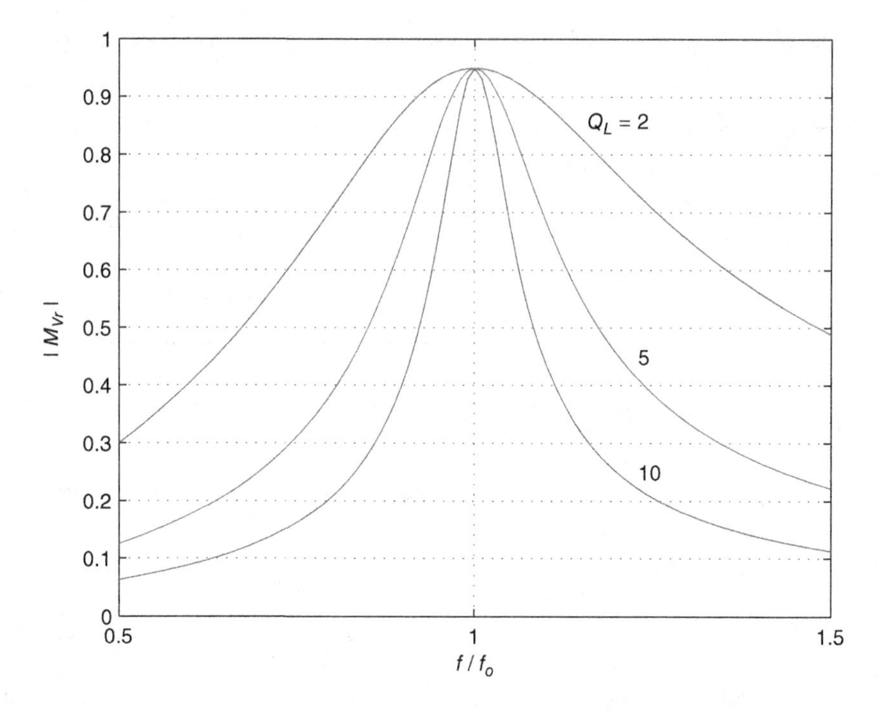

Figure 4.17 Transfer function $|M_{Vr}|$ as a function of f/f_o at fixed values of Q_L for $\eta_{Ir} = 0.95$.

Hence,

$$Q_L^2\left(\frac{f}{f_o} - \frac{f_o}{f}\right)^2 = 1 = (\pm 1)^2 \tag{4.65}$$

resulting in two second-order equations

$$Q_L\left(\frac{f}{f_o} - \frac{f_o}{f}\right) = \pm 1 \tag{4.66}$$

The first equation is

$$Q_L\left(\frac{f}{f_o} - \frac{f_o}{f}\right) = 1 \tag{4.67}$$

yielding the quadratic

$$\left(\frac{f}{f_o}\right)^2 - \frac{1}{Q_L}\left(\frac{f}{f_o}\right) - 1 = 0 \tag{4.68}$$

whose positive solution gives the upper 3-dB frequency

$$f_H = f_o\left[\sqrt{1 + \frac{1}{4Q_L^2}} + \frac{1}{2Q_L}\right] \approx f_o\left(1 + \frac{1}{2Q_L}\right) \tag{4.69}$$

Similarly, the second equation is

$$Q_L\left(\frac{f}{f_o} - \frac{f_o}{f}\right) = -1 \tag{4.70}$$

yielding the quadratic

$$\left(\frac{f}{f_o}\right)^2 + \frac{1}{Q_L}\left(\frac{f}{f_o}\right) - 1 = 0 \tag{4.71}$$

whose positive solution gives the lower 3-dB frequency

$$f_L = f_o \left[\sqrt{1 + \frac{1}{4Q_L^2}} - \frac{1}{2Q_L}\right] \approx f_o\left(1 - \frac{1}{2Q_L}\right) \tag{4.72}$$

Thus, the 3-dB bandwidth is obtained as

$$BW = \Delta f = f_H - f_L = f_o \left[\sqrt{1 + \frac{1}{4Q_L^2}} + \frac{1}{2Q_L}\right] - f_o \left[\sqrt{1 + \frac{1}{4Q_L^2}} - \frac{1}{2Q_L}\right] = \frac{f_o}{Q_L} \tag{4.73}$$

As the loaded quality factor Q_L decreases, the bandwidth BW increases.

4.8 Operation of Class D RF Power Amplifier at Resonance

4.8.1 Characteristics of Ideal Class D RF Power Amplifier

For frequencies lower than the resonant frequency f_o, the series-resonant circuit represents a high capacitive impedance. On the other hand, for frequencies higher than the resonant frequency f_o, the series-resonant circuit represents a high inductive impedance. The series-resonant circuit acts as a bandpass filter. If the loaded quality factor Q_L is high enough, the voltage across the resistance R_t at $f = f_o$ is sinusoidal

$$v = V_m \sin \omega t \tag{4.74}$$

and the current through the resonant circuit at $f = f_o$ is approximately sinusoidal and equal to the fundamental component

$$i = I_m \sin \omega t \tag{4.75}$$

where

$$I_m = \frac{V_m}{R_t} = \frac{2V_I}{\pi R_t} \tag{4.76}$$

At $f = f_o$, the current i_I drawn from the power supply V_I equals the current through the upper switch S_1 and is given by

$$i_I = i_{S1} = \begin{cases} I_m \sin \omega t & \text{for} \quad 0 < \omega t \le \pi \\ 0 & \text{for} \quad \pi < \omega t \le 2\pi \end{cases} \tag{4.77}$$

The dc component of the supply current is

$$I_I = \frac{1}{2\pi}\int_0^{2\pi} i_{S1} d(\omega t) = \frac{I_m}{2\pi}\int_0^{\pi} \sin \omega t d(\omega t) = \frac{I_m}{\pi} = \frac{V_m}{\pi R_t} \tag{4.78}$$

yielding

$$\gamma_1 = \frac{I_m}{I_I} = \pi \tag{4.79}$$

The dc supply power is

$$P_I = V_I I_I = \frac{2V_I^2}{\pi^2 R_t} \tag{4.80}$$

The output power is

$$P_O = \frac{RI_m^2}{2} = \frac{2V_I^2 R}{\pi^2 R_t^2} \tag{4.81}$$

Hence, the drain efficiency of the Class D RF power amplifier with ideal transistors ($r_{DS} = 0$ and switching times are zero) is

$$\eta_D = \frac{P_{DS}}{P_I} = \frac{1}{2}\left(\frac{I_m}{I_I}\right)\left(\frac{V_m}{V_I}\right) = \frac{1}{2}\gamma_1\xi_1 = \frac{1}{2} \times 1/2x\pi \times 2/\pi = 1 \tag{4.82}$$

4.8.2 Current and Voltage Stresses of Class D RF Power Amplifier

The voltage and current stresses of each MOSFET are

$$V_{SM} = V_I = \frac{\pi}{2}V_m \tag{4.83}$$

and

$$I_{SM} = I_m = \pi I_I \tag{4.84}$$

The amplitudes of the voltage across the resonant inductance and capacitance are

$$V_{Lm} = Z_o I_m = \omega_o L I_m \tag{4.85}$$

and

$$V_{Cm} = Z_o I_m = \frac{I_m}{\omega_o C} \tag{4.86}$$

4.8.3 Output-Power Capability of Class D RF Power Amplifier

The output-power capability is

$$c_p = \frac{1}{N}\frac{P_{Omax}}{I_{SM}V_{SM}} = \frac{1}{2N}\frac{I_m V_m}{I_{SM}V_{SM}} = \frac{1}{2N}\frac{I_m V_m}{I_{SM}V_{SM}} = \frac{1}{2N}\left(\frac{I_m}{I_{SM}}\right)\left(\frac{V_m}{V_{SM}}\right)$$
$$= \frac{1}{2\times 2} \times 1 \times \left(\frac{2}{\pi}\right) = \frac{1}{2\pi} \approx 0.159 \tag{4.87}$$

Another method to derive the output-power capability is

$$c_p = \frac{1}{N}\frac{P_{Omax}}{I_{SM}V_{SM}} = \frac{1}{N}\frac{\eta_D P_I}{I_{SM}V_{SM}} = \frac{1}{N}\eta_D\left(\frac{I_I}{I_{SM}}\right)\left(\frac{V_I}{V_{SM}}\right) = \frac{1}{2} \times 1 \times \left(\frac{1}{\pi}\right) \times 1$$
$$= \frac{1}{2\pi} \approx 0.159 \tag{4.88}$$

4.8.4 Power Losses and Efficiency of Class D RF Power Amplifier

The rms value of the current of each MOSFET is

$$I_{Srms} = \sqrt{\frac{1}{2\pi}\int_0^{2\pi} i_{S1}^2 d(\omega t)} = \sqrt{\frac{1}{2\pi}\int_0^{\pi} (I_m \sin\omega t)^2 d(\omega t)}$$
$$= I_m\sqrt{\frac{1}{2\pi}\int_0^{\pi} \frac{1}{2}(1 - \cos 2\omega t) d(\omega t)} = \frac{I_m}{2} \tag{4.89}$$

Hence, the conduction power loss in each MOSFET is

$$P_{rDS} = r_{DS} I_{Srms}^2 = \frac{r_{DS} I_m^2}{4} \tag{4.90}$$

The sinusoidal current of amplitude I_m flows effectively through the on-resistance r_{DS} during the entire cycle $T = 1/f$. Therefore, the conduction power loss in the on-resistances of both MOSFETs is

$$2P_{rDS} = \frac{1}{2} r_{DS} I_m^2 \tag{4.91}$$

Since

$$V_I = \frac{\pi}{2} V_m = \frac{\pi}{2} R_t I_m \tag{4.92}$$

the switching loss in each MOSFET is

$$P_{swQ1} = P_{swQ2} = \frac{1}{2} f C_o V_I^2 = \frac{\pi^2}{8} f C_o V_m^2 = \frac{\pi^2}{8} f C_o R_t^2 I_m^2 \tag{4.93}$$

The sum of the conduction and switching losses in each MOSFET is

$$P_{MOS} = P_{rDS} + P_{swQ1} = \frac{I_m^2}{4} \left(r_{DS} + \frac{\pi^2}{2} f C_o R_t^2 \right) \tag{4.94}$$

The total switching loss in both MOSFETs is

$$P_{swQ1Q2} = 2P_{swQ1} = f C_o V_I^2 = \frac{\pi^2}{4} f C_o V_m^2 = \frac{\pi^2}{4} f C_o R_t^2 I_m^2 \tag{4.95}$$

Charging a linear capacitor is a 50% efficient process. The same amount of energy is lost in the resistor present in the charging path as the amount of energy stored in the capacitor at the end of charging process. Therefore, the total switching loss in the Class D amplifier is

$$P_{sw} = 2P_{swQ1Q2} = 2 f C_o V_I^2 = \frac{\pi^2}{2} f C_o R_t^2 I_m^2 \tag{4.96}$$

The drain efficiency is

$$\eta_D = \frac{P_{DS}}{P_I} = \frac{P_O}{P_O + 2P_{rDS} + P_{sw}} = \frac{R}{R + r_{DS} + \frac{\pi^2}{2} f C_o R_t^2} \tag{4.97}$$

The power loss in the resonant circuit is

$$P_{rLrC} = \frac{1}{2}(r_L + r_C) I_m^2 \tag{4.98}$$

Hence, the efficiency of the resonant circuit is

$$\eta_r = \frac{P_O}{P_{DS}} = \frac{P_O}{P_O + P_{rLrC}} = \frac{R}{R + r_L + r_C} \tag{4.99}$$

The efficiency of the resonant circuit can be also expressed as

$$\eta_r = 1 - \frac{r_L + r_C}{R + r_L + r_C} = 1 - \frac{Q_L}{Q_o} \tag{4.100}$$

The overall efficiency is

$$\eta = \frac{P_O}{P_I} = \frac{P_O}{P_O + P_D + P_{rLrC} + P_{sw}} = \frac{R}{R + r_{DS} + r_L + r_C + \frac{\pi^2}{2} f C_o R_t^2} \tag{4.101}$$

4.8.5 Gate Drive Power

The gate-to-source capacitance C_{gs} is nearly linear. However, the gate-to-drain capacitance is highly nonlinear. Therefore, the short-circuit capacitance of a power MOSFET $C_{iss} = C_{gs} + C_{gd}$ at $v_{DS} = 0$ is also highly nonlinear. For this reason, the estimation of the gate drive power based on the input capacitance is difficult. Since $v_{DS} = v_{GS} + v_{DG}$, the voltage gain during the turn-on and turn-off transitions is

$$A_m = \frac{\Delta v_{DS}}{\Delta_{GS}} = \frac{\Delta v_{GS} + \Delta v_{DG}}{\Delta v_{GS}} = 1 + \frac{\Delta v_{DG}}{\Delta v_{GS}} \tag{4.102}$$

Hence, the Miller's effect capacitance reflected from the drain-to-gate terminals to the gate-to-source terminals is

$$C_m = (1 - A_m)C_{gd} = \left(1 - 1 - \frac{\Delta v_{DG}}{\Delta v_{GS}}\right) C_{gd} = \left(-\frac{\Delta v_{DG}}{\Delta v_{GS}}\right) C_{gd} \tag{4.103}$$

The input capacitance of the transistor during turn-on and turn-off transitions due to Miller's effect is

$$C_i = C_{gs} + C_m = C_{gs} + (1 - A_m)C_{gd} \tag{4.104}$$

A simpler method relies on the concept of the gate charge Q_g. The energy required to charge and discharge the MOSFET input capacitance is

$$W_G = Q_g V_{GSpp} \tag{4.105}$$

where Q_g is the gate charge given in data sheets of MOSFETs and V_{GSpp} is the peak-to-peak gate-to-source voltage. Hence, the drive power of each MOSFET associated with turning the device on and off is

$$P_G = \frac{W_G}{T} = fW_G = fQ_g V_{GSpp} \tag{4.106}$$

The power gain of the amplifier is

$$k_p = \frac{P_O}{2P_G} \tag{4.107}$$

The power-added efficiency is

$$\eta_{PAE} = \frac{P_O - 2P_G}{P_I} = \frac{P_O\left(1 - \frac{1}{k_p}\right)}{P_I} \tag{4.108}$$

Example 4.1

Design a Class D RF power amplifier to meet the following specifications: $V_I = 3.3$ V, $P_O = 1$ W, $f = f_o = 1$ GHz, $r_{DS} = 0.1\ \Omega$, $Q_L = 7$, $Q_{Lo} = 200$, and $Q_{Co} = 1000$.

Solution. Assume $\eta_{Ir} = 0.9$ and neglect switching losses. Hence, the dc input power is

$$P_I = \frac{P_O}{\eta} = \frac{1}{0.9} = 1.11 \text{ W} \tag{4.109}$$

The total resistance is

$$R_t = \frac{2V_I^2}{\pi^2 P_I} = \frac{2 \times 3.3^2}{\pi^2 \times 1.11} = 1.988\ \Omega \tag{4.110}$$

The load resistance is

$$R = \eta_{Ir} R_t = 0.9 \times 1.988 = 1.789\ \Omega \tag{4.111}$$

The maximum parasitic resistance is

$$r_{max} = R_t - R = 1.988 - 1.789 = 0.199\ \Omega \tag{4.112}$$

The inductance is

$$L = \frac{Q_L R_t}{\omega_o} = \frac{7 \times 1.988}{2\pi \times 10^9} = 2.215 \text{ nH} \tag{4.113}$$

The capacitance is

$$C = \frac{1}{\omega_o Q_L R_t} = \frac{1}{2\pi \times 10^9 \times 7 \times 1.988} = 11.437 \text{ pF} \tag{4.114}$$

The voltage stresses are

$$V_{DSM} = V_I = 3.3 \text{ V} \tag{4.115}$$

The amplitude of the output current is

$$I_m = I_{SM} = \sqrt{\frac{2P_O}{R}} = \sqrt{\frac{2 \times 1}{1.789}} = 1.057 \text{ A} \tag{4.116}$$

Assuming $K_n = \mu_{n0} C_{ox} = 0.142 \times 10^{-3}$ A/V^2, $V_t = 0.3$ V, $L = 0.18$ μm, $V_{GSH} = 3.3$ V, $a = 2$, and $I_{Dsat} = aI_{DM}$, we obtain the aspect ratio of n-channel Si transistor

$$\left(\frac{W}{L}\right)_N = \frac{2I_{Dsat}}{K_n (V_{GSH} - V_t)^2} = \frac{2I_{Dsat}}{K_n v_{DSsat}^2} = \frac{2aI_{DM}}{K_n v_{DSsat}^2}$$

$$= \frac{2 \times 2 \times 1.057}{0.142 \times 10^{-3} \times (3.3 - 0.3)^2} = 3308.29 \tag{4.117}$$

Pick

$$\left(\frac{W}{L}\right)_N = 3400 \tag{4.118}$$

The width of the n-channel transistor is

$$W_N = \left(\frac{W}{L}\right)_N L = 3400 \times 0.18 \times 10^{-6} = 612 \text{ μm} \tag{4.119}$$

The aspect ratio of p-channel Si transistor is

$$\left(\frac{W}{L}\right)_P = \frac{\mu_{n0}}{\mu_{p0}} \left(\frac{W}{L}\right)_N = 2.7 \times 3400 = 9,180. \tag{4.120}$$

The width of the p-channel transistor is

$$W_P = \left(\frac{W}{L}\right)_P L = 9,180 \times 0.18 = 1652.4 \text{ μm} = 1.652 \text{ mm}. \tag{4.121}$$

The peak voltages across the resonant components C and L are

$$V_{Cm} = \frac{I_m}{\omega_o C} = \frac{1.057}{2\pi \times 10^9 \times 11.437 \times 10^{-12}} = 14.709 \text{ V} \tag{4.122}$$

and

$$V_{Lm} = \omega_o L I_m = 2\pi \times 10^9 \times 2.215 \times 10^{-9} \times 1.057 = 14.71 \text{ V} \tag{4.123}$$

The conduction power loss in each MOSFET is

$$P_{rDS} = \frac{r_{DS} I_m^2}{4} = \frac{0.1 \times 1.057^2}{4} = 27.93 \text{ mW} \tag{4.124}$$

The switching loss in each MOSFET is

$$P_{swQ1} = \frac{1}{2} f C_o V_I^2 = \frac{1}{2} \times 10^9 \times 10 \times 10^{-12} \times 3.3^2 = 54.45 \text{ mW} \tag{4.125}$$

The power loss in each MOSFET is

$$P_{MOS} = P_{rDS} + P_{swQ1} = 27.93 + 54.45 = 82.38 \text{ mW} \tag{4.126}$$

The total switching loss in the amplifier is

$$P_{sw} = 2 f C_o V_I^2 = 2 \times 10^9 \times 10 \times 10^{-12} \times 3.3^2 = 217.8 \text{ mW} \tag{4.127}$$

The drain efficiency is

$$\eta_D = \frac{P_{DS}}{P_I} = \frac{P_I - 2P_{MOS}}{P_I} = 1 - \frac{2P_{MOS}}{P_I} = 1 - \frac{0.08238}{1.11} = 85.04\% \quad (4.128)$$

The ESR of the inductor is

$$r_L = \frac{\omega_o L}{Q_{Lo}} = \frac{2\pi \times 10^9 \times 2.215 \times 10^{-9}}{200} = 69.59\ \text{m}\Omega \quad (4.129)$$

and the ESR of the capacitor is

$$r_C = \frac{1}{\omega_o C Q_{Co}} = \frac{1}{2\pi \times 10^9 \times 11.437 \times 10^{-12} \times 1000} = 13.9\ \text{m}\Omega \quad (4.130)$$

The power loss in the inductor ESR is

$$P_{rL} = \frac{r_L I_m^2}{2} = \frac{0.06959 \times 1.057^2}{2} = 38.87\ \text{mW} \quad (4.131)$$

and the power loss in the capacitor ESR is

$$P_{rC} = \frac{r_C I_m^2}{2} = \frac{0.0139 \times 1.057^2}{2} = 7.765\ \text{mW} \quad (4.132)$$

The efficiency of the resonant circuit is

$$\eta_r = \frac{P_O}{P_O + P_{rL} + P_{rC}} = \frac{1}{1 + 0.03887 + 0.007765} = 95.53\% \quad (4.133)$$

The total power loss in the Class D amplifier (excluding the gate drive power) is

$$P_{LS} = 2P_{rDS} + P_{rL} + P_{rC} + P_{sw} = 2 \times 27.93 + 38.87 + 7.765 + 217.8 = 320.29\ \text{mW} \quad (4.134)$$

resulting in the total efficiency of the Class D amplifier

$$\eta = \frac{P_O}{P_O + P_{LS}} = \frac{1}{1 + 0.32029} = 75.74\% \quad (4.135)$$

4.9 Class D RF Power Amplifier with Amplitude Modulation

A circuit of the Class D power amplifier with AM is shown in Fig. 4.18. The modulating voltage source v_m is connected in series with the dc supply voltage V_I. AM can be accomplished in the Class D amplifier because the amplitude of the output voltage is directly proportional to the dc supply voltage V_I. Voltage waveforms that explain the AM process are depicted in Fig. 4.19. The modulating voltage waveform is given by

$$v_m(t) = V_m \sin \omega_m t \quad (4.136)$$

The AM voltage waveform across the bottom switch is

$$v = v_{DS2} = [V_I + v_m(t)] \left[\frac{1}{2} + \frac{2}{\pi} \sum_{n=1}^{\infty} \frac{1-(-1)^n}{2n} \sin(n\omega_c t) \right]$$

$$= (V_I + V_m \sin \omega_m t) \left[\frac{1}{2} + \frac{2}{\pi} \sum_{n=1}^{\infty} \frac{1-(-1)^n}{2n} \sin(n\omega_c t) \right]$$

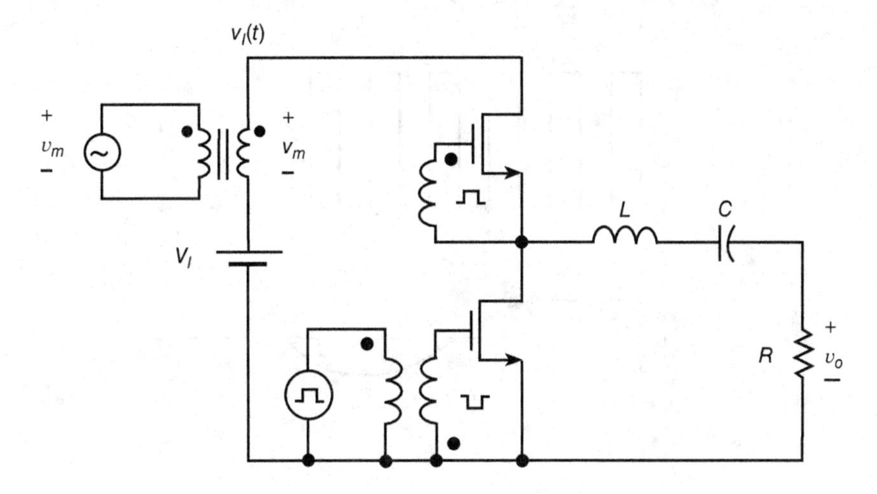

Figure 4.18 Class D RF power amplifier with amplitude modulation.

$$= V_I\left(1 + \frac{V_m}{V_I}\right)\sin\omega_m t\left[\frac{1}{2} + \frac{2}{\pi}\sum_{n=1}^{\infty}\frac{1-(-1)^n}{2n}\sin(n\omega_c t)\right]$$

$$= V_I(1+m)\sin\omega_m t\left[\frac{1}{2} + \frac{2}{\pi}\sum_{n=1}^{\infty}\frac{1-(-1)^n}{2n}\sin(n\omega_c t)\right] = \frac{V_I}{2} + \frac{mV_I}{2}\sin\omega_m t$$

$$+\frac{2V_I}{\pi}\sum_{n=1}^{\infty}\frac{1-(-1)^n}{2n}\sin(n\omega_c t) + \frac{2mV_I}{\pi}\sum_{n=1}^{\infty}\frac{1-(-1)^n}{2n}\sin(n\omega_c t)\sin\omega_m t \qquad (4.137)$$

Since

$$\sin x \sin y = \frac{1}{2}[\cos(x-y) - \cos(x+y)] \qquad (4.138)$$

the AM voltage waveform across the bottom switch is

$$v = v_{DS2} = \frac{V_I}{2} + \frac{mV_I}{2}\sin\omega_m t + \frac{2V_I}{\pi}\sum_{n=1}^{\infty}\frac{1-(-1)^n}{2n}\sin(n\omega_c t)$$

$$+\frac{mV_I}{\pi}\sum_{n=1}^{\infty}\frac{1-(-1)^n}{2n}\cos(n\omega_c - \omega_m)t - \frac{mV_I}{\pi}\sum_{n=1}^{\infty}\frac{1-(-1)^n}{2n}\cos(n\omega_c + \omega_m)t \qquad (4.139)$$

It can be seen that the fundamental component and the odd harmonics of the voltage across the bottom MOSFET (i.e., at the input of the resonant circuit) are modulated in amplitude. The amplitudes of all odd harmonics are directly proportional to the supply voltage V_I.

The AM fundamental component of the voltage at the input of the resonant circuit is

$$v_1 = v_{DS2(1)} = [V_I + v_m(t)]\left(\frac{2}{\pi}\sin\omega_c t\right)$$

$$= (V_I + V_m\sin\omega_m t)\left(\frac{2}{\pi}\sin\omega_c t\right) = \frac{2V_I}{\pi}\sin\omega_c t + \frac{2V_m}{\pi}\sin\omega_m t\sin\omega_c t$$

$$= \frac{2}{\pi}V_I\left(1 + \frac{V_m}{V_I}\sin\omega_m t\right)\sin\omega_c t = \frac{2V_I}{\pi}(1 + m\sin\omega_m t)\sin\omega_c t$$

$$= V_c(1 + m\sin\omega_m t)\sin\omega_c t = V_c\sin\omega_c t + \frac{mV_c}{2}\cos(\omega_c - \omega_m)t - \frac{mV_c}{2}\cos(\omega_c + \omega_m)t \qquad (4.140)$$

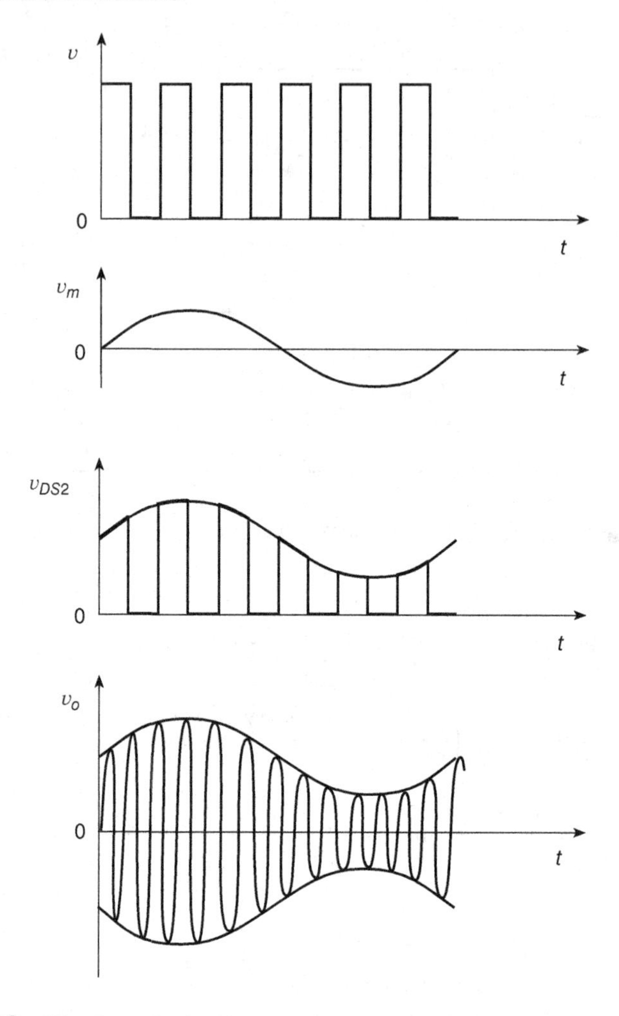

Figure 4.19 Waveforms in the Class D RF power amplifier with amplitude modulation.

where the amplitude of the carrier is

$$V_c = \frac{2}{\pi} V_I \tag{4.141}$$

and the modulation index is

$$m = \frac{V_m}{V_I} \tag{4.142}$$

Assume an ideal bandpass filter whose magnitude of the voltage gain is constantly equal to 1 and the phase is zero in the pass band, whereas the magnitude of the voltage gain is zero outside the pass band. Under this assumption, only AM fundamental component of the carrier is transmitted to the amplifier output. Hence, the AM output voltage is

$$\begin{aligned} v_o &= (V_I + V_m \sin \omega_m t)\left(\frac{2}{\pi} \sin \omega_c t\right) = \frac{2}{\pi} V_I \left(1 + \frac{V_m}{V_I} \sin \omega_m t\right) \sin \omega_c t \\ &= V_c(1 + m \sin \omega_m t) \sin \omega_c t = V_c \sin \omega_c t + \frac{mV_c}{2} \cos(\omega_c - \omega_m)t - \frac{mV_c}{2} \cos(\omega_c + \omega_m)t \end{aligned} \tag{4.143}$$

In reality, the side bands are attenuated by the nonideal bandpass filter

$$v_o = V_c \sin \omega_c t + \frac{mV_c|M_v|}{2} \cos[(\omega_c - \omega_m)t - \phi] \frac{mV_c|M_v|}{2} \cos[(\omega_c + \omega_m)t + \phi] \quad (4.144)$$

where $|M_v|$ and ϕ are the magnitude and phase of the voltage gain of the band pass filter at $\omega = \omega_c - \omega_m$ and $\omega = \omega_c + \omega_m$.

4.10 Operation of Class D RF Power Amplifier Outside Resonance

For frequencies lower than the resonant frequency f_o, the series-resonant circuit represents a high capacitive impedance. On the other hand, for frequencies higher than the resonant frequency f_o, the series-resonant circuit represents a high inductive impedance. If the operating frequency f is close to the resonant frequency f_o, the impedance of the resonant circuit is very high for higher harmonics, and therefore the current through the resonant circuit is approximately sinusoidal and equal to the fundamental component

$$i = I_m \sin(\omega t - \psi) \quad (4.145)$$

where from (4.57), (4.59), and (4.21)

$$I_m = \frac{V_m}{|Z|} = \frac{2V_I}{\pi|Z|} = \frac{2V_I \cos\psi}{\pi R_t} = \frac{2V_I}{\pi R_t \sqrt{1 + Q_L^2 \left(\frac{\omega}{\omega_o} - \frac{\omega_o}{\omega}\right)^2}}$$

$$= \frac{2V_I}{\pi Z_o \sqrt{\left(\frac{R_t}{Z_o}\right)^2 + \left(\frac{\omega}{\omega_o} - \frac{\omega_o}{\omega}\right)^2}}. \quad (4.146)$$

Figure 4.20 shows a three-dimensional representation of $I_m/(V_I/Z_o)$ as a function of f/f_o and Q_L. Plots of $I_m/(V_I/Z_o)$ as a function of f/f_o at fixed values of Q_L are depicted in Fig. 4.21. It can be seen that high values of $I_m/(V_I/Z_o)$ occur at the resonant frequency f_o and at low total resistance R_t. At $f = f_o$, the amplitude of the current through the resonant circuit and the transistors becomes

$$I_{mr} = \frac{2V_I}{\pi R_t} \quad (4.147)$$

The output voltage is sinusoidal

$$v_o = iR = V_m \sin(\omega t - \psi) \quad (4.148)$$

The input current of the amplifier i_I equals the current through the switch S_1 and is given by

$$i_I = i_{S1} = \begin{cases} I_m \sin(\omega t - \psi) & \text{for} \quad 0 < \omega t \le \pi \\ 0 & \text{for} \quad \pi < \omega t \le 2\pi \end{cases} \quad (4.149)$$

Hence, from (4.55), (4.57), and (4.21), one obtains the dc component of the input current

$$I_I = \frac{1}{2\pi} \int_0^{2\pi} i_{S1} d(\omega t) = \frac{I_m}{2\pi} \int_0^{\pi} \sin(\omega t - \psi) d(\omega t) = \frac{I_m \cos\psi}{\pi} = \frac{V_m \cos\psi}{\pi|Z|}$$

$$= \frac{2V_I \cos^2\psi}{\pi^2 R_t} = \frac{2V_I R_t}{\pi^2 |Z|^2} = \frac{I_m}{\pi\sqrt{1 + Q_L^2 \left(\frac{\omega}{\omega_o} - \frac{\omega_o}{\omega}\right)^2}} = \frac{2V_I}{\pi^2 R_t \left[1 + Q_L^2 \left(\frac{\omega}{\omega_o} - \frac{\omega_o}{\omega}\right)^2\right]}$$

(4.150)

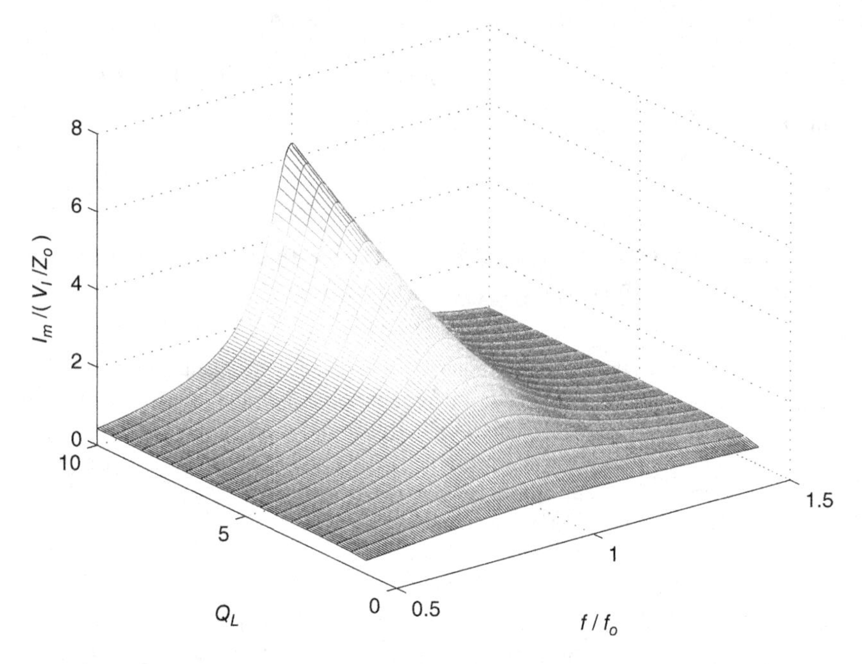

Figure 4.20 Normalized amplitude $I_m/(V_I/Z_o)$ of the current through the resonant circuit as a function of f/f_o and Q_L.

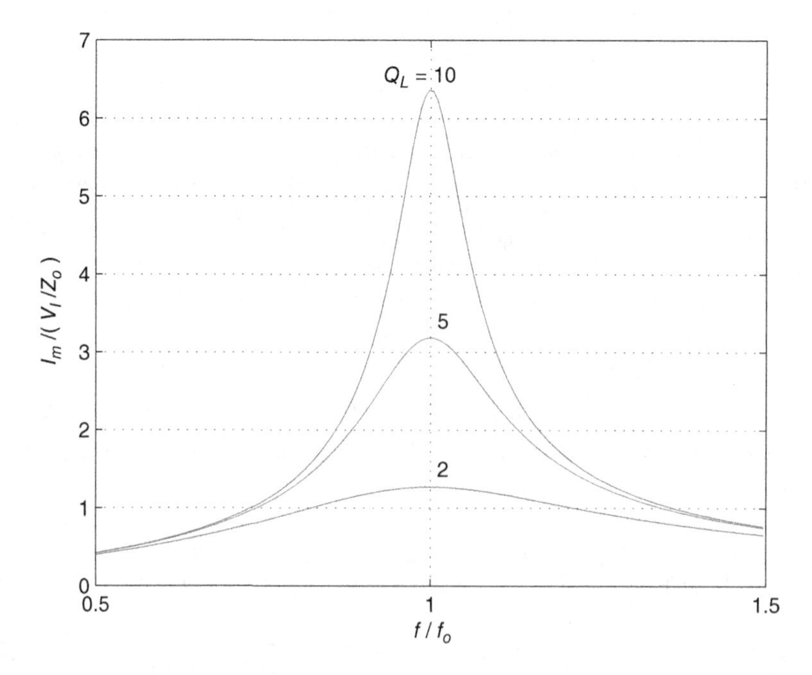

Figure 4.21 Normalized amplitude $I_m/(V_I/Z_o)$ of the current in the resonant circuit as a function of f/f_o at fixed values of Q_L.

At $f = f_o$,

$$I_I = \frac{I_m}{\pi} = \frac{2V_I}{\pi^2 R_t} \approx \frac{V_I}{5R_t} \tag{4.151}$$

Hence,

$$\gamma_1 = \frac{I_m}{I_I} = \pi\sqrt{1 + Q_L^2\left(\frac{\omega}{\omega_o} - \frac{\omega_o}{\omega}\right)^2} = \pi \cos\psi \tag{4.152}$$

At $f = f_o$,

$$\gamma_1 = \frac{I_m}{I_I} = \pi \tag{4.153}$$

The dc input power can be expressed as

$$P_I = I_I V_I = \frac{2V_I^2 \cos^2\psi}{\pi^2 R_t} = \frac{2V_I^2}{\pi^2 R_t\left[1 + Q_L^2\left(\frac{\omega}{\omega_o} - \frac{\omega_o}{\omega}\right)^2\right]} = \frac{2V_I^2 R_t}{\pi^2 Z_o^2\left[\left(\frac{R_t}{Z_o}\right)^2 + \left(\frac{\omega}{\omega_o} - \frac{\omega_o}{\omega}\right)^2\right]} \tag{4.154}$$

At $f = f_o$,

$$P_I = \frac{2V_I^2}{\pi^2 R_t} \approx \frac{V_I^2}{5R_t} \tag{4.155}$$

Using (4.146), one arrives at the output power

$$P_O = \frac{I_m^2 R}{2} = \frac{2V_I^2 R \cos^2\psi}{\pi^2 R_t^2} = \frac{2V_I^2 R}{\pi^2 R_t^2\left[1 + Q_L^2\left(\frac{\omega}{\omega_o} - \frac{\omega_o}{\omega}\right)^2\right]}$$

$$= \frac{2V_I^2 R}{\pi^2 Z_o^2\left[\left(\frac{R_t}{Z_o}\right)^2 + \left(\frac{\omega}{\omega_o} - \frac{\omega_o}{\omega}\right)^2\right]} \tag{4.156}$$

At $f = f_o$,

$$P_O = \frac{2V_I^2 R}{\pi^2 R_t^2} \approx \frac{2V_I^2}{\pi^2 R} \tag{4.157}$$

Figure 4.22 depicts $P_O/(V_I^2 R_t/Z_o^2)$ as a function of f/f_o and Q_L. The normalized output power $P_O/(V_I^2 R_t/Z_o^2)$ is plotted as a function of f/f_o at fixed values of Q_L in Fig. 4.23. The maximum output power occurs at the resonant frequency f_o and at low total resistance R_t.

The drain efficiency of the Class D RF power amplifier with ideal components is

$$\eta_D = \frac{P_O}{P_I} = \left(\frac{1}{2}\right)\left(\frac{I_m}{I_I}\right)\left(\frac{V_m}{V_I}\right)\cos\psi = \frac{1}{2}\gamma_1\xi_1\cos\psi$$

$$= \left(\frac{1}{2}\right)\left(\frac{2}{\pi}\right)(\pi)\sqrt{1 + Q_L^2\left(\frac{\omega}{\omega_o} - \frac{\omega_o}{\omega}\right)^2} \times \frac{1}{\sqrt{1 + Q_L^2\left(\frac{\omega}{\omega_o} - \frac{\omega_o}{\omega}\right)^2}} = 1 \tag{4.158}$$

The drain efficiency of the Class D RF power amplifier with ideal components at the resonant frequency f_o is

$$\eta_D = \frac{P_O}{P_I} = \left(\frac{1}{2}\right)\left(\frac{I_m}{I_I}\right)\left(\frac{V_m}{V_I}\right) = \frac{1}{2}\gamma_1\xi_1 = \left(\frac{1}{2}\right)\left(\frac{2}{\pi}\right)(\pi) = 1 \tag{4.159}$$

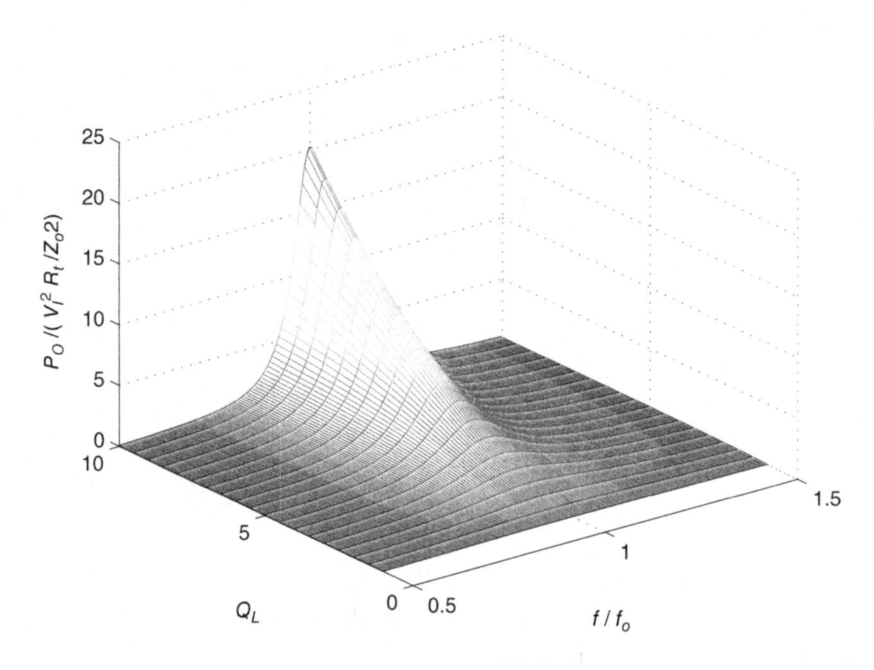

Figure 4.22 Normalized output power $P_O/(V_I^2 R_t/Z_o^2)$ as a function of f/f_o and Q_L.

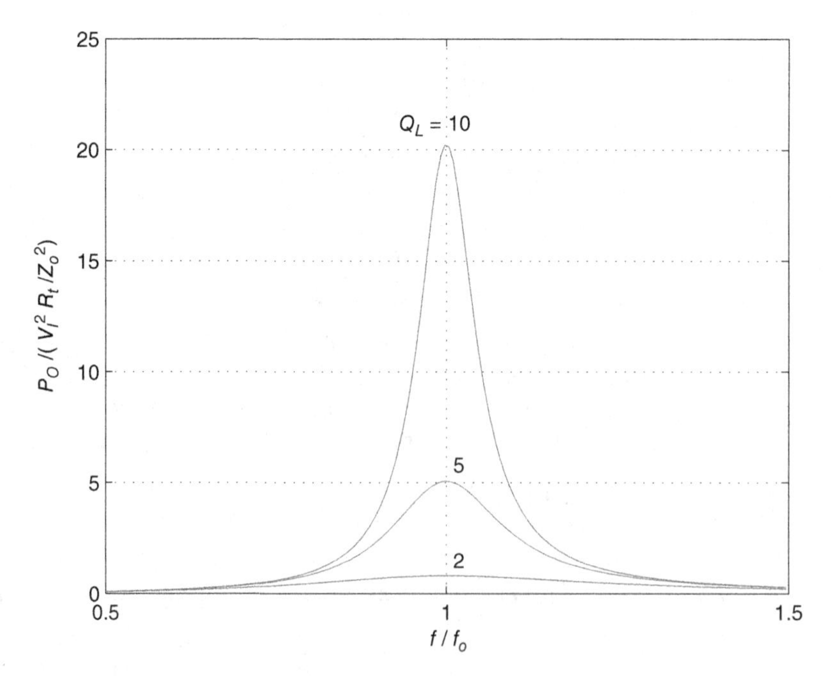

Figure 4.23 Normalized output power $P_O/(V_I^2 R_t/Z_o^2)$ as a function of f/f_o at fixed values of Q_L.

4.10.1 Current and Voltage Stresses

The peak voltage across each switch is equal to the dc input voltage

$$V_{SM} = V_I \tag{4.160}$$

The maximum value of the switch peak currents and the maximum amplitude of the current through the resonant circuit occurs at $f = f_o$. Hence, from (4.146)

$$I_{SM} = I_{mr} = \frac{V_m}{R_t} = \frac{2V_I}{\pi R_t} \tag{4.161}$$

The ratio of the input voltage of the resonant circuit to the voltage across the capacitor C is the voltage transfer function of the second-order low-pass filter. The amplitude of the voltage across the capacitor C is obtained from (4.146)

$$V_{Cm} = \frac{I_m}{\omega C} = \frac{2V_I}{\pi\left(\frac{\omega}{\omega_o}\right)\sqrt{\left(\frac{R_t}{Z_o}\right)^2 + \left(\frac{\omega}{\omega_o} - \frac{\omega_o}{\omega}\right)^2}} \tag{4.162}$$

A three-dimensional representation of V_{Cm}/V_I is shown in Fig. 4.24. Figure 4.25 shows plots of V_{Cm}/V_I as a function of f/f_o at fixed values of Q_L.

The ratio of the input voltage of the resonant circuit to the voltage across the inductor L is the transfer function of the second-order high-pass filter. The amplitude of the voltage across the inductor L is expressed as

$$V_{Lm} = \omega L I_m = \frac{2V_I\left(\frac{\omega}{\omega_o}\right)}{\pi\sqrt{\left(\frac{R_t}{Z_o}\right)^2 + \left(\frac{\omega}{\omega_o} - \frac{\omega_o}{\omega}\right)^2}} \tag{4.163}$$

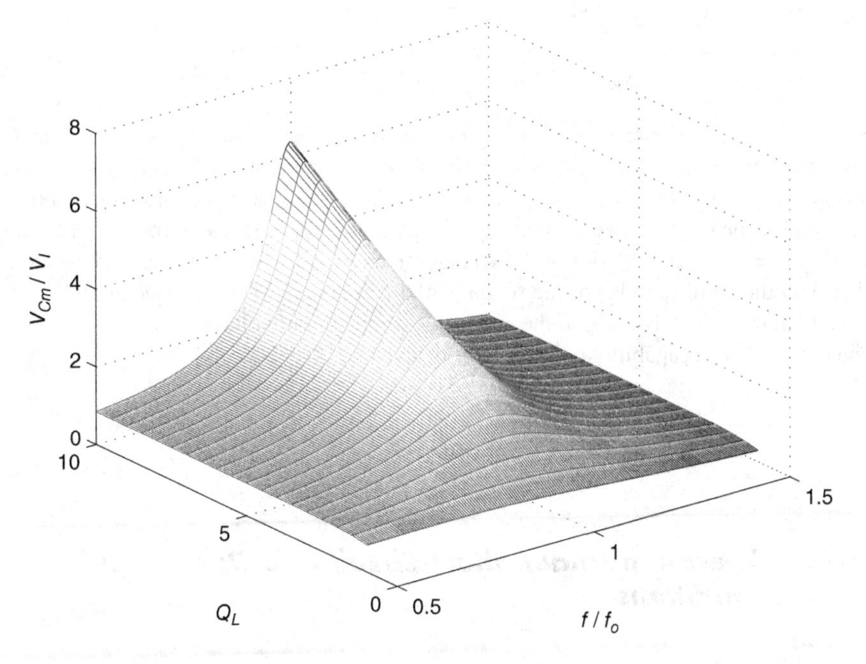

Figure 4.24 Normalized amplitude V_{Cm}/V_I of the voltage across the resonant capacitor C as a function of f/f_o and Q_L.

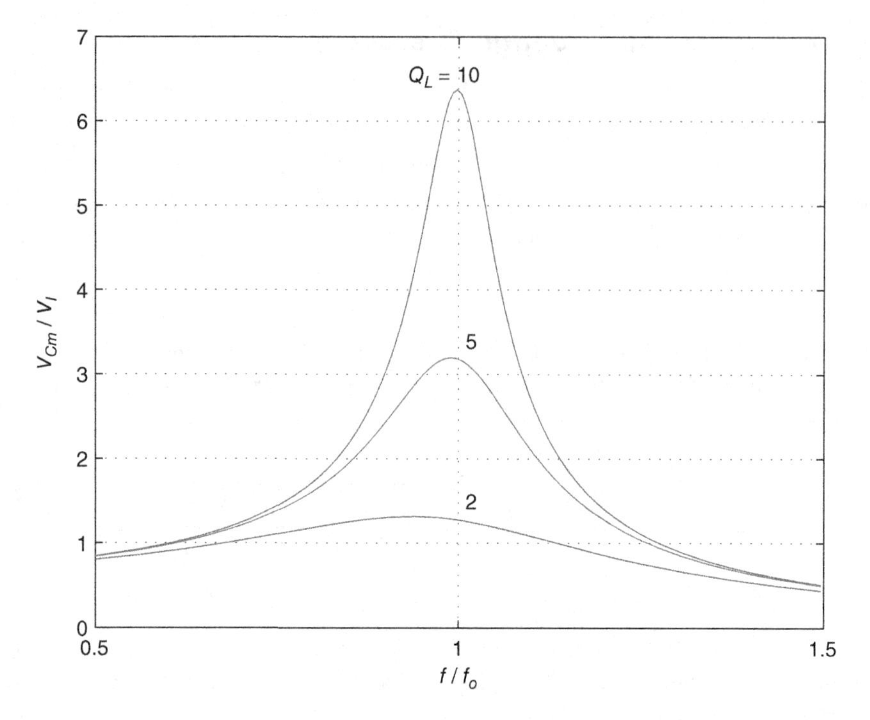

Figure 4.25 Normalized amplitude V_{Cm}/V_I of the voltage across the resonant capacitor C as a function of f/f_o at fixed values of Q_L.

Figure 4.26 shows V_{Lm}/V_I as a function of f/f_o and Q_L. Plots of V_{Lm}/V_I as a function of f/f_o at constant values of Q_L are displayed in Fig. 4.27. At $f = f_o$,

$$V_{Cm(\max)} = V_{Lm(\max)} = Z_o I_{mr} = Q_L V_m = \frac{2V_I Q_L}{\pi} \tag{4.164}$$

The maximum voltage stress of the resonant components occurs at the resonant frequency $f \approx f_o$, at a maximum dc input voltage $V_I = V_{Imax}$, and at a maximum loaded quality factor Q_L. Actually, the maximum value of V_{Lm} occurs slightly above f_o, and the maximum value of V_{Cm} occurs slightly below f_o. However, this effect is negligible for practical purposes. At the resonant frequency $f = f_o$, the amplitudes of the voltages across the inductor and capacitor are Q_L times higher than the amplitude V_m of the fundamental component of the voltage at the input of the resonant circuit, which is equal to the amplitude of the output voltage V_{om}.

The output-power capability per transistor is give by

$$c_p = \frac{P_O}{2I_{SM}V_{SM}} = \frac{1}{4}\left(\frac{I_m}{I_{SM}}\right)\left(\frac{V_m}{V_{SM}}\right) = \frac{1}{4} \times 1 \times \left(\frac{2}{\pi}\right) = \frac{1}{2\pi} = 0.159 \tag{4.165}$$

4.10.2 Operation Under Short-Circuit and Open-Circuit Conditions

The Class D amplifier with a series-resonant circuit can operate safely with an open circuit at the output. However, it is prone to catastrophic failure if the output is short-circuited at f close

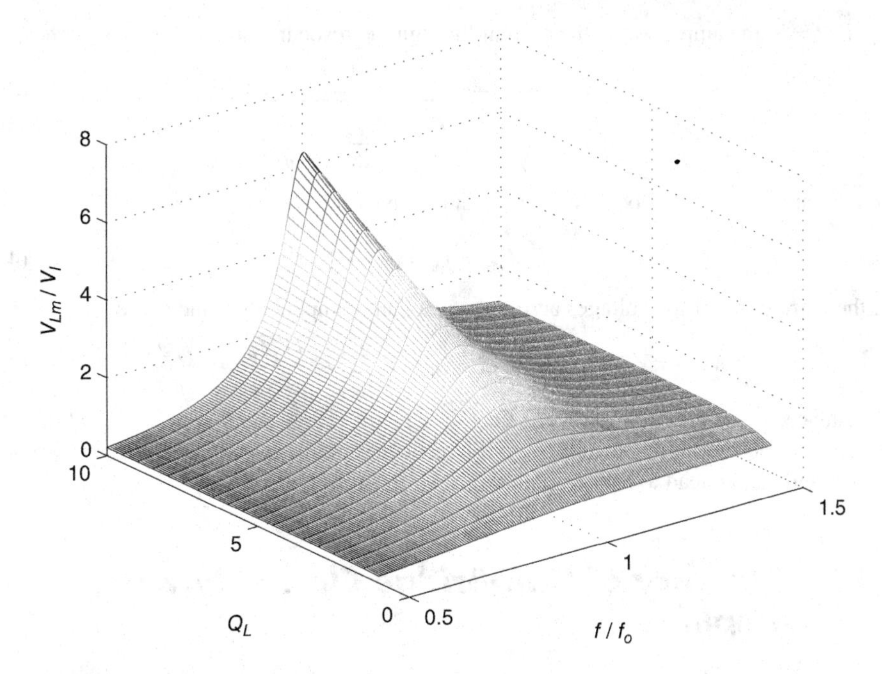

Figure 4.26 Normalized amplitude V_{Lm}/V_I of the voltage across the resonant inductor L as a function of f/f_o and Q_L.

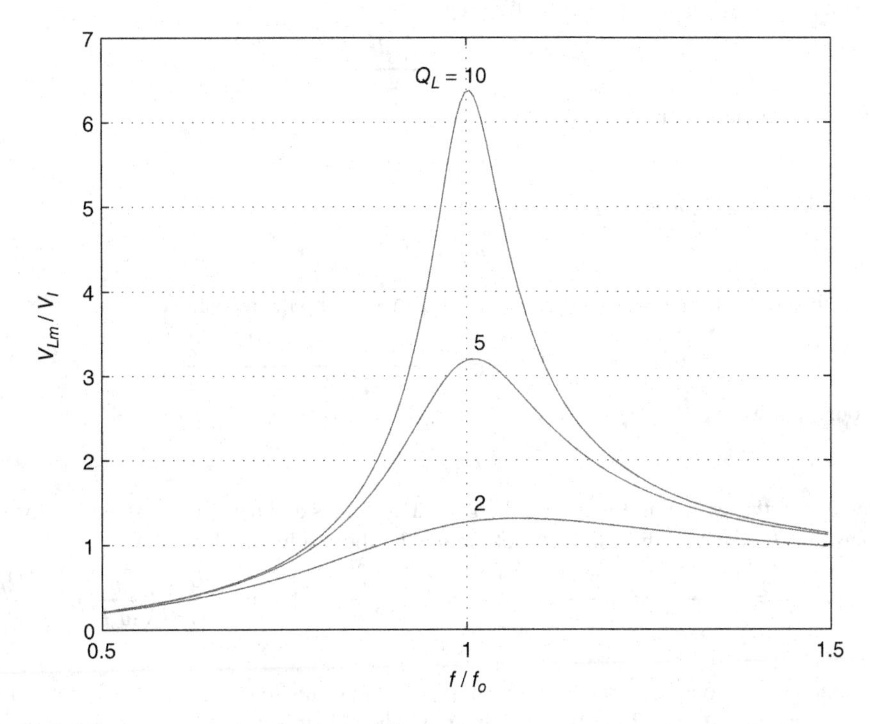

Figure 4.27 Normalized amplitude V_{Lm}/V_I of the voltage across the resonant inductor L as a function of f/f_o at fixed values of Q_L.

to f_o. If $R = 0$, the amplitude of the current through the resonant circuit and the switches is

$$I_m = \frac{2V_I}{\pi r\sqrt{1+\left(\frac{Z_o}{r}\right)^2\left(\frac{\omega}{\omega_o}-\frac{\omega_o}{\omega}\right)^2}} \tag{4.166}$$

The maximum value of I_m occurs at $f = f_o$ and is given by

$$I_{SM} = I_{mr} = \frac{2V_I}{\pi r} \tag{4.167}$$

and the amplitudes of the voltages across the resonant components L and C are

$$V_{Cm} = V_{Lm} = \frac{I_{mr}}{\omega_o C} = \omega_o L I_{mr} = Z_o I_{mr} = \frac{2V_I Z_o}{\pi r} = \frac{2V_I Q_o}{\pi} \tag{4.168}$$

For instance, if $V_I = 320$ V and $r = 2\ \Omega$, $I_{SM} = I_{mr} = 102$ A and $V_{Cm} = V_{Lm} = 80$ kV. Thus, the excessive current in the switches and the resonant circuit, as well as the excessive voltages across L and C, can lead to catastrophic failure of the amplifier.

4.11 Efficiency of Half-Bridge Class D Power Amplifier

4.11.1 Conduction Losses

The conduction loss for the power MOSFET is

$$P_{rDS} = \frac{r_{DS} I_m^2}{4} \tag{4.169}$$

for the resonant inductor is

$$P_{rL} = \frac{r_L I_m^2}{2} \tag{4.170}$$

and for the resonant capacitor is

$$P_{rC} = \frac{r_C I_m^2}{2} \tag{4.171}$$

Hence, the conduction power loss in both the transistors and the resonant circuit is

$$P_r = 2P_{rDS} + P_{rL} + P_{rC} = \frac{(r_{DS} + r_L + r_C) I_m^2}{2} = \frac{r I_m^2}{2} \tag{4.172}$$

The output power is

$$P_O = \frac{I_m^2 R}{2} \tag{4.173}$$

Neglecting the switching losses and the gate drive losses and using (4.154) and (4.156), one obtains the efficiency of the amplifier determined by the conduction losses only

$$\eta_{Ir} = \frac{P_O}{P_I} = \frac{P_O}{P_O + P_r} = \frac{R}{R+r} = \frac{1}{1+\frac{r}{R}} = 1 - \frac{r}{R+r} = 1 - \left(\frac{\omega_o L}{R+r}\right)\left(\frac{r}{\omega_o L}\right) 1 - \frac{Q_L}{Q_o} \tag{4.174}$$

Note that in order to achieve a high efficiency, the ratio of the load resistance R to the parasitic resistance r must be high. The efficiency is high, when Q_L is low and Q_o is high. For example, for $Q_L = 5$ and $Q_o = 200$, the efficiency is $\eta_{Ir} = 1 - 5/200 = 0.975$.

The turn-on loss for operation below resonance is given in the next section and the turn-off loss for operation above resonance is given in Section 4.11.3. Expressions for the efficiency for the two cases are given in those sections.

4.11.2 Turn-On Switching Loss

For operation below resonance, the turn-off switching loss is zero; however, there is a turn-on switching loss. This loss is associated with the charging and discharging of the output capacitances of the MOSFETs. The diode junction capacitance is

$$C_j(v_D) = \frac{C_{j0}}{\left(1 - \frac{v_D}{V_B}\right)^m} = \frac{C_{j0}V_B^m}{(V_B - v_D)^m}, \quad \text{for} \quad v_D \leq V_B \tag{4.175}$$

where C_{j0} is the junction capacitance at $v_D = 0$ and m is the grading coefficient; $m = 1/2$ for step junctions and $m = 1/3$ for graded junctions. The barrier potential is

$$V_B = V_T \ln\left(\frac{N_A N_D}{n_i^2}\right) \tag{4.176}$$

where n_i is the intrinsic carrier density (1.5×10^{10} cm^{-3} for silicon at 25 °C), N_A is the acceptor concentration, and N_D is the donor concentration. The thermal voltage is

$$V_T = \frac{kT}{q} = \frac{T}{11{,}609} \ \text{(V)} \tag{4.177}$$

where $k = 1.38 \times 10^{-23}$ J/K is Boltzmann's constant, $q = 1.602 \times 10^{-19}$ C is the charge per electron, and T is the absolute temperature in K. For p^+n diodes, a typical value of the acceptor concentration is $N_A = 10^{16}$ cm^{-3}, and a typical value of the donor concentration is $N_D = 10^{14}$ cm^{-3}, which gives $V_B = 0.57$ V. The zero-voltage junction capacitance is given by

$$C_{j0} = A\sqrt{\frac{\varepsilon_r\varepsilon_o q}{2V_B\left(\frac{1}{N_D} + \frac{1}{N_A}\right)}} \approx A\sqrt{\frac{\varepsilon_r\varepsilon_o q N_D}{2V_B}} \quad \text{for} \quad N_D \ll N_A \tag{4.178}$$

where A is the junction area in cm^2, $\varepsilon_r = 11.7$ for silicon, and $\varepsilon_o = 8.85 \times 10^{-14}$ (F/cm). Hence, $C_{j0}/A = 3.1234 \times 10^{-16}\sqrt{N_D}$ (F/cm^2). For instance, if $N_D = 10^{14}$ cm^{-3}, $C_{j0}/A \approx 3$ nF/cm^2. Typical values of C_{j0} are of the order of 1 nF for power diodes.

The MOSFET's drain–source capacitance C_{ds} is the capacitance of the body–drain *pn* step junction diode. Setting $v_D = -v_{DS}$ and $m = 1/2$, one obtains from (4.175)

$$C_{ds}(v_{DS}) = \frac{C_{j0}}{\sqrt{1 + \frac{v_{DS}}{V_B}}} = C_{j0}\sqrt{\frac{V_B}{v_{DS} + V_B}} \quad \text{for} \quad v_{DS} \geq -V_B \tag{4.179}$$

Hence,

$$\frac{C_{ds1}}{C_{ds2}} = \sqrt{\frac{v_{DS2} + V_B}{v_{DS1} + V_B}} \approx \sqrt{\frac{v_{DS2}}{v_{DS1}}} \tag{4.180}$$

where C_{ds1} is the drain–source capacitance at v_{DS1} and C_{ds2} is the drain–source capacitance at v_{DS2}. Manufacturers of power MOSFETs usually specify the capacitances $C_{oss} = C_{gd} + C_{ds}$ and $C_{rss} = C_{gd}$ at $V_{DS} = 25$ V, $V_{GS} = 0$ V, and $f = 1$ MHz. Thus, the drain–source capacitance at $V_{DS} = 25$ V can be found as $C_{ds(25V)} = C_{oss} - C_{rss}$. The interterminal capacitances of MOSFETs are essentially independent of frequency. From (4.180), the drain–source capacitance at the dc voltage V_I is

$$C_{ds(V_I)} = C_{ds(25V)}\sqrt{\frac{25 + V_B}{V_I + V_B}} \approx \frac{5C_{ds(25V)}}{\sqrt{V_I}} \ \text{(F)} \tag{4.181}$$

The drain–source capacitance at $v_{DS} = 0$ is

$$C_{j0} = C_{ds(25V)}\sqrt{\frac{25}{V_B} + 1} \approx 6.7C_{ds(25V)} \tag{4.182}$$

for $V_B = 0.57$ V. Also,

$$C_{ds}(v_{DS}) = C_{ds(V_I)}\sqrt{\frac{V_I + V_B}{v_{DS} + V_B}} \approx C_{ds(V_I)}\sqrt{\frac{V_I}{v_{DS}}} \tag{4.183}$$

Using (4.179) and $dQ_j = C_{ds}dv_{DS}$, the charge stored in the drain–source junction capacitance at v_{DS} can be found as

$$\begin{aligned} Q_j(v_{DS}) &= \int_{-V_B}^{v_{DS}} dQ_j = \int_{-V_B}^{v_{DS}} C_{ds}(v_{DS})dv_{DS} \\ &= C_{j0}\sqrt{V_B}\int_{-V_B}^{v_{DS}} \frac{dv_{DS}}{\sqrt{v_{DS} + V_B}} = 2C_{j0}\sqrt{V_B(v_{DS} + V_B)} \\ &= 2C_{j0}V_B\sqrt{1 + \frac{v_{DS}}{V_B}} = 2(v_{DS} + V_B)C_{ds}(v_{DS}) \approx 2v_{DS}C_{ds}(v_{DS}) \end{aligned} \tag{4.184}$$

which, by substituting (4.181) at $v_{DS} = V_I$, simplifies to

$$Q_j(V_I) = 2V_IC_{ds(V_I)} = 10C_{ds(25V)}\sqrt{V_I}(\mathrm{C}) \tag{4.185}$$

Hence, the energy transferred from the dc input source V_I to the output capacitance of the upper MOSFET after the upper transistor is turned off is

$$\begin{aligned} W_I &= \int_{-V_B}^{V_I} vidt = V_I\int_{-V_B}^{V_I} idt = V_IQ_j(V_I) \\ &= 2V_I^2C_{ds(V_I)} = 10\sqrt{V_I^3}C_{ds(25V)}(\mathrm{W}) \end{aligned} \tag{4.186}$$

Using $dW_j = (1/2)Q_jdv_{DS}$ and (4.184), the energy stored in the drain–source junction capacitance C_{ds} at v_{DS} is

$$\begin{aligned} W_j(v_{DS}) &= \frac{1}{2}\int_{-V_0}^{v_{DS}} Q_jdv_{DS} = C_{j0}\sqrt{V_B}\int_{-V_0}^{v_{DS}} \sqrt{v_{DS} + V_B}dv_{DS} \\ &= \frac{2}{3}C_{j0}\sqrt{V_B}(v_{DS} + V_B)^{\frac{3}{2}} = \frac{2}{3}C_{ds}(v_{DS})(v_{DS} + V_B)^2 \approx \frac{2}{3}C_{ds}(v_{DS})v_{DS}^2 \end{aligned} \tag{4.187}$$

Hence, from (4.181) the energy stored in the drain–source junction capacitance at $v_{DS} = V_I$ is

$$W_j(V_I) = \frac{2}{3}C_{ds(V_I)}V_I^2 = \frac{10}{3}C_{ds(25V)}\sqrt{V_I^3}(\mathrm{J}) \tag{4.188}$$

This energy is lost as heat when the transistor turns on and the capacitor is discharged through r_{DS}, resulting in the turn-on switching power loss per transistor

$$\begin{aligned} P_{tron} &= \frac{W_j(V_I)}{T} = fW_j(V_I) = \frac{2}{3}fC_{j0}\sqrt{V_B}(V_I + V_B)^{\frac{3}{2}} \\ &= \frac{2}{3}fC_{ds(V_I)}V_I^2 = \frac{10}{3}fC_{ds(25V)}\sqrt{V_I^3}(\mathrm{W}) \end{aligned} \tag{4.189}$$

Figure 4.28 shows C_{ds}, Q_j, and W_j as functions of v_{DS} given by (4.181), (4.184), and (4.187).

Using (4.186) and (4.188), one arrives at the energy lost in the resistances of the charging path during the charging process of the capacitance C_{ds}

$$W_{char}(V_I) = W_I(V_I) - W_j(V_I) = \frac{4}{3}C_{ds(V_I)}V_I^2 = \frac{20}{3}C_{ds(25V)}\sqrt{V_I^3}(\mathrm{W}) \tag{4.190}$$

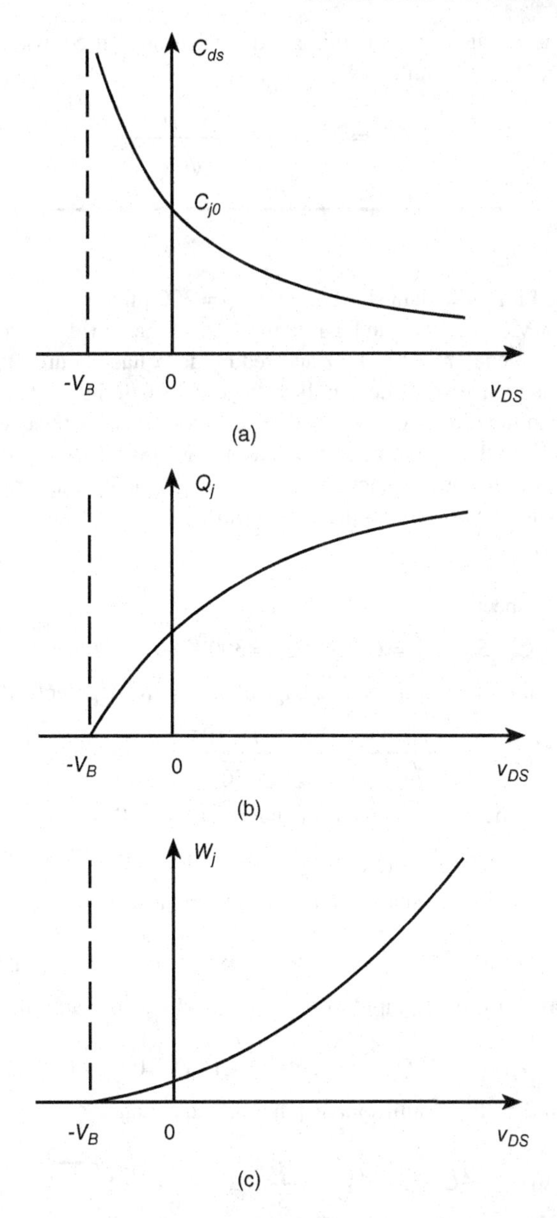

Figure 4.28 Plots of C_{ds}, Q_j, and W_j versus v_{DS}. (a) C_{ds} versus v_{DS}. (b) Q_j versus v_{DS}. (c) W_j versus v_{DS}.

and the corresponding power associated with the charging of the capacitance C_{ds} is

$$P_{char} = \frac{W_{char}(V_I)}{T} = fW_{char}(V_I) = \frac{4}{3}fC_{ds(V_I)}V_I^2 = \frac{20}{3}fC_{ds(25V)}\sqrt{V_I^3}(\mathrm{J}) \tag{4.191}$$

From (4.186), one arrives at the total switching power loss per transistor

$$P_{sw} = \frac{W(V_I)}{T} = fW_I(V_I) = 2fC_{j0}V_I\sqrt{V_B(V_I + V_B)}$$

$$= 2fC_{ds(V_I)}V_I^2 = 10fC_{ds(25V)}\sqrt{V_I^3}(\mathrm{W}) \tag{4.192}$$

The switching loss associated with charging and discharging an equivalent linear capacitance C_{eq} is $P_{sw} = fC_{eq}V_I^2$. Hence, from (4.192)

$$C_{eq} = 2C_{ds(V_I)} = \frac{10C_{ds(25V)}}{\sqrt{V_I}} \tag{4.193}$$

Example 4.2

For MTP5N40 MOSFETs, the data sheets give $C_{oss} = 300$ pF and $C_{rss} = 80$ pF at $V_{DS} = 25$ V and $V_{GS} = 0$ V. These MOSFETs are to be used in a Class D half-bridge series-resonant amplifier that is operated at frequency $f = 100$ kHz and fed by dc voltage source $V_I = 350$ V. Calculate the drain–source capacitance at the dc supply voltage V_I, the drain–source capacitance at $v_{DS} = 0$, the charge stored in the drain–source junction capacitance at V_I, the energy transferred from the dc input source V_I to the output capacitance of the MOSFET during turn-on transition, the energy stored in the drain–source junction capacitance C_{ds} at V_I, the turn-on switching power loss, and the total switching power loss per transistor in the amplifier operating below resonance. Assume $V_B = 0.57$ V.

Solution. Using data sheets,

$$C_{ds(25V)} = C_{oss} - C_{rss} = 300 - 80 = 220 \text{ pF} \tag{4.194}$$

From (4.181), one obtains the drain–source capacitance at the dc supply voltage $V_I = 350$ V

$$C_{ds(V_I)} = \frac{5C_{ds(25V)}}{\sqrt{V_I}} = \frac{5 \times 220 \times 10^{-12}}{\sqrt{350}} = 58.79 \approx 59 \text{ pF} \tag{4.195}$$

Equation (4.182) gives the drain–source capacitance at $v_{DS} = 0$

$$C_{j0} = 6.7C_{ds(25V)} = 6.7 \times 220 \times 10^{-12} = 1474 \text{ pF} \tag{4.196}$$

The charge stored in the drain–source junction capacitance at $V_I = 350$ V is obtained from (4.185)

$$Q_j(V_I) = 2V_I C_{ds(V_I)} = 2 \times 350 \times 59 \times 10^{-12} = 41.3 \text{ nC} \tag{4.197}$$

The energy transferred from the input voltage source V_I to the amplifier is calculated from (4.186) as

$$W_I(V_I) = V_I Q_j(V_I) = 350 \times 41.153 \times 10^{-9} = 14.4 \text{ μJ} \tag{4.198}$$

and the energy stored in the drain–source junction capacitance C_{ds} at V_I is calculated from (4.188) as

$$W_j(V_I) = \frac{10}{3}C_{ds(25V)}\sqrt{V_I^3} = \frac{10}{3} \times 220 \times 10^{-12}\sqrt{350^3} = 4.8 \text{ μJ} \tag{4.199}$$

Using (4.191), the power associated with charging the capacitance C_{ds} is calculated as

$$P_{char} = \frac{20}{3}fC_{ds(25V)}\sqrt{V_I^3} = \frac{20}{3} \times 10^5 \times 220 \times 10^{-12}\sqrt{350^3} = 0.96 \text{ W} \tag{4.200}$$

From (4.189), the turn-on switching power loss per transistor for operating below resonance is

$$P_{tron} = \frac{10}{3}fC_{ds(25V)}\sqrt{V_I^3} = \frac{10}{3} \times 10^5 \times 220 \times 10^{-12}\sqrt{350^3} = 0.48 \text{ W} \tag{4.201}$$

Using (4.192), one arrives at the total switching power loss per transistor for operating below resonance

$$P_{sw} = 10fC_{ds(25V)}\sqrt{V_I^3}(\text{W}) = 10 \times 10^5 \times 220 \times 10^{-12}\sqrt{350^3} = 1.44 \text{ W} \tag{4.202}$$

Note that $P_{tron} = \frac{1}{3}P_{sw}$ and $P_{char} = \frac{2}{3}P_{sw}$. The equivalent linear capacitance is $C_{eq} = 2C_{ds(V_I)} = 2 \times 59 = 118$ pF.

The overall power dissipation in the Class D amplifier is

$$P_T = P_r + 2P_{sw} + 2P_G = \frac{rI_m^2}{2} + 20fC_{ds(25V)}\sqrt{V_I^3} + 2fQ_gV_{GSpp} \tag{4.203}$$

Hence, the efficiency of the half-bridge amplifier for operating below resonance is

$$\eta_I = \frac{P_O}{P_O + P_T} = \frac{P_O}{P_O + P_r + 2P_{sw} + 2P_G} \tag{4.204}$$

4.11.3 Turn-Off Switching Loss

For operation above resonance, the turn-on switching loss is zero, but there is a turn-off switching loss. The switch current and voltage waveforms during turn-off for $f > f_o$ are shown in Fig. 4.29. These waveforms were observed in various Class D experimental circuits. Notice that the voltage

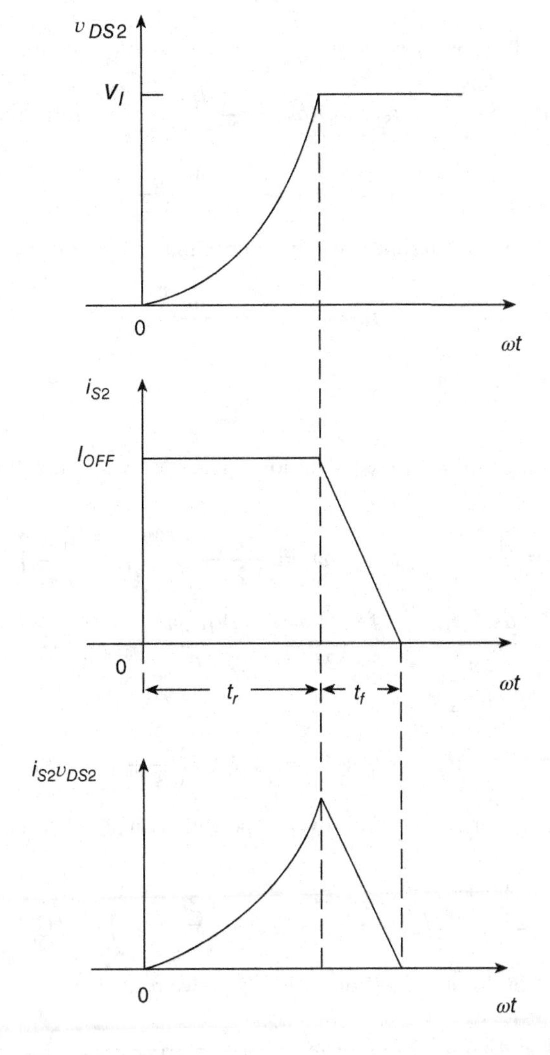

Figure 4.29 Waveforms of v_{DS2}, i_{S2}, and $i_{S2}v_{DS2}$ during turn-off for $f > f_o$.

v_{DS2} increases slowly at its lower values and much faster at its higher values. This is because the MOSFET output capacitance is highly nonlinear, and it is much higher at low voltage v_{DS2} than at high voltage v_{DS2}. The current that charges this capacitance is approximately constant. The drain-to-source voltage v_{DS2} during voltage rise time t_r can be approximated by a parabolic function

$$v_{DS2} = a(\omega t)^2 \tag{4.205}$$

Because $v_{DS2}(\omega t_r) = V_I$, one obtains

$$a = \frac{V_I}{(\omega t_r)^2} \tag{4.206}$$

Hence, (4.205) becomes

$$v_{DS2} = \frac{V_I(\omega t)^2}{(\omega t_r)^2} \tag{4.207}$$

The switch current during rise time t_r is a small portion of a sinusoid and can be approximated by a constant

$$i_{S2} = I_{OFF} \tag{4.208}$$

The average value of the power loss associated with the voltage rise time t_r is

$$\begin{aligned} P_{tr} &= \frac{1}{2\pi}\int_0^{2\pi} i_{S2} v_{DS2} d(\omega t) = \frac{V_I I_{OFF}}{2\pi(\omega t_r)^2}\int_0^{\omega t_r} (\omega t)^2 d(\omega t) \\ &= \frac{\omega t_r V_I I_{OFF}}{6\pi} = \frac{f t_r V_I I_{OFF}}{3} = \frac{t_r V_I I_{OFF}}{3T} \end{aligned} \tag{4.209}$$

The switch current during fall time t_f can be approximated by a ramp function

$$i_{S2} = I_{OFF}\left(1 - \frac{\omega t}{\omega t_f}\right) \tag{4.210}$$

and the drain-to-source voltage is

$$v_{DS2} = V_I \tag{4.211}$$

which yields the average value of the power loss associated with the fall time t_f of the current of the semiconductor device

$$\begin{aligned} P_{tf} &= \frac{1}{2\pi}\int_0^{2\pi} i_{S2} v_{DS2} d(\omega t) = \frac{V_I I_{OFF}}{2\pi}\int_0^{\omega t_f} \left(1 - \frac{\omega t}{\omega t_f}\right) d(\omega t) \\ &= \frac{\omega t_f V_I I_{OFF}}{4\pi} = \frac{f t_f V_I I_{OFF}}{2} = \frac{t_f V_I I_{OFF}}{2T} \end{aligned} \tag{4.212}$$

Hence, the turn-off switching loss is

$$P_{toff} = P_{tr} + P_{tf} = f V_I I_{OFF}\left(\frac{t_r}{3} + \frac{t_f}{2}\right). \tag{4.213}$$

Usually, t_r is much longer than t_f. The overall power dissipation in the Class D half-bridge amplifier is

$$P_T = P_r + 2P_{toff} + 2P_G = \frac{r I_m^2}{2} + f V_I I_{OFF}\left(\frac{2t_r}{3} + t_f\right) + 2f Q_g V_{GSpp} \tag{4.214}$$

Hence, the efficiency of the amplifier for operation above resonance is

$$\eta_I = \frac{P_O}{P_O + P_T} = \frac{P_O}{P_O + P_r + 2P_{toff} + 2P_G} \tag{4.215}$$

4.12 Design Example

A design procedure of the Class D voltage-switching power amplifier with a series-resonant circuit is illustrated by means of an example.

Example 4.3

Design a Class D half-bridge amplifier of Fig. 4.5 that meets the following specifications: $V_I = 100$ V, $P_O = 50$ W, and $f = 110$ kHz. Assume $Q_L = 5.5$, $\psi = 30°$ (that is, $\cos^2\psi = 0.75$), and the efficiency $\eta_{Ir} = 90\%i$. The converter employs IRF621 MOSFETs (International Rectifier) with $r_{DS} = 0.5\ \Omega$, $C_{ds(25V)} = 110$ pF, and $Q_g = 11$ nC. Check the initial assumption about η_{Ir} using $Q_{Lo} = 300$ and $Q_{Co} = 1200$. Estimate switching losses and gate drive power loss assuming $V_{GSpp} = 15$ V.

Solution. From (4.174), the dc input power of the amplifier is

$$P_I = \frac{P_O}{\eta_{Ir}} = \frac{50}{0.9} = 55.56 \text{ W} \tag{4.216}$$

Using (4.154), the overall resistance of the amplifier can be calculated as

$$R_t = \frac{2V_I^2}{\pi^2 P_I}\cos^2\psi = \frac{2\times 100^2}{\pi^2\times 55.56}\times 0.75 = 27.35\ \Omega \tag{4.217}$$

Relationships (4.174) and (4.29) give the load resistance

$$R = \eta_{Ir}R_t = 0.9\times 27.35 = 24.62\ \Omega \tag{4.218}$$

and the maximum total parasitic resistance of the amplifier

$$r = R_t - R = 27.35 - 24.62 = 2.73\ \Omega \tag{4.219}$$

The dc supply current is obtained from (4.154)

$$I_I = \frac{P_I}{V_I} = \frac{55.56}{100} = 0.556 \text{ A} \tag{4.220}$$

The peak value of the switch current is

$$I_m = \sqrt{\frac{2P_O}{R}} = \sqrt{\frac{2\times 50}{24.62}} = 2.02 \text{ A} \tag{4.221}$$

and from (4.160) the peak value of the switch voltage is equal to the input voltage

$$V_{SM} = V_I = 100 \text{ V} \tag{4.222}$$

Using (4.56), one arrives at the ratio f/f_o at full load

$$\frac{f}{f_o} = \frac{1}{2}\left(\frac{\tan\psi}{Q_L} + \sqrt{\frac{\tan^2\psi}{Q_L^2} + 4}\right) = \frac{1}{2}\left[\frac{\tan(30°)}{5.5} + \sqrt{\frac{\tan^2(30°)}{5.5^2} + 4}\right] = 1.054 \tag{4.223}$$

from which

$$f_o = \frac{f}{(f/f_o)} = \frac{110\times 10^3}{1.054} = 104.4 \text{ kHz} \tag{4.224}$$

The values of the reactive components of the resonant circuit are calculated from (4.27) as

$$L = \frac{Q_L R_t}{\omega_o} = \frac{5.5 \times 27.35}{2\pi \times 104.4 \times 10^3} = 229.3\ \mu\text{H} \tag{4.225}$$

$$C = \frac{1}{\omega_o Q_L R_t} = \frac{1}{2\pi \times 104.4 \times 10^3 \times 5.5 \times 27.35} = 10\ \text{nF} \tag{4.226}$$

From (4.26),

$$Z_o = \sqrt{\frac{L}{C}} = \sqrt{\frac{229.3 \times 10^{-6}}{10 \times 10^{-9}}} = 151.427\ \Omega \tag{4.227}$$

The maximum voltage stresses for the resonant components can be approximated using (4.164)

$$V_{Cm(max)} = V_{Lm(max)} = \frac{2V_I Q_L}{\pi} = \frac{2 \times 100 \times 5.5}{\pi} = 350\ \text{V} \tag{4.228}$$

Once the values of the resonant components are known, the parasitic resistance of the amplifier can be recalculated. From (4.52) and (4.53),

$$r_L = \frac{\omega L}{Q_{Lo}} = \frac{2\pi \times 110 \times 10^3 \times 229.3 \times 10^{-6}}{300} = 0.53\ \Omega \tag{4.229}$$

and

$$r_C = \frac{1}{\omega C Q_{Co}} = \frac{1}{2\pi \times 110 \times 10^3 \times 10 \times 10^{-9} \times 1200} = 0.12\ \Omega \tag{4.230}$$

Thus, the parasitic resistance is

$$r = r_{DS} + r_L + r_C = 0.5 + 0.53 + 0.12 = 1.15\ \Omega \tag{4.231}$$

From (4.169), the conduction loss in each MOSFET is

$$P_{rDS} = \frac{r_{DS} I_m^2}{4} = \frac{0.5 \times 2.02^2}{4} = 0.51\ \text{W} \tag{4.232}$$

Using (4.170), the conduction loss in the resonant inductor L is

$$P_{rL} = \frac{r_L I_m^2}{2} = \frac{0.53 \times 2.02^2}{2} = 1.08\ \text{W} \tag{4.233}$$

From (4.171), the conduction loss in the resonant capacitor C is

$$P_{rC} = \frac{r_C I_m^2}{2} = \frac{0.12 \times 2.02^2}{2} = 0.245\ \text{W} \tag{4.234}$$

Hence, one obtains the overall conduction loss

$$P_r = 2P_{rDS} + P_{rL} + P_{rC} = 2 \times 0.51 + 1.08 + 0.245 = 2.345\ \text{W} \tag{4.235}$$

The efficiency η_{Ir} associated with the conduction losses only at full power is

$$\eta_{Ir} = \frac{P_O}{P_O + P_r} = \frac{50}{50 + 2.345} = 95.52\% \tag{4.236}$$

Assuming the peak-to-peak gate–source voltage $V_{GSpp} = 15$ V, the gate drive power loss in both MOSFETs is

$$2P_G = 2fQ_g V_{GSpp} = 2 \times 110 \times 10^3 \times 11 \times 10^{-9} \times 15 = 0.036\ \text{W} \tag{4.237}$$

The sum of the conduction losses and the gate drive power loss is

$$P_{LS} = P_r + 2P_G = 2.345 + 0.036 = 2.28\ \text{W} \tag{4.238}$$

The turn-on conduction loss is zero because the amplifier is operated above resonance. The efficiency of the amplifier associated with the conduction loss and the gate drive power at full power is

$$\eta_I = \frac{P_O}{P_O + P_T} = \frac{50}{50 + 2.28} = 95.64\% \tag{4.239}$$

4.13 Transformer-Coupled Push–Pull Class D Voltage-Switching RF Power Amplifier

4.13.1 Waveforms

A push–pull Class D voltage-switching (voltage-source) RF power amplifier is shown in Fig. 4.30. Transistors are switched on and off alternately. Current and voltage waveforms are depicted in Fig. 4.31. The drain-to-source voltages v_{DS1} and v_{DS2} are square waves and the drain currents i_{D1} and i_{D2} are half-sine waves.

For the operating frequency equal to the resonant frequency, the series-resonant circuit forces a sinusoidal output current

$$i_o = I_m \sin \omega t \tag{4.240}$$

resulting in a sinusoidal output voltage

$$v_o = V_m \sin \omega t \tag{4.241}$$

where

$$V_m = R_L I_m \tag{4.242}$$

The resistance seen by each transistor across each primary winding with the other primary winding open is given by

$$R = n^2 R_L \tag{4.243}$$

where n is the transformer turn ratio of one primary winding to the secondary winding.

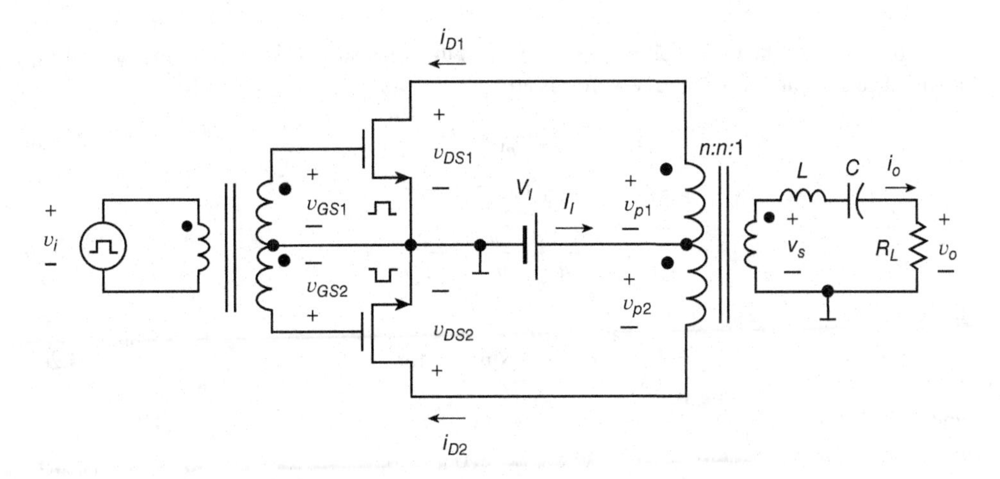

Figure 4.30 Push–pull Class D voltage-switching (voltage-source) RF power amplifier.

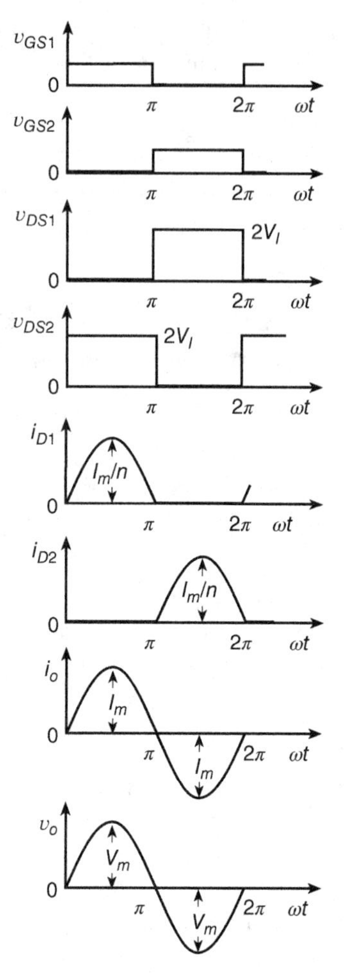

Figure 4.31 Waveforms in push–pull Class D voltage-switching (voltage-source) RF power amplifier.

For $0 < \omega t \leq \pi$, the drive voltage is positive and the transistor Q_1 is ON and the transistor Q_2 is OFF. The current and voltage waveforms are

$$v_{p2} = V_I = v_{p1} \tag{4.244}$$

$$v_{DS1} = V_I + v_{p1} = V_I + V_I = 2V_I \tag{4.245}$$

$$v_s = \frac{v_{p1}}{n} = \frac{v_{p2}}{n} = \frac{V_I}{n} \tag{4.246}$$

and

$$i_{D1} = I_{dm} \sin \omega t = \frac{I_m}{n} \sin \omega t \quad \text{for} \quad 0 < \omega t \leq \pi \tag{4.247}$$

and

$$i_{D2} = 0 \tag{4.248}$$

where I_{dm} is the peak drain current.

For $\pi < \omega t \leq 2\pi$, the drive voltage is negative and transistor Q_1 is OFF and transistor Q_2 is ON. The waveforms for this time interval are

$$v_{p1} = -V_I = v_{p1} \tag{4.249}$$

$$v_{DS2} = V_I - v_{p2} = V_I - (-V_I) = 2V_I \tag{4.250}$$

$$v_s = \frac{v_{p1}}{n} = \frac{v_{p2}}{n} = -\frac{V_I}{n} \tag{4.251}$$

$$i_{D1} = 0 \tag{4.252}$$

and

$$i_{D2} = -I_{dm} \sin \omega t = -\frac{I_m}{n} \sin \omega t \quad \text{for} \quad \pi < \omega t \leq 2\pi \tag{4.253}$$

The amplitude of the fundamental component of the drain-to-source voltage is

$$V_{dm} = \frac{4}{\pi} V_I \tag{4.254}$$

The amplitude of the output voltage is

$$V_m = \frac{V_{dm}}{n} = \frac{4}{\pi} \frac{V_I}{n} \tag{4.255}$$

The current through the dc voltage source V_I is a full-wave rectified sinusoid

$$i_I = i_{D1} + i_{D2} = I_{dm} |\sin \omega t| = \frac{I_m}{n} |\sin \omega t| \tag{4.256}$$

The dc supply current is

$$I_I = \frac{1}{2\pi} \int_0^{2\pi} \frac{I_m}{n} |\sin \omega t| d(\omega t) = \frac{2}{\pi} \frac{I_m}{n} = \frac{2}{\pi} I_{dm} = \frac{2}{\pi} \frac{V_m}{n^2 R_L} = \frac{8}{\pi^2} \frac{V_I}{n^2 R_L} \tag{4.257}$$

The dc resistance presented by the amplifier to the dc supply source V_I is

$$R_{DC} = \frac{V_I}{I_I} = \frac{\pi^2}{8} n^2 R_L = \frac{V_I}{I_I} = \frac{\pi^2}{8} R \tag{4.258}$$

4.13.2 Power

The output power is

$$P_O = \frac{V_m^2}{2R_L} = \frac{8}{\pi^2} \frac{V_I^2}{n^2 R_L} = \frac{8}{\pi^2} \frac{V_I^2}{R} \tag{4.259}$$

The dc supply power is

$$P_I = V_I I_I = \frac{8}{\pi^2} \frac{V_I^2}{n^2 R_L} \tag{4.260}$$

Neglecting conduction losses in the MOSFET on-resistances and switching losses, the drain efficiency is

$$\eta_D = \frac{P_O}{P_I} = 1 \tag{4.261}$$

4.13.3 Current and Voltage Stresses

The MOSFET current and voltage stresses are

$$I_{SM} = \frac{I_m}{n} \tag{4.262}$$

and

$$V_{SM} = 2V_I = \frac{\pi n}{2} V_m \tag{4.263}$$

The output-power capability is

$$c_p = \frac{P_{Omax}}{2I_{DM}V_{DSM}} = \frac{1}{4}\left(\frac{I_M}{I_{DM}}\right)\left(\frac{V_m}{V_{DSM}}\right) = \frac{1}{4} \times n \times \frac{2}{2\pi} = \frac{1}{2\pi} = 0.159 \tag{4.264}$$

4.13.4 Efficiency

The rms value of the drain current is

$$I_{Srms} = \sqrt{\frac{1}{2\pi}\int_{\pi}^{2\pi} \frac{I_m^2}{n^2} d(\omega t)} = \frac{I_m}{n\sqrt{2}} = \frac{I_{dm}}{\sqrt{2}} \tag{4.265}$$

Hence, the conduction power loss in the MOSFET on-resistance r_{DS} is

$$P_{rDS} = r_{DS} I_{Srms}^2 = \frac{r_{DS} I_{dm}^2}{2} = \frac{r_{DS} I_m^2}{2n^2} = \frac{r_{DS}}{4n^2 R_L} P_O \tag{4.266}$$

The drain efficiency is

$$\eta_D = \frac{P_O}{P_O + 2P_{rDS}} = \frac{1}{1 + \frac{2P_{rDS}}{P_O}} = \frac{1}{1 + \frac{r_{DS}}{2n^2 R_L}} \tag{4.267}$$

The power loss in the ESR of the resonant inductor r_L is

$$P_{rL} = \frac{r_L I_m^2 Q_L^2}{2} = \frac{r_L Q_L^2}{R_L} P_O \tag{4.268}$$

and the power loss in the ESR of the resonant capacitor r_C is

$$P_{rC} = \frac{r_C I_m^2 Q_L^2}{2} = \frac{r_C Q_L^2}{R_L} P_O \tag{4.269}$$

The total power loss is

$$P_{Loss} = 2P_{rDS} + P_{rL} + P_{rC} = P_O \left[\frac{r_{DS}}{n^2 R_L} + \frac{Q_L^2 (r_L + r_C)}{R_L}\right] \tag{4.270}$$

Hence, the overall efficiency is

$$\eta = \frac{P_O}{P_I} = \frac{P_O}{P_O + P_{Loss}} = \frac{1}{1 + \frac{r_{DS}}{n^2 R_L} + \frac{Q_L^2 (r_L + r_C)}{R_L}} \tag{4.271}$$

Example 4.4

Design a push–pull Class D voltage-switching RF power amplifier to meet the following specifications: $V_I = 28$ V, $P_O = 50$ W, $R_L = 50\ \Omega$, $BW = 240$ MHz, and $f = 2.4$ GHz. Neglect switching losses.

Solution. Assuming the efficiency of the resonant circuits $\eta_r = 0.96$, the drain power is

$$P_{DS} = \frac{P_O}{\eta_r} = \frac{50}{0.96} = 52.083 \text{ W} \tag{4.272}$$

Assume the minimum value of the drain-to-source voltage $V_{DSmin} = 1$ V. The resistance seen across one part of the primary winding is

$$R = \frac{8}{\pi^2}\frac{(V_I - V_{DSmin})^2}{P_{DS}} = \frac{8}{\pi^2}\frac{(28-1)^2}{52.083} = 11.345 \text{ W} \tag{4.273}$$

The transformer turns ratio is

$$n = \sqrt{\frac{R}{R_L}} = \sqrt{\frac{11.345}{50}} = 0.476 \tag{4.274}$$

Pick $n = \frac{1}{2}$.

The amplitude of the fundamental component of the drain-to-source voltage is

$$V_{dm} = \frac{4}{\pi}V_I = \frac{4}{\pi} \times 28 = 35.65 \text{ V} \tag{4.275}$$

The amplitude of the fundamental component of the drain current is

$$I_{dm} = \frac{V_{dm}}{R} = \frac{35.65}{11.345} = 3.142 \text{ A} \tag{4.276}$$

The dc supply current is

$$I_I = \frac{2}{\pi}I_{dm} = \frac{2}{\pi} \times 3.142 = 2 \text{ A} \tag{4.277}$$

The dc supply power is

$$P_I = V_I I_I = 28 \times 2 = 56 \text{ W} \tag{4.278}$$

The rms value of each MOSFET is

$$I_{Srms} = \frac{I_{dm}}{\sqrt{2}} = \frac{3.142}{\sqrt{2}} = 2.222 \text{ A} \tag{4.279}$$

Assuming $r_{DS} = 0.2\ \Omega$, we obtain the conduction loss in each MOSFET

$$P_{rDS} = r_{DS} I_{Srms}^2 = 0.2 \times 2.222^2 = 0.9875 \text{ W} \tag{4.280}$$

The drain efficiency is

$$\eta_D = \frac{P_O}{P_O + 2P_{rDS}} = \frac{50}{50 + 2 \times 0.9875} = 96.2\% \tag{4.281}$$

The maximum drain current is

$$I_{SM} = I_{dm} = 3.142 \text{ A} \tag{4.282}$$

The maximum drain-to-source voltage is

$$V_{SM} = 2V_I = 2 \times 28 = 56 \text{ V} \tag{4.283}$$

The loaded quality factor is

$$Q_L = \frac{f_c}{BW} = \frac{2400}{240} = 10 \tag{4.284}$$

The resonant inductance is

$$L = \frac{Q_L R_L}{\omega_c} = \frac{10 \times 11.345}{2\pi \times 2.4 \times 10^9} = 7.523 \text{ nH} \tag{4.285}$$

The resonant capacitance is

$$C = \frac{1}{\omega_c Q_L R_L} = \frac{1}{2\pi \times 2.4 \times 10^9 \times 10 \times 11.345} = 0.585 \text{ pF} \tag{4.286}$$

4.14 Class D Full-Bridge RF Power Amplifier

4.14.1 Currents, Voltages, and Powers

The circuit of a Class D full-bridge with a series-resonant amplifier is shown in Fig. 4.32. It consists of four controllable switches and a series-resonant circuit. Current and voltage waveforms in the amplifier are shown in Fig. 4.33. Notice that the voltage at the input of the resonant circuit is twice as high as that of the half-bridge amplifier. The average resistance of the on-resistances of power MOSFETs is $r_S = (r_{DS1} + r_{DS2} + r_{DS3} + r_{DS4})/4 \approx 2r_{DS}$. The total parasitic resistance is represented by

$$r \approx 2r_{DS} + r_L + r_C \tag{4.287}$$

which yields the overall resistance

$$R_t = R + r \approx R + 2r_{DS} + r_L + r_C \tag{4.288}$$

When the switches S_1 and S_3 are ON and the switches S_2 and S_4 are OFF, $v = V_I$. When the switches S_1 and S_3 are OFF and the switches S_2 and S_4 are ON, $v = -V_I$. Referring to Fig. 4.33, the input voltage of the series-resonant circuit is a square wave described by

$$v = \begin{cases} V_I & \text{for} \quad 0 < \omega t \leq \pi \\ -V_I & \text{for} \quad \pi < \omega t \leq 2\pi \end{cases} \tag{4.289}$$

The Fourier expansion of this voltage is

$$\begin{aligned} v &= \frac{4V_I}{\pi} \sum_{n=1}^{\infty} \frac{1-(-1)^n}{2n} \sin n\omega t = \frac{4V_I}{\pi} \sum_{k=1}^{\infty} \frac{\sin[(2k-1)\omega t]}{2k-1} \\ &= V_I \left(\frac{4}{\pi} \sin \omega t + \frac{4}{3\pi} \sin 3\omega t + \frac{4}{5\pi} \sin 5\omega t + \cdots \right) \end{aligned} \tag{4.290}$$

The fundamental component of voltage v is

$$v_{i1} = V_m \sin \omega t \tag{4.291}$$

where its amplitude is given by

$$V_m = \frac{4V_I}{\pi} \approx 1.273 V_I \tag{4.292}$$

Hence, one obtains the rms value of v_{i1}

$$V_{rms} = \frac{V_m}{\sqrt{2}} = \frac{2\sqrt{2}V_I}{\pi} \approx 0.9 V_I \tag{4.293}$$

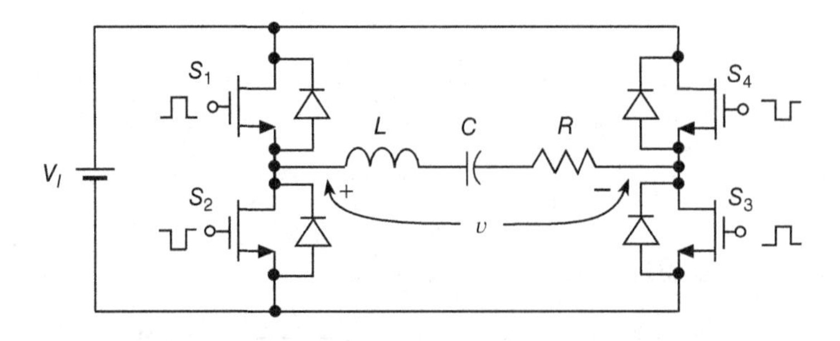

Figure 4.32 Full-bridge Class D power amplifier with a series-resonant circuit.

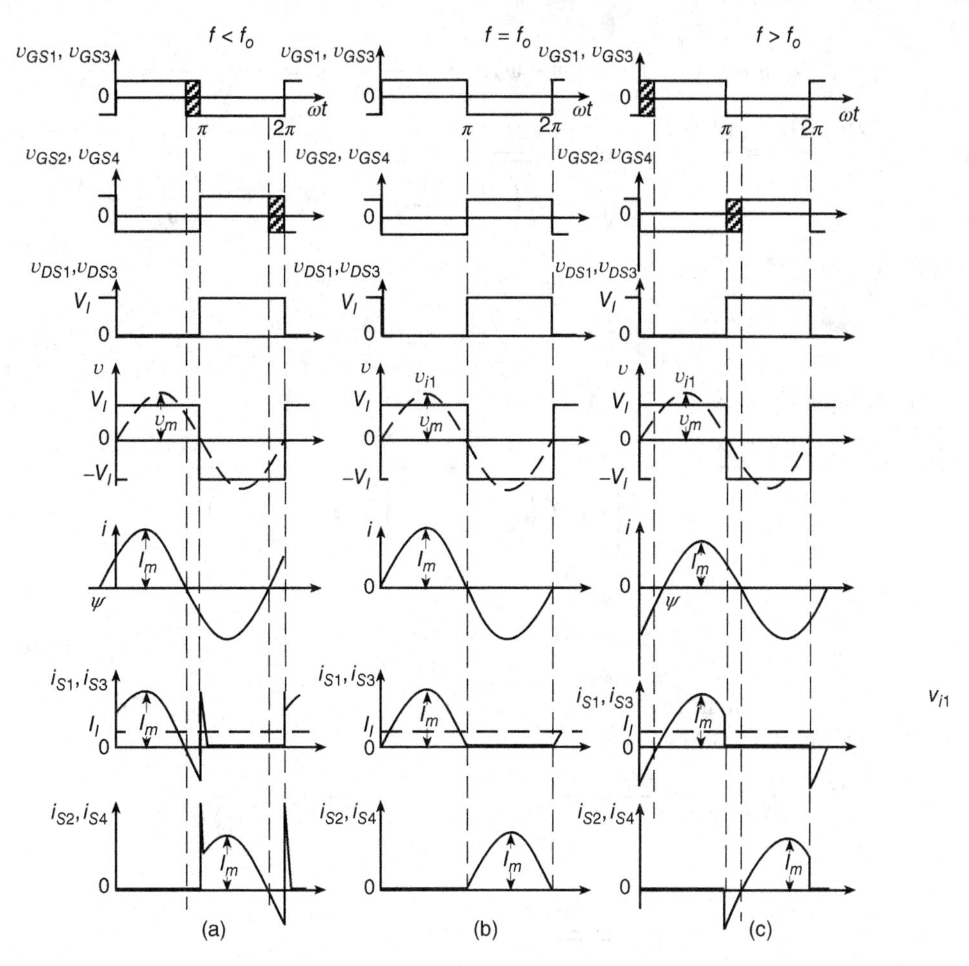

Figure 4.33 Waveforms in the Class D full-bridge power amplifier. (a) For $f < f_o$. (b) For $f = f_o$. (c) For $f > f_o$.

The current through the switches S_1 and S_3 is

$$i_{S1} = i_{S3} = \begin{cases} I_m \sin(\omega t - \psi) & \text{for} \quad 0 < \omega t \leq \pi \\ 0 & \text{for} \quad \pi < \omega t \leq 2\pi \end{cases} \tag{4.294}$$

and the current through the switches S_2 and S_4 is

$$i_{S2} = i_{S4} = \begin{cases} 0 & \text{for} \quad 0 < \omega t \leq \pi \\ -I_m \sin(\omega t - \psi) & \text{for} \quad \pi < \omega t \leq 2\pi \end{cases} \tag{4.295}$$

The input current of the amplifier is

$$i_I = i_{S1} + i_{S4} \tag{4.296}$$

The frequency of the input current is twice the operating frequency. Hence, from (4.55), (4.57), and (4.292), one obtains the dc component of the input current

$$I_I = \frac{1}{\pi}\int_0^{\pi} i_{S1} d(\omega t) = \frac{I_m}{\pi}\int_0^{\pi} \sin(\omega t - \psi) d(\omega t) = \frac{2I_m \cos\psi}{\pi} = \frac{2V_m \cos\psi}{\pi Z} = \frac{8V_I \cos\psi}{\pi^2 Z}$$
$$= \frac{8V_I \cos^2\psi}{\pi^2 R_t} = \frac{8V_I R_t}{\pi^2 Z^2} = \frac{2I_m}{\pi\sqrt{1 + Q_L^2\left(\frac{\omega}{\omega_o} - \frac{\omega_o}{\omega}\right)^2}} = \frac{8V_I}{\pi^2 R_t\left[1 + Q_L^2\left(\frac{\omega}{\omega_o} - \frac{\omega_o}{\omega}\right)^2\right]} \tag{4.297}$$

At $f = f_o$,

$$I_I = \frac{2I_m}{\pi} = \frac{8V_I}{\pi^2 R_t} \tag{4.298}$$

The dc input power is

$$P_I = I_I V_I = \frac{8V_I^2 \cos^2\psi}{\pi^2 R_t} = \frac{8V_I^2}{\pi^2 R_t\left[1 + Q_L^2\left(\frac{\omega}{\omega_o} - \frac{\omega_o}{\omega}\right)^2\right]} = \frac{8V_I^2 R_t}{\pi^2 Z_o^2\left[\left(\frac{R_t}{Z_o}\right)^2 + \left(\frac{\omega}{\omega_o} - \frac{\omega_o}{\omega}\right)^2\right]} \tag{4.299}$$

At $f = f_o$,

$$P_I = \frac{8V_I^2}{\pi^2 R_t} \tag{4.300}$$

The current through the series-resonant circuit is given by (4.145). From (4.57), (4.59), and (4.292), its amplitude can be found as

$$I_m = \frac{V_m}{Z} = \frac{4V_I}{\pi Z} = \frac{4V_I \cos\psi}{\pi R_t} = \frac{4V_I}{\pi R_t\sqrt{1 + Q_L^2\left(\frac{\omega}{\omega_o} - \frac{\omega_o}{\omega}\right)^2}}$$
$$= \frac{4V_I}{\pi Z_o\sqrt{\left(\frac{R_t}{Z_o}\right)^2 + \left(\frac{\omega}{\omega_o} - \frac{\omega_o}{\omega}\right)^2}} \tag{4.301}$$

At $f = f_o$,

$$I_{SM} = I_m(f_o) = \frac{4V_I}{\pi R_t} \tag{4.302}$$

The voltage stress of each switch is

$$V_{SM} = V_I \tag{4.303}$$

The output power is obtained from (4.301)

$$P_O = \frac{I_m^2 R}{2} = \frac{8V_I^2 R \cos^2\psi}{\pi^2 R_t^2} = \frac{8V_I^2 R}{\pi^2 R_t^2\left[1 + Q_L^2\left(\frac{\omega}{\omega_o} - \frac{\omega_o}{\omega}\right)^2\right]}$$
$$= \frac{8V_I^2 R}{\pi^2 Z_o^2\left[\left(\frac{R_t}{Z_o}\right)^2 + \left(\frac{\omega}{\omega_o} - \frac{\omega_o}{\omega}\right)^2\right]} \tag{4.304}$$

At $f = f_o$,

$$P_O = \frac{8V_I^2 R}{\pi^2 R_t^2} \approx \frac{8V_I^2}{\pi^2 R} \tag{4.305}$$

From (4.301), one obtains the amplitude of the voltage across the capacitor C

$$V_{Cm} = \frac{I_m}{\omega C} = \frac{4V_I}{\pi\left(\frac{\omega}{\omega_o}\right)\sqrt{\left(\frac{R_t}{Z_o}\right)^2 + \left(\frac{\omega}{\omega_o} - \frac{\omega_o}{\omega}\right)^2}} \tag{4.306}$$

Similarly, the amplitude of the voltage across the inductor L is

$$V_{Lm} = \omega L I_m = \frac{4V_I\left(\frac{\omega}{\omega_o}\right)}{\pi\sqrt{\left(\frac{R_t}{Z_o}\right)^2 + \left(\frac{\omega}{\omega_o} - \frac{\omega_o}{\omega}\right)^2}} \tag{4.307}$$

At $f = f_o$,

$$V_{Cm} = V_{Lm} = Z_o I_{mr} = Q_L V_m = \frac{4V_I Q_L}{\pi} \tag{4.308}$$

4.14.2 Efficiency of Full-Bridge Class D RF Power Amplifier

The conduction losses in each transistor, resonant inductor, and resonant capacitor are given by (4.169), (4.170), and (4.171), respectively. The conduction power loss in the four transistors and the resonant circuit is

$$P_r = \frac{rI_m^2}{2} = \frac{(2r_{DS} + r_L + r_C)I_m^2}{2} \tag{4.309}$$

The turn-on switching loss per transistor for operation below resonance P_{sw} is given by (4.192). The overall power dissipation in the Class D amplifier for operation below resonance is

$$P_T = P_r + 4P_{sw} + 4P_G = \frac{rI_m^2}{2} + 40C_{ds(25V)}\sqrt{V_I^3} + 4fQ_g V_{GSpp} \tag{4.310}$$

Hence, the efficiency of the full-bridge amplifier operating below resonance is

$$\eta_I = \frac{P_O}{P_O + P_T} = \frac{P_O}{P_O + P_r + 4P_{sw} + 4P_G} \tag{4.311}$$

The turn-off switching loss per transistor operating above resonance P_{toff} is given by (4.213). The overall power dissipation in the amplifier operating above resonance is

$$P_T = P_r + 4P_{toff} + 4P_G = \frac{rI_m^2}{2} + fV_I I_{OFF}\left(\frac{4t_r}{3} + 2t_f\right) + 4fQ_g V_{GSpp} \tag{4.312}$$

resulting in the efficiency of the Class D full-bridge series-resonant amplifier operating above resonance

$$\eta_I = \frac{P_O}{P_O + P_T} = \frac{P_O}{P_O + P_r + 4P_{toff} + 4P_G} \tag{4.313}$$

4.14.3 Operation Under Short-Circuit and Open-Circuit Conditions

The Class D amplifier with a series-resonant circuit can operate safely with an open circuit at the output. However, it is prone to catastrophic failure if the output is short-circuited at f close to f_o. If $R = 0$, the amplitude of the current through the resonant circuit and the switches is

$$I_m = \frac{4V_I}{\pi r\sqrt{1 + \left(\frac{Z_o}{r}\right)^2\left(\frac{\omega}{\omega_o} - \frac{\omega_o}{\omega}\right)^2}} \tag{4.314}$$

The maximum value of I_m occurs at $f = f_o$ and is given by

$$I_{mr} = \frac{4V_I}{\pi r} \tag{4.315}$$

The amplitudes of the voltages across the resonant components L and C are

$$V_{Cm} = V_{Lm} = \frac{I_{mr}}{\omega_o C} = \omega_o L I_{mr} = Z_o I_{mr} = \frac{4V_I Z_o}{\pi r} = \frac{4V_I Q_o}{\pi} \tag{4.316}$$

4.14.4 Voltage Transfer Function

The input of the amplifier to the input of the resonant circuit is

$$M_{Vs} = \frac{V_m}{V_I} = \frac{4}{\pi} \approx 1.273 \tag{4.317}$$

The magnitude of the dc-to-ac voltage transfer function for the Class D full-bridge amplifier with a series resonant is

$$M_{VI} = \frac{V_{om}}{V_I} = \frac{V_{om}}{V_m}\frac{V_m}{V_I} = M_{Vs}M_{Vr} = \frac{4\eta_{Ir}}{\pi\sqrt{1 + Q_L^2\left(\frac{\omega}{\omega_o} - \frac{\omega_o}{\omega}\right)^2}} \tag{4.318}$$

The maximum value of M_{VI} occurs at $f/f_o = 1$ and equals $M_{VImax} = 4\eta_{Ir}/\pi = 1.237\eta_{Ir}$. Thus, the values of M_{VI} range from 0 to $1.237\eta_{Ir}$.

Example 4.5

Design a Class D full-bridge amplifier of Fig. 4.32 to meet the following specifications: $V_I = 270$ V, $P_O = 500$ W, and $f = 110$ kHz. Assume $Q_L = 5.3$, $\psi = 30°$ (i.e., $\cos^2\psi = 0.75$), and the efficiency $\eta_{Ir} = 94\%$. Neglect switching losses.

Solution. The input power of the amplifier is

$$P_I = \frac{P_O}{\eta_{Ir}} = \frac{500}{0.94} = 531.9 \text{ W} \tag{4.319}$$

The overall resistance of the amplifier can be obtained from (4.304)

$$R_t = \frac{8V_I^2}{\pi^2 P_I}\cos^2\psi = \frac{8 \times 270^2}{\pi^2 \times 531.9} \times 0.75 = 83.3\ \Omega \tag{4.320}$$

Following the design procedure of Example 4.2, one obtains

$$R = \eta_{Ir} R_t = 0.94 \times 83.3 = 78.3\ \Omega \tag{4.321}$$

$$r = R_t - R = 83.3 - 78.3 = 5\ \Omega \tag{4.322}$$

$$I_I = \frac{P_I}{V_I} = \frac{531.9}{270} = 1.97 \text{ A} \tag{4.323}$$

$$I_m = \sqrt{\frac{2P_{R(rms)}}{R}} = \sqrt{\frac{2 \times 500}{78.3}} = 3.574 \text{ A} \tag{4.324}$$

$$\frac{f}{f_o} = \frac{1}{2}\left(\frac{\tan\psi}{Q_L} + \sqrt{\frac{\tan^2\psi}{Q_L^2} + 4}\right) = \frac{1}{2}\left[\frac{\tan(30^\circ)}{5.3} + \sqrt{\frac{\tan^2(30^\circ)}{5.3^2} + 4}\right] = 1.056 \tag{4.325}$$

$$f_o = \frac{f}{(f/f_o)} = \frac{110 \times 10^3}{1.056} = 104.2 \text{ kHz} \tag{4.326}$$

$$L = \frac{Q_L R_t}{\omega_o} = \frac{5.3 \times 83.3}{2\pi \times 104.2 \times 10^3} = 674 \text{ μH} \tag{4.327}$$

$$C = \frac{1}{\omega_o Q_L R_t} = \frac{1}{2\pi \times 104.2 \times 10^3 \times 5.3 \times 83.3} = 3.46 \text{ nF} \tag{4.328}$$

$$Z_o = \sqrt{\frac{L}{C}} = \sqrt{\frac{674 \times 10^{-6}}{3.46 \times 10^{-9}}} = 441.4\ \Omega \tag{4.329}$$

From (4.316), the maximum voltage stresses for the resonant components are

$$V_{Cm} = V_{Lm} = \frac{4V_I Q_L}{\pi} = \frac{4 \times 270 \times 5.3}{\pi} = 1822 \text{ V} \tag{4.330}$$

Referring to (4.303), the peak value of the switch voltage is equal to the input voltage

$$V_{SM} = V_I = 270 \text{ V} \tag{4.331}$$

4.15 Phase Control of Full-Bridge Class D Power Amplifier

Figure 4.34 shows a full-bridge Class D power amplifier with phase control. The output voltage, output current, and output power can be controlled by varying the phase shift $\Delta\phi$ between

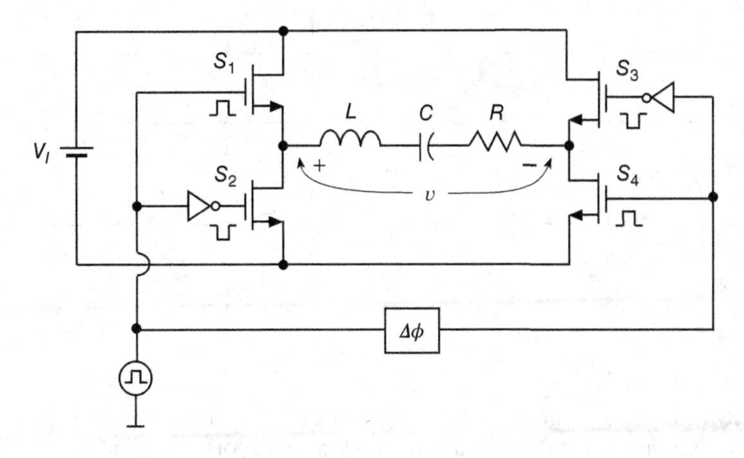

Figure 4.34 Full-bridge Class D power amplifier with phase control.

the gate-to-source voltages of the left leg and the right leg, while maintaining a constant operating frequency f. The operating frequency f can be equal to the resonant frequency f_o. Constant-frequency operation is preferred in most applications. Equivalent circuits of the amplifier are shown in Fig. 4.35. Figure 4.36 shows the waveforms of the gate-to-source voltages and the voltage across the resonant circuit.

Figure 4.35(a) shows the equivalent circuit when S_1 and S_3 are ON and S_2 and S_4 are OFF, producing the voltage across the resonant circuit $v = 0$. Figure 4.35(b) shows the equivalent circuit, when S_1 and S_4 are ON and S_2 and S_3 are OFF. In this case, the voltage across the resonant circuit is $v = V_I$. Figure 4.35(c) shows the equivalent circuit, when S_1 and S_3 are OFF and S_2 and S_4 are ON, generating the voltage across the resonant circuit $v = 0$. Figure 4.35(d) shows the equivalent circuit, when S_1 and S_4 are OFF and S_2 and S_3 are ON. In this case, the voltage across the resonant circuit is $v = -V_I$. The duty cycle corresponding to the width of the positive pulse or the negative pulse of the voltage across the resonant circuit v is given by

$$D = \frac{\pi - \Delta\phi}{2\pi} = \frac{1}{2} - \frac{\Delta\phi}{2\pi} \tag{4.332}$$

When the phase shift $\Delta\phi$ increases from 0 to π, the duty cycle D decreases from 50% to zero. The phase shift $\Delta\phi$ in terms of the duty cycle D is expressed by

$$\Delta\phi = \pi - 2\pi D = 2\pi\left(\frac{1}{2} - D\right) \quad \text{for} \quad 0 \le D \le 0.5 \tag{4.333}$$

The even harmonics of the voltage v are zero. The amplitudes of the odd harmonics of the voltage v are

$$V_{m(n)} = \frac{4V_I}{\pi n}\sin(n\pi D) \quad \text{for} \quad 0 \le D \le 0.5 \quad \text{and} \quad n = 1, 3, 5, \ldots \tag{4.334}$$

Higher harmonics are attenuated by the resonant circuit, which acts as a bandpass filter. The amplitude of the fundamental component V_m of the voltage v is given by

$$V_m = \frac{4V_I}{\pi}\sin(\pi D) = \frac{4V_I}{\pi}\cos\left(\frac{\Delta\phi}{2}\right) \quad \text{for} \quad 0 \le D \le 0.5 \tag{4.335}$$

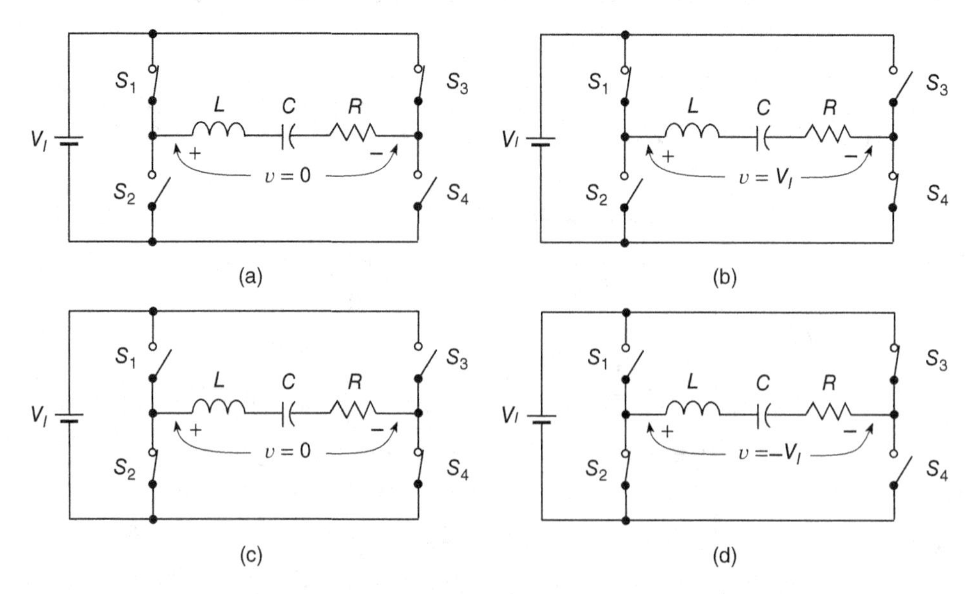

Figure 4.35 Equivalent circuits of full-bridge Class D power amplifier with phase control. (a) S_1 and S_3 ON, while S_2 and S_4 OFF. (b) S_1 and S_4 ON, while S_2 and S_3 OFF. (c) S_2 and S_4 ON, while S_1 and S_3 OFF. (d) S_2 and S_3 ON, while S_1 and S_4 OFF.

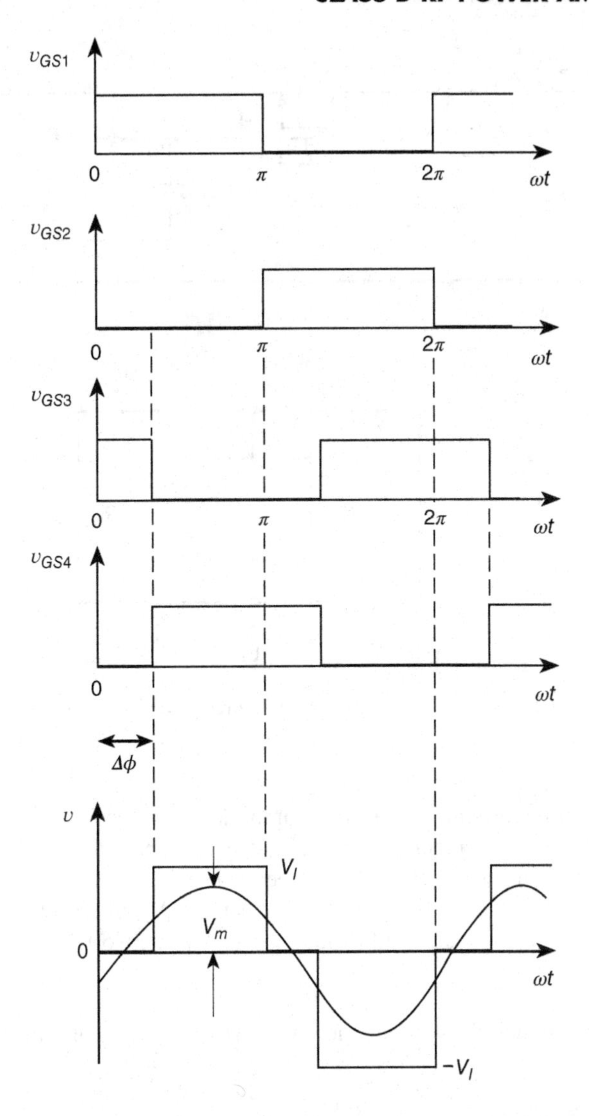

Figure 4.36 Waveforms of a full-bridge Class D power amplifier with phase control.

Thus, the amplitude of the fundamental component V_m of the voltage v decreases from $V_{m(\max)} = 4V_I/\pi$ to zero as D decreases from 0.5 to 0, or $\Delta\phi$ increases from 0 to π. Therefore, the output current, the output voltage, and the output power can be controlled by varying the phase shift $\Delta\phi$.

4.16 Class D Current-Switching RF Power Amplifier

4.16.1 Circuit and Waveforms

The Class D current-switching (current-source) RF power amplifier of in Fig. 4.37(a) was introduced in [12]. This circuit is dual to the Class D voltage-switching (voltage-source) RF

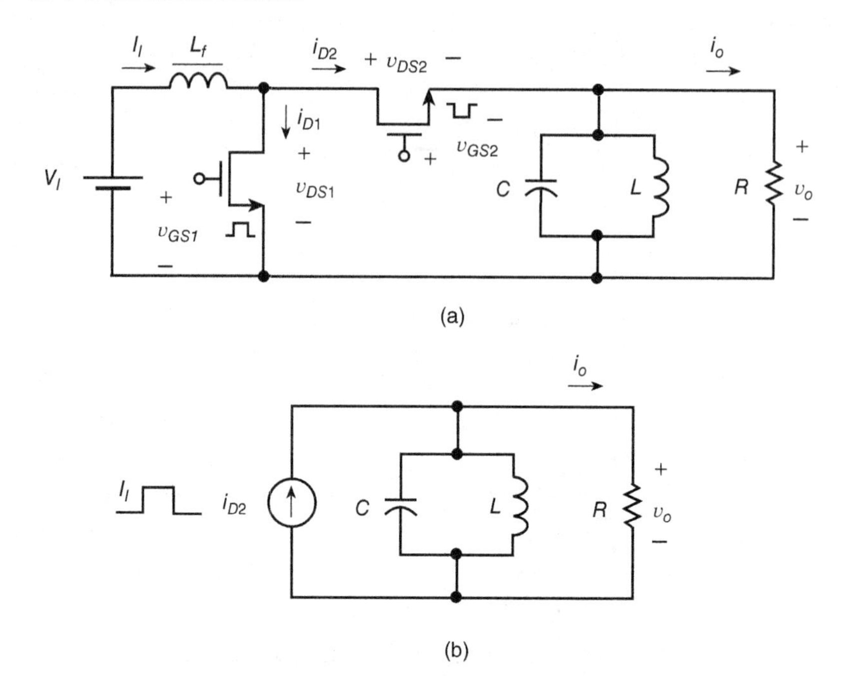

Figure 4.37 Class D current-switching (current-source) RF power amplifier [12]. (a) Circuit. (b) Equivalent circuit.

power amplifier. An equivalent circuit of the amplifier is shown in Fig. 4.37(b). Voltage and current waveforms, which explain the principle of operation of the amplifier, are depicted in Fig. 4.38. It can be seen the MOSFETs are operated under ZVS conditions. The dc voltage source V_I and the form a dc current source I_I. When the MOSFET switch S_1 is OFF and the MOSFET switch S_2 is ON, the current flowing into the load network is

$$i_{D2} = I_I \tag{4.336}$$

When the MOSFET switch S_1 is ON and the MOSFET switch S_2 is OFF, the current flowing into the load network is

$$i_{D2} = 0 \tag{4.337}$$

Thus, dc voltage source V_I, RFC, and two MOSFET switches form a square-wave current-source whose lower value is zero and the upper value is I_I. The parallel-resonant circuit behaves as a bandpass filter and filters out all harmonics, and only the fundamental component flows to the load resistance R. When the operating frequency is equal to the resonant frequency $f_0 = 1/(2\pi\sqrt{LC})$, the parallel-resonant circuit forces a sinusoidal voltage

$$v_o = V_m \sin \omega t \tag{4.338}$$

yielding a sinusoidal output current

$$i_o = I_m \sin \omega t \tag{4.339}$$

where the amplitude of the output current is

$$I_m = \frac{V_m}{R} \tag{4.340}$$

The dc input voltage is

$$V_I = \frac{1}{2\pi}\int_0^{2\pi} v_{DS1} d(\omega t) = \frac{1}{2\pi}\int_0^{2\pi} V_m \sin \omega t d(\omega t) = \frac{V_m}{\pi} = \frac{I_m R}{\pi} \tag{4.341}$$

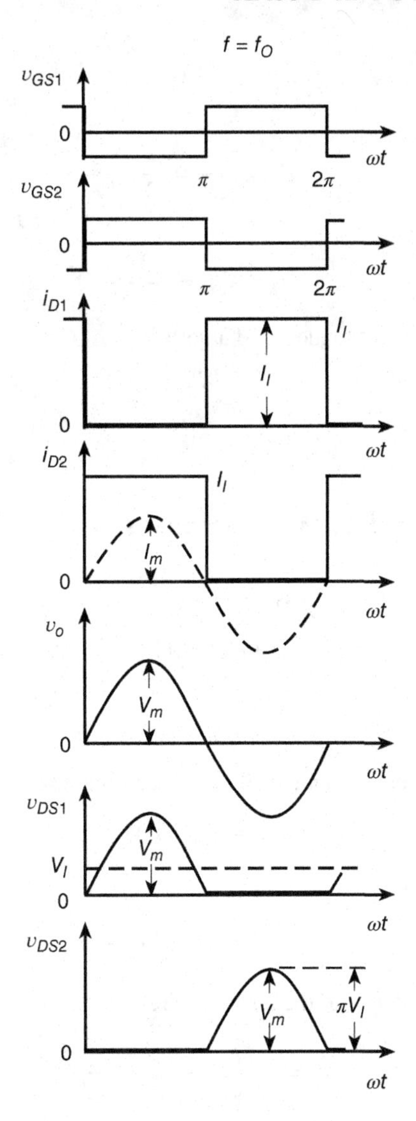

Figure 4.38 Waveforms in Class D current-switching (current-source) RF power amplifier.

The amplitude of the output current is

$$I_m = \frac{2}{\pi} I_I \tag{4.342}$$

Hence,

$$I_I = \frac{\pi}{2} I_m = \frac{\pi}{2} \frac{V_m}{R} = \frac{\pi^2}{2} \frac{V_I}{R} \tag{4.343}$$

The dc resistance is

$$R_{DC} = \frac{V_I}{I_I} = \frac{2}{\pi^2} \frac{V_m}{I_m} = \frac{2}{\pi^2} R = 0.2026R \tag{4.344}$$

4.16.2 Power

The output power is

$$P_O = \frac{V_m^2}{2R} = \frac{\pi^2}{2}\frac{V_I^2}{R} \tag{4.345}$$

The dc supply power is

$$P_I = V_I I_I = \frac{\pi^2}{2}\frac{V_I^2}{R} \tag{4.346}$$

Hence, the drain efficiency for the idealized amplifier is

$$\eta_D = \frac{P_O}{P_I} = 1 \tag{4.347}$$

4.16.3 Voltage and Current Stresses

The transistor current and voltage stresses are

$$V_{DSM} = \pi V_m \tag{4.348}$$

and

$$I_{DM} = I_I \tag{4.349}$$

The peak current in the resonant inductor L and capacitor are

$$I_{Lm} = I_{Cm} = Q_L I_m \tag{4.350}$$

where the loaded quality factor is

$$Q_L = \frac{R}{\omega_0 L} = \omega_0 CR = \frac{R}{Z_o} \tag{4.351}$$

and the characteristic impedance of the resonant circuit is

$$Z_o = \sqrt{\frac{L}{C}} = \omega_0 L = \frac{1}{\omega_0 C} \tag{4.352}$$

The output-power capability is

$$c_p = \frac{P_O}{2I_{DM}V_{DSM}} = \frac{1}{2}\left(\frac{I_I}{I_{DM}}\right)\left(\frac{V_I}{V_{DSM}}\right) = \frac{1}{2\pi} = 0.159 \tag{4.353}$$

4.16.4 Efficiency

The rms value of the MOSFET current is

$$I_{Srms} = \sqrt{\frac{1}{2\pi}\int_{\pi}^{2\pi} I_I^2 d(\omega t)} = \frac{I_I}{\sqrt{2}} \tag{4.354}$$

Hence, the conduction power loss in the MOSFET on-resistance r_{DS} is

$$P_{rDS} = r_{DS} I_{Srms}^2 = \frac{r_{DS} I_I^2}{2} = \frac{\pi^2}{4}\frac{r_{DS}}{R}P_O \tag{4.355}$$

The drain efficiency is

$$\eta_D = \frac{P_O}{P_O + 2P_{rDS}} = \frac{1}{1 + \frac{2P_{rDS}}{P_O}} = \frac{1}{1 + \frac{\pi^2}{2}\frac{r_{DS}}{R}} \tag{4.356}$$

The power loss in the ESR of the RFC r_{RFC} is

$$P_{RFC} = r_{RFC} I_I^2 = \frac{\pi^2 r_{RFC}}{4} I_m^2 = \frac{\pi^2}{2}\frac{r_{RFC}}{R} P_O \tag{4.357}$$

The power loss in the ESR of the resonant inductor r_L is

$$P_{rL} = \frac{r_L I_m^2 Q_L^2}{2} = \frac{r_L Q_L^2}{R} P_O \tag{4.358}$$

and the power loss in the ESR of the resonant capacitor r_C is

$$P_{rC} = \frac{r_C I_m^2 Q_L^2}{2} = \frac{r_C Q_L^2}{R} P_O \tag{4.359}$$

The total power loss is

$$P_{Loss} = 2P_{rDS} + P_{RFC} + P_{rL} + P_{rC} = P_O \left[\frac{\pi^2 r_{DS}}{2R} + \frac{\pi^2 r_{RFC}}{2R} + \frac{Q_L^2 (r_L + r_C)}{R} \right] \tag{4.360}$$

Hence, the overall efficiency is

$$\eta = \frac{P_O}{P_I} = \frac{P_O}{P_O + P_{Loss}} = \frac{1}{1 + \frac{\pi^2}{2}\frac{r_{DS} + r_{RFC}}{R} + \frac{Q_L^2 (r_L + r_C)}{R}} \tag{4.361}$$

Example 4.6

Design a push–pull Class D current-switching RF power amplifier to meet the following specifications: $V_I = 12$ V, $P_O = 10$ W, $BW = 240$ MHz, and $f_c = 2.4$ GHz.

Solution. Assuming the efficiency of the resonant circuit is $\eta_r = 0.94$, the drain power is

$$P_{DS} = \frac{P_O}{\eta_r} = \frac{10}{0.94} = 10.638 \text{ W} \tag{4.362}$$

The load resistance is

$$R = \frac{\pi^2}{2}\frac{V_I^2}{P_{DS}} = \frac{\pi^2}{2}\frac{12^2}{10.638} = 66.799 \text{ W} \tag{4.363}$$

The peak value of the drain-to-source voltage is

$$V_m = \pi V_I = \pi \times 12 = 37.7 \text{ V}. \tag{4.364}$$

The amplitude of the fundamental component of the drain current is

$$I_m = \frac{V_m}{R} = \frac{37.7}{66.799} = 0.564 \text{ A} \tag{4.365}$$

The dc supply current is

$$I_I = \frac{\pi}{2} I_m = \frac{\pi}{2} \times 0.564 = 0.8859 \text{ A} \tag{4.366}$$

The dc supply power is

$$P_I = V_I I_I = 12 \times 0.8859 = 10.631 \text{ W} \tag{4.367}$$

The rms value of the switch current is

$$I_{Srms} = \frac{I_I}{\sqrt{2}} = \frac{0.8859}{\sqrt{2}} = 0.626 \text{ A} \tag{4.368}$$

Assuming $r_{DS} = 0.1\ \Omega$, the conduction power loss of each MOSFET is

$$P_{rDS} = r_{DS} I_{Srms}^2 = 0.1 \times 0.626^2 = 0.0392 \text{ W} \tag{4.369}$$

The drain efficiency is

$$\eta_D = \frac{P_O}{P_O + 2P_{rDS}} = \frac{10}{10 + 2 \times 0.0392} = 99.22\% \tag{4.370}$$

The loaded quality factor is

$$Q_L = \frac{f_c}{BW} = \frac{2400}{240} = 10 \tag{4.371}$$

The resonant inductance is

$$L = \frac{R}{\omega_c Q_L} = \frac{66.799}{2\pi \times 2.4 \times 10^9 \times 10} = 0.443 \text{ nH} \tag{4.372}$$

The resonant capacitance is

$$C = \frac{Q_L}{\omega_c R} = \frac{10}{2\pi \times 2.4 \times 10^9 \times 66.799} = 9.93 \text{ pF} \tag{4.373}$$

4.17 Transformer-Coupled Push–pull Class D Current-Switching RF Power Amplifier

4.17.1 Waveforms

A circuit of a push–pull Class D current-switching power amplifier is shown in Fig. 4.39. It is the dual of the push–pull voltage-switching Class D amplifier. The impedance matching of the load to the transistors is achieved by the transformer. The magnetizing inductance on the transformer secondary side is absorbed into the resonant inductance L.

Current and voltage waveforms are shown in Fig. 4.40. The transistors turn on at zero voltage if the switching frequency is equal to the resonant frequency of the parallel-resonant circuit. The output current is

$$i_o = I_m \sin \omega t \tag{4.374}$$

and the output voltage is

$$v_o = V_m \sin \omega t \tag{4.375}$$

where

$$V_m = R_L I_m \tag{4.376}$$

When the drive voltage is positive, transistor Q_1 is ON and transistor Q_2 is OFF. The drain currents are

$$i_{D1} = 0 \quad \text{for} \quad 0 < \omega t \leq \pi \tag{4.377}$$

and

$$i_{D2} = I_I \quad \text{for} \quad 0 < \omega t \leq \pi \tag{4.378}$$

The transformer output current is

$$i = n i_{D1} = n I_I \quad \text{for} \quad 0 < \omega t \leq \pi \tag{4.379}$$

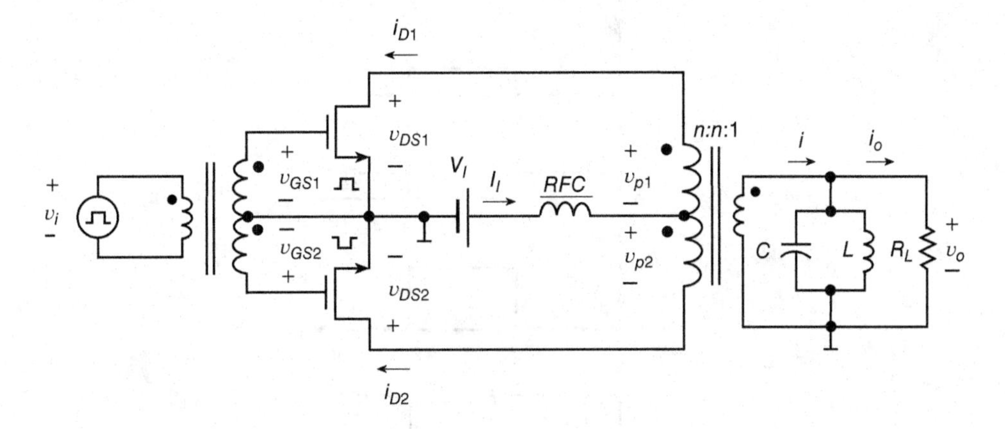

Figure 4.39 Push–pull Class D current-switching (current-source) RF power amplifier.

The voltage across the upper half of the primary winding is

$$v_{p1} = -nv_o = nV_m \sin \omega t = \frac{V_{dm}}{2} \sin \omega t \tag{4.380}$$

The drain-to-source voltages are

$$v_{DS1} = v_{p1} + v_{p2} = 2v_{p1} = V_{dm} \sin \omega t = 2nV_m \sin \omega t \quad \text{for} \quad 0 < \omega t \leq \pi \tag{4.381}$$

and

$$v_{DS2} = 0 \quad \text{for} \quad 0 < \omega t \leq \pi \tag{4.382}$$

where the peak value of the drain-to-source voltage is

$$V_{dm} = 2nV_m \tag{4.383}$$

When the drive voltage is negative, transistor Q_1 is OFF and transistor Q_2 is ON. The drain currents are

$$i_{D1} = I_I \quad \text{for} \quad \pi < \omega t \leq 2\pi \tag{4.384}$$

and

$$i_{D2} = 0 \quad \text{for} \quad \pi < \omega t \leq 2\pi \tag{4.385}$$

The transformer output current is

$$i = -ni_{D2} = -nI_I \quad \text{for} \quad \pi < \omega t \leq 2\pi \tag{4.386}$$

The voltage across the lower half of the primary winding is

$$v_{p2} = nv_o = -nV_m \sin \omega t = -\frac{V_{dm}}{2} \sin \omega t \quad \text{for} \quad \pi < \omega t \leq 2\pi \tag{4.387}$$

The drain-to-source voltages are

$$v_{DS1} = 0 \quad \text{for} \quad \pi < \omega t \leq 2\pi \tag{4.388}$$

and

$$v_{DS2} = v_{p1} = -V_{dm} \sin \omega t = -2nV_m \sin \omega t \quad \text{for} \quad \pi < \omega t \leq 2\pi \tag{4.389}$$

The output current of the transformer is

$$i = \frac{4nV_I}{\pi} \sum_{k=1}^{\infty} \frac{\sin[(2k-1)\omega t]}{2k-1} \tag{4.390}$$

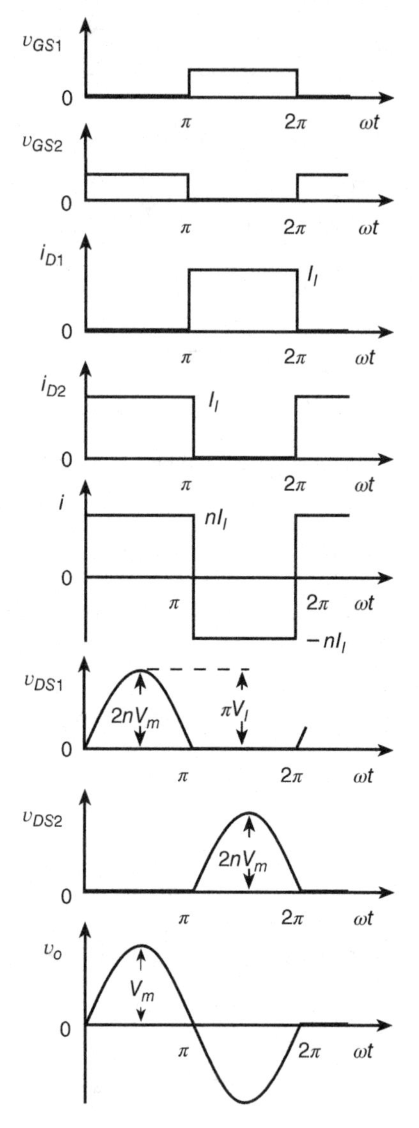

Figure 4.40 Waveforms in a push–pull Class D current-switching (current-source) RF power amplifier.

The parallel-resonant circuit exhibits a very low reactance for current harmonics and an infinite reactance for the fundamental frequency component. Therefore, only the fundamental component flows through the load resistance, resulting in sinusoidal output current and voltage waveforms.

The amplitude of the fundamental component of the drain current is

$$I_{dm} = \frac{2}{\pi} I_I \tag{4.391}$$

The amplitude of the output current is

$$I_m = \frac{4}{\pi} n I_I = 2nI_{dm} \tag{4.392}$$

The dc supply voltage is

$$V_I = \frac{V_{dm}}{\pi} = \frac{2nV_m}{\pi} \tag{4.393}$$

yielding

$$V_m = \frac{\pi}{2n} V_I \tag{4.394}$$

The resistance seen by each MOSFET is

$$R = n^2 R_L \tag{4.395}$$

4.17.2 Power

The dc supply current is

$$I_I = \frac{\pi}{2} I_{dm} = \frac{\pi}{4} \frac{I_m}{n} = \frac{\pi}{4n} \frac{V_m}{R_L} = \frac{\pi^2}{8} \frac{V_I}{n^2 R_L} = \frac{\pi^2}{8} \frac{V_I}{R} \tag{4.396}$$

The dc resistance seen by the dc power supply is

$$R_{DC} = \frac{V_I}{I_I} = \frac{8}{\pi^2} R \tag{4.397}$$

The output power is

$$P_O = \frac{V_m^2}{2R_L} = \frac{\pi^2}{8} \frac{V_I^2}{n^2 R_L} = \frac{\pi^2}{8} \frac{V_I^2}{R} \tag{4.398}$$

The dc supply current is

$$P_I = V_I I_I = \frac{\pi^2}{8} \frac{V_I^2}{n^2 R_L} = \frac{\pi^2}{8} \frac{V_I^2}{R} \tag{4.399}$$

Ideally, the drain efficiency is

$$\eta_D = \frac{P_O}{P_I} = 1 \tag{4.400}$$

4.17.3 Device Stresses

The current and voltage stresses of the MOSFETs are

$$I_{DM} = I_I \tag{4.401}$$

and

$$V_{DSM} = \pi V_I \tag{4.402}$$

The output-power capability is

$$c_p = \frac{P_O}{2I_{DM} V_{DSM}} = \frac{P_I}{2I_{DM} V_{DSM}} = \frac{I_I V_I}{2I_{DM} V_{DSM}} = \frac{1}{2\pi} = 0.159 \tag{4.403}$$

4.17.4 Efficiency

The rms drain current is

$$I_{Srms} = \sqrt{\frac{1}{2\pi} \int_{\pi}^{2\pi} I_I^2 d(\omega t)} = \frac{I_I}{\sqrt{2}} \tag{4.404}$$

Hence, the conduction power loss in the MOSFET on-resistance r_{DS} is

$$P_{rDS} = r_{DS} I_{Srms}^2 = \frac{r_{DS} I_I^2}{2} = \frac{\pi^2}{4} \frac{r_{DS}}{n^2 R_L} P_O \quad (4.405)$$

The drain efficiency is

$$\eta_D = \frac{P_O}{P_O + 2P_{rDS}} = \frac{1}{1 + \frac{2P_{rDS}}{P_O}} = \frac{1}{1 + \frac{\pi^2}{2} \frac{r_{DS}}{n^2 R_L}} \quad (4.406)$$

The power loss in the ESR of the RFC r_{RFC} is

$$P_{RFC} = r_{RFC} I_I^2 = \frac{\pi^2 r_{RFC}}{4} I_m^2 = \frac{\pi^2}{2} \frac{r_{RFC}}{n^2 R_L} P_O \quad (4.407)$$

The power loss in the ESR of the resonant inductor r_L is

$$P_{rL} = \frac{r_L I_m^2 Q_L^2}{2} = \frac{r_L Q_L^2}{R_L} P_O \quad (4.408)$$

and the power loss in the ESR of the resonant capacitor r_C is

$$P_{rC} = \frac{r_C I_m^2 Q_L^2}{2} = \frac{r_C Q_L^2}{R_L} P_O \quad (4.409)$$

The total power loss is

$$P_{Loss} = 2P_{rDS} + P_{RFC} + P_{rL} + P_{rC} = P_O \left[\frac{\pi^2 (r_{DS} + r_{RFC})}{2n^2 R_L} + \frac{Q_L^2 (r_L + r_C)}{R_L} \right] \quad (4.410)$$

Hence, the overall efficiency is

$$\eta = \frac{P_O}{P_I} = \frac{P_O}{P_O + P_{Loss}} = \frac{1}{1 + \frac{\pi^2}{2} \frac{r_{DS} + r_{RFC}}{n^2 R_L} + \frac{Q_L^2 (r_L + r_C)}{R_L}} \quad (4.411)$$

Example 4.7

Design a transformer-coupled push–pull Class D current-switching RF power amplifier to meet the following specifications: $V_I = 12$ V, $P_O = 12$ W, $BW = 180$ MHz, $R_L = 50\ \Omega$, and $f_c = 1.8$ GHz.

Solution. Assuming the efficiency of the resonant circuit is $\eta_r = 0.93$, the drain power is

$$P_{DS} = \frac{P_O}{\eta_r} = \frac{12}{0.93} = 12.903 \text{ W} \quad (4.412)$$

The resistance seen across one part of the primary winding is

$$R = \frac{\pi^2}{8} \frac{V_I^2}{P_{DS}} = \frac{\pi^2}{8} \frac{12^2}{12.903} = 13.768\ \Omega \quad (4.413)$$

The transformer turns ratio is

$$n = \sqrt{\frac{R}{R_L}} = \sqrt{\frac{13.768}{50}} = 0.5247 \quad (4.414)$$

Pick $n = \frac{1}{2}$. The resistance seen by each MOSFET is

$$R = n^2 R_L = \left(\frac{1}{2}\right)^2 \times 50 = 12.5\ \Omega \tag{4.415}$$

The peak value of the drain-to-source voltage is

$$V_{dm} = \pi V_I = \pi \times 12 = 37.7\ \text{V} \tag{4.416}$$

The amplitude of the output voltage is

$$V_m = \frac{\pi}{2n} V_I = \frac{\pi}{2 \times 0.5} \times 12 = 37.7\ \text{V} \tag{4.417}$$

The dc resistance seen by the dc power supply is

$$R_{DC} = \frac{V_I}{I_I} = \frac{8}{\pi^2} R = \frac{8}{\pi^2} \times 13.768 = 11.16\ \Omega \tag{4.418}$$

The amplitude of the fundamental component of the drain current is

$$I_{dm} = \frac{V_{dm}}{R} = \frac{37.7}{12.5} = 3.016\ \text{A} \tag{4.419}$$

The dc supply current is

$$I_I = \frac{\pi^2}{8} \frac{V_I}{R} = \frac{\pi^2}{8} \frac{12}{12.5} = 1.184\ \text{A} \tag{4.420}$$

The amplitude of the output current is

$$I_m = \frac{V_m}{R} = \frac{37.7}{12.5} = 3.016\ \text{A} \tag{4.421}$$

The output power is

$$P_O = \frac{V_m^2}{2R_L} = \frac{37.7^2}{2 \times 50} = 14.21\ \text{W} \tag{4.422}$$

The dc supply power is

$$P_I = V_I I_I = 12 \times 1.184 = 14.208\ \text{W} \tag{4.423}$$

The rms value of the switch current is

$$I_{Srms} = \frac{I_I}{\sqrt{2}} = \frac{1.184}{\sqrt{2}} = 0.8372\ \text{A} \tag{4.424}$$

Assuming $r_{DS} = 0.1\ \Omega$, the conduction power loss in each MOSFET is

$$P_{rDS} = r_{DS} I_{Srms}^2 = 0.1 \times 0.8372^2 = 0.07\ \text{W} \tag{4.425}$$

The drain efficiency is

$$\eta_D = \frac{P_O}{P_O + 2P_{rDS}} = \frac{1}{1 + \frac{\pi^2}{4n^2} \frac{r_{DS}}{R}} = \frac{1}{1 + \frac{\pi^2}{4 \times 0.5^2} \frac{0.1}{12.5}} = 92.68\% \tag{4.426}$$

Assuming the resistance $r_{RFC} = 0.11\ \Omega$, we have

$$P_{RFC} = r_{RFC} I^2 = 0.11 \times 1.184^2 = 0.154\ \Omega \tag{4.427}$$

The loaded quality factor is

$$Q_L = \frac{f_c}{BW} = \frac{1800}{180} = 10 \tag{4.428}$$

The resonant inductance is

$$L = \frac{R_L}{\omega_c Q_L} = \frac{50}{2\pi \times 1.8 \times 10^9 \times 10} = 0.442\ \text{nH} \tag{4.429}$$

The resonant capacitance is

$$C = \frac{Q_L}{\omega_c R_L} = \frac{10}{2\pi \times 1.8 \times 10^9 \times 50} = 17.68 \text{ pF} \tag{4.430}$$

Assuming $r_L = 0.009\ \Omega$ and $r_C = 0.01\ \Omega$, we get

$$P_{rL} = \frac{r_L I_m^2 Q_L^2}{2} = \frac{0.009 \times 3.016^2 \times 10^2}{2} = 4.09 \text{ W} \tag{4.431}$$

$$P_{rC} = \frac{r_C I_m^2 Q_L^2}{2} = \frac{0.01 \times 3.016^2 \times 10^2}{2} = 4.548 \text{ W} \tag{4.432}$$

Hence, the total power loss is

$$P_{Loss} = 2P_{rDS} + P_{RFC} + P_{rL} + P_{rC} = 2 \times 0.07 + 0.154 + 4.09 + 4.548 = 8.932 \text{ W} \tag{4.433}$$

The overall efficiency is

$$\eta = \frac{P_O}{P_O + P_{Loss}} = \frac{14.21}{14.21 + 8.932} = 61.4\% \tag{4.434}$$

4.18 Bridge Class D Current-Switching RF Power Amplifier

A circuit of a bridge Class D current-switching (current-source) RF power amplifier is shown in Fig. 4.41(a). When one transistor is ON, the other transistor is OFF. Current and voltage waveforms are shown in Fig. 4.42. Figure 4.41(c) shows an equivalent circuit when transistor Q_1 is OFF and transistor Q_2 is ON. For the switching frequency equal to the resonant frequency of the parallel-resonant circuit, the transistors turns on at zero voltage, reducing switching losses. The main disadvantage of the circuit is that the currents of both chokes flow through one of the transistors at all times, causing high conduction loss.

The current and voltage waveforms are given by

$$i = I_I \quad \text{for} \quad 0 < \omega t \leq \pi \tag{4.435}$$

$$i_{D1} = 0 \quad \text{for} \quad 0 < \omega t \leq \pi \tag{4.436}$$

$$i_{D2} = 2I_I \quad \text{for} \quad 0 < \omega t \leq \pi \tag{4.437}$$

and

$$v_{DS1} = V_m \sin \omega t \quad \text{for} \quad 0 < \omega t \leq \pi. \tag{4.438}$$

Figure 4.41(b) shows an equivalent circuit when transistor Q_1 is OFF and transistor Q_2 is ON. The waveforms are as follows:

$$i = -I_I \quad \text{for} \quad \pi < \omega t \leq 2\pi \tag{4.439}$$

$$i_{D1} = 2I_I \quad \text{for} \quad \pi < \omega t \leq 2\pi \tag{4.440}$$

$$i_{D2} = 0 \quad \text{for} \quad \pi < \omega t \leq 2\pi \tag{4.441}$$

and

$$v_{DS2} = -V_m \sin \omega t \quad \text{for} \quad \pi < \omega t \leq 2\pi \tag{4.442}$$

The output voltage is

$$v_o = V_m \sin \omega t \tag{4.443}$$

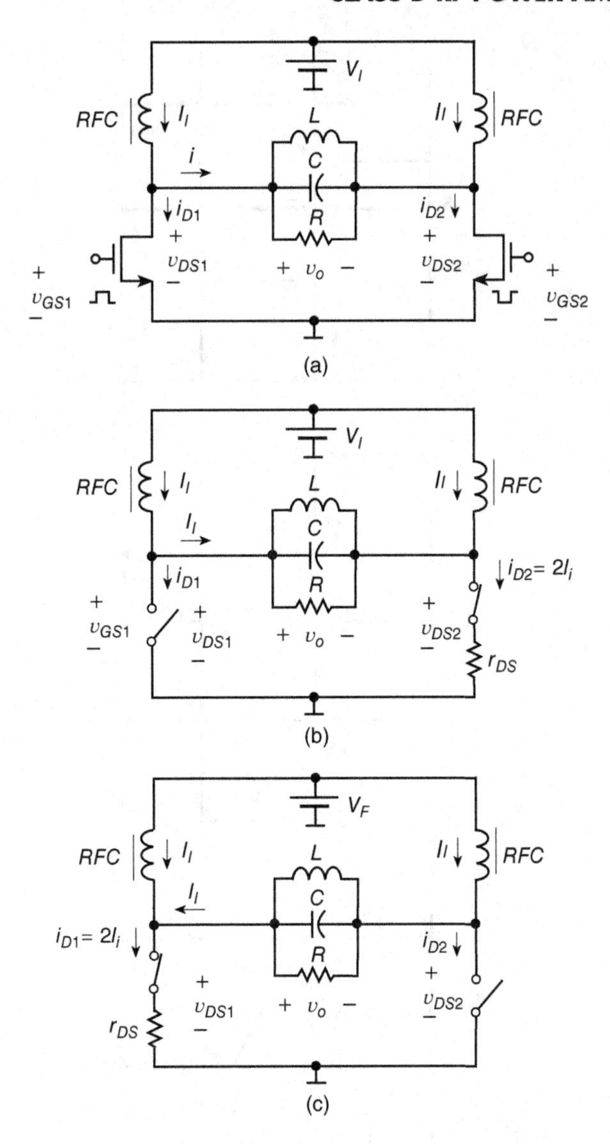

Figure 4.41 Bridge Class D current-switching (current-source) RF power amplifier. (a) Circuit. (b) Equivalent circuit when transistor Q_1 is OFF and transistor Q_2 is ON. (c) Equivalent circuit when transistor Q_1 is ON and transistor Q_2 is OFF.

where

$$V_m = RI_m = \pi V_I \tag{4.444}$$

The output current is

$$i_o = I_m \sin \omega t \tag{4.445}$$

where the amplitude of the output current is

$$I_m = \frac{4}{\pi} I_I \tag{4.446}$$

resulting in

$$I_I = \frac{\pi}{4} I_m = \frac{\pi}{4} \frac{V_m}{R} = \frac{\pi^2}{4} \frac{V_I}{R} \tag{4.447}$$

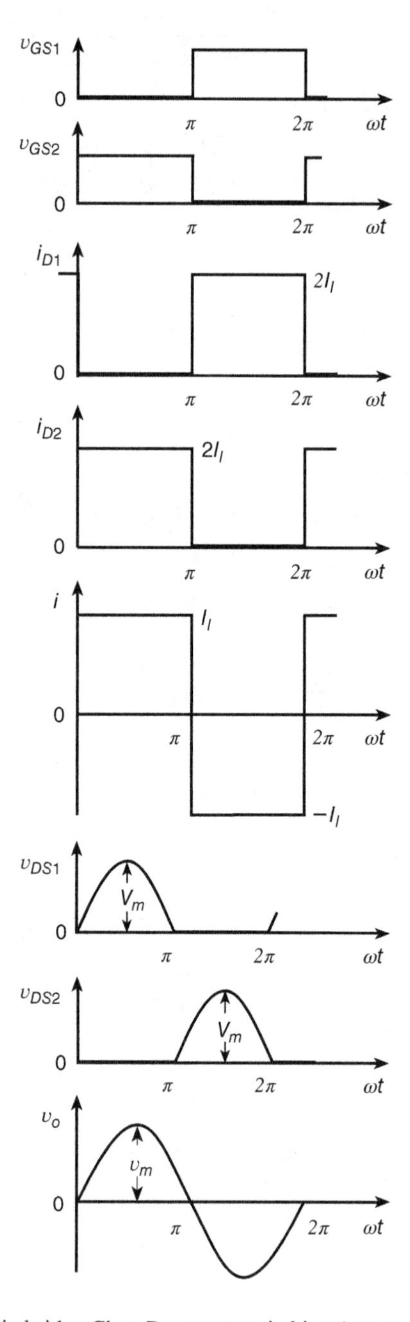

Figure 4.42 Waveforms in bridge Class D current-switching (current-source) RF power amplifier.

The dc resistance is

$$R_{DC} = \frac{V_I}{2I_I} = \frac{2}{\pi^2}R \tag{4.448}$$

The dc supply power is

$$P_I = 2I_I V_I = \frac{\pi^2}{2}\frac{V_I^2}{R} \tag{4.449}$$

The output power is

$$P_O = \frac{V_m^2}{2R} = \frac{RI_m^2}{2} = \frac{\pi^2}{2}\frac{V_I^2}{R} \tag{4.450}$$

The rms value of the switch current is

$$I_{Srms} = \sqrt{\frac{1}{2\pi}\int_0^{2\pi} i_{D2}^2 d(\omega t)} = \sqrt{\frac{1}{2\pi}\int_0^{\pi} (2I_I)^2 d(\omega t)} = \sqrt{2}I_I \tag{4.451}$$

Hence, the conduction loss in each transistor is

$$P_{rDS} = r_{DS} I_{Srms}^2 = 2r_{DS} I_I^2 = \frac{\pi^2}{8} r_{DS} I_m^2 = \frac{\pi^2}{4}\frac{r_{DS}}{R} P_O \tag{4.452}$$

The drain efficiency is

$$\eta_D = \frac{P_O}{P_O + 2P_{rDS}} = \frac{1}{1 + \frac{\pi^2}{2}\frac{r_{DS}}{R}} \tag{4.453}$$

The power loss in each RFC resistance r_{RFC} is

$$P_{RFC} = r_{RFC} I_I^2 = r_{RFC}\frac{\pi^2}{16}\frac{V_m^2}{R} = r_{RFC}\frac{\pi^2}{16}\frac{V_m^2}{R} = \frac{\pi^2}{8}\frac{r_{RFC}}{R} P_O \tag{4.454}$$

The power loss in the ESR of the resonant inductor r_L is

$$P_{rL} = \frac{r_L I_m^2 Q_L^2}{2} = \frac{r_L Q_L^2}{R_L} P_O \tag{4.455}$$

and the power loss in the ESR of the resonant capacitor r_C is

$$P_{rC} = \frac{r_C I_m^2 Q_L^2}{2} = \frac{r_C Q_L^2}{R_L} P_O \tag{4.456}$$

The total power loss is

$$P_{Loss} = 2P_{rDS} + 2P_{RFC} + P_{rL} + P_{rC} = P_O\left[\frac{\pi^2 r_{DS}}{2R} + \frac{\pi^2 r_{RFC}}{4R} + \frac{Q_L^2(r_L + r_C)}{R}\right] \tag{4.457}$$

Hence, the overall efficiency is

$$\eta = \frac{P_O}{P_I} = \frac{P_O}{P_O + P_{Loss}} = \frac{1}{1 + \frac{\pi^2}{2}\frac{r_{DS}}{R} + \frac{\pi^2}{4}\frac{r_{RFC}}{R} + \frac{Q_L^2(r_L + r_C)}{R}} \tag{4.458}$$

Example 4.8

Design a bridge Class D current-switching RF power amplifier to meet the following specifications: $V_I = 5$ V, $P_O = 6$ W, $BW = 240$ MHz, and $f_c = 2.4$ GHz.

Solution. Assuming the efficiency of the resonant circuit is $\eta_r = 0.92$, the drain power is

$$P_{DS} = \frac{P_O}{\eta_r} = \frac{6}{0.92} = 6.522 \text{ W} \tag{4.459}$$

The load resistance is

$$R = \frac{\pi^2}{2}\frac{V_I^2}{P_{DS}} = \frac{\pi^2}{2}\frac{5^2}{6.522} = 18.916\ \Omega \tag{4.460}$$

The amplitude of the output voltage and the peak value of the drain-to-source voltage is

$$V_m = \pi V_I = \pi \times 5 = 15.708 \text{ V} \tag{4.461}$$

The dc resistance seen by the dc power supply is

$$R_{DC} = \frac{V_I}{2I_I} = \frac{2}{\pi^2}R = \frac{2}{\pi^2} \times 18.916 = 3.833\ \Omega \tag{4.462}$$

The amplitude of the output current is

$$I_m = \frac{V_m}{R} = \frac{15.708}{18.916} = 0.83 \text{ A} \tag{4.463}$$

The dc supply current is

$$I_I = \frac{\pi^2}{4}\frac{V_I}{R} = \frac{\pi^2}{4}\frac{5}{18.916} = 0.652 \text{ A} \tag{4.464}$$

The dc supply power is

$$P_I = 2I_I V_I = 2 \times 0.652 \times 5 = 6.522 \text{ W} \tag{4.465}$$

Assuming $r_{DS} = 0.1\ \Omega$, the conduction power loss in each MOSFET is

$$P_{rDS} = r_{DS} I_I^2 = 0.1 \times 0.652^2 = 0.0425 \text{ W} \tag{4.466}$$

The drain efficiency is

$$\eta_D = \frac{P_O}{P_O + 2P_{rDS}} = \frac{1}{1 + \frac{\pi^2}{2}\frac{r_{DS}}{R}} = \frac{1}{1 + \frac{\pi^2}{2}\frac{0.1}{18.916}} = 97.46\% \tag{4.467}$$

Assuming the resistance $r_{RFC} = 0.12\ \Omega$,

$$P_{rRFC} = r_{RFC} I_I^2 = 0.12 \times 0.652^2 = 0.051\ \Omega \tag{4.468}$$

The loaded quality factor is

$$Q_L = \frac{f_c}{BW} = \frac{2400}{240} = 10 \tag{4.469}$$

The resonant inductance is

$$L = \frac{R_L}{\omega_c Q_L} = \frac{18.916}{2\pi \times 2.4 \times 10^9 \times 10} = 0.1254 \text{ nH} \tag{4.470}$$

The resonant capacitance is

$$C = \frac{Q_L}{\omega_c R_L} = \frac{10}{2\pi \times 2.4 \times 10^9 \times 18.916} = 35.06 \text{ pF} \tag{4.471}$$

Assuming $r_L = 0.08\ \Omega$ and $r_C = 0.05\ \Omega$, we get

$$P_{rL} = \frac{r_L I_m^2 Q_L^2}{2} = \frac{0.08 \times 0.83^2 \times 10^2}{2} = 2.7556 \text{ W} \tag{4.472}$$

$$P_{rC} = \frac{r_C I_m^2 Q_L^2}{2} = \frac{0.05 \times 0.83^2 \times 10^2}{2} = 1.722 \text{ W} \tag{4.473}$$

Hence, the total power loss is

$$P_{Loss} = 2P_{rDS} + 2P_{RFC} + P_{rL} + P_{rC} = 2 \times 0.0425 + 2 \times 0.051 + 2.7556 + 1.722 = 4.6646 \text{ W} \tag{4.474}$$

The overall efficiency is

$$\eta = \frac{P_O}{P_O + P_{Loss}} = \frac{5}{5 + 4.6646} = 51.74\% \tag{4.475}$$

4.19 Summary

- The maximum voltage across the switches in both Class D half-bridge and full-bridge amplifiers is low and equal to the dc input voltage V_I.
- Operation with a capacitive load (i.e., below resonance) is not recommended. The antiparallel diodes turn off at a high di/dt. If the MOSFET's body–drain *pn* junction diode (or any *pn* junction diode) is used as an antiparallel diode, it generates high reverse-recovery current spikes. These spikes occur in the switch current waveforms at both the switch turn-on and turn-off and may destroy the transistor. The reverse-recovery spikes may initiate the turn-on of parasitic BJT in the MOSFET structure and may cause the MOSFET to fail due to the second breakdown of parasitic BJT. The current spikes can be reduced by adding a Schottky antiparallel diode (if V_I is below 100 V), or a series diode and an antiparallel diode.
- For operation below resonance, the transistors are turned on at a high voltage equal to V_I and the transistor output capacitance is short-circuited by a low-transistor on-resistance, dissipating the energy stored in that capacitance. Therefore, the turn-on switching loss is high, Miller's effect is significant, the transistor input capacitance is high, the gate drive power is high, and the turn-on transition speed is reduced.
- Operation with an inductive load (i.e., above resonance) is preferred. The antiparallel diodes turn off at a low di/dt. Therefore, the MOSFET's body–drain *pn* junction diodes can be used as antiparallel diodes because these diodes do not generate reverse-recovery current spikes and are sufficiently fast.
- For operation above resonance, the transistors turn on at zero voltage. For this reason, the turn-on switching loss is reduced, Miller's effect is absent, the transistor input capacitance is low, the gate drive power is low, and turn-on speed is high. However, the turn-off is lossy.
- The efficiency is high at light loads because R/r increases when R increases [see Equation (4.174)].
- The amplifier can operate safely with an open circuit at the output.
- There is a risk of catastrophic failure if the output is short-circuited when the operating frequency f approaches the resonant frequency f_o.
- The input voltage of the resonant circuit in the Class D full-bridge amplifier is a square wave whose low level is $-V_I$ and whose high level is V_I. The peak-to-peak voltage across the resonant circuit in the full-bridge amplifier is twice as high as that of the half-bridge amplifier. Therefore, the output power of the full-bridge amplifier is four times higher than that in the half-bridge amplifier at the same load resistance R, at the same dc supply voltage V_I, and at the same ratio f/f_o.
- The dc voltage source V_I and the switches form an ideal square-wave voltage source; therefore, many loads can be connected between the two switches and ground, and can be operated without mutual interactions.
- Not only MOSFETs, but other power switches can be used, such as MESFETs, BJTs, thyristors, MOS-controlled thyristors (MCTs), gate turn-off thyristors (GTOs), and insulated gate bipolar transistors (IGBTs).
- The power losses in the parasitic resistances of the series-resonant circuit are Q_L^2 times lower than those in the parallel-resonant circuit.

4.20 Review Questions

4.1 Draw the inductive reactance X_L, capacitive reactance X_C, and total reactance $X_L - X_C$ versus frequency for the series-resonant circuit. What occurs at the resonance frequency?

4.2 What is the voltage across the switches in Class D half-bridge and full-bridge amplifiers?

4.3 What is the frequency range in which a series-resonant circuit represents a capacitive load to the switching part of the Class D series-resonant circuit amplifier?

4.4 What are the disadvantages of operation of the Class D series-resonant circuit amplifier with a capacitive load?

4.5 Is the turn-on switching loss of the power MOSFETs zero below resonance?

4.6 Is the turn-off switching loss of the power MOSFETs zero below resonance?

4.7 Is Miller's effect present at turn-on or turn-off below resonance?

4.8 What is the influence of zero-voltage switching on Miller's effect?

4.9 What is the frequency range in which a series-resonant circuit represents an inductive load to the switching part of the amplifier?

4.10 What are the merits of operation of the Class D amplifier with an inductive load?

4.11 Is the turn-on switching loss of the power MOSFETs zero above resonance?

4.12 Is the turn-off switching loss of the power MOSFETs zero above resonance?

4.13 What is the voltage stress of the resonant capacitor and inductor in half-bridge and full-bridge amplifiers?

4.14 What are the worst conditions for the voltage stresses of resonant components?

4.15 What happens when the output of the amplifier is short-circuited?

4.16 Is the part-load efficiency of the Class D amplifier with a series-resonant circuit high?

4.17 What kind of switching takes place in Class D current-switching power amplifiers when the operating frequency is equal to the resonant frequency?

4.18 What is the role of the transformer in impedance matching in push–pull power amplifiers?

4.21 Problems

4.1 Design a Class D RF power amplifier to meet the following specifications: $V_I = 5$ V, $P_O = 1$ W, $f = f_o = 1$ GHz, $r_{DS} = 0.1\ \Omega$, $Q_L = 7$, $Q_{Lo} = 200$, and $Q_{Co} = 1000$.

4.2 A series-resonant circuit consists of an inductor $L = 84\ \mu$H and a capacitor $C = 300$ pF. The ESRs of these components at the resonant frequency are $r_L = 1.4\ \Omega$ and $r_C = 50$ mΩ, respectively. The load resistance is $R = 200\ \Omega$. The resonant circuit is driven by a sinusoidal voltage source whose amplitude is $V_m = 100$ V. Find the resonant frequency f_o, characteristic impedance Z_o, loaded quality factor Q_L, unloaded quality factor Q_o, quality factor of the inductor Q_{Lo}, and quality factor of the capacitor Q_{Co}.

4.3 For the resonant circuit given in Problem 4.2, find the reactive power of the inductor Q and the total real power P_O.

4.4 For the resonant circuit given in Problem 4.2, find the voltage and current stresses for the resonant inductor and the resonant capacitor. Calculate also the reactive power of the resonant components.

4.5 Find the efficiency for the resonant circuit given in Problem 4.2. Is the efficiency dependent on the operating frequency?

4.6 Write general expressions for the instantaneous energy stored in the resonant inductor $w_L(t)$ and in the resonant capacitor $w_C(t)$, as well as the total instantaneous energy stored in the resonant circuit $w_t(t)$. Sketch these waveforms for $f = f_o$. Explain briefly how the energy is transferred between the resonant components.

4.7 A Class D half-bridge amplifier is supplied by a dc voltage source of 350–400 V. Find the voltage stresses of the switches. Repeat the same problem for the Class D full-bridge amplifier.

4.8 A series-resonant circuit, which consists of a resistance $R = 25\ \Omega$, inductance $L = 100\ \mu H$, and capacitance $C = 4.7$ nF, is driven by a sinusoidal voltage source $v = 100 \sin \omega t$ (V). The operating frequency can be changed over a wide range. Calculate exactly the maximum voltage stresses for the resonant components. Compare the results with voltages across the inductance and the capacitance at the resonant frequency.

4.9 Design a Class D half-bridge series-resonant amplifier that delivers power $P_O = 30$ W to the load resistance. The amplifier is supplied from input voltage source $V_I = 180$ V. It is required that the operating frequency is $f = 210$ kHz. Neglect switching and drive power losses.

4.10 Design a full-bridge Class D power amplifier with the following specifications: $V_I = 100$ V, $P_O = 80$ W, $f = f_o = 500$ kHz, and $Q_L = 5$.

4.11 Compare the power loss in the parallel and series resonant circuits at the same loaded quality factor Q_L and the amplitude of the load current I_m.

4.12 Design a Class D RF power amplifier to meet the following specifications: $P_O = 5$ W, $V_I = 12$ V, $V_{DSmin} = 0.5$ V, and $f = f_o = 5$ GHz.

References

[1] P. J. Baxandall, "Transistor sine-wave *LC* oscillators, some general considerations and new developments," *Proceedings of the IEE*, vol. 106, Pt. B, Suppl. 16, pp. 748–758, 1959.

[2] M. R. Osborne, "Design of tuned transistor power inverters," *Electron Engineering*, vol. 40, no. 486, pp. 436–443, 1968.

[3] W. J. Chudobiak and D. F. Page, "Frequency and power limitations of Class-D transistor inverter," *IEEE Journal of Solid-State Circuits*, vol. SC-4, pp. 25–37, 1969.

[4] H. L. Krauss, C. W. Bostian, and F. H. Raab, *Solid State Radio Engineering*, New York, NY: John Wiley & Sons, Ch. 14.1-2, pp. 432–448, 1980.

[5] M. K. Kazimerczuk and J. M. Modzelewski, "Drive-transformerless Class-D voltage switching tuned power amplifier," *Proceedings of IEEE*, vol. 68, pp. 740–741, 1980.

[6] F. H. Raab, "Class-D power inverter load impedance for maximum efficiency," *RF Technology Expo'85 Conference*, Anaheim, CA, January 23-25, 1985, pp. 287–295.

[7] N. Mohan, T. M. Undeland, and W. P. Robbins, *Power Electronics, Converters, Applications and Design*, New York, NY: John Wiley & Sons, 1989, Ch. 7.4.1, pp. 164–170.

[8] M. K. Kazimierczuk, "Class D voltage-switching MOSFET power inverter," *IEE Proceedings, Pt. B, Electric Power Applications*, vol. 138, pp. 286–296, 1991.

[9] J. G. Kassakian, M. F. Schlecht, and G. C. Verghese, *Principles of Power Electronics*, Reading, MA: Addison-Wesley, 1991, Ch. 9.2, pp. 202–212.

[10] M. K. Kazimierczuk and W. Szaraniec, "Class D voltage-switching inverter with only one shunt capacitor," *IEE Proceedings, Pt. B, Electric Power Applications*, vol. 139, pp. 449–456, 1992.

[11] S.-A. El-Hamamsy, "Design of high-efficiency RF Class D power amplifier," *IEEE Transactions on Power Electronics*, vol. 9, no. 3, pp. 297–308, 1994.

[12] M. K. Kazimierczuk and A. Abdulkarim, "Current-source parallel-resonant dc/dc converter," *IEEE Transactions on Industrial Electronics*, vol. 42, no. 2, pp. 199–208, 1995.

[13] M. K. Kazimierczuk and D. Czarkowski, *Resonant Power Converters*, New York, NY: John Wiley & Sons, 1995.

[14] L. R. Neorne, "Design of a 2.5-MHz, soft-switching, class-D converter for electrodless lighting," *IEEE Transactions on Power Electronics*, vol. 12, no. 3, pp. 507–516, 1997.

[15] A. J. Frazier and M. K. Kazimierczuk, "DC-AC power inversion using sigma-delta modulation," *IEEE Transactions on Circuits and Systems-I*, vol. 46, pp. 79–82, 2000.

[16] H. Kobaysashi, J. M. Hinriehs, and P. Asbeck, "Current mode Class-D power amplifiers for high efficiency RF applications," *IEEE Transactions on Microwave Theory and Technique*, vol. 49, no. 12, pp. 2480–2485, 2001.

[17] K. H. Abed, K. Y. Wong, and M. K. Kazimierczuk, "Implementations of novel low-power drivers for integrated buck converter," *IEEE Midwest Symposium on Circuits and Systems*, 2005.

[18] K. H. Abed, K. Y. Wong, and M. K. Kazimierczuk, "CMOS zero-cross-conduction low-power driver and power MOSFETs for integrated synchronous buck converter," *IEEE International Symposium on Circuits and Systems*, 2006, pp. 2745–2748.

[19] M. K. Kazimierczuk, *Pulse-Width Modulated DC-DC Power Converters*. New York, NY: John Wiley & Sons, 2008.

[20] C. Ekkaravarodome, K. Jirasereemornkul, and M. K. Kazimierczuk, "Class-D zero-current-switching rectifier as power-factor corrector for lighting applications," *IEEE Transactions on Power Electronics*, vol. 29, no. 9, pp. 4938–4948, September 2014.

5

Class E Zero-Voltage Switching RF Power Amplifiers

5.1 Introduction

There are two types of Class E power amplifiers [1–97], also called Class E dc–ac inverters: (1) Class E zero-voltage switching (ZVS) power amplifiers, which are the subject of this chapter, and (2) Class E zero-current switching (ZCS) power amplifiers. In Class E amplifiers, the transistor is operated as a switch. Class E ZVS power amplifiers [1–35] are the most efficient amplifiers known so far. The current and voltage waveforms of the switch are displaced with respect to time, yielding a very low-power dissipation in the transistor. In particular, the switch turns on at zero voltage if the component values of the resonant circuit are properly chosen. Since the switch current and voltage waveforms do not overlap during the switching time intervals, switching losses are virtually zero, yielding high efficiency.

We shall start by presenting a simple qualitative description of the operation of the Class E ZVS amplifier. Although simple, this description provides considerable insight into the performance of the amplifier as a basic power cell. Further, we shall quickly move to the quantitative description of the amplifier. Finally, we will present matching resonant circuits and give a design procedure for the amplifier. By the end of this chapter, the reader will be able to perform rapid first-order analysis as well as design a single-stage Class E ZVS amplifier.

5.2 Circuit Description

The basic circuit of the Class E ZVS power amplifier is shown in Fig. 5.1(a). It consists of the power MOSFET operating as a switch, *L*-*C*-*R* series-resonant circuit, shunt capacitor C_1,

RF Power Amplifiers, Second Edition. Marian K. Kazimierczuk.

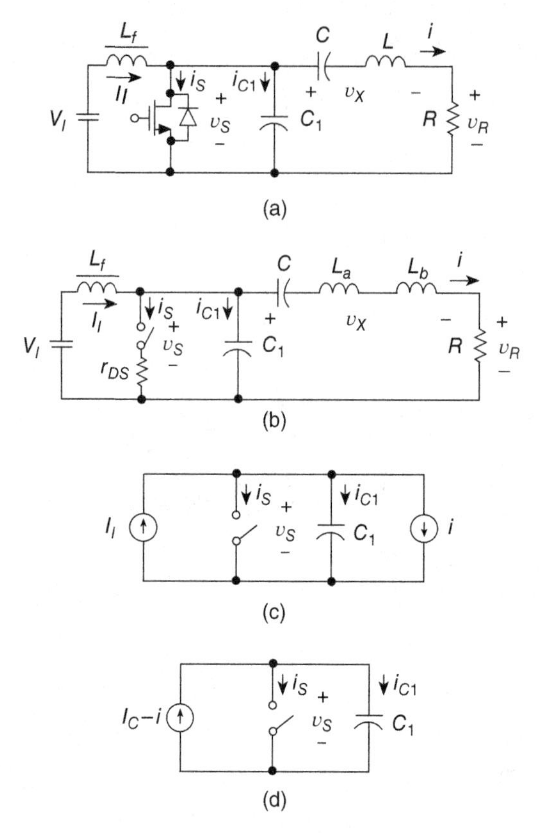

Figure 5.1 Class E zero-voltage switching RF power amplifier. (a) Circuit. (b) Equivalent circuit for operation above resonance. (c) Equivalent circuit with the dc voltage source V_I and the RF choke L_f replaced by a dc current source I_I and the series-resonant circuit replaced by an ac current source i. (d) Equivalent circuit with the two current sources combined into one current source $I_I - i$.

and choke inductor L_f. The switch turns on and off at the operating frequency $f = \omega/(2\pi)$ determined by a driver. The transistor output capacitance, the choke parasitic capacitance, and stray capacitances are included in the shunt capacitance C_1. For high operating frequencies, all of the capacitance C_1 can be supplied by the overall shunt parasitic capacitance. The resistor R is an ac load. The choke inductance L_f is assumed to be high enough so that the ac current ripple on the dc supply current I_I can be neglected. A small inductance with a large current ripple is also possible [41].

When the switch is ON, the resonant circuit consists of L, C, and R because the capacitance C_1 is short-circuited by the switch. However, when the switch is OFF, the resonant circuit consists of C_1, L, C, and R connected in series. Because C_1 and C are connected in series, the equivalent capacitance

$$C_{eq} = \frac{CC_1}{C + C_1} \tag{5.1}$$

is lower than C and C_1. The load network is characterized by two resonant frequencies and two loaded quality factors. When the switch is ON,

$$f_{o1} = \frac{1}{2\pi\sqrt{LC}} \tag{5.2}$$

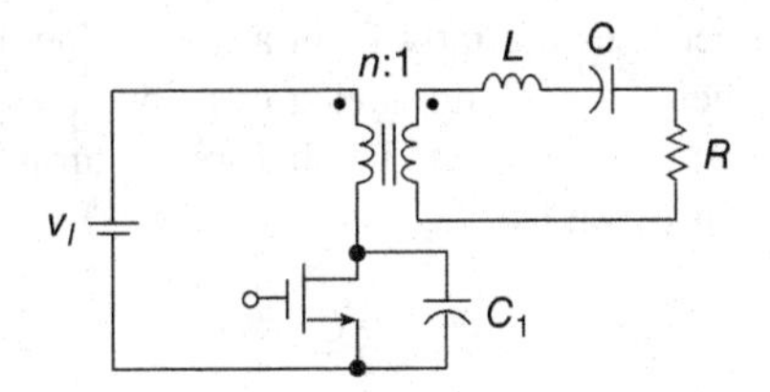

Figure 5.2 Class E zero-voltage switching RF power amplifier with a transformer.

and

$$Q_{L_1} = \frac{\omega_{o1} L}{R} = \frac{1}{\omega_{o1} CR} \tag{5.3}$$

When the switch is OFF,

$$f_{o2} = \frac{1}{2\pi\sqrt{\frac{LCC_1}{C + C_1}}} \tag{5.4}$$

and

$$Q_{L_2} = \frac{\omega_{o2} L}{R} = \frac{1}{\frac{\omega_{o2} RCC_1}{C + C_1}} \tag{5.5}$$

The ratio of the two resonant frequencies is

$$\frac{f_{o1}}{f_{o2}} = \frac{Q_{L_1}}{Q_{L_2}} = \sqrt{\frac{C_1}{C_1 + C}} \tag{5.6}$$

An equivalent circuit of the amplifier for operation above resonance is shown in Fig. 5.1(b). In Fig. 5.1(c), the dc source V_I and radio-frequency choke (RFC) L_f are replaced by a dc current source I_I and the series-resonant circuit is replaced by an ac current source i. Figure 5.1(d) shows an equivalent circuit of the Class E amplifier with the two current sources combined into one current source $I_I - i$.

If the operating frequency f is greater than the resonant frequency f_{o1}, the L-C-R series-resonant circuit represents an inductive load at the operating frequency f. Therefore, the inductance L can be divided into two inductances, L_a and L_b, connected in series such that $L = L_a + L_b$ and L_a resonates with C at the operating frequency f, that is,

$$\omega = \frac{1}{\sqrt{L_a C}} \tag{5.7}$$

The loaded quality factor is defined at the operating frequency as

$$Q_L = \frac{\omega L}{R} = \frac{\omega(L_a + L_b)}{R} = \frac{1}{\omega CR} + \frac{\omega L_b}{R} \tag{5.8}$$

A transformer version of the Class E amplifier is shown in Fig. 5.2. The transformer leakage inductance can be absorbed into the resonant inductance L.

5.3 Circuit Operation

Circuits with *hard-switching* operation of semiconductor components, such as pulse width modulation (PWM) power converters and digital gates, suffer from switching losses. The voltage waveform in these circuits decreases abruptly from a high value, often equal to the dc supply voltage V_I, to nearly zero, when a switching device turns on. The energy stored in the transistor output capacitance and load capacitance C just before the turn-on transitions (assuming that these capacitances are linear) is given by

$$W = \frac{1}{2}CV_I^2 \tag{5.9}$$

where V_I is the dc supply voltage. When the transistor is turned on, the current is circulating through the transistor on-resistance r_{DS}, and all the stored energy is lost in the on-resistance r_{DS} as heat. This energy is independent of the transistor on-resistance r_{DS}. The switching power loss of the transistor is given by

$$P_{sw} = \frac{1}{2}fCV_I^2 \tag{5.10}$$

The switching losses can be avoided if the voltage across the transistor v_S is zero, when the transistor turns on

$$v_S(t_{turn-on}) = 0 \tag{5.11}$$

Then the charge stored in the transistor output capacitance is zero, and the energy stored in this capacitance is zero. The main idea of the Class E radio-frequency (RF) power amplifier is that the transistor turns on as a switch at zero voltage, resulting in zero switching loss and high efficiency. The Class E power amplifier in its basic form contains a single switch. The switch turns on at zero voltage (ZVS), and the switch may also turn on at zero derivative (ZDS). In general, this type of operation is called *soft-switching*.

Figure 5.3 shows the current and voltage waveforms in the Class E ZVS amplifier for three cases: (1) $dv_S(\omega t)/d(\omega t) = 0$, (2) $dv_S(\omega t)/d(\omega t) < 0$, and (3) $dv_S(\omega t)/d(\omega t) > 0$ at $\omega t = 2\pi$ when the switch turns on. In all three cases, the voltage v_S across the switch and the shunt capacitance C_1 is zero when the switch turns on. Therefore, the energy stored in the shunt capacitance C_1 is zero when the switch turns on, yielding zero turn-on switching loss. Thus, the *ZVS condition* is expressed as

$$v_S(2\pi) = 0 \tag{5.12}$$

The choke inductor L_f forces a dc current I_I. To achieve ZVS turn-on of the switch, the operating frequency $f = \omega/(2\pi)$ should be greater than the resonant frequency $f_{o1} = 1/(2\pi\sqrt{LC})$, that is, $f > f_{o1}$. However, the operating frequency is usually lower than $f_{o2} = 1/(2\pi\sqrt{LC_{eq}})$, that is, $f < f_{o2}$. The shape of the waveform of current i depends on the loaded quality factor. If Q_L is high (i.e., $Q_L \geq 2.5$), the shape of the waveform of current i is approximately sinusoidal. If Q_L is low, the shape of the waveform of the current i becomes close to an exponential function [22]. The combination of the choke inductor L_f and the L-C-R series-resonant circuit acts as a current source whose current is I_I-i. When the switch is ON, the current I_I-i flows through the switch. When the switch is OFF, the current I_I-i flows through the capacitor C_1, producing the voltage across the shunt capacitor C_1 and the switch. Therefore, the shunt capacitor C_1 shapes the voltage across the switch.

5.4 ZVS and ZDS Operations of Class E Amplifier

When the transistor turns on at $\omega t = 2\pi$ in the Class E power amplifier, the ZVS condition and *zero-derivative switching* (ZDS) condition are satisfied

$$v_S(2\pi) = 0 \tag{5.13}$$

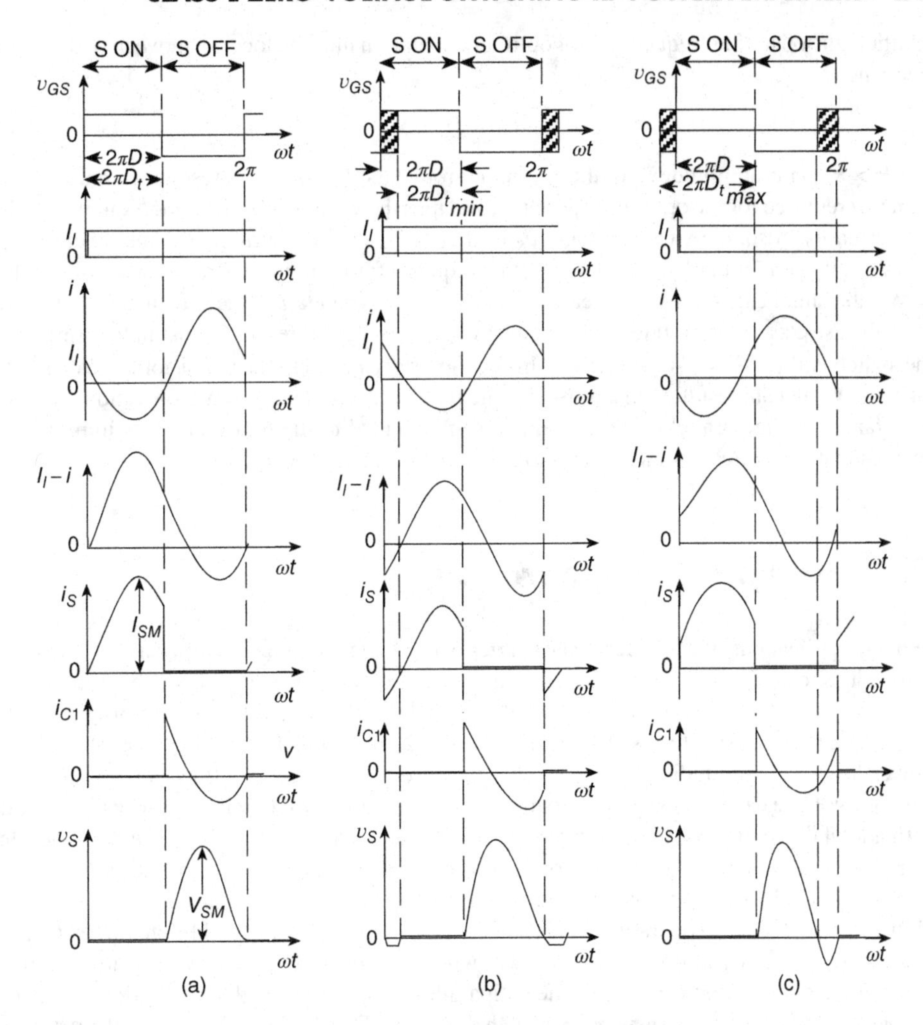

Figure 5.3 Waveforms in the Class E zero-voltage switching amplifier. (a) For optimum operation. (b) For suboptimum operation with $dv_S(\omega t)/d(\omega t) < 0$ at $\omega t = 2\pi$. (c) For suboptimum operation with $dv_S(\omega t)/d(\omega t) > 0$ at $\omega t = 2\pi$.

and

$$\frac{dv_S(\omega t)}{d(\omega t)}\Big|_{\omega t=2\pi} = 0 \tag{5.14}$$

Current and voltage waveforms for ZVS and ZDS operations are shown in Fig. 5.3(a). ZVS implies that the energy stored in the shunt capacitance C_1 is zero when the transistor turns on, yielding zero turn-on switching loss. Because the derivative of v_S is zero at the time when the switch turns on, the switch current i_S increases gradually from zero after the switch is closed. The operation for which both the ZVS and ZDS conditions are satisfied simultaneously is called the *nominal operation*or *optimum operation.* Both the switch voltage and the switch current waveforms are positive for the optimum operation. Therefore, there is no need to add any diode to the switch.

Relationships among C_1, L_b, R, f, and D must be satisfied to achieve optimum operation [22]. Therefore, optimum operation is achieved only for the optimum load resistance $R = R_{opt}$. In addition, the operating frequency f for optimum operation must be located between two resonant frequencies

$$f_{o1} < f < f_{o2} \tag{5.15}$$

If $R > R_{opt}$, the amplitude I_m of the current i through the L-C-R series-resonant circuit is lower than that required for the optimum operation, and thus the voltage drop across the shunt capacitor C_1 decreases. Also, the switch voltage v_S is greater than zero at turn-on. On the other hand, if $R < R_{opt}$, the amplitude I_m is higher than that required for optimum operation, the voltage drop across the shunt capacitor C_1 increases, and the switch voltage v_S is less than zero at turn-on. In both cases, assuming a linear capacitance C_1, the energy stored in C_1 just before turn-on of the switch is $W(2\pi-) = \frac{1}{2}C_1 v_S^2(2\pi-)$. This energy is dissipated in the transistor as heat after the switch is turned on, resulting in a turn-on switching loss. To obtain the ZVS operation at a wider load range, an antiparallel or a series diode can be added to the transistor. This improvement ensures that the switch automatically turns on at zero voltage for $R \leq R_{opt}$.

5.5 Suboptimum Operation

In many applications, the load resistance varies over a certain range. The turn-on of the switch at zero voltage can be achieved for suboptimum operation for $0 \leq R \leq R_{opt}$. For suboptimum operation, $v_S(2\pi) = 0$ and either $dv_S(\omega t)/d(\omega t) < 0$ or $dv_S(\omega t)/d(\omega t) > 0$. Figure 5.3(b) shows the current and voltage waveforms for the case when $v_S(2\pi) = 0$ and $dv_S(\omega t)/d(\omega t) < 0$ at $\omega t = 2\pi$. Power MOSFETs are bidirectional switches because their current can flow in both directions, but their voltage can only be greater than −0.7 V. When the switch voltage reaches −0.7 V, the antiparallel diode turns on and, therefore, the switch automatically turns on. The diode accelerates the time at which the switch turns on. This time is no longer determined by the gate-to-source voltage. Since the switch turns on at zero voltage, the turn-on switching loss is zero, yielding high efficiency. Such an operation can be achieved for $0 \leq R \leq R_{opt}$. In addition, if $R < R_{opt}$, the operating frequency f and the transistor ON switch duty cycle D_t can vary in bounded ranges. When the switch current is negative, the antiparallel diode is ON, but the transistor can either be ON or OFF. Therefore, the transistor ON switch duty cycle D_t is less than or equal to the ON switch duty cycle of the entire switch D. When the switch current is positive, the diode is OFF and the transistor must be ON. Hence, the range of D_t is $D_{tmin} \leq D_t \leq D$, as indicated in Fig. 5.3(b) by the shaded area.

Figure 5.3(c) depicts current and voltage waveforms for the case when $v_S(2\pi) = 0$ and $dv_S(\omega t)/d(\omega t) > 0$ at $\omega t = 2\pi$. Notice that the switch current i_S is always positive, but the switch voltage v_S has positive and negative values. Therefore, a unidirectional switch for current and a bidirectional switch for voltage is needed. Such a switch can be obtained by adding a diode in series with a MOSFET. When the switch voltage v_S is negative the diode is OFF and supports the switch voltage, regardless of the state of the MOSFET. The MOSFET is turned on during the time interval when the switch voltage is negative. Once the switch voltage reaches 0.7 V with a positive derivative, the diode turns on, turning the entire switch on. The series diode delays the time at which the switch turns on. The range of D_t is $D \leq D_t \leq D_{tmax}$, as shown in Fig. 5.3(c) by the shaded area. The disadvantages of the switch with a series diode are higher on-voltage and higher conduction loss. Another disadvantage is associated with the transistor output capacitance. When the switch voltage increases, the transistor output capacitance is charged through the series diode to the peak value of the switch voltage and then remains at this voltage until the transistor turns on (because the diode is OFF). At this time, the transistor output capacitance is discharged through the MOSFET on-resistance, dissipating the stored energy.

5.6 Analysis

5.6.1 Assumptions

The analysis of the Class E ZVS amplifier of Fig. 5.1(a) is carried out under the following assumptions:

1) The transistor and the antiparallel diode form an ideal switch whose on-resistance is zero, off-resistance is infinity, and switching times are zero.
2) The choke inductance is high enough so that the ac component is much lower than the dc component of the input current.
3) The loaded quality factor Q_L of the LCR series-resonant circuit is high enough so that the current i through the resonant circuit is sinusoidal.
4) The duty ratio D is 0.5.

5.6.2 Current and Voltage Waveforms

The current through the series-resonant circuit is sinusoidal and given by

$$i = I_m \sin(\omega t + \phi) \tag{5.16}$$

where I_m is the amplitude and ϕ is the initial phase of the current i. According to Fig. 5.1(a),

$$i_S + i_{C1} = I_I - i = I_I - I_m \sin(\omega t + \phi) \tag{5.17}$$

For the time interval $0 < \omega t \leq \pi$, the switch is ON and, therefore $i_{C1} = 0$. Consequently, the current through the MOSFET is given by

$$i_S = \begin{cases} I_I - I_m \sin(\omega t + \phi) & \text{for} \quad 0 < \omega t \leq \pi \\ 0 & \text{for} \quad \pi < \omega t \leq 2\pi \end{cases} \tag{5.18}$$

For the time interval $\pi < \omega t \leq 2\pi$, the switch is OFF which implies that $i_S = 0$. Hence, the current through the shunt capacitor C_1 is given by

$$i_{C1} = \begin{cases} 0 & \text{for} \quad 0 < \omega t \leq \pi \\ I_I - I_m \sin(\omega t + \phi) & \text{for} \quad \pi < \omega t \leq 2\pi \end{cases} \tag{5.19}$$

The voltage across the shunt capacitor and the switch is

$$\begin{aligned} v_S = v_{C1} &= \frac{1}{\omega C_1}\int_{\pi}^{\omega t} i_{C1}\, d(\omega t) = \frac{1}{\omega C_1}\int_{\pi}^{\omega t} [I_I - I_m \sin(\omega t + \phi)] d(\omega t) \\ &= \begin{cases} 0 & \text{for} \quad 0 < \omega t \leq \pi \\ \dfrac{1}{\omega C_1}\{I_I(\omega t - \pi) + I_m[\cos(\omega t + \phi) + \cos\phi]\} & \text{for} \quad \pi < \omega t \leq 2\pi \end{cases} \end{aligned} \tag{5.20}$$

Substitution of the ZVS condition $v_S(2\pi) = 0$ into (5.20) produces a relationship among I_I, I_m, and ϕ

$$I_m = -I_I \frac{\pi}{2\cos\phi} \tag{5.21}$$

Substitution of (5.21) into (5.18) yields the switch current waveform

$$\frac{i_S}{I_I} = \begin{cases} 1 + \dfrac{\pi}{2\cos\phi}\sin(\omega t + \phi) & \text{for} \quad 0 < \omega t \leq \pi \\ 0 & \text{for} \quad \pi < \omega t \leq 2\pi \end{cases} \tag{5.22}$$

Likewise, substituting (5.21) into (5.19), one obtains the current through the shunt capacitor C_1

$$\frac{i_{C1}}{I_I} = \begin{cases} 0 & \text{for} \quad 0 < \omega t \leq \pi \\ 1 + \dfrac{\pi}{2\cos\phi}\sin(\omega t + \phi) & \text{for} \quad \pi < \omega t \leq 2\pi \end{cases} \tag{5.23}$$

From (5.21), equation (5.20) becomes

$$v_S = \begin{cases} 0 & \text{for} \quad 0 < \omega t \leq \pi \\ \dfrac{I_I}{\omega C_1}\left\{\omega t - \dfrac{3\pi}{2} - \dfrac{\pi}{2\cos\phi}[\cos(\omega t + \phi)]\right\} & \text{for} \quad \pi < \omega t \leq 2\pi \end{cases} \tag{5.24}$$

Using the ZDS condition $dv_S/d(\omega t) = 0$ at $\omega t = 2\pi$, one obtains the phase of the output current waveform i

$$\tan\phi = -\frac{2}{\pi} \tag{5.25}$$

from which

$$\phi = \pi - \arctan\left(\frac{2}{\pi}\right) = 2.5747 \text{ rad} = 147.52^\circ \tag{5.26}$$

Using trigonometric relationships,

$$\sin\phi = \frac{2}{\sqrt{\pi^2 + 4}} \tag{5.27}$$

and

$$\cos\phi = -\frac{\pi}{\sqrt{\pi^2 + 4}} \tag{5.28}$$

Substitution of (5.28) into (5.21) yields the amplitude of the output current

$$I_m = \frac{\sqrt{\pi^2 + 4}}{2} I_I \approx 1.8621 I_I \tag{5.29}$$

yielding

$$\gamma_1 = \frac{I_m}{I_I} = \frac{\sqrt{\pi^2 + 4}}{2} \approx 1.8621 \tag{5.30}$$

Hence,

$$\frac{i_S}{I_I} = \begin{cases} 1 - \dfrac{\sqrt{\pi^2 + 4}}{2}\sin(\omega t + \phi) & \text{for} \quad 0 < \omega t \leq \pi \\ 0 & \text{for} \quad \pi < \omega t \leq 2\pi \end{cases} \tag{5.31}$$

$$\frac{i_{C1}}{I_I} = \begin{cases} 0 & \text{for} \quad 0 < \omega t \leq \pi \\ 1 - \dfrac{\sqrt{\pi^2 + 4}}{2}\sin(\omega t + \phi) & \text{for} \quad \pi < \omega t \leq 2\pi \end{cases} \tag{5.32}$$

and

$$v_S = \begin{cases} 0 & \text{for} \quad 0 < \omega t \leq \pi \\ \dfrac{I_I}{\omega C_1}\left(\omega t - \dfrac{3\pi}{2} - \dfrac{\pi}{2}\cos\omega t - \sin\omega t\right) & \text{for} \quad \pi < \omega t \leq 2\pi \end{cases} \tag{5.33}$$

The dc component of the voltage across an ideal choke inductor is zero. From (5.24), the dc input voltage is found as

$$V_I = \frac{1}{2\pi}\int_{\pi}^{2\pi} v_S \, d(\omega t) = \frac{I_I}{2\pi\omega C_1}\int_{\pi}^{2\pi}\left(\omega t - \frac{3\pi}{2} - \frac{\pi}{2}\cos\omega t - \sin\omega t\right) d(\omega t) = \frac{I_I}{\pi\omega C_1} \tag{5.34}$$

Rearrangement of this equation produces the dc input resistance of the Class E amplifier

$$R_{DC} \equiv \frac{V_I}{I_I} = \frac{1}{\pi\omega C_1} \tag{5.35}$$

Hence,

$$I_m = \frac{\sqrt{\pi^2+4}}{2}\pi\omega C_1 V_I \tag{5.36}$$

From (5.33) and (5.35), one arrives at the normalized switch voltage waveform

$$\frac{v_S}{V_I} = \begin{cases} 0 & \text{for} \quad 0 < \omega t \leq 2\pi \\ \pi\left(\omega t - \dfrac{3\pi}{2} - \dfrac{\pi}{2}\cos\omega t - \sin\omega t\right) & \text{for} \quad \pi < \omega t \leq 2\pi \end{cases} \tag{5.37}$$

5.6.3 Current and Voltage Stresses

Differentiating (5.22),

$$\frac{di_S}{d(\omega t)} = -I_I\frac{\sqrt{\pi^2+4}}{2}\cos(\omega t + \phi) = 0 \tag{5.38}$$

one obtains the value of ωt at which the peak value of the switch current occurs

$$\omega t_{im} = \frac{3\pi}{2} - \phi = 270^\circ - 147.52^\circ = 122.48^\circ \tag{5.39}$$

Substitution of this into (5.22) yields the switch peak current

$$I_{SM} = I_I\left(\frac{\sqrt{\pi^2+4}}{2} + 1\right) = 2.862 I_I \tag{5.40}$$

Differentiating the switch voltage waveform v_S in (5.24)

$$\frac{dv_S}{d(\omega t)} = \pi V_I\left[1 + \frac{\pi}{2\cos\phi}\sin(\omega t + \phi)\right] = 0 \tag{5.41}$$

yields the trigonometric equation

$$\sin(\omega t_{vm} + \phi) = -\frac{2\cos\phi}{\pi} = \frac{2}{\sqrt{\pi^2+4}} = \sin\phi = \sin(\pi - \phi) = \sin(2\pi + \pi - \phi) \tag{5.42}$$

the solution of which gives the value of ωt at which the peak value of the switch voltage occurs

$$\omega t_{vm} = 3\pi - 2\phi = 3 \times 180^\circ - 2 \times 147.52^\circ = 244.96^\circ \tag{5.43}$$

Hence,

$$V_{SM} = 2\pi(\pi - \phi)V_I = 2\pi(\pi - 2.5747)V_I = 3.562V_I \tag{5.44}$$

The amplitude of the voltage across the resonant capacitor C is

$$V_{Cm} = |X_C(f)|I_m = \frac{I_m}{\omega C} \tag{5.45}$$

and the amplitude of the voltage across the resonant inductor L is

$$V_{Lm} = X_L(f)I_m = \omega L I_m \tag{5.46}$$

Neglecting the power losses, the ac output power P_O is equal to the dc input power $P_I = V_I I_I$. Hence, using I_{SM}/I_I and V_{SM}/V_I, one obtains the power-output capability

$$c_p \equiv \frac{P_O}{I_{SM}V_{SM}} = \frac{I_I V_I}{I_{SM}V_{SM}} = \frac{1}{\pi(\pi - \phi)(\sqrt{\pi^2 + 4} + 2)} = \frac{1}{2.862} \times \frac{1}{3.562} = 0.0981 \tag{5.47}$$

5.6.4 Voltage Amplitudes Across the Series-Resonant Circuit

The current through the series-resonant circuit is sinusoidal. Consequently, higher harmonics of the input power are zero. Therefore, it is sufficient to consider the input impedance of the series-resonant circuit at the operating frequency f. Figure 5.4 shows an equivalent circuit of the series-resonant circuit above resonance at the operating frequency f. A phasor diagram of the voltages at the fundamental component is depicted in Fig. 5.5. The voltage across the load resistance R is

$$v_R = iR = V_{Rm}\sin(\omega t + \phi) \tag{5.48}$$

where $V_{Rm} = RI_m$ is the amplitude of the output voltage and ϕ is the initial phase. The voltage v_X across the L-C components is nonsinusoidal, equal to $v_X = v_S - v_R$. The fundamental component

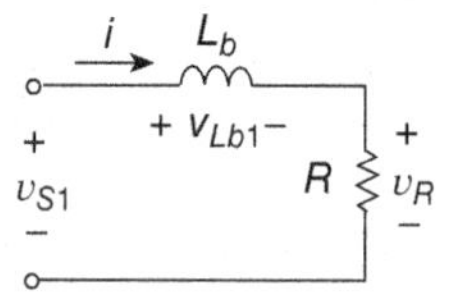

Figure 5.4 Equivalent circuit of the series-resonant circuit above resonance at the operating frequency f.

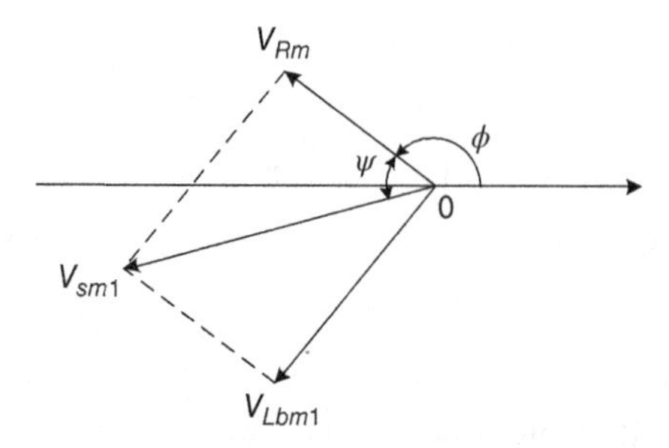

Figure 5.5 Phasor diagram of the voltages at the operating frequency f.

of the voltage across the C-L_a circuit is zero because the reactance of this circuit at the operating frequency f is zero. The fundamental component of the voltage across the inductance L_b is

$$v_{Lb1} = V_{Lbm1}\cos(\omega t + \phi) \tag{5.49}$$

where $V_{Lbm1} = \omega L_b I_m$. The fundamental component of the switch voltage, which is the input voltage of the series-resonant circuit at the operating frequency, is given by

$$v_{s1} = v_R + v_{Lb1} = V_{Rm}\sin(\omega t + \phi) + V_{Lbm1}\cos(\omega t + \phi) \tag{5.50}$$

Using (5.24) and the Fourier trigonometric series formula,

$$\begin{aligned} V_{Rm} &= \frac{1}{\pi}\int_{\pi}^{2\pi} v_S \sin(\omega t + \phi)d(\omega t) \\ &= \frac{1}{\pi}\int_{\pi}^{2\pi} V_I\pi\left[\omega t - \frac{3\pi}{2} - \frac{\pi}{2\cos\phi}\cos(\omega t + \phi)\right]\sin(\omega t + \phi)d(\omega t) \\ &= \frac{4}{\sqrt{\pi^2+4}}V_I \approx 1.074V_I \end{aligned} \tag{5.51}$$

The waveform of the output voltage is

$$v_o = V_{Rm}\sin(\omega t + \phi) = \frac{4}{\sqrt{\pi^2+4}}\sin(\omega t + 147.52^\circ) \tag{5.52}$$

Substituting (5.24) into the Fourier formula and using (5.34), the amplitude of the fundamental component of the voltage across the input reactance of the series-resonant circuit (equal to the reactance of the inductance L_b) is obtained as

$$\begin{aligned} V_{Lbm1} &= \omega L_b I_m = \frac{1}{\pi}\int_{\pi}^{2\pi} v_S \cos(\omega t + \phi)d(\omega t) \\ &= \frac{1}{\pi}\int_{\pi}^{2\pi} V_I\pi\left[\omega t - \frac{3\pi}{2} - \frac{\pi}{2\cos\phi}\cos(\omega t + \phi)\right]\cos(\omega t + \phi)d(\omega t) = \frac{\pi(\pi^2-4)}{4\sqrt{\pi^2+4}}V_I \\ &\approx 1.2378V_I \end{aligned} \tag{5.53}$$

The waveform of the fundamental component of the voltage across the inductance L_b is

$$v_{Lb1} = V_{Lbm1}\cos(\omega t + \phi) = \frac{\pi(\pi^2-4)}{\sqrt{4(\pi^2+4)}}\cos(\omega t + 147.52^\circ) \tag{5.54}$$

The amplitude of the fundamental component of the switch voltage is

$$V_{sm1} = \sqrt{V_{Rm}^2 + V_{Lbm1}^2} = V_I\sqrt{\frac{16}{\pi^2+4} + \frac{\pi^2(\pi^2-4)^2}{4(\pi^2+4)}} = \sqrt{\frac{256+\pi^2(\pi^2-4)^2}{16(\pi^2+4)^2}} = 1.7279V_I \tag{5.55}$$

yielding

$$\xi_1 = \frac{V_{sm1}}{V_I} = \sqrt{\frac{16}{\pi^2+4} + \frac{\pi^2(\pi^2-4)^2}{16(\pi^2+4)^2}} = \sqrt{\frac{256+\pi^2(\pi^2-4)^2}{16(\pi^2+4)^2}} = 1.7279 \tag{5.56}$$

The phase of the fundamental component of the switch voltage is

$$\theta = \phi + \psi = 147.52^\circ + 49.0525^\circ = 196.5725^\circ \tag{5.57}$$

The fundamental component of the switch voltage is

$$v_{s1} = V_{sm1}\sin(\omega t + \theta) = 1.7279V_I\sin(\omega t + 196.57^\circ) \tag{5.58}$$

5.6.5 Component Values of the Load Network

Combining (5.21), (5.34), and (5.51), we get

$$R = \frac{V_{Rm}}{I_m} = \frac{\dfrac{4}{\sqrt{\pi^2+4}}V_I}{\dfrac{\sqrt{\pi^2+4}}{2}\pi\omega C_1 V_I} = \frac{8}{\pi(\pi^2+4)\omega C_1} \tag{5.59}$$

resulting in

$$\omega C_1 R = \frac{8}{\pi(\pi^2+4)} \approx 0.1836 \tag{5.60}$$

and

$$\omega C_1 = \frac{P_O}{\pi V_I^2} \tag{5.61}$$

Similarly, using (5.21), (5.34), and (5.53),

$$X_{Lb} = \omega L_b = \frac{V_{Lbm1}}{I_m} = \frac{\dfrac{\pi(\pi^2-4)}{4\sqrt{\pi^2+4}}V_I}{\dfrac{\sqrt{\pi^2+4}}{2}\pi\omega C_1 V_I} = \frac{\pi^2-4}{2(\pi^2+4)\omega C_1} \tag{5.62}$$

producing

$$\omega^2 L_b C_1 = \frac{\pi^2-4}{2(\pi^2+4)} \approx 0.2116 \tag{5.63}$$

The ratio of (5.63) to (5.60) is

$$\tan\psi = \frac{\omega^2 L_b C_1}{\omega C_1 R} = \frac{\omega L_b}{R} = \frac{X_{Lb}}{R} = \frac{\pi(\pi^2-4)}{16} \approx 1.1525 \tag{5.64}$$

yielding

$$\psi = \arctan\left(\frac{V_{Lbm1}}{V_{Rm}}\right) = \arctan\left(\frac{\omega L_b}{R}\right) = \arctan\left[\frac{\pi(\pi^2-4)}{16}\right] = 49.05° \tag{5.65}$$

From (5.7), (5.8), and (5.64), the reactance of the resonant inductor is

$$\omega L = Q_L R \tag{5.66}$$

and the reactance of the resonant capacitor is

$$X_C = \frac{1}{\omega C} = \omega L_a = \omega(L - L_b) = Q_L R - \omega L_b = R\left(Q_L - \frac{\omega L_b}{R}\right)$$
$$= R\left[Q_L - \frac{\pi(\pi^2-4)}{16}\right] = R(Q_L - 1.1525) \tag{5.67}$$

The impedance of the series combination of components R and L_b at the operating frequency in terms of R is

$$Z = R + jX_{Lb} = R + j\omega L_b = R\left(1 + j\frac{\omega L_b}{R}\right) = R\left[1 + j\frac{\pi(\pi^2-4)}{16}\right]$$
$$\approx (1 + j1.1525)R = 1.52586^{j49.05°} R \ (\Omega) \tag{5.68}$$

and in terms of ωC_1 is

$$Z = R + j\omega L_b = \frac{8}{\pi(\pi^2+4)\omega C_1} + j\frac{\pi^2-4}{2(\pi^2+4)\omega C_1} = \frac{1}{(\pi^2+4)\omega C_1}\left(\frac{8}{\pi} + j\frac{\pi^2-4}{2}\right)$$
$$= \frac{1}{\pi^2+4}\sqrt{\frac{64}{\pi^2} + \frac{(\pi^2-4)^2}{4}}\,\frac{1}{\omega C_1}e^{j49.05^\circ} = \frac{0.28}{\omega C_1}e^{j49.05^\circ}\,(\Omega) \tag{5.69}$$

The reactance of the shunt capacitance at the operating frequency is

$$\frac{1}{j\omega C_1} = -jX_{C1} = -j\frac{\pi(\pi^2+4)R}{8} \approx -j5.44658R\,(\Omega) \tag{5.70}$$

The input impedance of the load network at the operating frequency is

$$Z_i = Z||\left(\frac{1}{j\omega C1}\right) = R\left[1 + j\frac{\pi(\pi^2-4)}{16}\right] || \left[-j\frac{\pi(\pi^2+4)R}{8}\right]$$
$$= \left[\frac{4\pi^6 + 32\pi^4 + 64\pi^2}{\pi^6 + 24\pi^4 + 144\pi^2 + 256} + j\frac{\pi(\pi^8 + 12\pi^6 - 16\pi^4 - 448\pi^2 - 1024)}{8\pi^6 + 192\pi^4 + 1152\pi^2 + 2048}\right]R$$
$$\approx (1.5261 + j1.1064)R \tag{5.71}$$

5.6.6 Output Power

From (5.51), one obtains the output power

$$P_O = \frac{V_{Rm}^2}{2R} = \frac{8}{\pi^2+4}\frac{V_I^2}{R} \approx 0.5768\frac{V_I^2}{R} \tag{5.72}$$

Since $R = 8/[\pi(\pi^2+4)\omega C_1]$, the output power is

$$P_O = \pi\omega C_1 V_I^2 = 2\pi^2 f C_1 V_I^2 \tag{5.73}$$

Because $1/R = \pi(\pi^2-4)/(16\omega L_b)$, the output power is

$$P_O = \frac{\pi(\pi^2-4)}{2(\pi^2+4)}\frac{V_I^2}{\omega L_b} \tag{5.74}$$

5.7 Drain Efficiency of Ideal Class E Amplifier

The drain efficiency of the Class E RF power amplifier with ideal components for optimum operation is

$$\eta_D = \frac{P_O}{P_I} = \frac{1}{2}\left(\frac{I_m}{I_I}\right)\left(\frac{V_m}{V_I}\right)\cos\psi = \frac{1}{2}\gamma_1\xi_1\cos\psi$$
$$= \frac{1}{2} \times \frac{\sqrt{\pi^2+4}}{2} \times \sqrt{\frac{256 + \pi^2(\pi^2-4)^2}{16(\pi^2+4)}} \times \frac{16}{\sqrt{256 + \pi^2(\pi^2-4)^2}} = 1 \tag{5.75}$$

5.8 RF Choke Inductance

In order to reduce the ripple input current, the RFC must have a large enough inductance $L_f > L_{fmin}$. When the switch is ON, the voltage across the choke inductor is

$$v_{L_f} = V_I \tag{5.76}$$

and the current through the choke inductance L_f is

$$i_{L_f} = \frac{1}{L_f}\int_0^t v_{L_f}\,dt + i_{L_f}(0) = \frac{1}{L_f}\int_0^t V_I\,dt + i_{L_f}(0) = \frac{V_I}{L_f}t + i_{L_f}(0) \tag{5.77}$$

Hence,

$$i_{L_f}\left(\frac{T}{2}\right) = \frac{V_I T}{2L_f} + i_{L_f}(0) = \frac{V_I}{2fL_f} + i_{L_f}(0) \tag{5.78}$$

The peak-to-peak current ripple through the choke inductor is

$$\Delta i_{L_f} = i_{L_f}\left(\frac{T}{2}\right) - i_{L_f}(0) = \frac{V_I}{2fL_f} \tag{5.79}$$

Thus, the minimum choke inductance required for the maximum peak-to-peak current ripple is

$$L_{fmin} = \frac{V_I}{2f\Delta i_{L_f max}} = \frac{V_I}{2fI_I\left(\frac{\Delta i_{L_f max}}{I_I}\right)} = \frac{R_{DC}}{2f\left(\frac{\Delta i_{L_f max}}{I_I}\right)} = \frac{\pi^2+4}{16}\frac{R}{f\left(\frac{\Delta i_{L_f max}}{I_I}\right)}$$

$$\approx 0.86685\frac{R}{f\left(\frac{\Delta i_{L_f max}}{I_I}\right)} \tag{5.80}$$

For $\Delta i_{L_f max}/I_I = 0.1$,

$$L_{fmin} = 8.6685\frac{R}{f} \tag{5.81}$$

An alternative method for determining L_{fmin} is by separating the low-frequency modulating voltage and the high-frequency carrier voltage in a Class E RF transmitter with amplitude modulation (AM). The minimum value of choke inductance L_{fmin} to obtain the peak-to-peak current ripple less than 10% of dc current I_I is given by Kazimierczuk [17]

$$L_{fmin} = 2\left(\frac{\pi^2}{4}+1\right)\frac{R}{f} \approx \frac{7R}{f} \tag{5.82}$$

5.9 Maximum Operating Frequency of Class E Amplifier

At high frequencies, the transistor output capacitance C_o becomes higher than the shunt capacitance C_1 required to achieve the ZVS and ZDS operations. The maximum frequency occurs when $C_1 = C_o$. Assume that the transistor output capacitance C_o is linear. For $D = 0.5$, the maximum operating frequency f_{max} at which Class E ZVS and ZDS operations are achievable is determined by the condition

$$2\pi f_{max} C_o R = \frac{8}{\pi(\pi^2+4)} \tag{5.83}$$

yielding

$$f_{max} = \frac{4}{\pi^2(\pi^2+4)C_o R} \approx \frac{0.02922}{C_o R} \tag{5.84}$$

Since

$$R = \frac{8}{\pi^2+4}\frac{V_I^2}{P_O} \tag{5.85}$$

we obtain the maximum operating frequency of the Class E amplifier that satisfies both the ZVS and ZDS conditions at $D = 0.5$

$$f_{max} = \frac{P_O}{2\pi^2 C_o V_I^2} \approx 0.05066 \frac{P_O}{C_o V_I^2} \tag{5.86}$$

A higher maximum operating frequency than that given by (5.86) for $D = 0.5$ can be obtained only when the ZVS condition is satisfied, but the ZDS condition is not. The maximum operating frequency of the Class E amplifier under the ZVS and ZDS conditions increases as the duty cycle D decreases [79]. However, at low values of D (usually for $D < 0.2$), the efficiency of the amplifier decreases [58]. For $f > f_{max}$, a Class CE operation can be obtained, which offers a reasonably high efficiency [27].

5.10 Summary of Parameters at D = 0.5

The parameters of the Class E ZVS amplifier for the duty cycle $D = 0.5$ are as follows:

$$\frac{i_S}{I_I} = \begin{cases} \frac{\pi}{2}\sin\omega t - \cos\omega t + 1 & \text{for} \quad 0 < \omega t \le \pi \\ 0 & \text{for} \quad \pi < \omega t \le 2\pi \end{cases} \tag{5.87}$$

$$\frac{v_S}{V_I} = \begin{cases} 0 & \text{for} \quad 0 < \omega t \le \pi \\ \pi\left(\omega t - \frac{3\pi}{2} - \frac{\pi}{2}\cos\omega t - \sin\omega t\right) & \text{for} \quad \pi < \omega t \le 2\pi \end{cases} \tag{5.88}$$

$$\frac{i_{C1}}{I_I} = \begin{cases} 0 & \text{for} \quad 0 < \omega t \le \pi \\ \frac{\pi}{2}\sin\omega t - \cos\omega t + 1 & \text{for} \quad \pi < \omega t \le 2\pi \end{cases} \tag{5.89}$$

$$\tan\phi = -\frac{2}{\pi} \tag{5.90}$$

$$\sin\phi = \frac{2}{\sqrt{\pi^2 + 4}} \tag{5.91}$$

$$\cos\phi = -\frac{\pi}{\sqrt{\pi^2 + 4}} \tag{5.92}$$

$$\phi = \pi - \arctan\left(\frac{2}{\pi}\right) = 2.5747 \quad \text{rad} = 147.52^\circ \tag{5.93}$$

$$R_{DC} \equiv \frac{V_I}{I_I} = \frac{1}{\pi\omega C_1} = \frac{\pi^2 + 4}{8} R = 1.7337R \tag{5.94}$$

$$\frac{I_{SM}}{I_I} = \frac{\sqrt{\pi^2 + 4}}{2} + 1 = 2.862 \tag{5.95}$$

$$\frac{V_{SM}}{V_I} = 2\pi(\pi - \phi) = 3.562 \tag{5.96}$$

$$c_p = \frac{I_I V_I}{I_{SM} V_{SM}} = \frac{1}{\pi(\pi - \phi)(2 + \sqrt{\pi^2 + 4})} = 0.0981 \tag{5.97}$$

$$\gamma_1 = \frac{I_m}{I_I} = \frac{\sqrt{\pi^2+4}}{2} = 1.8621 \tag{5.98}$$

$$\xi_1 = \frac{V_{sm1}}{V_I} = \sqrt{\frac{256+\pi^2(\pi^2-4)^2}{16(\pi^2+4)}} = 1.7279 \tag{5.99}$$

$$\frac{V_{Rm}}{V_I} = \frac{4}{\sqrt{\pi^2+4}} = 1.074 \tag{5.100}$$

$$\frac{V_{Lbm1}}{V_I} = \frac{\pi(\pi^2-4)}{4\sqrt{\pi^2+4}} = 1.2378 \tag{5.101}$$

$$P_O = \frac{V_{Rm}^2}{2R} = \frac{8}{\pi^2+4}\frac{V_I^2}{R} = 0.5768\frac{V_I^2}{R} = 2\pi^2 f C_1 V_I^2 \tag{5.102}$$

$$\omega C_1 R = \frac{8}{\pi(\pi^2+4)} = 0.1836 \tag{5.103}$$

$$\frac{\omega L_b}{R} = \frac{\pi(\pi^2-4)}{16} = 1.1525 \tag{5.104}$$

$$\omega^2 L_b C_1 = \frac{\pi^2-4}{2(\pi^2+4)} = 0.2116 \tag{5.105}$$

$$\frac{1}{\omega CR} = Q_L - \frac{\omega L_b}{R} = Q_L - \frac{\pi(\pi^2-4)}{16} \approx Q_L - 1.1525 \tag{5.106}$$

$$f_{max} = \frac{4}{\pi^2(\pi^2+4)C_o R} = \frac{P_O}{2\pi^2 C_o V_I^2} \tag{5.107}$$

$$Z = R + jX_{Lb} = \frac{0.28}{\omega C_1} e^{j49.05^\circ} \tag{5.108}$$

$$Z_i = Z || \left(\frac{1}{j\omega C_1}\right) = (1.5261 + j1.1064)R \tag{5.109}$$

5.11 Efficiency

The power losses and the efficiency of the Class E amplifier will be considered for the duty cycle $D = 0.5$. The current through the input choke inductor L_f is nearly constant. Hence, from (5.98) the rms value of the inductor current is

$$I_{L_{frms}} \approx I_I = \frac{2I_m}{\sqrt{\pi^2+4}} \tag{5.110}$$

The amplifier efficiency is defined as $\eta_I = P_O/P_I$ and $P_O = RI_m^2/2$. From (5.102) and (5.110), the power loss in the dc ESR r_{L_f} of the choke inductor L_f is

$$P_{rLf} = r_{L_f} I_{L_f rms}^2 = \frac{4I_m^2 r_{L_f}}{(\pi^2+4)} = \frac{8r_{L_f}}{(\pi^2+4)R} P_O. \tag{5.111}$$

For the duty cycle $D = 0.5$, the rms value of the switch current is found from (5.87)

$$I_{Srms} = \sqrt{\frac{1}{2\pi}\int_0^{\pi} i_S^2 \, d(\omega t)} = \frac{I_I\sqrt{\pi^2+28}}{4} = \frac{I_m}{2}\sqrt{\frac{\pi^2+28}{\pi^2+4}} \tag{5.112}$$

resulting in the switch conduction loss

$$P_{rDS} = r_{DS} I_{Srms}^2 = \frac{r_{DS} I_m^2(\pi^2+28)}{4(\pi^2+4)} = \frac{(\pi^2+28) r_{DS}}{2(\pi^2+4)R} P_O \tag{5.113}$$

Using (5.89), the rms value of the current through the shunt capacitor C_1 is

$$I_{C1rms} = \sqrt{\frac{1}{2\pi}\int_{\pi}^{2\pi} i_{C1}^2 \, d(\omega t)} = \frac{I_I\sqrt{\pi^2-4}}{4} = \frac{I_m}{2}\sqrt{\frac{\pi^2-4}{\pi^2+4}} \tag{5.114}$$

which leads to the power loss in the ESR r_{C1} of the shunt capacitor C_1

$$P_{rC1} = r_{C1} I_{C1rms}^2 = \frac{r_{C1} I_m^2(\pi^2-4)}{4(\pi^2+4)} = \frac{(\pi^2-4) r_{C1}}{2(\pi^2+4)R} P_O \tag{5.115}$$

The power loss in the ESR r_L of the resonant inductor L is

$$P_{r_L} = \frac{r_L I_m^2}{2} = \frac{r_L}{R} P_O \tag{5.116}$$

and in the ESR r_C of the resonant capacitor C is

$$P_{r_C} = \frac{r_C I_m^2}{2} = \frac{r_C}{R} P_O \tag{5.117}$$

The turn-on switching loss is zero if the ZVS condition is satisfied. The turn-off switching loss can be estimated as follows. Assume that the transistor current during the turn-off time t_f decreases linearly

$$i_S = 2I_I\left(1 - \frac{\omega t - \pi}{\omega t_f}\right) \quad \text{for} \quad \pi < \omega t \le \pi + \omega t_f \tag{5.118}$$

The sinusoidal current through the resonant circuit does not change significantly during the fall time t_f and is $i \approx 2I_I$. Hence, the current through the shunt capacitor C_1 can be approximated by

$$i_{C1} \approx \frac{2I_I(\omega t - \pi)}{\omega t_f} \quad \text{for} \quad \pi < \omega t \le \pi + \omega t_f \tag{5.119}$$

which gives the voltage across the shunt capacitor C_1 and the switch

$$v_S = \frac{1}{\omega C_1}\int_{\pi}^{\omega t} i_{C1} d(\omega t) = \frac{I_I}{\omega C_1}\frac{(\omega t)^2 - 2\pi\omega t + \pi^2}{\omega t_f} = \frac{V_I \pi[(\omega t)^2 - 2\pi\omega t + \pi^2]}{\omega t_f} \tag{5.120}$$

Thus, the average value of the power loss associated with the fall time t_f is

$$P_{t_f} = \frac{1}{2\pi}\int_{\pi}^{\pi+\omega t_f} i_S v_S \, d(\omega t) = \frac{(\omega t_f)^2}{12} P_I \approx \frac{(\omega t_f)^2}{12} P_O \tag{5.121}$$

From (5.111), (5.113), (5.115), (5.121), (5.116), and (5.117), one obtains the overall power loss

$$P_{LS} = P_{r_{Lf}} + P_{r_{DS}} + P_{r_{C1}} + P_{r_L} + P_{r_C} + P_{t_f}$$
$$= P_O\left[\frac{8 r_{L_f}}{(\pi^2+4)R} + \frac{(\pi^2+28) r_{DS}}{2(\pi^2+4)R} + \frac{r_{C1}(\pi^2-4)}{2(\pi^2+4)R} + \frac{r_L + r_C}{R} + \frac{(\omega t_f)^2}{12}\right] \tag{5.122}$$

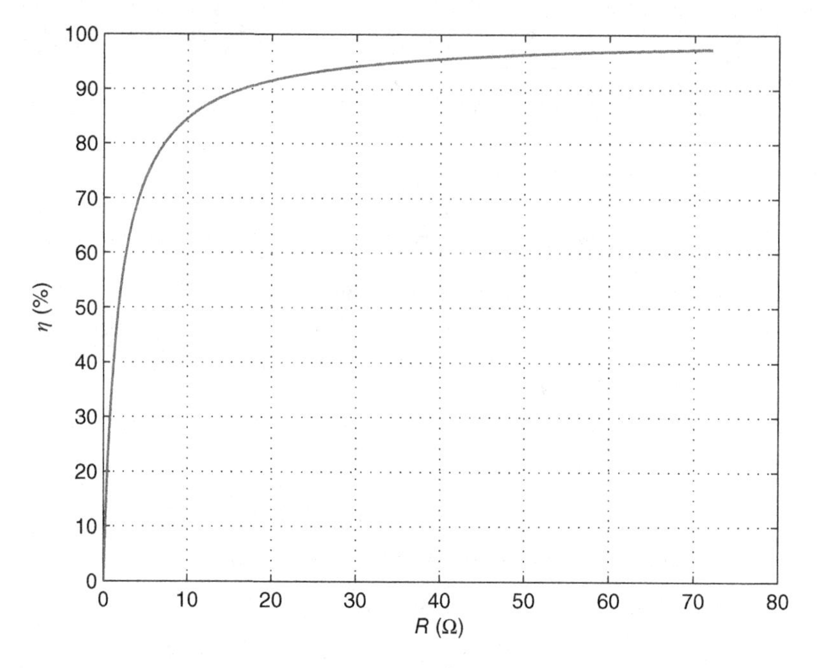

Figure 5.6 Efficiency of the Class E ZVS amplifier as a function of load resistance R for $r_{L_f} = 0.15\ \Omega$, $r_{DS} = 0.85\ \Omega$, $r_L = 0.5\ \Omega$, $r_C = 0.05\ \Omega$, $r_{C1} = 0.076\ \Omega$, and $t_f = 20$ ns.

The efficiency of the Class E amplifier is given by

$$\eta \equiv \frac{P_O}{P_I} = \frac{P_O}{P_O + P_{LS}} = \frac{1}{1 + \dfrac{P_{LS}}{P_O}}$$

$$= \frac{1}{1 + \dfrac{8r_{L_f}}{(\pi^2+4)R} + \dfrac{(\pi^2+28)r_{DS}}{2(\pi^2+4)R} + \dfrac{(\pi^2-4)r_{C1}}{2(\pi^2+4)R} + \dfrac{r_L + r_C}{R} + \dfrac{(\omega t_f)^2}{12}} \quad (5.123)$$

Figure 5.6 shows a plot of the efficiency η as a function of R.

The gate-drive power of each MOSFET that is required to charge and discharge a highly nonlinear MOSFET input capacitance is given by

$$P_G = fV_{GSm}Q_g \quad (5.124)$$

where V_{GSm} is the peak value of the gate-to-source voltage v_{GS} and Q_g is the gate charge at $v_{GS} = V_{GSm}$.

The power-added efficiency (PAE) is

$$\eta_{PAE} = \frac{P_O - P_G}{P_I} = \frac{P_O - P_G}{P_O + P_{LS}} \quad (5.125)$$

The power gain of the Class E ZVS amplifier at $D = 0.5$ is given by

$$k_p \equiv \frac{P_O}{P_G} = \frac{8}{\pi^2 + 4} \times \frac{V_I^2}{RfV_{GSm}Q_g} \quad (5.126)$$

The maximum efficiency of the Class E amplifier is achieved under ZVS and ZDS conditions only when all the components are ideal with zero parasitic resistances. However, real Class E amplifiers are built using real components. The transistor on-resistance is nonzero, and the inductors and capacitors have nonzero ESRs. The maximum efficiency of the Class E amplifier with real components is obtained when the ZVS and ZDS conditions are not satisfied. The switch voltage when the transistor turns on $V_{turn-on}$ is positive, usually

$$V_{turn-on} = (0.03 \text{ to } 0.15)V_I \tag{5.127}$$

The derivative of the switch voltage when the switch turns on is usually slightly positive. Under these conditions, there is a nonzero switching power loss at the transistor turn-on time

$$P_{sw} = \frac{1}{2} f_s C_1 V_{turn-on}^2 \tag{5.128}$$

However, the shapes of the current waveforms and their rms values are modified so that the conduction losses in different components are reduced, yielding the maximum overall efficiency.

5.12 Design of Basic Class E Amplifier

The component values of the resonant circuit of the basic Class E amplifier shown in Fig. 5.1(a) are obtained from (5.8), (5.102), (5.103), and (5.106) for optimum operation at $D = 0.5$

$$R = \frac{8}{\pi^2 + 4} \frac{V_I^2}{P_O} \approx 0.5768 \frac{V_I^2}{P_O} \tag{5.129}$$

$$X_{C1} = \frac{1}{\omega C_1} = \frac{\pi(\pi^2 + 4)R}{8} \approx 5.4466R \tag{5.130}$$

$$X_L = \omega L = Q_L R \tag{5.131}$$

$$X_C = \frac{1}{\omega C} = \left[Q_L - \frac{\pi(\pi^2 - 4)}{16}\right] R \approx (Q_L - 1.1525)R \tag{5.132}$$

Suboptimum operation (i.e., for only ZVS operation) occurs for load resistance $R_{(sub)}$ lower than that given in (5.129), that is,

$$0 \leq R_{(sub)} < R \tag{5.133}$$

The Class E ZVS amplifier with the basic resonant circuit shown in Fig. 5.1(a) operates safely under short-circuit conditions.

The basic resonant circuit of Fig. 5.1(a) does not have matching capability. In order to transfer a specified amount of power P_O at a specified dc voltage V_I, the load resistance R must be of the value determined by (5.129).

Example 5.1

Design a Class E ZVS amplifier of Fig. 5.1(a) to satisfy the following specifications: $V_I = 100$ V, $P_{Omax} = 80$ W, and $f = 1.2$ MHz. Assume $D = 0.5$.

Solution. Using (5.102), the full-load resistance is

$$R = \frac{8}{\pi^2 + 4} \frac{V_I^2}{P_O} = 0.5768 \times \frac{100^2}{80} = 72.1\ \Omega \tag{5.134}$$

From (5.94), the dc resistance of the amplifier is

$$R_{DC} = \frac{\pi^2 + 4}{8} R = 1.7337 \times 72.1 = 125\ \Omega \tag{5.135}$$

The amplitude of the output voltage is computed from (5.100) as

$$V_{Rm} = \frac{4}{\sqrt{\pi^2+4}}V_I = 1.074 \times 100 = 107.4\ \text{V} \tag{5.136}$$

The maximum voltage across the switch and the shunt capacitor C_1 can be calculated from (5.96) as

$$V_{SM} = V_{C1m} = 3.562V_I = 3.562 \times 100 = 356.2\ \text{V} \tag{5.137}$$

Assuming the amplifier efficiency $\eta = 0.9$, the dc input power is

$$P_I = \frac{P_O}{\eta} = \frac{80}{0.9} = 84.21\ \text{W} \tag{5.138}$$

yielding the dc input current

$$I_I = \frac{P_I}{V_I} = \frac{84.21}{100} = 0.8421\ \text{A} \tag{5.139}$$

The maximum switch current obtained using (5.95) is

$$I_{SM} = \left(\frac{\sqrt{\pi^2+4}}{2} + 1\right)I_I = 2.862 \times 0.8421 = 2.41\ \text{A} \tag{5.140}$$

The amplitude of the current through the resonant circuit computed from (5.98) is

$$I_m = \frac{\sqrt{\pi^2+4}}{2}I_I = 1.8621 \times 0.8421 = 1.568\ \text{A} \tag{5.141}$$

The amplitude of the current through the resonant circuit is

$$I_m = \sqrt{\frac{2P_O}{R}} = \sqrt{\frac{2 \times 80}{72.1}} = 1.49\ \text{A}. \tag{5.142}$$

Select the MOSFET process with $K_n = \mu_{n0}C_{ox} = 0.142 \times 10^{-3}\ \text{A/V}^2$, $V_t = 0.5$ V, and $L = 0.65$ μm. The peak value of a gate square voltage is $V_{GS} = 3.5$ V. Let us assume that $I_{Dsat}/I_{SM} = 2$. Then

$$I_{Dsat} = 2I_{SM} = 2 \times 2.41 = 4.82\ \text{A} \tag{5.143}$$

The aspect ratio of the n-channel MOSFET is

$$\frac{W}{L} = \frac{2I_{Dsat}}{K_n(V_{GS}-V_t)^2} = \frac{2 \times 4.82}{0.142 \times 10^{-3} \times (3.5-0.5)^2} = 7534 \tag{5.144}$$

Let $W/L = 7600$. The width of the channel is

$$W = \left(\frac{W}{L}\right)L = 7600 \times 0.65 \times 10^{-6} = 4940\ \mu\text{m} \tag{5.145}$$

Assuming $Q_L = 7$ and using (5.66), (5.103), and (5.106), the component values of the load network are

$$L = \frac{Q_L R}{\omega} = \frac{7 \times 72.1}{2\pi \times 1.2 \times 10^6} = 66.9\ \mu\text{H} \tag{5.146}$$

$$C = \frac{1}{\omega R\left[Q_L - \frac{\pi(\pi^2-4)}{16}\right]} = \frac{1}{2\pi \times 1.2 \times 10^6 \times 72.1 \times (7-1.1525)} = 314.6\ \text{pF} \tag{5.147}$$

The peak voltages across the resonant capacitor C and the peak voltage across the inductor L are

$$C_1 = \frac{8}{\pi(\pi^2+4)\omega R} = \frac{8}{2\pi^2(\pi^2+4) \times 1.2 \times 10^6 \times 72.1} = 337.4\ \text{pF} \tag{5.148}$$

Let the MOSFET output capacitance be $C_o = 37.4$ pF. Hence, the external shunt capacitance is

$$C_{1ext} = C_1 - C_o = 337.4 - 37.4 = 300 \text{ pF} \tag{5.149}$$

Pick $C_{1ext} = 300$ pF/400 V.

The peak voltages across the resonant capacitor C and the inductor L are

$$V_{Cm} = \frac{I_m}{\omega C} = \frac{1.49}{2\pi \times 1.2 \times 10^6 \times 314.6 \times 10^{-12}} = 628.07 \text{ V} \tag{5.150}$$

and

$$V_{Lm} = \omega L I_m = 2\pi \times 1.2 \times 10^6 \times 66.9 \times 10^{-6} \times 1.49 = 751.58 \text{ V} \tag{5.151}$$

Pick $C = 330$ pF/700 V.

It follows from (5.82) that in order to keep the current ripple in the choke inductor below 10% full-load dc input current I_I, the value of the choke inductance must be greater than

$$L_f = 2\left(\frac{\pi^2}{4} + 1\right)\frac{R}{f} = \frac{7 \times 72.1}{1.2 \times 10^6} = 420.58 \text{ μH} \tag{5.152}$$

The peak voltage across the RFC is

$$V_{Lfm} = V_{SM} - V_I = 356.2 - 100 = 256.2 \text{ V} \tag{5.153}$$

Assume that the dc ESR of the choke L_f is $r_{L_f} = 0.15\ \Omega$. Hence, from (5.111), the power loss in r_{L_f} is

$$P_{rLf} = r_{L_f} I_I^2 = 0.15 \times 0.8421^2 = 0.106 \text{ W} \tag{5.154}$$

From (5.112), the rms value of the switch current is

$$I_{Srms} = \frac{I_I\sqrt{\pi^2 + 28}}{4} = 0.8421 \times 1.5385 = 1.296 \text{ A} \tag{5.155}$$

Select an International Rectifier IRF840 power MOSFET, which has $V_{DSS} = 500$ V, $I_{Dmax} = 8$ A, $r_{DS} = 0.85\ \Omega$, $t_f = 20$ ns, and $Q_g = 63$ nC. The transistor conduction power loss is

$$P_{rDS} = r_{DS} I_{Srms}^2 = 0.85 \times 1.296^2 = 1.428 \text{ W} \tag{5.156}$$

Using (5.114), one obtains the rms current through the shunt capacitance C_1

$$I_{C1rms} = \frac{I_I\sqrt{\pi^2 - 4}}{4} = 0.8421 \times 0.6057 = 0.51 \text{ A}. \tag{5.157}$$

Assume the ESR of the capacitor C_1 to be $r_{C1} = 76$ mΩ, one arrives at the conduction power loss in r_{C1}

$$P_{rC1} = r_{C1} I_{C1rms}^2 = 0.076 \times 0.51^2 = 0.02 \text{ W} \tag{5.158}$$

Assuming the ESRs of the resonant inductor L and the resonant capacitor C to be $r_L = 0.5\ \Omega$ and $r_C = 50$ mΩ at $f = 1.2$ MHz. Hence, the power losses in the resonant components are

$$P_{rL} = \frac{r_L I_m^2}{2} = \frac{0.5 \times 1.49^2}{2} = 0.555 \text{ W} \tag{5.159}$$

and

$$P_{rC} = \frac{r_C I_m^2}{2} = \frac{0.05 \times 1.49^2}{2} = 0.056 \text{ W} \tag{5.160}$$

The drain current fall time is $t_f = 20$ ns. Hence, $\omega t_f = 2\pi \times 1.2 \times 10^6 \times 20 \times 10^{-9} = 0.151$ rad. From (5.121), the turn-off switching loss is

$$P_{tf} = \frac{(\omega t_f)^2 P_O}{12} = \frac{0.151^2 \times 80}{12} = 0.152 \text{ W}. \tag{5.161}$$

The power loss (excluding the gate-drive power) is

$$P_{LS} = P_{rLf} + P_{rDS} + P_{rC1} + P_{rL} + P_{rC} + P_{tf}$$
$$= 0.106 + 1.428 + 0.02 + 0.555 + 0.056 + 0.152 = 2.317 \text{ W} \quad (5.162)$$

The efficiency of the amplifier becomes

$$\eta = \frac{P_O}{P_O + P_{LS}} = \frac{80}{80 + 2.317} = 97.82\% \quad (5.163)$$

Assuming $V_{GSm} = 8$ V, one obtains the gate-drive power

$$P_G = f V_{GSm} Q_g = 1.2 \times 10^6 \times 8 \times 63 \times 10^{-9} = 0.605 \text{ W} \quad (5.164)$$

The PAE is

$$\eta_{PAE} = \frac{P_O - P_G}{P_I} = \frac{P_O - P_G}{P_O + P_{LS}} = \frac{80 - 0.605}{80 + 2.317} = 96.45\% \quad (5.165)$$

The power gain of the amplifier is

$$k_p = \frac{P_O}{P_G} = \frac{80}{0.605} = 132.23 = 21.21 \text{ dB} \quad (5.166)$$

The equivalent capacitance when the switch is OFF is

$$C_{eq} = \frac{CC_1}{C + C_1} = \frac{330 \times 337.4}{330 + 337.4} = 167 \text{ pF} \quad (5.167)$$

The resonant frequencies are

$$f_{o1} = \frac{1}{2\pi\sqrt{LC}} = \frac{1}{2\pi\sqrt{66.9 \times 10^{-6} \times 330 \times 10^{-12}}} = 1.071 \text{ MHz} \quad (5.168)$$

and

$$f_{o2} = \frac{1}{2\pi\sqrt{LC_{eq}}} = \frac{1}{2\pi\sqrt{66.9 \times 10^{-6} \times 167 \times 10^{-12}}} = 1.51 \text{ MHz} \quad (5.169)$$

Note that the operating frequency $f = 1.2$ MHz is between the resonant frequencies f_{o1} and f_{o2}.

5.13 Impedance Matching Resonant Circuits

The purpose of the impedance matching network is to convert the load resistance or impedance into the impedance required to produce the desired output power P_O at the specified supply voltage V_I and the operating frequency f. According to (5.129), V_I, P_O, and R are dependent quantities. In many applications, the load resistance is given and is different from that given in (5.129). Therefore, there is a need for a matching circuit that provides impedance transformation downward or upward. A block diagram of the Class E amplifier with an impedance matching circuit is shown in Fig. 5.7. Figure 5.8 shows various impedance matching resonant circuits. In the circuits shown in Fig. 5.8(a) and (c), impedance transformation is accomplished by tapping the resonant capacitance C, and in the circuits shown in Fig. 5.8(b) and (d) by tapping the resonant inductance L. All these circuits provide downward impedance transformation. The vertical inductance L_2 in Fig. 5.8(b) and the vertical inductance L_1 in Fig. 5.8(d) can be replaced by a step-up or step-down transformer, yielding additional impedance transformation. The topologies of these circuits are similar to the Greek letter π.

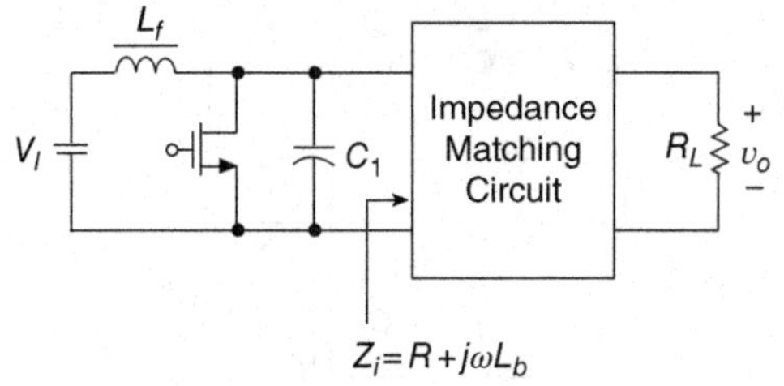

Figure 5.7 Block diagram of the Class E amplifier with impedance matching resonant circuit.

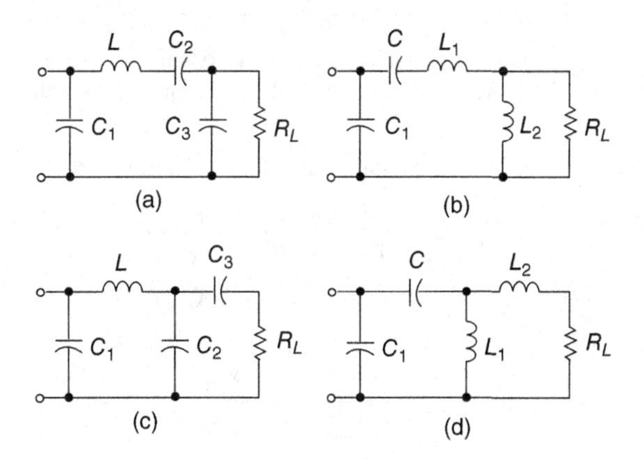

Figure 5.8 Matching resonant circuits. (a) Resonant circuit π1a. (b) Resonant circuit π2a. (c) Resonant circuit π1b. (d) Resonant circuit π2b.

5.13.1 Tapped Capacitor Downward Impedance Matching Resonant Circuit π 1a

Figure 5.9(a) shows the tapped capacitor matching circuit with downward impedance transformation ($R_L > R$). Its equivalent circuit is shown in Fig. 5.9(b). Let us assume that the load resistance R_L is given. Using (5.129), the series equivalent resistance for optimum operation at $D = 0.5$ is given by

$$R = R_s = \frac{8}{\pi^2 + 4} \frac{V_I^2}{P_O} \approx 0.5768 \frac{V_I^2}{P_O} \tag{5.170}$$

The components C_1 and L are given by (5.130) and (5.131)

$$X_{C1} = \frac{1}{\omega C_1} = \frac{\pi(\pi^2 + 4)R}{8} \approx 5.4466R \tag{5.171}$$

and

$$X_L = \omega L = Q_L R \tag{5.172}$$

The reactance factor of the R_L-C_3 and R_s-C_s equivalent two-terminal networks is

$$q = \frac{R_L}{X_{C_3}} = \frac{X_{Cs}}{R_s} = \frac{X_{Cs}}{R} \tag{5.173}$$

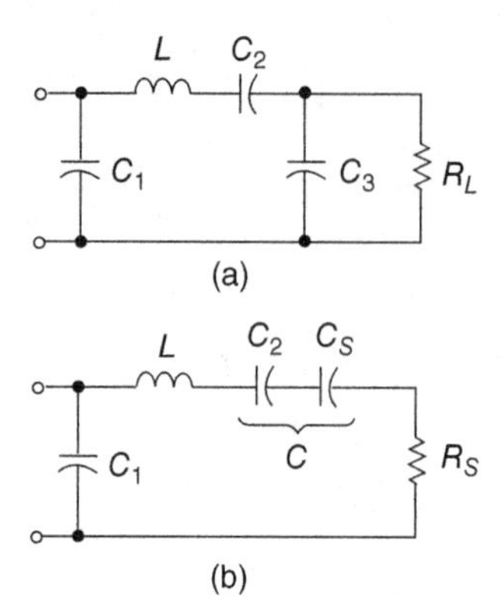

Figure 5.9 Tapped capacitor impedance matching resonant circuit π1a providing downward impedance transformation. (a) Matching circuit π1a. (b) Equivalent circuit of the matching circuit π1a.

Resistances R_s and R_L as well as the reactances X_{Cs} and X_{C3} are related by

$$R_s = R = \frac{R_L}{1+q^2} = \frac{R_L}{1+\left(\frac{R_L}{X_{C3}}\right)^2} \tag{5.174}$$

and

$$X_{Cs} = \frac{X_{C3}}{1+\frac{1}{q^2}} = \frac{X_{C3}}{1+(\frac{X_{C3}}{R_L})^2} \tag{5.175}$$

Hence,

$$C_s = C_3\left(1+\frac{1}{q^2}\right) = C_3\left[1+\left(\frac{X_{C3}}{R_L}\right)^2\right] \tag{5.176}$$

Rearrangement of (5.174) gives

$$q = \sqrt{\frac{R_L}{R_s} - 1} \tag{5.177}$$

From (5.173) and (5.177),

$$X_{C3} = \frac{1}{\omega C_3} = \frac{R_L}{q} = \frac{R_L}{\sqrt{\frac{R_L}{R_s} - 1}} \tag{5.178}$$

Substitution of (5.177) into (5.173) yields

$$X_{Cs} = qR_s = R_s\sqrt{\frac{R_L}{R_s} - 1} \tag{5.179}$$

Referring to Fig. 5.9 and using (5.132) and (5.179), one arrives at

$$X_{C2} = \frac{1}{\omega C_2} = X_C - X_{Cs} = \left[Q_L - \frac{\pi(\pi^2-4)}{16}\right]R_s - qR_s$$
$$= R_s\left[Q_L - \frac{\pi(\pi^2-4)}{16} - \sqrt{\frac{R_L}{R_s} - 1}\right] \tag{5.180}$$

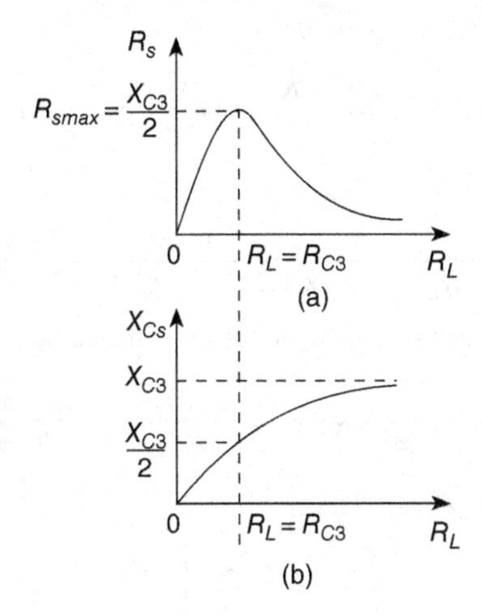

Figure 5.10 Series equivalent resistance R_s and reactance X_{Cs} as functions of load resistance R_L in the circuit π1a. (a) R_s versus R_L. (b) X_{Cs} versus R_L.

It follows from (5.178) that the circuit shown in Fig. 5.8(a) can match the resistances that satisfy the inequality

$$R_s < R_L \tag{5.181}$$

Suboptimum operation is obtained for

$$0 \le R_{s(sub)} < R_s \tag{5.182}$$

which corresponds to

$$R_L < R_{(sub)} < \infty \tag{5.183}$$

Expressions (5.174) and (5.175) are illustrated in Fig. 5.10. As R_L is increased from 0 to X_{C3}, R_s increases to $X_{C3}/2$, R_s reaches the maximum value $R_{smax} = X_{C3}/2$ at $R_L = X_{C3}$; as R_L is increased from X_{C3} to ∞, R_s decreases from $X_{C3}/2$ to 0. Thus, the R_L-C_3 circuit acts as an *impedance inverter* [20] for $R_L > X_{C3}$. If the optimum operation occurs at $R_L = X_{C3}$, then $R_{smax} = X_{C3}/2$ and the amplifier operates under ZVS conditions at any load resistance R_L [28]. This is because $R_s \le R_{smax} = X_{C3}/2$ at any values of R_L. As R_L increases from 0 to ∞, X_{Cs} increases from 0 to X_{C3} and C_s decreases from ∞ to C_3.

Example 5.2

Design a Class E ZVS amplifier of Fig. 5.1(a) to satisfy the following specifications: $V_I = 100$ V, $P_{Omax} = 80$ W, $R_L = 150\,\Omega$, and $f = 1.2$ MHz. Assume $D = 0.5$.

Solution. It is sufficient to design the amplifier for full power. The required series resistance is

$$R_s = R = \frac{8}{\pi^2 + 4} \frac{V_I^2}{P_O} = 0.5768 \times \frac{100^2}{80} = 72.1\,\Omega \tag{5.184}$$

Thus, the amplifier requires a matching circuit. The reactance factor is

$$q = \sqrt{\frac{R_L}{R_s} - 1} = \sqrt{\frac{150}{72.1} - 1} = 1.039 \tag{5.185}$$

Hence,

$$X_{C3} = \frac{1}{\omega C_3} = \frac{R_L}{q} = \frac{150}{1.039} = 144.37\ \Omega \tag{5.186}$$

resulting in the capacitance

$$C_3 = \frac{1}{\omega X_{C3}} = \frac{1}{2\pi \times 1.2 \times 10^6 \times 144.37} = 919\ \text{pF} \tag{5.187}$$

The amplitude of the voltage across the capacitor C_3 is the same as the amplitude of the voltage across the load resistance R_L. Thus,

$$V_{C3m} = V_{Rm} = \sqrt{2R_L P_O} = \sqrt{2 \times 150 \times 80} = 155\ \text{V} \tag{5.188}$$

Pick $C_3 = 910$ pF/200 V. Let $Q_L = 7$. Hence,

$$X_{C2} = \frac{1}{\omega C_2} = X_C - X_{C_s} = R_s \left[Q_L - \frac{\pi(\pi^2 - 4)}{16} - q\right]$$
$$= 72.1(7 - 1.1525 - 1.039) = 346.7\ \Omega \tag{5.189}$$

yielding

$$C_2 = \frac{1}{\omega X_{C2}} = \frac{1}{2\pi \times 1.2 \times 10^6 \times 346.7} = 383\ \text{pF} \tag{5.190}$$

The amplitude of the voltage across the capacitor C_2 is

$$V_{C2m} = X_{C2} I_m = 346.7 \times 1.49 = 517\ \text{V} \tag{5.191}$$

Select $C_2 = 390$ pF/600 V. All other parameters are the same as those in Example 5.1.

5.13.2 Tapped Inductor Downward Impedance Matching Resonant Circuit π2a

Figure 5.11(a) shows the tapped inductor matching resonant circuit of Fig. 5.8(b) that provides downward impedance transformation ($R_L > R$). Its equivalent circuit is depicted in Fig. 5.11(b).

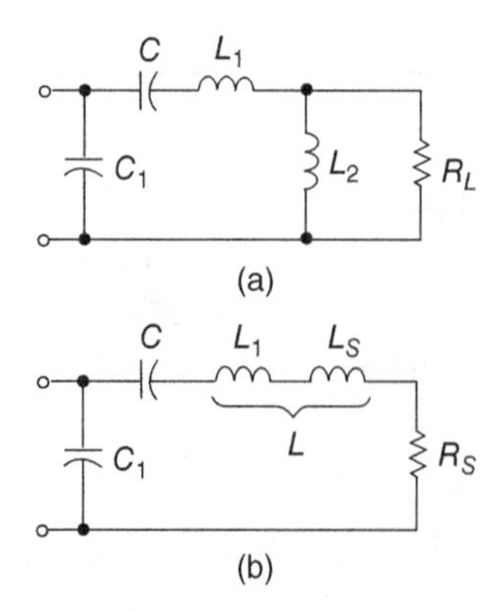

Figure 5.11 Tapped inductor downward impedance matching circuit π2a. (a) Matching circuit π2a. (b) The parallel impedance R_L–L_2 is converted into a series impedance R_s–L_s.

The values of R_s, X_{C1}, and X_C for the resonant circuit can be calculated for optimum operation at $D = 0.5$ from (5.170), (5.130), and (5.132), respectively. The series resistance is

$$R_s = R = \frac{8}{\pi^2 + 4}\frac{V_I^2}{P_O} \approx 0.5768\frac{V_I^2}{P_O} \tag{5.192}$$

$$X_{C1} = \frac{1}{\omega C_1} = \frac{\pi(\pi^2 + 4)R}{8} \approx 5.4466R \tag{5.193}$$

and

$$X_C = \frac{1}{\omega C} = \left[Q_L - \frac{\pi(\pi^2 - 4)}{16}\right]R \approx (Q_L - 1.1525)R \tag{5.194}$$

The reactance factor of the R_L-L_2 and R_s-L_s equivalent two-terminal networks is

$$q = \frac{R_L}{X_{L_2}} = \frac{X_{L_s}}{R_s} \tag{5.195}$$

Resistances R_s and R_L as well as the reactances X_{L_s} and X_{L_2} are related by

$$R_s = R = \frac{R_L}{1 + q^2} = \frac{R_L}{1 + \left(\frac{R_L}{X_{L_2}}\right)^2} \tag{5.196}$$

and

$$X_{L_s} = \frac{X_{L_2}}{1 + \frac{1}{q^2}} = \frac{X_{L_2}}{1 + \left(\frac{X_{L_2}}{R_L}\right)^2} \tag{5.197}$$

Hence,

$$L_s = \frac{L_2}{1 + \frac{1}{q^2}} = \frac{L_2}{1 + \left(\frac{X_{L3}}{R_L}\right)^2} \tag{5.198}$$

The reactance factor is

$$q = \sqrt{\frac{R_L}{R_s} - 1} \tag{5.199}$$

Hence,

$$X_{L_2} = \omega L_2 = \frac{R_L}{q} = \frac{R_L}{\sqrt{\frac{R_L}{R_s} - 1}} \tag{5.200}$$

Since

$$Q_L = \frac{\omega L}{R_s} = \frac{X_L}{R_s} \tag{5.201}$$

and

$$X_{L_s} = qR_s \tag{5.202}$$

one obtains

$$X_{L_1} = \omega L_1 = \omega(L - L_s) = X_L - X_{L_s} = (Q_L - q)R_s = \left(Q_L - \sqrt{\frac{R_L}{R_s} - 1}\right)R_s \tag{5.203}$$

The range of resistances that can be matched by the circuit shown in Fig. 5.8(b) is

$$R_s < R_L \tag{5.204}$$

Suboptimum operation takes place for

$$0 \leq R_{s(sub)} < R_s \tag{5.205}$$

and consequently for

$$R_L < R_{(sub)} < \infty \tag{5.206}$$

It follows from (5.196) that when R_L is increased from 0 to X_{L_2}, R_s increases from 0 to $R_{smax} = X_{L_2}/2$, and when R_L is increased from X_{L_2} to ∞, R_s decreases from $X_{L_2}/2$ to 0. It is clear that the R_L-L_2 circuit behaves like an *impedance inverter* for $R_L > X_{L_2}$. If the optimum operation occurs at $R_L = X_{L_2}$, then $R_s = X_{L_2}/2$ and the ZVS operation occurs at any load resistance [25, 30]. In this case, $R_s \leq R_{smax} = X_{L_2}/2$. As R_L increases from 0 to ∞, X_{L_s} increases from 0 to X_{L_2} and L_s increases from 0 to L_2.

The vertical inductance L_2 can be replaced by a step-up or step-down transformer, providing additional impedance transformation. The transformer leakage inductance can be absorbed into the horizontal inductance L_1. Figure 5.2 shows a Class E RF power amplifier with a transformer. The impedance transformation is increased by the square of the transformer turn ratio. The transformer leakage inductance L_{lp} is absorbed into inductance L. The magnetizing inductance L_m is used as an inductor L_2 connected in parallel with the load reflected to the primary side of the transformer. The load of this circuit can be a rectifier. The transformer may be used for wireless power transmission. The circuit can be used to charge a battery installed inside a person's body.

Example 5.3

Design a Class E ZVS amplifier with the tapped inductor downward impedance matching circuit π2a shown in Fig. 5.11(a) to satisfy the following specifications: $V_I = 100$ V, $P_{Omax} = 80$ W, $R_L = 150\ \Omega$, and $f = 1.2$ MHz. Assume $D = 0.5$.

Solution. The series resistance R_s is

$$R_s = R = \frac{8}{\pi^2 + 4}\frac{V_I^2}{P_O} = 0.5768 \times \frac{100^2}{80} = 72.1\ \Omega \tag{5.207}$$

A matching circuit is needed. The reactance factor is

$$q = \sqrt{\frac{R_L}{R_s} - 1} = \sqrt{\frac{150}{72.1} - 1} = 1.039 \tag{5.208}$$

The reactance X_{L_2} is

$$X_{L_2} = \omega L_2 = \frac{R_L}{q} = \frac{150}{1.039} = 144.37\ \Omega \tag{5.209}$$

resulting in the inductance

$$L_2 = \frac{X_{L_2}}{\omega} = \frac{144.37}{2\pi \times 1.2 \times 10^6} = 19\ \mu\text{H} \tag{5.210}$$

The amplitude of the voltage across the inductance L_2 is the same as the amplitude of the voltage across the load resistance R_L. Thus,

$$V_{L2m} = V_{Rm} = \sqrt{2R_L P_O} = \sqrt{2 \times 150 \times 80} = 155\ \text{V} \tag{5.211}$$

Let $Q_L = 7$. Hence,

$$X_{L_1} = \omega L_1 = X_L - X_{L_s} = R_s(Q_L - q) = 72.1(7 - 1.039) = 429.79\ \Omega \tag{5.212}$$

producing

$$L_1 = \frac{X_{L_1}}{\omega} = \frac{429.79}{2\pi \times 1.2 \times 10^6} = 57\ \mu\text{H} \tag{5.213}$$

The amplitude of the voltage across the inductor L_1 is

$$V_{L_1m} = X_{L_1} I_m = 429.79 \times 1.49 = 640.387\ \text{V} \tag{5.214}$$

All other parameters are identical to those in Example 5.1.

5.13.3 Matching Resonant Circuit π1b

Figure 5.12(a) shows the tapped capacitor downward and upward impedance matching circuit. The values of R_s, X_{C1}, and X_L for the circuit shown in Fig. 5.12(a) can be calculated for optimum operation at $D = 0.5$ from (5.170), (5.130), and (5.131), respectively. The series resistance R_s, and the reactances of the shunt capacitance C_1 and the resonant inductance L are

$$R_s = \frac{8}{\pi^2 + 4} \frac{V_I^2}{P_O} \approx 0.5768 \frac{V_I^2}{P_O} \tag{5.215}$$

$$X_{C1} = \frac{1}{\omega C_1} = \frac{\pi(\pi^2 + 4)R}{8} \approx 5.4466R \tag{5.216}$$

and

$$X_L = Q_L R \tag{5.217}$$

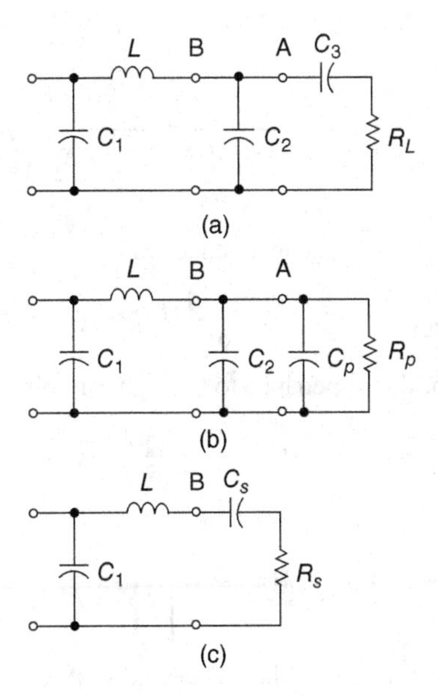

Figure 5.12 Tapped capacitor downward impedance matching circuit π1b. (a) Matching circuit π1b. (b) The series impedance R_L-C_3 is converted to a parallel impedance R_p-C_p. (c) The parallel impedance R_p-X_B is converted to the series impedance R_s-C_s.

The reactance factor of the series circuit R_L-C_3 located to the right of point A in Fig. 5.12(a) is

$$q_A = \frac{X_{C3}}{R_L} = \frac{R_p}{X_{Cp}} \tag{5.218}$$

The series circuit R_L–C_3 can be converted to the parallel circuit R_p–C_p as depicted in Fig. 5.12(b), using the following equations:

$$R_p = R_L(1 + q_A^2) = R_L\left[1 + \left(\frac{X_{C3}}{R_L}\right)^2\right] \tag{5.219}$$

$$X_{Cp} = X_{C3}\left(1 + \frac{1}{q_A^2}\right) = X_{C3}\left[1 + \left(\frac{R_L}{X_{C3}}\right)^2\right] \tag{5.220}$$

and

$$C_p = \frac{C_3}{1 + \frac{1}{q_A^2}} \tag{5.221}$$

The total capacitance to the right of point B is $C_B = C_2 + C_p$ and the total reactance to the right of point B is

$$\frac{1}{X_B} = \frac{1}{X_{C2}} + \frac{1}{X_{Cp}} \tag{5.222}$$

The parallel impedance R_p-X_B can be converted to the series impedance R_s-C_s as shown in Fig. 5.12(c). The conversion equations are

$$q_B = \frac{R_p}{X_B} = \frac{X_{Cs}}{R_s} \tag{5.223}$$

$$R_s = R = \frac{R_p}{1 + q_B^2} = \frac{R_p}{1 + \left(\frac{R_p}{X_B}\right)^2} \tag{5.224}$$

$$X_{Cs} = X_C = \frac{X_B}{1 + \frac{1}{q_B^2}} = \frac{X_B}{1 + \left(\frac{X_B}{R_p}\right)^2} \tag{5.225}$$

On the other hand, the reactance of the capacitance C is

$$X_C = X_{Cs} = \left[Q_L - \frac{\pi(\pi^2 - 4)}{16}\right] R \approx (Q_L - 1.1525)R \tag{5.226}$$

Hence, the reactance factor of the impedance to the right of point B is

$$q_B = \frac{X_{Cs}}{R_s} = \frac{X_C}{R_s} = Q_L - \frac{\pi(\pi^2 - 4)}{16} \approx Q_L - 1.1525 \tag{5.227}$$

The parallel resistance is

$$R_p = R(1 + q_B^2) = R\left\{1 + \left[Q_L - \frac{\pi(\pi^2 - 4)}{16}\right]^2\right\} = R[(Q_L - 1.1525)^2 + 1] \tag{5.228}$$

The reactance factor of the impedance to the right of point A is

$$q_A = \sqrt{\frac{R_p}{R_L} - 1} \tag{5.229}$$

Hence, the design equation for the reactance of the capacitor C_3 is

$$X_{C3} = \frac{1}{\omega C_3} = q_A R_L = R_L \sqrt{\frac{R[(Q_L - 1.1525)^2 + 1]}{R_L} - 1} \tag{5.230}$$

The parallel reactance is

$$X_{Cp} = X_{C3}\left(1 + \frac{1}{q_A^2}\right) = \frac{R[(Q_L - 1.1525)^2 + 1]}{\sqrt{\frac{R[(Q_L - 1.1525)^2 + 1]}{R_L} - 1}} \tag{5.231}$$

The parallel reactance is

$$X_B = \frac{R_p}{q_B} = \frac{R[(Q_L - 1.1525)^2 + 1]}{Q_L - 1.1525} \tag{5.232}$$

The reactance X_{C_2} is given by

$$\frac{1}{X_{C2}} = \frac{1}{X_B} - \frac{1}{X_p} \tag{5.233}$$

This gives the design equation for the reactance X_{C2} as

$$X_{C2} = \frac{1}{\omega C_2} = \frac{R[(Q_L - 1.1525)^2 + 1]}{Q_L - 1.1525 - \sqrt{\frac{R[(Q_L - 1.1525)^2 + 1]}{R_L} - 1}} \tag{5.234}$$

The resistances that can be matched by the above-mentioned circuit are

$$\frac{R_L}{(Q_L - 1.1525)^2 + 1} < R_s < R_L \tag{5.235}$$

Suboptimum operation takes place for

$$R_{s(sub)} < R_s \tag{5.236}$$

and therefore for

$$R_{s(sub)} < R_L \tag{5.237}$$

Example 5.4

Design a Class E ZVS amplifier with tapped capacitor downward impedance matching circuit π1b depicted in Fig. 5.12(a) to satisfy the following specifications: $V_I = 100$ V, $P_{Omax} = 80$ W, $R_L = 150\ \Omega$, and $f = 1.2$ MHz. Assume $D = 0.5$.

Solution. Assuming $Q_L = 7$, the reactance of the capacitor C_3 is calculated as

$$X_{C_3} = \frac{1}{\omega C_3} = R_L \sqrt{\frac{R[(Q_L - 1.1525)^2 + 1]}{R_L} - 1} = 150\sqrt{\frac{72.1[(7 - 1.1525)^2 + 1]}{150} - 1}$$
$$= 598.43\ \Omega \tag{5.238}$$

resulting

$$C_3 = \frac{1}{\omega X_{C3}} = \frac{1}{2\pi \times 1.2 \times 10^6 \times 598.43} = 221.6285 \text{ pF} \tag{5.239}$$

The amplitude of the current through the load resistance R_L and the capacitance C_3 is

$$I_{Rm} = \sqrt{\frac{2P_O}{R_L}} = \sqrt{\frac{2 \times 80}{150}} = 1.067 \text{ A} \tag{5.240}$$

Hence, the amplitude of the voltage across the capacitance C_3 is

$$V_{C3m} = X_{C3} I_{Rm} = 598.43 \times 1.067 = 638.32 \text{ V} \tag{5.241}$$

Pick $C_3 = 220$ pF/700 V.

The reactance of the capacitor C_2 is

$$X_{C_2} = \frac{1}{\omega C_2} = \frac{R[(Q_L - 1.1525)^2 + 1]}{Q_L - 1.1525 - \sqrt{\frac{R[(Q_L - 1.1525)^2 + 1]}{R_L} - 1}}$$

$$= \frac{72.1[(7 - 1.1525)^2 + 1]}{7 - 1.1525 - \sqrt{\frac{72.1[(7 - 1.1525)^2 + 1]}{150} - 1}} = 1365.69\ \Omega \tag{5.242}$$

yielding

$$C_2 = \frac{1}{\omega X_{C2}} = \frac{1}{2\pi \times 1.2 \times 10^6 \times 1365.69} = 97.115 \text{ pF} \tag{5.243}$$

The amplitude of the current through the capacitor C_2 is

$$V_{C_2m} = \sqrt{V_{Rm}^2 + V_{C_3m}^2} = \sqrt{160.05^2 + 638.32^2} = 658.08 \text{ V} \tag{5.244}$$

Pick $C_2 = 100$ pF/700 V. All other parameters are the same as those in Example 5.1.

5.13.4 Matching Resonant Circuit π2b

Figure 5.13(a) shows a downward and upward matching circuit π2b. The values of R_s, X_{C_1}, and X_C for the circuit shown in in Fig. 5.13(a) can be calculated for optimum operation at $D = 0.5$ from (5.170), (5.130), and (5.132), respectively. The series resistance R_s and the reactance of the shunt capacitance C_1 are

$$R_s = R = \frac{8}{\pi^2 + 4} \frac{V_I^2}{P_O} \approx 0.5768 \frac{V_I^2}{P_O} \tag{5.245}$$

and

$$X_{C1} = \frac{1}{\omega C_1} = \left[Q_L - \frac{\pi(\pi^2 - 4)}{16}\right] R \tag{5.246}$$

The reactance factor of the series circuit R_L-L_2 and the equivalent parallel circuit R_p-L_p is given by

$$q_A = \frac{X_{L_2}}{R_L} = \frac{R_p}{X_{L_p}} \tag{5.247}$$

The parallel components are

$$R_p = R_L(1 + q_A^2) = R_L \left[1 + \left(\frac{X_{L_2}}{R_L}\right)^2\right] \tag{5.248}$$

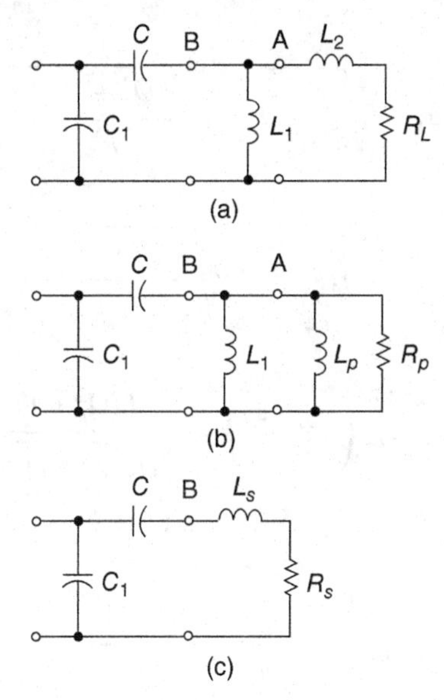

Figure 5.13 Tapped inductor downward impedance matching resonant circuit π2b. (a) Matching resonant circuit π2b. (b) Series impedance R_L–L_2 is converted into a parallel impedance R_p–L_p. (c) Parallel impedance R_p–L_p–L_1 is converted into a series impedance R_s–L_s.

$$X_p = X_{L_2}\left(1 + \frac{1}{q_A^2}\right) = X_{L_2}\left[1 + \left(\frac{R_L}{X_{L_2}}\right)^2\right] \tag{5.249}$$

and

$$L_p = L_2\left(1 + \frac{1}{q_A^2}\right) = L_2\left[1 + \left(\frac{R_L}{X_{L_2}}\right)^2\right] \tag{5.250}$$

The reactance of the components L_1-L_p-R_p is

$$\frac{1}{X_B} = \frac{1}{X_{L_1}} + \frac{1}{X_{Lp}} \tag{5.251}$$

The series components R_s-L_s can be described by

$$Q_L = q_B = \frac{R_p}{X_p} = \frac{X_{L_s}}{R_s} = \frac{X_L}{R} \tag{5.252}$$

$$R_s = R = \frac{R_p}{1 + Q_L^2} = \frac{R_p}{1 + \left(\frac{R_p}{X_B}\right)^2} \tag{5.253}$$

and

$$X_L = X_s = \frac{X_B}{1 + \frac{1}{Q_L^2}} = \frac{X_B}{1 + \left(\frac{X_B}{R_p}\right)^2} \tag{5.254}$$

The parallel resistance is

$$R_p = R_s(1 + q_B^2) = R(1 + Q_L^2) \tag{5.255}$$

Next,

$$q_A = \sqrt{\frac{R_p}{R_L} - 1} = \sqrt{\frac{R(Q_L^2 + 1)}{R_L} - 1} \tag{5.256}$$

Hence,

$$X_{L_2} = q_A R_L = R_L \sqrt{\frac{R(Q_L^2 + 1)}{R_L} - 1} \tag{5.257}$$

The parallel reactance is

$$X_{Lp} = X_{L_2}\left(1 + \frac{1}{q_A^2}\right) = \frac{R(Q_L^2 + 1)}{\sqrt{\frac{R(Q_L^2 + 1)}{R_L} - 1}} \tag{5.258}$$

The parallel reactance is

$$X_B = \frac{R_p}{q_B} = \frac{R(Q_L^2 + 1)}{Q_L} \tag{5.259}$$

Using

$$\frac{1}{X_{L_1}} = \frac{1}{X_B} - \frac{1}{X_{LP}} \tag{5.260}$$

we obtain

$$X_{L_1} = \omega L_1 = \frac{R(Q_L^2 + 1)}{Q_L - \sqrt{\frac{R(Q_L^2 + 1)}{R_L} - 1}} \tag{5.261}$$

This circuit can match resistances in the range

$$\frac{R_L}{Q_L^2 + 1} < R_s < R_L \tag{5.262}$$

Suboptimum operation takes place for

$$R_{s(sub)} < R_s \tag{5.263}$$

and therefore for

$$R_{s(sub)} < R_L \tag{5.264}$$

Example 5.5

Design a Class E ZVS amplifier with tapped inductor downward impedance matching circuit π2b shown in Fig. 5.13(a) to satisfy the following specifications: $V_I = 100$ V, $P_{Omax} = 80$ W, $R_L = 150\,\Omega$, and $f = 1.2$ MHz. Assume $D = 0.5$.

Solution. Assume that $Q_L = 7$. The reactance of the inductance L_1 is

$$X_{L_1} = \omega L_1 = \frac{R(Q_L^2 + 1)}{Q_L - \sqrt{\frac{R(Q_L^2 + 1)}{R_L} - 1}} = \frac{72.1(7^2 + 1)}{7 - \sqrt{\frac{72.1(7^2 + 1)}{150} - 1}} = 1638.1\,\Omega \tag{5.265}$$

Hence, the inductance L_1 is

$$L_1 = \frac{X_{L_1}}{\omega} = \frac{1638.1}{2\pi \times 1.2 \times 10^6} = 217.26\ \mu\text{H} \tag{5.266}$$

The reactance of L_2 is

$$X_{L_2} = R_L\sqrt{\frac{R(Q_L^2+1)}{R_L} - 1} = 150\sqrt{\frac{72.1(7^2+1)}{150} - 1} = 719.896\ \Omega \tag{5.267}$$

Hence, the inductance L_2 is

$$L_2 = \frac{X_{L_2}}{\omega} = \frac{719.896}{2\pi \times 1.2 \times 10^6} = 95.48\ \mu\text{H} \tag{5.268}$$

All other parameters are identical to those of Example 5.1.

5.13.5 Quarter-Wavelength Impedance Inverters

An impedance inverter can be realized using a quarter-wavelength transmission line inserted between the series-resonant circuit and a load impedance Z_L, as shown in Fig. 5.14. The length of the transmission line is $l = \lambda/4$ or an odd number multiple of a quarter wavelength $l = \lambda/[4(2n - 1)]$. The input impedance of the $\lambda/4$ transmission line loaded with impedance Z_L is given by

$$Z_i = Z_o \frac{Z_L + jZ_o \tan\left(\frac{2\pi}{\lambda}\frac{\lambda}{4}\right)}{Z_o + jZ_L \tan\left(\frac{2\pi}{\lambda}\frac{\lambda}{4}\right)} = \frac{Z_o^2}{Z_L} \tag{5.269}$$

Figure 5.15 shows two quarter-wavelength impedance inverters using π-configuration lumped-element resonant circuits. The characteristic impedance of both circuits is given by

$$Z_o = \omega L = \frac{1}{\omega C} \tag{5.270}$$

The input impedance of the impedance inverter of Fig. 5.15(a) is

$$Z_i = -jZ_o \| \left(jZ_o + \frac{-jZ_oZ_L}{Z_L - jZ_o}\right) = jZ_o \| \frac{Z_o^2}{Z_L - jZ_o} = \frac{Z_o^2}{Z_L} \tag{5.271}$$

Likewise, the input impedance of the impedance inverter of Fig. 5.15(b) is

$$Z_i = jZ_o \| \left(-jZ_o + \frac{jZ_oZ_L}{Z_L - jZ_o}\right) = jZ_o \| \frac{Z_o^2}{Z_L + jZ_o} = \frac{Z_o^2}{Z_L} \tag{5.272}$$

The expression for the input impedance Z_i of these circuits is the same as that for the input impedance of a quarter-wavelength impedance inverter using a transmission line. As $|Z_L|$ increases, $|Z_i|$ decreases.

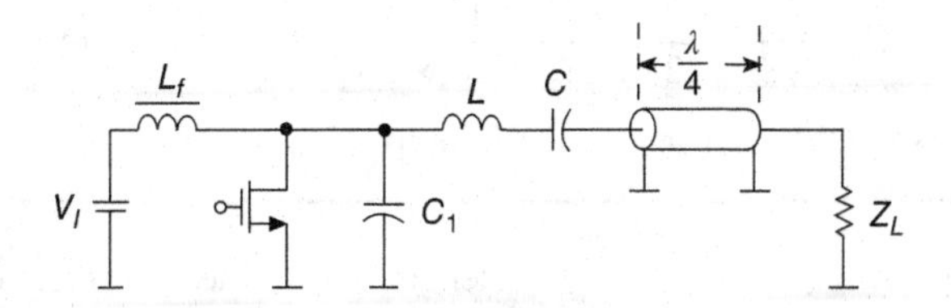

Figure 5.14 Class E RF power amplifier with a quarter-wavelength impedance inverter.

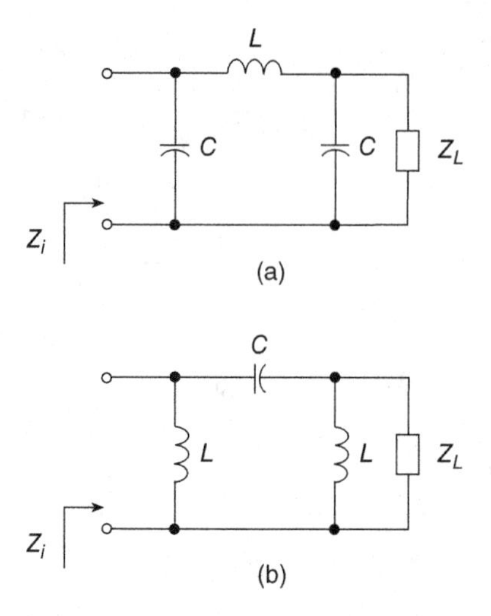

Figure 5.15 Quarter-wavelength impedance inverters using π networks consisting of lumped-element reactances. (a) Quarter-wavelength impedance inverter consisting of two capacitors and an inductor. (b) Quarter-wavelength impedance inverter consisting of two inductors and a capacitor.

If $Z_L = R_L$ and $Z_i = R$, then the required reactances of the quarter-wavelength matching circuits are given by

$$Z_o = X_L = \omega L = X_C = \frac{1}{\omega C} = \sqrt{RR_L} \tag{5.273}$$

Example 5.6

Design a Class E ZVS power amplifier with a quarter-wavelength impedance inverter using the lumped-element parameters shown in Fig. 5.15(a) to satisfy the following specifications: $V_I = 100$ V, $P_{Omax} = 80$ W, $R_L = 150\ \Omega$, and $f = 1.2$ MHz. Assume $D = 0.5$.

Solution. The reactances of the quarter-wavelength impedance inverter are given by

$$Z_o = X_L = \omega L = X_C = \frac{1}{\omega C} = \sqrt{RR_L} = \sqrt{72.1 \times 150} = 103.995\ \Omega x \tag{5.274}$$

Hence, the inductance is

$$L = \frac{Z_o}{\omega} = \frac{103.995}{2\pi \times 1.2 \times 10^6} = 13.79\ \mu\text{H} \tag{5.275}$$

and the capacitances are

$$C = \frac{1}{\omega Z_o} = \frac{1}{2\pi \times 1.2 \times 10^6 \times 103.995} = 1.2753\ \text{nF} \tag{5.276}$$

All other parameters are the same as those of Example 5.1.

Figure 5.16 shows two T-type quarter-wavelength transformers made up of lumped components. These transformers are introduced here. In the circuit of Fig. 5.16(a), the capacitances C and C_t can be combined into a single capacitor. In the circuit of Fig. 5.16(b), the inductances L

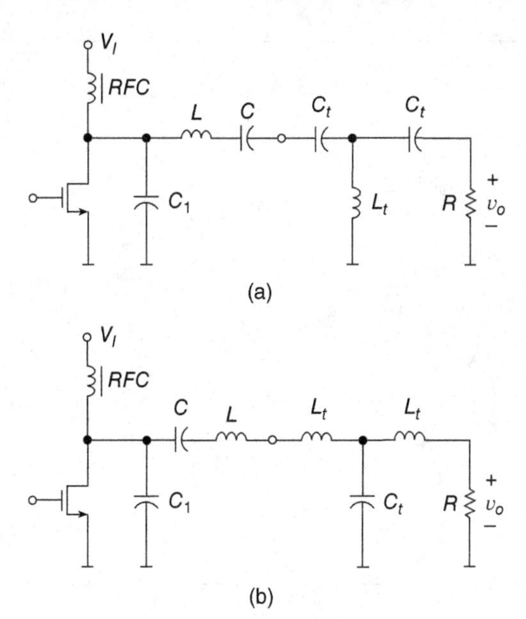

Figure 5.16 Quarter-wavelength impedance inverters using T-type networks consisting of lumped-element reactances. (a) Quarter-wavelength impedance inverter consisting of two capacitors and an inductor. (b) Quarter-wavelength impedance inverter consisting of two inductors and a capacitor.

and L_t can be combined into a single inductor. The characteristic impedance of both circuits is given by

$$Z_o = \omega L = \frac{1}{\omega C} \tag{5.277}$$

The input impedance of the transformer shown in Fig. 5.16(a) is

$$Z_i = -jZ_o + \frac{jZ_o(-jZ_o + Z_L)}{jZ_o - jZ_o + Z_L} = \frac{Z_o^2}{Z_L} \tag{5.278}$$

The input impedance of the transformer shown in Fig. 5.16(b) is

$$Z_i = jZ_o + \frac{(-jZ_o)(jZ_o + Z_L)}{-jZ_o + jZ_o + Z_L} = \frac{Z_o^2}{Z_L} \tag{5.279}$$

5.14 Class E ZVS RF Power Amplifier with Only Nonlinear Shunt Capacitance

In this section, we will consider the Class E ZVS RF power amplifier in which the MOSFET drain-to-source capacitance C_{ds} is the only shunt capacitance [40, 56]. The shunt capacitance is formed by the p-n junction capacitance of the MOSFET antiparallel diode

$$C_j = \frac{C_{j0}}{\left(1 - \dfrac{v_D}{V_{bi}}\right)^m} \quad \text{for} \quad v_D < V_{bi} \tag{5.280}$$

where v_D is the diode voltage, V_{bi} is the junction built-in potential, C_{j0} is the diode junction capacitance at $v_D = 0$, and m is the grading coefficient of the diode junction. Since

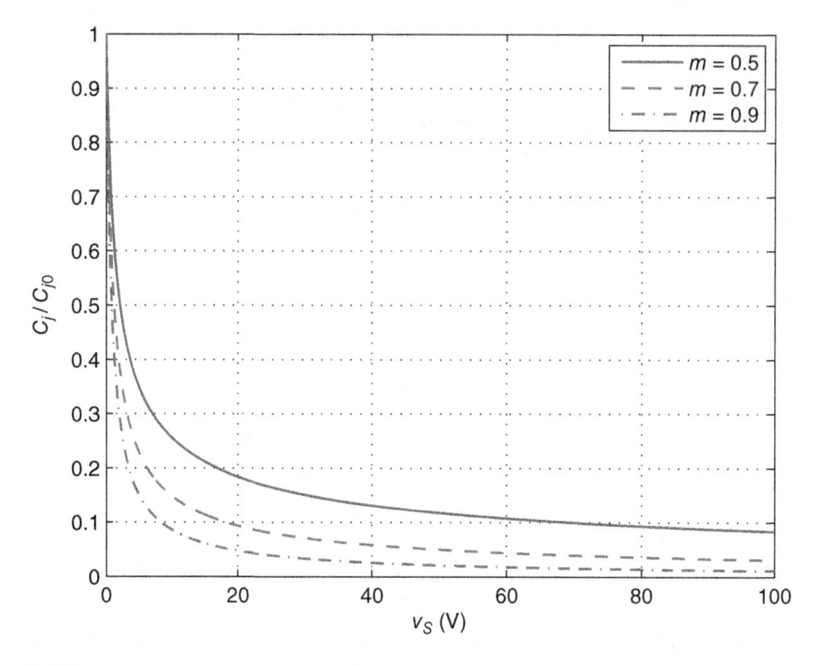

Figure 5.17 Ratio C_j/C_{j0} as a function of drain-to-source voltage v_S for $V_{bi} = 0.7$ V at selected values of grading coefficient m.

$v_D = -v_{DS} = -v_S$, the MOSFET drain-to-source capacitance is

$$C_{ds} = C_j = \frac{C_{j0}}{\left(1 + \frac{v_{DS}}{V_{bi}}\right)^m} = \frac{C_{j0}}{\left(1 + \frac{v_S}{V_{bi}}\right)^m} \quad \text{for} \quad v_S > -V_{bi} \tag{5.281}$$

Figures (5.17) and (5.18) illustrate C_j/C_{j0}.

The shunt capacitance consists of only the MOSFET nonlinear capacitance so that $C_1 = C_{ds}$. Assume that the RFC inductance is high enough to conduct only dc current and neglect its current ripple, that is, $i_{L_f} = I_I$. Assume also that the loaded quality factor of the series-resonant circuit is high enough to cause this circuit to conduct the sinusoidal current $i = I_m \sin(\omega t + \phi)$. Under these assumptions, all the currents, their harmonics, and the output voltage remain the same as those in the Class E amplifier with a linear shunt capacitance. Only the switch voltage waveform v_S is affected by the nonlinearity of the shunt capacitance. The current flowing through the nonlinear shunt capacitance C_{ds} when the switch is OFF is

$$i_{Cds} = I_I - i = I_I - I_m \sin(\omega t + \phi) \quad \text{for} \quad \pi \le \omega t \le 2\pi. \tag{5.282}$$

This current is related to the shunt capacitance C_{ds} and the switch voltage v_S by

$$i_{Cds} = \omega C_{ds}(v_S)\frac{dv_S}{d(\omega t)}, \tag{5.283}$$

yielding

$$C_{ds}(v_S)dv_S = \frac{1}{\omega} i_{Cds} d(\omega t). \tag{5.284}$$

Substitution of (5.281) and (5.282) into (5.284) produces

$$\frac{C_{j0}}{\left(1 + \frac{v_S}{V_{bi}}\right)^m} dv_S = \frac{1}{\omega}[I_I - I_m \sin(\omega t + \phi)]d(\omega t). \tag{5.285}$$

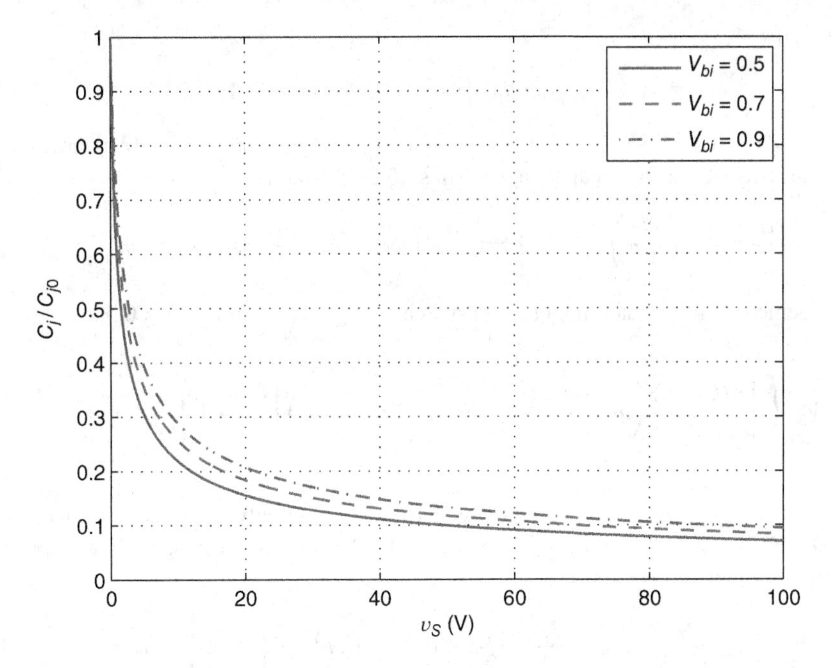

Figure 5.18 Ratio C_j/C_{j0} as a function of drain-to-source voltage v_S for $m = 0.5$ at selected values of V_{bi}.

Integrating both sides of this equation

$$\int_0^{v_S} \frac{C_{j0}}{\left(1 + \frac{v_S}{V_{bi}}\right)^m} dv_s = \frac{1}{\omega} \int_\pi^{\omega t} [I_I - I_m \sin(\omega t + \phi)] d(\omega t), \tag{5.286}$$

we obtain

$$\frac{V_{bi}}{1-m}\left[\left(1 + \frac{v_S}{V_{bi}}\right)^{1-m} - 1\right] = \frac{1}{\omega C_{j0}}\{I_I(\omega t - \pi) + I_m[\cos(\omega t + \phi) + \cos\phi]\}. \tag{5.287}$$

Under the ZVS condition $v_S(2\pi) = 0$, the left side of this equation is equal to zero, and therefore its right side must be also equal to zero, producing the relationship

$$I_m = -\frac{\pi}{2\cos\phi} I_I. \tag{5.288}$$

Hence, (5.287) becomes

$$\frac{V_{bi}}{1-m}\left[\left(1 + \frac{v_S}{V_{bi}}\right)^{1-m} - 1\right] = \frac{I_I}{\omega C_{j0}}\left\{\omega t - \frac{3\pi}{2} - \frac{\pi}{2\cos\phi}[\cos(\omega t + \phi)]\right\}, \tag{5.289}$$

yielding the switch voltage waveform under the ZVS condition at $D = 0.5$ and any value of $m \neq 1$

$$v_S = V_{bi}\left\{\left\langle \frac{I_I(1-m)}{\omega C_{j0} V_{bi}}\left\{\omega t - \frac{3\pi}{2} - \frac{\pi}{2\cos\phi}[\cos(\omega t + \phi)]\right\} + 1\right\rangle^{\frac{1}{1-m}} - 1\right\}$$

$$\text{for} \quad \pi \le \omega t \le 2\pi. \tag{5.290}$$

Imposing the ZDS condition $dv_S/d(\omega t) = 0$ at $\omega t = 2\pi$ on the switch voltage waveform, we obtain the optimum phase angle

$$\tan\phi_{opt} = -\frac{2}{\pi}, \tag{5.291}$$

resulting in

$$\phi_{opt} = \pi - \arctan\left(\frac{2}{\pi}\right) = 2.5747 \text{ rad} = 147.52^\circ. \tag{5.292}$$

Under the ZVS and ZDS conditions, the optimum phase angle $\phi_{opt} = -2/\pi$ is independent of the nonlinearity of the shunt capacitance. Thus, (5.289) becomes

$$\frac{V_{bi}}{1-m}\left[\left(1+\frac{v_S}{V_{bi}}\right)^{1-m} - 1\right] = \frac{I_I}{\omega C_{j0}}\left(\omega t - \frac{3\pi}{2} - \frac{\pi}{2}\cos\omega t - \sin\omega t\right). \tag{5.293}$$

Rearrangement of this equation yields the switch voltage waveform at $D = 0.5$ and any value of $m \neq 1$

$$v_S = V_{bi}\left\{\left[\frac{I_I(1-m)}{\omega C_{j0} V_{bi}}\left(\omega t - \frac{3\pi}{2} - \frac{\pi}{2}\cos\omega t - \sin\omega t\right) + 1\right]^{\frac{1}{1-m}} - 1\right\} \quad \text{for} \quad \pi \le \omega t \le 2\pi. \tag{5.294}$$

The switch voltage waveform of the Class E amplifier with only a nonlinear shunt capacitance C_{ds} under ZVS and ZDS conditions for duty cycle $D = 0.5$ and grading coefficient $m = 0.5$ is given by Suetsugu and Kazimierczuk [56]

$$v_S = V_{bi}\left\{\left[\frac{I_I}{2\omega C_{j0} V_{bi}}\left(\omega t - \frac{3\pi}{2} - \frac{\pi}{2}\cos\omega t - \sin\omega t\right) + 1\right]^2 - 1\right\} \quad \text{for} \quad \pi \le \omega t \le 2\pi \tag{5.295}$$

The dc supply voltage is

$$V_I = \frac{1}{2\pi}\int_0^{2\pi} v_S \, d(\omega t) \tag{5.296}$$

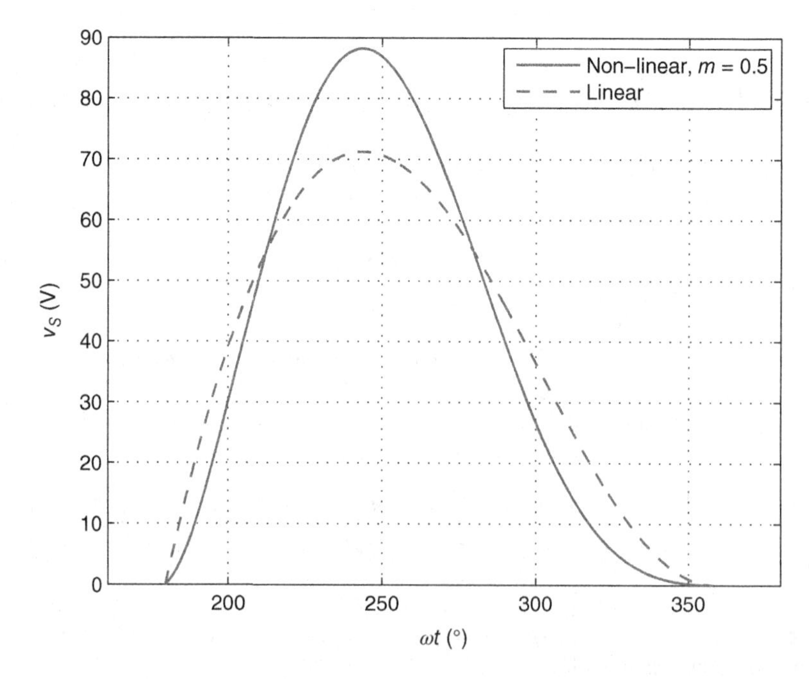

Figure 5.19 Waveforms of switch voltage for a linear shunt capacitance $C_1 = 298$ pF and a nonlinear shunt capacitance for $V_{bi} = 0.7$ V, $m = 0.5$, and $V_I = 20$ V.

This integral can be determined analytically only for $m = 0.5$. In this case,

$$R_{DC} = \frac{V_I}{I_I} = \frac{\pi^2 + 4}{8} R \tag{5.297}$$

Hence, the switch voltage waveform under both ZVS and ZDS conditions at $m = 0.5$ and $D = 0.5$ is given by

$$v_S = V_{bi} \left\{ \left[\frac{4V_I}{(\pi^2 + 4) V_{bi} \omega C_{j0} R} \left(\omega t - \frac{3\pi}{2} - \frac{\pi}{2} \cos \omega t - \sin \omega t \right) + 1 \right]^2 - 1 \right\}$$

$$\text{for} \quad \pi \leq \omega t \leq 2\pi \tag{5.298}$$

where the relationship among $\omega C_{j0} R$, V_{bi}, and V_I for optimum operation is [56]

$$\omega C_{j0} R = \frac{1}{3\pi(\pi^2 + 4) V_{bi}} \left[12V_{bi} + \sqrt{3V_{bi}(48V_{bi} - 36\pi^2 V_I + 5\pi^4 V_I)} \right] \tag{5.299}$$

Figure 5.19 shows the waveforms of switch voltage for a linear shunt capacitance $C_1 = 298$ pF and a nonlinear shunt capacitance for $V_{bi} = 0.7$ V, $m = 0.5$, and $V_I = 20$ V. It can be observed that the peak value of the switch voltage higher for the Class E amplifier with a nonlinear shunt capacitance than that with a linear shunt capacitance by a factor 1.25. For the Class E amplifier with a nonlinear shunt capacitance, $V_{SM}/V_I \approx 4.45$ at $m = 0.5$ and $D = 0.5$. The effect of V_{bi} on the switch waveform is small. As V_{bi} decreases, the switch peak voltage slightly increases. Using the equation for C_1, the design procedure of the Class E amplifier with a nonlinear shunt capacitance is the same as that with linear capacitance.

From (5.60), $\omega R = 8/[\pi(\pi^2 + 4)C_1]$ for the Class E amplifier with a linear shunt capacitance. We can also determine ωR from (5.299). Equating ωR for Class E amplifiers with linear and nonlinear shunt capacitance, the equivalent linear shunt capacitance for optimum operation at $m = 0.5$ and $D = 0.5$ is obtained as [56]

$$C_1 = \frac{24V_{bi} C_{j0}}{12V_{bi} + \sqrt{3V_{bi}(48V_{bi} - 36\pi^2 V_I + 5\pi^4 V_I)}} \tag{5.300}$$

The load resistance is $R = 8V_I^2/[(\pi^2 + 40)P_O]$. The maximum operating frequency at which both ZVS and ZDS conditions can be satisfied for the Class E amplifier with only nonlinear shunt capacitance at $D = 0.5$ and $m = 0.5$ is

$$f_{max} = \frac{1}{6\pi^2(\pi^2 + 4) V_{bi} C_{j0} R} \left[12V_{bi} + \sqrt{3V_{bi}(48V_{bi} - 36\pi^2 V_I + 5\pi^4 V_I)} \right]$$

$$= \frac{P_O}{48\pi^2 V_{bi} C_{j0} V_I^2} \left[12V_{bi} + \sqrt{3V_{bi}(48V_{bi} - 36\pi^2 V_I + 5\pi^4 V_I)} \right] \tag{5.301}$$

5.15 Push–Pull Class E ZVS RF Power Amplifier

A push–pull Class ZVS RF power amplifier with two RF chokes is depicted in Fig. 5.20 [5]. The circuit with the two RFCs can be simplified to the form depicted in Fig. 5.21. This amplifier consists of two transistors, two shunt capacitors C_1, RFC, center-tapped transformer, and series-resonant circuit driven by the secondary winding of the transformer. The transformer leakage inductance is absorbed into the resonant inductance L, and the transistor output capacitances are absorbed into shunt capacitances C_1.

Voltage and current waveforms, which explain the principle of operation of the amplifier of Fig. 5.21, are depicted in Fig. 5.22. The dc supply voltage source V_I is connected through an

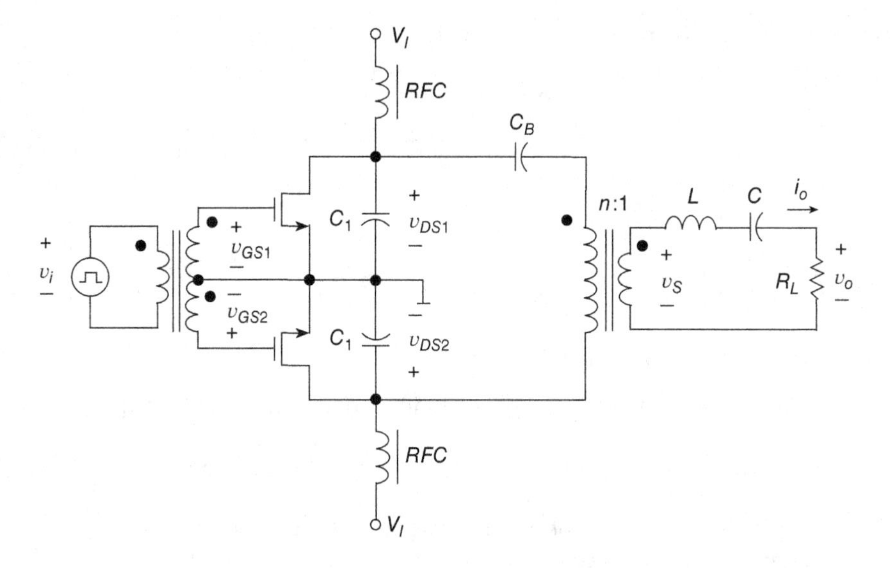

Figure 5.20 Push–pull Class E ZVS RF power amplifier with two RF chokes [5].

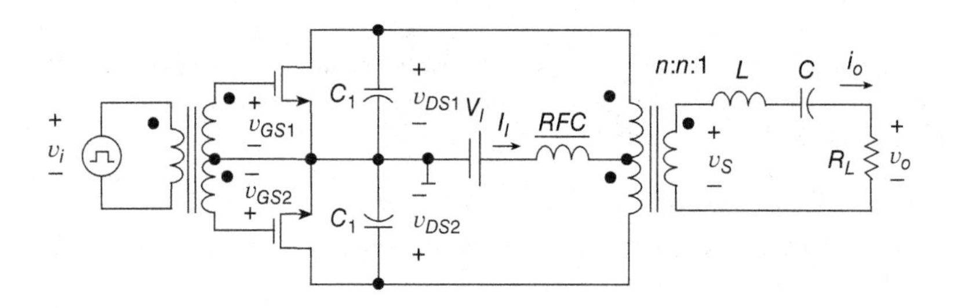

Figure 5.21 Push–pull Class E ZVS RF power amplifier.

RFC to the center tap of the primary winding of the output transformer. The switching transistors (MOSFETs) are driven on and off in opposite phase. The voltage across the secondary winding consists of positive and negative Class E pulses. The series-resonant circuit filters out all harmonics, and only the voltage of the fundamental component appears across the load resistance R_L. The odd harmonics are ideally zero in all push–pull amplifiers. The amplitude of the output voltage is twice the amplitude of the single-transistor Class E amplifier. As a result, the output power increases four times.

The load resistance seen by each transistor across each primary winding is given by

$$R = n^2 R_L \tag{5.302}$$

where n is the transformer turns ratio of the number of turns of one primary winding to the number of turns of the secondary winding. The relationships among acomponent values are

$$\omega C_1 = \frac{8}{\pi(\pi^2 + 4)R} = \frac{8}{\pi(\pi^2 + 4)n^2 R_L} \tag{5.303}$$

$$\frac{\omega L_b}{R_L} = \frac{\pi(\pi^2 - 4)}{16} \tag{5.304}$$

and

$$\frac{1}{\omega C R_L} = \left(Q_L - \frac{\omega L_b}{R_L}\right) = \left[Q_L - \frac{\pi(\pi^2 - 4)}{16}\right] \tag{5.305}$$

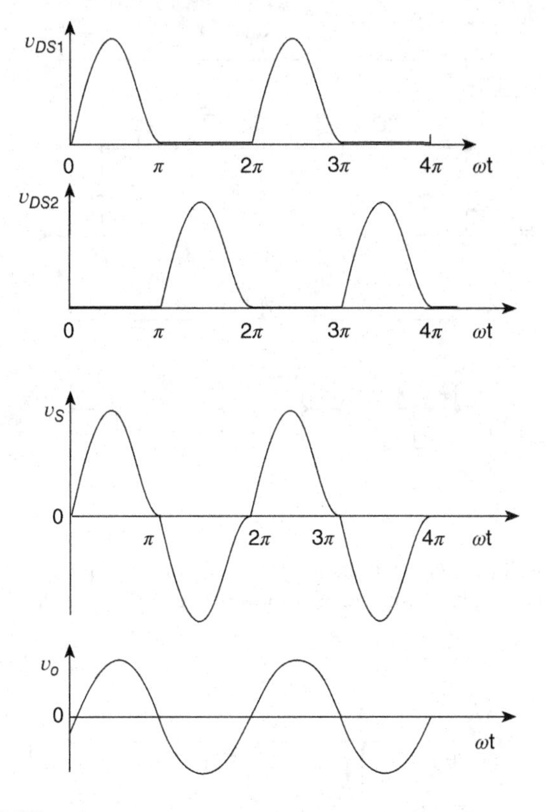

Figure 5.22 Waveforms in a push–pull Class E ZVS RF power amplifier.

The output power is

$$P_O = \frac{32}{\pi^2 + 4} \frac{V_I^2}{n^2 R_L} \tag{5.306}$$

Class E ZVS symmetrical RF power amplifiers are shown in Fig. 5.23 [53, 63]. The amplitudes of harmonics in the load are reduced in these amplifiers, such as in push–pull amplifiers.

5.16 Class E ZVS RF Power Amplifier with Finite DC-Feed Inductance

The Class E power amplifier can be operated with a finite dc-feed inductance L_f instead of an RF choke, as shown in Fig. 5.24 The output network is formed by a parallel-series resonant circuit. The inductance L_f and the shunt capacitance C_1 form a parallel-resonant Circuit, and the capacitor C and the inductor L form a series-resonant circuit. This circuit is easier to integrate because the dc-feed inductance is small. A small dc-feed inductance has a lower loss because of a smaller equivalent series resistance. In addition, if the amplifier is used as a radio transmitter with AM or any envelope modulation, it is easier to reduce distortion [17]. One such application of the Class E amplifier is in the envelope elimination and restoration (EER) system. The current and voltage waveforms of the Class E power amplifier with a dc-finite inductor are depicted in Fig. 5.25. The peak-to-peak current of the finite dc-feed inductance L_f increases as L_f decreases and becomes larger than the dc input current I_I. Table 5.1 gives the parameters for the Class E amplifier with finite dc-feed inductance [41]. As $\omega L_f / R_{DC}$

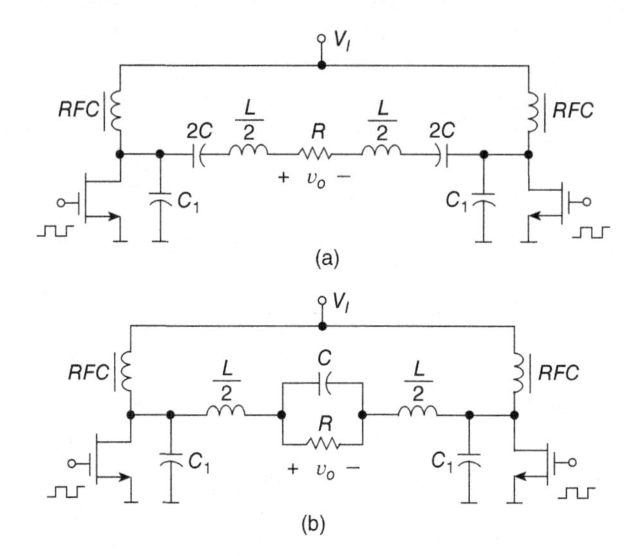

Figure 5.23 Class E ZVS symmetrical RF power amplifiers with reduced harmonics in the load. (a) Class E symmetrical amplifier with series-resonant circuit [53]. (b) Class E symmetrical amplifier with series-parallel circuit [63].

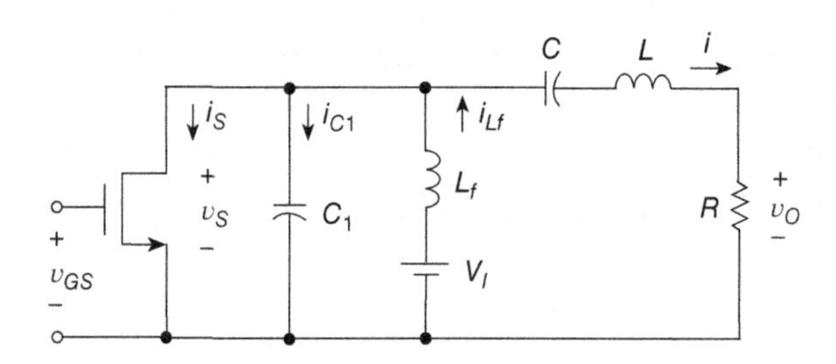

Figure 5.24 Class E power amplifier with finite dc-feed inductance L_f.

decreases, ωCR and $P_O R/V_I^2$ increase, and $\omega L_a/R$ decreases. For the last set of parameters in Table 5.1, $L_b = 0$, $L_a = L$, and the resonant frequency of the components L and C is equal to the operating frequency so that $\omega = 1/\sqrt{LC}$. Assuming the value of the loaded quality factor Q_L, these components can be calculated from the following relationship [54]:

$$\frac{\omega L}{R} = \frac{1}{\omega CR} = Q_L \tag{5.307}$$

Example 5.7

Design a Class E ZVS power amplifier to satisfy the following specifications: $V_I = 3.3$ V, $P_{Omax} = 0.25$ W, $f = 1$ GHz, and the bandwidth $BW = 0.2$ GHz. Assume $D = 0.5$.

Solution. Let us assume that $\omega L_f/R_{DC} = 1$. The load resistance is

$$R = 1.363\frac{V_I^2}{P_O} = 1.363 \times \frac{3.3^2}{0.25} = 59.372\,\Omega \tag{5.308}$$

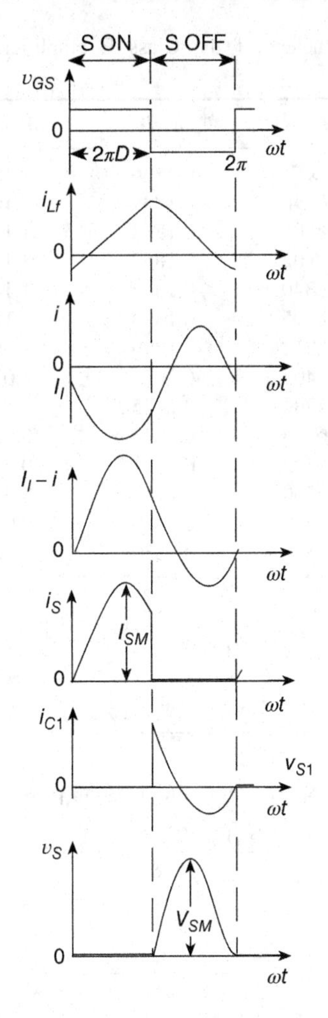

Figure 5.25 Waveforms in the Class E power amplifier with finite dc-feed inductance L_f.

The shunt capacitance is

$$C_1 = \frac{0.6839}{\omega R} = \frac{0.6839}{2\pi \times 10^9 \times 59.372} = 1.833 \text{ pF} \tag{5.309}$$

The dc-feed inductance is

$$L_f = \frac{1.363R}{\omega} = \frac{1.363 \times 59.372}{2\pi \times 10^9} = 12.879 \text{ nF} \tag{5.310}$$

The loaded quality factor is

$$Q_L = \frac{f_o}{BW} = \frac{10^9}{0.2 \times 10^9} = 5 \tag{5.311}$$

The inductance of the series-resonant circuit is

$$L = \frac{Q_L R}{\omega} = \frac{5 \times 59.372}{2\pi \times 10^9} = 47.246 \text{ nH} \tag{5.312}$$

The capacitance of the series-resonant circuit is

$$C = \frac{1}{\omega R Q_L} = \frac{1}{2\pi \times 10^9 \times 59.372 \times 5} = 0.537 \text{ pF} \tag{5.313}$$

Table 5.1 Parameters of the Class E amplifier with finite dc-feed inductance.

$\omega L_f/R_{DC}$	$\omega L_f/R$	$\omega C_1 R$	$\omega L_b/R$	RP_O/V_I^2
∞	∞	0.1836	1.152	0.5768
1000	574.40	0.1839	1.151	0.5774
500	289.05	0.1843	1.150	0.5781
200	116.02	0.1852	1.147	0.5801
100	58.340	0.1867	1.141	0.5834
50	29.505	0.1899	1.130	0.5901
20	12.212	0.1999	1.096	0.6106
15	9.3405	0.2056	1.077	0.6227
10	6.4700	0.2175	1.039	0.6470
5	3.6315	0.2573	0.9251	0.7263
3	2.5383	0.3201	0.7726	0.8461
2	2.0260	0.4142	0.5809	1.0130
1	1.3630	0.6839	0.0007	1.3630
0.9992	0.7320	0.6850	0.0000	1.3650

The dc input resistance is

$$R_{DC} = \frac{V_I}{I_I} = \frac{\omega L_f}{1} = \frac{2\pi \times 10^9 \times 12.879 \times 10^{-9}}{1} = 80.921\ \Omega \tag{5.314}$$

Assuming the efficiency $\eta_I = 0.8$, we obtain the dc input power

$$P_I = \frac{P_O}{\eta_I} = \frac{0.25}{0.8} = 0.3125\ \text{W} \tag{5.315}$$

and the dc input current

$$I_I = \frac{P_I}{V_I} = \frac{0.3125}{3.3} = 94.7\ \text{mA} \tag{5.316}$$

5.17 Class E ZVS Amplifier with Parallel-Series Resonant Circuit

A special case of a Class E ZVS amplifier with a finite dc inductance L_f is obtained, when the series components L and C are resonant at the operating frequency $f = \omega/(2\pi)$ [75]

$$\omega = \omega_s = \frac{1}{\sqrt{LC}} \tag{5.317}$$

The resonant frequency of the parallel-resonant circuit is

$$\omega_p = \frac{1}{\sqrt{L_f C_1}} \tag{5.318}$$

The ratio of the two resonant frequencies is

$$q = \frac{\omega_p}{\omega_s} = \frac{\omega_p}{\omega} = \frac{1}{\omega\sqrt{L_f C_1}} = \frac{\sqrt{LC}}{\sqrt{L_f C_1}} \tag{5.319}$$

The output current is assumed to be sinusoidal

$$i = I_m \sin(\omega t + \phi) \tag{5.320}$$

From KCL,

$$i_{L_f} = i_S + i_{C1} + i \tag{5.321}$$

When the switch is ON,

$$v_S = 0 \tag{5.322}$$

$$v_{L_f} = V_I \tag{5.323}$$

$$i_{C_1} = \omega C_1 \frac{dv_S}{d(\omega t)} = 0 \tag{5.324}$$

$$i_{L_f} = \frac{1}{\omega L_f} \int_0^{\omega t} v_{L_f} d(\omega t) + i_{L_f}(0) = \frac{1}{\omega L_f} \int_0^{\omega t} V_I \, d(\omega t) + i_{L_f}(0) = \frac{V_I}{\omega L_f} \omega t + i_{L_f}(0) \tag{5.325}$$

and the switch current is given by

$$i_S = i_{L_f} - i = \frac{V_I}{\omega L_f} \omega t + i_{L_f}(0) - I_m \sin(\omega t + \phi) \quad \text{for} \quad 0 < \omega t \leq \pi \tag{5.326}$$

Since

$$i_S(0) = i_{L_f}(0) - I_m \sin \phi = 0 \tag{5.327}$$

$$i_{L_f}(0) = I_m \sin \phi \tag{5.328}$$

Hence,

$$i_{L_f} = \frac{V_I}{\omega L_f} \omega t - I_m \sin \phi \tag{5.329}$$

and

$$i_S = \frac{V_I}{\omega L_f} \omega t + I_m [\sin \phi - \sin(\omega t + \phi)] \quad \text{for} \quad 0 < \omega t \leq \pi \tag{5.330}$$

When the switch is OFF, the switch current i_S is zero, $v_L = V_I - v_S$, and the current through the shunt capacitance C_1

$$i_{C_1} = i_{L_f} - i = \omega C_1 \frac{dv_S}{d(\omega t)} = \frac{1}{\omega L_f} \int_\pi^{\omega t} v_L \, d(\omega t) + i_{L_f}(\pi) - I_m \sin(\omega t + \phi)$$

$$= \frac{1}{\omega L_f} \int_\pi^{\omega t} (V_I - v_S) d(\omega t) + i_{L_f}(\pi) - I_m \sin(\omega t + \phi) \quad \text{for} \quad \pi < \omega t \leq 2\pi \tag{5.331}$$

where $v_S(\pi) = 0$ and

$$i_{L_f}(\pi) = \frac{\pi V_I}{\omega L_f} + I_m \sin \phi \tag{5.332}$$

Equation (5.331) can be differentiated to obtain a linear nonhomogeneous second-order differential equation

$$\omega^2 L_f C_1 \frac{d^2 v_S}{d(\omega t)^2} + v_S - V_I - \omega L_f I_m \cos(\omega + \phi) = 0 \tag{5.333}$$

The general solution of the differential equation is

$$\frac{v_S}{V_I} = A_1 \cos(q\omega t) + A_2 \sin(q\omega t) + 1 + \frac{q^2 p}{q^2 - 1} \cos(\omega t + \phi) \tag{5.334}$$

where

$$p = \frac{\omega L_f I_m}{V_I} \tag{5.335}$$

$$q = \frac{\omega_r}{\omega} = \frac{1}{\omega\sqrt{L_f C_1}} \tag{5.336}$$

$$A_1 = \frac{qp}{q^2 - 1}[q\cos\phi\cos\pi q + (2q^2 - 1)\sin\phi\sin\pi q - \cos\pi q - \pi q\sin\pi q \tag{5.337}$$

and

$$A_2 = \frac{qp}{q^2 - 1}[q\cos\phi\sin\pi q - (2q^2 - 1)\sin\phi\cos\pi q + \pi q\cos\pi q - \sin\pi q \tag{5.338}$$

The three unknowns in (5.334) are: p, q, and ϕ. Using the ZVS and ZDS conditions for v_S at $\omega t = 2\pi$ and the equation for the dc supply voltage

$$V_I = \frac{1}{2\pi}\int_o^{2\pi} v_S \, d(\omega t) \tag{5.339}$$

one obtains the numerical solution of (5.334)

$$q = \frac{\omega_r}{\omega} = \frac{1}{\omega\sqrt{L_f C_1}} \tag{5.340}$$

$$p = \frac{\omega L_f I_m}{V_I} = 1.21 \tag{5.341}$$

and

$$\phi = 195.155^\circ \tag{5.342}$$

The dc supply current is given by

$$I_I = \frac{1}{2\pi}\int_0^{2\pi} i_S \, d(\omega t) = \frac{I_m}{2\pi}\left(\frac{\pi^2}{2p} + 2\cos\phi - \pi\sin\phi\right) \tag{5.343}$$

The component of the switch voltage at the fundamental frequency is dropped across the load R and therefore its reactive component is zero

$$V_{X1} = \frac{1}{\pi}\int_0^{2\pi} v_S\cos(\omega t + \phi)d(\omega t) = 0 \tag{5.344}$$

The final results obtained from numerical solution are as follows:

$$\frac{\omega L_f}{R} = 0.732 \tag{5.345}$$

$$\omega C_1 R = 0.685 \tag{5.346}$$

$$\frac{\omega L}{R} = \frac{1}{\omega CR} = Q_L \tag{5.347}$$

$$\tan\psi = \frac{R}{\omega L_f} - \omega C_1 R \tag{5.348}$$

$$\psi = 145.856^\circ \tag{5.349}$$

$$V_m = 0.945 V_I \tag{5.350}$$

$$I_I = 0.826 I_m \tag{5.351}$$

$$I_m = 1.21 I_I \tag{5.352}$$

$$P_O = 1.365\frac{V_I^2}{R} \quad (5.353)$$

$$I_{SM} = 2.647I_I \quad (5.354)$$

$$V_{DSM} = 3.647V_I \quad (5.355)$$

$$c_p = 0.1036 \quad (5.356)$$

and

$$f_{max} = 0.0798\frac{P_O}{C_o V_I^2} \quad (5.357)$$

5.18 Class E ZVS Amplifier with Nonsinusoidal Output Voltage

Figure 5.26 shows a circuit of the Class E amplifier with nonsinusoidal output voltage [18]. This circuit is obtained by replacing the resonant capacitor C in the Class E amplifier of Fig. 5.1 with a blocking capacitor C_b. The loaded quality factor Q_L becomes low, and therefore the output voltage contains a lot of harmonics. The circuit of the Class E amplifier shown in Fig. 5.26 is able to operate under ZVS and ZDS at any duty cycle D. Waveforms in the Class E amplifier with a nonsinusoidal output voltage are shown in Fig. 5.27.

From the Kirchhoff current law (KCL),

$$i_S = I_I - i_{C1} - i \quad (5.358)$$

From the Kirchhoff voltage law (KVL),

$$v_S = V_I + v_L + v_o \quad (5.359)$$

When the switch is ON for $0 < \omega t \leq 2\pi D$ or $0 < t \leq t_1$,

$$v_S = 0 \quad (5.360)$$

$$i_{C1} = C_1\frac{dv_S}{d(\omega t)} = 0 \quad (5.361)$$

and

$$\frac{V_I}{s} = -sLI(s) + Li(0) - RI(s) \quad (5.362)$$

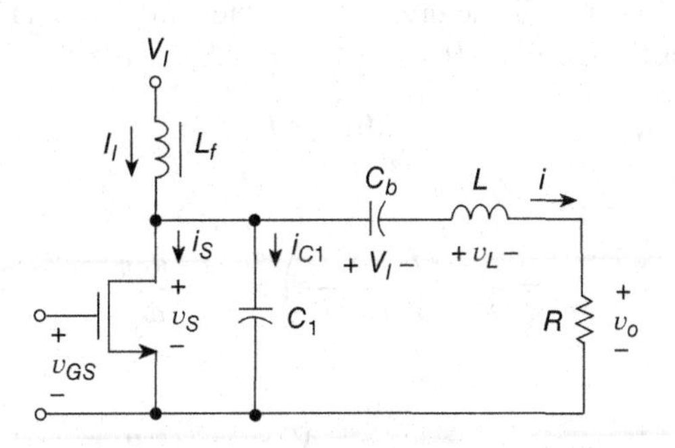

Figure 5.26 Class E amplifier with nonsinusoidal output voltage.

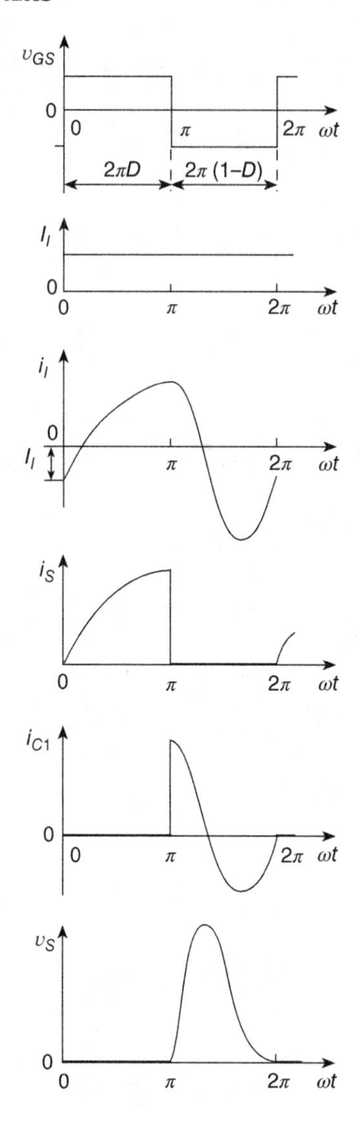

Figure 5.27 Waveforms of the Class E amplifier with nonsinusoidal output voltage.

where $i(0)$ is current through the inductor L at $t = 0$. Under the ZVS and ZDS conditions $v_S(0) = 0$ and $dv_S(0)/dt = 0$ at $t = 0$, $i_S(0) = 0$ and $i_{C1}(0) = 0$. Therefore,

$$i(0) = -I_I \tag{5.363}$$

From (5.362),

$$i = \frac{V_I}{R} - \left(I_I + \frac{V_I}{R}\right) \exp\left(-\frac{R\omega t}{\omega L}\right) \tag{5.364}$$

resulting in

$$i_S = I_I - i = \left(I_I + \frac{V_I}{R}\right) \exp\left(-\frac{R\omega t}{\omega L}\right) \tag{5.365}$$

The dc supply current is obtained as

$$I_I = \frac{1}{2\pi}\int_0^{2\pi} i_S \, d(\omega t) = \frac{a}{1-a}\frac{V_I}{R} \tag{5.366}$$

where

$$a = D + \frac{Q}{2\pi A}\left[\exp\left(-\frac{2\pi AD}{Q}\right)\right] \tag{5.367}$$

$$A = \frac{f_0}{f} = \frac{1}{\omega\sqrt{LC_1}} \tag{5.368}$$

$$Q = \frac{\omega_0 L}{R} = \frac{1}{\omega_0 C_1 R} \tag{5.369}$$

and

$$\omega_0 = \frac{1}{\sqrt{LC_1}} \tag{5.370}$$

Substitution of (5.366) into (5.365) yields the normalized switch current

$$\frac{i_S}{I_I} = \frac{1}{a}\left[1 - \exp\left(-\frac{A\omega t}{Q}\right)\right] \quad \text{for} \quad 0 < \omega t \leq 2\pi D \tag{5.371}$$

and

$$\frac{i_S}{I_I} = 0 \quad \text{for} \quad 2\pi D < \omega t \leq 2\pi \tag{5.372}$$

When the switch is OFF for $2\pi D < \omega t \leq 2\pi$ or $t_1 < t \leq T$,

$$i_S = 0 \tag{5.373}$$

$$V_S(s) = \frac{I_{C1}(s)}{sC_1} \tag{5.374}$$

$$I_{C1}(s) = I(s) + \frac{I_I}{sC_1}e^{-st_1} \tag{5.375}$$

and

$$V_S(s) = \frac{V_I}{s}e^{-st_1} - sLI(s) + Li(t_1)e^{-st_1} - RI(s) \tag{5.376}$$

where $i(t_1)$ is the initial condition of the inductor current at $t = t_1$. Thus,

$$V_S(s) = \left[\frac{1}{sC_1}\frac{\frac{V_I}{L} - \frac{I_I}{sLC_1} + si(t_1)}{s^2 + \frac{R}{L}s + \frac{1}{LC}} + \frac{I_I}{s^2 C}\right]e^{-st_1} \tag{5.377}$$

For the oscillatory case ($Q > \frac{1}{2}$),

$$V_S(s) = \left[\frac{1}{sC_1}\frac{\frac{V_I}{L} - \frac{I_I}{sLC_1} + si(t_1)}{(s+\alpha)^2 + \omega_n^2} + \frac{I_I}{s^2 C}\right]e^{-st_1} \tag{5.378}$$

where $\alpha = R/2L$ and $\omega_n = \omega_0\sqrt{1 - 1/4Q^2}$. Substituting (5.366) into (5.378) and taking the inverse Laplace transform, we obtain the normalized switch voltage

$$\frac{v_S}{V_I} = 0 \quad \text{for} \quad 0 < \omega t \leq 2\pi D \tag{5.379}$$

and

$$\frac{v_S}{V_I} = \frac{1}{1-a}\left\{1 - \exp\left[-\frac{A(\omega t - 2\pi D)}{2Q}\right]\left[\cos\left(\frac{A\sqrt{4Q^2-1}(\omega t - 2\pi D)}{2Q}\right)\right.\right.$$
$$\left.\left. - \frac{2Q^2\left(1-\exp\left(-\frac{2\pi AD}{Q}\right)\right)-1}{\sqrt{4Q^2-1}}\sin\left(\frac{A\sqrt{4Q^2-1}(\omega t - 2\pi D)}{2Q}\right)\right]\right\} \quad \text{for } 2\pi D < \omega t \le 2\pi \tag{5.380}$$

Imposing the ZVS and ZDS conditions on the switch voltage, we obtain the relationship among D, Q, and A in the form of a set of two equations:

$$\cos\left[\frac{\pi A(1-D)\sqrt{4Q^2-1}}{Q}\right] - \frac{2Q^2\left[1-\exp\left(-\frac{2\pi AD}{Q}\right)\right]-1}{\sqrt{4Q^2-1}}\sin\left[\frac{\pi A(1-D)\sqrt{4Q^2-1}}{Q}\right]$$
$$= \exp\left[\frac{\pi A(1-D)}{Q}\right] \tag{5.381}$$

and

$$\tan\left[\frac{\pi A(1-D)\sqrt{4Q^2-1}}{Q}\right] = \sqrt{4Q^2-1}\,\frac{\exp\left(-\frac{2\pi AD}{Q}\right)-1}{\exp\left(-\frac{2\pi AD}{Q}+1\right)} \tag{5.382}$$

These equations can be solved numerically. For $D = 0.5$, the solutions for which the ZVS and ZDS conditions are satisfied and the switch voltage is non-negative are

$$A = 1.6029 \tag{5.383}$$

and

$$Q = 2.856 \tag{5.384}$$

The output voltage is given by

$$\frac{v_o}{V_I} = \frac{a}{a-1} + \frac{1}{1-a}\left[1 - \exp\left(-\frac{A\omega t}{2Q}\right)\right] \quad \text{for} \quad 0 < \omega t \le 2\pi D \tag{5.385}$$

and

$$\frac{v_o}{V_I} = \frac{a}{a-1} + \frac{1}{Q(1-a)}\sqrt{Q^2\left[1-\exp\left(-\frac{2\pi AD}{Q}\right)\right]^2 + \exp\left(-\frac{2\pi AD}{Q}\right)}$$
$$\times \exp\left[-\frac{A(\omega t - 2\pi D)}{2Q}\right]\left\{\cos\left[\frac{A\sqrt{4Q^2-1}(\omega t - 2\pi D)}{2Q} - \psi\right]\right.$$
$$\left. - \frac{1}{\sqrt{4Q^2-1}}\sin\left[\frac{A\sqrt{4Q^2-1}(\omega t - 2\pi D)}{1Q} - \psi\right]\right\} \quad \text{for} \quad 2\pi D < \omega t \le 2\pi \tag{5.386}$$

where

$$\psi = \arctan\left\{\frac{\sqrt{4Q^2-1}}{2Q^2\left[1-\exp\left(-\frac{2\pi AD}{Q}\right)\right]-1}\right\} \tag{5.387}$$

The major parameters of the amplifier at the duty cycle $D = 0.5$ are [18]:

$$P_O = 0.1788\frac{V_I^2}{R} \tag{5.388}$$

$$R_{DC} = \frac{V_I}{I_I} = 2.7801R \tag{5.389}$$

$$V_{SM} = 3.1014V_I \tag{5.390}$$

$$I_{SM} = 4.2704I_I \tag{5.391}$$

$$Q = \frac{\omega_0 L}{R} = \frac{1}{\omega_0 C_1 R} = 2.856 \tag{5.392}$$

$$\omega C_1 R = 0.288 \tag{5.393}$$

$$\frac{\omega L}{R} = 2.4083 \tag{5.394}$$

$$c_p = \frac{P_O}{I_{SM}V_{SM}} = 0.0857 \tag{5.395}$$

and

$$f_{max} = 1.6108\frac{P_O}{C_o V_I^2} \tag{5.396}$$

where C_o is the transistor output capacitance. At $f = f_{max}$, $C_1 = C_o$.

Example 5.8

Design a Class E ZVS power amplifier with nonsinusoidal output voltage to satisfy the following specifications: $V_I = 100$ V, $P_{Omax} = 80$ W, and $f = 1.2$ MHz.

Solution. Let us assume the duty cycle $D = 0.5$. The load resistance is

$$R = 0.1788\frac{V_I^2}{P_O} = 0.1788\frac{100^2}{80} = 22.35\ \Omega \tag{5.397}$$

The dc input resistance seen by the power supply V_I is

$$R_{DC} = \frac{V_I}{I_I} = 2.7801R = 2.7801 \times 22.35 = 62.135\ \Omega \tag{5.398}$$

Assuming the amplifier efficiency $\eta = 0.95$, the dc supply power is

$$P_I = \frac{P_O}{\eta} = \frac{80}{0.95} = 84.21\ \text{W} \tag{5.399}$$

Hence, the dc supply current is

$$I_I = \frac{P_I}{V_I} = \frac{84.21}{100} = 0.8421\ \text{A} \tag{5.400}$$

The voltage stress of the switch is

$$V_{SM} = 3.1014V_I = 3.1014 \times 100 = 310.14 \text{ V} \tag{5.401}$$

and the current stress of the switch is

$$I_{SM} = 4.2704I_I = 4.2704 \times 0.8421 = 3.59 \text{ A} \tag{5.402}$$

The resonant frequency is

$$f_0 = Af = 1.6029 \times 1.2 \times 10^6 = 1.923 \text{ MHz} \tag{5.403}$$

The inductance is

$$L = \frac{QR}{\omega_0} = \frac{2.856 \times 22.35}{2\pi \times 1.923 \times 10^6} = 5.283 \text{ μH} \tag{5.404}$$

The shunt capacitance is

$$\frac{1}{Q\omega_0} = \frac{1}{2.856 \times 2\pi \times 1.923 \times 10^6} = 28.98 \text{ nF} \tag{5.405}$$

5.19 Class E ZVS Power Amplifier with Parallel-Resonant Circuit

Figure 5.28 shows a circuit of the Class E power amplifier with a parallel-resonant circuit. This circuit is also called a Class E amplifier with only one capacitor and one inductor [11–15]. Another version of this amplifier is depicted in Fig. 5.29 in which the dc supply voltage source V_I is connected in series with the inductor L and a blocking capacitor is connected in series with the load resistance R. The circuit of the amplifier is obtained from the conventional Class E ZVS power amplifier by replacing the RFC with a dc-feed inductance and the series-resonant circuit

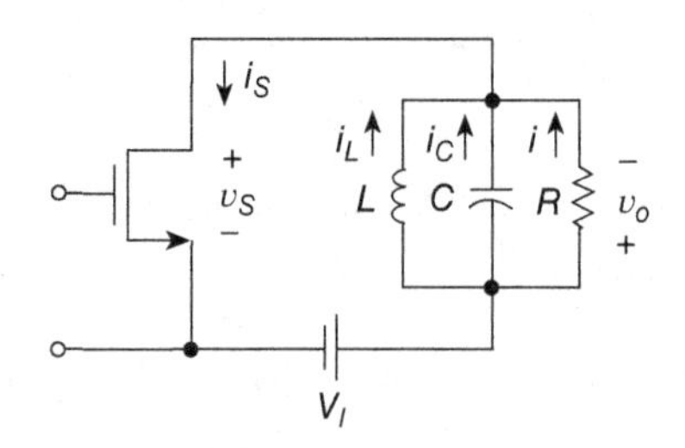

Figure 5.28 Class E amplifier with only one capacitor and one inductor.

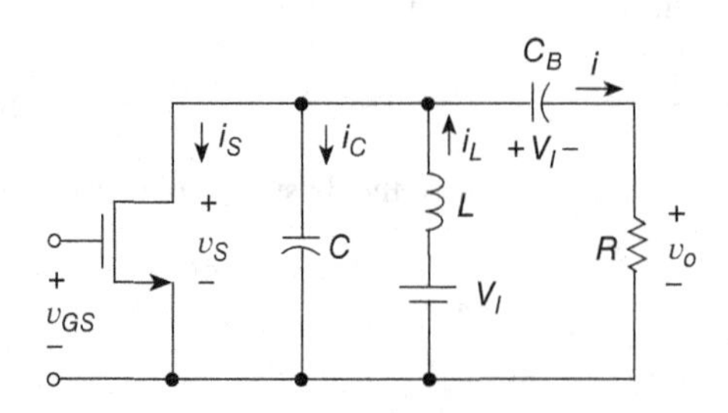

Figure 5.29 Another version of the Class E amplifier with only one capacitor and one inductor.

by a blocking capacitor C_B. Current and voltage waveforms are shown in Fig. 5.30. The drain current waveform is an increasing ramp, and the voltage waveform can satisfy both ZVS and ZDS conditions. The ZVS and ZDS conditions can be satisfied at any duty cycle D. From KCL and KVL,

$$i_S = i_L + i_C + i \tag{5.406}$$

and

$$v_S = V_I - v_o \tag{5.407}$$

Let us consider the amplifier circuit of Fig. 5.28. When the switch is ON for $0 < \omega t \leq 2\pi D$ or $0 < \omega t \leq t_1$,

$$v_S = 0 \tag{5.408}$$

$$v_o = V_I \tag{5.409}$$

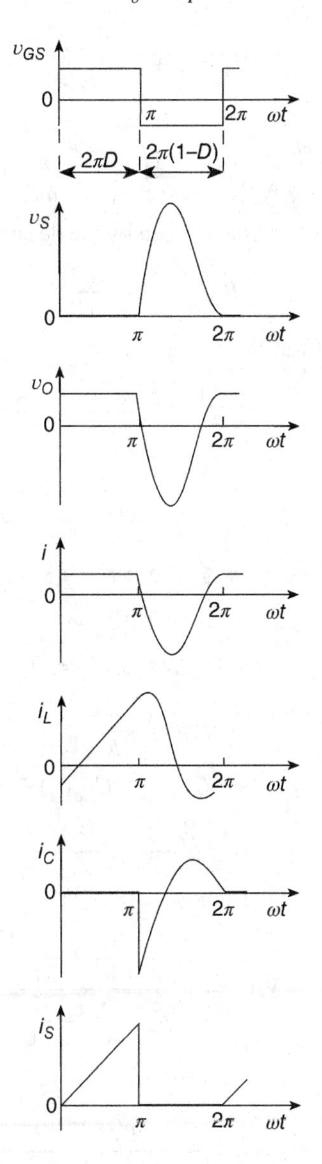

Figure 5.30 Waveforms of the Class E amplifier with only one capacitor and one inductor.

$$i = \frac{V_I}{R} \tag{5.410}$$

$$i_C = \omega C \frac{dv_o}{d(\omega t)} = 0 \tag{5.411}$$

$$i_L = \frac{1}{L}\int_0^{\omega t} v_o \, d(\omega t) + i_L(0) = \frac{1}{L}\int_0^{\omega t} V_I \, d(\omega t) + i_L(0) = \frac{V_I \omega t}{\omega L} + i_L(0) \tag{5.412}$$

$$i_S = \frac{V_I}{R} + \frac{V_I \omega t}{\omega L} + i_L(0) \tag{5.413}$$

Since $i_S(0) = 0$, $i_L(0) = -V_I/R$. Hence,

$$i_L = \frac{V_I \omega t}{\omega L} - \frac{V_I}{R} \tag{5.414}$$

and

$$i_S = i_L + i = \frac{V_I \omega t}{\omega L} \tag{5.415}$$

The dc supply voltage is

$$I_I = \frac{1}{2\pi}\int_0^{2\pi D} i_S \, d(\omega t) = \frac{1}{2\pi}\int_0^{2\pi D} \frac{V_I \omega t}{\omega L} d(\omega t) = \frac{\pi D^2 V_I}{\omega L} \tag{5.416}$$

where $2\pi D = \omega t_1$. Hence, the dc resistance seen by the dc supply voltage source is given by

$$R_{DC} = \frac{V_I}{I_I} = \frac{\omega L}{\pi D^2} \tag{5.417}$$

The normalized switch current waveform is given by

$$\frac{i_S}{I_I} = \frac{\omega t}{\pi D^2} \quad \text{for} \quad 0 < \omega t \leq 2\pi D \tag{5.418}$$

and

$$\frac{i_S}{I_I} = 0 \quad \text{for} \quad 2\pi D < \omega t \leq 2\pi \tag{5.419}$$

When the switch is OFF for $2\pi D < \omega t \leq 2\pi$ or $\omega t_1 < t \leq T$,

$$i_S = 0 \tag{5.420}$$

$$I_L(s) + I_C(s) + I(s) = 0 \tag{5.421}$$

$$I(s) = \frac{V_o(s)}{R} \tag{5.422}$$

$$I_C(s) = sCV_o(s) - Cv_o(t_1)e^{-st_1} \tag{5.423}$$

$$I_L(s) = \frac{V_o(s)}{sL} + \frac{i_L(t_1)}{s}e^{-st_1} \tag{5.424}$$

where $v_o(t_1) = V_I$ and $i_L(t_1) = V_I(t_1/L - 1/R)$. Hence,

$$V_o(s) = V_I \frac{s - \dfrac{t_1}{RC} + \dfrac{1}{RC}}{s^2 + \dfrac{s}{RC} + \dfrac{1}{LC}} \tag{5.425}$$

For the oscillatory case ($Q > \frac{1}{2}$),

$$V_o(s) = V_I \frac{(s^2 - \omega_0^2 t_1 + 2\alpha)e^{-st_1}}{(s + \alpha)^2 + \omega_n^2} \tag{5.426}$$

where

$$\alpha = \frac{1}{2RC} = \frac{\omega_0}{2Q} \tag{5.427}$$

and

$$\omega_n = \sqrt{\omega_0^2 - \alpha^2} = \omega_0\sqrt{1 - 1/4Q^2} \tag{5.428}$$

Thus, the normalized switch voltage is

$$\frac{v_S}{V_I} = 0 \quad \text{for} \quad 0 < \omega t \leq 2\pi D \tag{5.429}$$

and

$$\frac{v_S}{V_I} = 1 - \exp\left[-\frac{A}{2Q}(\omega t - 2\pi D)\right]\left\{\cos\left[\frac{A\sqrt{4Q^2-1}}{2Q}(\omega t - 2\pi D)\right]\right.$$
$$\left.-\frac{4\pi AQD - 1}{\sqrt{4Q^2-1}}\sin\left[\frac{A\sqrt{4Q^2-1}}{2Q}(\omega t - 2\pi D)\right]\right\} \quad \text{for} \quad 2\pi D < \omega t \leq 2\pi \tag{5.430}$$

where

$$\omega_0 = \frac{1}{\sqrt{LC}} \tag{5.431}$$

$$A = \frac{f_0}{f} = \frac{1}{\omega\sqrt{LC}} \tag{5.432}$$

and

$$Q = \omega_0 RC = \frac{R}{\omega_0 L} \tag{5.433}$$

Using the ZVS and ZDS conditions for the switch voltage at $\omega t = 2\pi$, we obtain a set of two equations:

$$\cos\left[\frac{\pi A(1-D)\sqrt{4Q^2-1}}{Q}\right] - \frac{4\pi AQD - 1}{\sqrt{4Q^2-1}}\sin\left[\frac{\pi A(1-D)\sqrt{4Q^2-1}}{Q}\right]$$
$$= \exp\left[\frac{\pi A(1-D)}{Q}\right] \tag{5.434}$$

and

$$\tan\left[\frac{\pi A(1-D)\sqrt{4Q^2-1}}{Q}\right] = \frac{\pi AD\sqrt{4Q^2-1}}{\pi AD - Q} \tag{5.435}$$

These sets of equations can be solved numerically. At $D = 0.5$, the solutions are

$$A = 1.5424 \tag{5.436}$$

and

$$Q = 1.5814 \tag{5.437}$$

The output voltage waveform is

$$\frac{v_o}{V_I} = 1 \quad \text{for} \quad 0 < \omega t \leq 2\pi D \tag{5.438}$$

and

$$\frac{v_o}{V_I} = \exp\left[-\frac{A}{2Q}(\omega t - 2\pi D)\right]\left\{\cos\left[\frac{A\sqrt{4Q^2-1}}{2Q}(\omega t - 2\pi D)\right]\right.$$

$$\left. -\frac{4\pi AQD - 1}{\sqrt{4Q^2-1}}\sin\left[\frac{A\sqrt{4Q^2-1}}{2Q}(\omega t - 2\pi D)\right]\right\} \quad \text{for} \quad 2\pi D < \omega t \leq 2\pi \tag{5.439}$$

The most important parameters of the amplifier at the duty cycle $D = 0.5$ are given by Kazimierczuk [15]

$$R_{DC} = \frac{V_I}{I_I} = 0.522R \tag{5.440}$$

$$I_{SM} = 4I_I \tag{5.441}$$

$$V_{SM} = 3.849V_I \tag{5.442}$$

$$P_O = \pi ADQ\frac{V_I^2}{R} = 1.9158\frac{V_I^2}{R} \tag{5.443}$$

$$A = \frac{f_0}{f} = 1.5424 \tag{5.444}$$

$$Q = \omega_0 CR = \frac{R}{\omega_0 L} = 1.5814 \tag{5.445}$$

$$\omega CR = 1.0253 \tag{5.446}$$

$$\frac{\omega L}{R} = 0.41 \tag{5.447}$$

$$c_p = \frac{P_O}{I_{SM}V_{SM}} = 0.0649 \tag{5.448}$$

and

$$f_{max} = 0.5318\frac{P_O}{C_o V_I^2} \tag{5.449}$$

Example 5.9

Design a Class E ZVS power amplifier with a parallel-resonant circuit to satisfy the following specifications: $V_I = 3.3$ V, $P_{Omax} = 1$ W, and $f = 2.4$ GHz.

Solution. We will assume the duty cycle $D = 0.5$. The load resistance is

$$R = 1.9158\frac{V_I^2}{P_O} = 1.9158 \times \frac{3.3^2}{1} = 20.86\ \Omega \tag{5.450}$$

The dc input resistance seen by the power supply V_I is

$$R_{DC} = \frac{V_I}{I_I} = 0.522R = 0.522 \times 20.86 = 10.89\ \Omega \tag{5.451}$$

Assuming the amplifier efficiency $\eta = 0.8$, the dc supply power is

$$P_I = \frac{P_O}{\eta} = \frac{1}{0.8} = 1.25\ \text{W} \tag{5.452}$$

Hence, the dc supply current is

$$I_I = \frac{P_I}{V_I} = \frac{1.25}{3.3} = 0.379 \text{ A}. \tag{5.453}$$

The voltage stress of the switch is

$$V_{SM} = 3.849V_I = 3.849 \times 3.3 = 12.7 \text{ V} \tag{5.454}$$

and the current stress of the switch is

$$I_{SM} = 4I_I = 4 \times 0.379 = 1.516 \text{ A} \tag{5.455}$$

The resonant frequency is

$$f_0 = Af = 1.5424 \times 2.4 \times 10^9 = 3.7 \text{ GHz} \tag{5.456}$$

The resonant inductance is

$$L = \frac{0.41R}{\omega} = \frac{0.41 \times 20.86}{2\pi \times 2.4 \times 10^9} = 0.567 \text{ nH} \tag{5.457}$$

The resonant capacitance is

$$C = \frac{1.0253}{\omega R} = \frac{1.0253}{2\pi \times 2.4 \times 10^9 \times 22.86} = 2.974 \text{ pF} \tag{5.458}$$

5.20 Amplitude Modulation of Class E ZVS RF Power Amplifier

The output voltage waveform in the Class E ZVS amplifier for continuous-wave (CW) operation, that is, without modulation, is given by

$$v_o = V_c \cos \omega_c t \tag{5.459}$$

where ω_c is the carrier angular frequency and V_c is the carrier amplitude. The amplitude of the output voltage of the Class E ZVS RF power amplifier V_m is directly proportional to the dc supply voltage V_I. For the duty cycle $D = 0.5$, the output voltage amplitude is

$$V_c = \frac{4}{\sqrt{\pi^2 + 4}} V_I \tag{5.460}$$

This property can be used for obtaining AM, as proposed in [17].

Figure 5.31 shows the circuit of a Class E ZVS RF power amplifier with drain AM [17]. The source of the modulating voltage v_m is connected in series with the dc supply voltage source V_I,

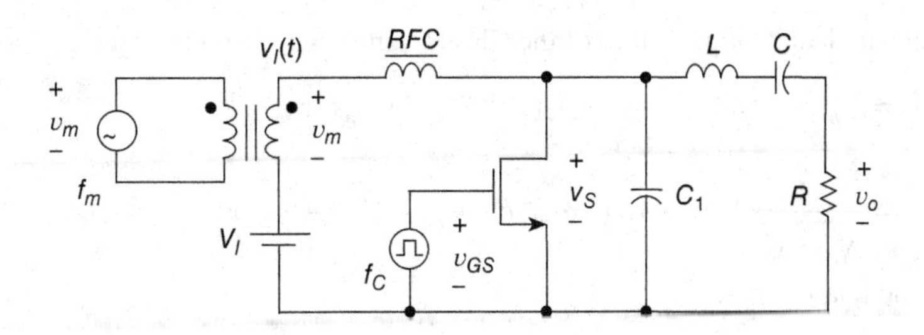

Figure 5.31 Class E ZVS RF power amplifier with drain amplitude modulation.

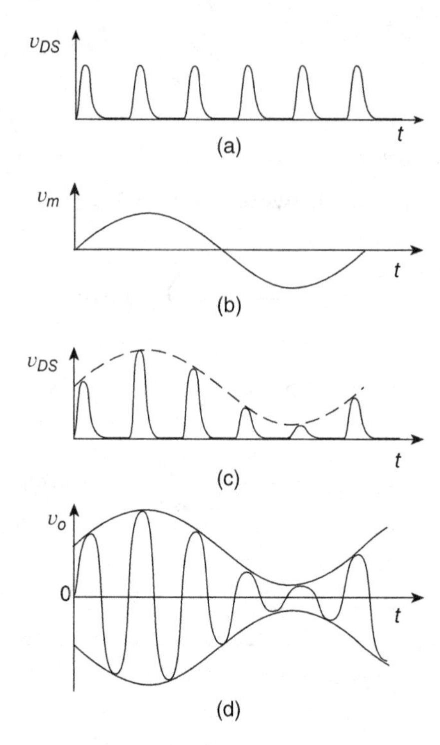

Figure 5.32 Waveforms in the Class E ZVS RF power amplifier with drain amplitude modulation. (a) CW drain-to-source voltage v_{DS}. (b) Modulating voltage v_m. (c) AM drain-to-source voltage v_{DS}. (d) AM output voltage v_o.

for example, via a transformer. The voltage waveforms in the AM Class E amplifier are shown in Fig. 5.32. A modulating voltage source v_m is connected in series with the dc voltage source V_I. The modulating voltage is given by

$$v_m = V_m \cos \omega_m t \tag{5.461}$$

Therefore, the supply voltage of the amplifier with AM is

$$v_I(t) = V_I + v_m(t) = V_I + V_m \cos \omega_m t \tag{5.462}$$

The amplitude of the output voltage is

$$V_m(t) = \frac{4}{\sqrt{\pi^2 + 4}}[V_I + v_m(t)] = \frac{4}{\sqrt{\pi^2 + 4}}(V_I + V_m \cos \omega_m t) \tag{5.463}$$

The switch voltage is $v_S = 0$ for $0 < \omega_c t \leq \pi$ and

$$v_S = (V_I + V_m \cos \omega_m t)\pi\left(\omega_c t - \frac{3\pi}{2} - \frac{\pi}{2}\cos \omega_c t - \sin \omega_c t\right) \quad \text{for} \quad \pi < \omega_c t \leq 2\pi \tag{5.464}$$

Assuming an ideal bandpass filter of the Class E amplifier, the output voltage waveform is

$$v_o = V_m(t) \cos \omega_c t = \frac{4}{\sqrt{\pi^2 + 4}}(V_I + V_m \cos \omega_m t) \cos \omega_c t$$

$$= \frac{4}{\sqrt{\pi^2 + 4}} V_I (1 + m \cos \omega_m t) \cos \omega_c t = V_c(1 + m \cos \omega_m t) \cos \omega_c t \tag{5.465}$$

where the modulation index is

$$m = \frac{V_m}{V_I} \tag{5.466}$$

The output voltage waveform can be rearranged to the form

$$v_o = V_c \cos \omega_c t + \frac{mV_c}{2} \cos(\omega_c - \omega_m)t + \frac{mV_c}{2} \cos(\omega_c + \omega_m)t \tag{5.467}$$

In reality, the voltage transfer function of the series-resonant circuit in the Class E amplifier under ZVS and ZDS conditions is not symmetrical with respect to the carrier frequency. The carrier frequency f_c is higher than the resonant frequency of the series-resonant circuit f_r (i.e., $f_c > f_r$). Therefore, the two sideband components are transmitted from the drain to the load resistance R with different magnitudes and different phase shifts (i.e., delays). The upper sideband is attenuated more than the lower sideband. In addition, the phase shift of the upper sideband is more negative than that of the lower sideband. The phase shift of the lower sideband may become even positive. These effects cause harmonic distortion of the envelope of the AM output voltage [17].

5.21 Summary

- The Class E ZVS RF power amplifier is defined as the circuit in which a single transistor is used and is operated as a switch, the transistor turns on at zero voltage, and the transistor may turn on at zero derivative.
- The transistor output capacitance, the choke parasitic capacitance, and the stray capacitance are absorbed into the shunt capacitance C_1 in the Class E ZVS power amplifier.
- The turn-on switching loss is zero.
- The operating frequency f is greater than the resonant frequency $f_0 = 1/(2\pi\sqrt{LC})$ of the series-resonant circuit. This results in an inductive load for the switch when it is ON.
- The antiparallel diode of the switch turns off at low di/dt and zero voltage, reducing reverse-recovery effects. Therefore, the MOSFET body diode can be used and there is no need for a fast diode.
- ZVS operation can be accomplished in the basic topology for load resistances ranging from zero to R_{opt}. Matching circuits can be used to match any impedance to the desired load resistance.
- The peak voltage across the transistor is about four times higher than the input dc voltage. Therefore, the circuit is suitable for low input voltage applications.
- The drive circuit is easy to build because the gate-to-source voltage of the transistor is referenced to ground.
- The circuit is very efficient and can be operated at high frequencies.
- The large choke inductance with a low-current ripple can be replaced by a low inductance with a large current ripple. In this case, the equations describing the amplifier operation will change [24].
- The loaded quality factor of the resonant circuit can be small. In the extreme case, the resonant capacitor becomes a large dc-blocking capacitor. The mathematical description will change accordingly [18].
- The maximum operating frequency of Class E operation with ZVS and ZDS or with ZVS only is limited by the output capacitance of the switch and is given by (5.84).

- The maximum operating frequency of Class E operation with ZVS and ZDS is inversely proportional to the shunt capacitance C_1 and to V_I^2, and is directly proportional to the output power P_O.
- The maximum operating frequency of the Class E ZVS power amplifier increases as the duty cycle decreases.

5.22 Review Questions

5.1 What is the zero-voltage switching (ZVS) operation?

5.2 What is the ZDS operation?

5.3 Is the transistor output capacitance absorbed into the Class E ZVS inverter topology?

5.4 Is it possible to obtain the ZVS operation at any load using the basic topology of the Class E ZVS inverter?

5.5 Is the turn-on switching loss zero in the Class E ZVS inverter?

5.6 Is the turn-off switching loss zero in the Class E ZVS inverter?

5.7 Is it possible to achieve the ZVS condition at any operating frequency?

5.8 Is the basic Class E ZVS inverter short-circuit proof?

5.9 Is the basic Class E ZVS inverter open-circuit proof?

5.10 Is it possible to use a finite dc-feed inductance in series with the dc input voltage source V_I?

5.11 Is it required to use a high value of the loaded quality factor of the resonant circuit in the Class E ZVS inverter?

5.23 Problems

5.1 Design a Class E RF power amplifier for wireless communication applications to meet the following specifications: $V_I = 3.3$ V, $P_O = 1$ W, $f = 1$ GHz, $C_o = 1$ pF, $Q_L = 5$, $r_{DS} = 0.01\ \Omega$, $r_{L_f} = 0.012\ \Omega$, $r_{C1} = 0.08\ \Omega$, $r_C = 0.05\ \Omega$, $r_L = 0.1\ \Omega$, and $t_f = 0$. Find the component values, reactances of reactive components, component stresses, and efficiency.

5.2 The Class E RF power amplifier given in Problem 5.1 had a load resistance $R_L = 50\ \Omega$. Design the impedance matching circuit.

5.3 Design an optimum Class E ZVS inverter to meet the following specifications: $P_O = 125$ W, $V_I = 48$ V, and $f = 2$ MHz. Assume $Q_L = 5$.

5.4 The rms value of the US utility voltage is from 92 to 132 V. This voltage is rectified by a bridge peak rectifier to supply a Class E ZVS inverter that is operated at a switch duty cycle of 0.5. What is the required value of the voltage rating of the switch?

5.5 Repeat Problem 5.4 for the European utility line, whose rms voltage is 220 ± 15.

5.6 Derive the design equations for the component values for the matching resonant circuit $\pi 2a$ shown in Fig. 5.8(b).

5.7 Find the maximum operating frequency at which pure Class E operation is still achievable for $V_I = 200$ V, $P_O = 75$ W, and $C_{out} = 100$ pF.

5.8 Design a Class E RF power amplifier for wireless communication applications to meet the following specifications: $V_I = 12$ V, $P_O = 10$ W, $f = 2.4$ GHz, $C_o = 1$ pF, $Q_L = 10$, $r_{DS} = 0.02\ \Omega$, $r_{L_f} = 0.01\ \Omega$, $r_{C1} = 0.09\ \Omega$, $r_C = 0.06\ \Omega$, $r_L = 0.2\ \Omega$, and $t_f = 0$. Find the component values, reactances of reactive components, component stresses, and efficiency.

5.9 The Class E power amplifier given in Problem 5.8 has $R_L = 50\ \Omega$. Design the impedance matching circuit.

5.10 Design a Class E RF power amplifier for wireless communication applications to meet the following specifications: $V_I = 3.3$ V, $P_O = 1$ W, $f = 1$ GHz, $C_o = 1$ pF, $Q_L = 5$, $r_{DS} = 0.01\ \Omega$, $r_{L_f} = 0.012\ \Omega$, $r_{C1} = 0.08\ \Omega$, $r_C = 0.05\ \Omega$, $r_L = 0.1\ \Omega$, and $t_f = 0$. Find the component values, reactances of reactive components, component stresses, and efficiency.

5.11 Find component values and their stresses for the Class E ZVS RF power amplifier with $V_I = 5$ V, $P_O = 0.25\ \Omega$, and $f_c = 2.4$ GHz.

References

[1] N. O. Sokal and A. D. Sokal, "Class E – a new class of high-efficiency tuned single-ended switching power amplifiers," *IEEE Journal of Solid-State Circuits*, vol. SC-10, pp. 168–176, 1975.

[2] N. O. Sokal and A. D. Sokal, "High efficiency tuned switching power amplifier," US Patent no. 3, 919, 656, November 11, 1975.

[3] J. Ebert and M. Kazimierczuk, "High efficiency RF power amplifier," *Bulletin of the Polish Academy of Sciences, Series Science Technical*, vol. 25, no. 2, pp. 13–16, 1977.

[4] N. O. Sokal, "Class E can boost the efficiency," *Electronic Design*, vol. 25, no. 20, pp. 96–102, 1977.

[5] F. H. Raab, "Idealized operation of the Class E tuned power amplifier," *IEEE Transactions on Circuits and Systems*, vol. CAS-24, pp. 725–735, 1977.

[6] N. O. Sokal and F. H. Raab, "Harmonic output of Class E RF power amplifiers and load coupling network design," *IEEE Journal of Solid-State Circuits*, vol. SC-12, pp. 86–88, 1977.

[7] F. H. Raab, "Effects of circuit variations on the Class E tuned power amplifier," *IEEE Journal of Solid-State Circuits*, vol. SC-13, pp. 239–247, 1978.

[8] F. H. Raab and N. O. Sokal, "Transistor power losses in the Class E tuned power amplifier," *IEEE Journal of Solid-State Circuits*, vol. SC-13, pp. 912–914, 1978.

[9] N. O. Sokal and A. D. Sokal, "Class E switching-mode RF power amplifiers – Low power dissipation, low sensitivity to component values (including transistors) and well-defined operation," *RF Design*, vol. 3, pp. 33–38, no. 41, 1980.

[10] J. Ebert and M. K. Kazimierczuk, "Class E high-efficiency tuned oscillator," *IEEE Journal of Solid-State Circuits*, vol. SC-16, pp. 62–66, 1981.

[11] N. O. Sokal, "Class E high-efficiency switching-mode tuned power amplifier with only one inductor and only one capacitor in load network – approximate analysis," *IEEE Journal of Solid–State Circuits*, vol. SC-16, pp. 380–384, 1981.

[12] M. K. Kazimierczuk, "Effects of the collector current fall time on the Class E tuned power amplifier," *IEEE Journal of Solid-State Circuits*, vol. SC-18, no. 2, pp. 181–193, 1983.

[13] M. K. Kazimierczuk, "Exact analysis of Class E tuned power amplifier with only one inductor and one capacitor in load network," *IEEE Journal of Solid-State Circuits*, vol. SC-18, no. 2, pp. 214–221, 1983.

[14] M. K. Kazimierczuk, "Parallel operation of power transistors in switching amplifiers," *Proceedings of the IEEE*, vol. 71, no. 12, pp. 1456–1457, 1983.

[15] M. K. Kazimierczuk, "Charge-control analysis of Class E tuned power amplifier," *IEEE Transactions on Electron Devices*, vol. ED-31, no. 3, pp. 366–373, 1984.

[16] B. Molnár, "Basic limitations of waveforms achievable in single-ended switching-mode (Class E) power amplifiers," *IEEE Journal of Solid-State Circuits*, vol. SC-19, no. 1, pp. 144–146, 1984.

[17] M. K. Kazimierczuk, "Collector amplitude modulation of the Class E tuned power amplifier," *IEEE Transactions on Circuits and Systems*, vol. CAS-31, no. 6, pp. 543–549, 1984.

[18] M. K. Kazimierczuk, "Class E tuned power amplifier with nonsinusoidal output voltage," *IEEE Journal of Solid-State Circuits*, vol. SC-21, no. 4, pp. 575–581, 1986.

[19] M. K. Kazimierczuk, "Generalization of conditions for 100-percent efficiency and nonzero output power in power amplifiers and frequency multipliers," *IEEE Transactions on Circuits and Systems*, vol. CAS-33, no. 8, pp. 805–506, 1986.

[20] M. K. Kazimierczuk and K. Puczko, "Impedance inverter for Class E dc/dc converters," *29th Midwest Symposium on Circuits and Systems*, Lincoln, Nebraska, August 10-12, 1986, pp. 707–710.

[21] G. Lüttke and H. C. Reats, "High voltage high frequency Class-E converter suitable for miniaturization," *IEEE Transactions on Power Electronics*, vol. PE-1, pp. 193–199, 1986.

[22] M. K. Kazimierczuk and K. Puczko, "Exact analysis of Class E tuned power amplifier at any Q and switch duty cycle," *IEEE Transactions on Circuits and Systems*, vol. CAS-34, no. 2, pp. 149–159, 1987.

[23] G. Lüttke and H. C. Reats, "220 V 500 kHz Class E converter using a BIMOS," *IEEE Transactions on Power Electronics*, vol. PE-2, pp. 186–193, 1987.

[24] R. E. Zulinski and J. W. Steadman, "Class E power amplifiers and frequency multipliers with finite dc-feed inductance," *IEEE Transactions on Circuits and Systems*, vol. CAS-34, no. 9, pp. 1074–1087, 1987.

[25] C. P. Avratoglou, N. C. Voulgaris, and F. I. Ioannidou, "Analysis and design of a generalized Class E tuned power amplifier," *IEEE Transactions on Circuits and Systems*, vol. CAS-36, no. 8, pp. 1068–1079, 1989.

[26] M. K. Kazimierczuk and X. T. Bui, "Class E dc-dc converters with a capacitive impedance inverter," *IEEE Transactions on Industrial Electronics*, vol. IE-36, pp. 425–433, 1989.

[27] M. K. Kazimierczuk and W. A. Tabisz, "Class C-E high-efficiency tuned power amplifier," *IEEE Transactions on Circuits and Systems*, vol. CAS-36, no. 3, pp. 421–428, 1989.

[28] M. K. Kazimierczuk and K. Puczko, "Power-output capability of Class E amplifier at any loaded Q and switch duty cycle," *IEEE Transactions on Circuits and Systems*, vol. CAS-36, no. 8, pp. 1142–1143, 1989.

[29] M. K. Kazimierczuk and X. T. Bui, "Class E dc/dc converters with an inductive impedance inverter," *IEEE Transactions on Power Electronics*, vol. PE-4, pp. 124–135, 1989.

[30] M. K. Kazimierczuk and K. Puczko, "Class E tuned power amplifier with antiparallel diode or series diode at switch, with any loaded Q and switch duty cycle," *IEEE Transactions on Circuits and Systems*, vol. CAS-36, no. 9, pp. 1201–1209, 1989.

[31] M. K. Kazimierczuk and X. T. Bui, "Class E amplifier with an inductive impedance inverter," *IEEE Transactions on Industrial Electronics*, vol. IE-37, pp. 160–166, 1990.

[32] G. H. Smith and R. E. Zulinski, "An exact analysis of Class E amplifiers with finite dc-feed inductance," *IEEE Transactions on Circuits and Systems*, vol. 37, no. 7, pp. 530–534, 1990.

[33] R. E. Zulinski and K. J. Grady, "Load-independent Class E power inverters: Part I – Theoretical development," *IEEE Transactions on Circuits and Systems*, vol. CAS-37, pp. 1010–1018, 1990.

[34] K. Thomas, S. Hinchliffe, and L. Hobson, "Class E switching-mode power amplifier for high-frequency electric process heating applications," *Electronics Letters*, vol. 23, no. 2, pp. 80–82, 1987.

[35] D. Collins, S. Hinchliffe, and L. Hobson, "Optimized Class-E amplifier with load variation," *Electronics Letters*, vol. 23, no. 18, pp. 973–974, 1987.

[36] D. Collins, S. Hinchliffe, and L. Hobson, "Computer control of a Class E amplifier," *International Journal of Electronics*, vol. 64, no. 3, pp. 493–506, 1988.

[37] S. Hinchliffe, L. Hobson, and R. W. Houston, "A high-power Class E amplifier for high frequency electric process heating," *International Journal of Electronics*, vol. 64, no. 4, pp. 667–675, 1988.

[38] M. K. Kazimierczuk, "Synthesis of phase-modulated dc/dc inverters an dc/dc converters," *IEE Proceedings, Part B, Electric Power Applications*, vol. 39, pp. 604–613, 1992.

[39] S. Ghandi, R. E. Zulinski, and J. C. Mandojana, "On the feasibility of load-independent output current in Class E amplifiers," *IEEE Transactions on Circuits and Systems*, vol. CAS-39, pp. 564–567, 1992.

[40] M. J. Chudobiak, "The use of parasitic nonlinear capacitors in Class-E amplifiers," *IEEE Transactions on Circuits and Systems I*, vol. CAS-41, no. 12, pp. 941–944, 1994.

[41] C.-H. Li and Y.-O. Yam, "Maximum frequency and optimum performance of class E power amplifier," *IEE Proceedings - Circuits Devices and Systems*, vol. 141, no. 3, pp. 174–184, 1994.

[42] M. K. Kazimierczuk and D. Czarkowski, *Resonant Power Converters*, 2nd Ed. New York, NY: John Wiley & Sons, 2011.

[43] T. Mader and Z. Popovic, "The transmission-line high-efficiency Class-E amplifier," *IEEE Transactions on Microwave Guided Wave Letters*, vol. 5, pp. 290–292, 1995.

[44] T. Sawlati, C. Andre, T. Salama, J. Stich, G. Robjohn, and D. Smith, "Low voltage high efficiency GaAs Class E power amplifiers," *IEEE Journal of Solid-State Circuits*, vol. 30, pp. 1074–1080, 1995.

[45] B. Grzesik, Z. Kaczmarczyk, and J. Janik, "A Class E inverter – the influence of inverter parameters on its characteristics," *27th IEEE Power Electronics Specialists Conference*, June 23-27, 1996, pp. 1832–1837.

[46] E. Bryetin, W. Shiroma, and Z. B. Popovic, "A 5-GHz high-efficiency Class-E oscillator," *IEEE Microwave and Guided Wave Letters*, vo. 6, no. 12, pp. 441–443, 1996.

[47] S. H.-L. Tu and C. Toumazou, "Low distortion CMOS complementary Class-E RF tuned power amplifiers," *IEEE Transactions on Circuits and Systems I*, vol. 47, pp. 774–779, 2000.

[48] W. H. Cantrell, "Tuning analysis for the high-Q Class-E power amplifier," *IEEE Transactions on Microwave Theory and Techniques*, vol. 48, no. 12, pp. 23-97-2402, 2000.

[49] A. J. Wilkinson and J. K. A. Everard, "Transmission-line load network topology for Class-E power amplifiers," *IEEE Transactions on Microwave Theory and Techniques*, vol. 49, no. 6, pp. 1202–1210, 2001.

[50] F. H. Raab, "Class-E, Class-C, and Class-F power amplifiers based upon a finite number of harmonics," *IEEE Transactions on Microwave Theory and Techniques*, vol. 49, no. 8, pp. 1462–1468, 2001.

[51] K. L. Martens and M. S. Steyaert, "A 700-MHz 1-W fully differential CMOS Class-E power amplifier," *IEEE Journal of Solid-State Circuits*, vol. 37, n. 2, pp. 137–141, 2002.

[52] F. H. Raab, P. Asbec, S. Cripps, P. B. Keningtopn, Z. B. Popovic, N. Potheary, J. Savic, and N. O. Sokal, "Power amplifiers and transistors for RF and microwaves," *IEEE Transactions on Microwave Theory and Techniques*, vol. 50, no. 3, pp. 814–826, 2002.

[53] S.-W. Ma, H. Wong, and Y.-O. Yam, "Optimal design of high output power Class E amplifier," *Proceedings of the 4th International Caracas Conference on Devices, Circuits and Systems*, pp. 012-1-012-3–, 2002.

[54] A. V. Grebennikov and H. J. Jaeger, "Class E amplifier with parallel circuit – A new challenge for high-efficiency RF and microwave power amplifiers," *IEEE MTT-S International Microwave Symposium Digest*, vol. 3, pp. 1627–1630, 2002.

[55] S. D. Kee, I. Aoki, A. Hajimiri, and D. Rutledge, "The Class-E/F family of ZVS switching amplifiers," *IEEE Transactions on Microwave Theory and Techniques*, vol. 51, no. 6, pp. 1677–1690, 2003.

[56] T. Suetsugu and M. K. Kazimierczuk, "Comparison of Class E amplifier with nonlinear and linear shunt capacitances," *IEEE Transactions on Circuits and Systems I, Fundamental Theory and Applications*, vol. 50, no. 8, pp. 1089–1097, 2003.

[57] T. Suetsugu and M. K. Kazimierczuk, "Analysis and design of Class E amplifier with shunt capacitance composed of linear and nonlinear capacitances," *IEEE Transactions on Circuits and Systems I: Regular Papers*, vol. 51, no. 7, pp. 1261–1268, 2004.

[58] D. Kessler and M. K. Kazimierczuk, "Power losses of Class E power amplifier at any duty cycle," *IEEE Transactions on Circuits and Systems I: Regular Papers*, vol. 51, no. 9, pp. 1675–1689, 2004.

[59] D. P. Kimber and P. Gardner, "Class E power amplifier steady-state solution as series in $1/Q$," *IEE Proceedings - Circuits Devices and Systems*, vol. 151, no. 6, pp. 557–564, 2004.

[60] T. Suetsugu and M. K. Kazimierczuk, "Design procedure of lossless voltage-clamped Class E amplifier with transformer and diode," *IEEE Transactions on Power Electronics*, vol. 20, no. 1, pp. 56–64, 2005.

[61] A. Grebennikov and N. O. Sokal, *Switchmode Power Amplifiers*. Amsterdam: Elsevier, 2005.

[62] M. K. Kazimierczuk, V. G. Krizhanovski, J. V. Rossokhina, and D. V. Chernov, "Class-E MOSFET tuned power oscillator design procedure," *IEEE Transactions on Circuits and Systems I: Regular Papers*, vol. 52, no. 6, pp. 1138–1147, 2005.

[63] S.-C. Wong and C. K. Tse, "Design of symmetrical Class E power amplifiers for low harmonic content applications," *IEEE Transactions on Circuits and Systems I: Regular Papers*, vol. 52, pp. 1684–1690, 2005.

[64] S. Jeon, A. Suarez, and D. B. Rutlege, "Global stability analysis and stabilization of a Class-E/F amplifiers with a distributed active transformer," *IEEE Transactions on Microwave Theory and Techniques*, vol. 53, no. 12, pp. 3712–3722, 2005.

[65] D. P. Kimber and P. Gardner, "Drain AM frequency response of the high-Q Class E power amplifier," *IEE Proceedings - Circuits Devices and Systems*, vol. 152, no. 6, pp. 752–756, 2005.

[66] D. P. Kimber and P. Gardner, "High Q Class E power amplifier analysis using energy conservation," *IEE Proceedings - Circuits Devices and Systems*, vol. 152, no. 6, pp. 592–597, 2005.

[67] A. Mazzanti, L. Larcher, R. Brama, and F. Svelto, "Analysis of reliability and power efficiency in cascade class-E PAs," *IEEE Journal of Solid-State Circuits*, vol. 41, no. 5, pp. 1222–1229, 2006.

[68] T. Suetsugu and M. K. Kazimierczuk, "Design procedure of Class E amplifier for off-nominal operation at 50% duty ratio," *IEEE Transactions on Circuits and Systems I: Regular Paper*, vol. 53, no. 7, pp. 1468–14, 2006.

[69] Z. Kaczmarczyk, "High-efficiency Class E, EF_2, and EF_3 inverters," *IEEE Transactions on Industrial Electronics*, vol. 53, no. 5, pp. 1584–1593, 2006.

[70] Z. Kaczmarczyk and W. Jurczyk, "Push-pull Class E inverter with improved efficiency," *IEEE Transactions on Industrial Electronics*, vol. 55, no. 4, pp. 1871–1874, 2008.

[71] Y. Y. Woo, Y. Yang, and B. Kim, "Analysis and experiments for high-frequency Class-F and inverse Class-F power amplifiers," *IEEE Transactions on Microwave Theory and Techniques*, vol. 54, no. 5, pp. 1969–1974, 2006.

[72] V. G. Krizhanovski, D. V. Chernov, and M. K. Kazimierczuk, "Low-voltage self-oscillating Class E electronic ballast for fluorescent lamps," IEEE International Symposium on Circuits and Systems, Island of Kos, Greece, May 21-24, 2006.

[73] K-C. Tsai and P. R. Gray, "A 1.9-GHz, 1-W CMOS Class-E power amplifier for wireless communications," *IEEE Journal of Solid-State Circuits*, vol. 34, no. 7, pp. 962–970, 1999.

[74] C. Yoo and Q. Huang, "A common-gate switched 0.9-W Class-E power amplifier with 41% PAE in 0.25-μm CMOS," *IEEE Journal of Solid-State Circuits*, vol. 36, no. 5, pp. 823–830, 2001.

[75] A. V. Grebennikov and H. Jaeger, "Class E with parallel circuit – A new challenges for high-efficiency RF and microwave power amplifiers," *IEEE MTT-S International Microwave Symposium Digest*, 2002, TJ2D-1, pp. 1627–1630.

[76] T. Suetsugu and M. K. Kazimierczuk, "Off-nominal operation of Class-E amplifier at any duty cycle," *IEEE Transactions on Circuits and Systems I: Regular Papers*, vol. 54, no. 6, pp. 1389–1397, 2007.

[77] A. V. Grebennikov and N. O. Sokal, *Switchmode RF Power Amplifiers*. Elsevier, Newnes, Oxford, UK, 2007.

[78] T. Suetsugu and M. K. Kazimierczuk, "ZVS operating frequency versus duty cycle of Class E amplifier with nonlinear capacitance," *IEEE International Symposium on Circuits and Systems*, Seattle, WA, May 23-26, 2008, pp. 3258–3226.

[79] T. Suetsugu and M. K. Kazimierczuk, "Maximum operating frequency of Class E power amplifier with any duty cycle," *IEEE Transactions on Circuits and Systems II: Express Briefs*, vol. 55, no. 8, pp. 768–770, 2008.

[80] F. You, S. He, X. Tang, and T. Cao, "Performance study of a Class-E power amplifier with tuned series-parallel resonance network," *IEEE Transactions on Microwave Theory and Techniques*, vol. 56, no. 10, pp. 2190–2200, 2008.

[81] Y. Abe, R. Ishikawa, and K. Hanjo, "Inverse Class-F AlGaN/GaN HEMT microwave amplifier based on lumped element circuit synthesis method," *IEEE Transactions on Microwave Theory and Techniques*, vol. 56, no. 12, pp. 2748–2753, 2008.

[82] A. Huhas and L. A. Novak, "Class-E, Class-C, and Class-F power amplifier based upon a finite number of harmonics," *IEEE Transactions on Microwave Theory and Techniques*, vol. 53, no. 6, pp. 1623–1625, 2009.

[83] N. Sagawa, H. Sekiya, and M. K. Kazimierczuk, "Computer-aided design of Class-E switching circuits taking into account optimized inductor design," *IEEE 25th Applied Power Electronics Conference, Palm Springs*, February 21-25, 2010, pp. 2212–2219.

[84] X. Wei, H. Sekiya, S. Kurokawa, T. Suetsugu, and M. K. Kazimierczuk, "Effect of MOSFET gate-to-drain parasitic capacitance on Class-E amplifier," *Proceedings IEEE International Symposium on Circuits and Systems*, Paris, France, May 31-June 2, 2010, pp. 2212–2219.

[85] T. Suetsugu and M. K. Kazimierczuk, "Power efficiency calculation of Class E amplifier with non-linear shunt capacitance," *Proceedings IEEE International Symposium on Circuits and Systems*, Paris, France, May 31-June 2, 2010, pp. 2714–2717.

[86] H. Sekiya, N. Sagawa, and M. K. Kazimierczuk, "Analysis of Class-DE amplifier with linear and nonlinear shunt capacitances at 25% duty cycle", *IEEE Transactions on Circuits and Systems I: Regular Papers*, vol. 57, no. 9, pp. 2334–2342, 2010

[87] R. Miyahara, H. Sekiya, and M. K. Kazimierczuk, "Novel design procedure of Class-E_Mi power amplifiers," *IEEE Transactions on Microwave Theory and Techniques*, vol. 58, no. 12, pp. 3607–3616, 2010.

[88] T. Nagashima, X. Wei, H. Sekiya, and M. K. Kazimierczuk, "Power conversion efficiency of Class-E amplifier ouside the nominal operations," *Proceedings IEEE International Symposium on Circuits and Systems*, Rio de Janeiro, Brazil, May 15-18, 2011, pp. 749–752.

[89] T. Suetsugu and M. K. Kazimierczuk, "Diode peak voltage clamping of class E amplifiers," *37th Annual Conference of the IEEE Industrial Electronics Society (IECON2011)*, Melbourne, Australia, November 7-10, 2011.

[90] X. Wei, H. Sekiya, S. Kurokawa, T. Suetsugu, and M. K. Kazimierczuk, "Effect of MOSFET parasitic capacitances on Class-E power amplifier," *IEEE Transactions on Circuits and Systems I: Regular Papers*, vol. 58, no. 10, pp. 2556–2564, 2011.

[91] X. Wei, S. Kurokawa, H. Sekiya, and M. K. Kazimierczuk, "Push-pull Class-E-M power amplifier for low harmonic-contents and high output-power applications," *IEEE Transactions on Circuits and Systems I: Regular Papers*, vol. 59, no. 9, pp. 21-37-2146–, 2012.

[92] T. Suetsugu, X. Wei, and M. K. Kazimierczuk, "Design equations for off-nominal operation of Class E amplifier with shunt capacitance at D = 0.5," *IEICE Transactions on Communications*, vol. E96B, no. 9, pp. 2198–2205, 2013.

[93] M. Hayati, A. Lofti, M. K. Kazimierczuk, and H. Sekiya, "Analysis and design of Class-E power amplifier with MOSFET parasitic linear and nonlinear capacitances at any duty cycle," *IEEE Transactions on Power Electronics*, vol. 28, no. 11, pp. 5222–5232, 2013.

[94] T. Nagashima, X. Wei, T. Suetsugu, M. K. Kazimierczuk, and H. Sekiya, "Wavefrom equations, output power and power conversion efficiency for Class-E inverter outside nominal operation," *IEEE Transactions on Circuits and Systems I: Regular Paper*, vol. 61, no. 4, pp. 1799–1810, 2014.

[95] X. Wei, T. Nagashima, M. K. Kazimierczuk, H. Sekiya, and T. Suetsugu, "Analysis and design of Class-E_M power amplifier," *IEEE Transactions on Circuits and Systems I: Regular Paper*, vol. 61, no. 4, pp. 976–986, April 2014.

[96] M. Hayati, A. Lofti, M. K. Kazimierczuk, and H. Sekiya, "Analysis, design, and implementation of Class-E ZVS amplifier with MOSFET nonlinear drain-to-source parasitic capacitance at any duty cycle," *IEEE Transactions on Power Electronics*, vol. 29, no. 9, pp. 4989–4999, September 2014.

[97] A. Mediano and N. O. Sokal, "A Class-E RF power amplifier with a flat-top transistor-voltage waveform," *IEEE Transactions on Microwave Theory and Techniques*, vol. 28, no. 11, pp. 5215–5220, 2013.

6

Class E Zero-Current Switching RF Power Amplifier

6.1 Introduction

In this chapter, a Class E RF zero-current switching (ZCS) power amplifier [1–6] is presented and analyzed. In this amplifier, the switch is turned off at zero current, yielding zero turn-off switching loss. Therefore, this circuit is also called an inverse Class E amplifier. A shortcoming of the Class E ZCS amplifier is that the switch output capacitance is not included in the basic amplifier topology. The switch turns on at a nonzero voltage, and the energy stored in the switch output capacitance is dissipated in the switching device, reducing the efficiency. Therefore, the upper operating frequency of the Class E ZCS amplifier is lower than that of the Class E ZVS amplifier.

6.2 Circuit Description

A circuit of a Class E ZCS RF power amplifier is depicted in Fig. 6.1(a). This circuit was introduced in [1]. It consists of a single transistor and a load network. The transistor operates cyclically as a switch at the desired operating frequency $f = \omega/(2\pi)$. The simplest type of load network consists of a resonant inductor L_1 connected in series with the dc source V_I, and an L-C-R series-resonant circuit. The resistance R is an ac load.

The equivalent circuit of the Class E ZCS amplifier is shown in Fig. 6.1(b). The capacitance C is divided into two series capacitances, C_a and C_b, so that capacitance C_a is series resonant with L at the operating frequency $f = \omega/2\pi$

$$\omega = \frac{1}{\sqrt{LC_a}} \tag{6.1}$$

RF Power Amplifiers, Second Edition. Marian K. Kazimierczuk.

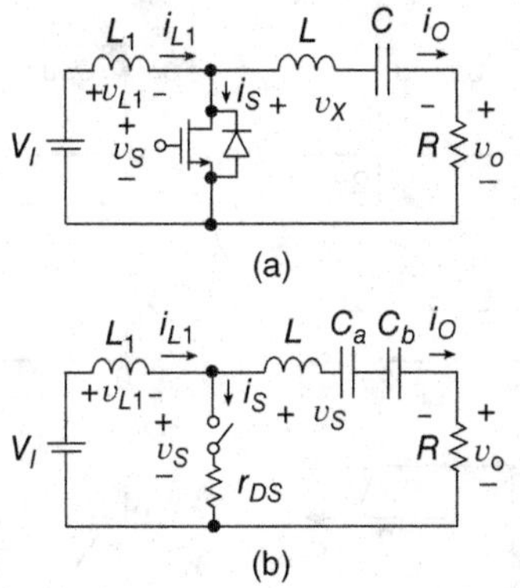

Figure 6.1 Class E RF zero-current switching amplifier. (a) Circuit. (b) Equivalent circuit.

The additional capacitance C_b signifies the fact that the operating frequency f is lower than the resonant frequency of the series-resonant circuit when the switch is ON $f_{o1} = 1/(2\pi\sqrt{LC})$. The loaded quality factor Q_L is defined by the expression

$$Q_L = \frac{X_{Cr}}{R} = \frac{C_a + C_b}{\omega R C_a C_b} \tag{6.2}$$

The choice of Q_L involves the usual tradeoff among (1) low harmonic content of the power delivered to R (high Q_L), (2) low change of amplifier performance with frequency (low Q_L), (3) high efficiency of the load network (low Q_L), and (4) high bandwidth (low Q_L).

6.3 Principle of Operation

The equivalent circuit of the amplifier is shown in Fig. 6.1(b). It is based on the following assumptions:

1) The elements of the load network are ideal.
2) The loaded quality factor Q_L of the series-resonant circuit is high enough that the output current is essentially a sinusoid at the operating frequency.
3) The switching action of the transistor is instantaneous and lossless; the transistor has zero output capacitance, zero saturation resistance, zero saturation voltage, and infinite "off" resistance.

It is assumed for simplicity that the switch duty ratio is 50%, that is, the switch is ON for half of the ac period and OFF for the remainder of the period. However, the duty ratio can be any arbitrarily chosen if the circuit component values are chosen to be appropriate for the chosen duty ratio. It will be explained in Section 6.4 that a duty ratio of 50% is one of the conditions for optimum amplifier operation.

The amplifier operation is determined by the switch when it is closed and by the transient response of the load network when the switch is open. The principle of operation of the amplifier is explained by the current and voltage waveforms, which are shown in Fig. 6.2. Figure 6.2(a) depicts the waveforms for optimum operation. When the switch is open, its current i_S is zero. Hence, the inductor current i_{L1} is equal to a nearly sinusoidal output current i_o. The current i_{L1}

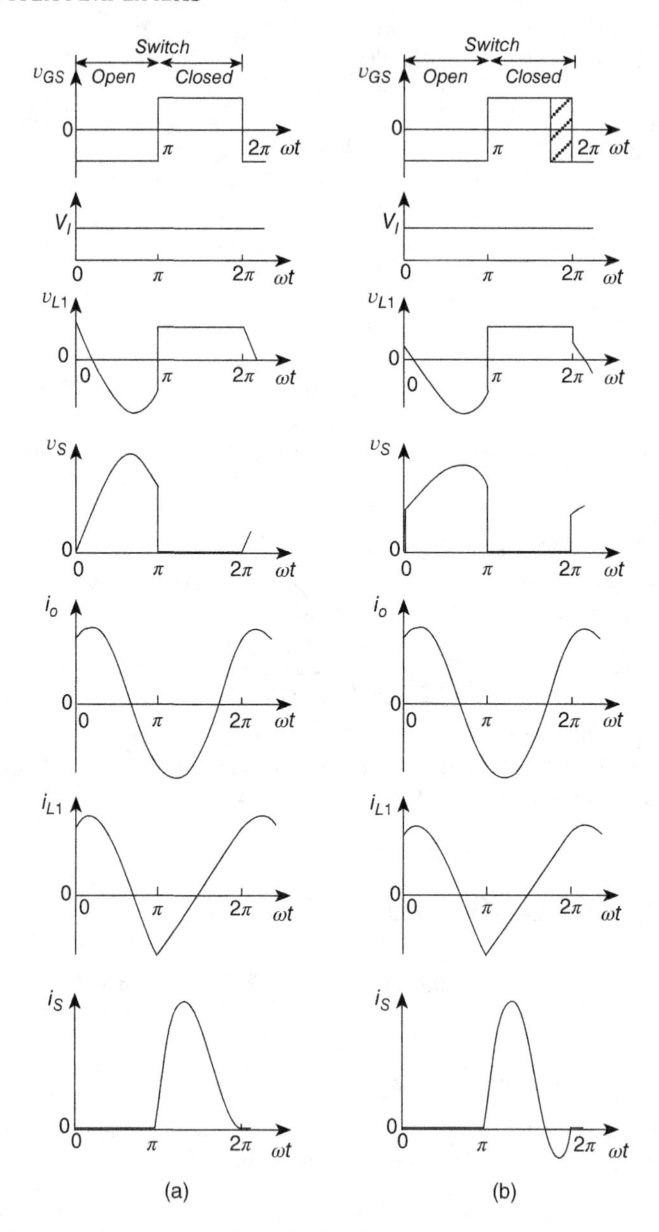

Figure 6.2 Current and voltage waveforms in the Class E ZCS amplifier. (a) For optimum operation. (b) For suboptimum operation.

produces the voltage drop v_{L1} across the inductor L_1. This voltage is approximately a section of a sine wave. The difference between the supply voltage V_I and the voltage v_{L1} is the voltage across the switch v_S. When the switch is closed, the voltage v_S is zero, and voltage v_{L1} equals the supply voltage V_I. This voltage produces the linearly increasing current i_{L1}. The difference between the current i_{L1} and current i flows through the switch.

In the Class E ZCS amplifier, it is possible to eliminate power losses due to on-to-off transition of the transistor, yielding high efficiency. Assuming that the transistor is turned off at $\omega t_{off} = 2\pi$, the ZCS condition at turn-off is

$$i_S(2\pi) = 0 \tag{6.3}$$

For optimum operation, the zero-derivative switching (ZDS) condition should also be satisfied:

$$\left.\frac{di_S}{d(\omega t)}\right|_{\omega t=2\pi} = 0 \tag{6.4}$$

If condition (6.3) is not satisfied, the transistor turns off at nonzero current. Consequently, there is a fall time of the drain (or collector) current during which the transistor acts as a current source. During the fall time, the drain current increases and the drain-source voltage increases. Since the transistor current and voltage overlap during the turn-off interval, there is a turn-off power loss. However, if the transistor current is already zero at turn-off, the transistor current fall time is also zero, there is no overlap of the transistor current and voltage, and the turn-off switching loss is zero.

Condition (6.3) eliminates dangerous voltage spikes at the output of the transistor. If this condition is not satisfied, the current i_S changes rapidly during turn-off of the transistor. Hence, the inductor current i_{L_1} also changes rapidly during turn-off. Therefore, inductive voltage spikes appear at the output of the transistor and device failure may occur. The rapid change of i_{L_1} during turn-off of the transistor causes a change of the energy stored in the inductor L_1. A part of this energy is dissipated in the transistor as heat, and the remainder is delivered to the series-resonant circuit L, C, and R. If condition (6.4) is satisfied, the switch current is always positive and the antiparallel diode never conducts. Furthermore, the voltage across the switch at the turn-off instant will be zero, that is, $v_S(2\pi) = 0$, and during the "off" state the voltage v_S will start to increase from zero only gradually. This zero starting voltage v_S is desirable in the case of the real transistor because the energy stored in the parasitic capacitance across the transistor is zero at the instant the transistor switches off. The parasitic capacitance comprises the transistor capacitances, the winding capacitance of L_1, and stray winding capacitance. The optimum operating conditions can be accomplished by a proper choice of the load-network components. The load resistance at which the ZCS condition is satisfied is $R = R_{opt}$.

Figure 6.2(b) shows the waveforms for suboptimum operation. This operation occurs only when the ZCS condition is satisfied. If the slope of the switch current at the time the switch current reaches zero is positive, the switch current will be negative during a portion of the period. If the transistor is OFF, the antiparallel diode conducts the negative switch current. If the transistor is ON, either only the transistor conducts or both the transistor and the antiparallel diode conduct. The transistor should be turned off during the time interval the switch current is negative. When the switch current reaches zero, the antiparallel diode turns off.

The voltage across the inductor L_1 is described by the expression

$$v_{L1} = \omega L_1 \frac{di_{L_1}}{d(\omega t)} \tag{6.5}$$

At the switch turn-on, the derivative of the inductor current i_{L_1} changes rapidly from a negative to a positive value. This causes a step change in the inductor voltage v_{L_1} and consequently in the switch voltage v_S.

According to assumption (3), the conduction power loss and the turn-on switching power loss are neglected. The conduction loss dominates at low frequencies, and the turn-on switching loss dominates at high frequencies. The off-to-on switching time is especially important in high-frequency operation. The parasitic capacitance across the transistor is discharged from the voltage $2V_I$ to zero when the transistor switches on. This discharge requires a nonzero length of time. The switch current i_S is increasing during this time. Since the switch voltage v_S and the switch current i_S are simultaneously nonzero, the power is dissipated in the transistor. The off-to-on switching loss becomes comparable to saturation loss at high frequencies. Moreover, the transient response of the load network depends on the parasitic capacitance when the switch is open. This influence is neglected in this analysis. According to assumption (1), the power losses in the parasitic resistances of the load network are also neglected.

6.4 Analysis

6.4.1 Steady-State Current and Voltage Waveforms

The basic equations of the equivalent amplifier circuit shown in Fig. 6.1(b) are

$$i_S = i_{L1} - i \tag{6.6}$$

$$v_S = V_I - v_{L1} \tag{6.7}$$

The series-resonant circuit forces a sinusoidal output current

$$i = I_m \sin(\omega t + \varphi) \tag{6.8}$$

The switch is OFF for the interval $0 < \omega t \leq \pi$. Therefore,

$$i_S = 0 \quad \text{for} \quad 0 < \omega \leq \pi \tag{6.9}$$

From (6.6), (6.8), and (6.9),

$$i_{L1} = i = I_m \sin(\omega t + \varphi) \quad \text{for} \quad 0 < \omega t \leq \pi \tag{6.10}$$

The voltage across the inductor L_1 is

$$v_{L1} = \omega L_1 \frac{di_{L1}}{d(\omega t)} = \omega L_1 I_m \cos(\omega t + \varphi) \quad \text{for} \quad 0 < \omega t \leq \pi \tag{6.11}$$

Hence, (6.7) becomes

$$v_S = V_I - v_{L1} = V_I - \omega L_1 I_m \cos(\omega t + \varphi) \quad \text{for} \quad 0 < \omega t \leq \pi \tag{6.12}$$

Using (6.10) and taking into account the fact that the inductor current i_{L1} is continuous,

$$i_{L1}(\pi+) = i_{L1}(\pi-) = I_m \sin(\pi + \varphi) = -I_m \sin \varphi \tag{6.13}$$

The switch is ON for the interval $\pi < \omega t \leq 2\pi$ during which

$$v_S = 0 \quad \text{for} \quad \pi < \omega t \leq 2\pi \tag{6.14}$$

Substitution of this into (6.7) then produces

$$v_{L1} = V_I \quad \text{for} \quad \pi < \omega t \leq 2\pi \tag{6.15}$$

Thus, from (6.13) and (6.15), the current through the inductor L_1 is

$$i_{L1} = \frac{1}{\omega L_1} \int_{\pi}^{\omega t} v_{L1}(u)du + i_{L1}(\pi+) = \frac{1}{\omega L_1} \int_{\pi}^{\omega t} V_I(u)du + i_{L1}(\pi+)$$

$$= \frac{V_I}{\omega L_1}(\omega t - \pi) - I_m \sin \varphi \quad \text{for} \quad \pi < \omega t \leq 2\pi \tag{6.16}$$

From (6.6) and (6.8), the switch current is obtained as

$$i_S = i_{L1} - i = \frac{V_I}{\omega L_1}(\omega t - \pi) - I_m[\sin(\omega t + \varphi) + \sin \varphi], \quad \text{for} \quad \pi < \omega t \leq 2\pi \tag{6.17}$$

Substituting the ZCS condition $i_S(2\pi) = 0$ into (6.17),

$$I_m = V_I \frac{\pi}{2\omega L_1 \sin \varphi} \tag{6.18}$$

Because $I_m > 0$,

$$0 < \varphi < \pi \tag{6.19}$$

From (6.9), (6.17), and (6.18),

$$i_S = \begin{cases} 0 & 0 < \omega t \leq \pi \\ \frac{V_I}{\omega L_1}\left[\omega t - \frac{3\pi}{2} - \frac{\pi}{2\sin\varphi}\sin(\omega t + \varphi)\right] & \pi < \omega t \leq 2\pi \end{cases} \tag{6.20}$$

Substitution of the condition of optimum operation given by (6.4) into (6.20) yields

$$\tan\varphi = \frac{\pi}{2} \tag{6.21}$$

From (6.19) and (6.21),

$$\varphi = \arctan\left(\frac{\pi}{2}\right) = 1.0039 \text{ rad} = 57.52^\circ \tag{6.22}$$

Consideration of trigonometric relationships shows that

$$\sin\varphi = \frac{\pi}{\sqrt{\pi^2+4}} \tag{6.23}$$

$$\cos\varphi = \frac{2}{\sqrt{\pi^2+4}} \tag{6.24}$$

From (6.20) and (6.21),

$$i_S = \begin{cases} 0 & 0 < \omega t \leq \pi \\ \frac{V_I}{\omega L_1}\left(\omega t - \frac{3\pi}{2} - \frac{\pi}{2}\cos\omega t - \sin\omega t\right) & \pi < \omega t \leq 2\pi \end{cases} \tag{6.25}$$

Using the Fourier formula, the supply dc current is

$$I_I = \frac{1}{2\pi}\int_{\pi}^{2\pi} i_S \, d(\omega t) = \frac{V_I}{2\pi\omega L_1}\int_{\pi}^{2\pi}\left(\omega t - \frac{3\pi}{2} - \frac{\pi}{2}\cos\omega t - \sin\omega t\right) d(\omega t) = \frac{V_I}{\pi\omega L_1} \tag{6.26}$$

The amplitude of the output current can be found from (6.18), (6.23), and (6.26)

$$I_m = \frac{\sqrt{\pi^2+4}}{2}\frac{V_I}{\omega L_1} = \frac{\pi\sqrt{\pi^2+4}}{2} I_I = 5.8499 I_I \tag{6.27}$$

Substitution of (6.27) into (6.25) yields the normalized steady-state switch current waveform

$$\frac{i_S}{I_I} = \begin{cases} 0 & 0 < \omega t \leq \pi \\ \pi\left(\omega t - \frac{3\pi}{2} - \frac{\pi}{2}\cos\omega t - \sin\omega t\right) & \pi < \omega t \leq 2\pi \end{cases} \tag{6.28}$$

From (6.12), (6.18), and (6.21), the normalized switch voltage waveform is found as

$$\frac{v_S}{V_I} = \begin{cases} \frac{\pi}{2}\sin\omega t - \cos\omega t + 1 & 0 < \omega t \leq \pi \\ 0 & \pi < \omega t \leq 2\pi \end{cases} \tag{6.29}$$

6.4.2 Peak Switch Current and Voltage

The peak switch current I_{SM} and voltage V_{SM} can be determined by differentiating waveforms (6.28) and (6.29), and by setting the results equal to zero. Finally, we obtain

$$I_{SM} = \pi(\pi - 2\varphi)I_I = 3.562 I_I \tag{6.30}$$

and

$$V_{SM} = \left(\frac{\sqrt{\pi^2+4}}{2} + 1\right)V_I = 2.8621 V_I. \tag{6.31}$$

Neglecting power losses, the output power equals the dc input power $P_I = I_I V_I$. Thus, the power-output capability c_p can be computed from the expression

$$c_p = \frac{P_O}{I_{SM} V_{SM}} = \frac{I_I V_I}{I_{SM} V_{SM}} = 0.0981 \tag{6.32}$$

It has the same value as the Class E ZVS amplifier with a shunt capacitor. It can be proved that the maximum power-output capability occurs at a duty ratio of 50%.

6.4.3 Fundamental-Frequency Components

The output voltage is sinusoidal and has the form

$$v_{R1} = V_m \sin(\omega t + \varphi) \tag{6.33}$$

where

$$V_m = RI_m \tag{6.34}$$

The voltage v_X across the elements L, C_a, and C_b is not sinusoidal. The fundamental-frequency component v_{X1} of the voltage v_X appears only across the capacitor C_b because the inductance L and the capacitance C_a are resonant at the operating frequency f and their reactance $\omega L - 1/(\omega C_a) = 0$. This component is

$$v_{X1} = V_{X1} \cos(\omega t + \varphi) \tag{6.35}$$

where

$$V_{X1} = -\frac{I_m}{\omega C_b} \tag{6.36}$$

The fundamental-frequency component of the switch voltage is

$$v_{S1} = v_{R1} + v_{X1} = V_m \sin(\omega t + \varphi) + V_{X1} \cos(\omega t + \varphi) \tag{6.37}$$

The phase shift between the voltages v_{R1} and v_{S1} is determined by the expression

$$\tan \psi = \frac{V_{X1}}{V_m} = -\frac{1}{\omega C_b R} \tag{6.38}$$

Using (6.29) and the Fourier formulas, we obtain

$$V_m = \frac{1}{\pi} \int_0^{2\pi} v_S \sin(\omega t + \varphi) d(\omega t) = \frac{4}{\pi\sqrt{\pi^2 + 4}} V_I = 0.3419 V_I \tag{6.39}$$

and

$$V_{X1} = \frac{1}{\pi} \int_0^{2\pi} v_S \cos(\omega t + \varphi) d(\omega t) = -\frac{\pi^2 + 12}{4\sqrt{\pi^2 + 4}} V_I = -1.4681 V_I \tag{6.40}$$

Substituting (6.26) and (6.27) into (6.39) and (6.40),

$$V_m = \frac{8}{\pi(\pi^2 + 4)} \omega L_1 I_m \tag{6.41}$$

$$V_{X1} = -\frac{\pi^2 + 12}{2(\pi^2 + 4)} \omega L_1 I_m \tag{6.42}$$

The fundamental-frequency components of the switch current

$$i_{s1} = I_{s1} \sin(\omega t + \gamma) \tag{6.43}$$

where

$$I_{s1} = I_I \sqrt{\left(\frac{\pi^2}{4} - 2\right)^2 + \frac{\pi^2}{2}} = 1.6389 I_I \tag{6.44}$$

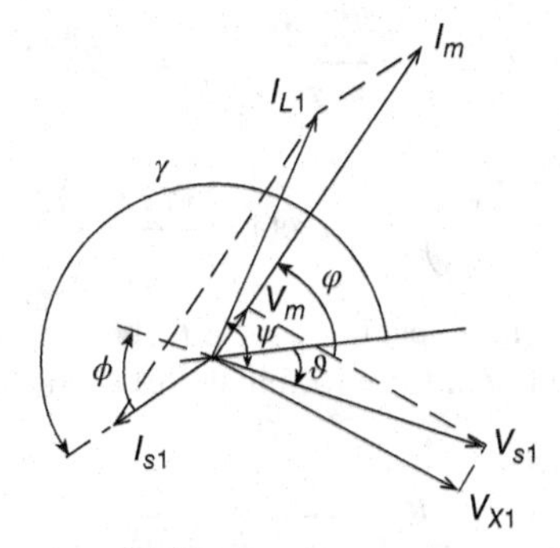

Figure 6.3 Phasor diagram of the fundamental-frequency components of the currents and voltages for optimum operation of Class E ZCS amplifier.

and

$$\gamma = 180^\circ + \arctan\left(\frac{\pi^2 - 8}{2\pi}\right) = 196.571^\circ \tag{6.45}$$

The switch voltage is

$$v_{s1} = V_{s1}\sin(\omega t + \vartheta) \tag{6.46}$$

where

$$V_{s1} = \sqrt{V_m^2 + V_{X1}^2} = V_I\sqrt{\frac{16}{\pi^2(\pi^2+4)} + \frac{(\pi^2+12)^2}{16(\pi^2+4)}} = 1.5074V_I \tag{6.47}$$

and

$$\vartheta = \varphi + \psi = -19.372^\circ \tag{6.48}$$

The phase ϕ of the input impedance of the load network at the operating frequency is

$$\phi = 180^\circ + \vartheta - \gamma = -35.945^\circ \tag{6.49}$$

This indicates that the input impedance of the load network is capacitive. Figure 6.3 shows a phasor diagram for the fundamental-frequency currents and voltages for optimum operation of the Class E ZCS amplifier.

6.5 Power Relationships

The dc input power P_I is

$$P_I = I_I V_I \tag{6.50}$$

From (6.39), the output power P_O is

$$P_O = \frac{V_m^2}{2R} = \frac{8}{\pi^2(\pi^2+4)}\frac{V_I^2}{R} = 0.05844\frac{V_I^2}{R} \tag{6.51}$$

6.6 Element Values of Load Network

From (6.34), (6.38), (6.41), and (6.42),

$$\frac{\omega L_1}{R} = \frac{\pi(\pi^2+4)}{8} = 5.4466 \tag{6.52}$$

$$\omega C_b R = \frac{16}{\pi(\pi^2 + 12)} = 0.2329 \tag{6.53}$$

and

$$\psi = \arctan\left(\frac{V_{X1}}{V_m}\right) = -\arctan\left[\frac{\pi(\pi^2 + 12)}{16}\right] = -76.89^\circ \tag{6.54}$$

Hence, according to Fig. 6.1(b), the capacitor C_b should be connected in series with C_a, L, and R. The values of L and C_a can be found from formulas (6.1) and (6.2).

From (6.27) and (6.52),

$$I_m = \frac{4}{\pi\sqrt{\pi^2 + 4}}\frac{V_I}{R}. \tag{6.55}$$

From (6.26) and (6.52),

$$I_I = \frac{8}{\pi^2(\pi^2 + 4)}\frac{V_I}{R} \tag{6.56}$$

The dc input resistance of the amplifier is obtained from (6.26) and (6.52)

$$R_{DC} \equiv \frac{V_I}{I_I} = \pi\omega L_1 = \frac{2\pi(\pi^2 + 4)}{(\pi^2 + 12)}\frac{1}{\omega C_b} = \frac{\pi^2(\pi^2 + 4)}{8}R = 17.11R \tag{6.57}$$

The element values of the load network can be computed from the following expressions:

$$R = \frac{8}{\pi^2(\pi^2 + 4)}\frac{V_I^2}{P_O} = 0.05844\frac{V_I^2}{P_O} \tag{6.58}$$

$$L_1 = \frac{\pi(\pi^2 + 4)}{8}\frac{R}{\omega} = 5.4466\frac{R}{\omega} \tag{6.59}$$

$$C_b = \frac{16}{\pi(\pi^2 + 12)}\frac{1}{\omega R} = \frac{0.2329}{\omega R} \tag{6.60}$$

$$C = \frac{1}{\omega R Q_L} \tag{6.61}$$

and

$$L = \left[Q_L - \frac{\pi(\pi^2 + 12)}{16}\right]\frac{R}{\omega} = (Q_L - 4.2941)\frac{R}{\omega} \tag{6.62}$$

It is apparent from (6.62) that the loaded quality factor Q_L must be greater than 4.2941.

6.7 Design Example

Example 6.1

Design the Class E ZCS RF power amplifier of Fig. 6.1(a) to meet the following specifications: $V_I = 5$ V, $P_{Omax} = 1$ W, and $f = 1$ GHz.

Solution. It is sufficient to design the amplifier for the full power. From (6.58), the full-load resistance is

$$R = \frac{8}{\pi^2(\pi^2 + 4)}\frac{V_I^2}{P_O} = 0.05844 \times \frac{5^2}{1} = 1.146\ \Omega \tag{6.63}$$

According to Section 6.6, the loaded quality factor Q_L must be greater than 4.2941. Let $Q_L = 8$. Using (6.59), (6.61), and (6.62), the values of the elements of the load network are

$$L_1 = \frac{\pi^2+4}{16}\frac{R}{f} = 0.8669 \times \frac{1.461}{10^9} = 1.2665 \text{ nH} \tag{6.64}$$

$$C = \frac{1}{\omega R Q_L} = \frac{1}{2\times\pi\times10^9\times1.461\times8} = 13.62 \text{ pF} \tag{6.65}$$

and

$$L = \left[Q_L - \frac{\pi(\pi^2+12)}{16}\right]\frac{R}{\omega} = \left[8 - \frac{\pi(\pi^2+12)}{16}\right] \times \frac{1.461}{2\times\pi\times10^9} = 0.862 \text{ nH} \tag{6.66}$$

The maximum voltage across the switch can be obtained using (6.31) as

$$V_{SM} = \left(\frac{\sqrt{\pi^2+4}}{2}+1\right)V_I = 2.8621\times5 = 14.311\ V \tag{6.67}$$

From (6.56), the dc input current is

$$I_I = \frac{8}{\pi^2(\pi^2+4)}\frac{V_I}{R} = 0.0584\times\frac{5}{1.461} = 0.2\ A \tag{6.68}$$

The maximum switch current is calculated using (6.30) as

$$I_{SM} = \pi(\pi - 2\varphi)I_I = 3.562\times0.2 = 0.7124\ A \tag{6.69}$$

and from (6.55) the maximum amplitude of the current through the resonant circuit is

$$I_m = \frac{4}{\pi\sqrt{\pi^2+4}}\frac{V_I}{R} = 0.3419\times\frac{5}{1.461} = 1.17\ A \tag{6.70}$$

The dc resistance seen by the dc power supply is

$$R_{DC} = \frac{\pi^2(\pi^2+4)}{8}R = 17.11\times1.461 = 25\ \Omega \tag{6.71}$$

The resonant frequency of the L-C series-resonant circuit when the switch is ON is

$$f_{o1} = \frac{1}{2\pi\sqrt{LC}} = \frac{1}{2\pi\sqrt{0.862\times10^{-9}\times13.62\times10^{-12}}} = 1.469 \text{ GHz} \tag{6.72}$$

and the resonant frequency of the L_1-L-C series-resonant circuit when the switch is OFF is

$$f_{o2} = \frac{1}{2\pi\sqrt{C(L+L_1)}} = \frac{1}{2\pi\sqrt{13.62\times10^{-12}(1.2665+0.862)\times10^{-9}}} = 0.9347 \text{ GHz} \tag{6.73}$$

6.8 Summary

- In the Class E ZCS RF power amplifier, the transistor turns off at zero current, reducing turn-off switching loss to zero, even if the transistor switching time is an appreciable fraction of the cycle of the operating frequency.
- The transistor output capacitance is not absorbed into the topology of the Class E ZCS amplifier.
- The transistor turns on at nonzero voltage, causing turn-on power loss.

- The efficiency of the Class E ZCS amplifier is lower than that of the Class E ZVS amplifier, using the same transistor and the same operating frequency.
- The voltage stress in the Class E ZCS amplifier is lower than that of the Class E ZVS amplifier.
- The ZCS condition can be satisfied for load resistances ranging from a minimum value of R to infinity.
- The load network of the amplifier can be modified for impedance transformation and harmonic suppression.

6.9 Review Questions

6.1 What is the ZCS technique?

6.2 What is the turn-off switching loss in the Class E ZCS amplifier?

6.3 What is the turn-on switching loss in the Class E ZCS amplifier?

6.4 Is the transistor output capacitance absorbed into the Class E ZCS amplifier topology?

6.5 Is the inductance connected in series with the dc input source V_I a large high-frequency choke in the Class E ZCS amplifier?

6.6 What are the switch voltage and current stresses for the Class E ZCS amplifier at $D = 0.5$?

6.7 Compare the voltage and current stresses for Class E ZVS and ZCS amplifiers at $D = 0.5$.

6.10 Problems

6.1 Design a Class E ZCS RF power amplifier to meet the following specifications: $V_I = 15$ V, $P_O = 10$ W, and $f = 900$ MHz.

6.2 A Class E ZCS amplifier is powered from a 340-V power supply. What is the required voltage rating of the switch if the switch duty cycle is 0.5?

6.3 Design a Class E ZCS power amplifier to meet the following specifications: $V_I = 180$ V, $P_O = 250$ W, and $f = 200$ kHz.

6.4 It has been found that a Class E ZCS amplifier has the following parameters: $D = 0.5, f =$ 400 kHz, $L_1 = 20$ μH, and $P_O = 100$ W. What is the maximum voltage across the switch in this amplifier?

6.5 Design the Class E ZCS power amplifier to meet the following specifications: $V_I = 100$ V, $P_{Omax} = 50$ W, and $f = 1$ MHz.

References

[1] M. K. Kazimierczuk, "Class E tuned power amplifier with shunt inductor," *IEEE Journal of Solid-State Circuits*, vol. SC-16, no. 1, pp. 2–7, 1981.

[2] N. C. Voulgaris and C. P. Avratoglou, "The use of a switching device in a Class E tuned power amplifier," *IEEE Transactions on Circuits and Systems*, vol. CAS-34, pp. 1248–1250, 1987.

[3] C. P. Avratoglou and N. C. Voulgaris, "A Class E tuned amplifier configuration with finite dc-feed inductance and no capacitance," *IEEE Transactions on Circuits and Systems*, vol. CAS-35, pp. 416–422, 1988.

[4] M. K. Kazimierczuk and D. Czarkowski, *Resonant Power Converters*, 2nd Ed., New York, NY: John Wiley & Sons, 2011.

[5] M. Hayati, A. Lofti, H. Sekiya, and M. K. Kazimierczuk, "Performance study of Class-E power amplifier with shunt inductor at sub-optimum condition," *IEEE Transactions on Power Electronics*, vol. 28, no. 8, pp. 3834–3844, 2013.

[6] M. Hayati, A. Lofti, M. K. Kazimierczuk, and H. Sekya, "Modeling and analysis of Class-E amplifier with a shunt inductor at sub-nominal condition for any duty ratio," *IEEE Transactions on Circuits and Systems I: Regular Papers*, vol. 61, no. 4, pp. 987–1000, April 2014.

7

Class DE RF Power Amplifier

7.1 Introduction

The Class DE RF switching-mode power amplifier [1–17], also called the Class D_{ZVS} RF power amplifier, consists of two transistors, series-resonant circuit, and shunt capacitors connected in parallel with the transistors. It combines the properties of low voltage stress of the Class D power amplifier and zero-voltage switching (ZVS) of the Class E power amplifier. Switching losses are zero in the Class DE power amplifier, yielding high efficiency. In the Class DE power amplifier, the transistors are driven in such a way that there are time intervals (dead times) when both transistors are OFF. In this chapter, we present the circuit of the Class DE amplifier and its principle of operation, analysis, and design procedure.

7.2 Analysis of Class DE RF Power Amplifier

A circuit of the Class DE RF power amplifier is depicted in Fig. 7.1. It consists of two transistors Q_1 and Q_2, series-resonant circuit LCR, and shunt capacitors C_1 and C_2 connected in parallel with the transistors. The transistor output capacitances C_{o1} and C_{o2} are absorbed into the shunt capacitances C_1 and C_2, respectively. At high frequencies, the shunt capacitances can be formed by the transistor output capacitances. We assume that the duty cycle of each transistor is fixed at $D = 0.25$. This means that the duty cycle of the normalized dead time t_D/T is also 0.25. In general, the duty cycle D can be in range from 0 to 0.5. Figure 7.2 shows equivalent circuits of the Class DE power amplifier for four time intervals during the cycle of the operating frequency f_s. Current and voltage waveforms in the Class DE RF power amplifier are shown in Fig. 7.3. During the dead-time intervals, the load current discharges one shunt capacitance and charges the other. It is also possible to use only one shunt capacitor [1]. For optimum operation, the Class E ZVS and zero-derivative switching (ZDS) conditions are satisfied, when the transistor turns on. Therefore, the efficiency of the Class DE power amplifier is high.

RF Power Amplifiers, Second Edition. Marian K. Kazimierczuk.

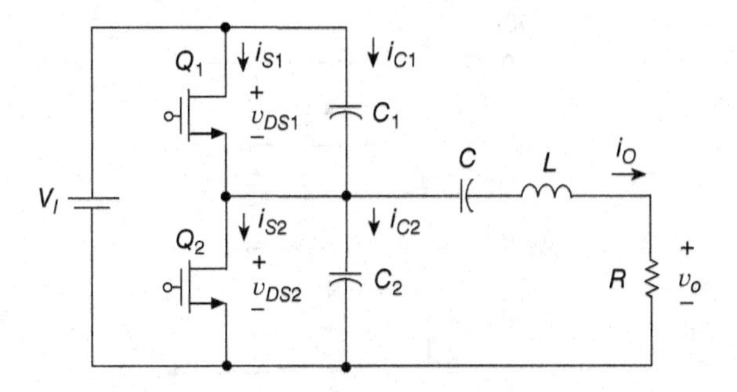

Figure 7.1 Class DE RF power amplifier with inductor L divided into two parts.

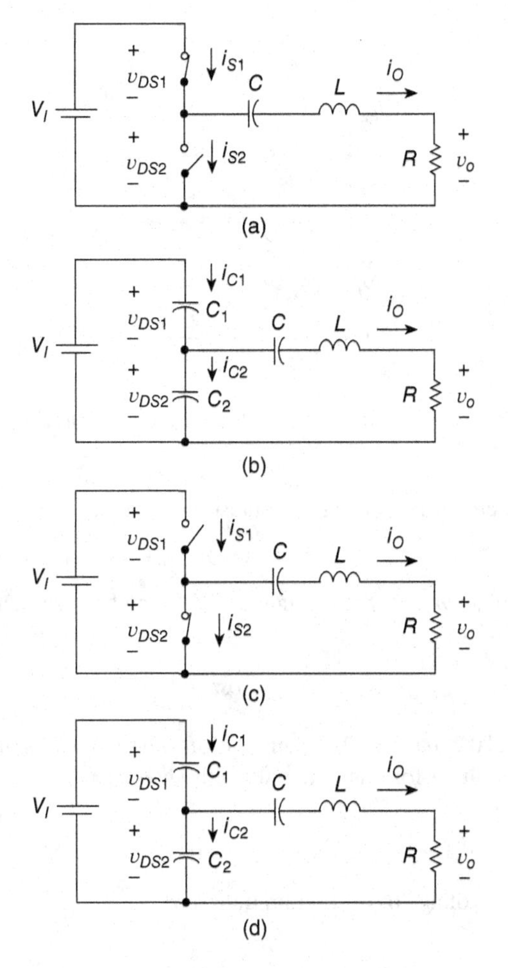

Figure 7.2 Equivalent circuits for Class DE RF power amplifier: (a) S_1 is ON and S_2 is OFF, (b) S_1 and S_2 are OFF, (c) S_1 is OFF and S_2 is ON, (d) S_1 and S_2 are OFF.

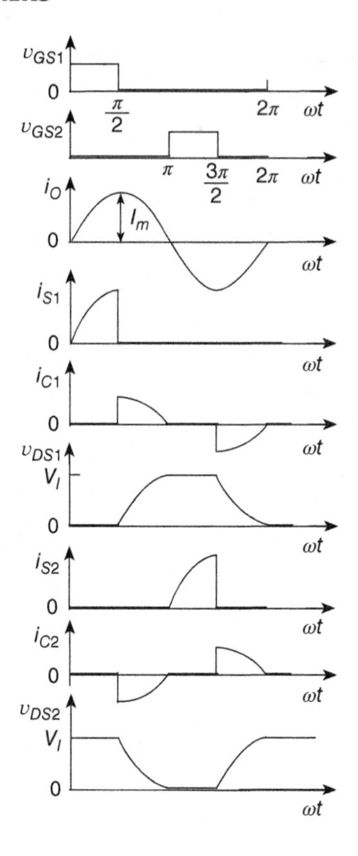

Figure 7.3 Voltage and current waveforms for Class DE RF power amplifier.

The series-resonant circuit forces nearly a sinusoidal current

$$i_o = I_m \sin(\omega t + \phi) \tag{7.1}$$

where I_m is the amplitude, $\omega = 2\pi f$ is the angular operating frequency, and ϕ is the phase of the output current. From KVL,

$$v_{DS1} + v_{DS2} = V_I \tag{7.2}$$

1) For the time interval $0 \le \omega t \le \pi/2$, switch S_1 is ON and switch S_2 is OFF. An equivalent circuit of the Class DE amplifier for this time interval is shown in Fig. 7.2(a). The drain current of the bottom transistor is

$$i_{S2} = 0 \tag{7.3}$$

The drain-to-source voltage of the bottom transistor is

$$v_{DS2} = V_I \tag{7.4}$$

resulting in

$$i_{C2} = \omega C_2 \frac{dv_{DS2}}{d(\omega t)} = \omega C_2 \frac{dV_I}{d(\omega t)} = 0 \tag{7.5}$$

The voltage across the upper transistor is

$$v_{DS1} = 0 \tag{7.6}$$

yielding

$$i_{C1} = \omega C_1 \frac{dv_{DS1}}{d(\omega t)} = 0 \tag{7.7}$$

The output current i_o flows through switch S_1

$$i_{S1} = i_o = I_m \sin(\omega t + \phi) \quad \text{for} \quad 0 \le \omega t \le \frac{\pi}{2} \tag{7.8}$$

2) For the time interval $\pi/2 \le \omega t \le \pi$, both switches S_1 and S_2 are OFF. An equivalent circuit of the Class DE amplifier for this time interval is shown in Fig. 7.2(b). During this time interval, capacitor C_1 is charged and capacitor C_2 is discharged. Therefore, the voltage v_{DS1} increases from zero to V_I and the voltage v_{DS2} decreases from V_I to zero. From KVL,

$$v_{DS1} = V_I - v_{DS2} \tag{7.9}$$

Note that

$$\frac{dv_{DS1}}{d(\omega t)} = -\frac{dv_{DS2}}{d(\omega t)} \tag{7.10}$$

From KCL,

$$-i_{C1} + i_{C2} = -i_o \tag{7.11}$$

This leads to

$$-\omega C_1 \frac{dv_{DS1}}{d(\omega t)} + \omega C_2 \frac{dv_{DS2}}{d(\omega t)} = -I_m \sin(\omega t + \phi) \tag{7.12}$$

resulting in

$$-\omega C_1 \frac{d(V_I - v_{DS2})}{d(\omega t)} + \omega C_2 \frac{dv_{DS2}}{d(\omega t)} = -I_m \sin(\omega t + \phi) \tag{7.13}$$

Hence,

$$\omega C_1 \frac{dv_{DS2}}{d(\omega t)} + \omega C_2 \frac{dv_{DS2}}{d(\omega t)} = -I_m \sin(\omega t + \phi) \tag{7.14}$$

Rearrangement of this equation gives

$$\frac{dv_{DS2}}{d(\omega t)} = -\frac{I_m}{\omega(C_1 + C_2)} \sin(\omega t + \phi) \tag{7.15}$$

The condition for ZDS of voltage v_{DS2} at $\omega t = \pi$ is given by

$$\frac{dv_{DS2}(\omega t)}{d(\omega t)}\Big|_{\omega t=\pi} = 0 \tag{7.16}$$

Imposing this condition on (7.15), we obtain

$$\sin(\pi + \phi) = 0 \tag{7.17}$$

The solutions of this equation are

$$\phi = 0 \tag{7.18}$$

and

$$\phi = \pi \tag{7.19}$$

Only $\phi = 0$ is the physical solution, which allows the charging of capacitor C_1 and discharging of capacitor C_2.
From (7.15),

$$dv_{DS2} = -\frac{I_m}{\omega(C_1 + C_2)} \sin \omega t \, d(\omega t) \tag{7.20}$$

Hence, the voltage v_{DS2} is

$$v_{DS2} = -\frac{I_m}{\omega(C_1+C_2)}\int_{\frac{\pi}{2}}^{\omega t} \sin \omega t \, d(\omega t) + v_{DS2}\left(\frac{\pi}{2}\right) = \frac{I_m}{\omega(C_1+C_2)} \cos \omega t + V_I \quad (7.21)$$

The ZVS condition of voltage v_{DS2} at $\omega t = \pi$ is given by

$$v_{DS2}(\pi) = 0 \quad (7.22)$$

Imposing this condition on (7.21), we get

$$\frac{I_m}{\omega(C_1+C_2)} = V_I \quad (7.23)$$

Hence, the voltage across the bottom switch is

$$v_{DS2} = V_I(\cos \omega t + 1) \quad (7.24)$$

and the voltage across the upper switch is

$$v_{DS1} = V_I - v_{DS2} = -V_I \cos \omega t \quad (7.25)$$

The output current is

$$i_o = I_m \sin \omega t \quad (7.26)$$

The current through the upper capacitor is

$$i_{C1} = \frac{i_o}{2} = \frac{1}{2} I_m \sin \omega t \quad (7.27)$$

and the current through the bottom capacitor is

$$i_{C2} = -\frac{i_o}{2} = -\frac{1}{2} I_m \sin \omega t \quad (7.28)$$

The voltage waveform across the upper shunt capacitor C_1 is

$$v_{DS1} = \frac{1}{\omega C_1}\int_{\frac{\pi}{2}}^{\omega t} i_{C1} \, d(\omega t) + v_{DS1}\left(\frac{\pi}{2}\right) = \frac{1}{\omega C_1}\int_{\frac{\pi}{2}}^{\omega t} \left(\frac{I_m}{2} \sin \omega t\right) d(\omega t)$$

$$= -\frac{I_m}{2\omega C_1} \cos \omega t \quad (7.29)$$

where $v_{DS1}(\pi/2) = 0$. Likewise, the voltage waveform across the bottom shunt capacitor C_2 is

$$v_{DS2} = \frac{1}{\omega C_2}\int_{\frac{\pi}{2}}^{\omega t} i_{C2} \, d(\omega t) + v_{DS2}\left(\frac{\pi}{2}\right) = \frac{1}{\omega C_2}\int_{\frac{\pi}{2}}^{\omega t} \left(-\frac{I_m}{2} \sin \omega t\right) d(\omega t) + V_I$$

$$= \frac{I_m}{2\omega C_1} \cos \omega t + V_I \quad (7.30)$$

where $v_{DS2}(\pi/2) = V_I$. Using the ZVS condition $v_{DS2}(\pi) = 0$, we obtain

$$\frac{I_m}{2\omega C_2} \cos \pi + V_I = 0 \quad (7.31)$$

yielding

$$\frac{I_m}{2\omega C_2} = V_I \quad (7.32)$$

Using (7.29) and $v_{DS1}(\pi) = V_I - v_{DS1}(2\pi) = V_I$,

$$-\frac{I_m}{2\omega C_1} \cos \pi = V_I \quad (7.33)$$

resulting in

$$\frac{I_m}{2\omega C_1} = V_I \tag{7.34}$$

Thus,

$$\frac{I_m}{2\omega C_1} = \frac{I_m}{2\omega C_2} \tag{7.35}$$

resulting in

$$C_1 = C_2 \tag{7.36}$$

In this analysis, we assume that the amplitude of the current through both shunt capacitances are equal to each other. Therefore, both shunt capacitances must be equal.

3) For the time interval $\pi \leq \omega t \leq 3\pi/2$, the switch S_1 is OFF and the switch S_2 is ON. An equivalent circuit of the Class DE amplifier for this time interval is shown in Fig. 7.2(c). The drain-to-source voltage of the bottom transistor is

$$v_{DS2} = 0 \tag{7.37}$$

which gives

$$i_{C2} = \omega C_2 \frac{dv_{DS2}}{d(\omega t)} = 0 \tag{7.38}$$

The drain-to-source voltage of the upper transistor is

$$v_{DS1} = V_I \tag{7.39}$$

producing

$$i_{C1} = \omega C_1 \frac{dv_{DS1}}{d(\omega t)} = \omega C_1 \frac{dV_I}{d(\omega t)} = 0 \tag{7.40}$$

The load current is

$$i_o = I_m \sin \omega t \quad \text{for} \quad \pi \leq \omega t \leq \frac{3\pi}{2} \tag{7.41}$$

Therefore, the current through the switch S_2 is

$$i_{S2} = -i_o = -I_m \sin \omega t \quad \text{for} \quad \pi \leq \omega t \leq \frac{3\pi}{2} \tag{7.42}$$

4) For the time interval $3\pi/2 \leq \omega t \leq 2\pi$, both switches are OFF. An equivalent circuit of the Class DE amplifier for this time interval is shown in Fig. 7.2(d). From KCL,

$$i_{C1} - i_{C2} = i_o = I_m \sin \omega t \tag{7.43}$$

Hence,

$$\omega C_1 \frac{dv_{DS1}}{d(\omega t)} - \omega C_2 \frac{dv_{DS2}}{d(\omega t)} = I_m \sin \omega t \tag{7.44}$$

Since

$$v_{DS2} = V_I - v_{DS1} \tag{7.45}$$

we have

$$\omega C_1 \frac{dv_{DS1}}{d(\omega t)} - \omega C_2 \frac{d(V_I - v_{DS1})}{d(\omega t)} = I_m \sin \omega t \tag{7.46}$$

which gives

$$\omega C_1 \frac{dv_{DS1}}{d(\omega t)} + \omega C_2 \frac{dv_{DS1}}{d(\omega t)} = I_m \sin \omega t \tag{7.47}$$

Thus,

$$\frac{dv_{DS1}}{d(\omega t)} = \frac{I_m}{\omega(C_1 + C_2)} \sin \omega t \tag{7.48}$$

resulting in

$$v_{DS1} = \frac{I_m}{\omega(C_1 + C_2)} \int_{3\pi/2}^{\omega t} \sin \omega t \, d(\omega t) + v_{DS1}\left(\frac{3\pi}{2}\right) = -\frac{I_m}{\omega(C_1 + C_2)} \cos \omega t + V_I \tag{7.49}$$

Applying the ZVS condition to the above equation at $\omega t = 2\pi$,

$$v_{DS1}(2\pi) = 0 \tag{7.50}$$

we obtain

$$\frac{I_m}{\omega(C_1 + C_2)} = V_I \tag{7.51}$$

Hence, the voltage across the switch S_1 is

$$v_{DS1} = V_I(1 - \cos \omega t) \tag{7.52}$$

and the voltage across the switch S_2 is

$$v_{DS2} = V_I - v_{DS1} = V_I \cos \omega t \tag{7.53}$$

The output current and voltage are

$$i_o = I_m \sin \omega t \tag{7.54}$$

and

$$v_o = V_m \sin \omega t \tag{7.55}$$

where $V_m = RI_m$.

The dc component of the current through the shunt capacitance C_1 is zero for steady-state operation. Therefore, the dc input current is

$$I_I = \frac{1}{2\pi} \int_0^{2\pi} i_{S1} \, d(\omega t) = \frac{1}{2\pi} \int_0^{\frac{\pi}{2}} I_m \sin \omega t \, d(\omega t) = \frac{I_m}{2\pi} = \frac{\omega(C_1 + C_2)}{2\pi} V_I \tag{7.56}$$

The dc input resistance of the amplifier is

$$R_{I(DC)} = \frac{V_I}{I_I} = \frac{2\pi}{\omega(C_1 + C_2)} = \frac{1}{f(C_1 + C_2)} \tag{7.57}$$

7.3 Components

Figure 7.4 shows the Class DE RF power amplifier with inductor L divided into two parts: L_a and L_b. The inductor L_a and the capacitor C are resonant at the operating frequency $f = \omega/(2\pi)$ so that

$$\omega = \frac{1}{\sqrt{L_a C}} \tag{7.58}$$

The net reactance of the components C and L_a is zero at the operating frequency f. Therefore, there is a voltage drop across inductance L_b and load resistance at the operating frequency f. Figure 7.5 shows the voltage waveforms of the fundamental components in the Class DE RF power amplifier under ZVS and ZDS conditions. Figure 7.6 shows the phasor diagram for the fundamental components in the Class DE RF power amplifier under ZVS and ZDS conditions.

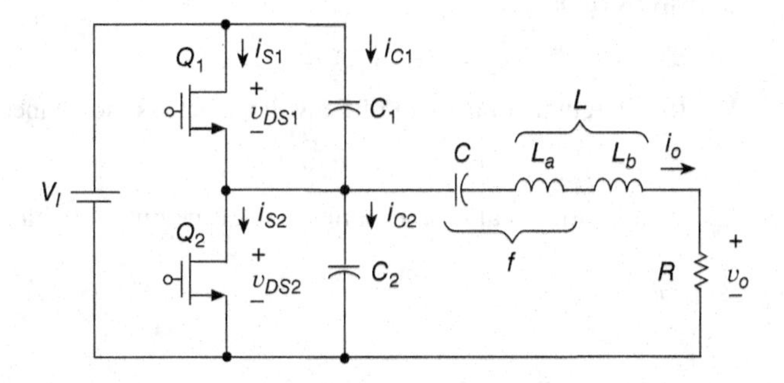

Figure 7.4 Class DE RF power amplifier with inductor L divided into two parts L_a and L_b.

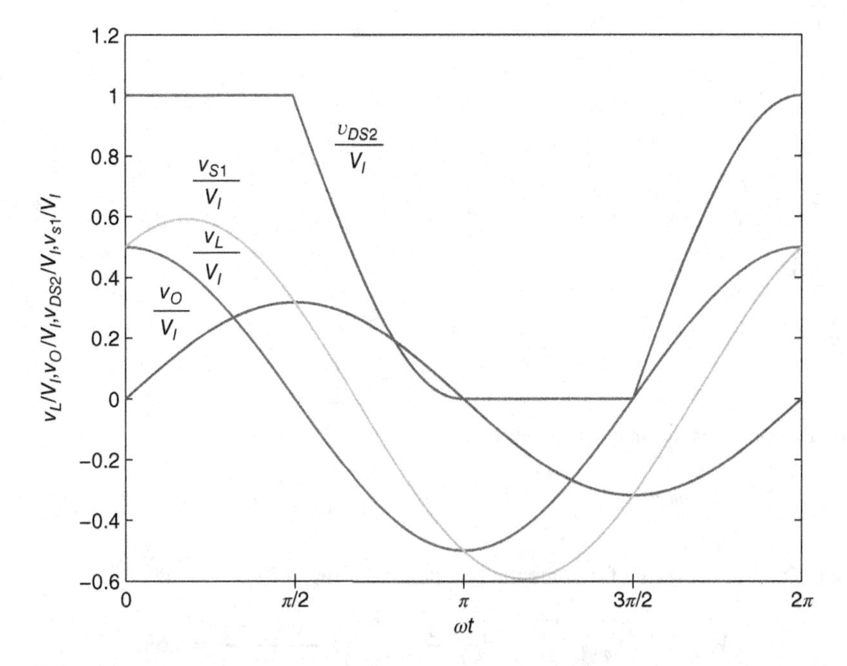

Figure 7.5 Voltage waveforms of the fundamental components in the Class DE RF power amplifier under ZVS and ZDS conditions.

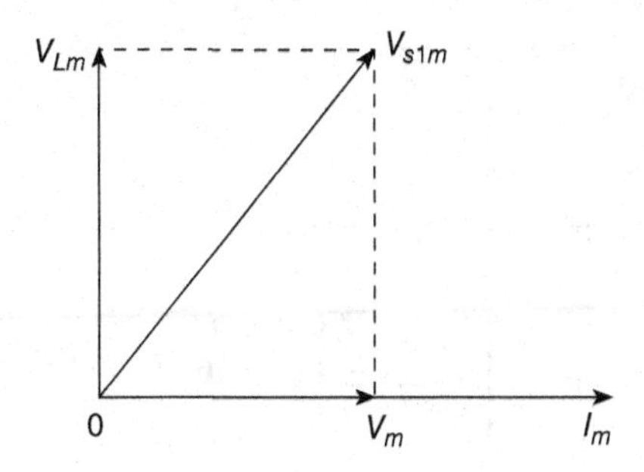

Figure 7.6 Phasor diagram for the fundamental components in the Class DE RF power amplifier under ZVS and ZDS conditions.

The output voltage waveform is

$$v_o = V_m \sin \omega t \tag{7.59}$$

where $V_m = RI_m$. The fundamental component of the voltage across inductance L is

$$v_{L1} = V_{Lm} \cos \omega t \tag{7.60}$$

where $V_{Lm} = \omega L_b I_m$. From Fourier analysis, the amplitude of the output voltage is

$$\begin{aligned} V_m &= \frac{1}{\pi}\int_0^{2\pi} v_{DS2} \sin \omega t \, d(\omega t) \\ &= \frac{1}{\pi}\left[\int_0^{\frac{\pi}{2}} V_I \sin \omega t \, d(\omega t) + \int_{\frac{\pi}{2}}^{\pi} V_I(\cos \omega t + 1) \sin \omega t d(\omega t) + \int_{\frac{3\pi}{2}}^{2\pi} V_I \cos \omega t \sin \omega t \, d(\omega t)\right] \\ &= \frac{V_I}{\pi} \end{aligned} \tag{7.61}$$

and

$$\begin{aligned} V_{Lm} &= \frac{1}{\pi}\int_0^{2\pi} v_{DS2} \cos \omega t d(\omega t) \\ &= \frac{1}{\pi}\left[\int_0^{\frac{\pi}{2}} V_I \cos^2 \omega t \, d(\omega t) + \int_{\frac{\pi}{2}}^{\pi} V_I(\cos \omega t + 1) \cos \omega t \, d(\omega t) + \int_{\frac{3\pi}{2}}^{2\pi} V_I \cos^2 \omega t (d\omega t)\right] \\ &= \frac{V_I}{2} \end{aligned} \tag{7.62}$$

Hence,

$$\frac{V_{Lm}}{V_m} = \frac{\omega L_b}{R} = \frac{\pi}{2} \tag{7.63}$$

The fundamental component of the voltage v_{DS2} is

$$v_{s1} = v_o + v_{L1} = V_m \sin \omega t + V_{Lm} \cos \omega t = V_I\left(\frac{1}{\pi} \sin \omega t + \frac{1}{2} \cos \omega t\right) = V_{s1m} \cos(\omega t + \psi) \tag{7.64}$$

where the amplitude of the fundamental component of the voltage v_{DS2} is

$$V_{s1m} = \sqrt{V_m^2 + V_{Lm}^2} = V_I\sqrt{\frac{1}{\pi^2} + \frac{1}{4}} = \frac{V_I\sqrt{\pi^2 + 4}}{2\pi} = 0.5927 V_I \tag{7.65}$$

$$\tan \psi = \frac{V_{Lm}}{V_m} = \frac{\omega L_b}{R} = \frac{2}{\pi} \tag{7.66}$$

$$\sin \psi = \frac{V_{Lm}}{V_{s1m}} = \frac{2}{\sqrt{\pi^2 + 4}} \tag{7.67}$$

and

$$\cos \psi = \frac{V_m}{V_{s1m}} = \frac{\pi}{\sqrt{\pi^2 + 4}} \tag{7.68}$$

$$\psi = \arcsin\left[\frac{\pi}{\sqrt{\pi^2 + 4}}\right] = 0.5669 \quad \text{rad} = 32.4816^\circ \tag{7.69}$$

The output power is

$$P_O = \frac{V_m^2}{2R} = \frac{V_I^2}{2\pi^2 R} \tag{7.70}$$

Assuming that the efficiency is 100%, the output power can also be expressed as

$$P_O = P_I = I_I V_I = \frac{\omega(C_1 + C_2)V_I^2}{2\pi} = f(C_1 + C_2)V_I^2 \tag{7.71}$$

Therefore,

$$\omega(C_1 + C_2)R = \frac{1}{\pi} \tag{7.72}$$

Assuming that $C_1 = C_2$,

$$\omega C_1 R = \omega C_2 R = \frac{1}{2\pi} \tag{7.73}$$

The loaded quality factor is defined at the operating frequency as

$$Q_L = \frac{\omega L}{R} \tag{7.74}$$

Hence,

$$L_a = L - L_b = \frac{Q_L R}{\omega} - \frac{\pi}{2}\frac{R}{\omega} = \left(Q_L - \frac{\pi}{2}\right)\frac{R}{\omega} \tag{7.75}$$

Since $\omega^2 = 1/(L_a C)$,

$$C = \frac{1}{\omega^2 L_a} = \frac{1}{\omega R\left(Q_L - \frac{\pi}{2}\right)} \tag{7.76}$$

7.4 Device Stresses

The maximum drain current is given by

$$I_{DMmax} = I_{m(max)} = \frac{V_m}{R} = \frac{V_I}{\pi R} \tag{7.77}$$

The maximum drain-to-source voltage is

$$V_{DSmax} = V_I \tag{7.78}$$

The maximum voltage across the series capacitor C is

$$V_{Cm(max)} = \frac{I_{m(max)}}{\omega C} \tag{7.79}$$

The maximum voltage across the series inductor L is

$$V_{Lm(max)} = \omega L I_{m(max)} \tag{7.80}$$

7.5 Design Equations

The design equations for the component values are:

$$R = \frac{V_I^2}{2\pi^2 P_O} \tag{7.81}$$

$$C_1 = C_2 = \frac{1}{2\pi\omega R} = \frac{\pi P_O}{\omega V_I^2} \tag{7.82}$$

$$L = \frac{Q_L R}{\omega} \tag{7.83}$$

and

$$C = \frac{1}{\omega R\left(Q_L - \frac{\pi}{2}\right)} \tag{7.84}$$

7.6 Maximum Operating Frequency

There is a maximum operating frequency f_{max} at which both the ZVS and ZDS conditions are satisfied. For $C_1 = C_2 = C_o, \omega C_o R = 1/(2\pi)$. Hence, the maximum operating frequency limited by the transistor output capacitance C_o is determined by

$$C_o = C_1 = C_2 = \frac{1}{4\pi^2 R f_{max}} \tag{7.85}$$

resulting in

$$f_{max} = \frac{1}{4\pi^2 R C_o} = \frac{P_O}{2 C_o V_I^2} \tag{7.86}$$

Example 7.1

Design a Class DE RF power amplifier to meet the following specifications: $P_O = 0.25$ W, $V_I = 3.3$ V, $BW = 100$ MHz, and $f = 1$ GHz. Assume that the transistor output capacitance C_o is linear and is equal to 1.5 pF.

Solution. The loaded quality factor is

$$Q_L = \frac{f_o}{BW} = \frac{10^9}{10^8} = 10 \tag{7.87}$$

We can calculate the component values as follows:

$$R = \frac{V_I^2}{2\pi^2 P_O} = \frac{3.3^2}{2\pi^2 \times 0.25} = 2.2\ \Omega \tag{7.88}$$

$$C_1 = C_2 = \frac{1}{2\pi\omega R} = \frac{\pi P_O}{\omega V_I^2} = \frac{\pi \times 0.25}{2\pi \times 10^9 \times 3.3^2} = 11.5\ \text{pF} \tag{7.89}$$

$$C_{1(ext)} = C_1 - C_o = 11.5 - 1.5 = 10\ \text{pF} \tag{7.90}$$

$$L = \frac{Q_L R}{\omega} = \frac{10 \times 2.2}{2\pi \times 10^9} = 3.5\ \text{nH} \tag{7.91}$$

and

$$C = \frac{1}{\omega R \left(Q_L - \frac{\pi}{2}\right)} = \frac{1}{2\pi \times 10^9 \times 2.2 \times \left(10 - \frac{\pi}{2}\right)} = 8.58\ \text{pF} \tag{7.92}$$

These components have low values and can be integrated. The output network may need a matching circuit.

The resonant frequency of the series-resonant circuit is

$$f_o = \frac{1}{2\pi\sqrt{LC}} = \frac{1}{2\pi\sqrt{3.5 \times 10^{-9} \times 8.58 \times 10^{-12}}} = 0.9184\ \text{GHz} \tag{7.93}$$

The ratio of the operating frequency to the resonant frequency is

$$\frac{f}{f_o} = \frac{1}{0.9184} = 1.09 \tag{7.94}$$

Thus, the operation is well above the resonance.

Assuming the amplifier efficiency $\eta = 0.94$, the dc supply power is

$$P_I = \frac{P_O}{\eta} = \frac{0.25}{0.94} = 0.266\ \text{W} \tag{7.95}$$

and the dc supply current is

$$I_I = \frac{P_I}{V_I} = \frac{0.266}{3.3} = 0.08 \text{ A} \tag{7.96}$$

The amplitude of the output voltage is

$$V_m = \sqrt{2P_O R} = \sqrt{2 \times 0.25 \times 2.2} = 1.0488 \text{ V} \tag{7.97}$$

The amplitude of the output current I_m and the MOSFET current stress I_{SMmax} are

$$I_{SMmax} = I_{m(max)} = \frac{V_m}{R} = \frac{1.0488}{2.2} = 0.4767 \text{ A} \tag{7.98}$$

The voltage stress of the MOSFETs is

$$V_{SM} = V_I = 3.3 \text{ V} \tag{7.99}$$

The MOSFET should have $V_{DSS} = 5$ V and $I_{DSmax} = 1$ A. The maximum voltage across the series capacitor C is

$$V_{Cmax} = \frac{I_{m(max)}}{\omega C} = \frac{0.4767}{2\pi \times 10^9 \times 8.58 \times 10^{-12}} = 8.84 \text{ V} \tag{7.100}$$

7.7 Class DE Amplifier with Only One Shunt Capacitor

Figure 7.7 shows the two circuits of Class DE power amplifier with only one shunt capacitor. A single-shunt capacitance C_s can be connected in parallel with the bottom transistor as in Fig. 7.7(a) or in parallel with the upper transistor as shown in Fig. 7.7(b). Figure 7.8 shows equivalent circuits for the Class DE power amplifier with only one capacitor connected in parallel with the bottom transistor. Voltage and current waveforms in the Class DE power amplifier with only one capacitor connected in parallel with the bottom transistor are depicted in Fig. 7.9.

1) For $0 < \omega t \leq \pi/2$, the switch S_1 is ON and the switch S_2 is OFF. The equivalent circuit for this time interval is depicted in Fig. 7.8(a). The analysis for this time interval is the same as that already presented for the case with two shunt capacitances C_1 and C_2.

2) For $\pi/2 < \omega t \leq \pi$, both switches S_1 and S_2 are OFF. The equivalent circuit for this time interval is depicted in Fig. 7.8(b). The current through the shunt capacitor is

$$i_{Cs} = \omega C_s \frac{dv_{DS2}}{d(\omega t)} = -i_o = -I_m \sin(\omega t + \phi) \tag{7.101}$$

Using the ZDS condition, $\phi = 0$. The output current waveform is

$$i_o = I_m \sin \omega t. \tag{7.102}$$

The current through the shunt capacitor is

$$i_{Cs} = -i_o = -I_m \sin \omega t \tag{7.103}$$

The drain-to-source voltage v_{DS2} is given by

$$v_{DS2} = \frac{1}{\omega C_s} \int_{\frac{\pi}{2}}^{\omega t} i_{Cs}\, d(\omega t) + V_I = -\frac{I_m}{\omega C_s} \int_{\frac{\pi}{2}}^{\omega t} \sin \omega t\, d(\omega t) + V_I = \frac{I_m}{\omega C_s} \cos \omega t + V_I \tag{7.104}$$

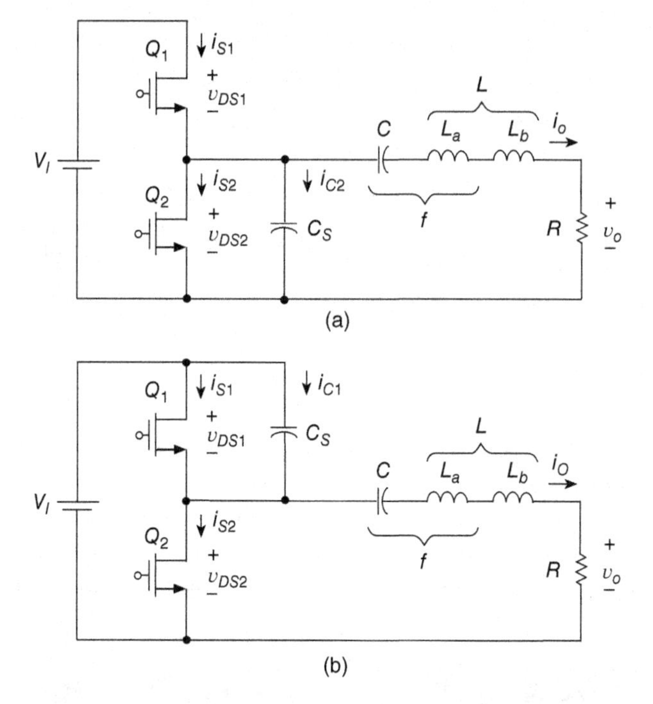

Figure 7.7 Class DE RF power amplifiers with only one shunt capacitance. (a) Shunt capacitance across the bottom transistor. (b) Shunt capacitance across the upper transistor.

Imposing the ZVS condition, we obtain

$$\frac{I_m}{\omega C_s} = V_I \tag{7.105}$$

Hence,

$$v_{DS2} = V_I(\cos\ \omega t + 1) \tag{7.106}$$

and

$$v_{DS1} = V_I - v_{DS2} = -V_I \cos\ \omega t \tag{7.107}$$

3) For $\pi < \omega t \leq 3\pi/2$, the switch S_1 is OFF and the switch S_2 is ON. The equivalent circuit for this time interval is depicted in Fig. 7.8(c). The analysis for this time interval is the same as that already presented for the case with two shunt capacitances C_1 and C_2.

4) For $3\pi/2 < \omega t \leq 2\pi$, switches S_1 and S_2 are OFF. The equivalent circuit for this time interval is depicted in Fig. 7.8(d). The current through the shunt capacitor is

$$i_{Cs} = \omega C_s \frac{dv_{DS2}}{d(\omega t)} = -i_o = -I_m \sin(\omega t + \phi) \tag{7.108}$$

The drain-to-source voltage v_{DS2} is given by

$$v_{DS2} = \frac{1}{\omega C_s} \int_{\frac{3\pi/2}{2}}^{\omega t} i_{Cs}\ d(\omega t) = -\frac{I_m}{\omega C_s} \int_{\frac{\pi}{2}}^{\omega t} \sin\ \omega t\ d(\omega t) = \frac{I_m}{\omega C_s} \cos\ \omega t \tag{7.109}$$

Imposing the ZVS condition, we obtain

$$\frac{I_m}{\omega C_s} = V_I \tag{7.110}$$

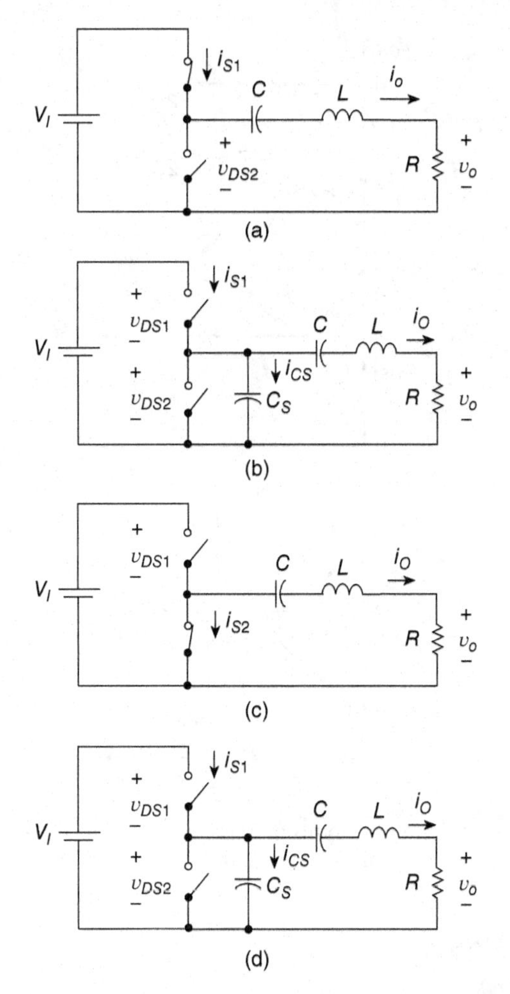

Figure 7.8 Equivalent circuits for Class DE RF power amplifiers with only one shunt capacitance connected in parallel with the bottom transistor: (a) S_1 is ON and S_2 is OFF, (b) Both transistors are OFF, (c) S_1 is OFF and S_2 is ON, (d) Both transistors are OFF.

Hence,

$$v_{DS2} = V_I \cos \omega t \tag{7.111}$$

and

$$v_{DS1} = V_I - v_{DS2} = V_I(1 - \cos \omega t) \tag{7.112}$$

The dc input current is

$$I_I = \frac{1}{2\pi} \int_0^{2\pi} i_{S1} \, d(\omega t) = \frac{1}{2\pi} \int_0^{\frac{\pi}{2}} I_m \sin \omega t \, d(\omega t) = \frac{I_m}{2\pi} = \frac{\omega C_s}{2\pi} V_I \tag{7.113}$$

The dc input resistance of the amplifier is

$$R_{I(DC)} = \frac{V_I}{I_I} = \frac{2\pi}{\omega C_s} = \frac{1}{f C_s} \tag{7.114}$$

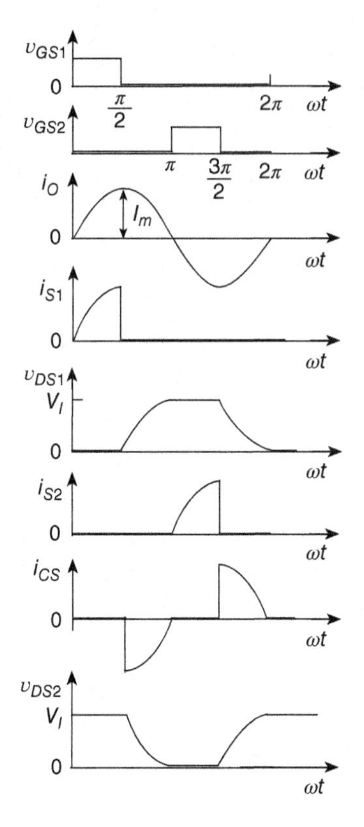

Figure 7.9 Voltage and current waveforms in Class DE RF power amplifiers with only one shunt capacitance connected in parallel with the bottom transistor.

7.8 Output Power

The output power is

$$P_O = \frac{V_m^2}{2R} = \frac{V_I^2}{2\pi^2 R} \tag{7.115}$$

Neglecting power losses, the efficiency is 100%. The output power can be expressed as

$$P_O = P_I = I_I V_I = \frac{\omega C_s V_I^2}{2\pi} = f C_s V_I^2 \tag{7.116}$$

Hence,

$$\omega C_s R = \frac{1}{\pi} \tag{7.117}$$

Thus,

$$C_s = 2C_1 = \frac{1}{\pi \omega R} \tag{7.118}$$

A single-shunt capacitance external to the transistors is given by

$$C_{s(ext)} = C_s - 2C_o = 2C_{1(ext)} \tag{7.119}$$

where C_o is the transistor output capacitance and $C_{s(ext)}$ is the external shunt capacitance. In Example 7.6, the two external shunt capacitances $C_{1(ext)} = 10\,\text{pF}$ can be replaced by a single-shunt capacitance $C_{s(ext)} = 2C_{1(ext)} = 20\,\text{pF}$.

7.9 Cancellation of Nonlinearities of Transistor Output Capacitances

The transistor output capacitance C_o is nonlinear. The output capacitance C_o is large when the drain-to-source voltage v_{DS} is low, and it is low when the drain-to-source voltage v_{DS} is high. The total voltage across both transistor output capacitances is constant

$$v_{DS1} + v_{DS2} = V_I \tag{7.120}$$

In the Class DE amplifier, the transistor capacitances are connected in parallel for the ac component [1]. When the voltage v_{DS2} across the lower output capacitance C_{o2} is low, the capacitance C_{o2} is high. At the same time, the voltage v_{DS1} is high and C_{o1} is low. Therefore, there is a partial cancellation of the nonlinearities of the total transistor output capacitance $C_{ot} = C_{o1} + C_{o2}$. The parallel combination of the shunt capacitance is approximately constant. For this reason, the expressions for load network components and the output power of the Class DE RF power amplifier are not affected by the nonlinearity of the transistor output capacitances [12, 16]. The shunt capacitances can be composed of only nonlinear transistor output capacitances or can be the combinations of nonlinear transistor output capacitances and linear external capacitances.

7.10 Amplitude Modulation of Class DE RF Power Amplifier

Assume that the modulating voltage is sinusoidal and given by

$$v_m(t) = V_m \sin \omega_m t \tag{7.121}$$

The amplitude-modulated fundamental component of the voltage waveform across the bottom MOSFET switch is

$$\begin{aligned} v_{DS2(1)}(t) &= [V_I + v_m(t)]\left(\frac{1}{\pi}\sin \omega_c t\right) = [V_I + V_m \sin \omega_m t]\left(\frac{1}{\pi}\sin \omega_c t\right) \\ &= \frac{V_I}{\pi}\sin \omega_c t + \frac{V_m}{\pi}\sin \omega_m t \sin \omega_c t = \frac{V_I}{\pi}\left(1 + \frac{V_m}{V_I}\sin \omega_m t\right) = \frac{V_I}{\pi}(1 + m \sin \omega_m t)\sin \omega_c t \\ &= V_c(1 + m \sin \omega_m t)\sin \omega_c t = V_c \sin \omega_c t + \frac{mV_c}{2}\cos(\omega_c - \omega_m)t - \frac{mV_c}{2}\cos(\omega_c + \omega_m)t \end{aligned} \tag{7.122}$$

where

$$V_c = \frac{V_I}{\pi} \tag{7.123}$$

and

$$m = \frac{V_m}{V_I} \tag{7.124}$$

7.11 Summary

- The Class DE RF amplifier consists of two transistors, a series-resonant circuit, and shunt capacitors.
- The transistors in the Class DE amplifier are operated as switches.

- There are two dead-time intervals in the gate-to-source voltages in each cycle during which the drain-to-source voltage of one transistor goes from high to low, and the drain-to-source voltage of the other transistor goes from low to high, and *vice versa*.
- The transistor output capacitances are absorbed into the shunt capacitances.
- Switching losses in the Class DE RF power amplifier are zero due to ZVS operation, like in the Class E amplifier.
- The voltage stress of power MOSFETs is low and equal to the supply voltage V_I, like in the Class D amplifier.
- The nonlinearity of the transistor output capacitances does affect the values of the load network components and the output power.
- The Class DE RF power amplifier can be implemented with only one shunt capacitor and the output capacitances of the transistors can be included in the circuit topology.

7.12 Review Questions

7.1 Compare Class D, Class E, and Class DE power amplifiers.

7.2 How large are the switching losses in the Class DE RF power amplifier?

7.3 How large are the current and voltage stresses in the Class DE power amplifier?

7.4 How many transistors are used in the Class DE amplifier?

7.5 How are the transistors driven in the Class DE amplifier?

7.6 Can the Class DE power amplifier operate under ZVS condition?

7.7 Can the Class DE power amplifier operate under ZDS condition?

7.8 Is there any limitation for the operating frequency of the Class DE power amplifier?

7.9 How does the nonlinearity of the transistor output capacitances affect the values of the load network components and the output power of the Class DE power amplifier?

7.13 Problems

7.1 Design a Class DE amplifier to meet the following specifications: $V_I = 5$ V, $P_O = 1$ W, and $f = 4$ GHz. Assume that the output capacitances of the transistors are linear and equal to 1 pF.

7.2 Design a Class DE amplifier with a single-shunt capacitance to meet the following specifications: $V_I = 5$ V, $P_O = 1$ W, $f = 4$ GHz, and $BW = 400$ MHz. Assume that the output capacitances of the transistors are linear and equal to 1 pF.

7.3 Design a Class DE RF power amplifier to meet the following specifications: $P_O = 5$ W, $V_I = 12$ V, $V_{DSmin} = 0.5$ V, and $f = 5$ GHz.

7.4 Design a Class DE RF power amplifier with $V_I = 5$ V, $P_O = 0.25$ W, $f_c = 2.4$ GHz, and $BW = 240$ MHz.

References

[1] M. K. Kazimierczuk and W. Szaraniec, "Class D zero-voltage switching inverter with only one shunt capacitor," *IEE Proceedings, Part B, Electric Power Applications*, vol. 139, pp. 449–456, 1992.

[2] M. K. Kazimierczuk and D. Czarkowski, *Resonant Power Converters*, 1st Ed, New York, NY: John Wiley & Sons, 1995, Ch. 10, pp. 295–308.

[3] H. Koizumi, T. Suetsugu, M. Fujii, K. Shinoda, S. Mori, and K. Iked, "Class DE high-efficiency tuned power amplifier," *IEEE Transactions on Circuits and Systems I: Theory and Applications*, vol. 43, no. 1, pp. 51–60, 1996.

[4] D. C. Hamil, "Class DE inverters and rectifiers," *IEEE Power Electronics Specialists Conference*, Baveno, Italy, June 23–27, 1996, pp. 854–860.

[5] K. Shinoda, T. Suetsugu, M. Matsuo, and S. Mori, "Idealized operation of the Class DE amplifies and frequency multipliers," *IEEE Transactions on Circuits and Systems I: Theory and Applications*, vol. 45, no. 1, pp. 34–40, 1998.

[6] I. D. de Vries, J. H. van Nierop, and J. R. Greence, "Solid state Class DE RF power source," *Proceedings of IEEE International Symposium on Industrial Electronics (ISIE'98)*, South Africa, July 1998, pp. 524–529.

[7] S. Hintea and I. P. Mihu, "Class DE amplifiers and their medical applications," *Proceedings of the 6th International Conference on Optimization of Electric and Electronic Equipment (OPTIM'98)*, Romania, May 1998, pp. 697–702.

[8] J. Modzelewski, "Optimum and suboptimum operation of high-frequency Class-D zero-voltage-switching tuned power amplifier," *Bulletin of the Polish Academy of Sciences, Technical Sciences*, vol. 46, no. 4., pp. 459–473, 1998.

[9] M. Albulet, "An exact analysis of Class DE amplifier at any Q", *IEEE Transactions on Circuits and Systems I: Theory and Applications*, vol. 46, no. 10, pp. 1228–1239, 1999.

[10] T. Suetsugu and M. K. Kazimierczuk, "Integration of Class DE inverter for on-chip power supplies," *IEEE International Symposium on Circuits and Systems*, May 2006, pp. 3133–3136.

[11] T. Suetsugu and M. K. Kazimierczuk, "Integration of Class DE dc-dc converter for on-chip power supplies," *37-th IEEE Power Electronics Specialists Conference (PESC'06)*, June 2006, pp. 1–5.

[12] H. Sekiya, T. Watanabe, T. Suetsugu, and M. K. Kazimierczuk, "Analysis and design of Class DE amplifier with nonlinear shunt capacitances," *The 7th IEEE International Conference on Power Electronics and Drive Systems (PEDS'07)*, Bangkok, Thailand, November 27–30, 2007, pp. 937–942.

[13] M. K. Kazimierczuk and D. Czarkowski, *Resonant Power Converters*, 2nd Ed. Hoboken, NJ: John Wiley & Sons, 2011.

[14] H. Sekiya, T. Watanabe, T. Suetsugu, and M. K. Kazimierczuk, "Analysis and design of Class DE amplifier with nonlinear shunt capacitances," *IEEE Transactions on Circuits and Systems I: Regular Papers*, vol. 56, no. 10, pp. 2363–2371, 2009.

[15] H. Sekiya, N. Sagawa, and M. K. Kazimierczuk, "Analysis of Class DE amplifier with nonlinear shunt capacitances at any grading coefficient for high Q and 25% duty ratio," *IEEE Transactions on Power Electronics*, vol. 25, no. 4, pp. 924–932, 2010.

[16] H. Sekiya, N. Sagawa, and M. K. Kazimierczuk, "Analysis of Class DE amplifier with nonlinear shunt capacitances at 25% duty ratio," *IEEE Transactions on Circuits and Systems I: Regular Papers*, vol. 57, no. 9, pp. 2334–2342, 2010.

[17] C. Ekkarayarodome, K. Jirsereeamornkul, and M. K. Kaszimierczuk, "Implementation of DC-side Class-DE low dv/dt rectifier as power-factor corrector for ballast applications," *IEEE Transactions on Industrial Electronics*, vol. 29, 2014.

8

Class F RF Power Amplifiers

8.1 Introduction

Class F radio-frequency (RF) power amplifiers [1–28] utilize multiple-harmonic resonators in the output network to shape the drain-to-source voltage such that the transistor loss is reduced and the efficiency is increased. These circuits are also called polyharmonic or multiresonant power amplifiers. The drain current flows when the drain-to-source voltage is flat and low, and the drain-to-source voltage is high when the drain current is zero. Therefore, the product of the drain current and the drain-to-source voltage is low, reducing the power dissipation in the transistor. This method of improving the efficiency is the oldest technique and was invented by Tyler in 1919 [1]. Class F power amplifiers with lumped-element resonant circuits tuned to the third harmonic or to the third and fifth harmonics have been widely used in high-power amplitude-modulated (AM) broadcast radio transmitters in the low-frequency (LF) range (30–300 kHz), the medium-frequency (MF) range (0.3–3 MHz), and the high-frequency (HF) range (3–30 MHz). Class F power amplifiers with a quarter-wavelength transmission line control all the odd harmonics and are used in very high-frequency (VHF) (30–300 MHz) frequency-modulated (FM) broadcast radio transmitters [1]. They are also used in ultrahigh frequency (UHF, 300 MHz–3 GHz) FM broadcast radio transmitters [3]. Dielectric resonators can be used in place of lumped-element resonant circuits. Output power of 40 W has been achieved at 11 GHz with an efficiency of 77%. In this chapter, we will present Class F power amplifier circuits, principle of operation, analysis, and design examples.

There are two groups of Class F RF power amplifiers:

- Odd harmonic Class F power amplifiers.
- Even harmonic Class F power amplifiers.

In Class F amplifiers with odd harmonics, the drain-to-source voltage contains only odd harmonics and the drain current contains only even harmonics. Therefore, the input impedance of the load network represents an open circuit at odd harmonics and a short circuit at even harmonics. The drain-to-source voltage v_{DS} of Class F amplifiers with odd harmonics is

RF Power Amplifiers, Second Edition. Marian K. Kazimierczuk.

symmetrical for the lower and upper halves of the cycle. In general, the drain-to-source voltage v_{DS} of Class F amplifiers with odd harmonics is given by

$$v_{DS} = V_I - V_m \cos \omega_o t + \sum_{n=3,5,7,\ldots}^{\infty} V_{mn} \cos n\omega_o t \tag{8.1}$$

and the drain current is

$$i_D = I_I + I_m \cos \omega_o t + \sum_{n=2,4,6,\ldots}^{\infty} I_{mn} \cos n\omega_o t \tag{8.2}$$

In Class F amplifiers with even harmonics, the drain-to-source voltage contains only even harmonics and the drain current contains only odd harmonics. Hence, the load network represents an open circuit at even harmonics and a short circuit at odd harmonics. The input impedance of the load network at each harmonic frequency is either zero or infinity. The drain-to-source voltage v_{DS} of Class F amplifiers with even harmonics is not symmetrical for the lower and upper halves of the cycle.

The drain-to-source voltage v_{DS} of Class F amplifiers with even harmonics is given by

$$v_{DS} = V_I - V_m \cos \omega_o t + \sum_{n=2,4,6,\ldots}^{\infty} V_{mn} \cos n\omega_o t \tag{8.3}$$

and the drain current is

$$i_D = I_I + I_m \cos \omega_o t + \sum_{n=3,5,7,\ldots}^{\infty} I_{mn} \cos n\omega_o t \tag{8.4}$$

No real power is generated at harmonics because there is either no current or no voltage present at each harmonic frequency. There are finite and infinite orders of Class F power amplifiers.

Two particular categories of high-efficiency Class F power amplifiers can be distinguished:

- Class F power amplifiers with maximally flat drain-to-source voltage waveform.
- Class F power amplifiers with maximum drain efficiency.

For Class F amplifiers with symmetrical drain-to-source voltage v_{DS}, the maximally flat voltage occurs at two values of $\omega_o t$, at the bottom part of the waveform, and at the top part of the waveform. For Class F amplifiers with unsymmetrical drain-to-source voltage v_{DS}, the maximally flat voltage occurs only at one value of $\omega_o t$ at the bottom part of the waveform. All the derivatives of voltage v_{DS} are zero at $\omega_o t$ at which the voltage v_{DS} is maximally flat.

8.2 Class F RF Power Amplifier with Third Harmonic

The circuit of the Class F RF power amplifier with a third-harmonic resonator [1] is shown in Fig. 8.1. This circuit is called the Class F_3 power amplifier. It consists of a transistor, load network, and RF choke (RFC). The load network consists of a parallel-resonant LCR circuit tuned to the operating frequency f_o and a parallel-resonant circuit $L_3C_3R_3$ tuned to the third harmonic $3f_o$, both connected in series. The ac power is delivered to the load resistance R.

Figure 8.2 shows the voltage, current, and power waveforms of the Class F_3 power amplifier with a third harmonic. The drain current waveform is

$$i_D = \begin{cases} I_{DM} \dfrac{\cos \omega_o t - \cos \theta}{1 - \cos \theta} & \text{for} \quad -\theta < \omega_o t \leq \theta \\ 0 & \text{for} \quad \theta \leq 2\pi - \theta \end{cases} \tag{8.5}$$

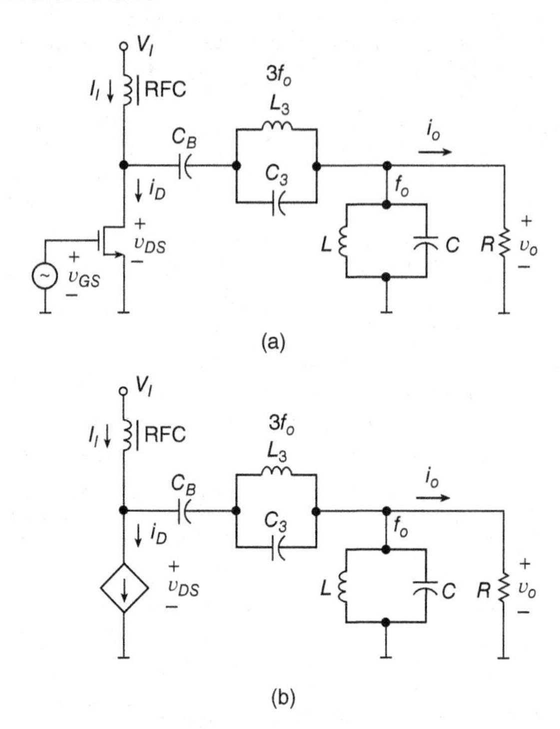

Figure 8.1 Class F_3 power amplifier with a third-harmonic resonator: (a) circuit and (b) equivalent circuit.

where I_{DM} is the peak value of the drain current, which occurs at $\omega_o t = 0$. The Fourier series of the drain current waveform is

$$i_D = I_I + \sum_{n=1}^{\infty} I_{mn} \cos n\omega_o t = I_{DM}\left(\alpha_0 + \sum_{n=1}^{\infty} \alpha_n \cos n\omega_o t\right)$$

$$= I_I\left(1 + \sum_{n=1}^{\infty} \frac{I_{mn}}{I_I} \cos n\omega_o t\right) = I_I\left(1 + \sum_{n=1}^{\infty} \gamma_n \cos n\omega_o t\right) \quad (8.6)$$

where

$$\gamma_1 = \frac{I_m}{I_I} = \frac{\theta - \sin\theta\cos\theta}{\sin\theta - \theta\cos\theta} \quad (8.7)$$

and

$$\gamma_n = \frac{I_{mn}}{I_I} = \frac{2[\sin n\theta\cos\theta - n\cos n\theta\sin\theta]}{n(n^2-1)\sin\theta - \theta\cos\theta} \quad \text{for} \quad n \geq 2 \quad (8.8)$$

Figure 8.3 shows plots of γ_n as functions of conduction angle θ.

The output resonant circuit is usually tuned to the fundamental frequency f_o and acts as a bandpass filter and filters out all the harmonics of the drain current. Therefore, the output voltage waveform is sinusoidal

$$v_o = -V_m \cos\omega_o t \quad (8.9)$$

The fundamental component of the drain-to-source voltage is

$$v_{ds1} = v_o = -V_m \cos\omega_o t \quad (8.10)$$

The voltage across the parallel-resonant circuit tuned to the third harmonic $3f_o$ is

$$v_{ds3} = V_{m3} \cos 3\omega_o t \quad (8.11)$$

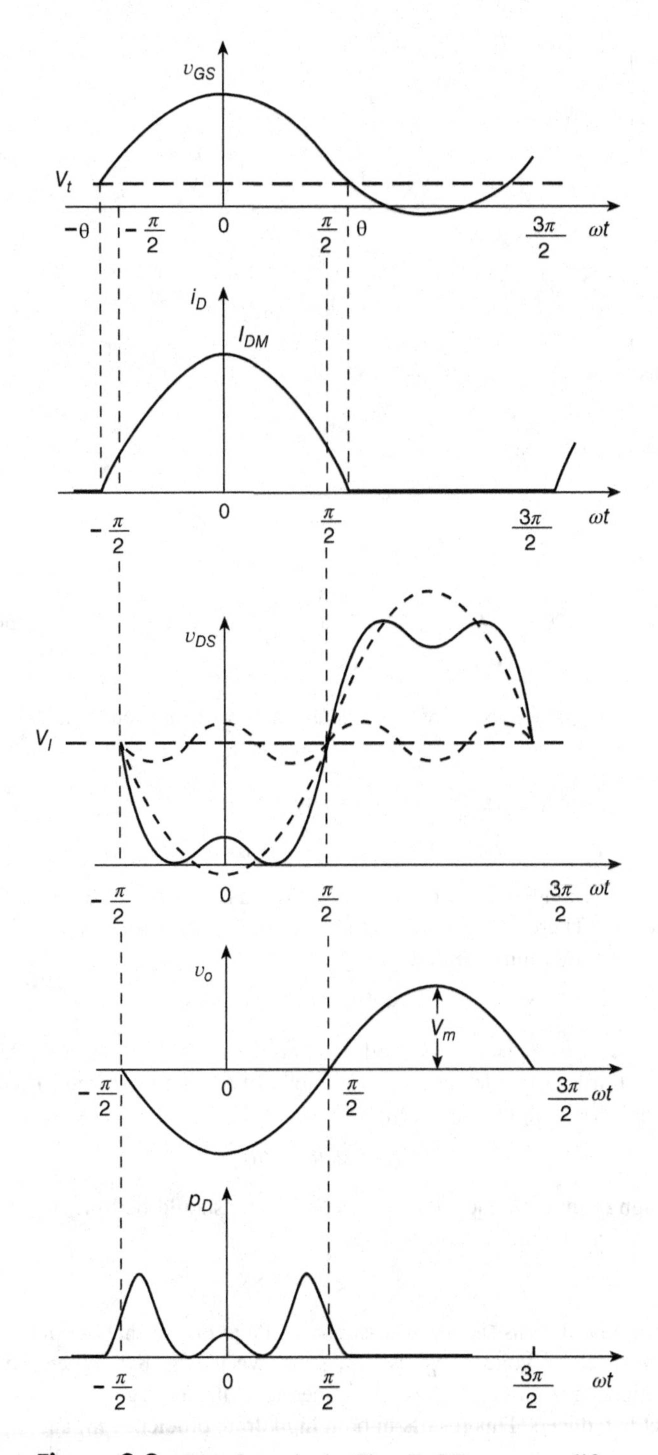

Figure 8.2 Waveforms in the Class F_3 RF power amplifier.

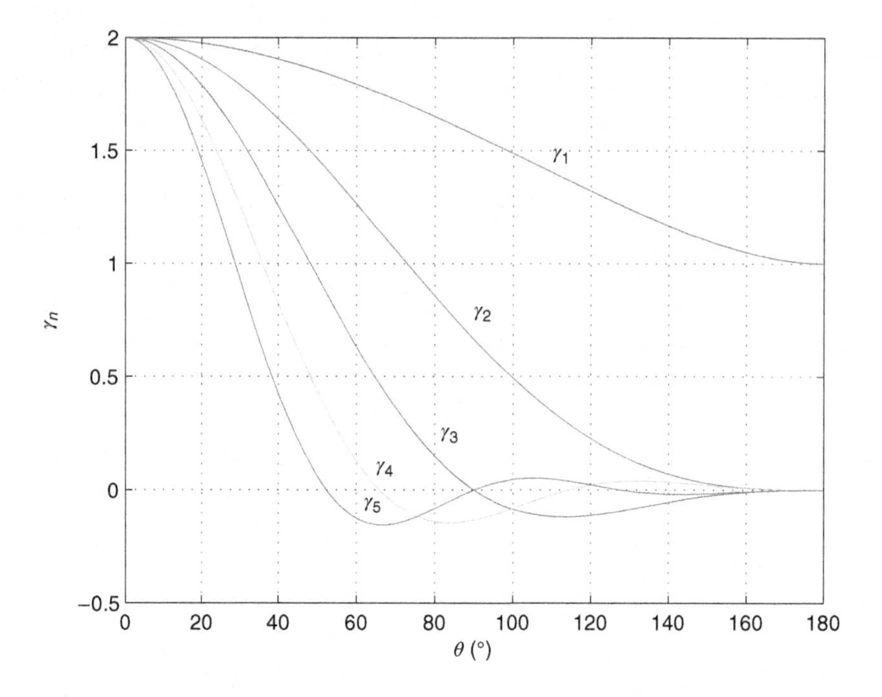

Figure 8.3 Plots of γ_n as functions of conduction angle θ.

The drain-to-source voltage waveform is given by

$$v_{DS} = V_I + v_{ds1} + v_{ds3} = V_I - V_m \cos \omega_o t + V_{m3} \cos 3\omega_o t \tag{8.12}$$

The third-harmonic voltage waveform is 180° out of phase with respect to the fundamental frequency voltage v_{ds1}. Therefore, according to Fig. 3.7, the conduction angle of the drain current for the Class F_3 amplifier must be in the range

$$90^\circ < \theta < 180^\circ \tag{8.13}$$

Only in this range of θ, α_1 is positive and α_3 is negative, as shown in Fig. 3.7. However, the amplitude of the third-harmonic current is very low for θ close to 90° and 180°. Therefore, the range of the conduction angle can be limited to

$$100^\circ < \theta < 170^\circ \tag{8.14}$$

To achieve a high drain efficiency, the conduction angle should be low. Hence, the conduction angle for Class F_3 can be limited to the range

$$100^\circ < \theta < 120^\circ \tag{8.15}$$

The third harmonic flattens the drain-to-source voltage v_{DS} so that the drain current i_D flows when the drain-to-source voltage v_{DS} is low, as shown in Fig. 8.2. When the drain-to-source voltage v_{DS} is high, the drain current is zero. Therefore, the average value of the drain voltage-current product is reduced. This results in both high drain efficiency η_D and high output-power capability c_p.

When the third harmonic is added to the fundamental component, the peak-to-peak swing $2V_{pk}$ of the ac component of the composed drain-to-source voltage waveform is reduced, and the peak value of the resulting ac component of the drain-to-source voltage V_{pk} is lower than the amplitude of the fundamental component V_m. The waveform of v_{DS} changes as the ratio V_{m3}/V_m increases.

1. For $0 < V_{m3}/V_m < 1/9$, the resulting waveform of v_{DS} has a single valley at $\omega_o t = 0$ and a single peak at $\omega_o t = \pi$. In other words, the minimum value of v_{DS} occurs at $\omega_o t = 0$ and the maximum value of v_{DS} occurs at $\omega_o t = \pi$. The peak value of the ac component of v_{DS} is

$$V_{pk} = V_I - v_{DSmin} = V_I - v_{DS}(0) = V_I - V_I + V_m - V_{m3} = V_m - V_{m3} \tag{8.16}$$

   For an ideal transistor, $v_{DSmin} = 0$ and $V_{pk} = V_I$. Hence,

$$\frac{V_m}{V_{pk}} = \frac{V_m}{V_I} = \frac{V_m}{V_m - V_{m3}} = \frac{1}{1 - \frac{V_{m3}}{V_m}} \tag{8.17}$$

   As V_{m3}/V_m increases, V_m/V_I also increases.

2. At $V_{m3}/V_m = 1/9$, the waveform of v_{DS} is maximally flat. The minimum value of v_{DS} occurs at $\omega_o t = 0$ and the maximum value of v_{DS} occurs at $\omega_o t = \pi$.

3. For $V_{m3}/V_m > 1/9$, the waveform of v_{DS} exhibits a ripple with double peaks, as shown in Fig. 8.2. There is a local maximum value of v_{DS} at $\omega_o t = 0$ and two local adjacent minimum values. In addition, there is a local minimum value of v_{DS} at $\omega_o t = \pi$ and two local adjacent maximum values.

4. As V_{m3}/V_m increases from 1/9, the ratio V_m/V_I reaches its maximum value at $V_{m3}/V_m = 1/6$ and is given by

$$\frac{V_m}{V_{pk}} = \frac{V_m}{V_I} = \frac{2}{\sqrt{3}}. \tag{8.18}$$

5. For $V_{m3}/V_m > 1/6$, the waveform of v_{DS} has double valleys and a peak at low voltages and double peaks and one valley at high voltages. The ratio V_m/V_{pk} decreases as V_{m3}/V_m increases.

It is possible to increase the ratio V_m/V_I by adding a small amount of third harmonic, while maintaining $v_{DS} \geq 0$. When V_I and I_I are held constant, P_I is constant. As V_m increases, $P_O = V_m I_m/2$ increases; therefore, $\eta_D = P_O/P_I$ also increases by the same factor by which V_m is increased.

8.2.1 Maximally Flat Class F$_3$ Amplifier

In order to find the maximally flat waveform of v_{DS}, we determine its derivative and set the result to zero

$$\frac{dv_{DS}}{d(\omega_o t)} = V_m \sin\omega_o t - 3V_{m3} \sin 3\omega_o t = V_m \sin\omega_o t - 3V_{m3}(3\sin\omega_o t - 4\sin^3\omega_o t)$$

$$= \sin\omega_o t(V_m - 9V_{m3} + 12V_{m3}\sin^2\omega_o t) = 0 \tag{8.19}$$

where $\sin 3x = 3\sin x - 4\sin^3 x$. Equation (8.19) has two solutions. One solution of this equation at nonzero values of V_m and V_{m3} is

$$\sin\omega_o t_m = 0 \tag{8.20}$$

which gives the location of one extremum (a minimum or a maximum) of v_{DS} at

$$\omega_o t_m = 0 \tag{8.21}$$

and the other extremum of v_{DS} at

$$\omega_o t_m = \pi \tag{8.22}$$

The other solutions are

$$\sin \omega_o t_m = \pm\sqrt{\frac{9V_{m3} - V_m}{12V_{m3}}} \quad \text{for} \quad V_{m3} \geq \frac{V_m}{9} \tag{8.23}$$

The real solutions of this equation do not exist for $V_{m3} < V_m/9$. For $V_{m3} > V_m/9$, there are two minimum values of v_{DS} at

$$\omega_o t_m = \pm \arcsin \sqrt{\frac{9V_{m3} - V_m}{12V_{m3}}} \tag{8.24}$$

and two maximum values of v_{DS} at

$$\omega_o t_m = \pi \pm \arcsin \sqrt{\frac{9V_{m3} - V_m}{12V_{m3}}} \tag{8.25}$$

For

$$\frac{V_{m3}}{V_m} = \frac{1}{9} \tag{8.26}$$

the location of three extrema of v_{DS} converges to zero and the location of the other three extrema converges to π. In this case, the waveform of v_{DS} is maximally flat at $\omega_o t = 0$ and at $\omega_o t = \pi$.

For the symmetrical and maximally flat drain-to-source voltage v_{DS}, the minimum value of v_{DS} occurs at $\omega_o t = 0$ and the maximum value occurs at $\omega_o t = \pi$. The Fourier coefficients for the maximally flat waveform of v_{DS} can be derived by setting all the derivatives of v_{DS} to zero. The first- and the second-order derivatives of voltage v_{DS} are

$$\frac{dv_{DS}}{d(\omega_o t)} = V_m \sin \omega_o t - 3V_{m3} \sin 3\omega_o t \tag{8.27}$$

and

$$\frac{d^2 v_{DS}}{d(\omega_o t)^2} = V_m \cos \omega_o t - 9V_{m3} \cos 3\omega_o t \tag{8.28}$$

The first-order derivative is zero at $\omega_o t = 0$ and $\omega_o t = \pi$, but it does not generate an equation because $\sin 0 = \sin \pi = 0$. From the second-order derivative of v_{DS} at $\omega_o t = 0$,

$$\left.\frac{d^2 v_{DS}}{d(\omega_o t)^2}\right|_{\omega_o t=0} = V_m - 9V_{m3} = 0 \tag{8.29}$$

Thus, a maximally flat waveform of the drain-to-source voltage occurs for

$$\frac{V_{m3}}{V_m} = \frac{1}{9} \tag{8.30}$$

Ideally, the minimum drain-to-source voltage is

$$v_{DSmin} = v_{DS}(0) = V_I - V_m + V_{m3} = V_I - V_m + \frac{V_m}{9} = V_I - \frac{8}{9}V_m = 0 \tag{8.31}$$

producing the maximum amplitude of the output voltage

$$V_m = \frac{9}{8}V_I \tag{8.32}$$

Similarly,

$$v_{DSmin} = v_{DS}(0) = V_I - V_m + V_{m3} = V_I - 9V_{m3} + V_{m3} = V_I - 8V_{m3} = 0 \tag{8.33}$$

yielding the amplitude of the third-harmonic voltage

$$V_{m3} = \frac{V_I}{8} \tag{8.34}$$

The maximum drain-to-source voltage is

$$V_{DSM} = v_{DS}(\pi) = V_I + V_m - V_{m3} = V_I + \frac{9}{8}V_I - \frac{1}{8}V_I = 2V_I \tag{8.35}$$

The normalized maximally flat waveform of v_{DS} is

$$\frac{v_{DS}}{V_I} = 1 - \frac{V_m}{V_I}\cos\omega_o t + \frac{V_{m3}}{V_I}\cos 3\omega_o t = 1 - \frac{9}{8}\cos\omega_o t + \frac{1}{8}\cos 3\omega_o t \tag{8.36}$$

Figure 8.4 shows the maximally flat waveform of the normalized drain-to-source voltage v_{DS}/V_I for the Class F_3 amplifier.

The normalized amplitude of the third harmonic is given by

$$\frac{V_{m3}}{V_m} = \frac{I_{m3}R_3}{I_m R} = \frac{|\alpha_3|I_{DM}R_3}{\alpha_1 I_{DM}R} = \frac{|\alpha_3|R_3}{\alpha_1 R} = \frac{1}{12}\left|\frac{\sin 3\theta\cos\theta - 3\cos 3\theta\sin\theta}{\theta - \sin\theta\cos\theta}\right|\left(\frac{R_3}{R}\right) = \frac{1}{9} \tag{8.37}$$

Hence,

$$R_3 = \frac{\alpha_1}{9|\alpha_3|}R = \frac{4}{3}\left|\frac{\theta - \sin\theta\cos\theta}{\sin 3\theta\cos\theta - 3\cos 3\theta\sin\theta}\right|R \tag{8.38}$$

Figure 8.5 shows the ratio R_2/R as a function of conduction angle θ for the Class F_3 amplifier with maximally flat voltage. The power loss in resistance R_3 is

$$P_{R3} = \frac{V_{m3}^2}{2R_3} = \frac{1}{2R_3}\left(\frac{V_m}{9}\right)^2 = \frac{V_m^2}{162R_3} \tag{8.39}$$

The load current is determined by the fundamental component of the drain current

$$i_o = \frac{v_o}{R} = -i_{d1} = -I_m\cos\omega_o t \tag{8.40}$$

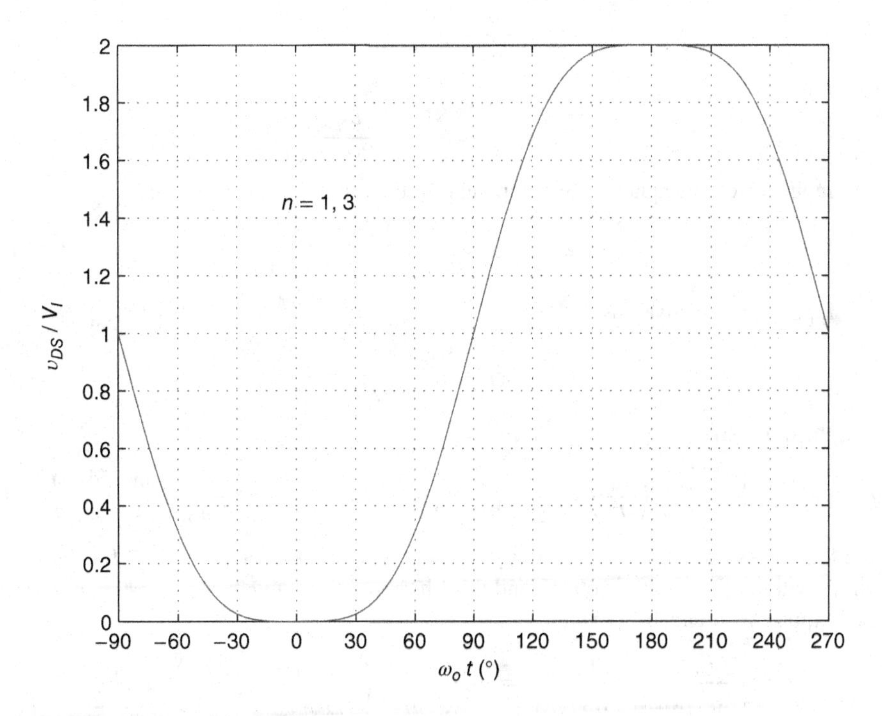

Figure 8.4 Maximally flat waveform of the normalized drain-to-source voltage v_{DS}/V_I for the Class F_3 RF power amplifier.

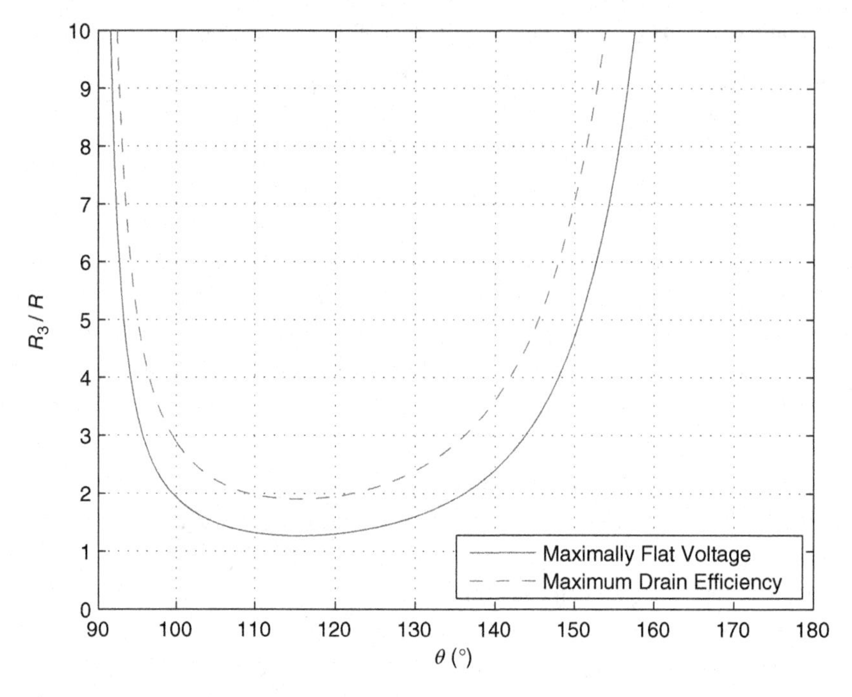

Figure 8.5 Ratio R_3/R as a function of conduction angle θ for the Class F_3 amplifier with maximally flat voltage and maximum drain efficiency.

where

$$I_m = \frac{V_m}{R} = \frac{9(V_I - v_{DSmin})}{8R} \tag{8.41}$$

The output power is

$$P_{DS} = \frac{V_m^2}{2R} = \frac{81(V_I - v_{DSmin})^2}{128R} \tag{8.42}$$

The dc input current is equal to the component of the drain current

$$I_I = \frac{I_m}{\gamma_1} = \frac{V_m}{\gamma_1 R} = \frac{9(V_I - v_{DSmin})}{8\gamma_1 R} \tag{8.43}$$

The dc input power is given by

$$P_I = V_I I_I = \frac{9V_I(V_I - v_{DSmin})}{8\gamma_1 R} \tag{8.44}$$

Hence, the drain efficiency of the amplifier is

$$\eta_D = \frac{P_{DS}}{P_I} = \frac{1}{2}\left(\frac{V_m}{V_I}\right)\left(\frac{I_m}{I_I}\right) = \frac{1}{2}\xi_1\gamma_1 = \left(\frac{1}{2}\right)\left(\frac{9}{8}\right)\gamma_1 = \frac{9\gamma_1}{16} = \frac{9(\theta - \sin\theta\cos\theta)}{16(\sin\theta - \theta\cos\theta)} \tag{8.45}$$

Figure 8.6 shows the drain efficiency η_D as a function of conduction angle θ for the Class F_3 amplifier with maximally flat voltage and third-harmonic peaking.

The output-power capability is

$$c_p = \frac{P_O}{I_{DM}V_{DSM}} = \frac{\eta_D P_I}{I_{DM}V_{DSM}} = \eta_D\left(\frac{I_I}{I_{DM}}\right)\left(\frac{V_I}{V_{DSM}}\right) = \left(\frac{9\gamma_1}{16}\right)(\alpha_0)\left(\frac{1}{2}\right)$$

$$= \frac{9}{32}\alpha_1 = \frac{9(\theta - \sin\theta\cos\theta)}{32\pi(1 - \cos\theta)} \tag{8.46}$$

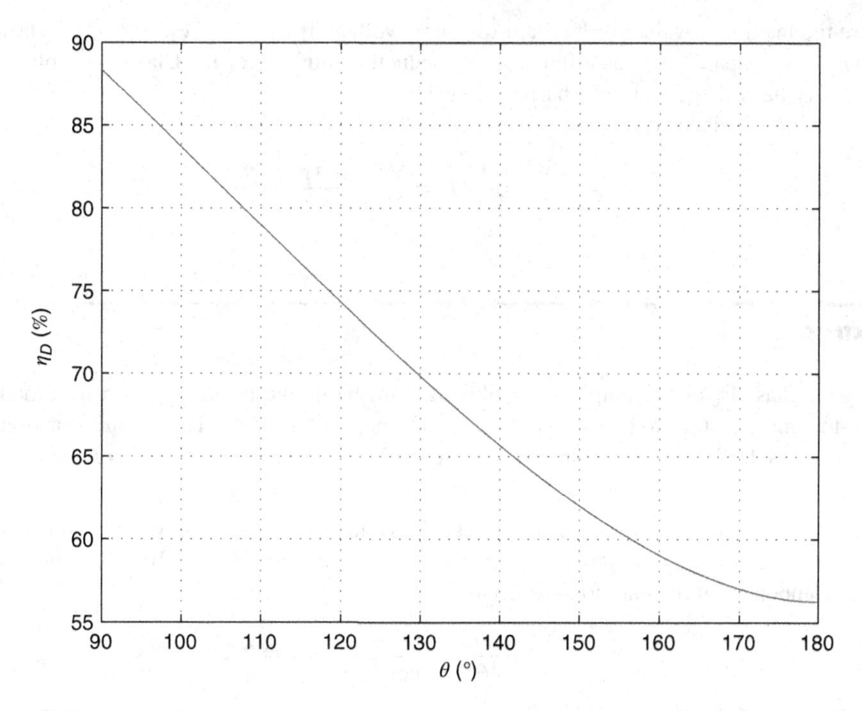

Figure 8.6 Drain efficiency η_D as a function of conduction angle θ for the Class F_3 amplifier with maximally flat voltage.

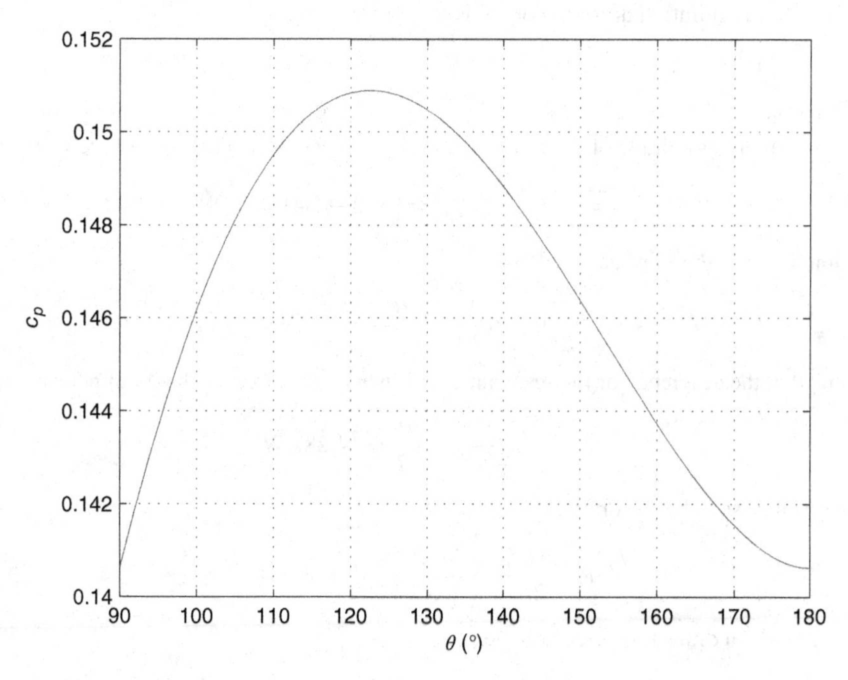

Figure 8.7 Output-power capability c_p as a function of conduction angle θ for the Class F_3 amplifier with maximally flat voltage.

where the maximum value of the drain-to-source voltage is $V_{DSM} = 2V_I$. Figure 8.7 shows the output-power capability c_p as a function of conduction angle θ for the Class F_3 amplifier with maximally flat voltage and third-harmonic peaking.

The dc input resistance is

$$R_{DC} = \frac{V_I}{I_I} = \frac{8\gamma_1}{9}R = \frac{8(\theta - \sin\theta\cos\theta)}{9(\sin\theta - \theta\cos\theta)}R \tag{8.47}$$

Example 8.1

Design a Class F_3 power amplifier employing third-harmonic peaking with a maximally flat drain-to-source voltage to deliver a power of 10 W at $f_c = 800$ MHz. The required bandwidth is $BW = 100$ MHz. The dc power supply voltage is 12 V. The gate drive power is $P_G = 0.4$ W.

Solution. Assume that the MOSFET threshold voltage is $V_t = 1$, the dc component of the gate-to-source voltage is $V_{GS} = 1.5$ V, and the conduction angle is $\theta = 110°$. The amplitude of the ac component of the gate-to-source voltage is

$$V_{gsm} = \frac{V_t - V_{GS}}{\cos\theta} = \frac{1 - 1.5}{\cos 110°} = 1.462 \text{ V} \tag{8.48}$$

The maximum gate-to-source voltage is

$$v_{GSmax} = V_{GS} + v_{gsm} \tag{8.49}$$

The maximum saturation drain-to-source voltage is

$$v_{DSsat} = v_{GSmax} - V_t = V_{GS} + V_{gsm} - V_t = 1.5 + 1.462 - 1 = 1.962 \text{ V} \tag{8.50}$$

Pick the minimum drain-to-source voltage $v_{DSmin} = 2.4$ V.

The maximum amplitude of the fundamental component of the drain-to-source voltage is

$$V_m = \frac{9}{8}(V_I - v_{DSmin}) = \frac{9}{8}(12 - 2.4) = 10.8 \text{ V} \tag{8.51}$$

The amplitude of the third harmonic is

$$V_{m3} = \frac{V_m}{9} = \frac{10.8}{9} = 1.2 \text{ V} \tag{8.52}$$

Assume that the efficiency of the resonant circuit is $\eta_r = 0.7$. Hence, the drain power is

$$P_{DS} = \frac{P_O}{\eta_r} = \frac{10}{0.7} = 14.289 \text{ W} \tag{8.53}$$

The resistance seen by the drain is

$$R = \frac{V_m^2}{2P_{DS}} = \frac{10.8^2}{2 \times 14.289} = 4.081\ \Omega \tag{8.54}$$

The maximum drain-to-source voltage is

$$V_{DSM} = V_I + V_I - v_{DSmin} = 12 + 12 - 2.4 = 21.6 \text{ V} \tag{8.55}$$

The amplitude of the fundamental component of the drain current is

$$I_m = \frac{V_m}{R} = \frac{10.8}{4.081} = 2.646 \text{ A} \tag{8.56}$$

Table 8.1 Fourier coefficients for drain current.

$\theta(°)$	α_0	α_1	γ_1	α_3	α_5	α_7
10	0.0370	0.0738	1.9939	0.0720	0.0686	0.0636
20	0.0739	0.1461	1.9756	0.1323	0.1075	0.0766
30	0.1106	0.2152	1.9460	0.1715	0.1029	0.0367
40	0.1469	0.2799	1.9051	0.1845	0.0625	−0.0124
45	0.1649	0.3102	1.8808	0.1811	0.0362	−0.0259
50	0.1828	0.3388	1.8540	0.1717	0.0105	−0.0286
60	0.2180	0.3910	1.7936	0.1378	−0.0276	−0.0098
70	0.2525	0.4356	1.7253	0.0915	−0.0378	0.0129
80	0.2860	0.4720	1.6505	0.0426	−0.0235	0.0147
90	0.3183	0.5000	1.5708	0.0000	0.0000	0.0000
100	0.3493	0.5197	1.4880	−0.0300	0.0165	−0.0104
110	0.3786	0.5316	1.4040	−0.0449	0.0185	−0.0063
120	0.4060	0.5363	1.3210	−0.0459	0.0092	0.0033
130	0.4310	0.5350	1.2414	−0.0373	−0.0023	0.0062
140	0.4532	0.5292	1.1675	−0.0244	−0.0083	0.0016
150	0.4720	0.5204	1.1025	−0.0123	−0.0074	−0.0026
160	0.4868	0.5110	1.0498	−0.0041	−0.0033	−0.0024
170	0.4965	0.5033	1.0137	−0.0006	−0.0005	−0.0005
180	0.5000	0.5000	1.0000	0.0000	0.0000	0.0000

Using Table 8.1, the coefficients at the conduction angle $\theta = 110°$ are as follows: $\alpha_1 = 0.5316$, $\alpha_3 = -0.0448754$, and $\gamma_1 = 1.404$. The maximum drain current is

$$I_{DM} = \frac{I_m}{\alpha_1} = \frac{2.646}{0.5316} = 4.977 \text{ A} \tag{8.57}$$

The dc supply current is

$$I_I = \frac{I_m}{\gamma_1} = \frac{2.646}{1.404} = 1.885 \text{ A} \tag{8.58}$$

The dc supply power is

$$P_I = V_I I_I = 12 \times 1.885 = 22.62 \text{ W} \tag{8.59}$$

The drain power loss of the transistor is

$$P_D = P_I - P_{DS} = 22.62 - 14.289 = 8.331 \text{ W} \tag{8.60}$$

The drain efficiency is

$$\eta_D = \frac{P_{DS}}{P_I} = \frac{14.289}{22.62} = 63.17\% \tag{8.61}$$

The amplifier efficiency is

$$\eta = \frac{P_O}{P_I} = \frac{10}{22.62} = 44.2\% \tag{8.62}$$

The power-added efficiency is

$$\eta_{PAE} = \frac{P_O - P_G}{P_I} = \frac{10 - 0.4}{22.62} = 42.44\% \tag{8.63}$$

The dc resistance that the amplifier presents to the dc source is

$$R_{DC} = \frac{V_I}{I_I} = \frac{12}{1.884} = 6.3685\ \Omega \tag{8.64}$$

The MOSFET parameter is

$$K = \frac{I_{DM}}{v_{DSsat}^2} = \frac{4.977}{1.962^2} = 1.2929 \text{ A/V} \quad (8.65)$$

yielding

$$\frac{W}{L} = \frac{2K}{\mu_{n0}C_{ox}} = \frac{2 \times 1.2929}{0.142 \times 10^{-3}} = 18,210 \quad (8.66)$$

The loaded quality factor is

$$Q_L = \frac{f_c}{BW} = \frac{800}{100} = 8 \quad (8.67)$$

The definition of the loaded quality factor for a parallel-resonant circuit is

$$Q_L = \frac{R}{\omega_c L} = \omega_c CR \quad (8.68)$$

The inductance of the resonant circuit tuned to the fundamental is

$$L = \frac{R}{\omega_c Q_L} = \frac{4.081}{2\pi \times 0.8 \times 10^9 \times 8} = 101.49 \text{ pH} \quad (8.69)$$

The capacitance of the resonant circuit tuned to the fundamental is

$$C = \frac{Q_L}{\omega_c R} = \frac{8}{2\pi \times 0.8 \times 10^9 \times 4.081} = 390 \text{ pF} \quad (8.70)$$

The resistance connected in parallel with the resonant circuit tuned to the third harmonic is

$$R_3 = \frac{\alpha_1}{9|\alpha_3|}R = \frac{0.5316}{9|-0.0448754|} \times 4.081 = 1.31623 \times 4.081 = 5.372\ \Omega \quad (8.71)$$

The power loss in resistor R_3 is

$$P_{R3} = \frac{V_{m3}^2}{2R_3} = \frac{1.2^2}{2 \times 5.372} = 0.134 \text{ W} \quad (8.72)$$

Assuming the loaded quality factor of the resonant circuit tuned to the third harmonic to be $Q_{L3} = 15$, we obtain

$$L_3 = \frac{R_3}{3\omega_c Q_{L3}} = \frac{5.372}{3 \times 2\pi \times 0.8 \times 10^9 \times 15} = 23.75 \text{ pH} \quad (8.73)$$

and

$$C_3 = \frac{Q_{L3}}{3\omega_c R_3} = \frac{15}{3 \times 2\pi \times 0.8 \times 10^9 \times 5.372} = 185.17 \text{ pF} \quad (8.74)$$

The reactance of the RFC inductance is

$$X_{Lf} = 10R = 10 \times 4.081 = 40.81\ \Omega \quad (8.75)$$

yielding

$$L_f = \frac{X_{Lf}}{\omega_0} = \frac{40.81}{2\pi \times 0.8 \times 10^9} = 8.12 \text{ nH} \quad (8.76)$$

The reactance of the coupling capacitor is

$$X_{Cc} = \frac{R}{10} = \frac{4.081}{10} = 0.4081\ \Omega \quad (8.77)$$

producing

$$C_c = \frac{1}{\omega_0 X_{Cc}} = \frac{1}{2\pi \times 0.8 \times 10^9 \times 0.4081} = 487.49 \text{ pF} \quad (8.78)$$

8.2.2 Maximum Drain Efficiency Class F$_3$ Amplifier

The maximum drain efficiency η_D and the maximum output-power capability c_p do not occur at the maximally flat drain-to-source voltage v_{DS}, but when the waveform v_{DS} exhibits slight ripple. The drain efficiency is given by

$$\eta_D = \frac{P_{DS}}{P_I} = \frac{\frac{1}{2} I_m V_m}{I_I V_I} = \frac{1}{2}\left(\frac{I_m}{I_I}\right)\left(\frac{V_m}{V_I}\right) = \frac{1}{2}\gamma_1(\theta)\xi_1 \tag{8.79}$$

At a fixed value of conduction angle θ, γ_1 is constant. Therefore, the maximum drain efficiency η_{Dmax} at a given conduction angle θ occurs at the maximum value of $\xi_{1max} = V_{m(max)}/V_I$

$$\eta_{Dmax} = \frac{P_{DS(max)}}{P_I} = \frac{1}{2}\gamma_1(\theta)\xi_{1max} \tag{8.80}$$

From (8.23),

$$\cos\omega_o t_m = \sqrt{1 - \sin^2\omega_o t_m} = \sqrt{1 - \frac{9V_{m3} - V_m}{12V_{m3}}} = \sqrt{\frac{1}{4} + \frac{V_m}{12V_{m3}}} \quad \text{for} \quad \frac{V_{m3}}{V_m} \geq \frac{1}{9} \tag{8.81}$$

The minimum value of v_{DS} occurs at $\omega_o t_m$ and is equal to zero

$$v_{DS}(\omega_o t_m) = V_I - V_m\cos\omega_o t_m + V_{m3}\cos 3\omega_o t_m = 0 \tag{8.82}$$

which gives

$$\begin{aligned} V_I &= V_m\cos\omega_o t_m - V_{m3}\cos 3\omega_o t_m = V_m\cos\omega_o t_m - V_{m3}(4\cos^3\omega_o t_m - 3\cos\omega_o t_m) \\ &= \cos\omega_o t_m(V_m - 4V_{m3}\cos^2\omega_o t_m + 3V_{m3}) = \sqrt{\frac{1}{4} + \frac{V_m}{12V_{m3}}}\left(\frac{2V_m}{3} + 2V_{m3}\right) \\ &= V_m\left[\sqrt{\frac{1}{4} + \frac{V_m}{12V_{m3}}}\left(\frac{2}{3} + \frac{2V_{m3}}{V_m}\right)\right] = V_m\left[\left(\frac{1}{3} + \frac{V_{m3}}{V_m}\right)\sqrt{1 + \frac{V_m}{3V_{m3}}}\right] \end{aligned} \tag{8.83}$$

Hence,

$$\frac{V_m}{V_I} = \frac{1}{\left(\frac{1}{3} + \frac{V_{m3}}{V_m}\right)\sqrt{1 + \frac{1}{\frac{3V_{m3}}{V_m}}}} \quad \text{for} \quad \frac{V_{m3}}{V_m} \geq \frac{1}{9} \tag{8.84}$$

Figure 8.8 shows V_m/V_I as a function of V_{m3}/V_m computed from (8.17) and (8.84).

In order to maximize the drain efficiency

$$\eta_D = \frac{P_O}{P_I} = \frac{1}{2}\left(\frac{I_m}{I_I}\right)\left(\frac{V_m}{V_I}\right) \tag{8.85}$$

at a given waveform of drain current i_D, that is, at a given ratio I_m/I_I, we will maximize the ratio V_m/V_I. To maximize V_m/V_I, we take the derivative and set it to zero

$$\frac{d\left(\frac{V_m}{V_I}\right)}{d\left(\frac{V_{m3}}{V_m}\right)} = \frac{1}{\left(\frac{1}{3} + \frac{V_{m3}}{V_m}\right)\sqrt{1 + \frac{1}{\frac{3V_{m3}}{V_m}}}}\left[\frac{1}{6\left(\frac{V_{m3}}{V_m}\right)^2 + 2\frac{V_{m3}}{V_m}} - \frac{1}{\frac{1}{3} + \frac{V_{m3}}{V_m}}\right] = 0 \tag{8.86}$$

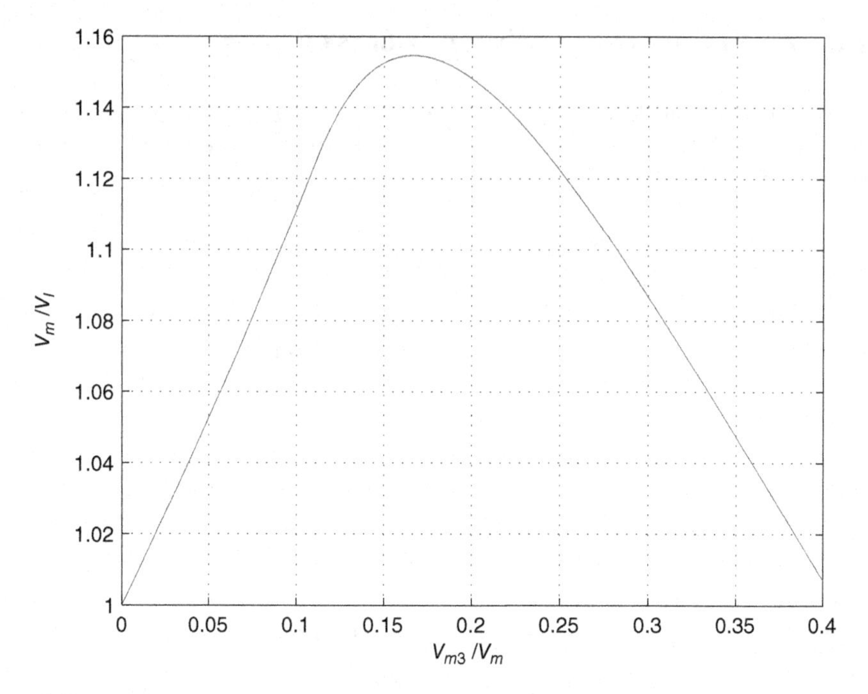

Figure 8.8 Ratio V_m/V_I as a function of V_{m3}/V_m for the Class F_3 RF power amplifier with maximum drain efficiency.

producing

$$18\left(\frac{V_{m3}}{V_m}\right)^2 + 3\left(\frac{V_{m3}}{V_m}\right) - 1 = 0 \tag{8.87}$$

Solution of this equation yields

$$\frac{V_{m3}}{V_m} = \frac{1}{6} \approx 0.1667 \tag{8.88}$$

Substitution of this equation into (8.84) yields

$$\frac{V_m}{V_I} = \frac{2}{\sqrt{3}} \approx 1.1547 \tag{8.89}$$

Finally,

$$V_{m3} = \frac{V_m}{6} = \frac{V_I}{3\sqrt{3}} \tag{8.90}$$

yielding

$$\frac{V_{m3}}{V_I} = \frac{1}{3\sqrt{3}} \approx 0.19245 \tag{8.91}$$

The minimum values of v_{DS} occur at

$$\omega_o t_m = \pm \arcsin\sqrt{\frac{9V_{m3} - V_m}{12V_{m3}}} = \pm \arcsin\left(\frac{1}{2}\right) = \pm 30^\circ \tag{8.92}$$

The normalized waveform v_{DS}/V_I for achieving the maximum drain efficiency is given by

$$\frac{v_{DS}}{V_I} = 1 - \frac{V_m}{V_I}\cos\omega_o t + \frac{V_{m3}}{V_I}\cos 3\omega_o t = 1 - \frac{2}{\sqrt{3}}\cos\omega_o t + \frac{1}{3\sqrt{3}}\cos 3\omega_o t \tag{8.93}$$

The normalized drain-to-source voltage waveform v_{DS}/V_I for the Class F_3 amplifier with the maximum drain efficiency is shown in Fig. 8.9.

The normalized amplitude of the third-harmonic voltage is

$$\frac{V_{m3}}{V_m} = \frac{I_{m3}R_3}{I_m R} = \frac{|\alpha_3| I_{DM} R_3}{\alpha_1 I_{DM} R} = \frac{|\alpha_3| R_3}{\alpha_1 R} = \frac{1}{12}\left|\frac{\sin 3\theta \cos\theta - 3\cos 3\theta \sin\theta}{\theta - \sin\theta\cos\theta}\right|\left(\frac{R_3}{R}\right) = \frac{1}{6} \tag{8.94}$$

Hence,

$$R_3 = \frac{\alpha_1}{6|\alpha_3|}R = 2\left|\frac{\theta - \sin\theta\cos\theta}{\sin 3\theta\cos\theta - 3\cos 3\theta\sin\theta}\right| R \tag{8.95}$$

Figure 8.5 shows a plot of R_3/R as a function of conduction angle θ for the Class F_3 amplifier with maximum drain efficiency.

The power loss in resistor R_3 is

$$P_{R3} = \frac{V_{m3}^2}{2R_3} = \frac{1}{2R_3}\left(\frac{V_m}{6}\right)^2 = \frac{V_m^2}{72R_3} \tag{8.96}$$

The amplitude of the output current is

$$I_m = \frac{V_m}{R} \tag{8.97}$$

The output power is

$$P_O = \frac{V_m^2}{2R} = \frac{2V_I^2}{3R} \tag{8.98}$$

The dc supply current is

$$I_I = \frac{I_m}{\gamma_1} = \frac{2V_I}{\sqrt{3}\gamma_1 R} \tag{8.99}$$

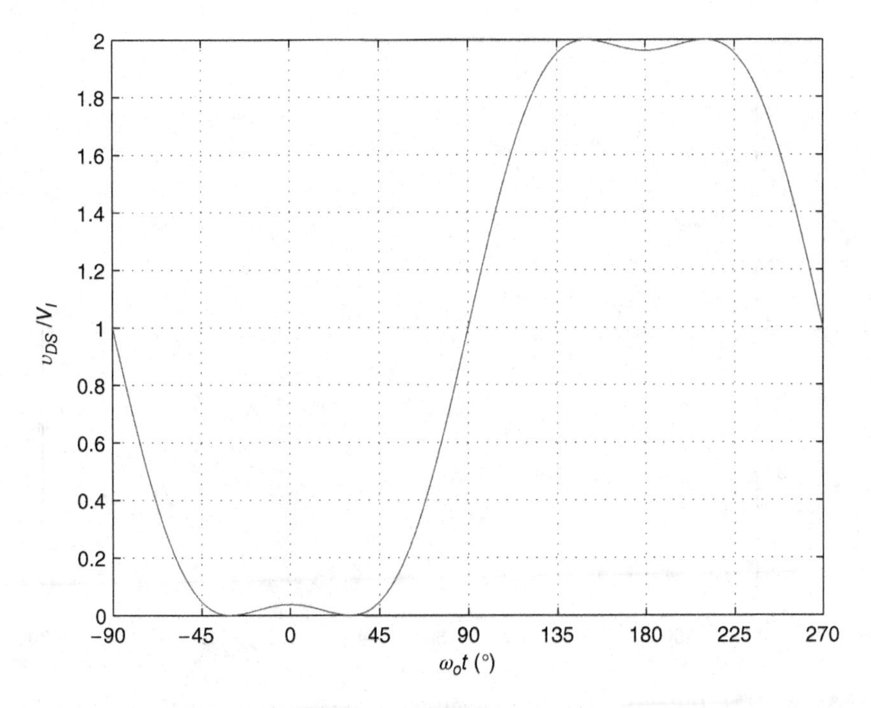

Figure 8.9 Waveform of the normalized drain-to-source voltage v_{DS}/V_I for the Class F_3 RF power amplifier with the maximum drain efficiency η_{Dmax} and output-power capability c_{pmax}.

The dc supply power is

$$P_I = V_I I_I = \frac{2V_I^2}{\sqrt{3}\gamma_1 R} \tag{8.100}$$

The drain efficiency is

$$\eta_D = \frac{P_{DS}}{P_I} = \frac{1}{2}\left(\frac{V_m}{V_I}\right)\left(\frac{I_m}{I_I}\right) = \frac{1}{2} \times \frac{2}{\sqrt{3}}\gamma_1 = \frac{\gamma_1}{\sqrt{3}} = \frac{\theta - \sin\theta\cos\theta}{\sqrt{3}(\sin\theta - \theta\cos\theta)} \tag{8.101}$$

Figure 8.10 shows the drain efficiency η_D as a function of the conduction angle θ at $V_m/V_I = 2/\sqrt{3}$ for the Class F_3 amplifier with maximum drain efficiency.

The efficiency of the Class AB amplifier for $V_m = V_I$, that is, $\xi_{1(AB)} = 1$, is given by

$$\eta_{D(AB)} = \frac{1}{2}\gamma_1 \xi_{1(AB)} = \frac{1}{2}\gamma_1 \tag{8.102}$$

The efficiency of the Class F_3 amplifier is

$$\eta_{D(F_3)} = \frac{1}{2}\gamma_1 \xi_{1(F_3)} \tag{8.103}$$

The ratio of the efficiency of any Class F_3 amplifier to the efficiency of the Class AB amplifier at the same conduction angle θ, that is, at the same value of γ_1, is expressed as

$$\frac{\eta_{D(F_3)}}{\eta_{D(AB)}} = \xi_{1(F_3)} = \frac{V_m}{V_I} \tag{8.104}$$

where

$$\frac{\eta_{D(F_3)}}{\eta_{D(AB)}} = \frac{1}{1 - \frac{V_{m3}}{V_m}} \quad \text{for} \quad 0 \le \frac{V_{m3}}{V_m} \le \frac{1}{9} \tag{8.105}$$

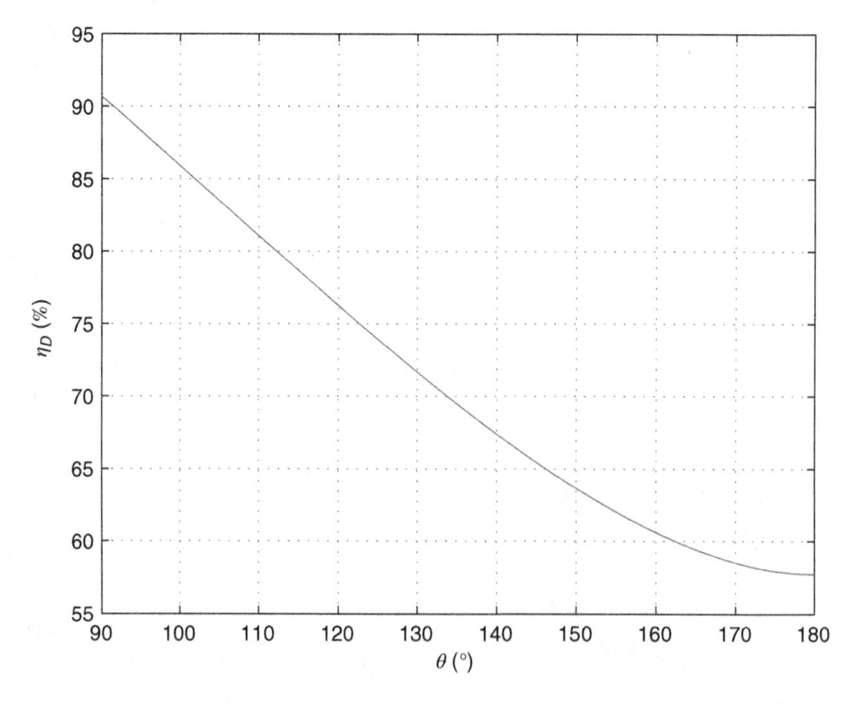

Figure 8.10 Drain efficiency η_D as a function of conduction angle θ at $V_m/V_I = 2/\sqrt{3}$ for the Class F_3 RF power amplifier with maximum drain efficiency.

and

$$\frac{\eta_{D(F_3)}}{\eta_{D(AB)}} = \frac{1}{\left(\frac{1}{3} + \frac{V_{m3}}{V_m}\right)\sqrt{1 + \frac{V_m}{3V_{m3}}}} \quad \text{for} \quad \frac{V_{m3}}{V_m} \geq \frac{1}{9} \tag{8.106}$$

The ratio of the drain efficiency of any Class F_3 amplifier to the drain efficiency of the Class AB amplifier $\eta_{D(F_3)}/\eta_{D(AB)}$ as a function of V_{m3}/V_m at the same conduction angle θ is shown in Fig. 8.11.

The ratio of the drain efficiency of the Class F_3 amplifier with maximum drain efficiency $\eta_{DME(F3)}$ to the drain efficiency of the Class F_3 amplifier with maximally flat voltage $\eta_{DMF(F3)}$ at $v_{DSmin} = 0$ is

$$\frac{\eta_{DME(F3)}}{\eta_{DMF(F3)}} = \frac{16}{9\sqrt{3}} \approx 1.0264 \tag{8.107}$$

Thus, the improvement in the efficiency is by 2.64%. The maximum drain-to-source voltage is

$$V_{DSM} = 2V_I \tag{8.108}$$

The maximum output-power capability is

$$c_p = \frac{P_{DS}}{I_{DM}V_{DSM}} = \frac{\eta_D P_I}{I_{DM}V_{DSM}} = \eta_D\left(\frac{V_I}{V_{DSM}}\right)\left(\frac{I_I}{I_{DM}}\right) = \frac{\gamma_1}{\sqrt{3}} \times \left(\frac{1}{2}\right)\alpha_0 = \frac{\alpha_1}{2\sqrt{3}}$$
$$= \frac{\theta - \sin\theta\cos\theta}{2\sqrt{3}\pi(1 - \cos\theta)} \tag{8.109}$$

Figure 8.12 shows the output-power capability c_p as a function of θ for the Class F_3 RF power amplifier with maximum drain efficiency.

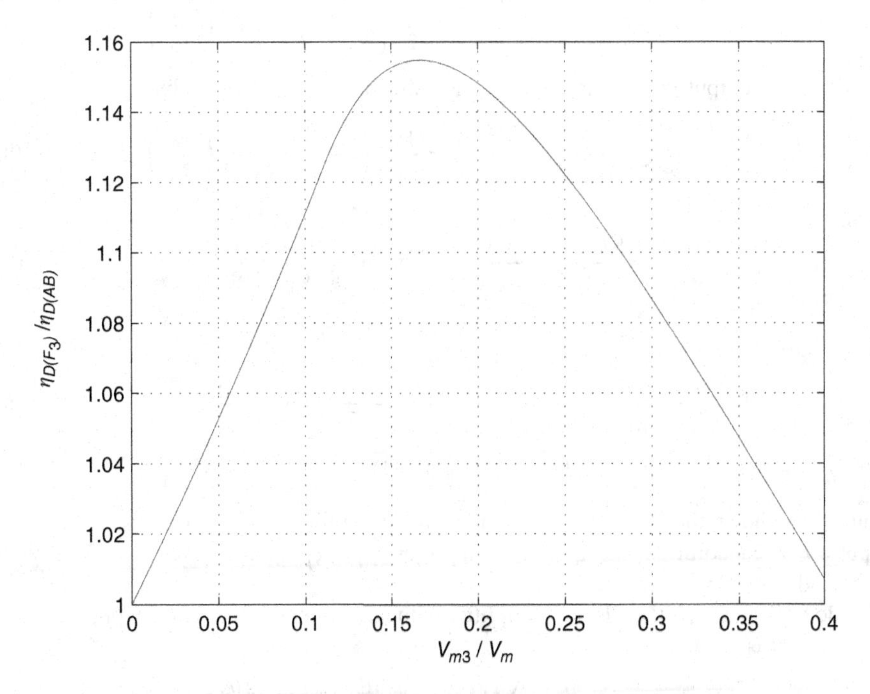

Figure 8.11 Ratio of drain efficiency of any Class F_3 amplifier to the efficiency of the Class AB amplifier $\eta_{D(F_3)}/\eta_{D(AB)}$ as a function of V_{m3}/V_m at the same conduction angle θ.

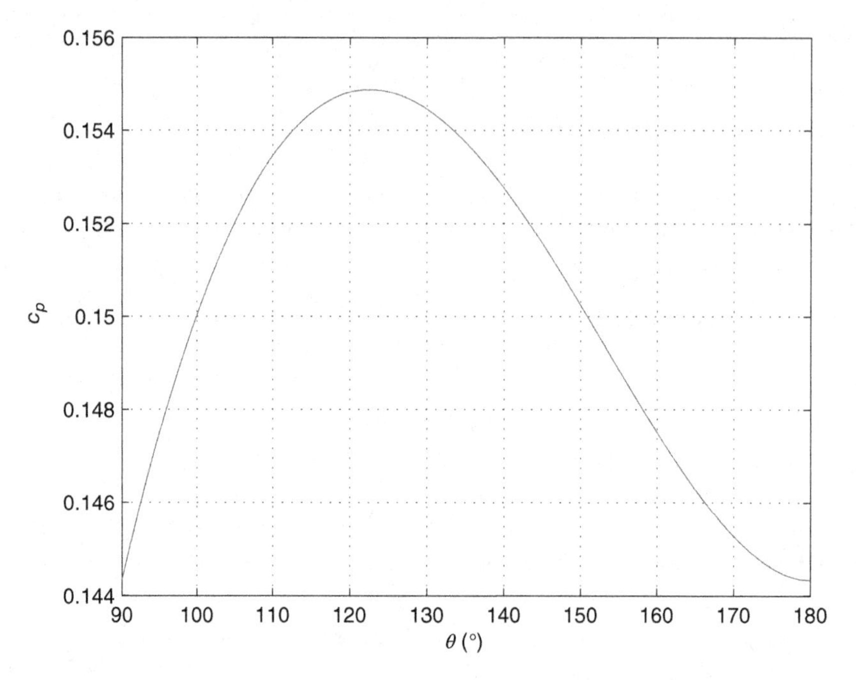

Figure 8.12 Output-power capability c_p as a function of θ for the Class F_3 RF power amplifier with maximum drain efficiency.

The output-power capability of the Class AB amplifier is

$$c_p = \frac{1}{2}\left(\frac{I_m}{I_{DM}}\right)\left(\frac{V_m}{V_{DSM}}\right) = \frac{1}{2}\alpha_1\left(\frac{V_m}{2V_I}\right) = \frac{1}{4}\alpha_1\left(\frac{V_m}{V_I}\right) \tag{8.110}$$

The maximum output-power capability at any value of V_{m3}/V_m is given by

$$c_p = \frac{1}{2}\left(\frac{I_m}{I_{DM}}\right)\left(\frac{V_m}{V_{DSM}}\right) = \frac{1}{2}\alpha_1\left(\frac{V_m}{2V_I}\right) = \frac{1}{4}\alpha_1\left(\frac{V_m}{V_I}\right) \tag{8.111}$$

Thus,

$$\frac{c_{p(F_3)}}{c_{p(AB)}} = \frac{1}{1 - \dfrac{V_{m3}}{V_m}} \quad \text{for} \quad 0 \le \frac{V_{m3}}{V_m} \le \frac{1}{9} \tag{8.112}$$

and

$$\frac{c_{p(F_3)}}{c_{p(AB)}} = \frac{1}{\left(\dfrac{1}{3} + \dfrac{V_{m3}}{V_m}\right)\sqrt{1 + \dfrac{V_m}{3V_{m3}}}} \quad \text{for} \quad \frac{V_{m3}}{V_m} \ge \frac{1}{9} \tag{8.113}$$

Figure 8.13 shows the ratio of the output-power capability of any Class F_3 amplifier to the output-power capability of the Class AB amplifier $c_{p(F3)}/c_{p(AB)}$ as a function of V_{m3}/V_m at the same conduction angle θ.

The dc resistance that the Class F_3 amplifier with maximum drain efficiency presents to the dc power supply is

$$R_{DC} = \frac{V_I}{I_I} = \frac{\sqrt{3}\gamma_1}{2}R = \frac{\sqrt{3}(\theta - \sin\theta\cos\theta)}{2(\sin\theta - \theta\cos\theta)}R. \tag{8.114}$$

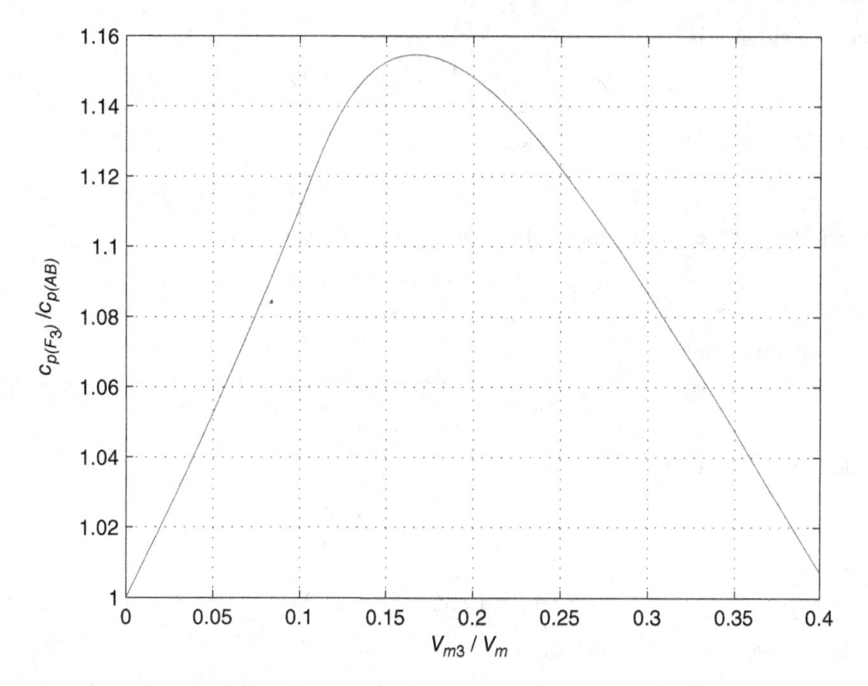

Figure 8.13 Ratio of the output-power capability of any Class F_3 amplifier to the output-power capability of the Class AB amplifier $c_{p(F_3)}/c_{p(AB)}$ as a function of V_{m3}/V_m at the same conduction angle θ.

Example 8.2

Design a Class F_3 power amplifier employing third-harmonic peaking and having a maximum drain efficiency to deliver a power of 10 W at $f_c = 800$ MHz with $BW = 100$ MHz. The dc power supply voltage is 12 V. The efficiency related to all passive components is $\eta_r = 0.7$. The gate drive power is $P_G = 0.4$ W.

Solution. Assume that the MOSFET threshold voltage is $V_t = 1$. The dc component of the gate-to-source voltage is $V_{GS} = 1.5$ V, and the conduction angle is $\theta = 110°$. The amplitude of the ac component of the gate-to-source voltage is

$$V_{gsm} = \frac{V_t - V_{GS}}{\cos\theta} = \frac{1 - 1.5}{\cos 110°} = 1.462 \text{ V} \tag{8.115}$$

The saturation drain-to-source voltage is

$$v_{DSsat} = v_{GS} - V_t = V_{GS} + V_{gsm} - V_t = 1.5 + 1.462 - 1 = 1.962 \text{ V} \tag{8.116}$$

Pick the minimum drain-to-source voltage $v_{DSmin} = 2.4$ V. The maximum amplitude of the fundamental component of the drain-to-source voltage is

$$V_m = \frac{2}{\sqrt{3}}(V_I - v_{DSmin}) = 1.1547(12 - 2.4) = 11.085 \text{ V} \tag{8.117}$$

The amplitude of the third harmonic is

$$V_{m3} = \frac{V_m}{6} = \frac{11.085}{6} = 1.848 \text{ V} \tag{8.118}$$

The drain power is

$$P_{DS} = \frac{P_O}{\eta_r} = \frac{10}{0.7} = 14.289 \text{ W} \tag{8.119}$$

The load resistance is

$$R = \frac{V_m^2}{2P_{DS}} = \frac{11.085^2}{2 \times 14.289} = 4.2998\ \Omega \tag{8.120}$$

The maximum drain-to-source voltage is

$$v_{DSmax} = 2V_I - v_{DSmin} = 2 \times 12 - 2.4 = 21.6\ \text{V} \tag{8.121}$$

The amplitude of the fundamental component of the drain current is

$$I_m = \frac{V_m}{R} = \frac{11.085}{4.2998} = 2.578\ \text{A} \tag{8.122}$$

The maximum drain current is

$$I_{DM} = \frac{I_m}{\alpha_1} = \frac{2.578}{0.5316} = 4.8496\ \text{A} \tag{8.123}$$

The dc supply current is

$$I_I = \frac{I_m}{\gamma_1} = \frac{2.578}{1.404} = 1.836\ \text{A} \tag{8.124}$$

The dc supply power is

$$P_I = I_I V_I = 1.836 \times 12 = 22.0344\ \text{W} \tag{8.125}$$

The drain power loss of the transistor is

$$P_D = P_I - P_{DS} = 22.0344 - 14.289 = 7.745\ \text{W} \tag{8.126}$$

The drain efficiency is

$$\eta_D = \frac{P_{DS}}{P_I} = \frac{14.289}{22.0344} = 64.85\% \tag{8.127}$$

The amplifier efficiency is

$$\eta = \frac{P_O}{P_I} = \frac{10}{22.0344} = 45.38\% \tag{8.128}$$

The power-added efficiency is

$$\eta_{PAE} = \frac{P_O - P_G}{P_I} = \frac{10 - 0.4}{22.0344} = 43.56\% \tag{8.129}$$

The dc resistance presented by the amplifier to the dc power supply is

$$R_{DC} = \frac{V_I}{I_I} = \frac{12}{1.836} = 6.5352\ \Omega \tag{8.130}$$

The loaded quality factor of the resonant circuit tuned to the fundamental of the carrier frequency is

$$Q_L = \frac{f_c}{BW} = \frac{800}{100} = 8 \tag{8.131}$$

The resonant circuit components of the output circuit are

$$L = \frac{R}{\omega_c Q_L} = \frac{4.2998}{2\pi \times 0.8 \times 10^9 \times 8} = 106.92\ \text{pH} \tag{8.132}$$

and

$$C = \frac{Q_L}{\omega_c R} = \frac{8}{2\pi \times 0.8 \times 10^9 \times 4.2998} = 370.14\ \text{pF} \tag{8.133}$$

The resistance connected in parallel with the resonant circuit tuned to the third harmonic is

$$R_3 = \frac{\alpha_1}{6|\alpha_3|} R = \frac{0.5316}{6|-0.0448754|} \times 4.2998 = 1.97436 \times 4.2998 = 8.4894\ \Omega \tag{8.134}$$

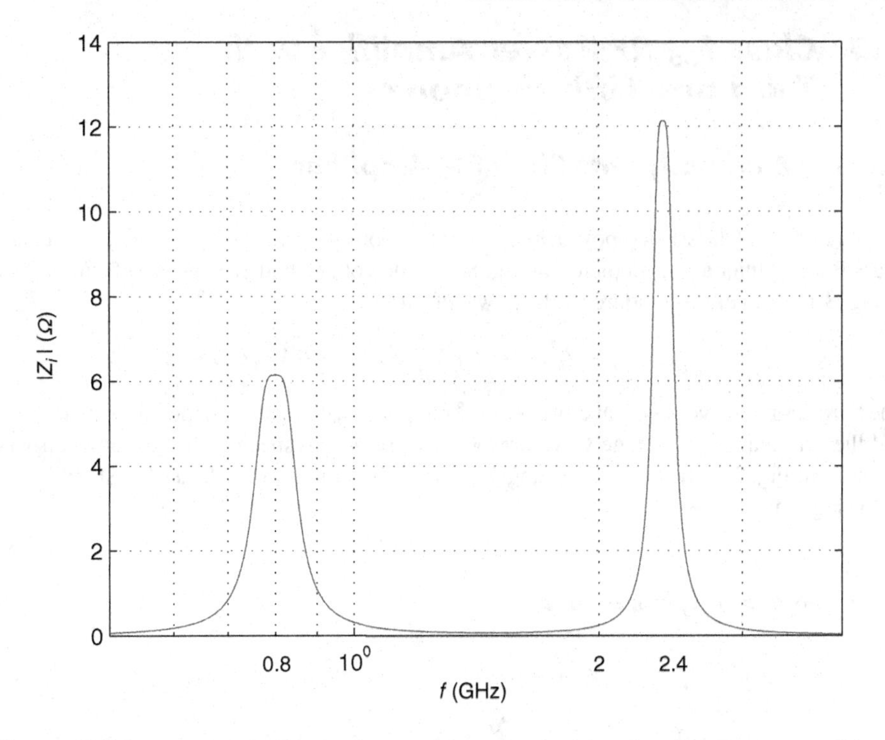

Figure 8.14 Magnitude of the load network impedance of the Class F_3 RF power amplifier.

Assuming the loaded quality factor of the resonant circuit tuned to the third harmonic to be $Q_{L3} = 15$, we obtain

$$L_3 = \frac{R_3}{3\omega_c Q_{L3}} = \frac{8.4894}{3 \times 2\pi \times 0.8 \times 10^9 \times 15} = 37.53 \text{ pH} \tag{8.135}$$

and

$$C_3 = \frac{Q_{L3}}{3\omega_c R_3} = \frac{15}{3 \times 2\pi \times 0.8 \times 10^9 \times 8.4894} = 117.17 \text{ pF} \tag{8.136}$$

The reactance of the RFC inductance is

$$X_{Lf} = 10R = 10 \times 4.2998 = 42.998 \ \Omega \tag{8.137}$$

yielding

$$L_f = \frac{X_{Lf}}{\omega_0} = \frac{42.998}{2\pi \times 0.8 \times 10^9} = 8.55 \text{ nH} \tag{8.138}$$

The reactance of the coupling capacitor is

$$X_{Cc} = \frac{R}{10} = \frac{4.2998}{10} = 0.42998 \ \Omega \tag{8.139}$$

producing

$$C_c = \frac{1}{\omega_0 X_{Cc}} = \frac{1}{2\pi \times 0.8 \times 10^9 \times 0.42998} = 462.68 \text{ pF} \tag{8.140}$$

Figure 8.14 shows the magnitude of the impedance of the load network for the designed Class F_3 RF power amplifier.

8.3 Class F$_{35}$ RF Power Amplifier with Third and Fifth Harmonics

8.3.1 Maximally Flat Class F$_{35}$ Amplifier

The circuit of a Class F RF power amplifier with both third and fifth harmonics is called the Class F$_{35}$ amplifier and is depicted in Fig. 8.15. The voltage and current waveforms are shown in Fig. 8.16. The drain-to-source voltage waveform is

$$v_{DS} = V_I - V_m \cos \omega_o t + V_{m3} \cos 3\omega_o t - V_{m5} \cos 5\omega_o t \tag{8.141}$$

The third-harmonic voltage waveform is 180° out phase with respect to the fundamental voltage, and the fifth-harmonic voltage waveform is in phase with respect to the fundamental voltage. Thus, from Fig. 3.7, the conduction angle of the drain current of the Class F$_{35}$ should be in the following range:

$$90^\circ \leq \theta \leq 127.76^\circ \tag{8.142}$$

In this case, $\alpha_1 > 0$, $\alpha_3 < 0$, and $\alpha_5 > 0$.

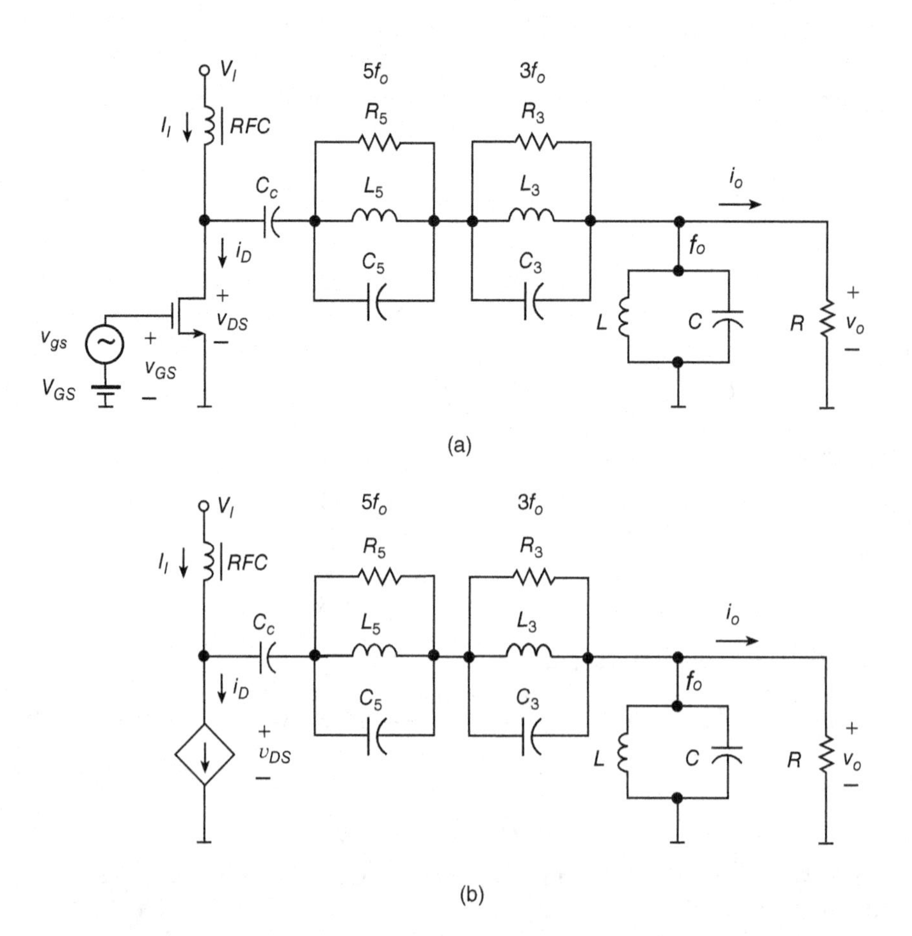

Figure 8.15 Class F$_{35}$ power amplifier: (a) circuit and (b) equivalent circuit.

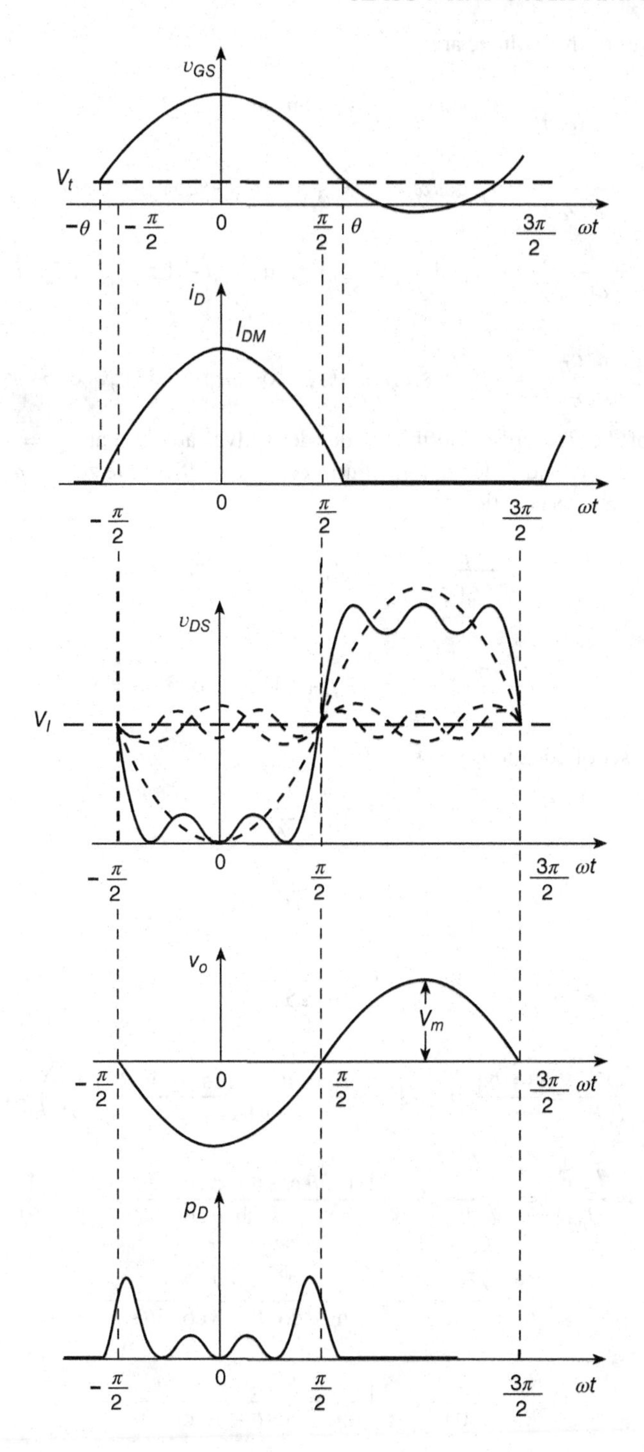

Figure 8.16 Waveforms in the Class F_{35} RF power amplifier with third- and fifth- harmonic peaking.

The derivatives of this voltage are

$$\frac{dv_{DS}}{d(\omega_o t)} = V_m \sin\omega_o t - 3V_{m3}\sin 3\omega_o t + 5V_{m5}\sin 5\omega_o t \tag{8.143}$$

$$\frac{d^2 v_{DS}}{d(\omega_o t)^2} = V_m \cos\omega_o t - 9V_{m3}\cos 3\omega_o t + 25V_{m5}\cos 5\omega_o t \tag{8.144}$$

$$\frac{d^3 v_{DS}}{d(\omega_o t)^3} = -V_m \sin\omega_o t + 27V_{m3}\sin 3\omega_o t - 125V_{m5}\sin 5\omega_o t \tag{8.145}$$

and

$$\frac{d^4 v_{DS}}{d(\omega_o t)^4} = -V_m \cos\omega_o t + 81V_{m3}\cos 3\omega_o t - 625V_{m5}\cos 5\omega_o t \tag{8.146}$$

All the terms of the first-order and third-order derivatives are zero at $\omega_o t = 0$ and do not give any equations. The second-order and the fourth-order derivatives are zero at $\omega_o t = 0$, yielding a set of two simultaneous equations

$$\left.\frac{d^2 v_{DS}}{d(\omega_o t)^2}\right|_{\omega_o t=0} = V_m - 9V_{m3} + 25V_{m5} = 0 \tag{8.147}$$

and

$$\left.\frac{d^4 v_{DS}}{d(\omega_o t)^4}\right|_{\omega_o t=0} = -V_m + 81V_{m3} - 625V_{m5} = 0 \tag{8.148}$$

Solution of this set of equations gives

$$V_{m3} = \frac{V_m}{6} \tag{8.149}$$

$$V_{m5} = \frac{V_m}{50} \tag{8.150}$$

and

$$V_{m5} = \frac{3}{25}V_{m3} \tag{8.151}$$

Hence,

$$\frac{V_{m3}}{V_m} = \frac{I_{m3}R_3}{I_m R} = \frac{|\alpha_3|R_3}{\alpha_1 R} = \frac{1}{12}\left|\frac{\sin 3\theta\cos\theta - 3\cos 3\theta\sin\theta}{\theta - \sin\theta\cos\theta}\right|\left(\frac{R_3}{R}\right) = \frac{1}{6} \tag{8.152}$$

and

$$\frac{V_{m5}}{V_m} = \frac{I_{m5}R_5}{I_m R} = \frac{|\alpha_5|R_5}{\alpha_1 R} = \frac{1}{60}\left|\frac{\sin 5\theta\cos\theta - 5\cos 5\theta\sin\theta}{\theta - \sin\theta\cos\theta}\right| = \frac{1}{50} \tag{8.153}$$

The resistances are

$$\frac{R_3}{R} = \frac{\alpha_1}{6|\alpha_3|} = 2\left|\frac{\theta - \sin\theta\cos\theta}{\sin 3\theta\cos\theta - 3\cos 3\theta\sin\theta}\right| \tag{8.154}$$

and

$$\frac{R_5}{R} = \frac{\alpha_1}{50\alpha_5} = \frac{5}{6}\left|\frac{\theta - \sin\theta\cos\theta}{\sin 5\theta\cos\theta - 5\cos 5\theta\sin\theta}\right| \tag{8.155}$$

Now

$$v_{DS}(0) = V_I - V_m + V_{m3} - V_{m5} = V_I - V_m + \frac{V_m}{6} - \frac{V_m}{50} = V_I - \frac{64}{75}V_m = 0 \tag{8.156}$$

producing

$$V_m = \frac{75}{64}V_I \approx 1.1719V_I \tag{8.157}$$

Similarly,

$$v_{DS}(0) = V_I - V_m + V_{m3} - V_{m5} = V_I - 6V_{m3} + V_{m3} - \frac{3V_{m3}}{25} = V_I - \frac{128}{25}V_{m3} = 0 \quad (8.158)$$

yielding

$$V_{m3} = \frac{25}{128}V_I \approx 0.1953V_I \quad (8.159)$$

Finally,

$$v_{DS}(0) = V_I - V_m + V_{m3} - V_{m5} = V_I - 50V_{m5} + \frac{25}{3}V_{m5} - V_{m5} = V_I - \frac{128}{3}V_{m5} = 0 \quad (8.160)$$

leading to

$$V_{m5} = \frac{3}{128}V_I \approx 0.02344V_I \quad (8.161)$$

The maximum drain-to-source voltage is

$$V_{DSM} = v_{DS}(\pi) = V_I + V_m - V_{m3} + V_{m5} = V_I + \frac{75}{64}V_I - \frac{25}{128}V_I + \frac{3}{128}V_I = 2V_I \quad (8.162)$$

The normalized maximally flat voltage is given by

$$\frac{v_{DS}}{V_I} = 1 - \frac{V_m}{V_I}\cos\omega_o t + \frac{V_{m3}}{V_m}\cos 3\omega_o t - \frac{V_{m5}}{V_m}\cos 5\omega_o t$$
$$= 1 - \frac{75}{64}\cos\omega_o t + \frac{25}{128}\cos 3\omega_o t - \frac{3}{128}\cos 5\omega_o t \quad (8.163)$$

Figure 8.17 shows the maximally flat waveform of the normalized drain-to-source voltage v_{DS}/V_I. It can be seen that the flat part of the drain-to-source voltage waveform for the Class F_{35} amplifier is wider than that of the Class F_3 amplifier. Table 8.2 lists the normalized voltage amplitudes in the odd harmonic maximally flat the Class F_3 and Class F_{35} amplifiers.

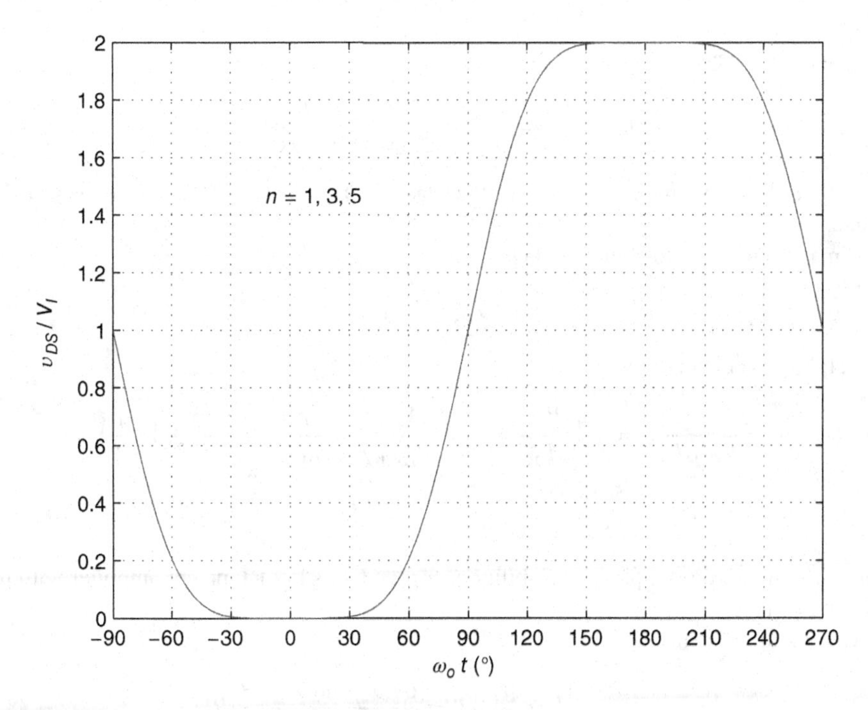

Figure 8.17 Maximally flat waveform of the normalized drain-to-source voltage v_{DS}/V_I for the Class F_{35} RF power amplifier.

Table 8.2 Voltage amplitudes in odd harmonic maximally flat Class F_3 and F_{35} power amplifiers.

Class		$\frac{V_{m3}}{V_I}$	$\frac{V_{m5}}{V_I}$	$\frac{V_{m3}}{V_m}$	$\frac{V_{m5}}{V_m}$
F_3	$\frac{9}{8}$	$\frac{1}{8} = 0.125$	0	$\frac{1}{9} = 0.1111$	0
F_{35}	$\frac{75}{64}$	$\frac{25}{128} = 0.1953$	$\frac{3}{128} = 0.02344$	$\frac{1}{6} = 1.667$	$\frac{1}{50} = 0.02$

The amplitude of the fundamental component of the current is

$$I_m = \frac{V_m}{R} = \frac{75}{64}\frac{V_I}{R} \tag{8.164}$$

The dc input current is

$$I_I = \frac{I_m}{\gamma_1} = \frac{75}{64}\frac{V_I}{\gamma_1 R} \tag{8.165}$$

The dc input power is

$$P_I = I_I V_I = \frac{75}{64}\frac{V_I^2}{\gamma_1 R} \tag{8.166}$$

The output power is

$$P_O = \frac{V_m^2}{2R} = \frac{5625}{8192}\frac{V_I^2}{R} \approx 0.6866\frac{V_I^2}{R} \tag{8.167}$$

The drain efficiency is

$$\eta_D = \frac{P_O}{P_I} = \frac{75\gamma_1}{128} = \frac{75(\theta - \sin\theta\cos\theta)}{128(\sin\theta - \theta\cos\theta)} \tag{8.168}$$

Figure 8.18 shows drain efficiency η_D for the Class F_{35} RF power amplifier with maximally flat voltage.

The maximum drain-to-source voltage is

$$V_{DSM} = 2V_I \tag{8.169}$$

The output-power capability is

$$c_p = \frac{P_O}{V_{DSM}I_{DM}} = \frac{\eta_D P_I}{V_{DSM}I_{DM}} = \eta_D\left(\frac{V_I}{V_{DSM}}\right)\left(\frac{I_I}{I_{DM}}\right) = \left(\frac{75\gamma_1}{128}\right)\left(\frac{1}{2}\right)\alpha_0$$
$$= \frac{75}{256}\alpha_1 = \frac{75(\theta - \sin\theta\cos\theta)}{256\pi(1 - \cos\theta)} \tag{8.170}$$

Figure 8.19 shows output-power capability c_p for the Class F_{35} RF power amplifier with maximally flat voltage.

The dc input resistance is

$$R_{DC} = \frac{V_I}{I_I} = \frac{64\gamma_1}{75}R = \frac{64(\theta - \sin\theta\cos\theta)}{75(\sin\theta - \theta\cos\theta)}R. \tag{8.171}$$

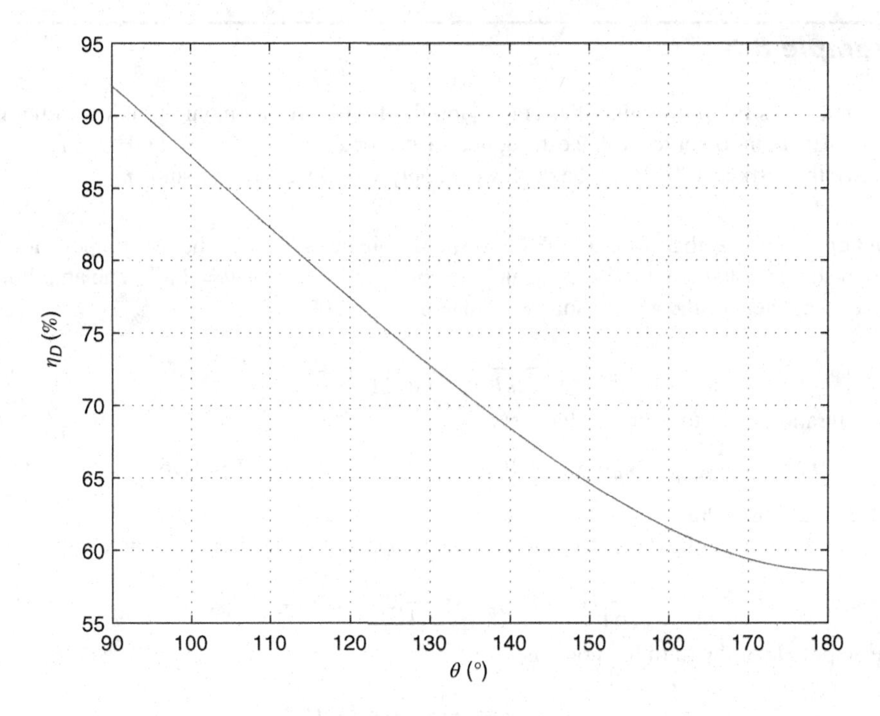

Figure 8.18 Drain efficiency η_D for the Class F_{35} RF power amplifier with maximally flat voltage.

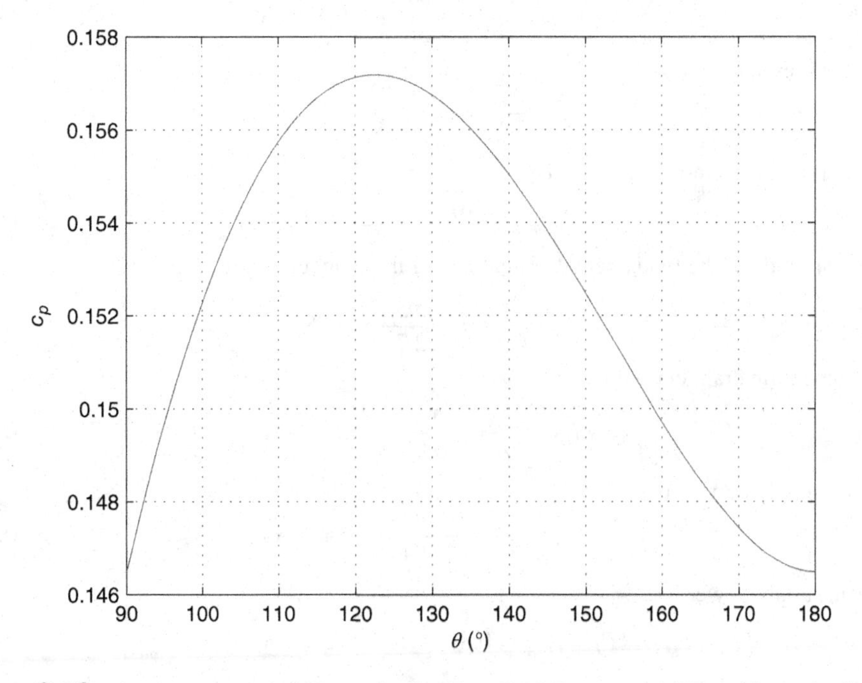

Figure 8.19 Output-power capability c_p for the Class F_{35} RF power amplifier with maximally flat voltage.

Example 8.3

Design a Class F_{35} power amplifier employing third- and fifth-harmonic peaking with a maximally flat drain-to-source voltage to deliver a power of 16 W at $f_c = 1$ GHz. The required bandwidth is $BW = 100$ MHz. The dc power supply voltage is 24 V. Assume $\eta_r = 1$.

Solution. Assume that the MOSFET threshold voltage is $V_t = 1$, the dc component of the gate-to-source voltage is $V_{GS} = 1.5$ V, and the conduction angle is $\theta = 110°$. The amplitude of the ac component of the gate-to-source voltage is

$$V_{gsm} = \frac{V_t - V_{GS}}{\cos\theta} = \frac{1 - 1.5}{\cos 110°} = 1.462 \text{ V} \tag{8.172}$$

The saturation drain-to-source voltage is

$$v_{DSsat} = v_{GSmax} - V_t = V_{GS} + V_{gsm} - V_t = 1.5 + 1.462 - 1 = 1.962 \text{ V} \tag{8.173}$$

Pick the minimum drain-to-source voltage $v_{DSmin} = 2.4$ V.

The maximum amplitude of the fundamental component of the drain-to-source voltage is

$$V_m = \frac{75}{64}(V_I - v_{DSmin}) = \frac{75}{64}(24 - 2.4) = 26.953 \text{ V} \tag{8.174}$$

The amplitude of the third harmonic is

$$V_{m3} = \frac{V_m}{6} = \frac{26.953}{6} = 4.492 \text{ V} \tag{8.175}$$

$$V_{m5} = \frac{V_m}{50} = \frac{26.953}{50} = 0.5391 \text{ V} \tag{8.176}$$

The load resistance is

$$R = \frac{V_m^2}{2P_O} = \frac{26.953^2}{2 \times 16} = 22.7 \ \Omega \tag{8.177}$$

The maximum drain-to-source voltage is

$$v_{DSmax} = 2V_I = 2 \times 24 = 48 \text{ V} \tag{8.178}$$

The amplitude of the fundamental component of the drain current is

$$I_m = \frac{V_m}{R} = \frac{26.953}{22.7} = 1.187 \text{ A} \tag{8.179}$$

The maximum drain current is

$$I_{DM} = \frac{I_m}{\alpha_1} = \frac{1.187}{0.5316} = 2.233 \text{ A} \tag{8.180}$$

The dc supply current is

$$I_I = \frac{I_m}{\gamma_1} = \frac{1.187}{1.404} = 0.8454 \text{ A} \tag{8.181}$$

The dc supply power is

$$P_I = I_I V_I = 0.8454 \times 24 = 20.2896 \text{ W} \tag{8.182}$$

The drain power loss of the transistor is

$$P_D = P_I - P_O = 20.2896 - 16 = 4.2896 \text{ W} \tag{8.183}$$

The drain efficiency is

$$\eta_D = \frac{P_O}{P_I} = \frac{16}{20.2896} = 78.85\% \tag{8.184}$$

The dc resistance presented by the amplifier to the dc source is

$$R_{DC} = \frac{32\pi}{75}R = 1.34 \times 22.7 = 30.418\ \Omega \tag{8.185}$$

The loaded quality factor is

$$Q_L = \frac{f_c}{BW} = \frac{1000}{100} = 10 \tag{8.186}$$

The inductance of the resonant circuit tuned to the fundamental frequency is

$$L = \frac{R}{\omega_c Q_L} = \frac{22.7}{2\pi \times 10^9 \times 10} = 0.36128\ \text{nH} \tag{8.187}$$

The capacitance of the resonant circuit tuned to the fundamental frequency is

$$C = \frac{Q_L}{\omega_c R} = \frac{10}{2\pi \times 10^9 \times 22.7} = 70.11\ \text{pF} \tag{8.188}$$

The resistance connected in parallel with the resonant circuit tuned to the third harmonic is

$$R_3 = \frac{\alpha_1}{9|\alpha_3|}R = \frac{0.5316}{6|-0.0448754|} \times 22.7 = 1.9744 \times 22.7 = 44.819\ \Omega \tag{8.189}$$

The resistance connected in parallel with the resonant circuit tuned to the fifth harmonic is

$$R_5 = \frac{\alpha_1}{9|\alpha_5|}R = \frac{0.5316}{50|0.01855|} \times 22.7 = 0.5732 \times 22.7 = 13.012\ \Omega \tag{8.190}$$

Assuming the loaded quality factor of the resonant circuit tuned to the third harmonic to be $Q_{L3} = 12$, we obtain

$$L_3 = \frac{R_3}{3\omega_c Q_{L3}} = \frac{44.819}{3 \times 2\pi \times 0.8 \times 10^9 \times 12} = 247.68\ \text{pH} \tag{8.191}$$

and

$$C_3 = \frac{Q_{L3}}{3\omega_c R_3} = \frac{12}{3 \times 2\pi \times 0.8 \times 10^9 \times 44.819} = 66.31\ \text{pF} \tag{8.192}$$

Assuming the loaded quality factor of the resonant circuit tuned to the fifth harmonic to be $Q_{L5} = 15$, we obtain

$$L_5 = \frac{R_5}{3\omega_c Q_{L5}} = \frac{13.012}{3 \times 2\pi \times 0.8 \times 10^9 \times 15} = 83.39\ \text{pH} \tag{8.193}$$

and

$$C_5 = \frac{Q_{L5}}{3\omega_c R_5} = \frac{15}{3 \times 2\pi \times 0.8 \times 10^9 \times 13.02} = 76.446\ \text{pF} \tag{8.194}$$

The reactance of the RFC inductance is

$$X_{Lf} = 10R = 10 \times 22.7 = 227\ \Omega \tag{8.195}$$

yielding

$$L_f = \frac{X_{Lf}}{\omega_0} = \frac{227}{2\pi \times 0.8 \times 10^9} = 45.18\ \text{nH} \tag{8.196}$$

The reactance of the coupling capacitor is

$$X_{Cc} = \frac{R}{10} = \frac{22.7}{10} = 2.27\ \Omega \tag{8.197}$$

producing

$$C_c = \frac{1}{\omega_0 X_{Cc}} = \frac{1}{2\pi \times 0.8 \times 10^9 \times 2.27} = 87.68\ \text{pF} \tag{8.198}$$

8.3.2 Maximum Drain Efficiency Class F_{35} Amplifier

The maximum drain efficiency and the output-power capability of the Class F amplifier with the third and fifth harmonics are obtained for [23]

$$\frac{V_{m3}}{V_m} = 0.2323 \tag{8.199}$$

$$\frac{V_{m5}}{V_m} = 0.0607 \tag{8.200}$$

$$\frac{V_{m5}}{V_{m3}} = 0.2613 \tag{8.201}$$

$$\frac{V_m}{V_I} = 1.2071 \tag{8.202}$$

$$\frac{V_{m3}}{V_I} = 0.2804 \tag{8.203}$$

and

$$\frac{V_{m5}}{V_I} = 0.07326 \tag{8.204}$$

The normalized waveform v_{DS}/V_I for the maximum drain efficiency is given by

$$\frac{v_{DS}}{V_I} = 1 - \frac{V_m}{V_I}\cos\omega_o t + \frac{V_{m3}}{V_m}\cos\omega_o t + \frac{V_{m5}}{V_I}\cos 5\omega_o t$$
$$= 1 - 1.2071\cos\omega_o t + 0.2804V_m\cos\omega_o t + 0.07326\cos 5\omega_o t \tag{8.205}$$

The waveform of normalized drain-to-source voltage v_{DS}/V_I in the Class F amplifier with the third and fifth harmonics, yielding the maximum drain efficiency and the maximum output-power capability is shown in Fig. 8.20.

Table 8.3 gives the normalized voltage amplitudes V_{mn}/V_I for the odd harmonic maximum drain efficiency Class F_3 and F_{35} amplifiers.

The amplitude of the fundamental component of the drain current is

$$I_m = \frac{V_m}{R} = \frac{1.207V_I}{R} \tag{8.206}$$

The dc supply current is

$$I_I = \frac{I_m}{\gamma_1} = \frac{1.207}{\gamma_1}\frac{V_I}{R}. \tag{8.207}$$

The dc supply current is

$$P_I = I_I V_I = \frac{1.207}{\gamma_1}\frac{V_I^2}{R} \tag{8.208}$$

The output power is

$$P_O = \frac{V_m^2}{2R} = \frac{1}{2}\frac{(1.207V_I)^2}{R} \approx 0.7284\frac{V_I^2}{R} \tag{8.209}$$

The maximum drain efficiency is

$$\eta_D = \frac{P_O}{P_I} = \frac{1}{2}\left(\frac{I_m}{I_I}\right)\left(\frac{V_m}{V_I}\right) = \frac{1}{2}\gamma_1(1.2071) = 0.6035\gamma_1 = \frac{0.60355(\theta - \sin\theta\cos\theta)}{\sin\theta - \theta\cos\theta} \tag{8.210}$$

Figure 8.21 shows drain efficiency η_D for the Class F_{35} RF power amplifier with third and fifth harmonics, yielding the maximum drain efficiency and output-power capability.

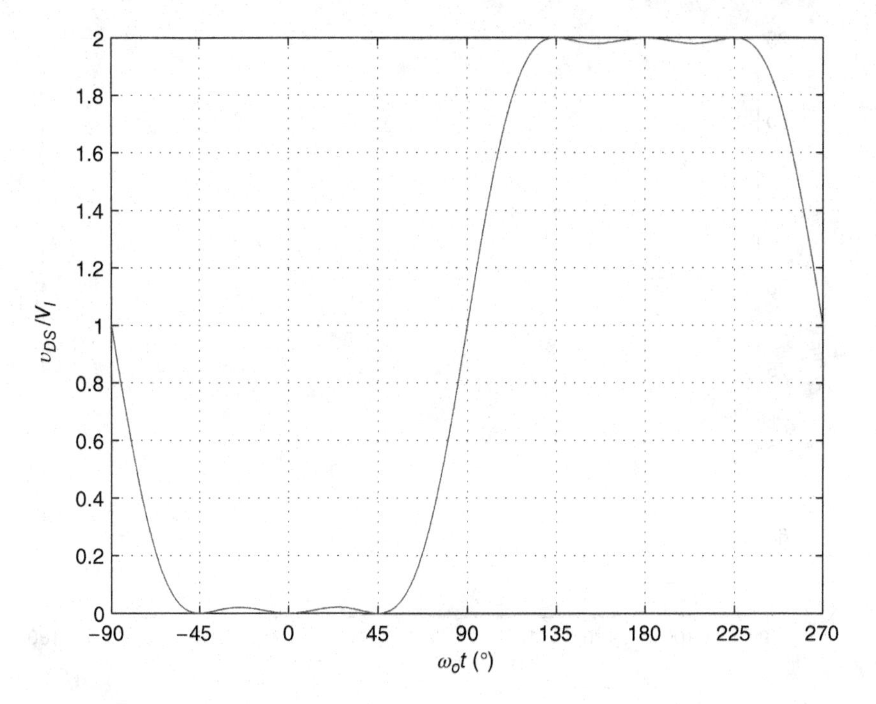

Figure 8.20 Waveform of the normalized drain-to-source voltage υ_{DS}/V_I for the Class F_{35} RF power amplifier with maximum drain efficiency and output-power capability.

Table 8.3 Voltage amplitudes in odd harmonic maximum drain efficiency Class F_3 and F_{35} power amplifiers.

Class	$\frac{V_m}{V_I}$	$\frac{V_{m3}}{V_I}$	$\frac{V_{m5}}{V_I}$	$\frac{V_{m3}}{V_m}$	$\frac{V_{m5}}{V_m}$
F_3	$\frac{3}{\sqrt{3}}$	$\frac{1}{3\sqrt{3}} = 0.1925$	0	$\frac{1}{6} = 0.1667$	0
F_{35}	1.2071	0.2804	0.07326	0.2323	0.0607

The maximum value of υ_{DS} is

$$V_{DSM} = 2V_I = \frac{2}{1.1719}V_m = 1.7066V_m \tag{8.211}$$

The output-power capability is

$$c_p = \frac{P_O}{I_{DM}V_{DSM}} = \frac{1}{2}\left(\frac{I_m}{I_{DM}}\right)\left(\frac{V_m}{V_{DSM}}\right) = \frac{1}{2}\alpha_1\left(\frac{1}{1.7066}\right) = 0.2929\alpha_1$$
$$= \frac{0.2929(\theta - \sin\theta\cos\theta)}{\pi(1 - \cos\theta)} \tag{8.212}$$

The dc input resistance is

$$R_{DC} = \frac{V_I}{I_I} = 0.8284\gamma_1 R = 0.8284\frac{\theta - \sin\theta\cos\theta}{\sin\theta - \theta\cos\theta}R \tag{8.213}$$

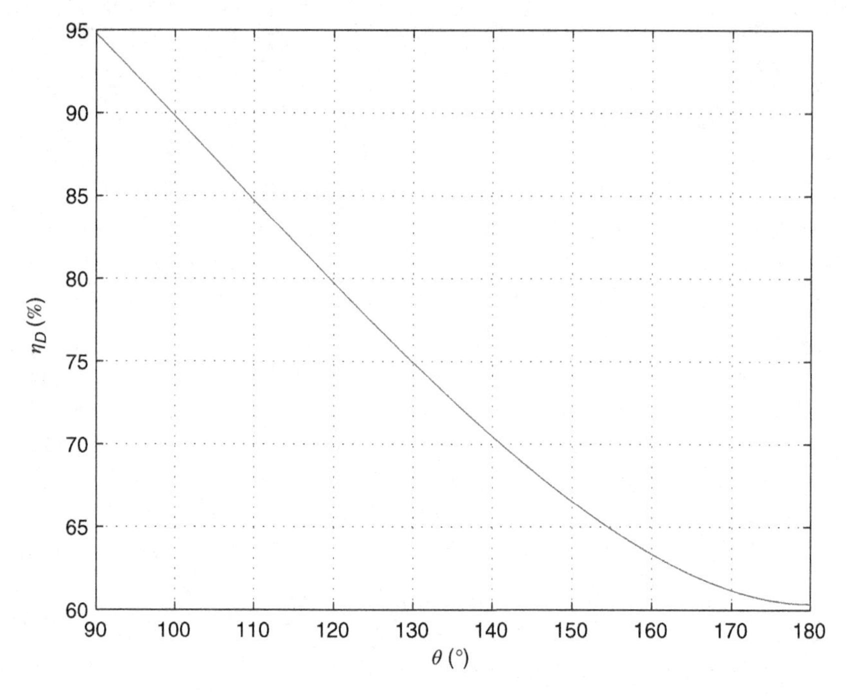

Figure 8.21 Drain efficiency η_D for the Class F_{35} RF power amplifier with third and fifth harmonics, yielding the maximum drain efficiency and output-power capability.

8.4 Class F_{357} RF Power Amplifier with Third, Fifth, and Seventh Harmonics

For the maximally flat Class F_{357} amplifier,

$$V_{m3} = \frac{V_{m1}}{5} \tag{8.214}$$

$$V_{m5} = \frac{V_{m1}}{25} \tag{8.215}$$

$$V_{m7} = \frac{V_{m1}}{245} \tag{8.216}$$

$$V_{m1} = \frac{1225}{1024} V_I \tag{8.217}$$

$$V_{m3} = \frac{245}{2024} V_I \tag{8.218}$$

$$V_{m5} = \frac{49}{1024} V_I \tag{8.219}$$

and

$$V_{m7} = \frac{5}{1024} V_I \tag{8.220}$$

8.5 Class F_T RF Power Amplifier with Parallel-Resonant Circuit and Quarter-Wavelength Transmission Line

Figure 8.22(a) shows the circuit of a Class F power amplifier with a quarter-wavelength transmission line and a parallel-resonant circuit, called the Class F_T amplifier. This circuit is also referred to as *Class F_∞ amplifier*. The input impedance of the quarter-wavelength transmission line is $Z_i = Z_o^2/Z_L$, where Z_o is the characteristic impedance of the transmission line. The quarter-wavelength transmission line is equivalent to an infinite number of parallel-resonant circuits, which behave as an open circuit for odd harmonics and as a short circuit for even harmonics, as shown in Fig. 8.22(b). The characteristic impedance of the transmission line is Z_o. At the fundamental frequency, $R_i = Z_o^2/R_L$, which becomes $R = Z_o^2/R = R^2/R$ at $Z_o = R$. For all harmonics, the load of the transmission line is nearly zero, $Z_L = 1/(n\omega_o C) \approx 0$ for $n \geq 2$. For odd harmonics, $Z_i = Z_o^2/0 = \infty$. For even harmonics, the transmission line acts as a half-wavelength transformer and its input impedance is $Z_i = Z_L = 0$.

The Class F_T amplifier is widely used in VHF and UHF FM radio transmitters [3]. Current and voltage waveforms in the Class F_T amplifier are shown in Fig. 8.23. The drain current waveform is a half-sine wave. The conduction angle of the drain current is $\theta = \pi$. The drain-to-source voltage is a square wave. These current and voltage waveforms constitute the waveforms of an

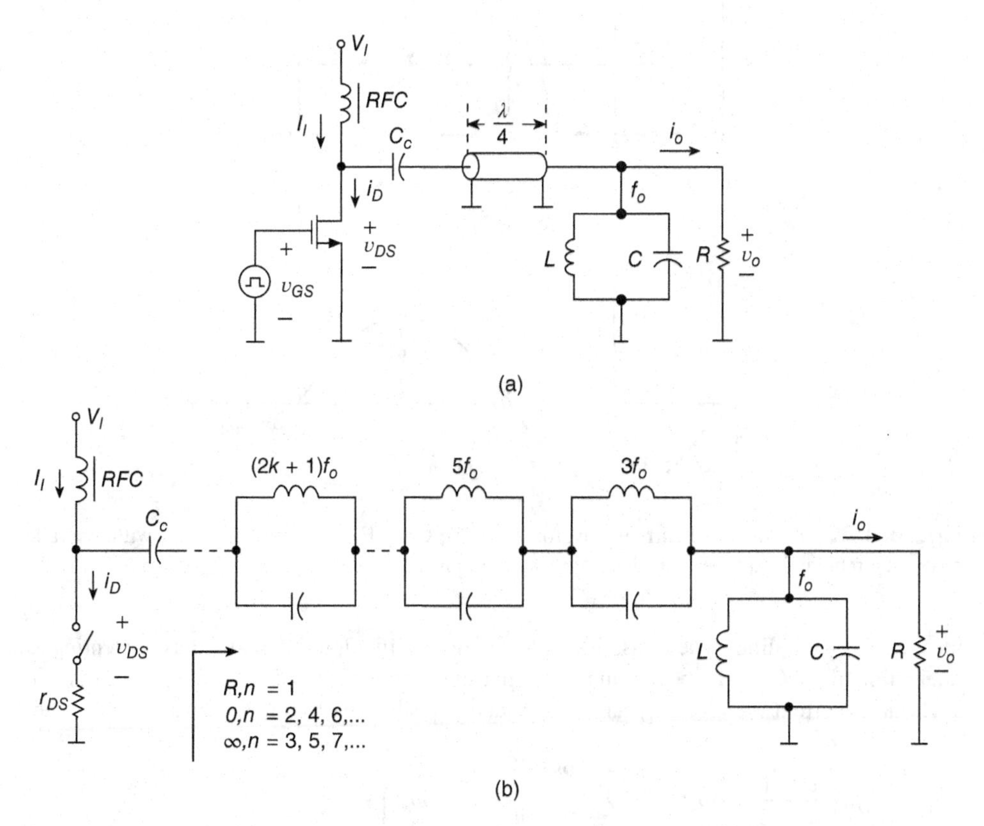

Figure 8.22 Class F_T power amplifier employing a parallel-resonant circuit and a quarter-wavelength transmission line, where $n = (2k + 1)$ with $k = 0, 1, 2, 3, \ldots$: (a) circuit and (b) equivalent circuit at $Z_o = R$.

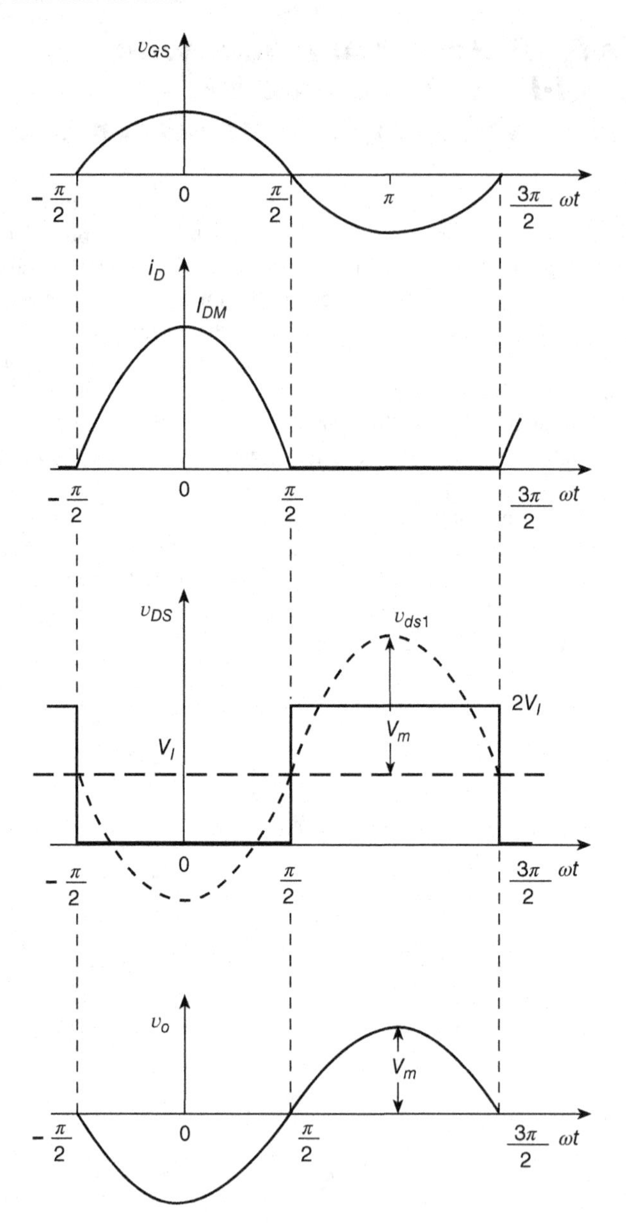

Figure 8.23 Voltage and current waveforms for the Class F_T RF power amplifier with a parallel-resonant circuit and a quarter-wavelength transmission line.

ideal Class F_T amplifier. The transmission line may be difficult to integrate, thus preventing full integration of the Class F_T power amplifier on a single chip. At $f = 2.4$ GHz, $\lambda/4 \approx 2$ cm.

The drain current waveform i_D can be expanded into a Fourier trigonometric series at $\theta = 90°$

$$i_D = I_{DM}\left[\frac{1}{\pi} + \frac{1}{2}\cos\omega_o t + \frac{2}{\pi}\sum_{n=2}^{\infty}\frac{\cos\left(\frac{n\pi}{2}\right)}{1-n^2}\cos n\omega_o t\right]$$
$$= I_{DM}\left(\frac{1}{\pi} + \frac{1}{2}\cos\omega_o t + \frac{2}{3\pi}\cos 2\omega_o t - \frac{2}{15\pi}\cos 4\omega_o t + \frac{2}{35\pi}\cos 6\omega_o t + \cdots\right) \quad (8.221)$$

Thus, the drain current waveform i_D contains only the dc component, fundamental component, and even harmonic components.

The drain-to-source voltage v_{DS} is a square wave, and it can be expanded into a Fourier series at $\theta = 90°$

$$v_{DS} = V_I \left[1 + \frac{4}{\pi}\left(\cos\omega_o t - \frac{1}{3}\cos 3\omega_o t + \frac{1}{5}\cos 5\omega_o t - \frac{1}{7}\cos 7\omega_o t + \cdots\right)\right] \tag{8.222}$$

The drain-to-source voltage v_{DS} contains only the dc component, fundamental component, and odd harmonic components.

The peak value of the drain-to-source voltage is

$$V_{DSM} = 2V_I \tag{8.223}$$

The amplitude of the drain-to-source voltage is

$$V_m = \frac{4}{\pi}(V_I - v_{DSmax}) \tag{8.224}$$

and the amplitude of the drain current is

$$I_m = \frac{I_{DM}}{2} \tag{8.225}$$

The input impedance of the load network seen by the drain and source terminals at the fundamental frequency f_o is

$$Z(f_o) = \frac{V_m}{I_m} = \frac{\frac{4}{\pi}(V_I - v_{DSmin})}{\frac{1}{2}I_{DM}} = \frac{8}{\pi}\frac{(V_I - v_{DSmax})}{I_{DM}} = R_i = \frac{Z_o^2}{R} \tag{8.226}$$

The impedance at even harmonics is

$$Z(nf_o) = Z(2kf_o) = \frac{V_{m(n)}}{I_{m(n)}} = \frac{0}{\textit{finite value}} = 0 \quad \text{for} \quad n = 2, 4, 6, \ldots \tag{8.227}$$

and the impedance at odd harmonics is

$$Z(nf_o) = Z[(2k+1)f_o] = \frac{V_{m(n)}}{I_{m(n)}} = \frac{\textit{finite value}}{0} = \infty \quad \text{for} \quad n = 3, 5, 7, \ldots \tag{8.228}$$

where $k = 1, 2, 3, \ldots$

The input impedance of the transmission line is given by

$$Z_i(l) = Z_o \frac{Z_L + jZ_o \tan\left(\frac{2\pi}{\lambda}l\right)}{Z_o + jZ_L \tan\left(\frac{2\pi}{\lambda}l\right)} \tag{8.229}$$

where l is the length of the transmission line, the wavelength in the transmission line is

$$\lambda = \frac{v_p}{f} = \frac{c}{f\sqrt{\epsilon_r}} \tag{8.230}$$

where $v_p = c/\sqrt{\epsilon_r}$ is the phase velocity, c is the speed of light, ϵ_r is the relative dielectric constant, and λ_o is the wavelength in free space.

The input impedance of the transmission line for $l = \lambda/4$ is

$$Z_i\left(\frac{\lambda}{4}\right) = Z_o \frac{Z_L + jZ_o \tan\left(\frac{2\pi}{\lambda}\frac{\lambda}{4}\right)}{Z_o + jZ_L \tan\left(\frac{2\pi}{\lambda}\frac{\lambda}{4}\right)} = \frac{Z_o^2}{Z_L}\,\frac{\frac{Z_L}{Z_o} + j\tan\left(\frac{\pi}{2}\right)}{\frac{Z_o}{Z_L} + j\tan\left(\frac{\pi}{2}\right)} = \frac{Z_o^2}{Z_L} \tag{8.231}$$

In this case, the transmission line acts as an *impedance inverter*. The input impedance of the transmission line for $l = \lambda/2$ is

$$Z_i\left(\frac{\lambda}{2}\right) = Z_o \frac{Z_L + jZ_o \tan\left(\frac{2\pi}{\lambda}\frac{\lambda}{2}\right)}{Z_o + jZ_L \tan\left(\frac{2\pi}{\lambda}\frac{\lambda}{2}\right)} = Z_L \quad (8.232)$$

The transmission line under this condition acts as an *impedance repeater*. The capacitor in the parallel-resonant circuit represents a short circuit for all harmonics. The quarter-wavelength transmission line converts short circuits at its output to open circuits at its input at odd harmonics, and it converts short circuits to short circuits at even harmonics. The high impedances at odd harmonics shape a square wave of the drain-to-source voltage v_{DS} shown in Fig. 8.23 because this voltage waveform contains only odd harmonics. The low impedances at even harmonics shape a half-sine wave of the drain current i_D since this current waveform contains only even harmonics. Ideally, the voltage and current waveforms in Class F_T and Class D amplifiers are identical.

The fundamental component of the drain-to-source voltage is given by

$$V_m = \frac{4}{\pi}(V_I - v_{DSmin}) \quad (8.233)$$

The fundamental component of the drain current is

$$i_{d1} = I_m \cos \omega_o t \quad (8.234)$$

where

$$I_m = \frac{1}{\pi}\int_{-\frac{\pi}{2}}^{\frac{\pi}{2}} i_D \cos \omega_o t \, d(\omega_o t) = \frac{1}{\pi}\int_{-\frac{\pi}{2}}^{\frac{\pi}{2}} I_{DM}\cos^2 \omega_o t \, d(\omega_o t) = \frac{I_{DM}}{2} \quad (8.235)$$

or

$$I_m = \frac{V_m}{R_i} = \frac{4(V_I - v_{DSmin})}{\pi R_i} \quad (8.236)$$

The dc input current is equal to the dc component of the drain current

$$I_I = \frac{1}{2\pi}\int_{-\frac{\pi}{2}}^{\frac{\pi}{2}} i_D \, d(\omega_o t) = \frac{1}{2\pi}\int_{-\frac{\pi}{2}}^{\frac{\pi}{2}} I_{DM} \cos \omega_o t \, d(\omega_o t) = \frac{I_{DM}}{\pi} = \frac{2}{\pi} I_m = \frac{8(V_I - v_{DSmin})}{\pi^2 R_i} \quad (8.237)$$

The dc input power is given by

$$P_I = I_I V_I = \frac{I_{DM} V_I}{\pi} \quad (8.238)$$

The output power is

$$P_O = \frac{1}{2} I_m V_m = \frac{1}{2} \times \frac{I_{DM}}{2} \times \frac{4}{\pi}(V_I - v_{DSmin}) = \frac{1}{\pi} I_{DM}(V_I - v_{DSmin}) \quad (8.239)$$

or

$$P_O = \frac{V_m^2}{2R_i} = \frac{8(V_I - v_{DSmin})^2}{\pi^2 R_i} \quad (8.240)$$

The drain efficiency of the amplifier is

$$\eta_D = \frac{P_O}{P_I} = 1 - \frac{v_{DSmin}}{V_I} \quad (8.241)$$

The output-power capability is

$$c_p = \frac{P_O}{I_{DM} V_{DSM}} = \frac{1}{2\pi}\left(1 - \frac{v_{DSmin}}{V_I}\right) = 0.159\left(1 - \frac{v_{DSmin}}{V_I}\right) \quad (8.242)$$

The dc resistance is

$$R_{DC} = \frac{V_I}{I_I} = \frac{\pi^2}{8} R_i \tag{8.243}$$

The rms value of the drain current is

$$I_{DSrms} = \sqrt{\frac{1}{2\pi} \int_{\pi/2}^{\pi/2} I_m^2 \cos^2 \omega_o t \, d(\omega_o t)} = \frac{I_m}{2} \tag{8.244}$$

If the transistor is operated as a switch, the conduction loss in the MOSFET on-resistance r_{DS} is given by

$$P_{rDS} = r_{DS} I_{DSrms}^2 = \frac{r_{DS} I_m^2}{4} \tag{8.245}$$

The output power is

$$P_O = \frac{R_i I_m^2}{2} \tag{8.246}$$

Hence, the drain efficiency is

$$\eta_D = \frac{P_O}{P_I} = \frac{P_O}{P_O + P_{rDS}} = \frac{1}{1 + \frac{P_{rDS}}{P_O}} = \frac{1}{1 + \frac{r_{DS}}{2R_i}} = \frac{R_i}{R_i + \frac{r_{DS}}{2}} = 1 - \frac{R_i}{R_i + \frac{r_{DS}}{2}} \tag{8.247}$$

Example 8.4

Design a Class F_T power amplifier with a quarter-wave transmission line to deliver a power of 16 W to the load of $R = 50\ \Omega$ at $f_c = 5$ GHz. The dc power supply voltage is 28 V and $v_{DSmin} = 1$ V. The MOSFET on-resistance is $r_{DS} = 0.2\ \Omega$.

Solution. The maximum amplitude of the fundamental component of the drain-to-source voltage is

$$V_m = \frac{4}{\pi}(V_I - v_{DSmin}) = \frac{4}{\pi}(28 - 1) = 34.377 \text{ V} \tag{8.248}$$

The input resistance into the transmission line is

$$R_i = \frac{V_m^2}{2P_O} = \frac{34.377^2}{2 \times 16} = 36.93\ \Omega \tag{8.249}$$

The maximum drain-to-source voltage is

$$v_{DSmax} = 2V_I = 2 \times 28 = 56 \text{ V} \tag{8.250}$$

The amplitude of the fundamental component of the drain current is

$$I_m = \frac{V_m}{R_i} = \frac{34.377}{36.93} = 0.9308 \text{ A} \tag{8.251}$$

The maximum drain current is

$$I_{DM} = 2I_m = 2 \times 0.9308 = 1.8616 \text{ A} \tag{8.252}$$

The dc supply current is

$$I_I = \frac{I_{DM}}{\pi} = \frac{1.8616}{\pi} = 0.5926 \text{ A} \tag{8.253}$$

The dc supply power is

$$P_I = I_I V_I = 0.5926 \times 28 = 16.5928 \text{ W} \tag{8.254}$$

The drain power loss of the transistor is

$$P_D = P_I - P_O = 16.5928 - 16 = 0.5928 \text{ W} \tag{8.255}$$

The drain efficiency is

$$\eta_D = \frac{P_O}{P_I} = \frac{16}{16.5928} = 96.42\% \tag{8.256}$$

The dc resistance presented by the amplifier to the dc source is

$$R_{DC} = \frac{\pi^2}{8} R_i = 1.2337 \times 36.93 = 45.56\ \Omega \tag{8.257}$$

The characteristic impedance of the transmission line is

$$Z_o = \sqrt{R_i R} = \sqrt{36.93 \times 50} = 42.97\ \Omega \tag{8.258}$$

Assume the dielectric constant used for construction of the transmission line to be $\epsilon_r = 2.1$. The wavelength in the transmission line is

$$\lambda = \frac{c}{\sqrt{\epsilon_r} f_c} = \frac{3 \times 10^8}{\sqrt{2.1} \times 5 \times 10^9} = 4.14\ \text{cm} \tag{8.259}$$

Hence, the length of the transmission line is

$$l_{TL} = \frac{\lambda}{4} = \frac{4.14}{4} = 1.035\ \text{cm} \tag{8.260}$$

Assume that the loaded quality factor $Q_L = 7$. The loaded quality factor for a parallel-resonant circuit is defined as

$$Q_L = \frac{R}{\omega L} = \omega C R_L \tag{8.261}$$

Hence, the inductance of the resonant circuit tuned to the fundamental is

$$L = \frac{R}{\omega_c Q_L} = \frac{50}{2\pi \times 5 \times 10^9 \times 7} = 0.227\ \text{nH} \tag{8.262}$$

The capacitance of the resonant circuit tuned to the fundamental is

$$C = \frac{Q_L}{\omega_c R} = \frac{7}{2\pi \times 5 \times 10^9 \times 50} = 4.456\ \text{pF} \tag{8.263}$$

Neglecting switching losses, the drain efficiency is

$$\eta_D = \frac{R_i}{R_i + \frac{r_{DS}}{2}} = \frac{36.93}{36.93 + \frac{0.2}{2}} = 99.73\% \tag{8.264}$$

8.6 Class F_2 RF Power Amplifier with Second Harmonic

A circuit of the Class F RF power amplifier with a second-harmonic peaking, called the *Class F_2 amplifier*, is shown in Fig. 8.24. Figure 8.25 shows the voltage and current waveforms for this amplifier. The gate-to-source voltage waveform is rectangular.

8.6.1 Fourier Series of Rectangular Drain Current Waveform

The drain current waveform is a rectangular wave given by

$$i_D = \begin{cases} I_{DM} & \text{for} \quad -\theta \le \omega t \le \theta \\ 0 & \text{for} \quad \theta \le \omega t \le 2\pi - \theta \end{cases} \tag{8.265}$$

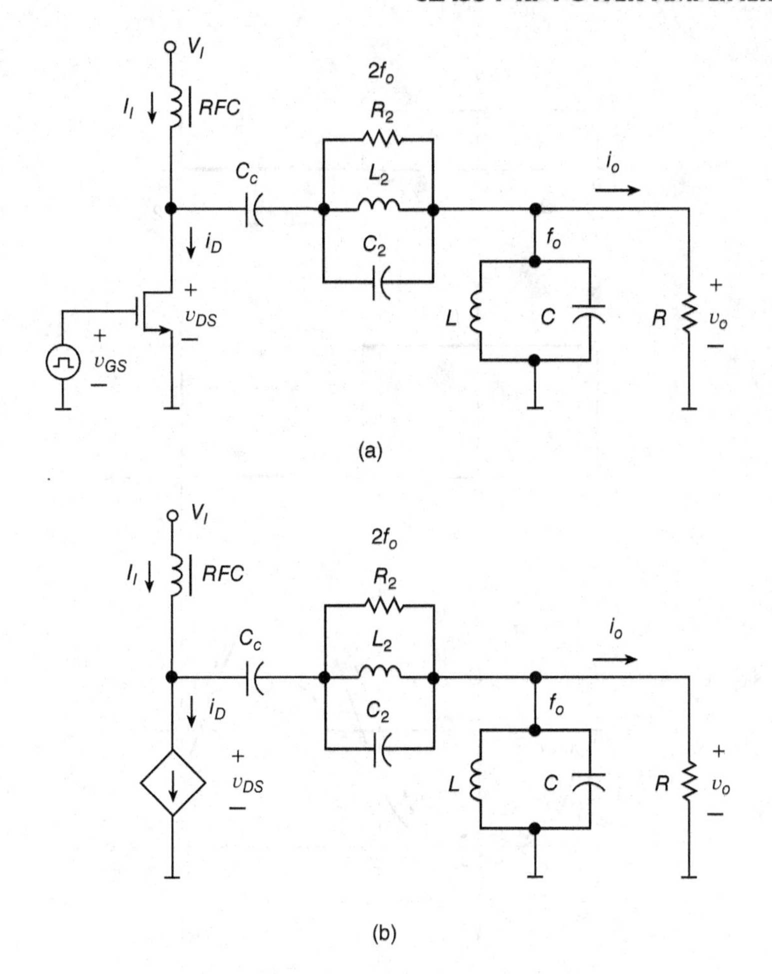

Figure 8.24 Class F_2 power amplifier: (a) circuit and (b) equivalent circuit.

The rectangular drain current can be expanded into Fourier series

$$i_D = I_I + \sum_{n=1}^{n=\infty} I_{mn} \cos n\omega t = I_{DM} \left(\alpha_0 + \sum_{n=1}^{n=\infty} \alpha_n \cos n\omega t \right) \tag{8.266}$$

where

$$\alpha_0 = \frac{I_I}{I_{DM}} = \frac{\theta}{\pi} \tag{8.267}$$

$$\alpha_n = \frac{I_{mn}}{I_{DM}} = \frac{2 \sin n\theta}{n\pi} \tag{8.268}$$

and

$$\gamma_n = \frac{I_{mn}}{I_I} = \frac{\alpha_n}{\alpha_0} = \frac{2 \sin n\theta}{n\theta} \tag{8.269}$$

Figure 8.26 shows Fourier coefficients α_n as functions of conduction angle θ for rectangular drain current. Figure 8.27 shows Fourier coefficients γ_n as functions of conduction angle θ for rectangular drain current waveform. Table 8.4 gives the Fourier coefficients for a rectangular drain current waveform as a function of θ.

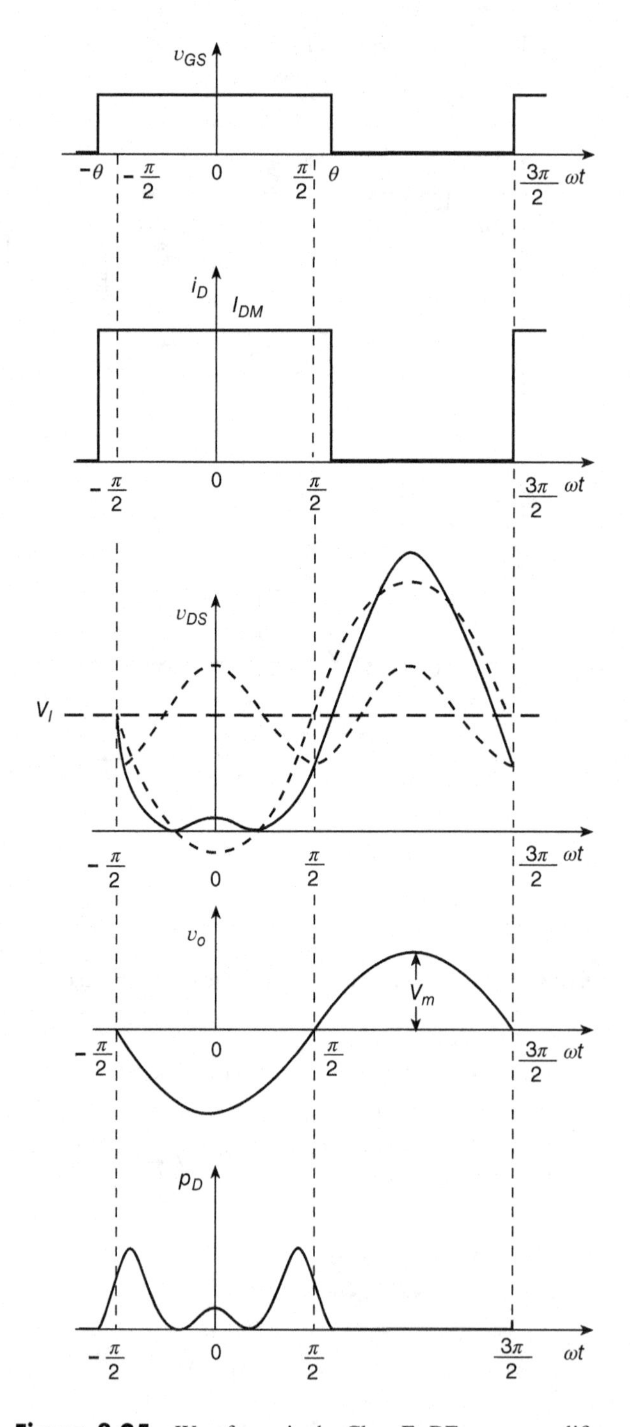

Figure 8.25 Waveforms in the Class F_2 RF power amplifier.

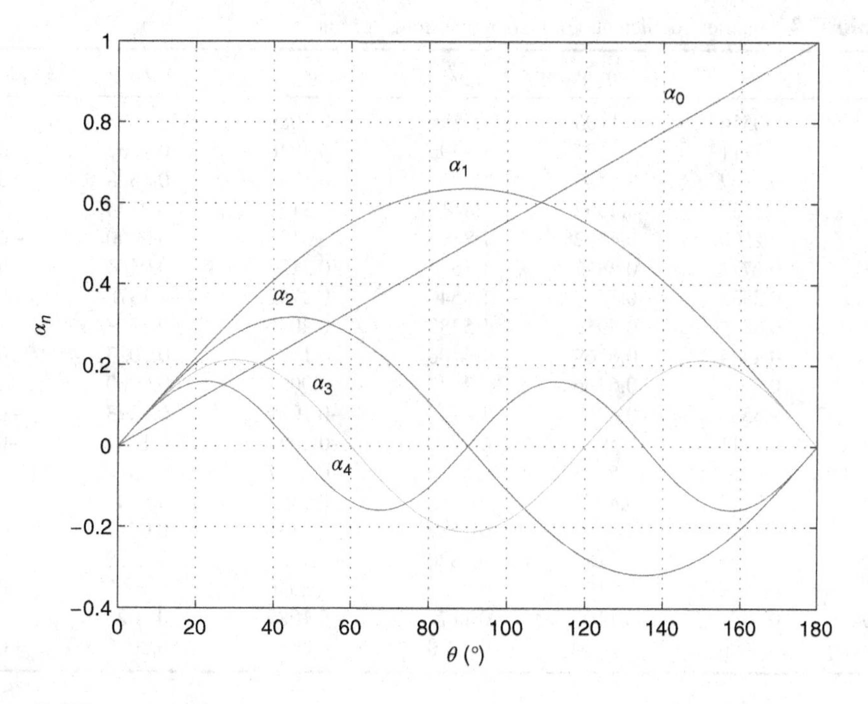

Figure 8.26 Fourier coefficients α_n as functions of conduction angle θ for rectangular drain current.

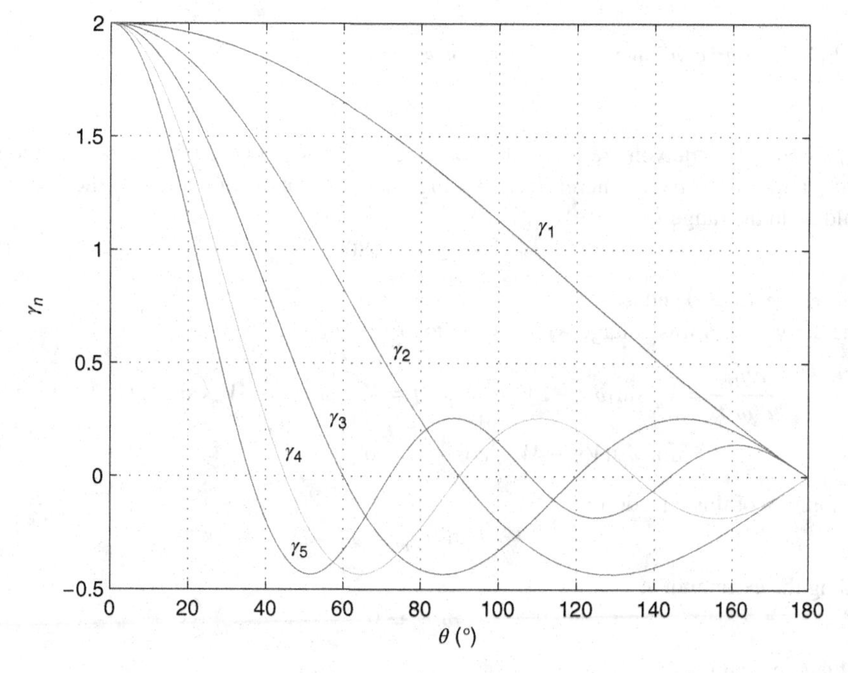

Figure 8.27 Fourier coefficients γ_n as functions of conduction angle θ for rectangular drain current waveform.

Table 8.4 Fourier coefficients for rectangular drain current.

$\theta(°)$	α_0	α_1	γ_1	α_2	α_4	α_6
10	0.0556	0.1105	1.9899	0.1089	0.1023	0.0919
20	0.1111	0.2177	1.9596	0.2046	0.1567	0.0919
30	0.1667	0.3183	1.9099	0.2757	0.1378	0.0000
40	0.2222	0.4092	1.8415	0.3135	0.0544	−0.0919
45	0.2500	0.4502	1.8006	0.3183	0.0000	−0.1061
50	0.2778	0.4877	1.7556	0.3135	−0.0544	−0.0919
60	0.3333	0.5513	1.6540	0.2757	−0.1378	0.0000
70	0.3889	0.5982	1.5383	0.2046	−0.1567	0.0919
80	0.4444	0.6269	1.4106	0.1089	−0.1023	0.0919
90	0.5000	0.6366	1.2732	0.0000	0.0000	0.0000
100	0.5556	0.6269	1.1285	−0.1089	0.1023	−0.0919
110	0.6111	0.5982	0.9789	−0.2046	0.1567	−0.0919
120	0.6667	0.5513	0.8270	−0.2757	0.1378	0.0000
130	0.7222	0.4877	0.6752	−0.3135	0.0544	0.0919
140	0.7778	0.4092	0.5261	−0.3135	−0.0544	0.0919
150	0.8333	0.3183	0.3820	−0.2757	−0.1378	0.0000
160	0.8889	0.2177	0.2450	−0.2046	−0.1567	−0.0919
170	0.9444	0.1105	0.1171	−0.1089	−0.1023	−0.0919
180	1.0000	0.0000	0.0000	0.0000	0.0000	0.0000

8.6.2 Maximally Flat Class F$_2$ Amplifier

The drain-to-source voltage waveform v_{DS} is given by

$$v_{DS} = V_I - V_m \cos \omega_o t + V_{m2} \cos 2\omega_o t \tag{8.270}$$

The second-harmonic voltage waveform must be 180° out of phase with respect to the fundamental frequency voltage. Therefore, using Fig. 8.26, the conduction angle of the drain current should be in the range

$$90° \leq \theta \leq 180° \tag{8.271}$$

In this range, $\alpha_1 > 0$ and $\alpha_2 < 0$.

The derivative of this voltage with respect to $\omega_o t$ is

$$\frac{dv_{DS}}{d(\omega_o t)} = V_m \sin \omega_o t - 2V_{m2} \sin 2\omega_o t = V_m \sin \omega_o t - 4V_{m2} \sin \omega_o t \cos \omega_o t$$

$$= \sin \omega_o t(V_m - 4V_{m2} \cos \omega_o t) = 0 \tag{8.272}$$

One solution of this equation is

$$\sin \omega_o t_m = 0 \tag{8.273}$$

yielding the extremum at

$$\omega_o t_m = 0 \tag{8.274}$$

and the maximum at

$$\omega_o t_m = \pi \tag{8.275}$$

The other solution is

$$\cos \omega_o t_m = \frac{V_m}{4V_{m2}} \leq 1 \tag{8.276}$$

For $0 \le V_{m2}/V_m \le 1/4$, the waveform of v_{DS} has only one minimum value at $\omega_o t = 0$ and only one maximum value at $\omega_o t = \pi$. For $v_{DSmin} = 0$,

$$\frac{V_m}{V_{pk}} = \frac{V_m}{V_I} = \frac{V_m}{V_m - V_{m2}} = \frac{1}{1 - \frac{V_{m2}}{V_m}} \tag{8.277}$$

The waveform of v_{DS} is maximally flat for

$$\frac{V_{m2}}{V_m} = \frac{1}{4} \tag{8.278}$$

producing

$$\omega t_m = \pm \arccos\left(\frac{V_m}{4V_{m2}}\right) \quad \text{for} \quad V_{m2} > \frac{V_m}{4} \tag{8.279}$$

The solutions do not exist for $V_{m2} > V_m/4$.

Using the trigonometric identity $\sin^2\omega_o t + \cos^2\omega_o t = 1$, we get

$$\sin^2\omega_o t = 1 - \cos^2\omega_o t = 1 - \frac{V_m^2}{16V_{m2}^2} \tag{8.280}$$

Hence,

$$\sin\omega_o t_m = \pm\sqrt{1 - \frac{V_m^2}{16V_{m2}^2}} = \pm\sqrt{\frac{16V_{m2}^2 - V_m^2}{16V_{m2}^2}} \quad \text{for} \quad V_{m2} \ge \frac{V_m}{4} \tag{8.281}$$

For $V_{m2} < V_m/4$, the real solutions of this equation do not exist. For $V_{m2} > V_m/4$, the waveform of v_{DS} has two maxima at

$$\omega_o t_m = \pm \arcsin\sqrt{1 - \frac{V_m^2}{16V_{m2}^2}} \tag{8.282}$$

The derivatives of the voltage v_{DS} are

$$\frac{dv_{DS}}{d(\omega_o t)} = V_m \sin\omega_o t - 2V_{m2}\sin 2\omega_o t \tag{8.283}$$

and

$$\frac{d^2 v_{DS}}{d(\omega_o t)^2} = V_m \cos\omega_o t - 4V_{m2}\cos 2\omega_o t \tag{8.284}$$

Both terms of the first-order derivative at $\omega_o t = 0$ are zero. Using the second derivative, we get

$$\left.\frac{d^2 v_{DS}}{d(\omega_o t)^2}\right|_{\omega_o t = 0} = V_m - 4V_{m2} = 0 \tag{8.285}$$

yielding

$$V_{m2} = \frac{V_m}{4} \tag{8.286}$$

and

$$\frac{V_{m2}}{V_m} = \frac{I_{m2}R_2}{I_m R} = \frac{|\alpha_2| I_{DM} R_2}{\alpha_1 I_{DM} R} = \frac{|\alpha_2| R_2}{\alpha_1 R} = \frac{R_2}{R}\cos\theta = \frac{1}{4} \tag{8.287}$$

yielding the ratio of the resistances

$$\frac{R_2}{R} = \frac{\alpha_1}{|\alpha_2|} = \frac{\cos\theta}{4} \tag{8.288}$$

Next,

$$v_{DS}(0) = V_{m2} = V_I - V_m + V_{m2} = V_I - V_m + \frac{V_m}{4} = V_I - \frac{3}{4}V_m = 0 \tag{8.289}$$

producing

$$V_m = \frac{4}{3} V_I \tag{8.290}$$

Similarly,

$$v_{DS}(0) = V_{m2} = V_I - V_m + V_{m2} = V_I - 4V_{m2} + V_{m2} = V_I - 3V_{m2} = 0 \tag{8.291}$$

which gives

$$V_{m2} = \frac{V_I}{3} \tag{8.292}$$

The maximum value of the drain-to-source voltage is

$$V_{DSM} = v_{DS}(\pi) = V_I + V_m + V_{m2} = V_I + \frac{4}{3}V_I + \frac{1}{3}V_I = \frac{8}{3}V_I. \tag{8.293}$$

The normalized maximally flat waveform v_{DS}/V_I is given by

$$\frac{v_{DS}}{V_I} = 1 - \frac{V_m}{V_I}\cos\omega_o t + \frac{V_{m2}}{V_m}\cos 2\omega_o t = 1 - \frac{4}{3}\cos\omega_o t + \frac{1}{3}\cos 2\omega_o t \tag{8.294}$$

Figure 8.28 depicts the maximally flat waveform of the normalized drain-to-source voltage v_{DS}/V_I for a Class F_2 amplifier.

The amplitude of the drain current is

$$I_m = \frac{V_m}{R} = \frac{4}{3}\frac{V_I}{R} \tag{8.295}$$

or

$$I_m = \alpha_1 I_{DM} \tag{8.296}$$

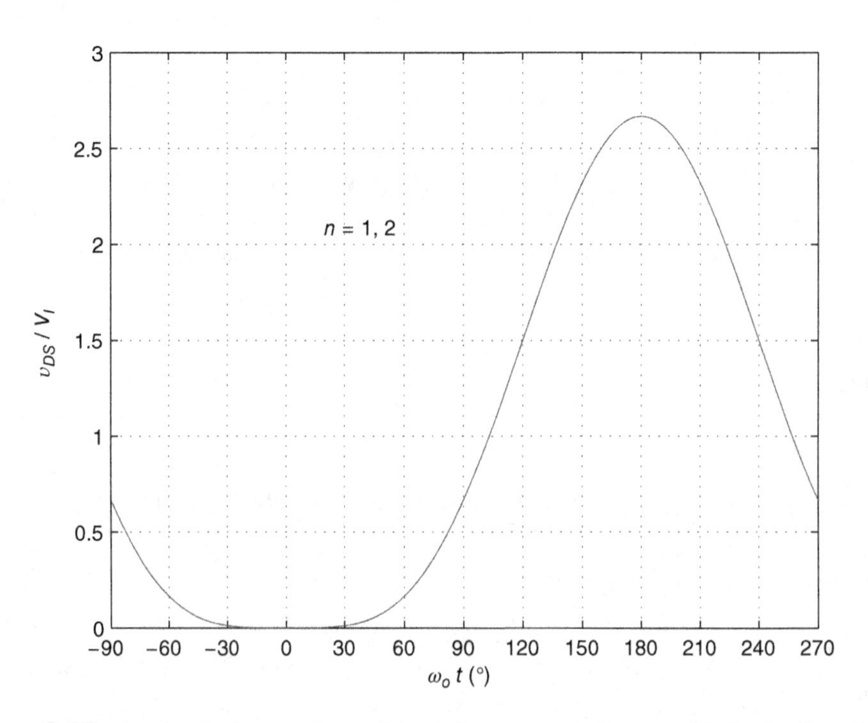

Figure 8.28 Maximally flat waveform of the drain-to-source voltage v_{DS}/V_I for the Class F_2 RF power amplifier.

Hence, the maximum drain current is

$$I_{DM} = \frac{I_m}{\alpha_1} = \frac{V_m}{\alpha_1 R} = \frac{4}{3}\frac{V_I}{\alpha_1 R} \tag{8.297}$$

The dc input current is

$$I_I = \frac{I_m}{\gamma_1} = \frac{4}{3}\frac{V_I}{\gamma_1 R} \tag{8.298}$$

The dc input power is

$$P_I = I_I V_I = \frac{4}{3}\frac{V_I^2}{\gamma_1 R} \tag{8.299}$$

The output power is

$$P_O = \frac{V_m^2}{2R} = \frac{8}{9}\frac{V_I^2}{R} \tag{8.300}$$

The drain efficiency is

$$\eta_D = \frac{P_O}{P_I} = \frac{1}{2}\left(\frac{I_m}{I_I}\right)\left(\frac{V_m}{V_I}\right) = \frac{1}{2} \times \gamma_1 \times \frac{4}{3} = \frac{2}{3}\gamma_1 = \frac{4\sin\theta}{3\theta} \tag{8.301}$$

Figure 8.29 shows drain efficiency η_D for the Class F_2 RF power amplifier with the second-harmonic peaking with maximally flat voltage.

The output-power capability is

$$c_p = \frac{P_O}{I_{DM}V_{DSM}} = \frac{\eta_D P_I}{I_{DM}V_{DSM}} = \eta_D\left(\frac{I_I}{I_{DM}}\right)\left(\frac{V_I}{V_{DSM}}\right) = \left(\frac{2\gamma_1}{3}\right)(\alpha_0)\left(\frac{3}{8}\right) = \frac{\alpha_1}{4}$$
$$= \frac{\sin\theta}{2\pi} \tag{8.302}$$

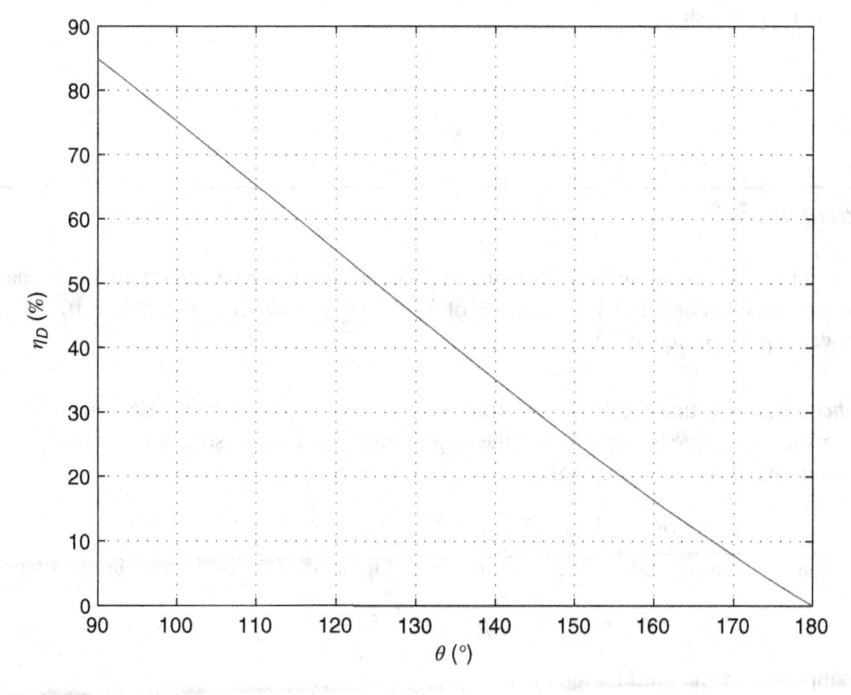

Figure 8.29 Drain efficiency η_D for the Class F_2 RF power amplifier with the second-harmonic peaking with maximally flat voltage.

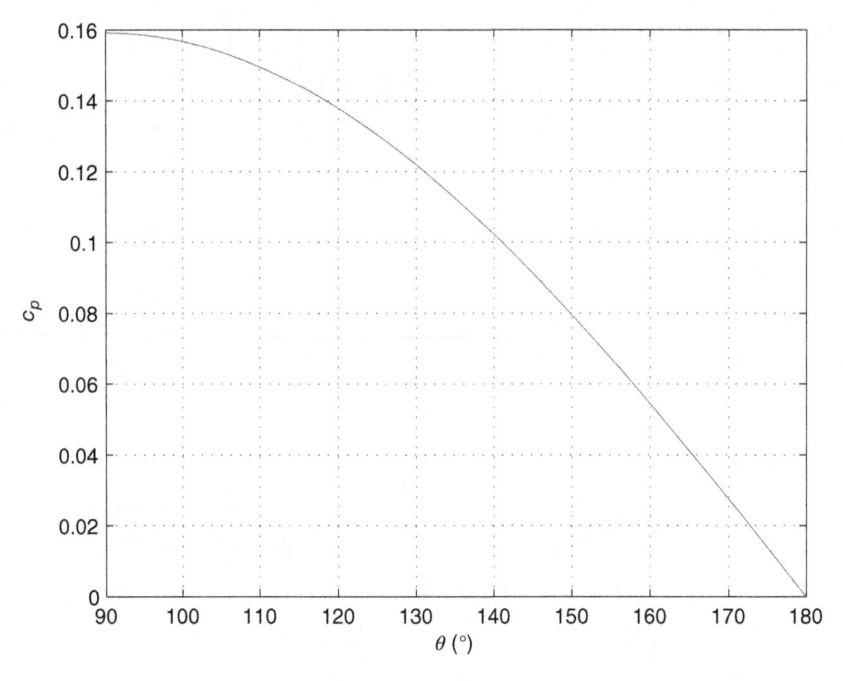

Figure 8.30 Output-power capability c_p for the Class F_2 RF power amplifier with maximally flat voltage.

Figure 8.30 shows output-power capability c_p for the Class F_2 RF power amplifier with the second-harmonic peaking with maximally flat voltage.

The dc input resistance is

$$R_{DC} = \frac{V_I}{I_I} = \frac{3\gamma_1}{4}R = \frac{3\sin\theta}{2\theta}R \tag{8.303}$$

Example 8.5

Design a Class F_2 power amplifier employing second-harmonic peaking with a maximally flat drain-to-source voltage to deliver a power of 25 W at $f_c = 900$ MHz with $BW = 100$ MHz. The dc power supply voltage is 32 V.

Solution. Assume that the MOSFET threshold voltage is $V_t = 1$ V, the dc component of the gate-to-source voltage is $V_{GS} = 0$, and the magnitude of the gate-to-source voltage is $V_{gsm} = 2$ V. The saturation drain-to-source voltage is

$$v_{DSsat} = v_{DSmin} = v_{GS} - V_t = V_{gsm} - V_t = 2 - 1 = 1 \text{ V} \tag{8.304}$$

The maximum amplitude of the fundamental component of the drain-to-source voltage is

$$V_m = \frac{4}{3}(V_I - v_{DSmin}) = \frac{4}{3}(32 - 1) = 41.333 \text{ V} \tag{8.305}$$

The amplitude of the third harmonic is

$$V_{m2} = \frac{V_m}{4} = \frac{41.333}{4} = 10.332 \text{ V} \tag{8.306}$$

The load resistance is

$$R = \frac{V_m^2}{2P_O} = \frac{41.333^2}{2 \times 25} = 34.168\ \Omega \tag{8.307}$$

The maximum drain-to-source voltage is

$$v_{DSmax} = \frac{8V_I}{3} = \frac{8 \times 32}{3} = 85.333\ \text{V} \tag{8.308}$$

The amplitude of the fundamental component of the drain current is

$$I_m = \frac{V_m}{R} = \frac{41.333}{34.168} = 1.2097\ \text{A} \tag{8.309}$$

Assuming the conduction angle of the drain current $\theta = 120°$, the maximum drain current is

$$I_{DM} = \frac{I_m}{\alpha_1} = \frac{1.2097}{0.5513} = 0.6669\ \text{A} \tag{8.310}$$

The dc supply current is

$$I_I = \alpha_0 I_{DM} = 0.6667 \times 0.6669 = 0.4446\ \text{A} \tag{8.311}$$

The dc supply power is

$$P_I = I_I V_I = 0.4446 \times 32 = 30.4\ \text{W} \tag{8.312}$$

The drain power loss of the transistor is

$$P_D = P_I - P_O = 30.4 - 25 = 5.4\ \text{W} \tag{8.313}$$

The drain efficiency is

$$\eta_D = \frac{P_O}{P_I} = \frac{25}{30.4} = 82.23\% \tag{8.314}$$

The dc resistance presented by the amplifier to the dc source is

$$R_{DC} = \frac{V_I}{I_I} = \frac{32}{0.4446} = 71.974\ \Omega \tag{8.315}$$

The loaded quality factor of the resonant circuit tuned to the fundamental is

$$Q_L = \frac{f_c}{BW} = \frac{900}{100} = 9 \tag{8.316}$$

Hence, the components of this circuit are

$$L = \frac{R}{\omega_c Q_L} = \frac{34.168}{2\pi \times 0.9 \times 10^9 \times 9} = 0.671\ \text{nH} \tag{8.317}$$

and

$$C = \frac{Q_L}{\omega_c R} = \frac{9}{2\pi \times 0.9 \times 10^9 \times 34.168} = 46.58\ \text{pF} \tag{8.318}$$

The resistance R_2 is

$$R_2 = \frac{\alpha_1}{4|\alpha_2|} R = \frac{R}{4|\cos\theta|} = \frac{R}{4|\cos 120°|} = \frac{R}{2} = \frac{34.168}{2} = 17.084\ \Omega \tag{8.319}$$

Assuming the loaded quality factor $Q_{L2} = 30$, the components of the resonant circuit tuned to the second harmonic are

$$L_2 = \frac{R}{2\omega_c Q_{L2}} = \frac{17.084}{2 \times 2\pi \times 0.9 \times 10^9 \times 30} = 50.37\ \text{pH} \tag{8.320}$$

and

$$C_2 = \frac{Q_L}{2\omega_c R} = \frac{30}{2 \times 2\pi \times 0.9 \times 10^9 \times 17.084} = 155.34\ \text{pF} \tag{8.321}$$

The reactance of the RFC inductance is

$$X_{Lf} = 10R = 10 \times 34.168 = 341.68\ \Omega \tag{8.322}$$

yielding

$$L_f = \frac{X_{Lf}}{\omega_0} = \frac{341.68}{2\pi \times 0.9 \times 10^9} = 60.45\ \text{nH} \tag{8.323}$$

The reactance of the coupling capacitor is

$$X_{Cc} = \frac{R}{10} = \frac{34.168}{10} = 3.4168\ \Omega \tag{8.324}$$

producing

$$C_c = \frac{1}{\omega_0 X_{Cc}} = \frac{1}{2\pi \times 0.9 \times 10^9 \times 3.4168} = 56.22\ \text{pF} \tag{8.325}$$

8.6.3 Maximum Drain Efficiency Class F_2 Amplifier

The optimum amplitudes of the fundamental component and the second harmonic for achieving the maximum drain efficiency for the Class F amplifier with the second harmonic are derived later. The minimum value of the drain-to-source voltage with ripple is given by

$$v_{DS}(\omega_o t_m) = V_I - V_m \cos\omega_o t_m + V_{m2}\cos 2\omega_o t_m = 0 \tag{8.326}$$

Using (8.276),

$$\begin{aligned} V_I &= V_m \cos\omega_o t_m - V_{m2}\cos 2\omega_o t_m = V_m\cos\omega_o t_m - V_{m2}(2\cos^2\omega_o t_m - 1) \\ &= V_m \times \frac{V_m}{4V_{m2}} - V_{m2}\left(2 \times \frac{V_m^2}{16V_{m2}^2} - 1\right) = \frac{V_m^2}{8V_{m2}} + V_{m2} \end{aligned} \tag{8.327}$$

producing

$$\frac{V_m}{V_I} = \frac{1}{\dfrac{V_{m2}}{V_m} + \dfrac{V_m}{8V_{m2}}} \tag{8.328}$$

Figure 8.31 shows a plot of V_m/V_I as function of V_{m2}/V_m.

To maximize the ratio V_m/V_I, we take the derivative and set it equal to zero

$$\frac{d\left(\dfrac{V_m}{V_I}\right)}{d\left(\dfrac{V_{m2}}{V_m}\right)} = \frac{-1 + \dfrac{1}{8}\left(\dfrac{V_m}{V_{m2}}\right)^2}{\left(\dfrac{V_{m2}}{V_m} + \dfrac{V_m}{8V_{m2}}\right)^2} = 0 \tag{8.329}$$

yielding the optimum ratio of the amplitudes of the second harmonic and the fundamental component

$$\frac{V_{m2}}{V_m} = \frac{1}{2\sqrt{2}} \approx 0.3536 \tag{8.330}$$

Hence,

$$\frac{V_{m2}}{V_m} = \frac{I_{m2}R_2}{I_m R} = \frac{|\alpha_2|R_2}{\alpha_1 R} = \frac{R_2}{R}\cos\theta = \frac{1}{2\sqrt{2}} \tag{8.331}$$

yielding the normalized resistance

$$\frac{R_2}{R} = 2\sqrt{2}/\cos\theta \tag{8.332}$$

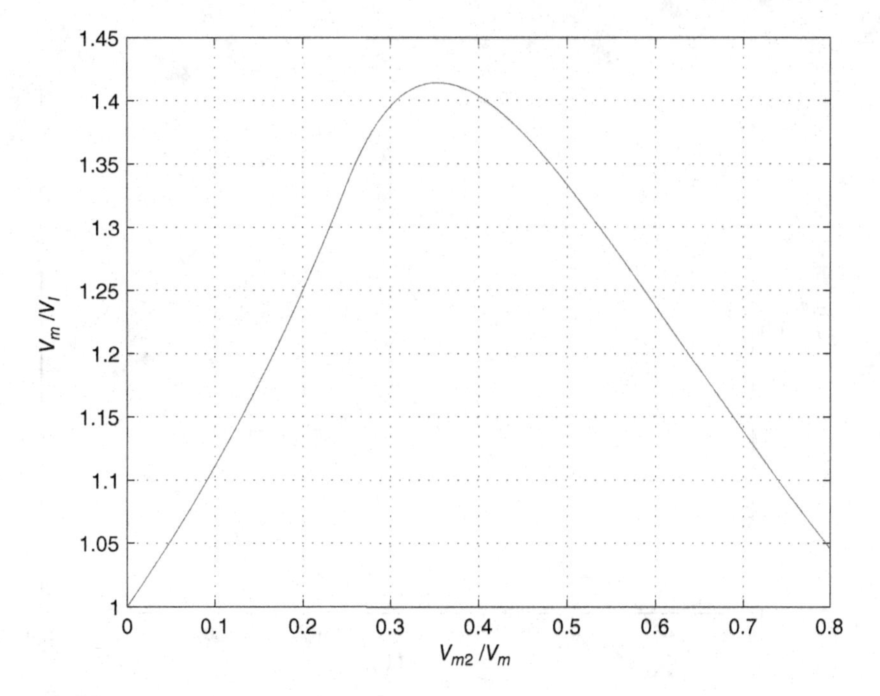

Figure 8.31 Ratio V_m/V_I as a function of V_{m2}/V_m for the Class F_2 power amplifier with second harmonic.

The dc supply voltage V_I can be expressed in terms of V_m to give

$$V_I = V_m \cos \omega_o t_m - V_{m2}(2\cos^2 \omega_o t_m - 1) = \frac{V_m}{\sqrt{2}} - 2\sqrt{2}\left(2 \times \frac{1}{2} - 1\right) = \frac{V_m}{\sqrt{2}} \tag{8.333}$$

producing

$$\frac{V_m}{V_I} = \sqrt{2} \tag{8.334}$$

Finally,

$$V_{m2} = \frac{V_m}{2\sqrt{2}} = \frac{V_I}{2} \tag{8.335}$$

yielding

$$\frac{V_{m2}}{V_I} = \frac{1}{2} \tag{8.336}$$

When V_I reaches its minimum value, the ratio V_m/V_I reaches the maximum value, and therefore the drain efficiency also reaches the maximum value. The maximum drain-to-source voltage is

$$V_{DSM} = v_{DS}(\pi) = V_I + V_m + V_{m2} = V_I + \sqrt{2}V_I + 0.5V_I = (1.5 + \sqrt{2})V_I \approx 2.914V_I \tag{8.337}$$

The two minimum values of v_{DS} for the maximum efficiency are located at

$$\omega_o t_m = \pm \arcsin \sqrt{1 - \frac{1}{16}\left(\frac{V_m}{V_{m3}}\right)^2} = \pm \arcsin \sqrt{1 - \frac{1}{16}\left(2\sqrt{2}\right)^2} = \pm \arcsin \frac{1}{\sqrt{2}} = \pm 45^\circ \tag{8.338}$$

The normalized waveform v_{DS}/V_I at the maximum drain efficiency is given by

$$\frac{v_{DS}}{V_I} = 1 - \frac{V_m}{V_I} \cos \omega_o t + \frac{V_{m2}}{V_I} \cos 2\omega_o t = 1 - \sqrt{2} \cos \omega_o t + \frac{1}{2} \cos 2\, \omega_o t \tag{8.339}$$

and is illustrated in Fig. 8.32.

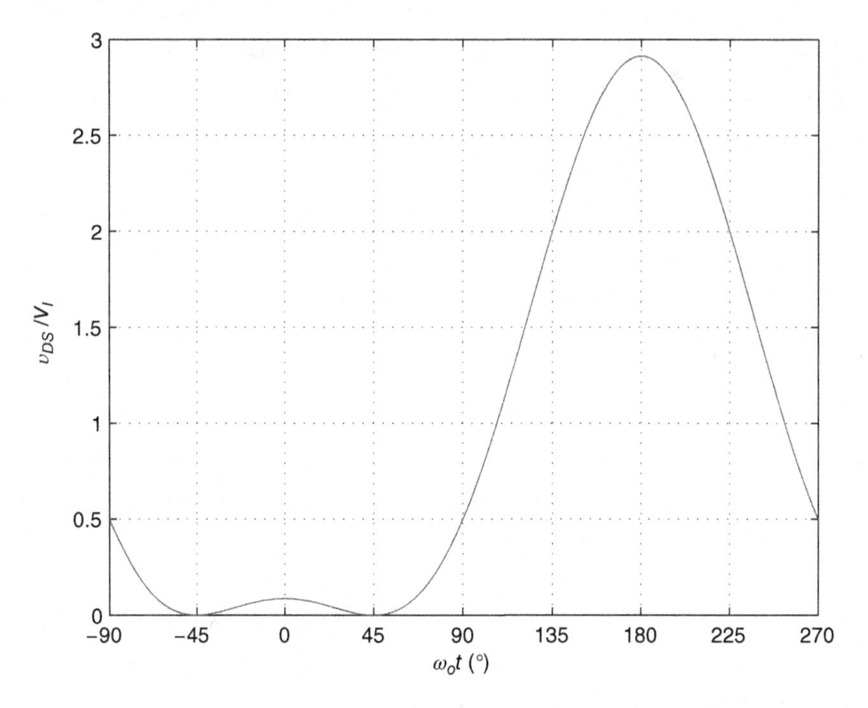

Figure 8.32 Waveform of the drain-to-source voltage v_{DS}/V_I for the Class F_2 RF power amplifier with the second-harmonic peaking at the maximum drain efficiency.

The amplitude of the fundamental component of the drain current is

$$I_m = \frac{V_m}{R} = \frac{\sqrt{2}V_I}{R} \tag{8.340}$$

The dc supply current is

$$I_I = \frac{I_m}{\gamma_1} = \frac{\sqrt{2}V_I}{\gamma_1 R} \tag{8.341}$$

The dc supply power is

$$P_I = I_I V_I = \frac{\sqrt{2}V_I^2}{\gamma_1 R} \tag{8.342}$$

The output power is

$$P_O = \frac{V_m^2}{2R} = \frac{(\sqrt{2}V_I)^2}{2R} = \frac{V_I^2}{R} \tag{8.343}$$

The drain efficiency is

$$\eta_D = \frac{P_O}{P_I} = \frac{1}{2}\left(\frac{I_m}{I_I}\right)\left(\frac{V_m}{V_I}\right) = \frac{1}{2} \times \gamma_1 \times \sqrt{2} = \frac{\gamma_1}{\sqrt{2}} = \frac{\sqrt{2}\sin\theta}{\theta} \tag{8.344}$$

Figure 8.33 shows drain efficiency η_D for the Class F_2 RF power amplifier with the second-harmonic peaking with the maximum drain efficiency.

The ratio of the maximum drain efficiency $\eta_{Dmax(F2)}$ to the drain efficiency with maximally flat voltage is

$$\frac{\eta_{Dmax(F2)}}{\eta_{D(F2)}} = \frac{3\sqrt{2}}{4} \approx 1.0607 \tag{8.345}$$

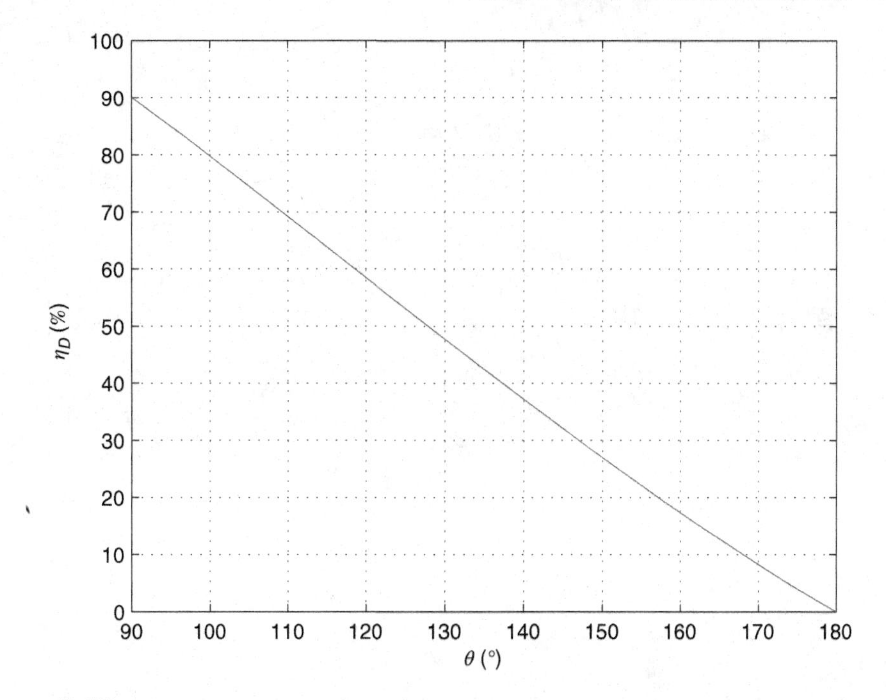

Figure 8.33 Drain efficiency η_D for the Class F_2 RF power amplifier with the second-harmonic peaking with the maximum drain efficiency.

The drain efficiency of the Class B amplifier with a square-wave drain current and a sinusoidal drain-to-source voltage is given by

$$\eta_B = \frac{P_O}{P_I} = \frac{2}{\pi} \tag{8.346}$$

The drain efficiency of the Class AB amplifier with a square-wave drain current and a sinusoidal drain-to-source voltage is given by

$$\eta_{AB} = \frac{P_O}{P_I} = \frac{1}{2}\gamma_1 \xi_{1(AB)} \tag{8.347}$$

The drain efficiency of the Class F_2 amplifier with the second harmonic is expressed as

$$\eta_D = \frac{P_O}{P_I} = \frac{1}{2}\gamma_1 \xi_{1(F_2)} \tag{8.348}$$

The ratio of the efficiencies is

$$\frac{\eta_{D(F_2)}}{\eta_{D(AB)}} = \xi_{1(F_2)} = \frac{V_m}{V_I} = \frac{1}{1 - \frac{V_{m2}}{V_m}} \quad \text{for} \quad 0 \leq \frac{V_{m2}}{V_m} \leq \frac{1}{4} \tag{8.349}$$

and

$$\frac{\eta_{D(F_2)}}{\eta_{D(AB)}} = \frac{1}{\frac{V_{m2}}{V_m} + \frac{V_m}{8V_{m2}}} \quad \text{for} \quad \frac{V_{m2}}{V_m} \geq \frac{1}{4} \tag{8.350}$$

Figure 8.34 shows ratio $\eta_{D(F_2)}/\eta_{D(AB)}$ for the Class F_2 RF power amplifier with the second-harmonic peaking with the maximum drain efficiency.

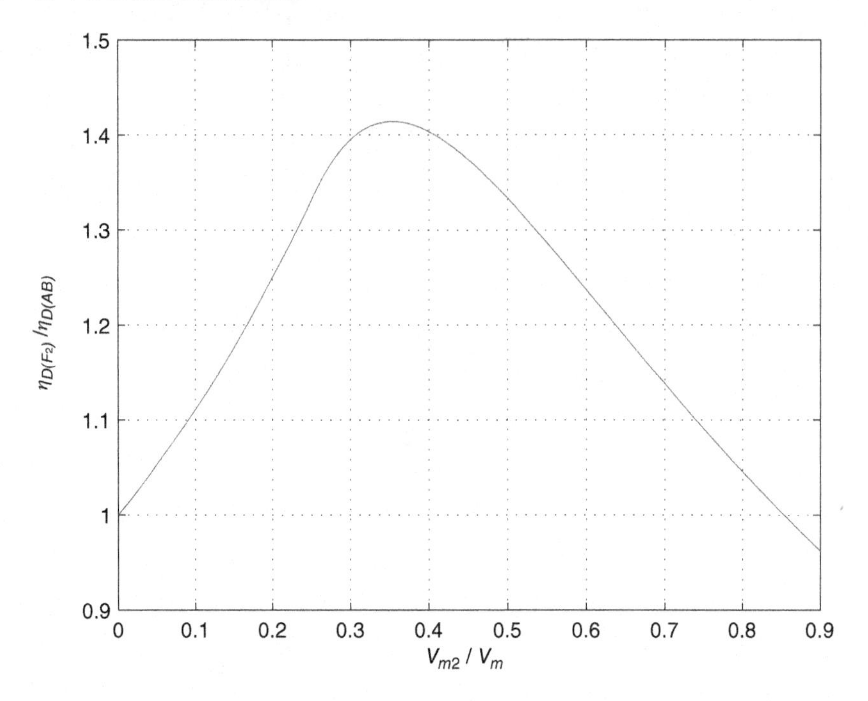

Figure 8.34 Ratio $\eta_{D(F_2)}/\eta_{D(AB)}$ for the Class F_2 RF power amplifier with the maximum drain efficiency.

The maximum output-power capability at any value of V_{m2}/V_m is given by

$$c_p = \frac{P_{O(max)}}{I_{DM}V_{DSM}} = \frac{1}{2}\left(\frac{I_m}{I_{DM}}\right)\left(\frac{V_m}{V_{DSM}}\right) = \frac{1}{2}(\alpha_1)\left(\frac{\sqrt{2}}{(\sqrt{2}+1.5)}\right) = \frac{\sqrt{2}}{\sqrt{2}+1.5}\frac{\sin\theta}{\pi} \quad (8.351)$$

Figure 8.35 shows output-power capability c_p as a function of θ for Class F_2 power amplifier with second harmonic and maximum drain efficiency.

The dc input resistance is

$$R_{DC} = \frac{V_I}{I_I} = \frac{\gamma_1}{\sqrt{2}}R = \frac{\theta - \sin\theta\cos\theta}{\sqrt{2}(\sin\theta - \theta\cos\theta)}R \quad (8.352)$$

8.7 Class F_{24} RF Power Amplifier with Second and Fourth Harmonics

8.7.1 Maximally Flat Class F_{24} Amplifier

Figure 8.36 shows the circuit of a Class F RF power amplifier with the second and fourth harmonics. This circuit is called the Class F_{24} amplifier. The voltage and current waveforms are depicted in Fig. 8.37. The drain-to-source voltage is

$$v_{DS} = V_I - V_m\cos\omega_o t + V_{m2}\cos 2\omega_o t - V_{m4}\cos 4\omega_o t \quad (8.353)$$

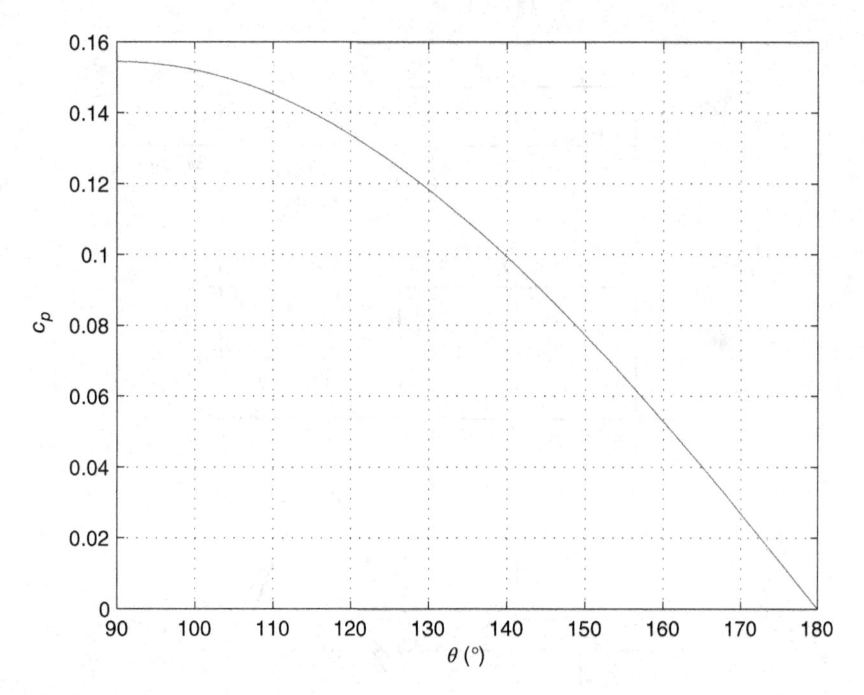

Figure 8.35 Output-power capability c_p as a function of θ for the Class F power amplifier with second harmonic and maximum drain efficiency.

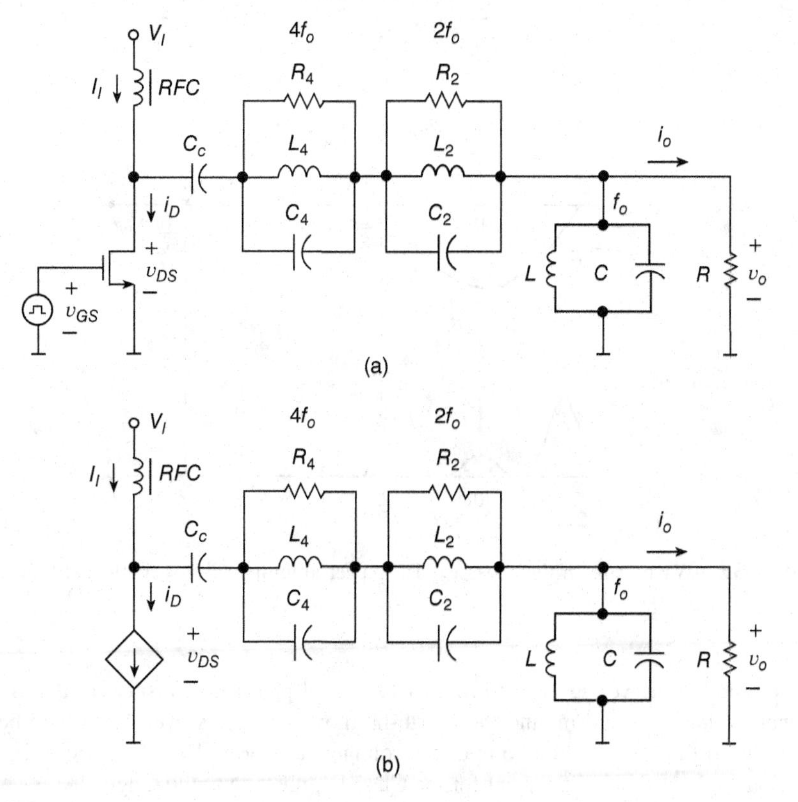

Figure 8.36 Class F_{24} power amplifier with second and fourth harmonics: (a) circuit and (b) equivalent circuit.

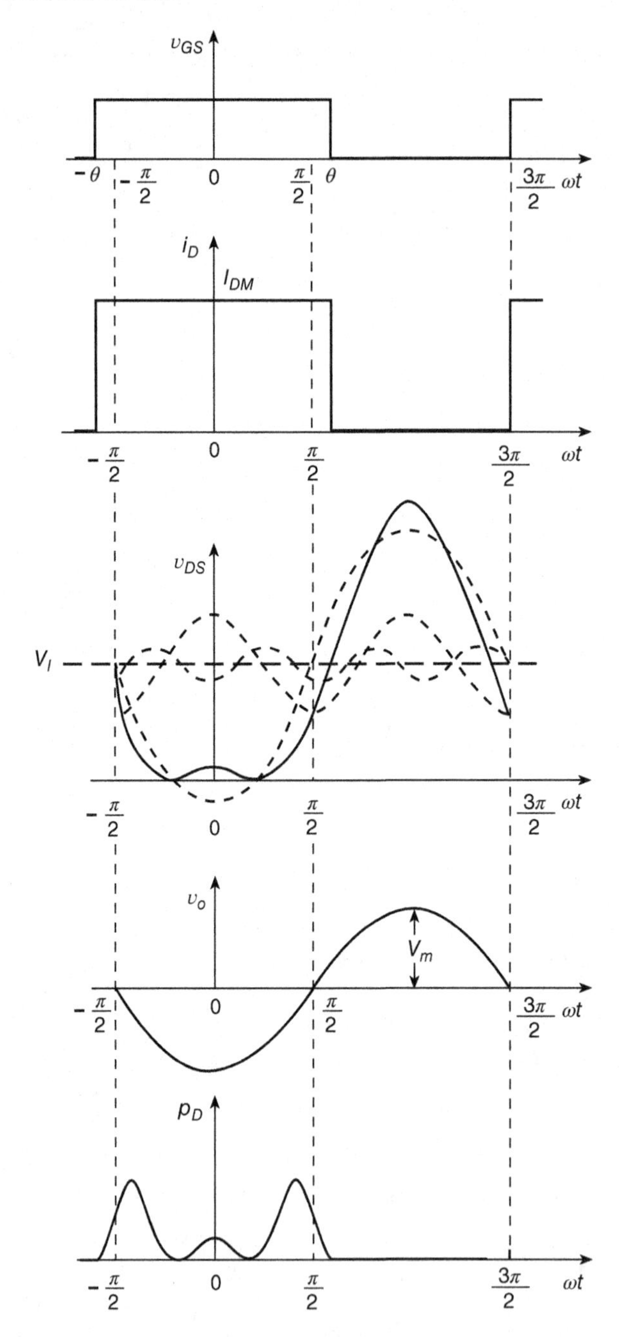

Figure 8.37 Waveforms in the Class F_{24} RF power amplifier with second- and fourth-harmonic peaking.

The second-harmonic voltage waveform is 180° out of phase with respect to the fundamental component voltage waveform, and the fourth-harmonic voltage waveform should be in phase with respect to the fundamental component voltage waveform. Therefore, from Fig. 8.26 the conduction angle of the drain current for the Class F_{24} must be in the range

$$90^\circ \leq \theta \leq 135^\circ \tag{8.354}$$

The derivatives of this voltage are

$$\frac{dv_{DS}}{d(\omega_o t)} = V_m \sin \omega_o t - 2V_{m2} \sin 2\omega_o t + 4V_{m4} \sin 4\omega_o t \tag{8.355}$$

$$\frac{d^2 v_{DS}}{d(\omega_o t)^2} = V_m \cos \omega_o t - 4V_{m2} \cos 2\omega_o t + 16V_{m4} \cos 4\omega_o t \tag{8.356}$$

$$\frac{d^3 v_{DS}}{d(\omega_o t)^3} = -V_m \sin \omega_o t + 8V_{m2} \sin 2\omega_o t - 64V_{m4} \sin 4\omega_o t \tag{8.357}$$

and

$$\frac{d^4 v_{DS}}{d(\omega_o t)^4} = -V_m \cos \omega_o t + 16V_{m2} \cos 2\omega_o t - 256V_{m4} \cos 4\omega_o t \tag{8.358}$$

The first-order and third-order derivatives are zero at $\omega_o t = 0$ and do not generate any equations. Using the second-order and forth-order derivatives, we obtain

$$\left.\frac{d^2 v_{DS}}{d(\omega_o t)^2}\right|_{\omega_o t=0} = V_m - 4V_{m2} + 16V_{m4} = 0 \tag{8.359}$$

and

$$\left.\frac{d^4 v_{DS}}{d(\omega_o t)^4}\right|_{\omega_o t=0} = V_m - 16V_{m2} + 256V_{m4} = 0 \tag{8.360}$$

Solution for this set of equations yields

$$V_{m2} = \frac{5}{16}V_m \tag{8.361}$$

$$V_{m4} = \frac{1}{64}V_m \tag{8.362}$$

and

$$V_{m2} = 20V_{m4} \tag{8.363}$$

Thus,

$$\frac{V_{m2}}{V_m} = \frac{I_{m2}R_2}{I_m R} = \frac{|\alpha_2|R_2}{\alpha_1 R} = \frac{R_2}{R}\cos\theta = \frac{5}{16} \tag{8.364}$$

and

$$\frac{V_{m4}}{V_m} = \frac{I_{m4}R_4}{I_m R} = \frac{|\alpha_4|R_2}{\alpha_1 R} = \frac{R_4}{R}\cos\theta\cos 2\theta = \frac{1}{64} \tag{8.365}$$

Hence, the ratios of resistances are

$$\frac{R_2}{R} = \frac{5\alpha_1}{16|\alpha_2|} = \frac{5}{16\cos\theta} \tag{8.366}$$

and

$$\frac{R_4}{R} = \frac{\alpha_1}{64\alpha_4} = \frac{1}{64\cos\theta\cos 2\theta} \tag{8.367}$$

The following equation holds true:

$$v_{DS}(0) = V_I - V_m + V_{m2} - V_{m4} = V_I - V_m + \frac{5}{16}V_m - \frac{1}{64}V_m = V_I - \frac{45}{64}V_m = 0 \tag{8.368}$$

producing

$$V_m = \frac{64}{45}V_I \tag{8.369}$$

Now

$$v_{DS}(0) = V_I - V_m + V_{m2} - V_{m4} = V_I - \frac{16}{5}V_{m2} + V_{m2} - \frac{1}{20}V_{m2} = V_I - \frac{9}{4}V_{m2} = 0 \tag{8.370}$$

yielding

$$V_{m2} = \frac{4}{9}V_I \tag{8.371}$$

Finally,

$$v_{DS}(0) = V_I - V_m + V_{m2} - V_{m4} = V_I - 64V_{m4} + 20V_{m4} - V_{m4} = 0 \tag{8.372}$$

giving

$$V_{m4} = \frac{1}{45}V_I \tag{8.373}$$

The maximum drain-to-source voltage is

$$V_{DSM} = v_{DS}(\pi) = V_I + V_m + V_{m2} - V_{m4} = V_I + \frac{64}{45}V_I + \frac{4}{9}V_I - \frac{1}{45}V_I = \frac{128}{45}V_I \approx 2.844V_I \tag{8.374}$$

The normalized maximally flat waveform v_{DS}/V_I is given by

$$\frac{v_{DS}}{V_I} = 1 - \frac{V_m}{V_I}\cos\omega_o t + \frac{V_{m2}}{V_I}\cos 2\omega_o t - \frac{V_{m4}}{V_I}\cos 4\omega_o t$$

$$= 1 - \frac{64}{45}\cos\omega_o t + \frac{4}{9}\cos 2\omega_o t - \frac{1}{45}\cos 4\omega_o t \tag{8.375}$$

and is shown in Fig. 8.38. Table 8.5 gives the normalized voltage amplitudes in the even harmonic maximally flat Class F_2 and Class F_{24} amplifiers.

The amplitude of the fundamental component of the drain current is

$$I_m = \frac{V_m}{R} = \frac{64}{45}\frac{V_I}{R} \tag{8.376}$$

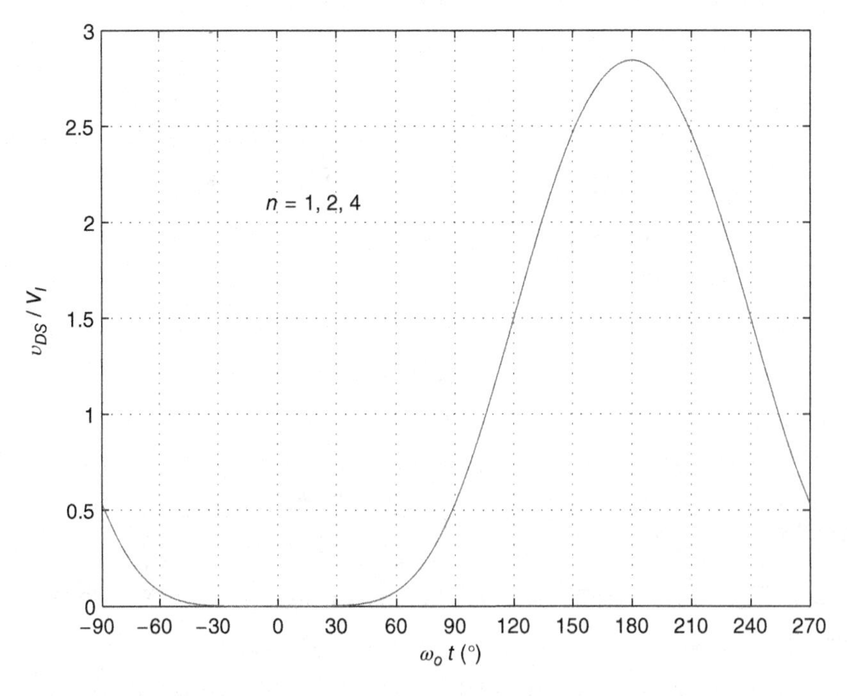

Figure 8.38 Maximally flat waveform of the drain-to-source voltage v_{DS}/V_I for the Class F_{24} RF power amplifier with the second- and fourth-harmonic peaking.

Table 8.5 Voltage amplitudes in even harmonic maximally flat Class F_2 and F_{24} power amplifiers.

Class	$\frac{V_m}{V_I}$	$\frac{V_{m2}}{V_I}$	$\frac{V_{m4}}{V_I}$	$\frac{V_{m2}}{V_m}$	$\frac{V_{m4}}{V_m}$
F_2	$\frac{4}{3}$	$\frac{1}{3} = 0.333$	0	$\frac{1}{4} = 0.25$	0
F_{24}	$\frac{64}{45}$	$\frac{4}{9} = 0.4444$	$\frac{1}{45} = 0.02222$	$\frac{5}{16} = 0.3125$	$\frac{1}{64} = 0.01563$

or

$$I_m = \alpha_1 I_{DM} \tag{8.377}$$

Thus,

$$I_{DM} = \frac{I_m}{\alpha_1} = \frac{64}{45}\frac{V_I}{\alpha_1 R} \tag{8.378}$$

The dc input current is

$$I_I = \alpha_0 I_{DM} = \frac{I_m}{\gamma_1} = \frac{64}{45}\frac{V_I}{\gamma_1 R} \tag{8.379}$$

The dc input power is

$$P_I = I_I V_I = \frac{64}{45}\frac{V_I^2}{\gamma_1 R} \tag{8.380}$$

The output power is

$$P_O = \frac{V_m^2}{2R} = \frac{2048}{2025}\frac{V_I^2}{R} \tag{8.381}$$

The drain efficiency of the amplifier is

$$\eta_D = \frac{P_O}{P_I} = \frac{32\gamma_1}{45} = \frac{64 \sin\theta}{45\theta} \tag{8.382}$$

Figure 8.39 shows drain efficiency η_D for the Class F_{24} RF power amplifier with the second- and fourth-harmonic peaking.

The maximum drain-to-source voltage is

$$V_{DSM} = v_{DS}(\pi)V_I + V_m + V_{m2} - V_{m4} = V_I\left(1 + \frac{64}{45} + \frac{4}{9} - \frac{1}{45}\right) = \frac{128}{45}V_I \approx 2.844V_I \tag{8.383}$$

The output-power capability is

$$c_p = \frac{P_O}{V_{DSM}I_{DM}} = \frac{\eta_D P_I}{V_{DSM}I_{DM}} = \eta_D\left(\frac{I_I}{I_{DM}}\right)\left(\frac{V_I}{V_{DSM}}\right) = \left(\frac{32\gamma_1}{45}\right)(\alpha_0)\left(\frac{45}{128}\right)$$
$$= \frac{\alpha_1}{4} = \frac{\sin\theta}{2\pi} \tag{8.384}$$

Figure 8.40 shows output-power capability c_p for the Class F_{24} RF power amplifier with the second- and fourth-harmonic peaking.

The dc input resistance is

$$R_{DC} = \frac{V_I}{I_I} = \frac{45\gamma_1}{64}R = \frac{45\sin\theta}{32\theta}R \tag{8.385}$$

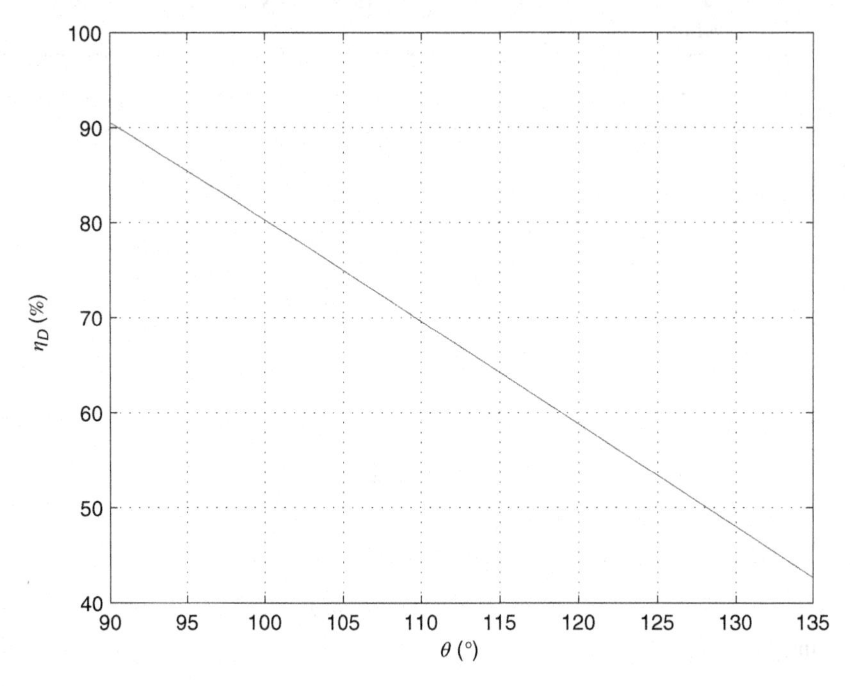

Figure 8.39 Drain efficiency η_D for the Class F_{24} RF power amplifier with the second- and fourth-harmonic peaking with maximally flat voltage.

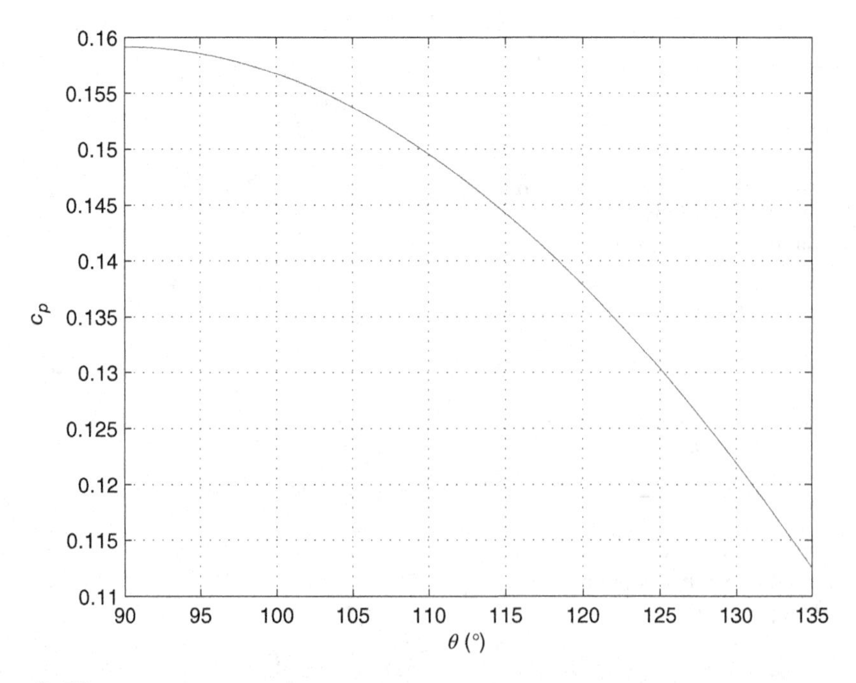

Figure 8.40 Output-power capability c_p for the Class F_{24} RF power amplifier with the second- and fourth-harmonic peaking with maximally flat voltage.

8.7.2 Maximum Drain Efficiency Class F_{24} Amplifier

To find the extrema of the drain-to-source voltage v_{DS}, we take the derivative of this voltage and set the result equal to zero

$$\begin{aligned}\frac{dv_{DS}}{d(\omega t)} &= V_m \sin\omega_o t - 2V_{m2}\sin 2\omega_o t + 4V_{m4}\sin 4\omega_o t\\ &= V_m \sin\omega_o t - 4V_{m2}\sin\omega_o t\cos\omega_o t + 8V_{m4}\sin 2\omega_o t\cos 2\omega t\\ &= V_m \sin\omega_o t - 4V_{m2}\sin\omega_o t\cos\omega_o t + 16V_{m4}\sin\omega_o t\cos\omega_o t\cos 2\omega t\\ &= \sin\omega_o t(V_m - 4V_{m2}\cos\omega_o t + 16V_{m4}\cos\omega_o t\cos 2\omega_o t) = 0\\ &= \sin\omega_o t[V_m - 4V_{m2}\cos\omega_o t + 16V_{m4}\cos\omega_o t(2\cos^2\omega_o t - 1)] = 0\end{aligned} \tag{8.386}$$

Hence, one solution is

$$\sin\omega_o t_m = 0 \tag{8.387}$$

which gives

$$\omega_o t_m = 0 \tag{8.388}$$

and

$$\omega_o t_m = \pi \tag{8.389}$$

The second equation is

$$32V_{m4}\cos 3\omega_o t - (4V_{m2} + 16V_{m4})\cos\omega_o t + V_m = 0 \tag{8.390}$$

which can be rearranged to the form

$$32\frac{V_{m4}}{V_m}\cos^3\omega_o t - \left(4\frac{V_{m2}}{V_m} + 16\frac{V_{m4}}{V_m}\right)\cos\omega_o t + 1 = 0 \tag{8.391}$$

This equation can only be solved numerically. The Fourier coefficients of the drain-to-source voltage waveform v_{DS} for achieving the maximum drain efficiency are [23]

$$\frac{V_m}{V_I} = 1.5 \tag{8.392}$$

$$\frac{V_{m2}}{V_I} = 0.5835 \tag{8.393}$$

and

$$\frac{V_{m4}}{V_I} = 0.0834 \tag{8.394}$$

$$\frac{V_{m2}}{V_m} = 0.389 \tag{8.395}$$

and

$$\frac{V_{m4}}{V_m} = 0.0556 \tag{8.396}$$

Table 8.5 gives the normalized voltage amplitudes V_{mn}/V_I and V_{mn}/V_m in the even harmonic maximum drain efficiency Class F_2 and F_{24} amplifiers. The normalized waveform v_{DS}/V_I is given by

$$\begin{aligned}\frac{v_{DS}}{V_I} &= 1 - \frac{V_m}{V_I}\cos\omega_o t + \frac{V_{m2}}{V_I}\cos 2\omega_o t - \frac{V_{m4}}{V_I}\cos 4\omega_o t\\ &= 1 - 1.5\cos\omega_o t + 0.5835\cos 2\omega_o t - 0.0834\cos 4\omega_o t\end{aligned} \tag{8.397}$$

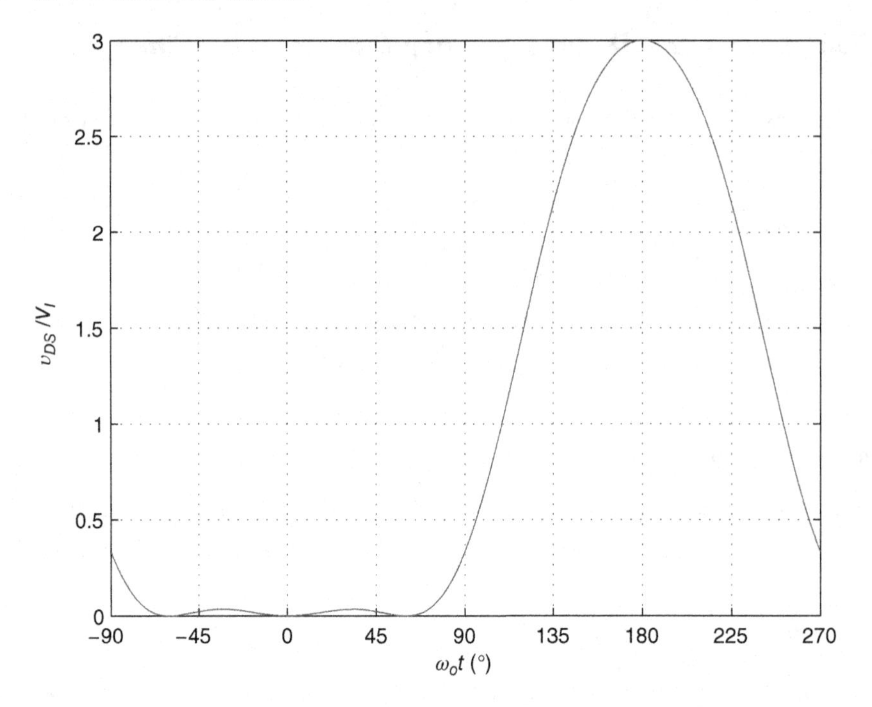

Figure 8.41 Waveform of the normalized drain-to-source voltage v_{DS}/V_I for the Class F_{24} RF power amplifier with the second- and fourth-harmonic peaking for the maximum drain efficiency.

The waveform of v_{DS}/V_I for the maximum drain efficiency is shown in Fig. 8.41. Table 8.6 gives the normalized voltage amplitudes V_{mn}/V_I for the even harmonic maximally flat Class F_2 and F_{24} power amplifiers.

The amplitude of the fundamental component of the drain current is

$$I_m = \frac{V_m}{R} = \frac{1.5V_I}{R} \tag{8.398}$$

The maximum drain current is

$$I_{DM} = \frac{I_m}{\alpha_1} = 1.5 \times \frac{V_I}{\alpha_1 R} \tag{8.399}$$

The dc supply current is

$$I_I = \frac{I_m}{\gamma_1} = 1.5 \times \frac{V_I}{\gamma_1 R} \tag{8.400}$$

The dc input current is

$$P_I = I_I V_I = 1.5 \frac{V_I^2}{\gamma_1 R} \tag{8.401}$$

Table 8.6 Voltage amplitudes in even harmonic maximum drain efficiency Class F power amplifiers.

Class	$\frac{V_m}{V_I}$	$\frac{V_{m2}}{V_I}$	$\frac{V_{m4}}{V_I}$	$\frac{V_{m2}}{V_m}$	$\frac{V_{m4}}{V_m}$
F_2	$\sqrt{2} = 1.4142$	$\frac{1}{2} = 0.5$	0	$\frac{1}{2}\sqrt{2} = 0.3536$	0
F_{24}	1.5	0.5835	0.0834	0.389	0.0556

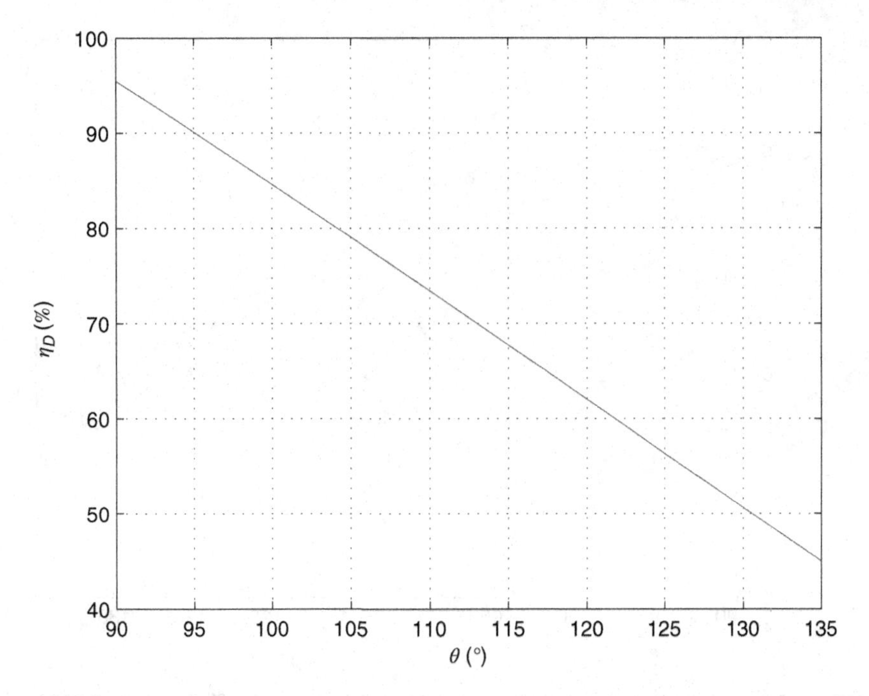

Figure 8.42 Drain efficiency η_D for the Class F_{24} RF power amplifier with the second- and fourth-harmonic peaking for the maximum drain efficiency.

The output power is

$$P_O = \frac{V_m^2}{2R} = \frac{(1.5V_I)^2}{2R} = 1.125\frac{V_I^2}{R} \tag{8.402}$$

The drain efficiency is

$$\eta_D = \frac{P_O}{P_I} = 0.75\gamma_1 = \frac{3\sin\theta}{2\theta} \tag{8.403}$$

Figure 8.42 shows drain efficiency η_D for the Class F_{24} RF power amplifier with the second- and fourth-harmonic peaking for the maximum drain efficiency.

The maximum drain-to-source voltage is

$$V_{DSM} = v_{DS}(\pi) = V_I(1 + 1.5 + 0.5835 - 0.0834) = 3V_I = 2V_m \tag{8.404}$$

The output-power capability is

$$c_p = \frac{P_O}{P_I} = \frac{1}{2}\left(\frac{I_m}{I_{DM}}\right)\left(\frac{V_m}{V_{DSM}}\right) = \frac{1}{2}(\alpha_1)\left(\frac{1}{2}\right) = \frac{\alpha_1}{4} = \frac{\sin\theta}{2\pi} \tag{8.405}$$

Figure 8.43 shows output-power capability c_p for the Class F_{24} RF power amplifier with the second- and fourth-harmonic peaking for the maximum drain efficiency.

The dc input resistance is

$$R_{DC} = \frac{V_I}{I_I} = \frac{2\gamma_1}{3}R = \frac{4\sin\theta}{3\theta}R \tag{8.406}$$

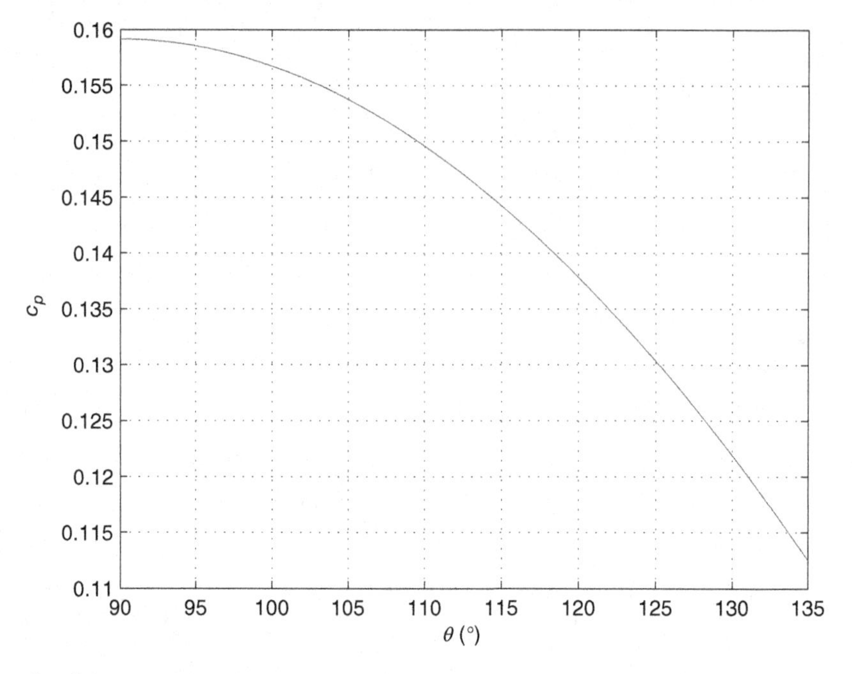

Figure 8.43 Output-power capability c_p for the Class F_{24} RF power amplifier with the second- and fourth-harmonic peaking for the maximum drain efficiency.

8.8 Class F_{246} RF Power Amplifier with Second, Fourth, and Sixth Harmonics

For the maximally flat Class F_{246} amplifier,

$$V_{m2} = \frac{175}{512} V_m \tag{8.407}$$

$$V_{m4} = \frac{7}{256} V_m \tag{8.408}$$

$$V_{m6} = \frac{1}{512} V_m \tag{8.409}$$

$$V_{m1} = \frac{64}{45} V_I \tag{8.410}$$

$$V_{m2} = \frac{4}{9} V_I \tag{8.411}$$

and

$$V_{m4} = \frac{1}{45} V_I \tag{8.412}$$

8.9 Class F_K RF Power Amplifier with Series-Resonant Circuit and Quarter-Wavelength Transmission Line

A Class F RF power amplifier with a series-resonant circuit and a quarter-wavelength transmission line was introduced in [8] and is shown in Fig. 8.44(a). This circuit is referred to as Class F_K

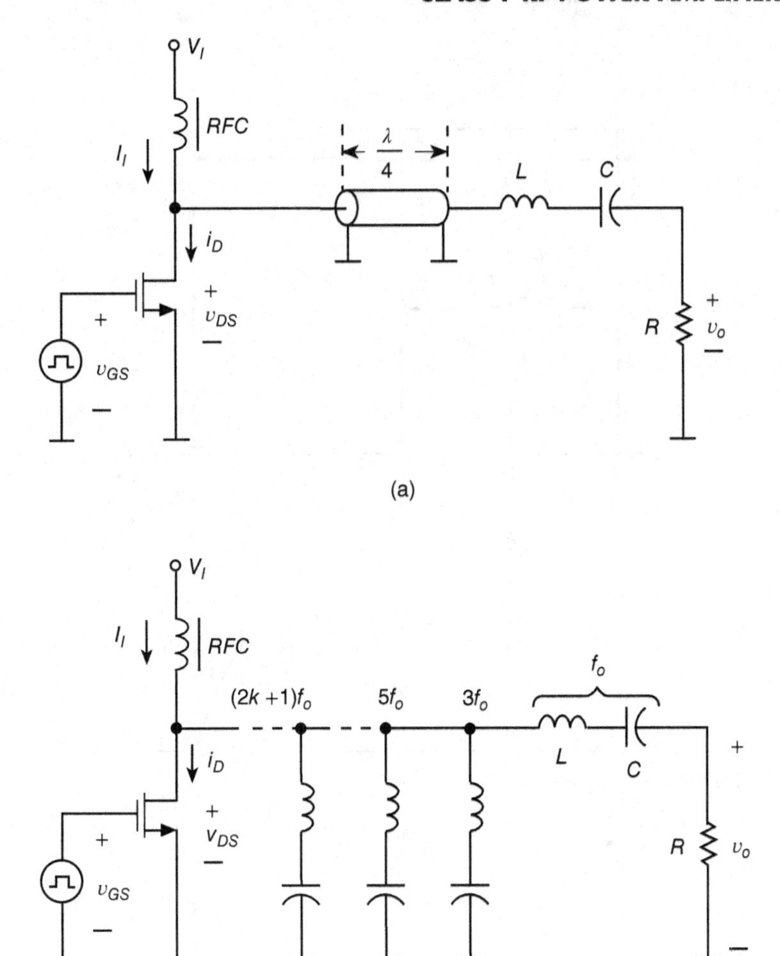

Figure 8.44 Class F_K power amplifier employing a series-resonant circuit and a quarter-wavelength transmission line, where $n = 2k + 1$ with $k = 0, 1, 2, \ldots$.: (a) circuit and (b) equivalent circuit at $Z_o = R$.

amplifier or Class F inverse amplifier. The quarter-wavelength transmission line is equivalent to an infinite number of series-resonant circuits, which behaves as short circuits for odd harmonics and as open circuits for even harmonics. The equivalent circuit of the amplifier is depicted in Fig. 8.44(b). Voltage and current waveforms of the amplifier are shown in Fig. 8.45. The drain current is a square wave, and the drain-to-source voltage is a half-sine wave. The circuit and its waveforms are dual to those of the Class F_K amplifier with a parallel-resonant circuit. The drain current and the drain-to-source voltage waveforms are interchanged.

The drain current is a square wave, and it can be expanded into a Fourier series

$$i_D = I_{DM}\left\{\frac{1}{2} + \frac{2}{\pi}\left[\cos\omega_o t - \frac{1}{3}\cos 3\omega_o t + \frac{1}{5}\cos 5\omega_o t - \frac{1}{7}\cos 7\omega_o t + \cdots\right]\right\} \tag{8.413}$$

The drain current i_D contains only the dc component, fundamental component, and odd harmonic components.

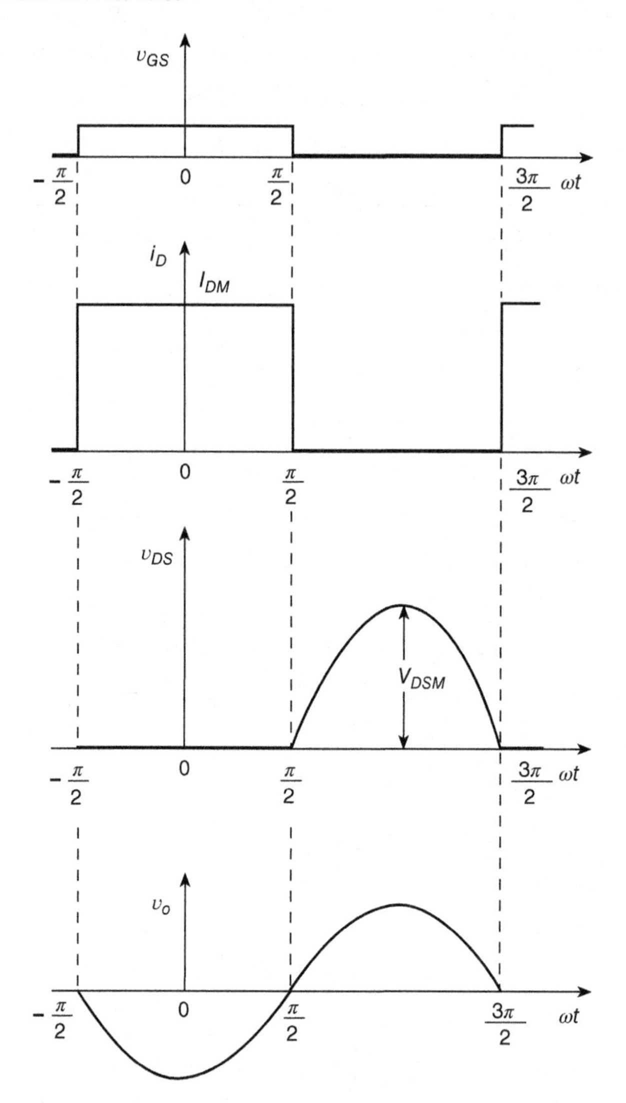

Figure 8.45 Voltage and current waveforms in the Class F_K RF power amplifier with a series-resonant circuit and a quarter-wavelength transmission line [8].

The drain-to-source voltage waveform v_{DS} can be expanded into a Fourier trigonometric series

$$v_{DS} = V_{DSM}\left[\frac{1}{\pi} + \frac{1}{2}\cos\omega_o t + \frac{2}{\pi}\sum_{n=2}^{\infty}\frac{\cos\left(\frac{n\pi}{2}\right)}{1-n^2}\cos n\omega_o t\right]$$

$$= V_{DSM}\left[\frac{1}{\pi} + \frac{1}{2}\cos\omega_o t + \frac{2}{3\pi}\cos 2\omega_o t - \frac{2}{15\pi}\cos 4\omega_o t + \frac{2}{35\pi}\cos 6\omega_o t + \cdots\right] \tag{8.414}$$

Thus, the drain current waveform i_D contains only the dc component, fundamental component, and the even harmonic components.

The output voltage is equal to the fundamental component of the drain-to-source voltage

$$v_o = V_m \cos\omega_o t \tag{8.415}$$

where

$$V_m = \frac{V_{DSM}}{2} = \frac{\pi}{2} V_I \tag{8.416}$$

The output current is equal to the fundamental component of the drain current

$$i_o = i_{d1} = I_m \cos \omega_o t \tag{8.417}$$

where

$$I_m = \frac{2}{\pi} I_{DM} = \frac{4}{\pi} I_I \tag{8.418}$$

or

$$I_m = \frac{V_m}{R} = \frac{\pi V_I}{2R} \tag{8.419}$$

The dc supply current is equal to the dc component of the drain current

$$I_I = \frac{I_{DM}}{2} = \frac{\pi}{4} I_m = \frac{\pi^2 V_I}{8R} \tag{8.420}$$

and dc supply voltage is

$$V_I = \frac{V_{DSM}}{\pi} \tag{8.421}$$

Hence, the dc supply power is

$$P_I = I_I V_I = \frac{I_{DM} V_I}{2} = \frac{I_{DM} V_{DSM}}{2\pi} \tag{8.422}$$

The output power is

$$P_O = \frac{1}{2} I_m V_m = \frac{1}{2} \times \frac{2}{\pi} \times I_{DM} \frac{V_{DSM}}{2} = \frac{1}{2\pi} I_{DM} V_{DSM} \tag{8.423}$$

or

$$P_O = \frac{V_m^2}{2R} = \frac{\pi^2 V_I^2}{8R} \tag{8.424}$$

The efficiency of the amplifier is

$$\eta = \frac{P_O}{P_I} = 1 \tag{8.425}$$

The output-power capability is

$$c_p = \frac{P_O}{I_{DM} V_{DSM}} = \frac{1}{2\pi} = 0.159 \tag{8.426}$$

The dc resistance is

$$R_{DC} = \frac{V_I}{I_I} = \frac{8}{\pi^2} R \approx 0.811R \tag{8.427}$$

The input impedance of the load network seen by the drain and source terminals at the fundamental frequency f_o is

$$Z(f_o) = \frac{V_m}{I_m} = R_i = \frac{Z_o^2}{R} \tag{8.428}$$

the input impedance at even harmonics is

$$Z(nf_o) = Z(2kf_o) = \frac{V_{m(n)}}{I_{m(n)}} = \frac{\textit{finite value}}{0} = \infty \quad \text{for} \quad n = 2, 4, 6, \ldots \tag{8.429}$$

and the input impedance at odd harmonics is

$$Z(nf_o) = Z[(2k+1)f_o] = \frac{V_{m(n)}}{I_{m(n)}} = \frac{0}{\textit{finite value}} = 0 \quad \text{for} \quad n = 3, 5, 7, \ldots \tag{8.430}$$

where $k = 1, 2, 3, \ldots$

The rms value of the drain current is

$$I_{DSrms} = \sqrt{\frac{1}{2\pi}\int_{\pi/2}^{\pi/2} I_{DM}^2 \, d(\omega_o t)} = \frac{I_{DM}}{\sqrt{2}} \tag{8.431}$$

If the transistor is operated as a switch, the conduction loss in the MOSFET on-resistance r_{DS} is given by

$$P_{rDS} = r_{DS} I_{DSrms}^2 = \frac{r_{DS} I_m^2}{2} = \frac{\pi^2}{8} r_{DS} I_m^2 \tag{8.432}$$

The output power is

$$P_O = \frac{r_{DS} I_m^2}{2} \tag{8.433}$$

Thus, the drain efficiency is

$$\eta_D = \frac{P_O}{P_I} = \frac{P_O}{P_O + P_{rDS}} = \frac{1}{1 + \frac{P_{rDS}}{P_O}} = \frac{1}{1 + \frac{\pi^2}{4}\frac{r_{DS}}{R_i}} \tag{8.434}$$

Example 8.6

Design a Class F_K power amplifier with a quarter-wave transmission line to deliver a power of 10 W at $f_c = 5$ GHz. The required bandwidth is $BW = 500$ MHz. The dc power supply voltage is 28 V and $v_{DSmin} = 1$ V. The load resistance is $R = 50\ \Omega$. The MOSFET on-resistance is $r_{DS} = 0.2\ \Omega$.

Solution. The maximum amplitude of the fundamental component of the drain-to-source voltage is

$$V_m = \frac{\pi}{2}(V_I - v_{DSmin}) = \frac{\pi}{2}(28 - 1) = 42.41 \text{ V} \tag{8.435}$$

The input resistance of the load circuit is

$$R_i = \frac{V_m^2}{2P_O} = \frac{42.41^2}{2 \times 10} = 89.93\ \Omega \tag{8.436}$$

The maximum drain-to-source voltage is

$$v_{DSmax} = \pi V_I = \pi \times 28 = 87.965 \text{ V} \tag{8.437}$$

The amplitude of the fundamental component of the drain current is

$$I_m = \frac{V_m}{R_i} = \frac{42.41}{89.93} = 0.472 \text{ A} \tag{8.438}$$

The maximum drain current is

$$I_{DM} = \frac{\pi}{2} I_m = \frac{\pi}{2} \times 0.472 = 0.7414 \text{ A} \tag{8.439}$$

The dc supply current is

$$I_I = \frac{I_{DM}}{2} = \frac{0.7414}{2} = 0.37 \text{ A} \tag{8.440}$$

The dc supply power is

$$P_I = I_I V_I = 0.37 \times 28 = 10.36 \text{ W} \tag{8.441}$$

The drain power loss of the transistor is

$$P_D = P_I - P_O = 10.36 - 10 = 0.36 \text{ W} \tag{8.442}$$

The drain efficiency is

$$\eta_D = \frac{P_O}{P_I} = \frac{10}{10.36} = 96.525\% \tag{8.443}$$

The dc resistance presented by the amplifier to the dc source is

$$R_{DC} = \frac{8}{\pi^2} = 0.81 \times 89.93 = 72.84\ \Omega \tag{8.444}$$

The characteristic impedance of the transmission line is

$$Z_o = \sqrt{R_i R} = \sqrt{89.93 \times 50} = 67\ \Omega \tag{8.445}$$

Assume the dielectric constant used for construction of the transmission line to be $\epsilon_r = 2.1$. The wavelength in the transmission line is

$$\lambda = \frac{c}{\sqrt{\epsilon_r} f_c} = \frac{3 \times 10^8}{\sqrt{2.1} \times 5 \times 10^9} = 4.14\ \text{cm} \tag{8.446}$$

Hence, the length of the transmission line is

$$l_{TL} = \frac{\lambda}{4} = \frac{4.14}{4} = 1.035\ \text{cm} \tag{8.447}$$

The loaded quality factor is

$$Q_L = \frac{f_c}{BW} = \frac{5 \times 10^9}{500 \times 10^6} = 10 \tag{8.448}$$

The loaded quality factor for a series-resonant circuit is defined as

$$Q_L = \frac{\omega_c L}{R} = \omega_c C R_L \tag{8.449}$$

Hence, the inductance of the resonant circuit tuned to the fundamental is

$$L = \frac{Q_L R}{\omega_c} = \frac{10 \times 50}{2\pi \times 5 \times 10^9} = 15.915\ \text{nH} \tag{8.450}$$

The capacitance of the resonant circuit tuned to the fundamental is

$$C = \frac{1}{\omega_c Q_L R} = \frac{1}{2\pi \times 5 \times 10^9 \times 10 \times 50} = 0.06366\ \text{pF} \tag{8.451}$$

The drain efficiency is

$$\eta_D = \frac{1}{1 + \frac{\pi^2}{4}\frac{r_{DS}}{R_i}} = \frac{1}{1 + \frac{\pi^2}{4}\frac{0.2}{89.93}} = 99.45\% \tag{8.452}$$

8.10 Summary

- The transistor is operated as a dependent current source in Class F RF power amplifiers.
- The Class F RF power amplifier circuit consists of a transistor and a parallel or series-resonant circuit, tuned to the fundamental frequency, and resonant circuits, tuned to the harmonics.
- The circuit diagrams of Class F power amplifiers are complex.
- The addition of harmonics of the correct amplitudes and phases into the drain-to-source voltage flattens the v_{DS} voltage, reducing the transistor power loss and improving the efficiency.

- The Class F power amplifier with a parallel-resonant circuit and a quarter-wavelength transmission line ideally has a square-wave drain-to-source voltage v_{DS} and a half-sinusoidal drain current i_D.
- The Class F power amplifier with a series-resonant circuit and a quarter-wavelength transmission line forms a square-wave drain current i_D and a half-sinusoidal drain-to-source voltage v_{DS}.
- Transmission lines can be used to create higher order harmonics in Class F power amplifiers.
- There is a trade-off between the efficiency and circuit complexity of Class F amplifiers.
- The transistor output capacitance can dominate the total impedance of the output network and hamper the synthesis of the desired output network with high impedances at harmonics in Class F power amplifiers.

8.11 Review Questions

8.1 What is the principle of operation of Class F RF power amplifiers?

8.2 What is the topology of the Class F_3 power amplifier?

8.3 Are the topologies of Class F power amplifiers simple or complex?

8.4 What is the region of operation of the transistor in Class F power amplifiers?

8.5 What is a Class F amplifier with a maximally flat voltage?

8.6 What is a Class F amplifier with a maximum drain efficiency?

8.7 How high is the efficiency of Class F power amplifiers?

8.8 What is the amplitude and phase of the third-harmonic voltage in the Class F_3 amplifier with the third-harmonic peaking for maximally flat drain-to-source voltage?

8.9 What is the amplitude and phase of the third-harmonic voltage in the Class F_3 amplifier with the third-harmonic peaking for maximum drain efficiency?

8.10 What is the range of the drain current conduction angle in the Class F_3 amplifier?

8.11 What are the amplitudes and phases of the third- and fifth-harmonic voltages in the Class F_{35} amplifier with the third- and fifth-harmonic peaking for maximum drain efficiency?

8.12 What are the amplitudes and phases of the third- and fifth-harmonic voltages in the Class F_{35} amplifier with the third- and fifth-harmonic peaking for maximum drain efficiency?

8.13 What is the range of the drain current conduction angle in the Class F_{35} amplifier?

8.14 How can the transmission lines be used to control the impedances at harmonics in Class F power amplifiers?

8.15 What is the range of the drain current conduction angle in the Class F_2 amplifier?

8.16 What is the range of the drain current conduction angle in the Class F_{24} amplifier?

8.17 What are the applications of Class F RF power amplifiers?

8.18 Does the transistors output capacitance affect the operation of Class F RF power amplifiers?

8.19 List the applications of Class F RF power amplifiers.

8.12 Problems

8.1 Design a Class F power amplifier employing third-harmonic peaking with a maximally flat drain-to-source voltage to deliver a power of 100 W at $f = 2.4$ MHz. The dc power supply voltage is 120 V and $v_{DSmin} = 1$ V.

8.2 Design a Class F power amplifier employing third-harmonic peaking with a maximally drain efficiency to deliver a power of 100 W at $f = 2.4$ MHz. The dc power supply voltage is 120 V and $v_{DSmin} = 1$ V.

8.3 Design a Class F power amplifier employing second-harmonic peaking with a maximally flat drain-to-source voltage to deliver a power of 1 W at $f = 2.4$ MHz. The dc power supply voltage is 5 V and $v_{DSmin} = 0.3$ V.

8.4 Design a Class F_3 RF power amplifier to meet the following specifications: $V_I = 48$ V, $v_{DSmin} = 2$ V, $P_O = 100$ W, $R_L = 50\ \Omega$, $\theta = 60°$, $f_c = 88$ MHz, and $BW = 10$ MHz.

8.5 Design a Class F RF power amplifier with a quarter-wave transmission line and a square-wave drain current to meet the following specifications: $P_O = 50$ W, $V_I = 100$ V, $v_{DSmin} = 2$ V, $R_L = 50\ \Omega$, and $f = 250$ kHz.

References

[1] V. J. Tyler, "A new high-efficiency high-power amplifier," *Marconi Review*, vol. 21, no. 130, pp. 96–109, 1958.

[2] L. B. Hallman, "A Fourier analysis of radio-frequency power amplifier waveforms," *Proceedings of the IRE*, vol. 20, no. 10, pp. 1640–1659, 1932.

[3] D. M. Snider, "A theoretical analysis and experimental confirmation of the optimally loaded and over-driven RF power amplifier," *IEEE Transactions on Electron Devices*, vol. 14, pp. 851–857, 1967.

[4] V. O. Stocks, *Radio Transmitters: RF Power Amplification*. London, England: Van Nostrand, 1970, pp. 38–48.

[5] N. S. Fuzik, "Biharmonic modes of tuned power amplifier," *Telecommunications and Radio Engineering, Part 2*, vol. 25, pp. 117–124, 1970.

[6] S. R. Mazumber, A. Azizi, and F. E. Gardiol, "Improvement of a Class C transistor power amplifiers by second-harmonic tuning," *IEEE Transactions on Microwave Theory and Technique*, vol. 27, no, 5, pp. 430–433, 1979.

[7] H. L. Krauss, C. V. Bostian, and F. H. Raab, *Solid State Radio Engineering*. New York, NY: John Wiley & Sons, 1980.

[8] M. K. Kazimierczuk, "A new concept of Class F tuned power amplifier," *Proceedings of the 27th Midwest Symposium on Circuits and Systems*, Morgantown, WV, June 11-12, 1984, pp. 425–428.

[9] Z. Zivkovic and A. Markovic, "Third harmonic injection increasing the efficiency of high power RF amplifiers," *IEEE Transactions on Broadcasting*, vol. 31, no. 2, pp. 34–39, 1985.

[10] Z. Zivkovic and A. Markovic, "Increasing the efficiency of high power triode RF amplifier. Why not with the second harmonic?," *IEEE Transactions on Broadcasting*, vol. 32, no. 1, pp. 5–10, 1986.

[11] X. Lu, "An alternative approach to improving the efficiency of the high power radio frequency amplifiers," *IEEE Transactions on Broadcasting*, vol. 38, pp. 85–89, 1992.

[12] F. H. Raab, "An introduction to Class-F power amplifiers," *RF Design*, vol. 19, no. 5, pp. 79–84, 1996.

[13] F. H. Raab, "Class-F power amplifiers with maximally flat waveforms," *IEEE Transactions on Microwave Theory and Technique*, vol. 45, no. 11, pp. 2007–2012, 1997.

[14] B. Tugruber, W. Pritzl, D. Smely, M. Wachutka, and G. Margiel, "High-efficiency harmonic control amplifier," *IEEE Transactions on Microwave Theory and Technique*, vol. 46, no. 6, pp. 857–862, 1998.

[15] F. H. Raab, "Class-F power amplifiers with reduced conduction angle," *IEEE Transactions on Broadcasting*, vol. 44, no. 4, pp. 455–459, 1998.

[16] C. Trask, "Class-F amplifier loading networks: a unified design approach," IEEE MTT-S International Microwave Symposium Digest, vol. 1, pp. 351–354, 1999.

[17] S. C. Cripps, *RF Power Amplifiers for Wireless Communications*, Norwood, MA: Artech House, 1999.

[18] P. Colantonio, F. Giannini, G. Leuzzi, and E. Limiti, "On the class-F power amplifier design," *RF Microwave Computer-Aided Engineering*, vol. 32, no. 2, pp. 129–149, 1999.

[19] A. N. Rudiakova, V. G. Krizhanovski, and M. K. Kazimierczuk, "Phase tuning approach to polyharmonic power amplifiers," Proc. European Microwave Week Conference, London, UK, September 10-14, 2001, pp. 105–107.

[20] F. Fortes and M. J. Rosario, "A second harmonic Class-F power amplifier in standard CMOS technology," *IEEE Transactions on Microwave Theory and Technique*, vol. 49, no. 6, pp. 1216–1220, 2001.

[21] M. Weiss, F. H. Raab, and Z. Popovic "Linearity of X-band Class F power amplifiers in high-mode transmitters," *IEEE Microwave Theory and Techniques*, vol. 49, no. 6, pp. 1174–1179, 2001.

[22] F. H. Raab, "Class-E, Class-C, and Class-F power amplifiers based upon a finite number of harmonics," *IEEE Transactions on Microwave Theory and Technique*, vol. 49, no. 8, pp. 1462–1468, 2001.

[23] F. H. Raab, "Maximum efficiency and output of Class-F power amplifiers," *IEEE Transactions on Microwave Theory and Technique*, vol. 49, no. 6, pp. 1162–1166, 2001.

[24] F. Lepine, A. Adahl, and H. Zirath, "L-band LDMOS power amplifier based on an inverse Class F architecture," *IEEE Transactions on Microwave Theory and Technique*, vol. 53, no. 6, pp. 2007–2012, 2005.

[25] A. Grebennikov, *RF and Microwave Power Amplifier Design*. New York, NY: McGraw-Hill, 2005.

[26] A. N. Rudiakova and V. G. Krizhanovski, *Advanced Design Techniques for RF Power Amplifiers*. Berlin, Springer, 2006.

[27] M. Roberg and Z. Popović, "Analysis of high-efficiency power amplifiers with arbitrary output harmonic terminations," *IEEE Transactions on Microwave Theory and Technique*, vol. 59, no. 5, pp. 2037–2048, 2011.

[28] M. K. Kazimierczuk and R. Wojda, "Maximum drain efficiency Class F_3 RF power amplifier," *IEEE International Symposium on Circuits and Systems*, Rio de Janeiro, Brazil, May 18-21, pp. 2785–2788, 2011.

9

Linearization and Efficiency Improvements of RF Power Amplifiers

9.1 Introduction

High efficiency and high linearity of power amplifiers [1–85] are of primary importance in wireless communication systems. Wireless communication systems transmit voice, video, and data with high data rates. High efficiency is required for low energy consumption, a longer battery lifetime, and thermal management. Linearity is required for achieving low distortion of the amplified signals. The radio-frequency (RF) power amplifier specifications call for intermodulation (IM) levels of −60 dBc. Linearity and efficiency enhancement techniques of RF power amplifiers, which are used in transmitters, are studied in this chapter.

Signals used in modern digital wireless communications systems have a time-varying envelope (amplitude modulation; AM) and time-varying angle (phase modulation; PM)

$$v_{AM/PM} = V_m(t)\cos[\cos\omega_c t + \phi(t)] \tag{9.1}$$

A nonconstant envelope signal requires linear power amplifiers in the transmitters. The output power in CDMA2000 and WCDMA transmitters may vary in a wide dynamic range of 80 dB. The average output power is usually lower than the peak power by 15–25 dB. The transmitters must be designed for the maximum output power. A high peak-to-average ratio (PAR) drives many power amplifiers into saturation, causing signal distortion and generating out-band interference. The maximum efficiency of linear power amplifiers such as Class A, AB, and B amplifiers occurs at the maximum output power. In a linear Class A power amplifier, the average power is usually set at a back-off (BO) of 10–12 dB, resulting in a very low average efficiency. In this case, the maximum amplitude of the drain-to-source voltage V_m is nearly equal to the supply voltage V_I. This corresponds to the AM index equal to 1. However, the average AM index is in the range of 0.2–0.3, when the amplitude of the drain-to-source voltage V_m is much lower than

RF Power Amplifiers, Second Edition. Marian K. Kazimierczuk.

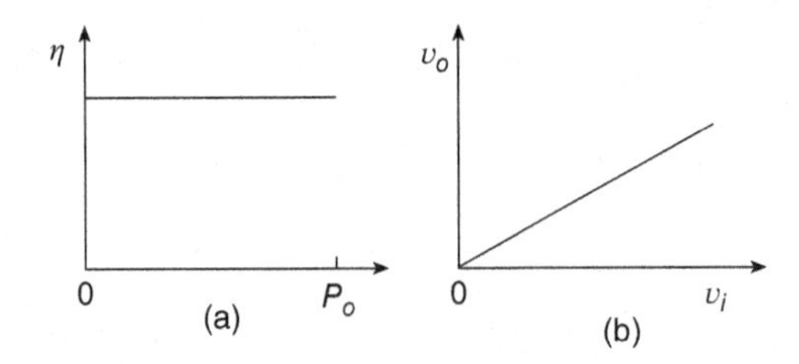

Figure 9.1 Ideal characteristics of a power amplifier. (a) Efficiency η as a function of the output power. (b) Output voltage v_o as a function of the input voltage v_i.

the supply voltage V_I. Therefore, the statistical average efficiency of linear power amplifiers with AM is much lower than their maximum efficiency. In wireless communication systems, there are two types of transmitters: base-station transmitters and handset transmitters.

Ideal characteristics of a power amplifier are depicted in Fig. 9.1. The efficiency η should be very high over a wide range of the output power, as shown in Fig. 9.1(a). The output voltage v_o should be a linear function of the input voltage v_i, as shown in Fig. 9.1(b). In this case, the voltage gain of the amplifier $A_v = v_o/v_i$ is constant over a wide range of the input voltage.

Power control of RF transmitters is required in modern digital wireless communications. In CDMA, power control is used in both the base-station transmitters and handset transmitters. In base-station transmitters, a higher power is required to transmit the signal to the edges of a cell. In handset transmitters, the output power should be transmitted at variable levels so that the power levels of the signals received at the base station are similar to all users.

This chapter presents an overview of basic linearization and efficiency- enhancement techniques of power amplifiers invented over the years. The purpose of some of these techniques is to improve either the linearity or the efficiency only, while some other techniques address both the linearization and efficiency enhancement together.

9.2 Predistortion

Nonlinear distortion in the output signal of a power amplifier is caused by changes in the slope of the transfer function $v_o = f(v_i)$. These changes are caused by the nonlinear properties of transistors (MOSFETs or BJTs) used in the power amplifier. The changes in the slope of the transfer function cause changes in the gain at different values of the input voltage or power. There are several techniques for reduction of nonlinear distortion.

Predistortion is a technique that seeks to linearize a power amplifier by making suitable modifications to the amplitude and phase of the power amplifier input signal v_i [1–3]. This technique includes analog predistortion and digital predistortion. A block diagram of a power amplifier with predistortion is depicted in Fig. 9.2. A nonlinear block is connected in the signal path to compensate for the nonlinearity of the power amplifier. This block is called a *predistorter* or a *predistortion linearizer*. Predistortion can be performed either at RF or baseband frequencies. Predistortion at baseband frequencies is favorable because of digital signal processing

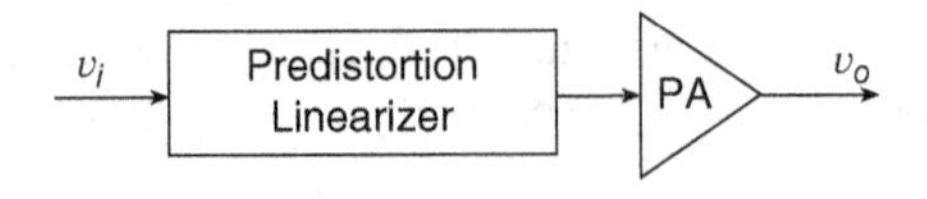

Figure 9.2 Block diagram of power amplifier with a predistortion system.

(DSP) capabilities. This is called "digital predistortion." DSP ICs can be used to perform digital predistortion. It is an open-loop system, which is inherently stable. The disadvantage of the system is the difficulty in generating the transfer function of the predistorter such that the transfer function of the overall system is linear.

Example 9.1

A power amplifier voltage transfer function has the gain $A_1 = 100$ for small signals and the gain $A_2 = 50$ for large signals, as depicted in Fig. 9.3(a). Find the transfer function of the predistorter to achieve a linear transfer function of the overall system.

Solution. Let us assume the gain of the predistorter for small signals to be $A_3 = 1$. The ratio of the two gains of the power amplifiers is

$$\frac{A_1}{A_2} = \frac{100}{50} = 2 \tag{9.2}$$

Hence, the gain of the predistorter for large signals is

$$A_4 = 2 \tag{9.3}$$

The gain of the overall system for low signals is

$$A_5 = A_1 A_3 = 100 \times 1 = 100 \tag{9.4}$$

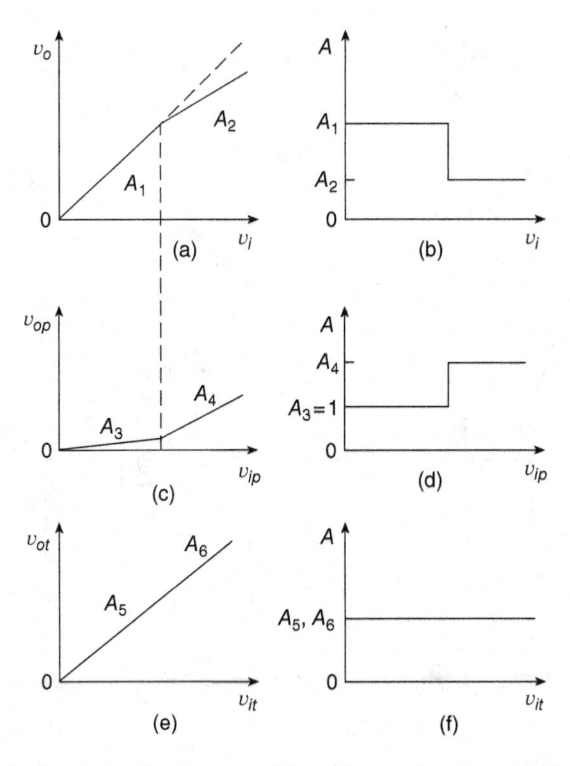

Figure 9.3 Transfer functions of a power amplifier using predistortion. (a) Nonlinear transfer function of a power amplifier $v_o = f(v_i)$. (b) Transfer function of the predistorter A. (c) Transfer function of the predistorter $v_{op} = f(v_{ip})$. (d) Transfer function of a predistortion amplifier A_p. (e) Transfer function of the total amplifier $v_{ot} = f(v_{it})$. (f) Linearized transfer function of the overall system A.

and for large signals is

$$A_6 = A_2 A_4 = 50 \times 2 = 100. \tag{9.5}$$

Thus, $A_6 = A_5$, resulting in a linear transfer function over a wide range of the input signal. Figure 9.3 illustrates the linearization of a power amplifier using the predistortion technique.

9.3 Feedforward Linearization Technique

A block diagram of a basic power amplifier system with feedforward linearization technique [4–9] is shown in Fig. 9.4(a). The feedforward linearization technique relies on the cancellation of the distortion signal. This is done by generating a proper error voltage and subtracting it from the distorted output voltage of a nonlinear power amplifier. The input voltage v_i is applied to two channels. The voltage in one channel is applied to the input of the main power amplifier with the voltage gain A_v and it is amplified. The other channel uses the original input voltage v_i as a reference signal for future comparison. The output voltage of the main power amplifier contains the undistorted and amplified input voltage $A_v v_i$ and the distortion voltage v_d produced by the amplifier nonlinearity

$$v_p = A_v v_i + v_d \tag{9.6}$$

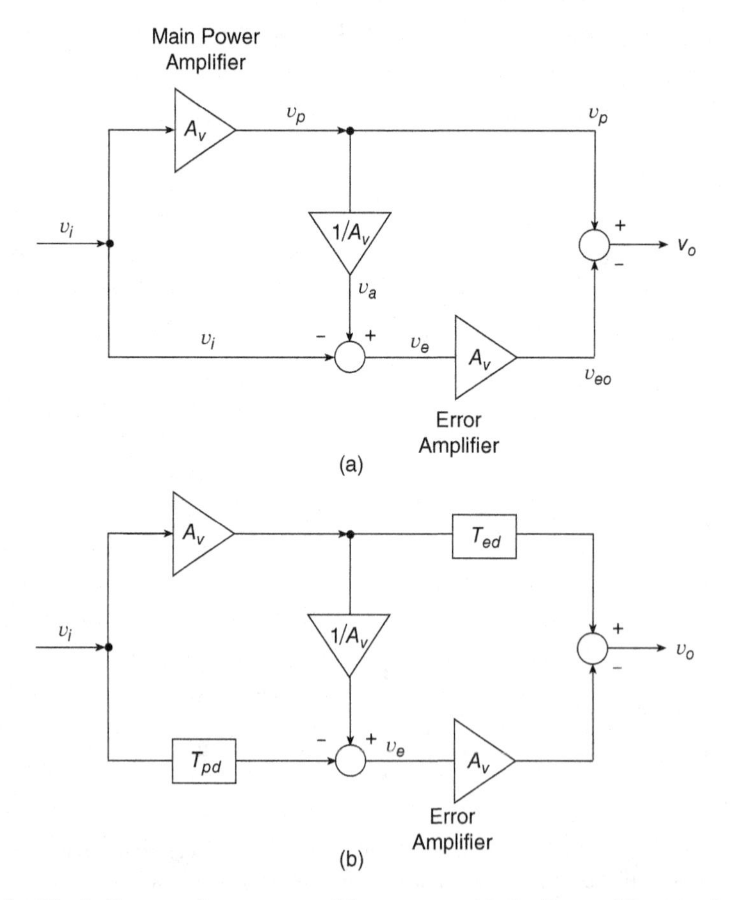

Figure 9.4 Block diagram of a power amplifier system with feedforward linearization. (a) Basic feedforward system. (b) Feedforward system with delay blocks.

The total output voltage of the main power amplifier is attenuated by a circuit with the transfer function equal to $1/A_v$

$$v_a = \frac{v_p}{A_v} = v_i + \frac{v_d}{A_v} \tag{9.7}$$

The attenuated voltage is compared with the original voltage v_i in the subtractor to produce an error voltage v_e

$$v_e = v_a - v_i = v_i + \frac{v_d}{A_v} - v_i = \frac{v_d}{A_v} \tag{9.8}$$

The error voltage v_e is amplified by the error amplifier with the voltage gain A_v, yielding

$$v_{eo} = A_v v_e = A_v \frac{v_d}{A_v} = v_d \tag{9.9}$$

Next, the output voltage of the main power amplifier v_p is compared with the error amplifier output voltage v_{eo} by a subtractor, producing the output voltage of the entire system

$$v_o = v_p - v_{eo} = A_v v_i + v_d - v_d = A_v v_i \tag{9.10}$$

It can be seen that the distortion signal v_d is cancelled out for perfect amplitude and phase matching at each subtractor. In addition to the cancellation of the nonlinear component, the system suppresses the noise added to the signal in the main power amplifier. The first loop is called the signal cancellation loop and the second loop is called the error cancellation loop.

The feedforward linearization system offers a wide bandwidth, reduced noise level, and is stable in spite of large phase shifts at RF because the output signal is not applied to the input. Wideband operation is useful in multicarrier wireless communications such as wireless base stations.

The main disadvantage of the feedforward linearization system is the gain mismatch and the phase (delay) mismatch in the signal channels. The amount of linearization depends on the amplitude and phase matching at each subtractor. If the first loop from the input voltage v_i to the first subtractor has a relative gain mismatch $\Delta A/A$ and the phase mismatch $\Delta\phi$, the attenuation of the magnitude of the IM terms in the output voltage is [5, 7]

$$A_{IM} = \sqrt{1 - 2\left(1 + \frac{\Delta A}{A}\right)\cos\Delta\phi + \left(1 + \frac{\Delta A}{A}\right)^2} \tag{9.11}$$

To improve the signal quality, delay blocks can be added as shown in Fig. 9.4(b). A delay block T_{pd} is added in the bottom path to compensate for the delay introduced by the main power amplifier and the attenuator. The other delay block T_{ed} is added in the upper path to compensate for the delay introduced by the error amplifier. The delay blocks can be implemented using passive lumped-element networks or transmission lines. However, these blocks dissipate power and reduce the amplifier efficiency. In addition, the design of wideband delay blocks is difficult.

9.4 Negative Feedback Linearization Technique

In general, negative feedback reduces nonlinearity [12, 13]. A block diagram of a power amplifier with a negative feedback is depicted in Fig. 9.5. The transfer function $v_o = f(v_s)$ of a power amplifier can be considerably linearized through the application of negative feedback, reducing nonlinear distortion. Large changes in the open-loop gain cause much smaller changes in the closed-loop gain.

Let us assume that the output voltage of a power amplifier without negative feedback contains a distortion component (an IM term of a harmonic) given by

$$v_d = V_d \sin \omega_d t \tag{9.12}$$

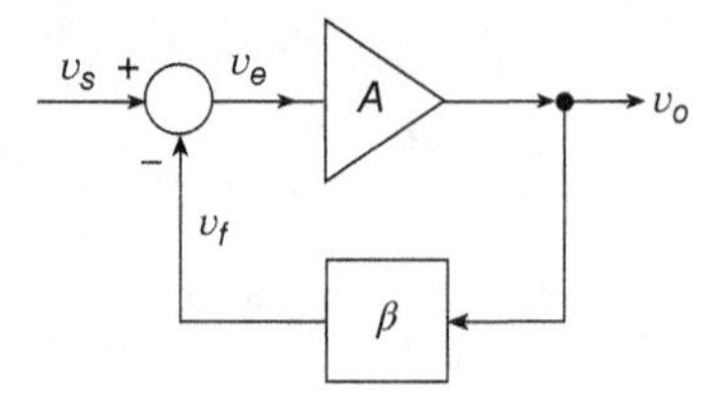

Figure 9.5 Block diagram of a power amplifier system with negative feedback.

The distortion component of the output voltage of the power amplifier with negative feedback is

$$v_{df} = V_{df} \sin \omega_d t \tag{9.13}$$

We wish to find the relationship between v_d and v_{df}. The feedback voltage is

$$v_f = \beta v_{df} \tag{9.14}$$

The reference voltage for the distortion component is zero. Hence, the input voltage of the power amplifier is

$$v_e = -v_f = -\beta v_{df} \tag{9.15}$$

The distortion component at the power amplifier output is

$$v_{od} = Av_e = -Av_f = -\beta A v_{df} \tag{9.16}$$

Thus, the output voltage contains two terms: the original distortion component generated by the power amplifier v_d and the component v_{od}, which represents the effect of negative feedback. Hence, the overall output voltage is

$$v_{df} = v_d - v_{od} = v_d - \beta A v_{df} \tag{9.17}$$

yielding

$$v_{df} = \frac{v_d}{1 + \beta A} = \frac{v_d}{1 + T} \tag{9.18}$$

Since A is generally a function of frequency, the amplitude of the distortion component must be evaluated at the frequency of the distortion component $f_d = \omega_d/(2\pi)$.

It follows from (9.18) that the reduction of nonlinear distortion is large when the loop gain $T = \beta A$ is high. However, the voltage gain of RF power amplifiers is low, and therefore the loop gain T is also low at RF. In addition, stability of the loop is of great concern due to a large number of poles introduced by various parasitic components. A high-order power amplifier may have an excessive phase shift, causing oscillations. For a large loop gain $T = \beta A$,

$$A_f \approx \frac{1}{\beta} \tag{9.19}$$

This equation indicates that the gain depends only on the linear feedback network β. The output voltage of a power amplifier is given by

$$v_o = A_f v_s \approx \frac{v_i}{\beta} \tag{9.20}$$

Thus, the transfer function is nearly linear. However, the feedback amplifier requires a larger input voltage in order to produce the same output as the amplifier without feedback.

Figure 9.6 shows a block diagram of a power amplifier with negative feedback and frequency translation. The forward path consists of a low-frequency high-gain error amplifier, an upconversion mixer, and a power amplifier. The mixer converts the frequency of the input signal f_i to an RF frequency $f_{RF} = f_i + f_{LO}$, where f_{LO} is the frequency of the local oscillator. The feedback

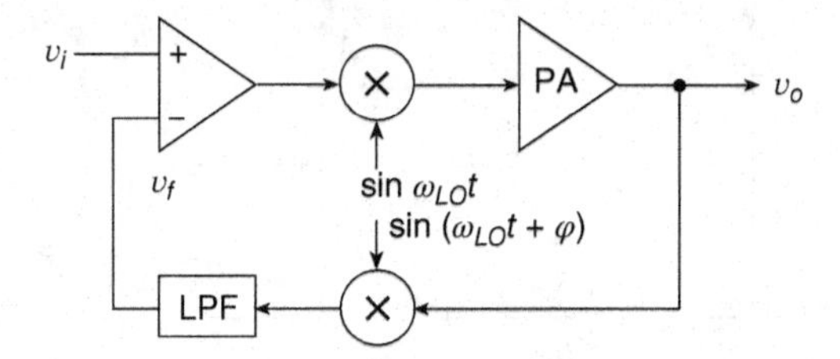

Figure 9.6 Block diagram of a power amplifier system with negative feedback linearization.

path consists of a downconversion mixer and a low-pass filter (LPF). The mixer converts the frequency of the RF signal f_{RF} to the input frequency $f_i = f_{RF} - f_{LO}$. In this system, a high loop gain T is obtained at low frequencies. Therefore, the negative feedback is very effective in reducing the nonlinear distortion. The total phase shift in the upper path by the error amplifier, the mixer, and the power amplifier usually exceeds 180°. Therefore, a phase shift is added to the local oscillator to ensure loop stability.

Example 9.2

An open-loop power amplifier has a gain $A_1 = 100$ for small signals and a gain $A_2 = 50$. The transfer function of the feedback network is $\beta = 0.1$. Find the closed-loop gains of the power amplifier with negative feedback.

Solution. The ratio of the open-loop gains is

$$\frac{A_1}{A_2} = \frac{100}{50} = 2 \tag{9.21}$$

The relative change in the gain of the open-loop power amplifier is

$$\frac{A_1 - A_2}{A_1} = \frac{100 - 50}{100} = 50\% \tag{9.22}$$

The loop gain for small signals is

$$T_1 = \beta A_1 = 0.1 \times 100 = 10 \tag{9.23}$$

and the closed-loop gain for small signals is

$$A_{f1} = \frac{A_1}{1 + T_1} = \frac{100}{1 + 10} = 9.091 \tag{9.24}$$

The loop gain for large signals is

$$T_2 = \beta A_2 = 0.1 \times 50 = 5 \tag{9.25}$$

and the closed-loop gain for large signals is

$$A_{f2} = \frac{A_2}{1 + T_2} = \frac{50}{1 + 5} = 8.333 \tag{9.26}$$

The ratio of the closed-loop gain is

$$\frac{A_{f1}}{A_{f2}} = \frac{9.091}{8.333} = 1.091 \tag{9.27}$$

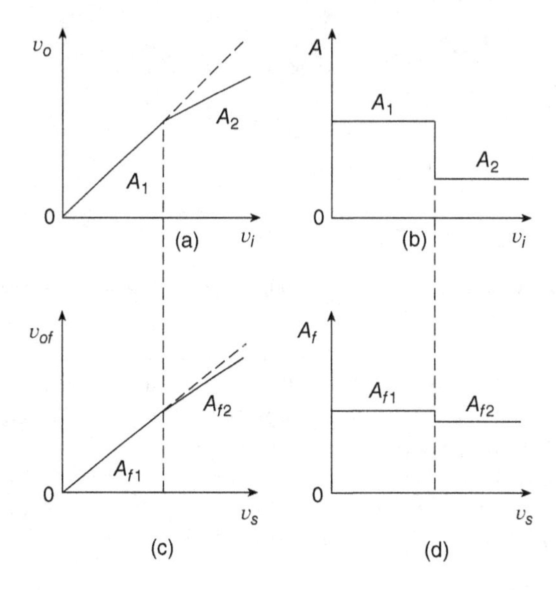

Figure 9.7 Transfer functions of a power amplifier without and with negative feedback. (a) Input–output voltage characteristic of amplifier without feedback. (b) Voltage gain of amplifier without feedback. (c) Input–output voltage characteristic of amplifier with feedback. (d) Voltage gain of amplifier with feedback.

The relative change in the gain of the closed-loop power amplifier is

$$\frac{A_{f1} - A_{f2}}{A_{f1}} = \frac{9.091 - 8.333}{9.091} = 8.333\% \tag{9.28}$$

Thus, the open-loop transfer function changes by a factor of 2, whereas the closed-loop transfer function changes by a factor of 1.091. Figure 9.7 illustrates the linearization of a power amplifier using the negative feedback technique.

Example 9.3

The relationship between the input voltage v_e and the output voltage v_o of a power stage of an RF amplifier is given by

$$v_o = Av_e^2 \tag{9.29}$$

Find the relationship between the input voltage v_s and the output voltage v_o for the RF amplifier with negative feedback. Draw the plots of $v_o = f(v_e)$ and $v_o = f(v_s)$ for $A = 100$ and $\beta = 1/10$.

Solution. The error voltage is

$$v_e = v_s - \beta v_o \tag{9.30}$$

Hence, the output voltage with negative feedback is

$$v_o = A(v_s - \beta v_o)^2 \tag{9.31}$$

Solving for v_s, we get

$$v_s = \beta v_o + \sqrt{\frac{v_o}{A}} \tag{9.32}$$

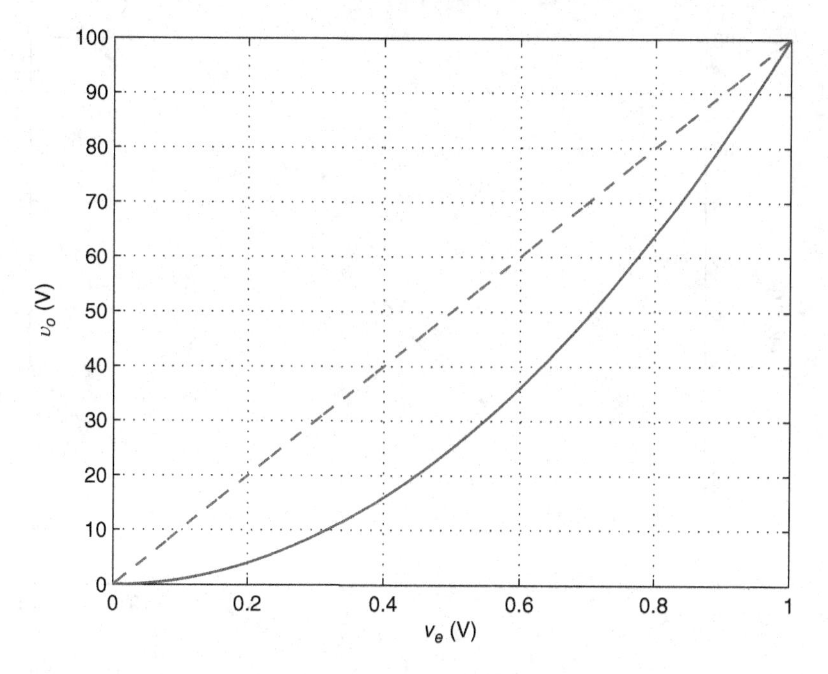

Figure 9.8 Output voltage v_o as functions of input voltage for a power amplifier without feedback.

or

$$v_o = \frac{v_s}{\beta} - \frac{1}{\beta}\sqrt{\frac{v_o}{A}} \tag{9.33}$$

Figures 9.8 and 9.9 illustrates the linearization of a power amplifier using the negative feedback technique. It can be seen that the plot of the amplifier with negative feedback shown in Fig. 9.9 is much more linear than that of the amplifier without negative feedback shown in Fig. 9.8.

9.5 Envelope Elimination and Restoration

A block diagram of a power amplifier with an envelope elimination and restoration system (EER) is shown in Fig. 9.10. The concept of the EER system was first presented by Kahn [14]. It has been used to increase linearity and efficiency of radio transmitters [14–23]. One path of the EER system consists of an RF limiter and an RF nonlinear high-efficiency switching-mode power amplifier. The other path consists of an envelope detector and an AM, which is a low-frequency power amplifier.

An AM and PM voltage applied to the amplifier input can be represented by

$$v_i = V(t)\cos[\omega_c t + \theta(t)] \tag{9.34}$$

This voltage is applied to both an RF limiter and an envelope detector. The output voltage of an RF limiter is a constant-envelope phase-modulated (CEPM) voltage

$$v_{PM} = V_c \cos[\omega_c t + \theta(t)] \tag{9.35}$$

The RF limiter should be designed to minimize amplitude-to-phase conversion. The output voltage of an envelope detector is

$$v_m(t) = V(t) \tag{9.36}$$

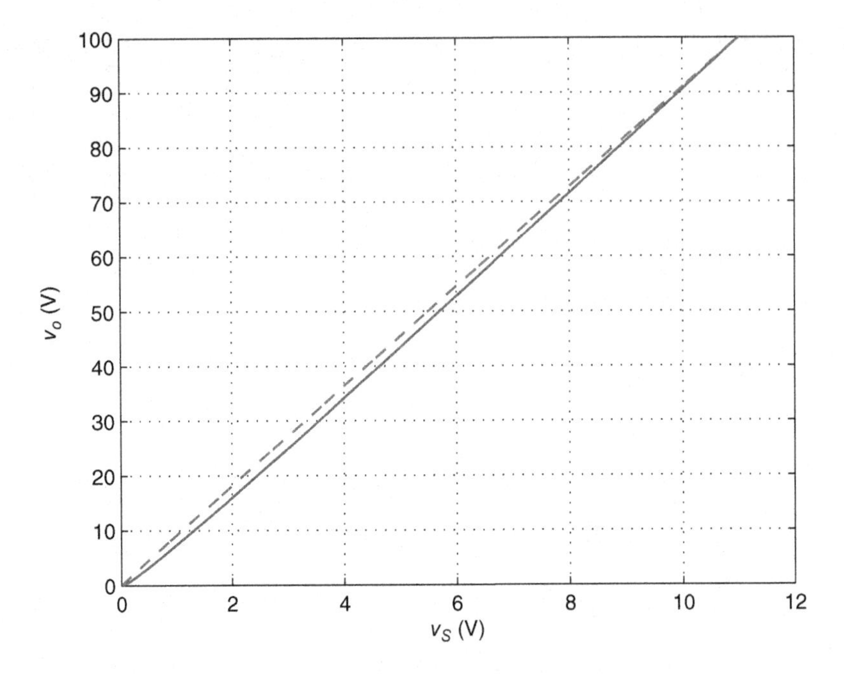

Figure 9.9 Output voltage as function of input voltage for a power amplifier with negative feedback.

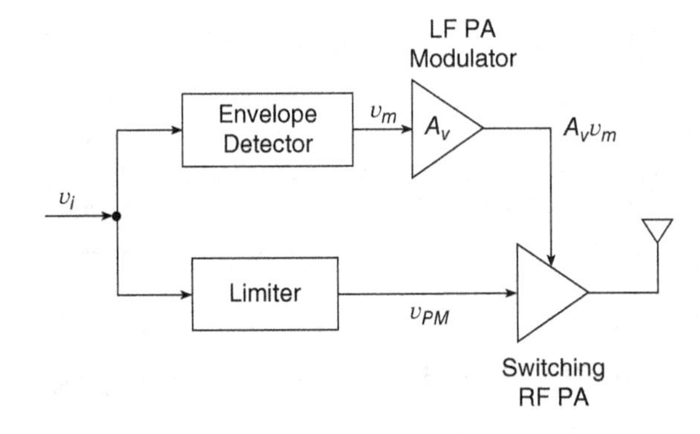

Figure 9.10 Block diagram of power amplifier with the envelope elimination and restoration (EER).

The envelope of the input voltage is extracted using an envelope detector, usually a diode AM detector to obtain the AM voltage $v_m(t)$. Next, it is amplified by a low-frequency power amplifier with the voltage gain A_v, producing a modulating voltage $A_v v_m(t)$. The envelope is eliminated from the AM–PM voltage using a limiter. The PM voltage v_{PM} is simultaneously amplified and amplitude modulated by an RF power amplifier to restore the original amplitude and phase information contained in the input signal v_i. The PM voltage v_{PM} is applied to the gate of the transistor in the RF power amplifier, and the modulating voltage v_m is used to modulate the supply voltage of the RF power amplifier, producing $v_{DD} = v_m(t)A_v V_{DD}$. In switching-mode RF power amplifiers, the amplitude of the output voltage is directly proportional to the supply voltage V_{DD}. A very important advantage of the EER system is that a linear RF power amplifier is not required. Linear RF power amplifiers exhibit very low efficiency, which decreases when the AM index decreases. The average AM index is in the range from 0.2 to 0.3. A highly efficient nonlinear RF power amplifier can be used in the EER system. Switching-mode RF power amplifiers, such as Class D, E, DE, and F amplifiers, exhibit a very high efficiency, greater than 90%. Their efficiency is approximately independent of the AM index. An AM high-efficiency Class E power amplifier was studied in [24].

The EER system suffers from a couple of shortcomings. First, there is mismatch between the total phase shift (delay) and gain in the two paths. It is a hard task because the phase shift in the low-frequency path is much larger than that in the high-frequency path. The IM term due to the phase mismatch is given by Raab [16]

$$IDM \approx 2\pi BW_{RF}^2 \Delta\tau^2 \tag{9.37}$$

where BW_{RF} is the bandwidth of the RF signal and $\Delta\tau$ is the delay mismatch.

Second, phase distortion is introduced by the limiter. Limiters are built by using nonlinear devices and exhibit a considerable AM–PM conversion at high frequencies, corrupting the phase $\theta(t)$ of the PM signal v_{PM}. Third, the phase distortion is introduced by the nonlinear transistor capacitance because of the AM of the supply voltage of the RF power amplifier.

9.6 Envelope Tracking

The main purpose of the envelope tracking (ET) technique is to maintain a high efficiency over a wide range of the output power P_O. A block diagram of an RF power amplifier with ET is shown in Fig. 9.11 [29–48]. This system is also called the power amplifier with dynamic power supply (DPS), the power amplifier with variable voltage power supply, the power amplifier with adaptive biasing, the power amplifier with dynamic control of supply voltage, or the dynamic drain bias control.

There are two reasons of applications of the DPS in RF power amplifiers:

1. To increase the efficiency of linear RF power amplifiers.
2. To obtain AM in nonlinear switching-mode RF power amplifiers.

All power amplifiers are nonlinear because the transistors are nonlinear and the signals are large. Amplification of AM signals requires linear amplifiers. Some power amplifiers are nearly linear, such as Class A power amplifier, and therefore are called linear power amplifiers. However, linear RF power amplifiers have very low efficiency, especially at a low ratio V_m/V_I. In these amplifiers, the transistor is operated as a dependent current source. Ideally, the amplitude of the ac component of the drain-to-source voltage V_m is independent of the supply voltage V_I when the transistor is operated as a dependent current source. In reality, the transistor cannot be

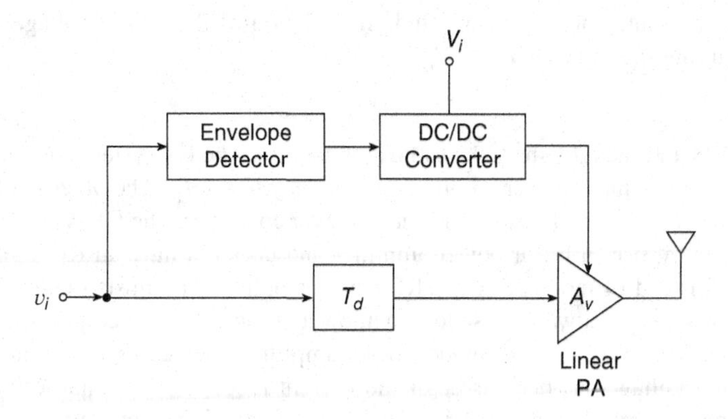

Figure 9.11 Block diagram of RF power amplifier with envelope tracking (ET).

operated as an ideal dependent current source because of the channel-length modulation and the resulting finite MOSFET output resistance. The Class A RF power amplifier is the most commonly used linear amplifier. The drain efficiency of the Class A amplifier with a fixed supply voltage V_I is low and is given by

$$\eta_D = \frac{P_{DS}}{P_I} = \frac{1}{2}\left(\frac{V_m}{V_I}\right)^2 \tag{9.38}$$

It can be seen that the drain efficiency η_D is low when V_m/V_I is low. When the supply voltage V_I and the dc supply current are fixed, such as in the Class A RF power amplifier, the supply power $P_I = I_I V_I$ is also fixed and independent of the output power P_O. In AM systems, the amplitude of the ac component of the drain-to-source voltage is a function of time $V_m(t)$. The instantaneous drain efficiency of Class A RF amplifier applied to amplify AM signals at a fixed supply voltage V_I is

$$\eta_D(t) = \frac{P_{DS}(t)}{P_I} = \frac{1}{2}\left(\frac{V_m(t)}{V_I}\right)^2 \tag{9.39}$$

The average value of the drain voltage amplitude V_m is usually much lower than the supply voltage V_I. Therefore, the long-term average drain efficiency of the Class A amplifier is very low, for example, 5%.

The efficiency of the Class A RF power amplifier is high, close to 50%, when the ratio V_m/V_I is close to 1. The instantaneous drain efficiency of the Class A RF power amplifier with a variable supply voltage $V_I(t) = f[V_m(t)]$ is

$$\eta_D(t) = \frac{P_{DS}(t)}{P_I(t)} = \frac{1}{2}\left(\frac{V_m(t)}{V_I(t)}\right)^2 \tag{9.40}$$

If the supply voltage V_I tracks the amplitude $V_m(t)$ according to the envelope voltage so that the ratio $V_m(t)/V_I(t)$ remains close to 1, the instantaneous drain efficiency η_D is high over a wide range of output power, for example, 45%. Therefore, the long-term average drain efficiency is also high. The envelope voltage of the RF input signal is detected by an envelope detector or is derived from the baseband I/Q waveform. Then, the output voltage of the DPS $V_I(t)$ is used to supply the RF power amplifier, which is the output stage of the RF transmitter.

In nonlinear amplifiers, the transistors are operated as switches. The efficiency of switch-mode amplifier is high because the transistor voltage is low when the transistor current is high, and the transistor current is zero when the transistor voltage is high. Examples of nonlinear amplifiers are Class D, Class E, and Class DE RF power amplifiers. In these amplifiers, the amplitude of the output voltage V_m is proportional to the supply voltage V_I

$$V_m = kV_I \tag{9.41}$$

where k is a constant. Therefore, amplitude modulation of the output voltage can be accomplished by varying the supply voltage V_I

$$V_m(t) = kV_I(t) \tag{9.42}$$

Both the DPS and the AM share the same architecture. The ET system consists of two paths. One path contains a linear power amplifier and a delay block T_d. The other path consists of an envelope detector and a dc–dc switching-mode power converter. The ET system is similar to the EER system, but it uses a linear power amplifier and does not have an RF limiter as the EER system does. Instead of modulating the RF power amplifier, it adjusts the level of the supply voltage so that V_m/V_I is always close to 1 to allow an efficient linear amplification.

The drain efficiency of current-mode power amplifiers decreases as the amplitude of the drain-to-source voltage V_m decreases. In the current-mode power amplifiers, such as Class A, AB, B, and C amplifiers, the transistor is operated as a dependent current source. In these amplifiers, the amplitude of the drain-to-source voltage V_m is proportional to the amplitude of the input voltage V_{im}. The drain efficiency of the power amplifiers at the resonant frequency is given by

$$\eta_D = \frac{P_{DS}}{P_I} = \frac{1}{2}\left(\frac{I_m}{I_I}\right)\left(\frac{V_m}{V_I}\right) \tag{9.43}$$

It can be seen from this equation that the drain efficiency η_D is proportional to the ratio V_m/V_I. As V_m decreases due to an increase in V_{im} at fixed value of the supply voltage V_I, the ratio V_m/V_I also decreases, reducing the drain efficiency. If the supply voltage V_I is increased proportionally to the amplitude of the input voltage V_{im}, the ratio V_{im}/V_I, then the ratio V_m/V_I can be maintained at a fixed value close to 1 (e.g., 0.9), resulting in high efficiency at any amplitude V_m. When V_{im} is varied in current-mode power amplifiers, the conduction angle of the drain current remains approximately constant and the ratio I_m/I_I is also constant. For instance, for the Class B amplifier, $I_m/I_I = \pi/2$. The efficiency is improved, while low distortion of the RF linear power amplifier is preserved.

9.7 The Doherty Amplifier

The Doherty power amplifier was first proposed in 1936 [52]. The main purpose of this system is to maintain high efficiency over a wide range of the input voltage of a linear power amplifier with AM signals [52–62]. The average efficiency of linear power amplifiers with AM is very low because the average AM index is very low, usually from 0.2 to 0.3. The Doherty power amplifier architecture delivers high efficiency with input signals that have high peak-to-average power ratio (PAPR) in the range of 6–10 dB. This system has the potential to deliver high efficiency in base station transmitters. Current and emerging wireless systems produce high PAR signals, including WCDMA, CDMA2000, and orthogonal frequency division multiplexing (OFDM). For example, at 3-dB output power BO (i.e., at 50% of full power), the efficiency of the Class B power amplifier decreases to $\pi/4 = 39\%$. The transmitter accounts for a high percentage of the overall power consumption. Increased efficiency can lower the electricity cost and minimize cooling requirements.

A block diagram of Doherty power amplifier is shown in Fig. 9.12. It consists of a *main power amplifier* and an *auxiliary power amplifier*. The main power amplifier is called a *carrier amplifier* and an auxiliary power amplifier is called *peak amplifier*. The main power amplifier is usually a Class B (or a Class AB) amplifier and the auxiliary amplifier is usually a Class C amplifier. A quarter-wave transformer is connected at the output of the main power amplifier. Also, a

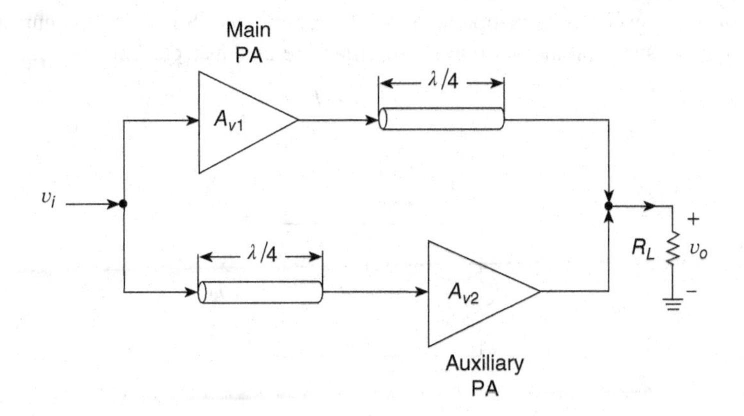

Figure 9.12 Block diagram of Doherty power amplifier.

quarter-wave transformer is connected at the input of the auxiliary amplifier to compensate for a 90° phase shift introduced by the transformer in the main amplifier. The transistors in both amplifiers are operated as dependent current sources. The main amplifier saturates at high input power, and therefore its voltage gain decreases. The auxiliary amplifier turns on at high input power when the main power amplifier reaches saturation. As a result, the linearity of the overall system is improved at high input power levels. The system maintains good efficiency under power BO conditions. It is common for power amplifiers in mobile transmitters to operate at output power levels of 10–40 dB BO from peak power.

9.7.1 Condition for High Efficiency Over Wide Power Range

The efficiency of power amplifiers at the resonant frequency is given by

$$\eta_D = \frac{P_{DS}}{P_I} = \frac{1}{2}\left(\frac{I_m}{I_I}\right)\left(\frac{V_m}{V_I}\right) = \frac{1}{2}\left(\frac{I_m}{I_I}\right)\left(\frac{RI_m}{V_I}\right) \tag{9.44}$$

When V_{im} is varied in current-mode power amplifiers (such as in the Class B and C amplifiers), the peak drain current is proportional to V_{im}, the conduction angle of the drain current remains approximately constant, and therefore the ratio I_m/I_I also remains constant. The ratio V_m/V_I should be held close to 1 (e.g., equal to 0.9) to achieve high efficiency at any value of the output voltage amplitude. This can be accomplished if $V_m/V_I = RI_m/V_I$ is maintained at a fixed value close to 1. At a fixed supply voltage V_I, the following condition should be satisfied:

$$V_m = RI_m = Const \tag{9.45}$$

where R is the load resistance. If I_m is decreased, R should be increased so that the product RI_m remains constant. This idea is partially realized in the Doherty amplifier because the carrier power amplifier is presented with a *modulated load impedance*, *dynamic load variation*, *drive-dependent load resistance* or *load-pulling effect*, which enables high efficiency and a constant voltage gain over a wide input power range. The main shortcoming of the Doherty power amplifier is its poor linearity. However, recent advancements in digital and analog predistortion and feedforward techniques can dramatically reduce nonlinear distortion.

9.7.2 Impedance Modulation Concept

The load impedance of a current source can be modified by applying a current from another current source. Consider the circuit depicted in Fig. 9.13. The current source I_1 represents the transistor of one power amplifier operated as a current source. Similarly, the current source I_2 represents the transistor of another power amplifier operated as a current source. From KCL,

$$I_L = I_1 + I_2 \tag{9.46}$$

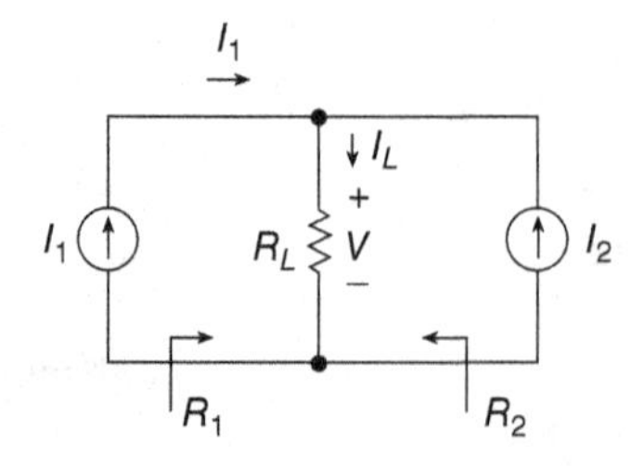

Figure 9.13 Circuit with load impedance modulation.

and from Ohm's law,

$$R_L = \frac{V_L}{I_L} \tag{9.47}$$

The load impedance seen by the current sources I_1 is

$$R_1 = \frac{V}{I_1} = \frac{V}{I_L - I_2} = \frac{V}{I_L\left(1 - \frac{I_2}{I_L}\right)} = \frac{V_L}{I_L\left(1 - \frac{I_2}{I_1 + I_2}\right)} = R_L\left(1 + \frac{I_2}{I_1}\right) \tag{9.48}$$

Thus, the load resistance seen by the current source I_1 is dependent on the current I_2. If the currents I_1 and I_2 are in phase, the load voltage V is increased, and therefore the current source I_1 sees a higher load resistance. For $I_2 = I_1$,

$$R_1 = 2R_L \tag{9.49}$$

For $I_2 = 0$,

$$R_1 = R_L \tag{9.50}$$

The changes in the load resistance R_1 in the circuit shown in Fig. 9.13 are not compatible with power amplifier requirements because R_1 decreases as I_2/I_1 decreases.

Similarly, the load resistance of the current source I_2 is expressed as

$$R_2 = \frac{V}{I_2} = R_L\left(1 + \frac{I_1}{I_2}\right) \tag{9.51}$$

9.7.3 Equivalent Circuit of Doherty Amplifier

Both the main and auxiliary power amplifiers should deliver the maximum powers at the maximum overall output power. As the overall power is decreased, the power of each amplifier should be reduced. The peak drain current is proportional to the peak gate-to-source voltage in Class A, AB, B, and C amplifiers. The maximum output power occurs at the maximum value of the fundamental component of the drain current I_{1max}. The amplitude of the fundamental component of the drain-to-source voltage also has the maximum value $V_{m(max)}$ at the maximum output power $P_{O(max)}$. The load resistance seen by the drain and source terminal at the operating frequency for $P_{o(max)}$ is

$$R_1 = R_{1(opt)} = \frac{V_{m(max)}}{I_{1max}} \tag{9.52}$$

As the drive power is decreased, the output power decreases, and the fundamental component of the drain current decreases. In order to maintain a constant value of the fundamental component of the drain-to-source voltage V_m when I_{d1m} is decreased, the load resistance R_1 should be increased. Therefore, an *impedance inverter* is needed between the load R_L and the current source I_1, as shown in Fig. 9.14. A quarter-wave transmission line transformer acts as an impedance inverter

$$R_1 = \frac{Z_o^2}{R_3} \tag{9.53}$$

where Z_o is the characteristic impedance of the transmission line. As R_3 increases, R_1 decreases. Lumped-component impedance inverters are depicted in Fig. 9.15. The components of these inverters are given by

$$\omega L = \frac{1}{\omega C} = Z_o \tag{9.54}$$

The phase shift of the impedance inverter shown in Fig. 9.15(a) is −90° and the phase shift of impedance inverter shown in Fig. 9.15(b) is 90°.

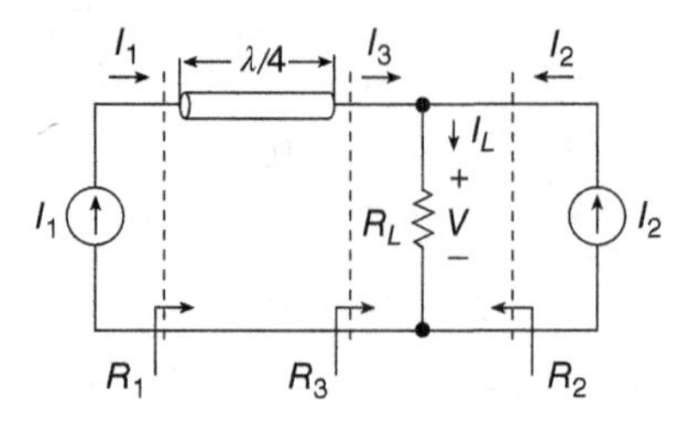

Figure 9.14 Equivalent circuit of Doherty power amplifier.

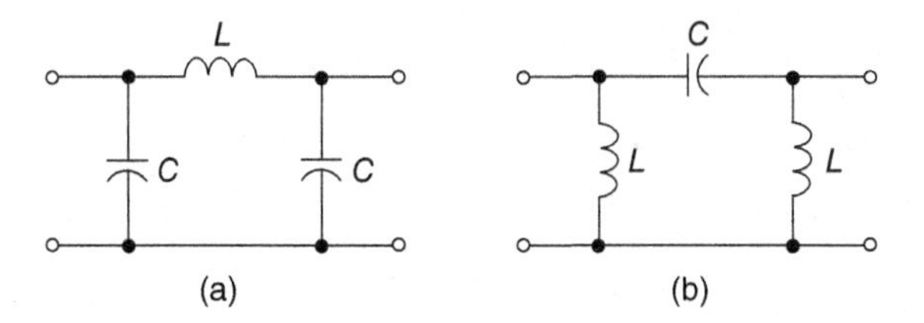

Figure 9.15 Lumped-component impedance inverters. (a) Two-capacitor single-inductor π impedance inverter. (b) Two-inductor single-capacitor π impedance inverter.

An equivalent circuit of the most basic form of the Doherty amplifier is shown in Fig. 9.14. The load resistance of the quarter-wave transformer is given by

$$R_3 = \frac{V}{I_3} = \frac{V}{I_L - I_2} = \frac{V}{I_L\left(1 - \frac{I_2}{I_L}\right)} = R_L \frac{1}{\frac{I_3}{I_2 + I_3}} = R_L\left(1 + \frac{I_2}{I_3}\right) \tag{9.55}$$

The load resistance of the current source I_1 is

$$R_1 = \frac{Z_o^2}{R_3} = \frac{Z_o^2}{R_L\left(1 + \frac{I_2}{I_3}\right)} \tag{9.56}$$

As I_2 decreases, R_1 increases. Therefore, as I_1 decreases, V_m remains constant. For $I_2 = I_3$,

$$R_1 = \frac{Z_o^2}{2R_L} \tag{9.57}$$

and for $I_2 = 0$,

$$R_1 = \frac{Z_o^2}{R_L} \tag{9.58}$$

As I_2 decreases from its maximum value to zero, the load resistance of the current source I_1 increases from $Z_o^2/(2R_L)$ to Z_o^2/R_L.

The load resistance of the current source I_2 is given by

$$R_2 = R_L\left(1 + \frac{I_3}{I_2}\right) \tag{9.59}$$

9.7.4 Power and Efficiency of Doherty Amplifier

The output power averaged over a cycle of the modulating frequency of the main power amplifier with AM single-frequency signal is

$$P_M = \left(1 - \frac{m}{\pi} + \frac{m^2}{4}\right) P_C \tag{9.60}$$

where P_C is the power of the carrier and m is the modulation index. The average output power of the auxiliary amplifier is

$$P_A = \left(\frac{m}{\pi} + \frac{m^2}{4}\right) P_C \tag{9.61}$$

For $m = 1$,

$$P_M = \left(1 - \frac{1}{\pi} + \frac{1}{4}\right) = 0.9317 P_C \tag{9.62}$$

$$P_A = \left(\frac{1}{\pi} + \frac{1}{4}\right) = 0.5683 P_C \tag{9.63}$$

and the total average output power is

$$P_T = P_M + P_A = 1.5 P_C \tag{9.64}$$

For $m = 0$,

$$P_T = P_M = P_C \tag{9.65}$$

and

$$P_A = 0 \tag{9.66}$$

The drain efficiency averaged over a cycle of the modulating frequency of the main power amplifier with AM single-frequency signal is

$$\eta_M = \left(1 - \frac{m}{\pi} + \frac{m^2}{4}\right) \eta_C \tag{9.67}$$

where η_C is the drain efficiency for the operation at carrier only ($m = 0$). The average drain efficiency of the auxiliary amplifier is

$$\eta_A = 1.15 \left(\frac{1}{2} + \frac{\pi}{8} m\right) \eta_C \tag{9.68}$$

The average drain efficiency of the overall amplifier is

$$\eta = \frac{1 + \frac{m^2}{2}}{1 + 1.15 \frac{2}{\pi} m} \eta_C \tag{9.69}$$

The total average efficiency can be expressed in terms of the amplitude of the input voltage V_i normalized with respect to its maximum value V_{imax}

$$\eta = \frac{\pi}{2} \left(\frac{V_i}{V_{imax}}\right) \quad \text{for} \quad 0 \leq V_i \leq \frac{V_{imax}}{2} \tag{9.70}$$

and

$$\eta = \frac{\pi}{2} \frac{\left(\frac{V_i}{V_{im}}\right)^2}{3\left(\frac{V_i}{V_{imax}}\right) - 1} \quad \text{for} \quad \frac{V_{imax}}{2} \leq V_i \leq V_{imax} \tag{9.71}$$

Figure 9.16 depicts the overall efficiency of the Doherty power amplifier as a function of the normalized input voltage amplitude V_i/V_{imax}. Figure 9.17 shows the normalized voltage and current amplitudes of the main power amplifier (M) and the auxiliary power amplifier (A) as functions of the normalized amplitude of the input voltage V_i/V_{imax}.

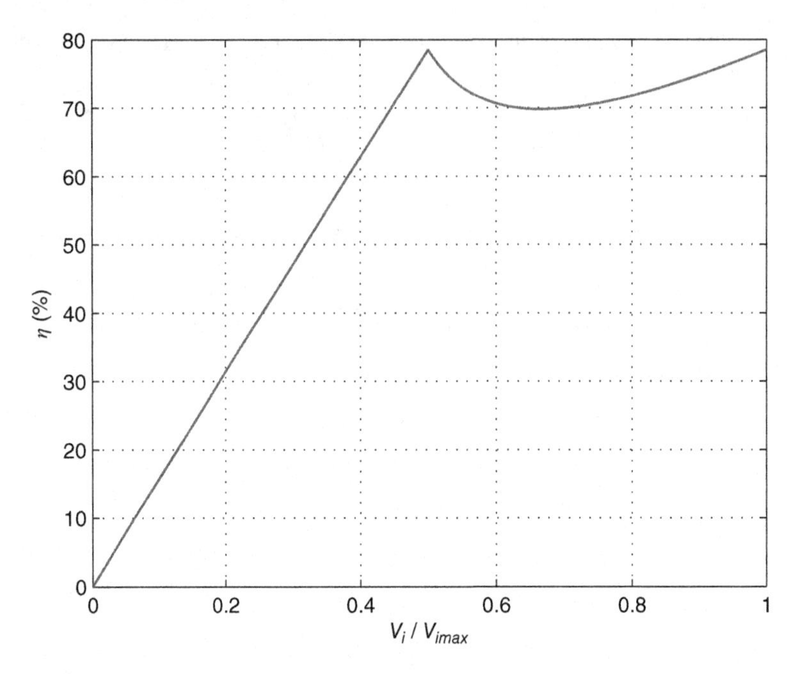

Figure 9.16 Efficiency of the Doherty power amplifier as a function of the normalized input voltage V_i/V_{imax}.

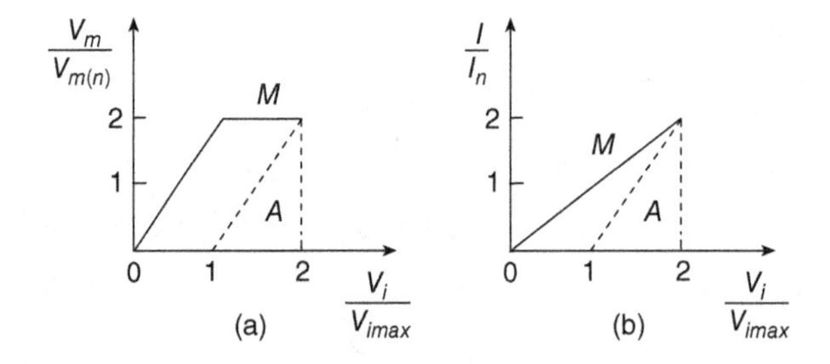

Figure 9.17 Voltage and current amplitudes of the main power amplifier (M) and the auxiliary power amplifier (A) as functions of V_i/V_{imax} in the Doherty system. (a) $V_m/V_{m(n)}$ versus V_i/V_{imax}. (b) I/I_n versus V_i/V_{imax}.

9.8 Outphasing Power Amplifier

An outphasing power amplifier system was invented by Chireix [64] to improve the linearity and the efficiency of AM signal power amplifiers [64–79]. The system is depicted in Fig. 9.18. It consists of a signal component separator (SCS) and PM, two identical nonlinear high-efficiency power amplifiers, and RF power adder (combiner). Power amplifiers can be Class D, E, or DE switching-mode constant-envelope amplifiers. The signal separator and phase modulator is a complex stage. The outphasing power amplifier was studied in [65–82]. The subsequent analysis shows the possibility of a linear amplification with nonlinear components (LINC) [65].

A bandpass signal with AM and PM can be described by

$$v_i = V(t)\cos[\omega_c t + \phi(t)] = \cos[\omega_c t + \phi(t)]\cos[\arccos V(t)] \tag{9.72}$$

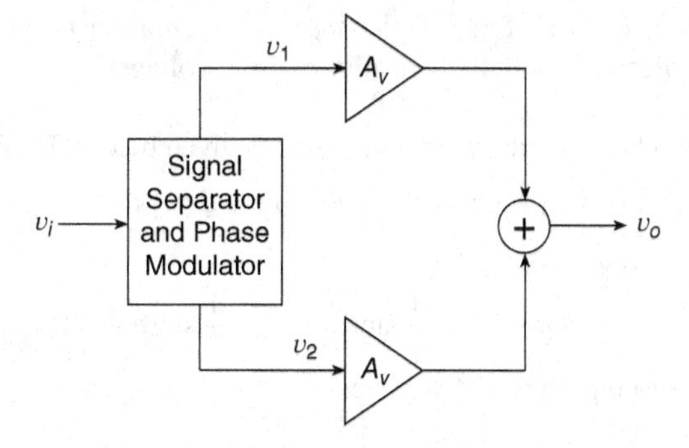

Figure 9.18 Block diagram of outphasing power amplifier system.

Using the trigonometric relationship,

$$\cos\alpha\cos\beta = \frac{1}{2}\cos(\alpha+\beta) + \frac{1}{2}\cos(\alpha-\beta) \tag{9.73}$$

the AM signal can be represented as the sum of two CEPM signals

$$v_i = v_1 + v_2 \tag{9.74}$$

where

$$v_1 = \frac{1}{2}\cos[\omega_c t + \phi(t) + \arccos V(t)] \tag{9.75}$$

and

$$v_2 = \frac{1}{2}\cos[\omega_c t + \phi(t) - \arccos V(t)]. \tag{9.76}$$

Both the amplitude and phase information of the original input signal v_i is included in the PM component signals v_1 and v_2. The waveforms of voltages v_1 and v_2 have constant amplitudes. The constant-envelope signals v_1 and v_2 can be amplified by nonlinear high-efficiency switching-mode power amplifiers, such as Class D, E, and DE circuits.

Using the trigonometric formula,

$$\cos\alpha + \cos\beta = 2\cos\frac{\alpha+\beta}{2}\cos\frac{\alpha-\beta}{2} \tag{9.77}$$

the two amplified signals with the same voltage gain A_v can be added up to form the output voltage waveform

$$\begin{aligned} v_o &= A_v(v_1+v_2) = A_v\{\frac{1}{2}\cos[\omega_c t + \phi(t) + \arccos V(t)] + \frac{1}{2}\cos[\omega_c t + \phi(t) - \arccos V(t)] \\ &= A_v\cos\frac{\omega_c t + \phi(t) + \arccos V(t) + \omega_c t + \phi(t) - \arccos V(t)}{2} \\ &\quad \times \cos\frac{\omega_c t + \phi(t) + \arccos V(t) - \omega_c t - \phi(t) + \arccos V(t)}{2} \\ &= A_v\cos[\omega_c t + \phi(t)]\cos[\arccos V(t)] = A_v V(t)\cos[\omega_c t + \phi(t)] \\ &= V_{om}(t)\cos[\omega_c t + \phi(t)] \end{aligned} \tag{9.78}$$

where the amplitude of the output voltage is

$$V_{om}(t) = A_v V(t) \tag{9.79}$$

The output signal v_o contains the same AM and PM information as v_i does.

The practical implementation of the outphasing power amplifier is very difficult. The SCS is not an easy circuit to design and build. The phase of the voltages v_1 and v_2 must be modulated by a highly nonlinear functions $\pm$ arccos $V(t)$.

An alternative method of splitting the input signal is given below. The input signal is

$$v_i = V(t)\cos[\omega_c t + \phi(t)] = \cos[\omega_c t + \phi(t)]\sin[\arcsin V(t)] \tag{9.80}$$

From the trigonometric relationship,

$$\cos\alpha\sin\beta = \frac{1}{2}\sin(\alpha+\beta) - \frac{1}{2}\sin(\alpha-\beta) \tag{9.81}$$

we obtain two constant-amplitude PM voltages

$$v_1 = \frac{1}{2}\sin[\omega_c t + \phi(t) + \arcsin V(t)] \tag{9.82}$$

and

$$v_2 = -\frac{1}{2}\sin[\omega_c t + \phi(t) - \arcsin V(t)] \tag{9.83}$$

From the relationship,

$$\sin\alpha - \sin\beta = 2\cos\frac{\alpha+\beta}{2}\sin\frac{\alpha-\beta}{2} \tag{9.84}$$

we obtain the output voltage waveform

$$v_o = A_v(v_1 + v_2) = \cos[\omega_c t + \phi(t)]\sin[\arcsin V(t)] = A_v V(t)\cos[\omega_c t + \phi(t)] \tag{9.85}$$

There are two more combinations of splitting and adding the signals.

A complex representation of the input signal is

$$v_i(t) = A(t)e^{j\phi(t)}\ 0 \le A(t) \le A_{max} \tag{9.86}$$

This signal can be split into two components

$$v_1(t) = v_i(t) - v(t) \tag{9.87}$$

and

$$v_2(t) = v_i(t) + v(t) \tag{9.88}$$

where $v(t)$ is the quadrature voltage described by

$$v(t) = jv_i(t)\sqrt{\frac{A_{max}^2}{A^2(t)} - 1} \tag{9.89}$$

9.9 Summary

- The predistortion technique linearizes power amplifiers by making suitable modifications to the amplitude and phase of its input signal so that the overall transfer function of the system is linear.
- The principle of operation of the feedforward linearization system of power amplifiers is based on cancellation of the distortion signal produced by the nonlinearity of the power amplifier.
- The feedforward linearization system of power amplifiers is wideband, low noise, and stable.
- The feedforward linearization system of power amplifiers suffers from gain and phase (delay) mismatches of signals in two channels.
- The negative feedback reduces nonlinear distortion by a factor $1 + T$.
- The EER system improves the linearity and efficiency of RF power transmitters with AM and PM.

- In the EER system, the AM–PM signal is first decomposed into a CEPM signal by a limiter and an envelope signal by an envelope detector. Next, the two signals are recombined by an AM of an RF power amplifier to restore the amplitude and phase of the original signal.
- Highly nonlinear RF power amplifiers can be used in the EER system, such as Class D, E, DE, and F amplifiers.
- A phase shift and gain mismatch in two signal paths is a drawback of the EER system.
- The phase distortion, introduced by the limiter and the nonlinearity of the transistor output capacitance, is a shortcoming of the EER system.
- The ET technique adjusts the supply voltage of a power amplifier so that $V_m/V_I \approx 1$, yielding high efficiency of a linear power amplifier.
- The load impedance of power amplifiers can be modulated by driving the load with another current-source power amplifier.
- A Doherty power amplifier requires an impedance inverter to increase its load resistance when its output power decreases. Therefore, the amplitude of the drain-to-source voltage $V_m = R_1 I_1$ remains constant when I_1 decreases.
- The outphasing power amplifier separates AM input signal into two constant-envelope signals, amplifies them by high-efficiency nonlinear amplifiers, and combines the amplified signals into one output signal.
- The outphasing power amplifier offers wideband high linearity and high efficiency simultaneously.

9.10 Review Questions

9.1 Explain the principle of reduction of nonlinear distortion using the predistortion technique.

9.2 Explain the principle of operation of the feedforward linearization system of power amplifiers.

9.3 List advantages of the feedforward linearization system of power amplifiers.

9.4 List disadvantages of the feedforward linearization system of power amplifiers.

9.5 How large is the reduction of nonlinear distortion using the negative feedback technique?

9.6 Explain the principle of operation of the EER system.

9.7 What are advantages of the EER system?

9.8 What are disadvantages of the EER system?

9.9 What is the principle of operation of envelope tracking?

9.10 What is the difference between the EER and envelope tracking techniques?

9.11 What is the main goal of using the Doherty power amplifier?

9.12 Why is an impedance inverter used in a Doherty power amplifier?

9.13 Is the linearity of the Doherty amplifier good?

9.14 Explain the principle of operation of the outphasing power amplifier.

9.11 Problems

9.1 An RF power amplifier has the voltage gain $A_1 = 180$ for $v_s = 0$ to 1 V and the voltage gain $A_2 = 60$ for $v_s = 1$ to 2 V. Find the voltage transfer function of predistorter to obtain a linear function of the entire amplifier.

9.2 The voltage gains of the power stage of an RF power amplifier are $A_1 = 900$ for $v_i = 0$ to 1 V, $A_2 = 600$ for $v_i = 1$ to 2 V, and $A_3 = 300$ for $v_i = 2$ to 3 V. Negative feedback with $\beta = 0.1$ is used to linearize the amplifier. Find the gains of the amplifier with negative feedback.

9.3 The relationship between the input voltage v_e and the output voltage v_o of an power stage of an RF amplifier is given by

$$v_o = Av_e^2$$

Find the relationship between the input voltage v_s and v_o of the amplifier with negative feedback. Draw the plots $v_o = f(v_e)$ and $v_o = f(v_s)$ at $A = 50$ and $\beta = 1/5$. Compare the nonlinearity of both plots.

9.4 The relationship between the input voltage v_e and the output voltage v_o of an power stage of an RF amplifier is given by

$$v_o = 100\left(v_e + \frac{v_e^2}{2}\right)$$

Find the relationship between the input voltage v_s and v_o of the amplifier with negative feedback. Draw the plots $v_o = f(v_e)$ and $v_o = f(v_s)$ at $\beta = 1/20$. Compare the nonlinearity of both plots.

References

[1] C. Haskins, T. Winslow, and S. Raman, "FET diode linearizer optimization for amplifier predistortion in digital radios," *IEEE Microwave Guided Wave Letters*, vol. 10, no. 1, pp. 21–23, 2000.

[2] J. Yi, Y. Yang, M. Park, W. Kong, and B. Kim, "Analog predistortion linearizer for high-power RF amplifiers," *IEEE Transactions on Microwave Theory and Techniques*, vol. 48, no. 12, pp. 2709–2713, 2000.

[3] S. Y. Lee, Y. S. Lee, K. I. Jeon, and Y. H. Jeong, "High linear predistortion power amplifiers with phase-controlled error generator," *IEEE Microwave and Wireless Components Letters*, vol. 16, no. 12, pp. 690–692, 2006.

[4] H. Seidel, "A microwave feedforward experiments," *Bell System Technical Journal*, vol 50, pp. 2879–2916, 1994.

[5] R. G. Mayer, R. Eschenbach, and W. M. Edgerley, "A wideband feedforward amplifier," *IEEE Journal of Solid-State Circuits*, vol. 9, pp. 422–488, 1974.

[6] D. P. Myer, "A multicarrier feedforward amplifier design," *Microwave Journal*, pp. 78–88, 1994.

[7] E. E. Eid, F. M. Ghannouchi, and F. Beauregard, "A wideband feedforward linearization system design," *Microwave Journal*, pp. 78–86, 1995.

[8] N. Pothecary, *Feedforward Linear Power Amplifiers*. Norwood, MA: Artech House, 1999.

[9] A. Shrinivani and B. A. Wooley, *Design and Control of RF Power Amplifiers*. Norwell, MA: Kluwer, 2003.

[10] S. C. Cripps, *Advanced Techniques in RF Power Amplifiers Design*. Norwell, MA: Kluwer, 2002.

[11] S. C. Cripps, *RF Power Amplifiers for Wireless Communications*, 2nd Ed. Norwell, MA: Kluwer, 2006.

[12] M. Johnson and T. Mattson, "Transmitter linearization using Cartesian feedback for linear TDMA modulation," *Proceedings IEEE Transactions on Vehicular Technology Conference*, pp. 439–44, May 1991.

[13] J. L. Dawson and T. H. Lee, *Feedback Linearization of RF Power Amplifiers.* Norwell, MA: Kluwer, 2004.

[14] L. R. Kahn, "Single-sideband transmission by envelope elimination and restoration," *Proceedings of the IRE*, vol. 40, pp. 803–806, 1952.

[15] H. L. Krauss, C. W. Bostian, and F. H. Raab, *Solid State Radio Engineering.* New York, NY: John Wiley & Sons, 1980.

[16] F. H. Raab, "Envelope elimination and restoration system requirements," *Proceedings RF Technology Expo*, February 1988, pp. 499–512.

[17] F. H. Raab and D. Rupp, "High-efficiency single-sideband HF/VHF transmitter based upon envelope elimination and restoration," *Proceedings of International Conference on HF systems and Techniques*, July 1994, pp. 21–25.

[18] F. H. Raab. B. E. Sigmon, R. G. Myers, and R. M. Jackson, "L-bend transmitter using Kahn EER technique," *IEEE Transactions on Microwave Theory and Techniques*, vol. 44, pp. 2220–2225, 1996.

[19] F. H. Raab, "Intermodulation distortion in Kahn-technique transmitters," *IEEE Transactions on Microwave Theory and Techniques*, vol. 44, no. 12, pp. 2273–2278, 1996.

[20] D. K. Su and W. McFarland, "An IC for linearizing RF power amplifier using envelope elimination and restoration," *IEEE Journal of Solid-State Circuits*, vo. 33, pp. 2252–2258, 1998.

[21] H. Rabb, B. E. Sigmon, R. G. Meyer, and R. M. Jackson, "L-band transmitter using Khan EER technique," *IEEE Transactions on Microwave Theory and Techniques*, vol. 46, no. 12, pp. 2273–2278, 1998.

[22] I. Kim, J. Kim, J. Moon, and B. kim, "Optimized envelope shaping for hybrid EER transmitter of mobile WiMAX optimized ET operation," *IEEE Microwave Wireless Component Letters*, vol. 14, no. 8, pp. 389–391, 2004.

[23] I. Kim, Y. Woo, J. Kim, J. Woo, J. Kim, and B. Kim, "High-efficiency hybrid transmitter using optimized power amplifier," *IEEE Transactions on Microwave Theory and Techniques*, vol. 56, no. 11, pp. 2582–2593, 2008.

[24] M. Kazimierczuk, "Collector amplitude modulation of Class E tuned power amplifier," *IEEE Transactions on Circuits and Systems*, vol. 31, pp. 543–549, 1984.

[25] F. Wang, D. F. Kimball, J. D. Popp, A. H. Yang, D. Y. C. Lie, P. M. Asbeck, and L. E. Larsen, "An improved power-aided efficiency 19-dBm hybrid envelope elimination and restoration power amplifier for 802.11g WLAN applications<" *IEEE Transactions on Microwave Theory and Technique*, vol. 54, no. 12, pp. 4086–4099, 2006.

[26] J Groe, "Polar transmitters for wireless communications," *IEEE Communications Magazine*, vol. 45, no. 9, pp. 58–63, 2007.

[27] I. Kim, Y. Y. Woo, J. Kim, J. Moon, J. Kimi, and B. Kim, "High-efficiency EER transmitter using optimized power amplifier," *IEEE Transactions on Microwave Theory and Technique*, vol. 56, no. 11, pp. 3848–3856, 2008.

[28] A. A. M. Saleh and D. C. Cox, "Improving the power-added efficiency of FET amplifiers with operating with varying envelope signals," *IEEE Transactions on Microwave Theory and Techniques*, vol. 31, no. 1, pp. 51–55, 1983.

[29] G. Hannington, P.-F. Chen, V. Radisic, T. Itoch, and P. M. Asbeck, "Microwave power amplifier efficiency improvement with a 10 MHz HBT dc-dc converter," *1998 IEEE MTT-S Microwave Symposium Digest*, pp. 313–316, 1998.

[30] G. Hannington, P.-F. Chen, P. M. Asbeck, and L. E. Larson, "High-efficiency power amplifier using dynamic power supply voltage for CDMA applications," *IEEE Transactions on Microwave Theory and Techniques*, vol. 47, no. 8, pp. 1471–1476, 1999.

[31] Y.-S. Joen, J. Cha, and S. Nam, "High-efficiency power amplifier using novel dynamic bias switching," *IEEE Transactions on Microwave Theory and Techniques*, vol. 55, no. 4, pp. 690–696, 2000.

[32] M. Ranjan, K. H. Koo, G. Hannington, C. Fallesen, and P. M. Asbeck, "Microwave power amplifier with digitally-controlled power supply voltage for high efficiency and high linearity," *2000 IEEE MTT-S Microwave Symposium Digest*, vol. 1, pp. 493–496, 2000.

[33] J. Straudinger, B. Glisdorf, D. Neumen, G. Sadowniczak, and R. Sherman, "High-efficiency CDMA RF power amplifier using dynamic envelope-tracking technique," *2000 IEEE MTT-S Microwave Symposium Digest*, vol. 2, pp. 873–876, 2000.

[34] P. Midya, K. Haddad, L. Connel, S. Bergstedt, and B. Roekner, "Tracking power converter for supply modulation of RF power amplifiers," *IEEE Power Electronics Specialists Conference*, vol. 3, pp. 1540–1545, 2001.

[35] S. Abedinpour, I. Deligoz, J. Desai, M. Figel, and S. Kiaei, "Monolithic supply modulated RF power amplifier and dc-dc power converter IC," 2003 IEEE MTT-S Microwave Symposium Digest, vol. 1, pp. 89–92, 2003.

[36] F. H. Raab, P. Asbeck, S. Cripps, P. B. Kenington, A. B. Popović, N. Pothecary, J. F. Sevic, and N. O. Sokal, "Power amplifiers and transmitters for RF and microwave," *IEEE Transactions on Microwave Theory and Techniques*, vol. 50, no. 3, pp. 814–826, 2003.

[37] A. Sato, J. A. Oliver, J. A. Cobos, J. Cerzon, and F. Arevalo, "Power supply for a radio transmitter with modulated supply voltage," *IEEE Applied Power Electronics Conference*, vol. 1, pp. 392–398, 2004.

[38] V. Yousefzadeh, E. Alarcon, D. Maksimović, "Efficiency optimization in linear assisted switching power converters for envelope tracking in RF power amplifiers," IEEE International Symposium on Circuits and Systems, vol. 2, pp. 1302–1305, 2005.

[39] E. McCune, "High-efficiency, multi-mode, multi-band terminal power amplifiers," *IEEE Microwave Magazine*, vol. 6, no, 1, pp. 44–55, 2005.

[40] M. C. W. Hoyerby and M. A. E. Anderson, "Envelope tracking power supply with 4-th order filter," IEEE Applied Power Electronic Conference, March 2006.

[41] V. Yousefzadeh, E. Alarcon, D. Maksimović, "Three-level buck converter foe envelope tracking in RF power amplifiers," *IEEE Transactions on Power Electronics*, vol. 21 no. 2, pp. 549–552, March 2006.

[42] D. F. Kimball, J. Jeong, C. Hsia, P. Draxler, S. Lanfranco, W. Nagy, K. Linthicum, L. E. Larson, and P. M. Asbeck, "Envelope tracking W-CDMA base-station amplifier using GaN HFETs," *IEEE Transactions on Microwave Theory and Techniques*, vol. 54, no. 11, pp. 3848–3856, 2006.

[43] V. Vasić, O. Garcia, J. A. Oliver, P. Alou, D. Diaz, and J. A. Cobos, "Power supply for high efficiency RF amplifier," IEEE Applied Power Electronic Conference, February 2009.

[44] J. Choi, D. Kang, D. Kim, and B. Kim, "A polar transmitter with CMOS programmable hysteretic-controlled hybrid- switching supply modulator for multistandard applications," *IEEE Transactions on Microwave Theory and Techniques*, vol. 57, no. 7, pp. 1675–1686, 2009.

[45] M. McCunne, "Envelope tracking or polar - Which is it?," *IEEE Microwave Magazine*, pp. 54–56, 2012.

[46] J. KIm J. Son, J. Jee, S. KIm, and B. Kim, "Optimization of envelope tracking power amplifier for base-station applications," *IEEE Transactions on Microwave Theory and Techniques*, vol. 61, no. 4, pp. 1620–1627, 2013.

[47] A. Ayachit and M. K. Kazimierczuk, "Two-phase buck converter as a dynamic power supply for RF power amplifier applications," *IEEE Midwest Symposium on Circuits and Systems*, Columbus, OH, pp. 493–496. August 3-7, 2013.

[48] T. R. Salvatierra and M. K. Kazimierczuk, "Inductor design for PWM buck converter operated as dynamic supply or amplitude modulator for RF transmitters," IEEE *Midwest Symposium on Circuits and Systems*, Columbus, OH, pp. 37–40. August 3-7, 2013.

[49] O. Garcia, M. Vlasić, J. A. Oliver, P. Alou, and J. A. Cobos, "An overview of fast dc-dc converters for envelope amplifier in RF transmitters," *IEEE Transactions on Power Electronics*, vol. 28, no. 10, pp. 4712–4720, 2014.

[50] M. Vlasić, O. Garcia, J. A. Oliver, P. Alou, and J. A. Cobos, "Theoretical limits of a serial and parallel linear-assisted switching converter as an envelope amplifier," *IEEE Transactions on Power Electronics*, vol. 29 no. 2, pp. 719–728, 2014.

[51] P. F. Miaja, J. Sebastián, R. Marante, and J. A. Garcia, "A linear assisted switched envelope amplifier for a UHF polar transmitter," *IEEE Transactions on Power Electronics*, vol. 29 no. 4, pp. 1850–1861, 2014.

[52] W. H. Doherty, "A new high-efficiency power amplifier for modulated waves," *Proceedings of the IRE*, vol. 24, pp. 1163–1182, 1936.

[53] F. H. Raab, "Efficiency of Doherty RF power amplifier systems," *IEEE Transactions on Broadcasting*, vol. 33, no. 9, pp. 77–83, 1987.

[54] N. Srirattana, A. Raghavan, D. Heo, P. E. Allen, and J. Laskar, "Analysis and design of a high-efficiency multistage Doherty power amplifier for wireless communications," *IEEE Transactions on Microwave Theory and Techniques*, vol. 22, pp. 852–860, 2005.
[55] P. B. Kenington, *High-Linearity RF Amplifier Design*. Norwood, MA: Artech House, 2000.
[56] C. P. Campball, "A fully integrated Ku-band Doherty amplifier MMIC," *IEEE Microwave and Guided Letters*, vol. 9, no. 3, pp. 114–116, 1999.
[57] C. P. Carrol, G. D. Alley, S. Yates, and R. Matreci, "A 20 GHz Doherty power amplifier MMIC with high efficiency and low distortion designed for broad band digital communications," *2000 IEEE MTT-S International Microwave Symposium Digest*, vol. 1, pp. 537–540, 2000.
[58] B. Kim, J. Kim, I Kim, and J. Cha, "Efficiency enhancement of linear power amplifier using load modulation techniques," IEEE International Microwave and Optical Technology Symposium Digest, pp. 505–508, June 2001.
[59] Y. Yang, J. Yi, Y. Y. Woo, and B. Kim, "Optimum design for linearity and efficiency of microwave Doherty amplifier using a new load matching technique," *Microwave Journal*, vol. 44, no. 12, pp 20–36, 2001.
[60] B. Kim, J. Kim, I. Kim, and J. Cha, "The Doherty power amplifier," *IEEE Microwave Magazine*, vol. 7, no. 5, pp. 42–50, 2006.
[61] M. Pelk, W. Neo, J. Gajadharsing, R. Pengelly, and L de Vreede, "A high-efficiency 100-W GaN three-way Doherty amplifier for base-station applications," *IEEE Transactions on Microwave Theory and Techniques*, vol. 56, no. 7, pp. 1582–1591, 2008.
[62] C. Burns, A. Chang, and D. Runton, "A 900 MHz, 500 W Doherty power amplifier using optimized output matched SiIDMOS power transistors," IEEE International Microwave and Optical Technology Symposium Digest, pp. 1577–1591, July 2008.
[63] J. Choi, D. Kong, D. Kim, and B. Kim, "Optimized envelope tracking operation of Doherty power amplifier for high efficiency over an extended dynamic range," *IEEE Transactions on Microwave Theory and Techniques*, vol. 57, no. 6, pp. 1508–1515, 2009.
[64] H. Chireix, "High-power outphasing modulation," *Proceedings of the IRE*, vol. 23, pp. 1370–1392, 1935.
[65] D. C. Cox, "Linear amplification with nonlinear components," *IEEE Transactions on Communications*, vol. 22, pp. 1942–1945, 1974.
[66] D. C. Cox and R. P. Leck, "Component signal separation and recombination for linear amplification with nonlinear components," *IEEE Transactions on Communications*, vol. 23, pp. 1281–1187, 1975.
[67] D. C. Cox and R. P. Leck, "A VHF implementation of a LINC amplifier," *IEEE Transactions on Communications*, vol. 23, pp. 1018–1022, 1976.
[68] F. H. Raab, "Average efficiency of outphasing power-amplifier systems," *IEEE Transactions on Communications*, vol. 33, no. 9, pp. 1094–1099, 1985.
[69] S. Tomatso, K. Chiba, and K. Murota, "Phase error free LINC modulator," *Electronics Letters*, vol. 25, pp. 576–577, 1989.
[70] F. J. Casedevall, "The LINC transmitter," *RF Design*, pp. 41–48, February 1990.
[71] S. A. Hetzel, A. Bateman, and J. P. McGeehan, "LINC transmitter," *Electronics Letters*, vol. 27. no. 10. pp. 844–846, 1991.
[72] X. Zhang and L. E. Larson, "Gain and phase error-free LINC transmitter," *IEEE Transaction on Vehicular Technology*, vol. 49, no. 5, pp. 1986–1994, 2000.
[73] X. Zhang, L. E. Larson, and P.M. Asbeck, *Design of Linear RF Outphasing Power Amplifiers*. Norwood, MA: Artech House, 2003.
[74] A. Birafane and A. B. Kouki, "On the linearity and efficiency of outphasing microwave amplifier," *IEEE Transactions on Microwave Theory and Techniques*, vol. 52, no. 7, pp. 1702–1708, 2004.
[75] A. Birafane and A. B. Kouki, "Phase-only predistortion for LINC amplifiers with Chireix-outphasing combiners," *IEEE Transactions on Microwave Theory and Techniques*, vol. 53, no. 6, pp. 2240–2250, 2005.
[76] M. K. Kazimierczuk, "Synthesis of phase-modulated dc/ac inverters and dc/dc converters," *IEE Proceedings, Part B, Electric Power Applications*, vol. 139, pp. 387–394, 1992.

[77] R. Langrindge, T. Thorton, P. M. Asbeck, and L. E. Larson, "Average efficiency of outphasing power re-use technique for improved efficiency of outphasing microwave power amplifiers," *IEEE Transactions on Microwave Theory and Techniques*, vol. 47, pp. 1467–1470, 2001.

[78] X. Zhang, L. E. Larson, P. M. Asbeck, and P. Nanawa, "Gain/phase imbalance minimization techniques for LINC transmitters," *IEEE Transactions on Microwave Theory and Techniques*, vol. 49, no. 12, pp. 2507–2515, 2001.

[79] M. P. van der Heijden, M. Axer, J. S. Vromans, and D. A. Calvillo-Corts, "A 19 W high-efficiency wide-bandCMOS-gan Class-E Chireix RF outphasing power amplifier," IEEE MTT-S International Microwave Symposium Digest (MTT), June 2011, pp. 1–4.

[80] A. S. Tripathi and C. L. Delano, "Method and apparatus for oversampled, nois-shaping, mixed signal processing," US Patent 5, 777, 512, July 7, 1998.

[81] M. Weiss, F. Raab, and Z. Popović, "Linerity of X-band Class-F power amplifiers in high-efficiency transmitters," *IEEE Transactions on Microwave Theory and Techniques*, vol. 49, no. 6, pp. 1174–1179, 2001.

[82] S. Hamedi-High and C. A. T. Salama, "Wideband CMOS integrated RF combiner for LINC transmitters," 2003 IEEE MTT-S Microwave Symposium Digest, vol. 1, pp. 41–43, 2003.

[83] B. Kim. J. Moon, and I. Kim. "Efficiently amplified," *IEEE Microwave Magazine*, pp. 87–100, August 2010.

[84] J. Kim, J. Kim, J. Son, I. Kim, and B. Kim,"Saturated power amplifier optimized for efficiency using self-generated harmonic current and voltages," *IEEE Transactions on Microwave Theory and Techniques*, vol. 59, no. 8, pp. 2049–2058, 2011.

[85] I. Aoki, S. Kee, D. Rutledge, and A. Hajimiri, "Fully integrated CMOS power amplifier design using the distributed active-transformer architecture," *IEEE Journal of Solid-State Circuits*, vol. 37, pp. 371–383, 2002.

10

Integrated Inductors

10.1 Introduction

There is a great demand for on-chip integrated inductors with a high quality factor. Capacitors can easily be integrated in CMOS technology using polysilicon layers, resulting in high capacitances per unit area. One of the largest challenges in making completely monolithic integrated circuits is to make high-performance integrated inductors [1–56]. Lack of a good integrated inductor is the most important disadvantage of standard integrated circuit (IC) processing for many applications. Front-end radio-frequency (RF) ICs require monolithic integrated inductors to construct resonant circuits, power amplifiers, impedance matching networks, bandpass filters, low-pass filters, and high-impedance chokes. RF IC inductors are essential elements of wireless communication circuit blocks, such as low-noise amplifiers (LNAs), mixers, intermediate frequency filters (IFFs), and voltage-controlled oscillators (VCOs) that drive cellular phone transmitters. Monolithic inductors are also used to bias RF amplifiers, RF oscillators, tuning varactors, PIN diodes, transistors, and monolithic circuits. The structure of planar spiral inductors is based on the standard CMOS and BiCMOS technologies as the cost is one of the major factors. Applications of RF IC inductors include cell phone, wireless local Network (WLN), TV tuners, and radars. On-chip and off-chip inductors are used in RF circuits. There are four major types of integrated RF inductors: planar spiral inductors, meander inductors, bondwire inductors, and micro-electromechanical system (MEMS) inductors. In this chapter, integrated inductors will be discussed.

10.2 Skin Effect

At low frequencies, the current density J in a conductor is uniform. In this case, the ac resistance of the conductor R_{ac} is equal to the dc resistance R_{dc}, and the conduction power loss is identical to ac and dc currents if $I_{rms} = I_{dc}$. In contrast, the current density J at high frequencies is not uniform because of eddy currents. Eddy currents are present in any conducting material, which is subjected to a time-varying magnetic field. The skin effect is the tendency of high-frequency

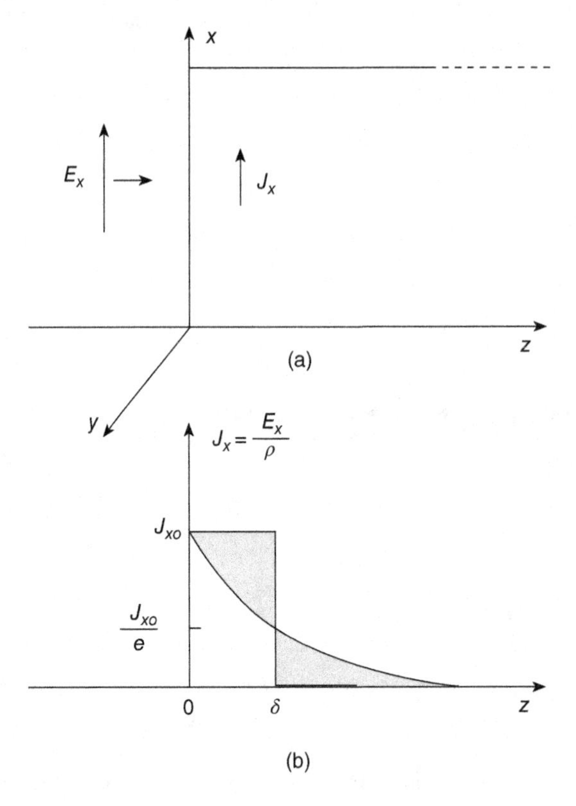

Figure 10.1 Semi-infinite slab. (a) Conductor. (b) Current density distribution.

current to flow mostly near the surface of a conductor. This causes an increase in the effective ac resistance of a conductor R_{ac} above its dc resistance R_{dc}, resulting in an increase in conduction power loss and heat.

Consider a semi-infinite slab of a conductor with conductivity $\sigma = 1/\rho$ that occupies $z \geq 0$, where ρ is the resistivity of the conductor. The semi-infinite conductor is shown in Fig. 10.1(a). The harmonic electric field incident on the conductor is given by

$$E_x(z) = E_{xo}e^{-\frac{z}{\delta}} \tag{10.1}$$

where E_{xo} is the amplitude of the electric field just on the surface of the conductor and the skin depth is

$$\delta = \frac{1}{\sqrt{\pi\mu\sigma f}} = \sqrt{\frac{\rho}{\pi\mu f}} \tag{10.2}$$

The current density in the conductor is

$$J_x(z) = \sigma E_x(z) = \frac{E_x(z)}{\rho} = J_{xo}e^{-\frac{z}{\delta}} \tag{10.3}$$

where the $J_{xo} = \sigma E_{xo} = E_{xo}/\rho$ is the amplitude of the current density at the surface. The current density distribution in the conductor is shown in Fig. 10.1(b). The current density J_x exponentially decays as a function of depth z into the conductor.

The current flows in the x-direction through the surface extending into the z-direction from $z = 0$ to $z = \infty$. The surface has a width w in the y-direction. The total current is obtained as

$$I = \int_{z=0}^{z=\infty} \int_{y=0}^{w} J_{xo}e^{-\frac{z}{\delta}}\, dy\, dz = J_{xo}w\delta \tag{10.4}$$

Thus, the equivalent current is obtained if we assume that the current density J_x is constant equal to J_{xo} from the surface down to skin depth δ and is zero for $z > \delta$. The area of a rectangle of sides J_{xo} and δ is equivalent to the area of the exponential curve. It can be shown that 95% of the total current flows in thickness of 3δ and 99.3% of the total current flows in the thickness of 5δ.

The voltage drop across the distance l in the x-direction is

$$V = E_{xo}l = \rho J_{xo}l \tag{10.5}$$

Hence, the ac resistance of the semi-infinite conductor of width w and length l that extends from $z = 0$ to $z = \infty$ is

$$R_{ac} = \frac{V}{I} = \frac{\rho}{\delta}\left(\frac{l}{w}\right) = \frac{1}{\sigma\delta}\left(\frac{l}{w}\right) = \left(\frac{l}{w}\right)\sqrt{\pi\rho\mu f} \tag{10.6}$$

As the skin depth δ decreases with increasing frequency, the ac resistance R_{ac} increases.

Example 10.1

Calculate the skin depth at $f = 10\,\text{GHz}$ for (a) copper, (b) aluminum, (c) silver, and (d) gold.

Solution.

(a) The resistivity of copper at $T = 20\,°\text{C}$ is $\rho_{Cu} = 1.724 \times 10^{-8}\,\Omega\text{m}$. The skin depth of the copper at $f = 10\,\text{GHz}$ is

$$\delta_{Cu} = \sqrt{\frac{\rho_{Cu}}{\pi\mu_0 f}} = \sqrt{\frac{1.724 \times 10^{-8}}{\pi \times 4\pi \times 10^{-7} \times 10 \times 10^9}} = 0.6608\,\mu\text{m} \tag{10.7}$$

(b) The resistivity of aluminum at $T = 20\,°\text{C}$ is $\rho_{Al} = 2.65 \times 10^{-8}\,\Omega\text{m}$. The skin depth of the aluminum at $f = 10\,\text{GHz}$ is

$$\delta_{Al} = \sqrt{\frac{\rho_{Al}}{\pi\mu_0 f}} = \sqrt{\frac{2.65 \times 10^{-8}}{\pi \times 4\pi \times 10^{-7} \times 10 \times 10^9}} = 0.819\,\mu\text{m} \tag{10.8}$$

(c) The resistivity of silver at $T = 20\,°\text{C}$ is $\rho_{Ag} = 1.59 \times 10^{-8}\,\Omega\text{m}$. The skin depth of the silver at $f = 10\,\text{GHz}$ is

$$\delta_{Ag} = \sqrt{\frac{\rho_{Ag}}{\pi\mu_0 f}} = \sqrt{\frac{1.59 \times 10^{-8}}{\pi \times 4\pi \times 10^{-7} \times 10 \times 10^9}} = 0.6346\,\mu\text{m} \tag{10.9}$$

(d) The resistivity of gold at $T = 20\,°\text{C}$ is $\rho_{Au} = 2.44 \times 10^{-8}\,\Omega\text{m}$. The skin depth of the gold at $f = 10\,\text{GHz}$ is

$$\delta_{Au} = \sqrt{\frac{\rho_{Au}}{\pi\mu_0 f}} = \sqrt{\frac{2.44 \times 10^{-8}}{\pi \times 4\pi \times 10^{-7} \times 10 \times 10^9}} = 0.786\,\mu\text{m} \tag{10.10}$$

10.3 Resistance of Rectangular Trace

Consider a trace (a slab) of a conductor of thickness h, length l, and width w, as shown in Fig. 10.2(a). The dc resistance of the trace is

$$R_{dc} = \rho\frac{l}{wh} \tag{10.11}$$

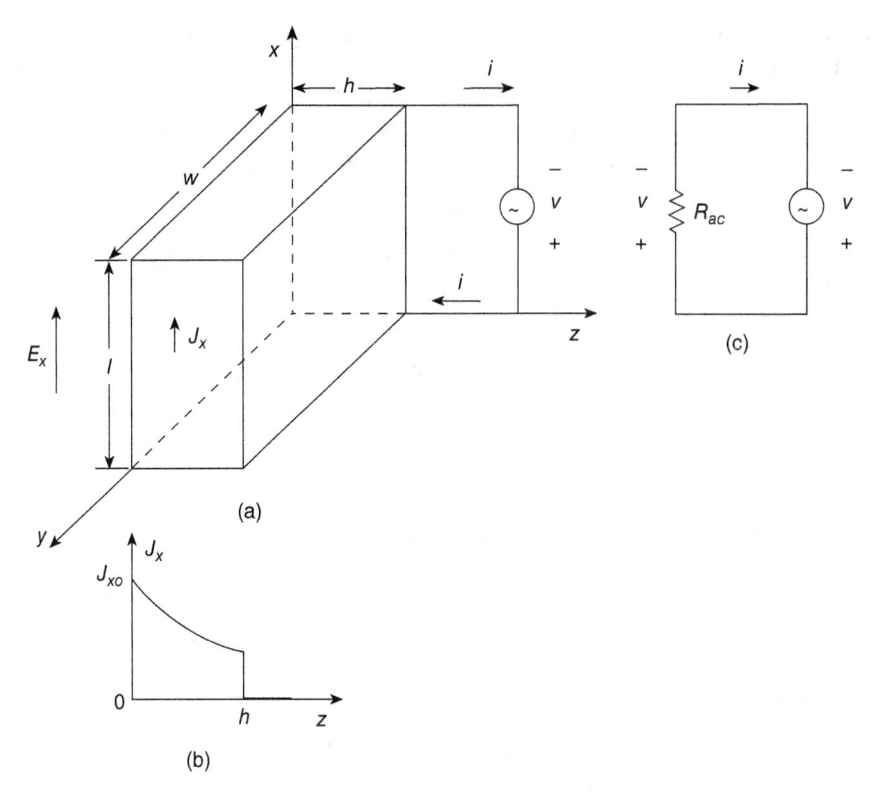

Figure 10.2 Slab conductor. (a) Slab. (b) Current density distribution. (c) Equivalent circuit.

The harmonic electric field incident on the conductor is given by

$$E_x(z) = E_{xo}e^{-\frac{z}{\delta}} \tag{10.12}$$

The current density at high frequencies is given by

$$J_x = J_{xo}e^{-\frac{z}{\delta}} \quad \text{for} \quad 0 \leq z \leq h \tag{10.13}$$

and

$$J_x = 0 \quad \text{for} \quad z > h \tag{10.14}$$

The current density distribution in the trace conductor is shown in Fig. 10.2(b). Hence, the total current flowing in the conductor is

$$I = \int_{z=0}^{h} \int_{y=0}^{w} J_{xo}e^{-\frac{z}{\delta}}\, dy\, dz = w\delta J_{xo}\left(1 - e^{-\frac{h}{\delta}}\right) \tag{10.15}$$

Since $E_{xo} = \rho J_{xo}$, the voltage drop across the conductor in the x-direction of the current flow at $z = 0$ is

$$V = E_{xo}l = \frac{J_{xo}}{\sigma}l = \rho J_{xo}l \tag{10.16}$$

The ac trace resistance is given by

$$R_{ac} = \frac{V}{I} = \frac{\rho l}{w\delta(1 - e^{-h/\delta})} = \frac{l}{w(1 - e^{-h/\delta})}\sqrt{\pi\rho\mu f} \quad \text{for} \quad \delta < h \tag{10.17}$$

At high frequencies, the ac resistance of the metal trace R_{ac} is increased by the skin effect. The current density in the metal trace increases on the side of the substrate. The equivalent circuit of the trace conductor for high frequencies is shown in Fig. 10.2(c).

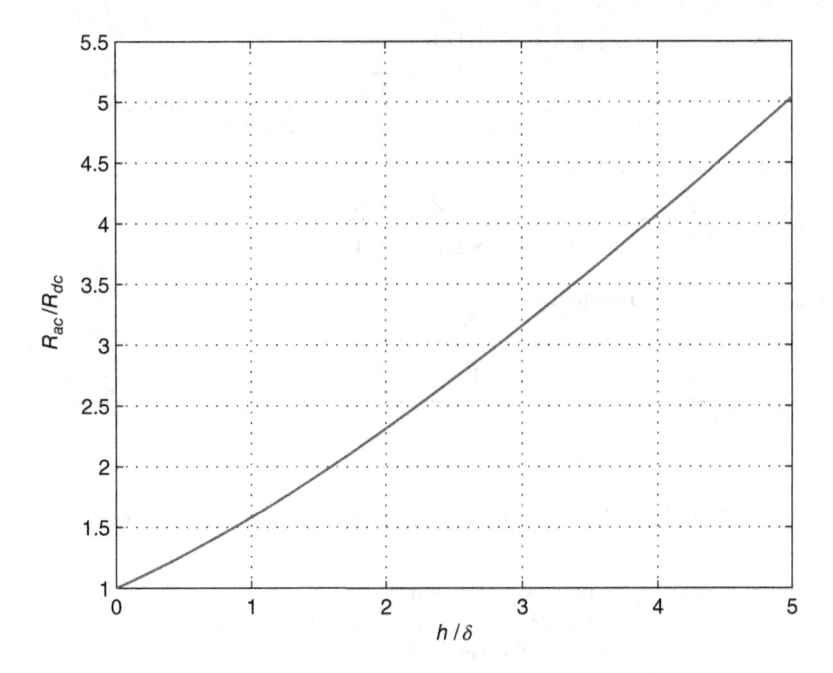

Figure 10.3 R_{ac}/R_{dc} as a function of h/δ.

The ratio of the ac-to-dc resistances is

$$F_R = \frac{R_{ac}}{R_{dc}} = \frac{h}{\delta(1 - e^{-\frac{h}{\delta}})} \tag{10.18}$$

The ratio R_{ac}/R_{dc} as a function of h/δ is shown in Fig. 10.3. As h/δ increases with increasing frequency, R_{ac}/R_{dc} increases.

For very high frequencies, this expression can be approximated by

$$R_{ac} \approx \frac{h}{\delta} = h\sqrt{\frac{\pi\mu f}{\rho}} \quad \text{for} \quad \delta < \frac{h}{5} \tag{10.19}$$

The power loss in the metal trace resistance because of the flow of a sinusoidal current of amplitude I_m is given by

$$P_{Rac} = \frac{1}{2}R_{ac}I_m^2 \tag{10.20}$$

Example 10.2

Calculate the dc resistance of a copper trace with $l = 50\,\mu\text{m}$, $w = 1\,\mu\text{m}$, and $h = 1\,\mu\text{m}$. Find the ac resistance of the trace at 10 GHz. Calculate F_R.

Solution. The resistivity of copper at $T = 20\,°\text{C}$ is $\rho_{Cu} = 1.724 \times 10^{-8}\,\Omega\text{m}$. The dc resistance of the trace is

$$R_{dc} = \rho_{Cu}\frac{l}{wh} = \frac{1.724 \times 10^{-8} \times 50 \times 10^{-6}}{1 \times 10^{-6} \times 1 \times 10^{-6}} = 0.862\,\Omega \tag{10.21}$$

The skin depth of the copper at $f = 10\,\text{GHz}$ is

$$\delta_{Cu} = \sqrt{\frac{\rho_{Cu}}{\pi\mu_0 f}} = \sqrt{\frac{1.724\times10^{-8}}{\pi\times4\pi\times10^{-7}\times10\times10^{9}}} = 0.6608\,\mu\text{m} \tag{10.22}$$

The ac resistance of the trace is

$$R_{ac} = \frac{\rho l}{w\delta(1-e^{-h/\delta})} = \frac{1.724\times10^{-8}\times50\times10^{-6}}{1\times10^{-6}\times0.6608\times10^{-6}\left(1-e^{-\frac{1}{0.6608}}\right)} = 1.672\,\Omega \tag{10.23}$$

The ratio of the ac-to-dc resistances is

$$F_R = \frac{R_{ac}}{R_{dc}} = \frac{1.672}{0.862} = 1.93 \tag{10.24}$$

10.4 Inductance of Straight Rectangular Trace

A straight rectangular trace inductor is shown in Fig. 10.4. The self-inductance of a solitary rectangular straight conductor at low frequencies, where the current density is uniform and skin effect can be neglected, is given by Grover's formula [8]

$$L = \frac{\mu_0 l}{2\pi}\left[\ln\left(\frac{2l}{w+h}\right) + \frac{w+h}{3l} + 0.50049\right]\ (\text{H}) \quad \text{for} \quad w \le 2l \quad \text{and} \quad h \le 2l \tag{10.25}$$

where l, w, and h are the conductor length, width, and thickness in meters, respectively. Since $\mu_0 = 2\pi\times10^{-7}$, the trace inductance is

$$L = 2\times10^{-7} l\left[\ln\left(\frac{2l}{w+h}\right) + \frac{w+h}{3l} + 0.50049\right]\ (\text{H}) \quad \text{for} \quad w \le 2l \quad \text{and} \quad h \le 2l \tag{10.26}$$

where all dimensions are in meters. Grover's formula can also be expressed as

$$L = 0.0002 l\left[\ln\left(\frac{2l}{w+h}\right) + \frac{w+h}{3l} + 0.50049\right]\ (\text{nH}) \quad \text{for} \quad w \le 2l \quad \text{and} \quad h \le 2l \tag{10.27}$$

where l, w, and h are the conductor length, width, and thickness in micrometers, respectively.

Figures 10.5 through 10.7 show the inductance L versus w, h, and l, respectively. The inductance L of a conducting trace increases as l increases and as w and h decrease. Straight traces are segments of inductors of more complex structures of integrated planar inductors, such as square-integrated planar inductors. They are also used as interconnections in ICs.

The quality factor of the straight rectangular trace at frequency $f = \omega/(2\pi)$ is defined as

$$Q = 2\pi\frac{\text{Peak magnetic energy stored in } L}{\text{Energy dissipated per cycle}} = \frac{\omega L}{R_{ac}} \tag{10.28}$$

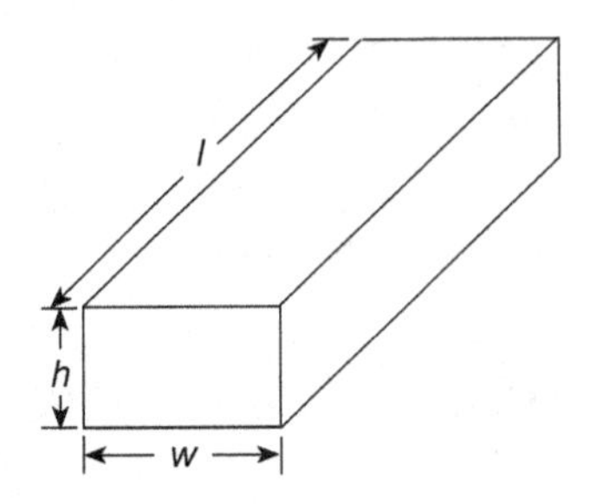

Figure 10.4 Straight rectangular trace inductor.

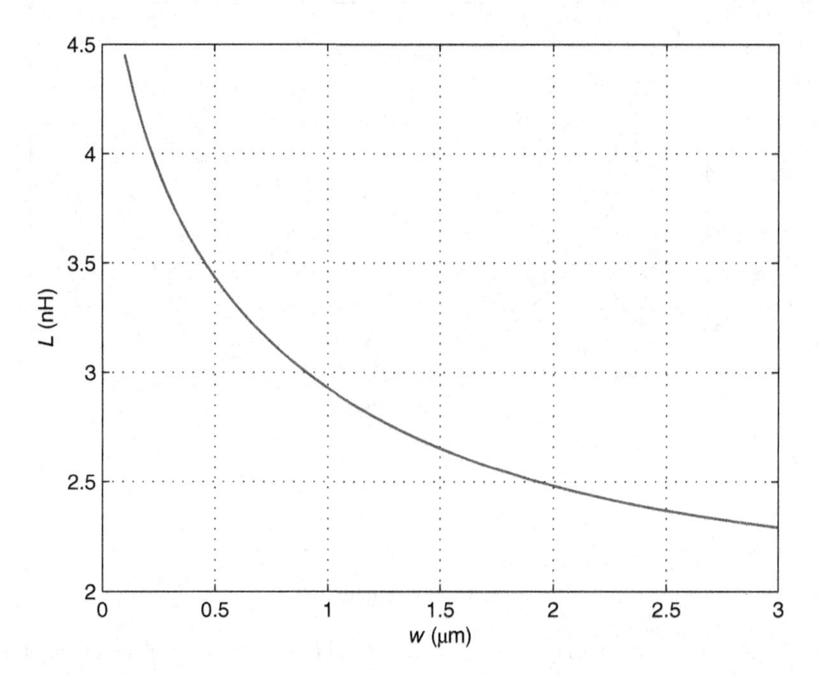

Figure 10.5 Inductance L as a function of trace width w for straight rectangular trace at $h = 1$ μm and $l = 50$ μm.

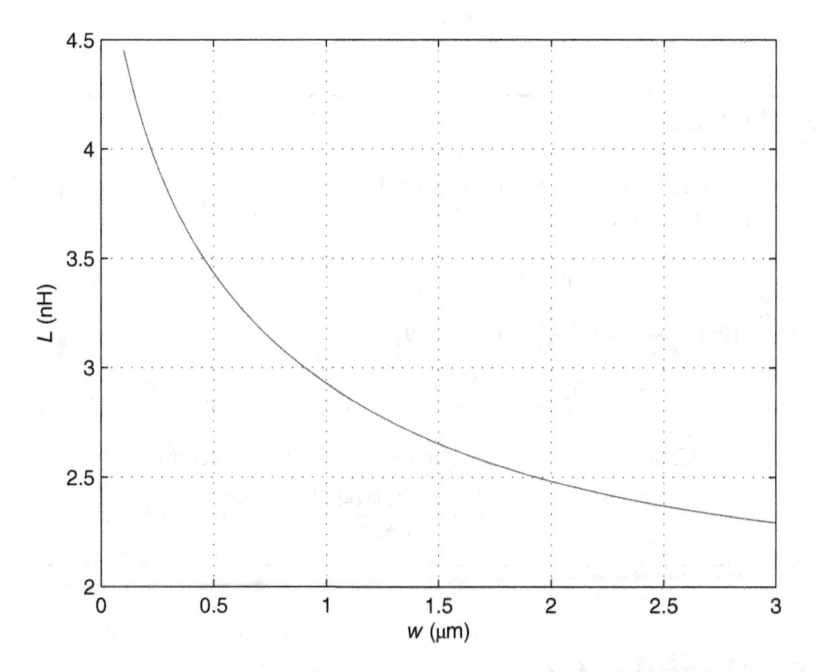

Figure 10.6 Inductance L as a function of trace height h for straight rectangular trace at $w = 1$ μm and $l = 50$ μm.

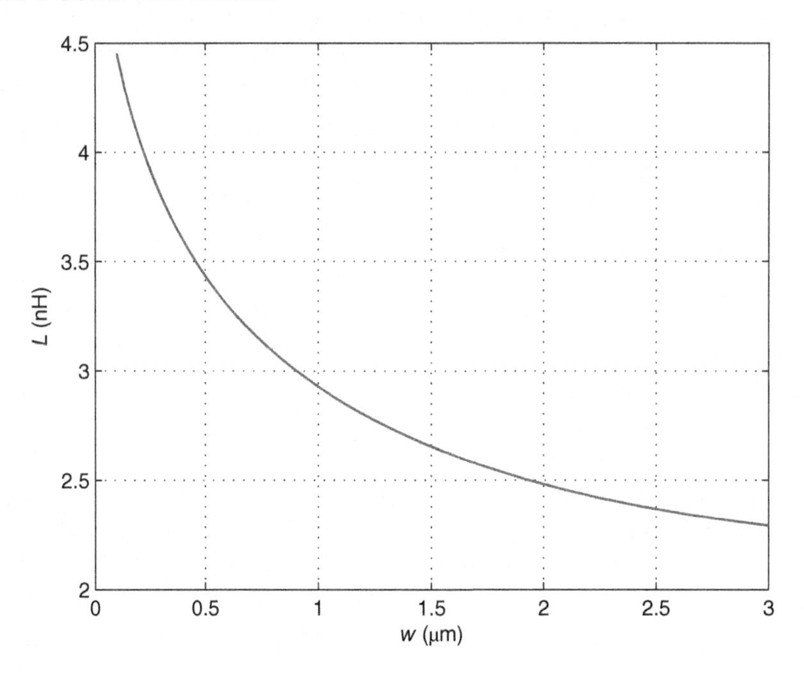

Figure 10.7 Inductance L as a function of trace length l for straight rectangular trace at $w = 1$ μm and $h = 1$ μm.

Another definition of the quality factor is given by

$$Q = 2\pi \frac{\text{Peak magnetic energy} - \text{Peak electric energy}}{\text{Energy dissipated per cycle}} \tag{10.29}$$

Example 10.3

Calculate the inductance of a straight rectangular trace with $l = 50$ μm, $w = 1$ μm, $h = 1$ μm, and $\mu_r = 1$. Find its quality factor Q at a frequency of 10 GHz.

Solution. The inductance of the straight trace is

$$L = \frac{\mu_0 l}{2\pi}\left[\ln\left(\frac{2l}{w+h}\right) + \frac{w+h}{3l} + 0.50049\right]$$
$$= \frac{4\pi \times 10^{-7} \times 50 \times 10^{-6}}{2\pi}\left[\ln\left(\frac{2 \times 50}{1+1}\right) + \frac{1+1}{3 \times 50} + 0.50049\right] = 0.04426\,\text{nH} \tag{10.30}$$

From Example 10.2, $R_{ac} = 1.672\,\Omega$. The quality factor of the inductor is

$$Q = \frac{\omega L}{R_{ac}} = \frac{2\pi \times 10 \times 10^9 \times 0.04426 \times 10^{-9}}{1.672} = 1.7092 \tag{10.31}$$

10.5 Meander Inductors

Meander inductors have a simple layout as shown in Fig. 10.8. It is sufficient to use only one metal layer because both metal contacts are on the same metal level. This simplifies the

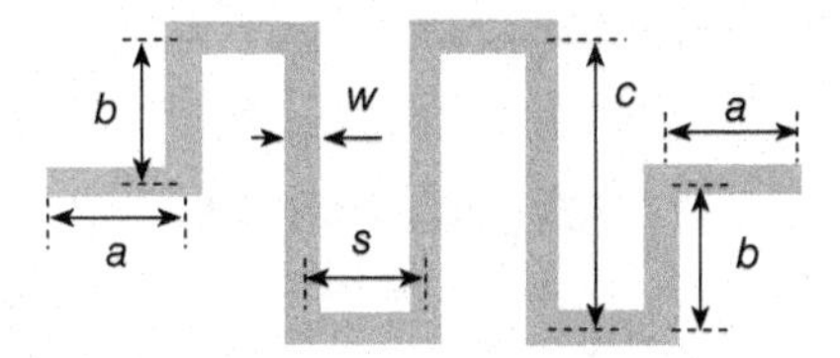

Figure 10.8 Meander inductor.

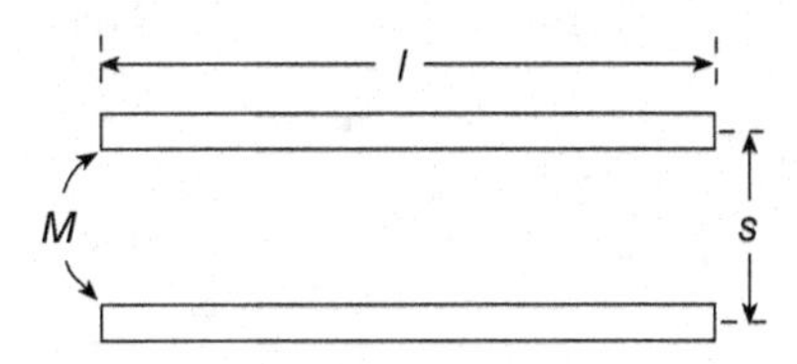

Figure 10.9 Two equal parallel straight rectangular conductors.

technological process by avoiding two levels of photolithography. They usually have a low inductance per unit area and a large ac resistance because the length of the trace is large, causing a large dc resistance. This results in a low quality factor. The meander inductor can be divided into straight segments. The self-inductance of each straight segment is given by (10.27). The total self-inductance is equal to the sum of the self-inductances of all straight segments and is given by Stojanovic et al. [52]

$$L_{self} = 2L_a + 2L_b + NL_c + (N+1)L_s \tag{10.32}$$

where

$$L_a = \frac{\mu_0 a}{2\pi}\left[\ln\left(\frac{2a}{w+c}\right) + \frac{w+c}{3a} + 0.50049\right] \ \text{(H)} \tag{10.33}$$

$$L_b = \frac{\mu_0 b}{2\pi}\left[\ln\left(\frac{2b}{w+c}\right) + \frac{w+c}{3b} + 0.50049\right] \ \text{(H)} \tag{10.34}$$

$$L_c = \frac{\mu_0 c}{2\pi}\left[\ln\left(\frac{2c}{w+c}\right) + \frac{w+c}{3c} + 0.50049\right] \ \text{(H)} \tag{10.35}$$

$$L_s = \frac{\mu_0 s}{2\pi}\left[\ln\left(\frac{2s}{w+c}\right) + \frac{w+c}{3s} + 0.50049\right] \ \text{(H)} \tag{10.36}$$

N is the number of segments of length c and L_a, L_b, L_c, and L_s are self-inductances of the segments. All dimensions are in meters.

The mutual inductance between two equal parallel conductors (segments) shown in Fig. 10.9 is given by Grover [8]

$$M = \pm\frac{\mu_0 l}{2\pi}\left\{\ln\left[\frac{l}{s} + \sqrt{1+\left(\frac{l}{s}\right)^2}\right] - \sqrt{1+\left(\frac{s}{l}\right)^2} + \frac{s}{l}\right\} \ \text{(H)} \tag{10.37}$$

where l is the length of the parallel conductors and s is the center-to-center separation of the conductors. All dimensions are in meters. If the currents flow in both parallel segments in the same direction, the mutual inductance is positive. If the currents flow in the parallel segments in the opposite directions, the mutual inductance is negative. The mutual inductance of perpendicular conductors is zero.

It can be shown that the input inductance of a transformer with the inductance of each winding equal to L and both currents flowing in the same direction is

$$L_i = L + M \tag{10.38}$$

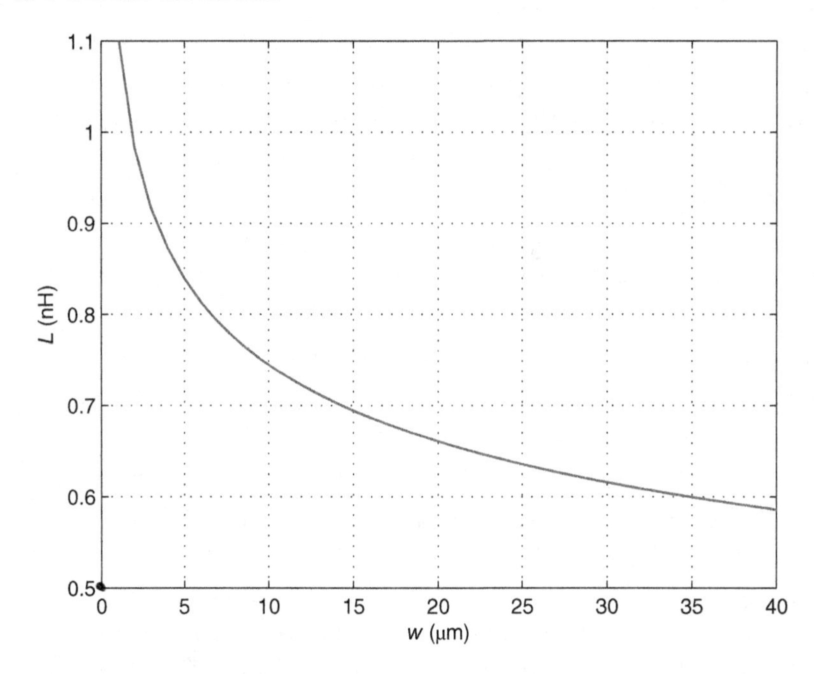

Figure 10.10 Inductance L as a function of trace width w for meander inductor at $N = 5$, $a = 40$ μm, $s = 20$ μm, and $c = 100$ μm.

The input inductance of a transformer with the inductance of each winding equal to L and both currents flowing in the opposite directions is

$$L_i = L - M \tag{10.39}$$

The total inductance of the meander inductor is the sum of the self-inductances of all segments and the positive and negative mutual inductances.

The monomial equation for the total inductance of the meander inductor is given by Stojanovic et al. [52]

$$L = 0.00266a^{0.0603}c^{0.4429}N^{0.954}s^{0.606}w^{-0.173} \text{ (nH)} \tag{10.40}$$

where $b = c/2$. All dimensions are in micrometers. The accuracy of this equation is better than 12%. Figures 10.10 through 10.14 show plots of the inductance L as a function of w, s, c, a, and N for the meander inductor.

Example 10.4

Calculate the inductance of a meander inductor with $N = 5$, $w = 40\,\mu$m, $s = 40\,\mu$m, $a = 40\,\mu$m, $c = 100\,\mu$m, and $\mu_r = 1$.

Solution. The inductance of the trace is

$$L = 0.00266a^{0.0603}c^{0.4429}N^{0.954}s^{0.606}w^{-0.173}$$
$$= 0.00266 \times 40^{0.0603}100^{0.4429}5^{0.954}40^{0.606}40^{-0.173} = 0.585\,\text{nH} \tag{10.41}$$

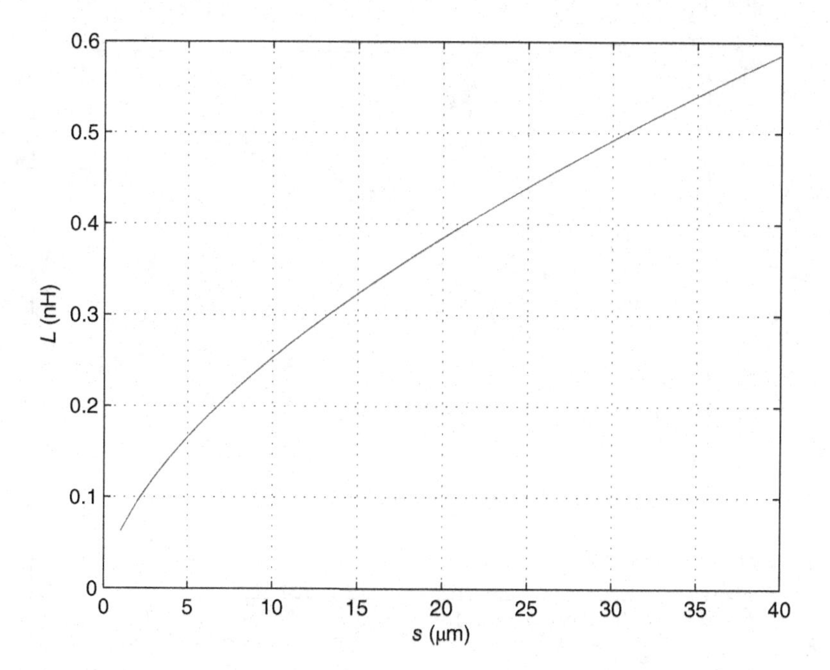

Figure 10.11 Inductance L as a function of separation s for meander inductor at $N = 5$, $a = 40$ μm, $w = 40$ μm, and $c = 100$ μm.

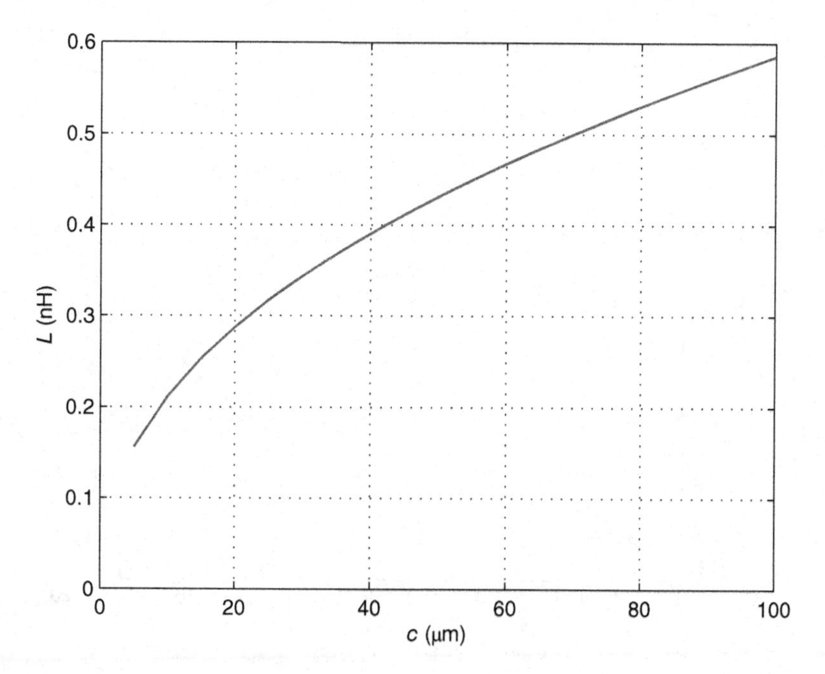

Figure 10.12 Inductance L as a function of segment length c for meander inductor at $N = 5$, $s = 40$ μm, $w = 40$ μm, and $a = 40$ μm.

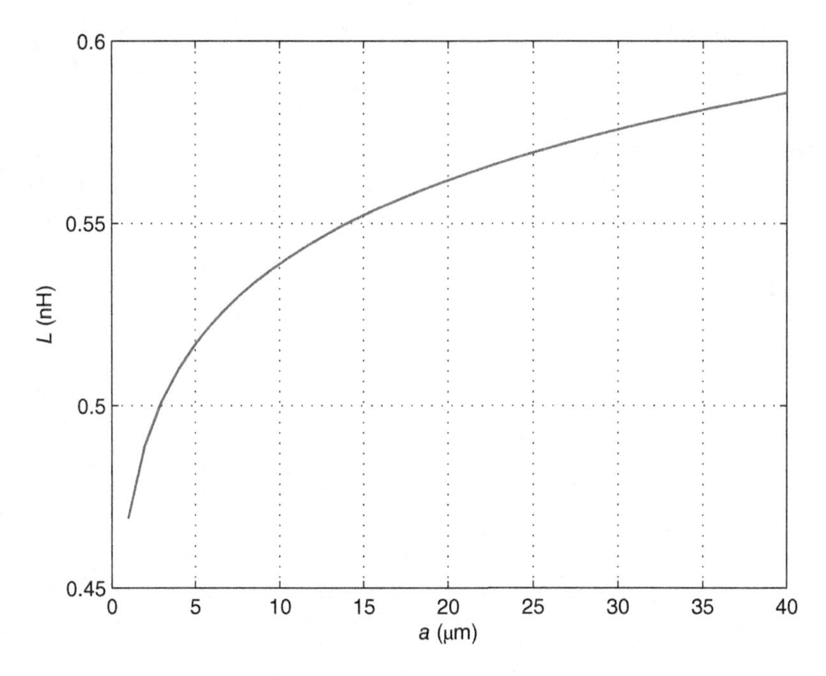

Figure 10.13 Inductance L as a function of a for meander inductor at $N = 5$, $s = 40$ μm, $w = 40$ μm, and $c = 100$ μm.

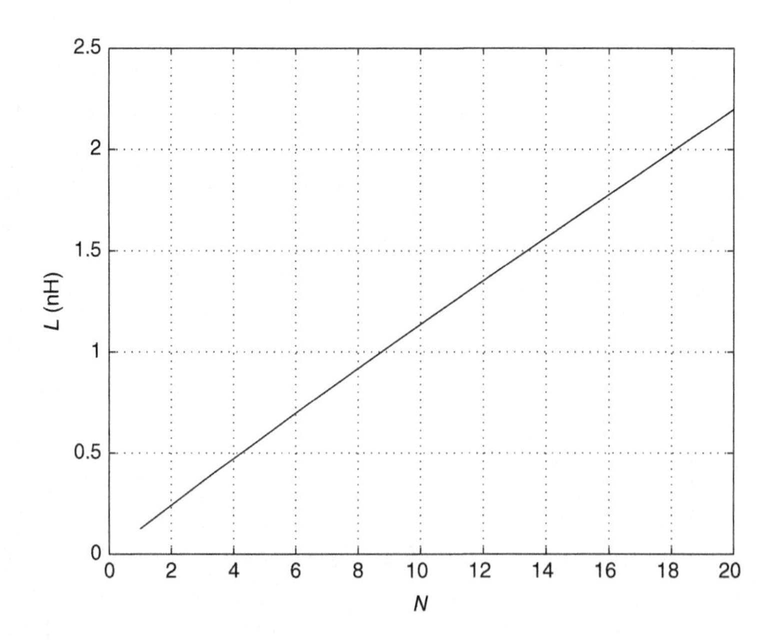

Figure 10.14 Inductance L as a function of N for meander inductor at $s = 40$ μm, $w = 40$ μm, $c = 100$ μm, and $a = 40$ μm.

10.6 Inductance of Straight Round Conductor

The inductance of a round straight inductor of length l and conductor radius a, when the effects of nearby conductors (including the return current) and the skin effect can be neglected, is given by Grover [8]

$$L = \frac{\mu_0 l}{2\pi}\left[\ln\left(\frac{2l}{a}\right) + \frac{a}{l} - \frac{3}{4}\right] \text{ (H)} \tag{10.42}$$

where the dimensions are in meters. This equation can be simplified to the form

$$L = 0.0002l\left[\ln\left(\frac{2l}{a}\right) + \frac{a}{l} - 0.75\right] \text{ (nH)} \tag{10.43}$$

where the dimensions are in micrometers.

Another expression for a round straight conductor is

$$L = \frac{\mu l}{8\pi} \quad \text{for} \quad \delta > a \tag{10.44}$$

The inductance at high frequencies decreases with frequency. The high-frequency inductance of a round straight inductor when the skin effect cannot be neglected is given by

$$L_{HF} = \frac{l}{4\pi a}\sqrt{\frac{\mu\rho}{\pi f}} \quad \text{for} \quad \delta < a. \tag{10.45}$$

Example 10.5

Calculate the inductance of a copper round straight inductor with $l = 2\,\text{mm}$, $a = 0.1$ mm, and $\mu_r = 1$ for $\delta > h$. Also, calculate the inductance of a copper round straight inductor at $f = 1$ GHz.

Solution. The inductance of the round straight inductor for $\delta > a$ (low-frequency operation) is given by

$$\begin{aligned} L &= \frac{\mu_0 l}{2\pi}\left[\ln\left(\frac{2l}{a}\right) + \frac{a}{l} - 0.75\right] \text{ (H)} \\ &= \frac{4\pi\times 10^{-7}\times 2\times 10^{-3}}{2\pi}\left[\ln\left(\frac{2\times 2}{0.1}\right) + \frac{0.1}{2} - 0.75\right] = 1.1956\,\text{nH} \end{aligned} \tag{10.46}$$

The skin depth in copper at $f = 1$ GHz is

$$\delta = \sqrt{\frac{\rho}{\pi\mu_0 f}} = \sqrt{\frac{1.724\times 10^{-8}}{\pi\times 4\pi\times 10^{-7}\times 1\times 10^{9}}} = 2.089\,\mu\text{m} \tag{10.47}$$

Since $a = 0.1\,\text{mm} = 100\,\mu\text{m}$, $\delta \ll a$. Therefore, the operation at $f = 1$ GHz is the high-frequency operation. The inductance of the round straight inductor at $f = 1$ GHz is

$$L_{HF} = \frac{l}{4\pi a}\sqrt{\frac{\mu_0\rho}{\pi f}} = \frac{2\times 10^{-3}}{4\pi\times 0.1\times 10^{-3}}\sqrt{\frac{4\pi\times 10^{-7}\times 1.724\times 10^{-8}}{\pi\times 1\times 10^{9}}} = 4.179\,\text{pH} \tag{10.48}$$

Hence,

$$F_R = \frac{L_{HF}}{L_{LF}} = \frac{4.179\times 10^{-12}}{1.1956\times 10^{-9}} = 3.493\times 10^{-3} = \frac{1}{286} \tag{10.49}$$

10.7 Inductance of Circular Round Wire Loop

The inductance of a circular loop of round wire with loop radius r and wire radius a is [8]

$$L = \mu_0 a \left[\ln\left(\frac{8r}{a}\right) - 1.75\right] \text{ (H)} \tag{10.50}$$

All dimensions are in meters. The dependence of the inductance on wire radius is weak.

The inductance of a round wire square loop is

$$L = \frac{\mu_0 l}{2\pi}\left[\ln\left(\frac{l}{4a}\right) - 0.52401\right] \text{ (H)} \tag{10.51}$$

where l is the loop length and a is the wire radius. All dimensions are in meters.

10.8 Inductance of Two-Parallel Wire Loop

Consider a loop formed by two parallel conductors whose length is l, radius is r, and the distance between the conductors is s, where $l \gg s$. The conductors carry equal currents in opposite directions. The inductance of the two-parallel wire loop is [8]

$$L = \frac{\mu_0 l}{\pi}\left[\ln\left(\frac{s}{r}\right) - \frac{s}{l} + 0.25\right] \text{ (H)} \tag{10.52}$$

10.9 Inductance of Rectangle of Round Wire

The inductance of a rectangle of round wire of radius r with rectangle side lengths x and y is given by Grover [8]

$$L = \frac{\mu_0}{\pi}[x ln\left(\frac{2x}{a}\right) + y ln\left(\frac{2y}{a}\right) + 2\sqrt{x^2+y^2} - x\, arcsinh\left(\frac{x}{y}\right) - y\, arcsinh\left(\frac{y}{x}\right) - 1.75(x+y) \tag{10.53}$$

10.10 Inductance of Polygon Round Wire Loop

The inductance of a single-loop polygon can be approximated by [8]

$$L \approx \frac{\mu p}{2\pi}\left[\ln\left(\frac{2p}{a}\right) - \ln\left(\frac{p^2}{A}\right) + 0.25\right] \text{ (H)} \tag{10.54}$$

where p is the perimeter of the coil, A is the area enclosed by the coil, and a is the wire radius. All dimensions are in meters. The inductance is strongly dependent on the perimeter and weakly dependent on the loop area and wire radius. Inductors enclosing the same perimeter with similar shape have approximately the same inductance.

The inductance of a triangle is [8]

$$L = \frac{\mu_0 l}{2\pi}\left[\ln\left(\frac{l}{3a}\right) - 1.15546\right] \text{ (H)} \tag{10.55}$$

where a is the wire radius. The inductance of a square is

$$L = \frac{\mu_0 l}{2\pi}\left[\ln\left(\frac{l}{4a}\right) - 0.52401\right] \text{ (H)} \tag{10.56}$$

The inductance of a pentagon is

$$L = \frac{\mu_0 l}{2\pi}\left[\ln\left(\frac{l}{5a}\right) - 0.15914\right] \quad \text{(H)} \tag{10.57}$$

The inductance of a hexagon is

$$L = \frac{\mu_0 l}{2\pi}\left[\ln\left(\frac{l}{6a}\right) + 0.09848\right] \quad \text{(H)} \tag{10.58}$$

The inductance of a octagon is

$$L = \frac{\mu_0 l}{2\pi}\left[\ln\left(\frac{l}{8a}\right) + 0.46198\right] \quad \text{(H)} \tag{10.59}$$

All dimensions are in meters.

10.11 Bondwire Inductors

Bondwire inductors are frequently used in RF ICs. Figure 10.15 shows a bondwire inductor. It can be regarded as a fraction of a round turn. Bondwires are often used for chip connections. The main advantage of bondwire inductors is a very small series resistance. Standard bondwires have a relatively large diameter of about 25 μm and can handle substantial currents with low loss. They may be placed well above any conductive planes to reduce parasitic capacitances and hence increase the self-resonant frequency (SRF) f_r. Typical inductances range from 2 to 5 nH. Since bondwires have much larger cross-sectional area than the traces of planar spiral inductors, these inductors have lower resistances and hence a higher quality factor Q_{Lo}, typically $Q_{Lo} = 20$ to 50 at 1 GHz. These inductors may be placed well above any conductive planes to reduce parasitic capacitances, yielding a higher SRF f_r. The low-frequency inductance of bondwire inductors is given by

$$L \approx \frac{\mu_0 l}{2\pi}\left[\ln\left(\frac{2l}{a}\right) - 0.75\right] \quad \text{(H)} \tag{10.60}$$

where l is the length of the bondwire and a is the radius of the bondwire. A standard bondwire inductance with $l = 1$ mm gives $L = 1$ nH or 1 nH/mm. The resistance is

$$R = \frac{l}{2\pi a\delta\sigma} \tag{10.61}$$

where σ is the conductivity and δ is the skin depth. For aluminum, $\sigma = 4 \times 10^7$ S/m. The skin depth for aluminum is $\delta_{Al} = 2.5$ μm at 1 GHz. The resistance per unit length is $R_{Al}/l \approx 0.2\,\Omega$/mm at 2 GHz.

Bondwires are often used as interconnects and package leads in ICs. A large number of closely spaced bondwires are often connected in parallel. The mutual inductance of two parallel bond wires is

$$M = \frac{\mu_0 l}{2\pi}\left(\ln\frac{2l}{s} + \frac{s}{l} - 1\right) \tag{10.62}$$

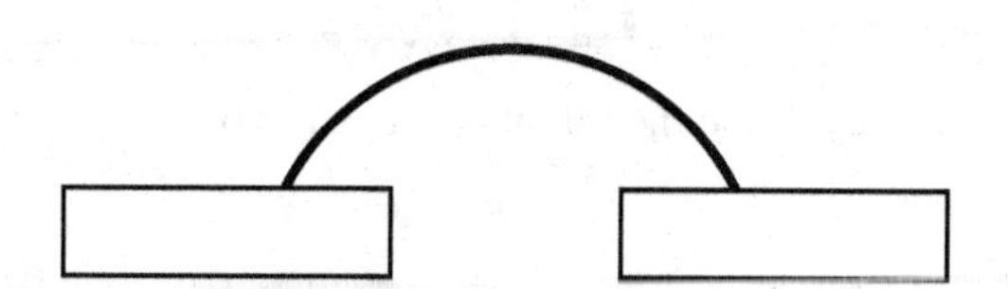

Figure 10.15 Bondwire inductor.

where s is the distance (separation) between the two bondwires. Typically, $M = 0.3\,\text{nH/mm}$ at $s = 0.2\,\text{mm}$. A typical value of the coupling coefficient k is 0.3.

The major disadvantage of bondwire inductors is low predictability because of possible variations in length and spacing. Bondwire inductance depends on bonding geometry and the existence of neighboring bondwires, making accurate prediction of the inductance value difficult. However, once the configuration for a particular inductance is known, it is rather repeatable in subsequent bondings. The sensitivity of the circuit performance to the variations in the bondwire inductances and mutual inductances should be examined. Common bondwire metals include gold and aluminum. Gold is generally preferred because of its higher conductivity and flexibility. This allows higher quality factor Q_{Lo} with shorter physical lengths to be bonded for a given die height.

Example 10.6

Calculate the inductance of a round bondwire inductor with $l = 2\,\text{mm}$, $a = 0.2\,\text{mm}$, and $\mu_r = 1$ for $\delta > a$.

Solution. The inductance of the bondwire inductor for $\delta > a$ is

$$L \approx \frac{\mu_0 l}{2\pi}\left[\ln\left(\frac{2l}{a}\right) - 0.75\right] = \frac{4\pi \times 10^{-7} \times 2 \times 10^{-3}}{2\pi}\left[\ln\left(\frac{2 \times 2}{0.2}\right) - 0.75\right] = 0.8982\,\text{nH} \tag{10.63}$$

10.12 Single-Turn Planar Inductor

The self-inductance is a special case of mutual inductance. Therefore, an expression for the self-inductance of an inductor can be derived using the concept of the mutual inductance. This is demonstrated by deriving an expression for the inductance of a single-turn planar inductor.

Figure 10.16 shows a single-turn strip inductor of width w, inner radius b, and outer radius a. The concept of mutual inductance can be used to determine a self-inductance. Consider two circuits: a line circuit along the inner surface of the strip and a line current along the outer surface of the strip. The vector magnetic potential caused by circuit 2 is

$$\mathbf{A}_2 = I_2 \oint_{C_2} \frac{\mu d\mathbf{l}_2}{4\pi h} \tag{10.64}$$

The integration of the vector magnetic potential $\mathbf{A}_2$ along circuit 1 resulting from the current I_2 in circuit 2 gives the mutual inductance between the two circuits

$$M = \frac{1}{I_2} \oint_{C_1} \mathbf{A}_2 \cdot d\mathbf{l}_1 \tag{10.65}$$

Hence, the mutual inductance of two conductors is given by

$$M = \frac{\mu}{4\pi} \oint_{C_1} \oint_{C_2} \frac{d\mathbf{l}_1 \cdot d\mathbf{l}_2}{h} \tag{10.66}$$

From geometrical consideration and the following substitution

$$\theta = \pi - 2\phi \tag{10.67}$$

we have

$$dl_1 = |d\mathbf{l}_1| = a d\theta = -2a d\phi \tag{10.68}$$

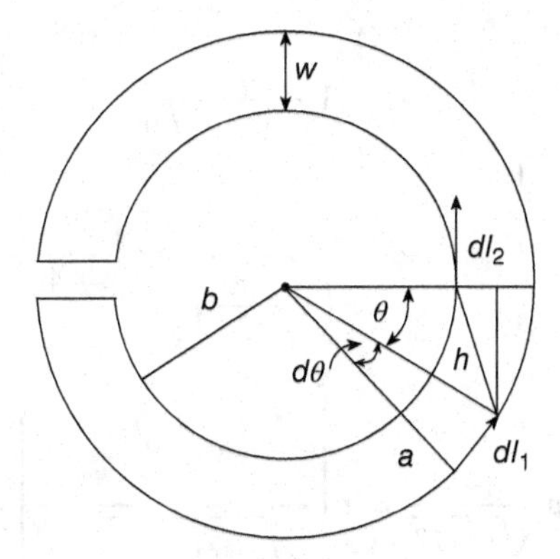

Figure 10.16 Single-turn planar inductor.

yielding

$$d\mathbf{l}_1 d\mathbf{l}_2 = dl_1\, dl_2 \cos\theta = 2a\cos 2\phi dl_2\, d\phi \tag{10.69}$$

and

$$\begin{aligned} h &= \sqrt{(a\cos\theta - b)^2 + (a\sin\theta)^2} = \sqrt{a^2 + b^2 - 2ab\cos\theta} = \sqrt{a^2 + b^2 + 2ab\cos 2\phi} \\ &= \sqrt{a^2 + b^2 + 2ab(1 - 2\sin^2\phi)} = \sqrt{(a+b)^2 - 4ab\sin^2\phi} = (a+b)\sqrt{1 - k^2\sin^2\phi} \end{aligned} \tag{10.70}$$

where

$$k^2 = \frac{4ab}{(a+b)^2} = 1 - \frac{(a-b)^2}{(a+b)^2} \tag{10.71}$$

Also,

$$\oint dl_2 = 2\pi b \tag{10.72}$$

The mutual inductance is

$$\begin{aligned} M &= \frac{\mu}{4\pi}\oint dl_2 \int_0^{2\pi} \frac{a\cos\theta\, d\theta}{\sqrt{a^2 + b^2 - 2ab\cos\theta}} = \frac{2\mu ab}{a+b}\int_0^{\pi/2} \frac{(2\sin^2\phi - 1)d\phi}{\sqrt{1 - k^2\sin^2\phi}} \\ &= \mu\sqrt{ab}k\int_0^{\pi/2} \frac{(2\sin^2\phi - 1)d\phi}{\sqrt{1 - k^2\sin^2\phi}} = \mu\sqrt{ab}\left[\left(\frac{2}{k} - k\right)K(k) - \frac{2}{k}E(k)\right] \end{aligned} \tag{10.73}$$

where the complete elliptic integrals are

$$E(k) = \int_0^{\pi/2} \sqrt{1 - k^2, \sin^2\phi}\, d\phi \tag{10.74}$$

and

$$K(k) = \int_0^{\pi/2} \frac{d\phi}{\sqrt{1 - k^2\sin^2\phi}} \tag{10.75}$$

The width of the strip is

$$w = a - b \tag{10.76}$$

resulting in

$$a + b = 2a - w \tag{10.77}$$

$$ab = a(a - w) = a^2\left(1 - \frac{w}{a}\right) \tag{10.78}$$

and

$$k^2 = 1 - \frac{w^2}{(2a - w)^2} \tag{10.79}$$

If $w/a \ll 1$, $k \approx 1$, $E(k) \approx 1$,

$$k^2 = 1 - \frac{w^2}{4a^2\left(1 - \frac{w}{2a}\right)^2} \approx 1 - \frac{w^2}{4a^2} \tag{10.80}$$

and

$$K(k) \approx \ln \frac{4}{\sqrt{1-k^2}} \approx \ln \left[\frac{4}{\sqrt{1 - \left(1 - \frac{w^2}{4a^2}\right)}} \right] = \ln\left(\frac{8a}{w}\right) \tag{10.81}$$

Hence, the self-inductance of a single-turn inductor is given by

$$L \approx \mu a \left[\ln\left(\frac{8a}{w}\right) - 2\right] = \frac{\mu l}{2\pi}\left[\ln\left(\frac{8a}{w}\right) - 2\right] \quad \text{for} \quad a \gg w \tag{10.82}$$

where $l = 2\pi a$.

Example 10.7

Calculate the inductance of a single-turn round planar inductor with $a = 100\,\mu\text{m}$, $w = 1\,\mu\text{m}$, and $\mu_r = 1$ for $\delta > a$.

Solution. The inductance of the single-turn round planar inductor for $\delta < a$ is

$$L \approx \mu a \left[\ln\left(\frac{8a}{w}\right) - 2\right] = 4\pi \times 10^{-7} \times 100 \times 10^{-6}\left[\ln\left(\frac{8 \times 100}{1}\right) - 2\right] = 0.58868\,\text{nH} \tag{10.83}$$

10.13 Inductance of Planar Square Loop

The self-inductance of a square coil made of rectangular wire of length $l \gg w$ and width w is

$$L \approx \frac{\mu_0 l}{\pi}\left[arcsinh\left(\frac{l}{2w}\right) - 1\right] \ (\text{H}) \tag{10.84}$$

Both l and w are in meters.

10.14 Planar Spiral Inductors

10.14.1 Geometries of Planar Spiral Inductors

Planar spiral inductors are the most widely used RF IC inductors. Figure 10.17 shows square, octagonal, hexagonal, and circular planar spiral inductors. A planar spiral inductor consists of a low-resistivity metal trace (aluminum, copper, gold, or silver), silicon dioxide (SiO_2) of thickness t_{ox}, and a silicon substrate, as shown in Fig. 10.18. A metal layer embedded in silicon dioxide is used to form the metal spiral. The topmost metal layer is usually the thickest and thus most conductive. In addition, a larger distance of the topmost metal layer to the substrate

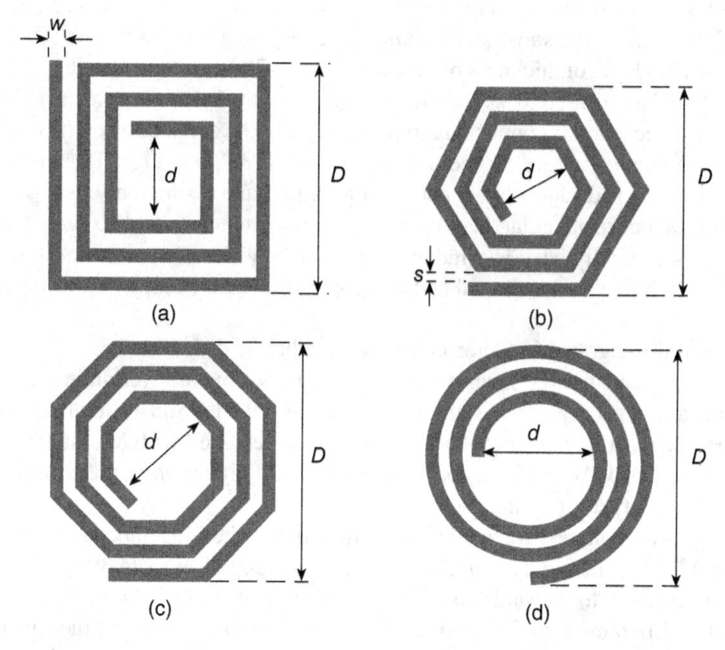

Figure 10.17 Planar spiral integrated inductors. (a) Square inductor. (b) Hexagonal inductor. (c) Octagonal inductor. (d) Circular inductor.

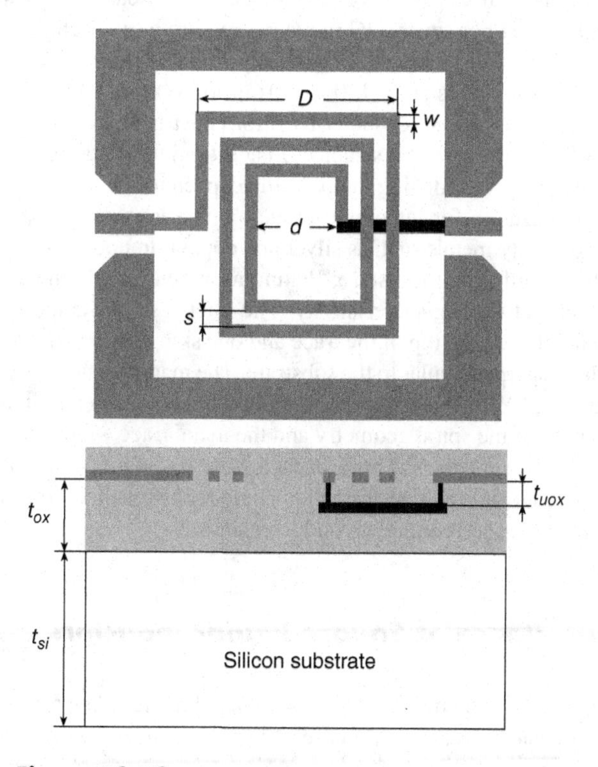

Figure 10.18 Cross section of a planar RF IC inductor.

reduces the parasitic capacitance $C = \epsilon A_m/t_{ox}$, increasing the SRF $f_r = 1/(2\pi\sqrt{LC})$, where A_m is the metal trace area. The substrate is a thick silicon (Si), gallium arsenide (GaAs), or silicon germanium (SiGe) layer of thickness of the order of 500–700 μm. The thin silicon dioxide (SiO_2) of thickness 0.4–3 μm is used to isolate the metal strips of the inductor from the silicon substrate. The outer end of the spiral is connected directly to a port. The connection of the innermost turn is made through an *underpass* by another metal layer or via an *air bridge*. The whole spiral structure is connected to pads and surrounded by a ground plane. Commonly used shapes of spiral inductors are square, rectangular, hexagonal, octagonal, and circular. Hexagonal and octagonal spiral inductors generally have less inductance and less series resistance per turn as compared to square spirals. Since the hexagonal and octagonal structures occupy larger chip area, they are rarely used.

A planar spiral integrated inductor can only be fabricated with technologies having two or more metal layers because the inner connection requires a metal layer different from that used for the spiral to connect the inside turn of the coil to outside. Square inductor shapes are the most compatible with IC layout tools. They are easily designed with Manhattan style physical layout tools, such as MAGIC. Hexagonal, octagonal, and higher order polygon spiral inductors have higher quality factor Q_{Lo} than square spirals.

The parameters of interest for integrated inductors are the inductance L_s, the quality factor Q_{Lo}, and the SRF f_r. The typical inductances are in the range from 1 to 30 nH, the typical quality factors are from 5 to 20, and the typical self-resonant frequencies are from 2 to 20 GHz. The geometrical parameters of an inductor are the number of turns N, the metal strip width w, the metal strip height h, the turn-to-turn space s, the inner diameter d, the outer diameter D, the thickness of the silicon substrate t_{Si}, the thickness of the oxide layer t_{ox}, and the thickness of oxide between the metal strip and the underpass t_{uox}. The typical metal strip width w is 30 μm, the typical turn-to-turn spacing s is 20 μm, and the typical metal strip sheet resistance is $R_{sheet} = \rho/h = 0.03$–$0.1\,\Omega$ per square. IC inductors require a lot of chip area. Therefore, practical IC inductors have small inductances, typically $L \leq 10\,\text{nH}$, but they can be as high as 30 nH. The typical inductor size is from $130 \times 130\,\mu\text{m}$ to $1000 \times 1000\,\mu\text{m}$. The inductor area is $A = D^2$. The topmost metal layer is usually used for construction of IC inductors because it is the thickest and thus has the lowest resistance. In addition, the distance from the topmost layer to the substrate is the largest, reducing the parasitic capacitances.

The preferred metalization for integrated inductors is a low-resistivity, inert metal, such as gold. Other low-resistivity metals such as silver and copper do not offer the same level of resistance to atmospheric sulfur and moisture. Platinum, another noble metal, is two times more expensive than gold and has higher resistivity. The thickness of the metal h should be higher than 2δ, one skin depth δ on the top of the trace and one skin depth on the bottom. The direction of the magnetic flux is perpendicular to the substrate. The magnetic field penetrates the substrate, inducing eddy currents. A high substrate conductivity tends to lower the inductor quality factor Q_{Lo}. The optimization of the spiral geometry and the metal trace width leads to minimization of the trace ohmic resistance and the substrate capacitance. To reduce power losses in the substrate, the substrate can be made of high-resistive silicon oxide such as silicon-on-insulator (SOI), thick dielectric layers, or thick and multilayer conductor lines.

10.14.2 Inductance of Square Planar Inductors

Several expressions have been developed to estimate RF spiral planar inductances. In general, the inductance of planar inductors L increases when the number of turns N and the area of an inductor A increase.

The inductance of a single loop of area A and any shape (e.g., square, rectangular, hexagonal, octagonal, or circular) is given by

$$L \approx \mu_0 \sqrt{\pi A} \tag{10.85}$$

Hence, the inductance of a single-turn round inductor of radius r is

$$L \approx \pi \mu_0 r = 4\pi^2 \times 10^{-7} r = 4 \times 10^{-6} r \ \text{(H)} \tag{10.86}$$

For example, $L = 4\,\text{nH}$ for $r = 1$ mm.

The inductance of an arbitrary planar spiral inductor with N turns, often used in ICs, is given by

$$L \approx \pi \mu_0 r N^2 = 4\pi^2 \times 10^{-7} r N^2 = 4 \times 10^{-6} r N^2 \ \text{(H)} \tag{10.87}$$

where N is the number of turns and r is the spiral radius.

Bryan's Formula. The inductance of the square planar spiral inductor is given by empirical Bryan's equation [8]

$$L = 6.025 \times 10^{-7} (D + d) N^{\frac{5}{3}} \ln \left[4 \left(\frac{D+d}{D-d} \right) \right] \ \text{(H)} \tag{10.88}$$

where the outermost diameter is

$$D = d + 2Nw + 2(N-1)s \tag{10.89}$$

N is the number of turns, and D and d are the outermost and innermost diameters of the inductor in meters, respectively. Figures 10.19 through 10.22 illustrate equations developed by various authors for the inductance L as a function of d, w, s, a, and N.

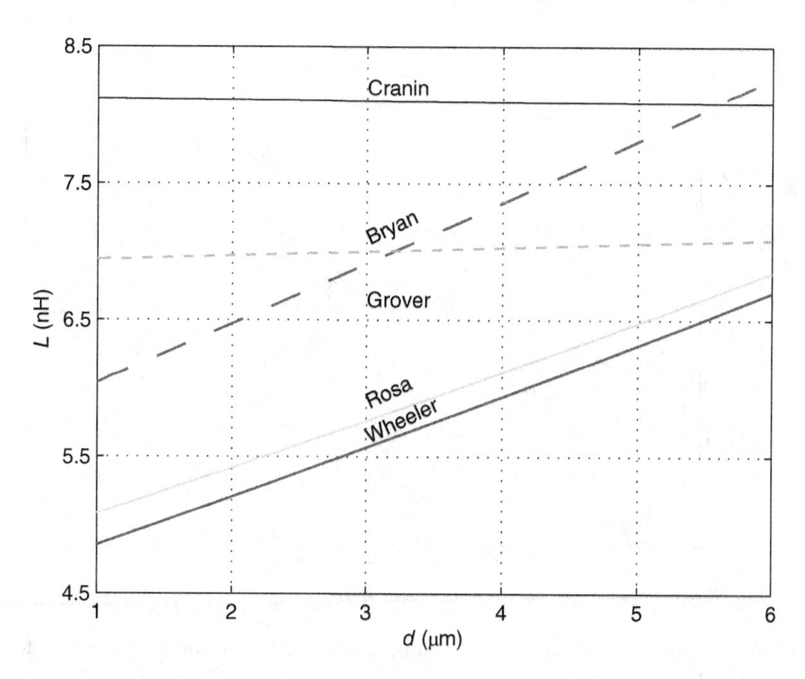

Figure 10.19 Inductance L as a function of inner diameter d for square spiral inductor at $N = 5$, $w = 30$ μm, $s = 20$ μm, and $h = 20$ μm.

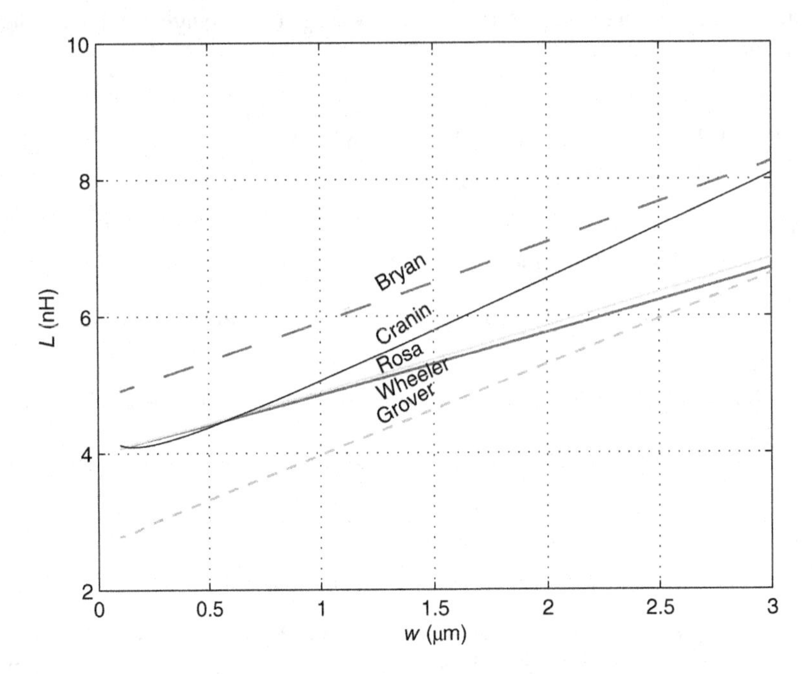

Figure 10.20 Inductance L as a function of trace width w at $N = 5$, $d = 60$ μm, $s = 20$ μm, and $h = 3$ μm.

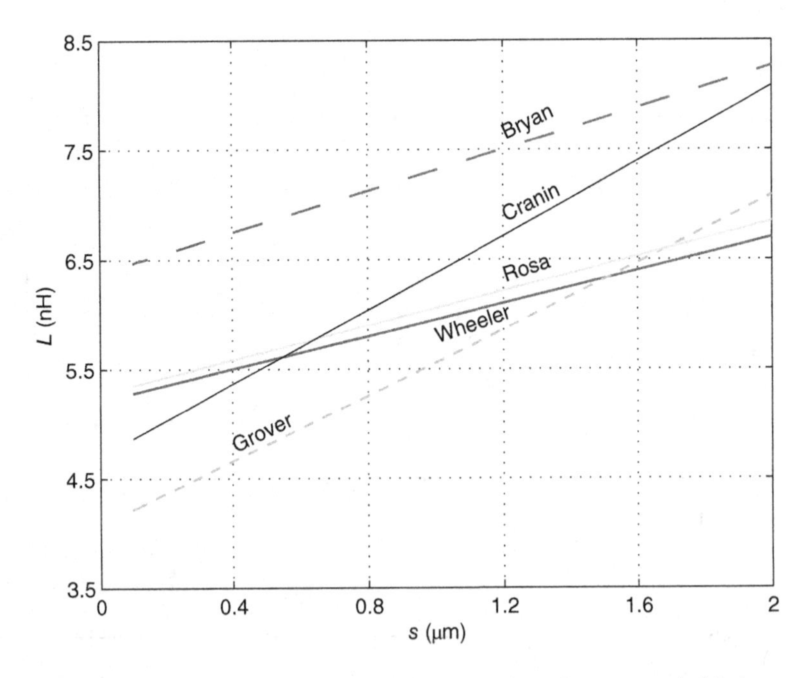

Figure 10.21 Inductance L as a function of trace separation s for square spiral inductor at $N = 5$, $d = 60$ μm, $w = 30$ μm, and $h = 3$ μm.

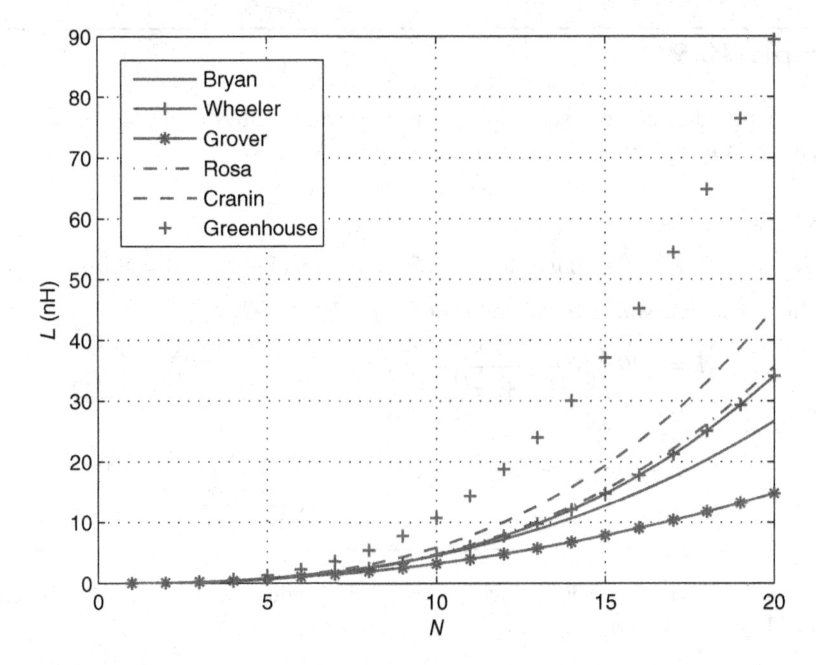

Figure 10.22 Inductance L as a function of number of turns N for square spiral inductor at $d = 60$ μm, $w = 30$ μm, $s = 20$ μm, and $h = 3$ μm.

Example 10.8

Calculate the inductance of a square planar spiral inductor with $N = 5$, $d = 60\,\mu\text{m}$, $w = 30\,\mu\text{m}$, $s = 20\,\mu\text{m}$, and $\mu_r = 1$ for $\delta > h$. Use Bryan's formula.

Solution. The outermost diameter is

$$D = d + 2Nw + 2(N-1)s = 60 + 2 \times 5 \times 30 + 2 \times (5-1) \times 20 = 520\,\mu\text{m} \tag{10.90}$$

The inductance of the round planar inductor for $\delta > h$ is

$$L = 6.025 \times 10^{-7}(D+d)N^{\frac{5}{3}} \ln\left[4\left(\frac{D+d}{D-d}\right)\right] \text{ (H)}$$
$$= 6.025 \times 10^{-7} \times (520+60) \times 10^{-6} \times 5^{\frac{5}{3}} \ln\left[4\left(\frac{520+60}{520-60}\right)\right] = 8.266\,\text{nH} \tag{10.91}$$

Wheeler's Formula. The inductance of a square spiral inductor is given by modified Wheeler's formula [4, 33]

$$L = 1.17\mu_0 N^2 \frac{D+d}{1 + 2.75\frac{D-d}{D+d}} \text{ (H)} \tag{10.92}$$

where

$$D = d + 2Nw + 2(N-1)s \tag{10.93}$$

All dimensions are in meters. The inductance in terms of N, w, and s is

$$L = \frac{2.34\mu_0 N^2}{3.75N(w+s) - 3.75s + d} \text{ (H)} \tag{10.94}$$

Example 10.9

Calculate the inductance of a square planar spiral inductor with $N = 5$, $d = 60\,\mu m$, $w = 30\,\mu m$, $s = 20\,\mu m$, and $\mu_r = 1$ for $\delta > h$. Use Wheeler's formula.

Solution. The outermost diameter is

$$D = d + 2Nw + 2(N-1)s = 60 + 2 \times 5 \times 30 + 2 \times (5-1) \times 20 = 520\,\mu m \quad (10.95)$$

The inductance of the square planar inductor for $\delta > h$ is

$$L = 1.17\mu_0 N^2 \frac{D+d}{1 + 2.75\dfrac{D-d}{D+d}} \text{ (H)}$$

$$= 1.17 \times 4\pi \times 10^{-7} \times 5^2 \times \frac{(520+60) \times 10^{-6}}{1 + 2.75 \times \dfrac{520-60}{520+60}} = 6.7\,\text{nH} \quad (10.96)$$

Greenhouse Formula. The inductance of spiral inductors is given by Greenhouse's equation [14]

$$L = 10^{-9} DN^2 \left\{ 180 \left[\log\left(\frac{D}{4.75(w+h)} \right) + 5 \right] + \frac{1.76(w+h)}{D} + 7.312 \right\} \text{ (nH)} \quad (10.97)$$

All dimensions are in meters.

Example 10.10

Calculate the inductance of a square planar spiral inductor with $N = 5$, $d = 60\,\mu m$, $w = 30\,\mu m$, $s = 20\,\mu m$, $h = 20\,\mu m$, and $\mu_r = 1$ for $\delta > h$. Use Greenhouse's formula.

Solution. The outermost diameter is

$$D = d + 2Nw + 2(N-1)s = 60 + 2 \times 5 \times 30 + 2 \times (5-1) \times 20 = 520\,\mu m. \quad (10.98)$$

The inductance of the single-turn round planar inductor for $\delta > h$ is

$$L = 10^{-9} DN^2 \left\{ 180 \left[\log\left(\frac{D}{4.75(w+h)} \right) + 5 \right] + \frac{1.76(w+h)}{D} + 7.312 \right\} \text{ (nH)}$$

$$= 10^{-9} \times 520 \times 10^{-6} \times 5^2 \left\{ 180 \left[\log\left(\frac{520}{4.75(30+20)} \right) + 5 \right] + \frac{1.76(30+20)}{520} + 7.312 \right\}$$

$$= 12.59\,\text{nH} \quad (10.99)$$

Grover's Formula. The inductance of a square or rectangular spiral planar inductor with a rectangular cross section of the trace is expressed as [8, 31]

$$L = \frac{\mu_0 l}{2\pi} \left[\ln\left(\frac{2l}{w+h} \right) + \frac{w+h}{3l} + 0.50049 \right] \text{ (H)} \quad (10.100)$$

where the length of the metal trace is

$$l = 2(D+w) + 2N(2N-1)(w+s) \quad (10.101)$$

the outer diameter is

$$D = d + 2Nw + 2(N-1)s \tag{10.102}$$

and h is the metal trace height. All dimensions l, w, s, and h are in meters. This equation is similar to that for a straight trace inductor. It accounts for the self-inductance only and neglects mutual inductances between the traces.

Example 10.11

Calculate the inductance of a square planar spiral inductor with $N = 5$, $d = 60\,\mu\text{m}$, $w = 30\,\mu\text{m}$, $s = 20\,\mu\text{m}$, $h = 20\,\mu\text{m}$, and $\mu_r = 1$ for $\delta > h$. Use Grover's formula.

Solution. The outer diameter is

$$D = d + 2Nw + 2(N-1)s = 60 + 2 \times 5 \times 30 + 2 \times (5-1) \times 20 = 520\,\mu\text{m} \tag{10.103}$$

The length of the metal trace is

$$l = 2(D + w) + 2N(2N-1)(w + s) = 2(520 + 30) + 2 \times 5(2 \times 5 - 1)(30 + 20) = 5600\,\mu\text{m} \tag{10.104}$$

The inductance of the square spiral planar inductor for $\delta > h$ is

$$\begin{aligned} L &= \frac{\mu_0 l}{2\pi}\left[\ln\left(\frac{2l}{w+h}\right) + \frac{w+h}{3l} + 0.50049\right] \ (\text{H}) \\ &= \frac{4\pi \times 10^{-7} \times 5600 \times 10^{-6}}{2\pi}\left[\ln\left(\frac{2 \times 5600}{30+20}\right) + \frac{30+20}{3 \times 5600} + 0.50049\right] \\ &= 6.608\,\text{nH} \end{aligned} \tag{10.105}$$

Rosa's Formula. The equation for the inductance of square planar spiral inductors is expressed as [1, 9, 33]

$$L = 0.3175\mu N^2(D+d)\left\{\ln\left[\frac{2.07(D+d)}{D-d}\right] + \frac{0.18(D-d)}{D+d} + 0.13\left(\frac{D-d}{D+d}\right)^2\right\} \ (\text{H}) \tag{10.106}$$

All dimensions are in meters.

Example 10.12

Calculate the inductance of a square planar spiral inductor with $N = 5$, $d = 60\,\mu\text{m}$, $w = 30\,\mu\text{m}$, $s = 20\,\mu\text{m}$, $h = 20\,\mu\text{m}$, and $\mu_r = 1$ for $\delta > h$. Use Rosa's formula.

Solution. The outer diameter is

$$D = d + 2Nw + 2(N-1)s = 60 + 2 \times 5 \times 30 + 2 \times (5-1) \times 20 = 520\,\mu\text{m} \tag{10.107}$$

The inductance of the square spiral planar inductor for $\delta > h$ is

$$L = 0.3175\mu_0 N^2(D+d)\left\{\ln\left[\frac{2.07(D+d)}{D-d}\right] + \frac{0.18(D-d)}{D+d} + 0.13\left(\frac{D-d}{D+d}\right)^2\right\}$$

$$= 0.3175 \times 4\pi \times 10^{-7} \times 5^2 \times (520+60) \times 10^{-6}$$

$$\times\left\{\ln\left[\frac{2.07(520+60)}{520-60}\right] + \frac{0.18 \times (520-60)}{520+60} + 0.13\left(\frac{520-60}{520+60}\right)^2\right\}$$

$$= 6.849 \text{ nH} \quad (10.108)$$

Cranin's Formula. An empirical expression for the inductance of spiral planar inductors that has less than 10% error for inductors in the range from 5 to 50 nH is given by Craninckx and Steyeart [22]

$$L \approx 1.3 \times 10^{-7} \frac{A_m^{5/3}}{A_{tot}^{1/6} w^{1.75}(w+s)^{0.25}} \quad \text{(nH)} \quad (10.109)$$

where A_m is the metal area, A_{tot} is the total inductor area, w is the metal trace width, and s is the spacing between metal traces. All dimensions are in meters.

Example 10.13

Calculate the inductance of a square planar spiral inductor with $N = 5$, $d = 60\,\mu$m, $w = 30\,\mu$m, $s = 20\,\mu$m, $h = 20\,\mu$m, and $\mu_r = 1$ for $\delta > h$. Use Cranin's formula.

Solution. The outer diameter is

$$D = d + 2Nw + 2(N-1)s = 60 + 2 \times 5 \times 30 + 2 \times (5-1) \times 20 = 520\,\mu\text{m} \quad (10.110)$$

The length of the metal trace is

$$l = 2(D+w) + 2N(2N-1)(w+s) = 2(520+30) + 2 \times 5(2 \times 5 - 1)(30+20) = 5600\,\mu\text{m} \quad (10.111)$$

Hence, the total area of the inductor is

$$A_{tot} = D^2 = (520 \times 10^{-6})^2 = 0.2704 \times 10^{-6}\,\text{m}^2 \quad (10.112)$$

The metal area is

$$A_m = wl = 30 \times 10^{-6} \times 5600 \times 10^{-6} = 0.168 \times 10^{-6}\,\text{m}^2 \quad (10.113)$$

The inductance of the square spiral planar inductor for $\delta > h$ is

$$L \approx 1.3 \times 10^{-7} \frac{A_m^{5/3}}{A_{tot}^{1/6} w^{1.75}(w+s)^{0.25}} \quad \text{(H)}$$

$$= 1.3 \times 10^{-7} \frac{(0.168 \times 10^{-6})^{5/3}}{(0.2704 \times 10^{-6})^{1/6}(30 \times 10^{-6})^{1.75}[(30+20) \times 10^{-6}]^{0.25}} = 8.0864 \text{ nH} \quad (10.114)$$

Monomial Formula. The data-fitted monomial empirical expression for the inductance of the square spiral inductor is [33]

$$L = 0.00162 D^{-1.21} w^{-0.147}\left(\frac{D+d}{2}\right)^{2.4} N^{1.78} s^{-0.03} \quad \text{(nH)} \quad (10.115)$$

All dimensions are in micrometers.

Example 10.14

Calculate the inductance of a square planar spiral inductor with $N = 5$, $d = 60\,\mu\text{m}$, $w = 30\,\mu\text{m}$, $s = 20\,\mu\text{m}$, $h = 20\,\mu\text{m}$, and $\mu_r = 1$ for $\delta > h$. Use monomial formula.

Solution. The outer diameter is

$$D = d + 2Nw + 2(N-1)s = 60 + 2 \times 5 \times 30 + 2 \times (5-1) \times 20 = 520\,\mu\text{m} \tag{10.116}$$

The inductance of the square spiral planar inductor for $\delta < a$ is

$$L = 0.00162 D^{-1.21} w^{-0.147} \left(\frac{D+d}{2}\right)^{2.4} N^{1.78} s^{-0.03} \text{ (nH)}$$

$$= 0.00162 \times (520)^{-1.21} 30^{-0.147} \left(\frac{520+60}{2}\right)^{2.4} 5^{1.78} 20^{-0.03} = 6.6207 \text{ nH} \tag{10.117}$$

Jenei's Formula. The total inductance consists of the self-inductance L_{self} and the sum of the positive mutual inductances M^+ and the sum of the negative mutual inductances M^-. The derivation of an expression for the total inductance is given in [37]. The self-inductance of one straight segment is [8]

$$L_{self1} = \frac{\mu_0 l_{seg}}{2\pi}\left[\ln\left(\frac{2l_{seg}}{w+h}\right) + 0.5\right] \tag{10.118}$$

where l_{seg} is the segment length, w is the metal trace width, and h is the metal trace thickness. The total length of the conductor is

$$l = (4N+1)d + (4N_i+1)N_i(w+s) \tag{10.119}$$

where N is the number of turns, N_i is the integer part of N, and s is the spacing between the segments. The total self-inductance of a square planar inductor is equal to the sum of $4N$ self-inductances of all the segments.

$$L_{self} = 4NL_{self1} = \frac{\mu_0 l}{2\pi}\left[\ln\left(\frac{2l}{N(w+h)}\right) - 0.2\right] \tag{10.120}$$

The anti-parallel segments contribute to negative mutual inductance M^-. The sum of all interactions is approximately equal to $4N^2$ average interactions between segments of an average length and an average distance. The negative mutual inductance is

$$M^- = \frac{0.47\mu_0 Nl}{2\pi} \tag{10.121}$$

The positive mutual inductance is caused by interactions between parallel segments on the same side of a square. The average distance is

$$b = (w+s)\frac{(3N - 2N_i - 1)(N_i + 1)}{3(2N - N_i - 1)} \tag{10.122}$$

For $N_i = N$,

$$b = \frac{(w+s)(N+1)}{3} \tag{10.123}$$

The total positive mutual inductance is

$$M^+ = \frac{\mu_0 l(N-1)}{2\pi}\left\{\ln\left[\sqrt{1+\left(\frac{l}{4Nb}\right)^2} + \frac{l}{4Nb}\right] - \sqrt{1+\left(\frac{4Nb}{l}\right)^2} + \frac{4Nb}{l}\right\} \tag{10.124}$$

The inductance of a square planar inductors is

$$L = L_{self} + M^- + M^+ = \frac{\mu_0 l}{2\pi}\left\langle \ln\left[\frac{l}{N(w+h)}\right] - 0.2 - 0.47N \right.$$
$$\left. +(N-1)\left\{\ln\left[\sqrt{1+\left(\frac{l}{4Nb}\right)^2}+\frac{l}{4Nb}\right] - \sqrt{1+\left(\frac{4Nb}{l}\right)^2}+\frac{4Nb}{l}\right\}\right\rangle \quad (10.125)$$

All dimensions are in meters.

Example 10.15

Calculate the inductance of a square planar spiral inductor with $N = N_i = 5$, $d = 60\,\mu m$, $w = 30\,\mu m$, $s = 20\,\mu m$, $h = 20\,\mu m$, and $\mu_r = 1$ for $\delta > h$. Use Jenei's formula.

Solution. The total length of the conductor is

$$l = (4N+1)d + (4N_i+1)N_i(w+s) = (4\times 5+1)\times 60\times 10^{-6} + (4\times 5+1)5(30+20)\times 10^{-6}$$
$$= 6500\,\mu m \quad (10.126)$$

The average distance is

$$b = (w+s)\frac{(3N-2N_i-1)(N_i+1)}{3(2N-N_i-1)}$$
$$= (30+20)\times 10^{-6}\frac{(3\times 5-2\times 5-1)(5+1)}{3(2\times 5-5-1)} = 100\,\mu m \quad (10.127)$$

The inductance of the square spiral planar inductor for $\delta > h$ is

$$L = \frac{\mu_0 l}{2\pi}\left\langle \ln\left[\frac{l}{N(w+h)}\right] - 0.2 - 0.47N \right.$$
$$\left. + (N-1)\left\{\ln\left[\sqrt{1+\left(\frac{l}{4Nb}\right)^2}+\frac{l}{4Nb}\right] - \sqrt{1+\left(\frac{4Nb}{l}\right)^2}+\frac{4Nb}{l}\right\}\right\rangle$$
$$= \frac{4\pi\times 10^{-7}\times 6500\times 10^{-6}}{2\pi}\left\langle \ln\left[\frac{6500}{5(30+20)}\right] - 0.2 - 0.47\times 5 + (5-1)\right.$$
$$\times\left\{\ln\left[\sqrt{1+\left(\frac{6500}{4\times 5\times 100}\right)^2}+\frac{6500}{4\times 5\times 100}\right]\right.$$
$$\left.\left. -\sqrt{1+\left(\frac{4\times 5\times 100}{6500}\right)^2}+\frac{4\times 5\times 100}{6500}\right\}\right\rangle$$
$$= 6.95\ \text{nH} \quad (10.128)$$

Dill's Formula. The inductance of a square planar inductor is [10]

$$L = 8.5\times 10^{-10} N^{5/3}\ \text{(H)} \quad (10.129)$$

For $N = 5, L = 12.427\,\text{nH}$.

Terman's Formula. The inductance of a square planar spiral single-turn inductor is

$$L = 18.4173 \times 10^{-4} D \left[\log \left(\frac{0.7874 \times 10^{-4} D^2}{w+h} \right) - \log(0.95048 \times 10^{-4} D) \right] + 10^{-4} \times [7.3137D + 1.788(w+h)] \text{ (nH)}. \quad (10.130)$$

All dimensions are in micrometers.

The inductance of a square planar spiral multiple-turn inductor is

$$L = 18.4173 \times 10^{-4} DN^2 \left[\log \left(\frac{0.7874 \times 10^{-4} D^2}{w+h} \right) - \log(0.95048 \times 10^{-4} D) \right] + 8 \times 10^{-4} N^2 [0.914D + 0.2235(w+h)] \text{ (nH)} \quad (10.131)$$

where the outer diameter is

$$D = d + 2Nw + 2(N-1)s \quad (10.132)$$

All dimensions are in meters.

Example 10.16

Calculate the inductance of a square planar spiral inductor with $N = 5$, $D = 520\,\mu\text{m}$, $w = 30\,\mu\text{m}$, $h = 20\,\mu\text{m}$, and $\mu_r = 1$ for $\delta > h$. Use Terman's formula.

Solution. The inductance of a square planar spiral multiple-turn inductor is

$$L = 18.4173 \times 10^{-4} DN^2 \left[\log \left(\frac{0.7874 \times 10^{-4} D^2}{w+h} \right) - \log(0.95048 \times 10^{-4} D) \right] + 8 \times 10^{-4} N^2 [0.914D + 0.2235(w+h)] \text{ (nH)}$$

$$= 18.4173 \times 10^{-4} \times 520 \times 5^2 \times \left[\log \left(\frac{0.7874 \times 10^{-4} \times 520^2}{30+20} \right) - \log(0.95048 \times 10^{-4} \times 520) \right] + 8 \times 10^{-4} 5^2 [0.914 \times 520 + 0.2235(30+20)] = 32.123\,\text{nH} \quad (10.133)$$

Rosa's, Wheeler's, Grover's, Bryan's, Jenei's, and monomial formulas give similar values of inductance. Greenhouse's, Cranin's, and Terman's formulas give higher values of inductance than those calculated from the former group of formulas.

10.14.3 Inductance of Hexagonal Spiral Inductors

Wheeler's Formula. The inductance of a hexagonal spiral inductor is given by modified Wheeler's formula [4, 33]

$$L = 1.165 \mu_0 N^2 \frac{D+d}{1 + 3.82 \dfrac{D-d}{D+d}} \text{ (H)} \quad (10.134)$$

where

$$D = d + 2Nw + 2(N-1)s \quad (10.135)$$

All dimensions are in meters.

Rosa's Formula. The equation for the inductance of hexagonal planar spiral inductors is expressed as [1, 9, 33]

$$L = 0.2725\mu N^2(D+d)\left\{\ln\left[\frac{2.23(D+d)}{D-d}\right] + 0.17\left(\frac{D-d}{D+d}\right)^2\right\} \quad \text{(H)} \tag{10.136}$$

All dimensions are in meters.

Example 10.17

Calculate the inductance of a hexagonal planar spiral inductor with $N = 5$, $d = 60\,\mu\text{m}$, $w = 30\,\mu\text{m}$, $s = 20\,\mu\text{m}$, $h = 20\,\mu\text{m}$, and $\mu_r = 1$ for $\delta > h$. Use Rosa's formula.

Solution. The outer diameter is

$$D = d + 2Nw + 2(N-1)s = 60 + 2\times 5\times 30 + 2\times(5-1)\times 20 = 520\,\mu\text{m} \tag{10.137}$$

The inductance of the square spiral planar inductor for $\delta > h$ is

$$\begin{aligned} L &= 0.2725\mu N^2(D+d)\left\{\ln\left[\frac{2.23(D+d)}{D-d}\right] + 0.17\left(\frac{D-d}{D+d}\right)^2\right\} \quad \text{(H)} \\ &= 0.2725\times 4\pi\times 10^{-7}\times 5^2(520+60)\times 10^{-6} \\ &\quad \times\left\{\ln\left[\frac{2.23(520+60)}{520-60}\right] + 0.17\left(\frac{520-60}{520+60}\right)^2\right\} \\ &= 5.6641\ \text{nH} \end{aligned} \tag{10.138}$$

Grover's Formula. The inductance of a hexagonal planar inductor is [8]

$$L = \frac{2\mu l}{\pi}\left[\left(\frac{l}{6r}\right) + 0.09848\right] \quad \text{(H)} \tag{10.139}$$

All dimensions are in micrometers.

Monomial Formula. The data-fitted monomial empirical expression for the inductance of the hexagonal spiral inductor is [33]

$$L = 0.00128D^{-1.24}w^{-0.174}\left(\frac{D+d}{2}\right)^{2.47}N^{1.77}s^{-0.049} \quad \text{(nH)} \tag{10.140}$$

All dimensions are in micrometers.

10.14.4 Inductance of Octagonal Spiral Inductors

Wheeler's Formula. The inductance of an octagonal spiral inductor is given by modified Wheeler's formula [4, 33]

$$L = 1.125\mu_0 N^2\frac{D+d}{1+3.55\frac{D-d}{D+d}} \quad \text{(H)} \tag{10.141}$$

where

$$D = d + 2Nw + 2(N-1)s \tag{10.142}$$

All dimensions are in meters.

Rosa's Formula. The inductance of an octagonal planar spiral inductors is expressed as [1, 9, 33]

$$L = 0.2675\mu N^2(D+d)\left\{\ln\left[\frac{2.29(D+d)}{D-d}\right] + 0.19\left(\frac{D-d}{D+d}\right)^2\right\} \quad \text{(H)} \tag{10.143}$$

All dimensions are in meters.

Grover's Formula. The inductance of an octagonal planar inductor is [8]

$$L = \frac{2\mu l}{\pi}\left[\left(\frac{l}{8r}\right) - 0.03802\right] \quad \text{(H)} \tag{10.144}$$

All dimensions are in meters.

Monomial Formula. The data-fitted monomial empirical expression for the inductance of an octagonal spiral inductor is [33]

$$L = 0.00132D^{-1.21}w^{-0.163}\left(\frac{D+d}{2}\right)^{2.43} N^{1.75}s^{-0.049} \quad \text{(nH)} \tag{10.145}$$

All dimensions are in micrometers.

The inductance of octagonal inductors is almost the same as that of hexagonal inductors.

10.14.5 Inductance of Circular Spiral Inductors

Rosa's Formula. The inductance of circle planar spiral inductor is expressed as [1, 9, 33]

$$L = 0.25\mu N^2(D+d)\left\{\ln\left[\frac{2.46(D+d)}{D-d}\right] + 0.19\left(\frac{D-d}{D+d}\right)^2\right\} \quad \text{(H)} \tag{10.146}$$

All dimensions are in meters.

Example 10.18

Calculate the inductance of a circular planar spiral inductor with $N = 5$, $d = 60\,\mu\text{m}$, $w = 30\,\mu\text{m}$, $s = 20\,\mu\text{m}$, and $\mu_r = 1$ for $\delta > h$. Use Rosa's formula.

Solution. The outer diameter is

$$D = d + 2Nw + 2(N-1)s = 60 + 2\times 5\times 30 + 2\times(5-1)\times 20 = 520\,\mu\text{m} \tag{10.147}$$

The inductance of the circular spiral planar inductor for $\delta < a$ is

$$L = 0.25\mu N^2(D+d)\left\{\ln\left[\frac{2.46(D+d)}{D-d}\right] + 0.19\left(\frac{D-d}{D+d}\right)^2\right\} \quad \text{(H)}$$
$$= 0.25\times 4\pi\times 10^{-7}\times 5^2(520+60)\times 10^{-6}$$
$$\times\left\{\ln\left[\frac{2.46(520+60)}{520-60}\right] + 0.19\left(\frac{520-60}{520+60}\right)^2\right\}$$
$$= 5.7\ \text{nH} \tag{10.148}$$

Wheeler's Formula. The inductance of a circular planar inductor is [3]

$$L = 31.33\mu N^2\frac{a^2}{8a+11h} \quad \text{(H)} \tag{10.149}$$

All dimensions are in meters.

Schieber Formula. The inductance of a circular planar inductor is [16]

$$L = 0.874\pi \times 10^{-5} DN^2 \text{ (H)} \tag{10.150}$$

All dimensions are in meters.

Spiral IC inductors have high parasitic resistances and high shunt capacitances, resulting in a low Q_{Lo} and low self-resonant frequencies. It is difficult to achieve $Q_{Lo} > 10$ and f_r greater than a few gigahertzs for planar inductors because of the loss in the substrate and metalization. The RF MEMS technology has a potential to improve the performance of RF IC inductors by removing the substrate under the planar spiral via top-side etching, which decouples the RF IC inductor performance from substrate characteristics.

10.15 Multi-Metal Spiral Inductors

Multi-metal planar spiral inductors (called stacked inductors) are also used to achieve compact high-inductance magnetic devices. A double-layer spiral inductor can be implemented using metal 1 and metal 2 layers, as shown in Fig. 10.23 [46, 47]. An equivalent circuit of a two-layer inductor is shown in Fig. 10.24.

The impedance of a two-layer inductor is given by

$$Z = j\omega(L_1 + L_2 + 2M) = j\omega(L_1 + L_2 + 2k\sqrt{L_1L_2}) \approx j\omega(L_1 + L_2 + 2\sqrt{L_1L_2}) \tag{10.151}$$

where the coupling coefficient $k \approx 1$. If both inductors are equal,

$$Z \approx j\omega(L + L + 2L) = j\omega(4L) = j\omega L_s \tag{10.152}$$

Thus, the total inductance L_s for a two-layer inductor increases nearly four times due to mutual inductance. For an m-layer inductor, the total inductance is increased by a factor of m^2 as compared to a self-inductance of a single-layer spiral inductor. Modern CMOS technologies provide more than five metal layers, and stacking inductors or transformers give large inductances in a small chip area.

A patterned ground shield reduces eddy currents. Many RF IC designs incorporate several inductors on the same die. Since these structures are physically large, substrate coupling can cause significant problems, such as feedback and oscillations.

The spiral inductors have several drawbacks. First, the size is large compared with other inductors for the same number of turns N. Second, the spiral inductor requires a lead wire to connect

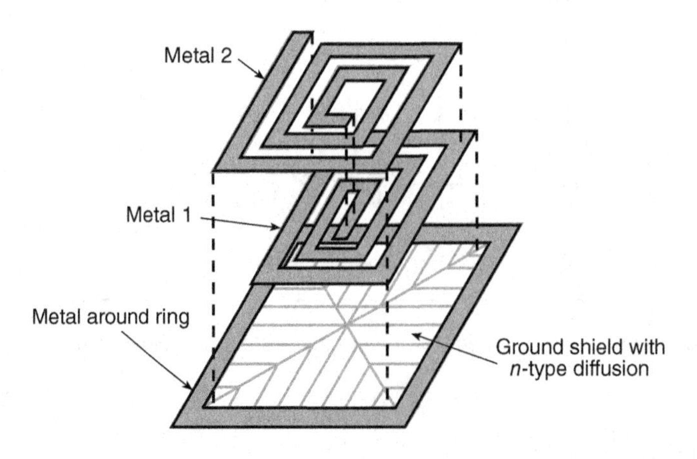

Figure 10.23 Multi-metal spiral inductor that uses metal 1 and metal 2.

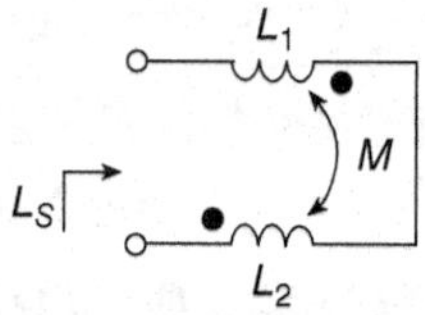

Figure 10.24 Equivalent circuit of two-layer inductor.

the inside end of the coil to the outside, which introduces a capacitance between the conductor and the lead wire, and this capacitance is the dominant component of the overall stray capacitance. Third, the direction of the magnetic flux is perpendicular to the substrate, which can interfere with the underlying circuit. Fourth, the quality factor is very low. Fifth, the self-resonant frequency is low.

10.16 Planar Transformers

Monolithic transformers are required in many RF designs. The principle of planar transformers relies on lateral magnetic coupling. They can be used as narrowband or wideband transformers. An interleaved planar spiral transformer is shown in Fig. 10.25. The coupling coefficient of these transformers is $k \approx 0.7$. Figure 10.26 shows a planar spiral transformer, in which the turns ratio $n = N_p : N_s$ is not equal to 1. The coupling coefficient k is low in this transformer, typically $k = 0.4$. Stacked transformers use multiple metal layers.

10.17 MEMS Inductors

MEMS technology can improve the performance of integrated inductors. MEMS inductors usually are solenoids fabricated using surface micromachining techniques and polymer/metal

Figure 10.25 Interleaved planar spiral transformer.

Figure 10.26 Planar spiral transformer in which turns ratio $n = N_p : N_s$ is not equal to 1.

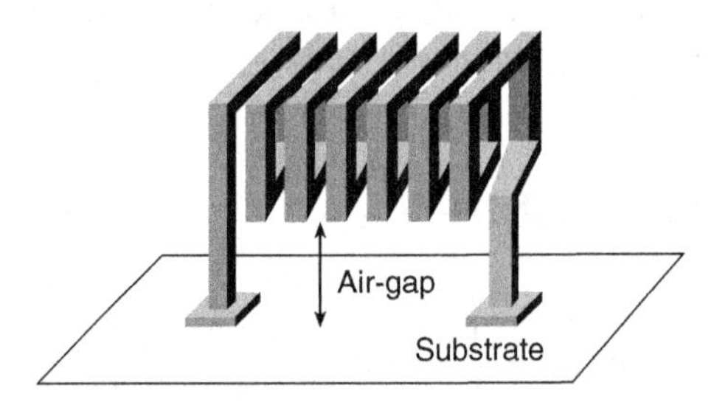

Figure 10.27 Integrated MEMS solenoid inductor [30].

multilayer process techniques. Figure 10.27 shows an integrated solenoid inductor [30]. These inductors may have air cores or electroplated Ni–Fe Permalloy cores. The winding is made up of electroplated copper layers. There is an air gap between the metal winding and the substrate. This geometry gives small inductor size, low stray capacitance C_S, high SRF f_r, low power loss, and high quality factor. However, it requires a 2D design approach.

The magnetic field intensity inside the solenoid is uniform and given by

$$B = \frac{\mu NI}{l_c} \tag{10.153}$$

where l_c is the length of the core and N is the number of turns. The magnetic flux is

$$\phi = A_c B = \frac{\mu NIA_c}{l_c} \tag{10.154}$$

where A_c is the cross-sectional area of the core. The magnetic linkage is

$$\lambda = N\phi = \frac{\mu N^2 IA_c}{l_c} \tag{10.155}$$

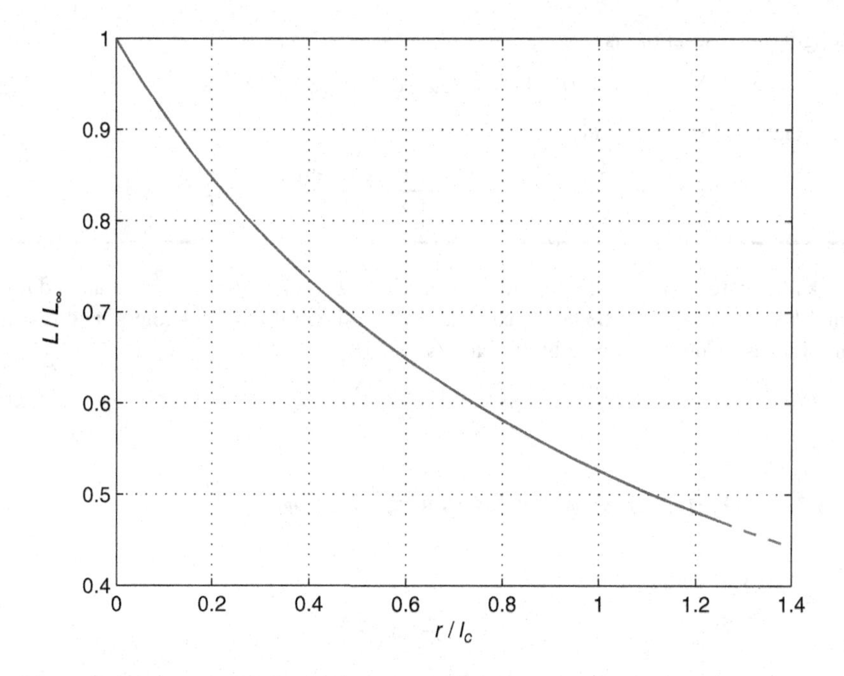

Figure 10.28 Ratio L/L_∞ versus r/l_c.

The inductance of the infinitely long MEMS solenoid inductor is given by

$$L = \frac{\lambda}{I} = \frac{\mu N^2 A_c}{l_c} \tag{10.156}$$

The inductance of a short solenoid is lower than that of an infinitely long solenoid. The inductance of a round solenoid of radius r and length l_c can be determined from Wheeler equation: [3]

$$L = \frac{L_\infty}{1 + 0.9\frac{r}{l_c}} = \frac{\mu A_c N^2}{l_c\left(1 + 0.9\frac{r}{l_c}\right)} \tag{10.157}$$

For a square solenoid, we can use an equivalent cross-sectional area of the solenoid

$$A = h^2 = \pi r^2 \tag{10.158}$$

yielding

$$r = \frac{h}{\sqrt{\pi}} \tag{10.159}$$

Figure 10.28 shows a plot of L/L_∞ as a function of r/l_c.

Example 10.19

Calculate the inductance of an MEMS solenoid made of $N = 10$ turns of square wire with thickness $h = 8\,\mu m$ and space between the turns $s = 8\,\mu m$. The solenoid has a square shape with width $w = 50\,\mu m$.

Solution. The cross-sectional area of the MEMS solenoid is

$$A_c = w^2 = (50 \times 10^{-6})^2 = 25 \times 10^{-10}\ \text{m}^2 \tag{10.160}$$

The length of the solenoid is

$$l_c = (N-1)(h+s) = (10-1)(8+8)\times 10^{-6} = 144\times 10^{-6}\,\text{m} = 144\,\mu\text{m} \tag{10.161}$$

The inductance of the solenoid is

$$L = \frac{\mu N^2 A_c}{l_c} = \frac{4\pi\times 10^{-7}\times 10^2\times 25\times 10^{-10}}{144\times 10^{-6}} = 2.182\,\text{nH} \tag{10.162}$$

Brooks Inductor. The maximum inductance with a given length of wire is obtained for $a/c = 3/2$ and $b = c$, where a is the coil mean radius, b is the coil axial thickness, and c is the coil radial thickness. The inductance in this case is [17, 18]

$$L = 1.353\mu a N^2 \tag{10.163}$$

10.18 Inductance of Coaxial Cable

The inductance of a coaxial cable is

$$L = \frac{\mu l}{2\pi}\ln\left(\frac{b}{a}\right) \tag{10.164}$$

where a is the inner radius of the dielectric, b is the outer radius of the dielectric, and l is the length of the coaxial cable.

10.19 Inductance of Two-Wire Transmission Line

The inductance of a two-parallel-wire transmission line is

$$L = \frac{\mu l}{\pi}\left[\ln\left(\frac{d}{2a}\right) + \sqrt{\left(\frac{d}{2a}\right)^2 - 1}\right] \tag{10.165}$$

where a is the radius of the wires, d is the distance between the centers of the wires, and l is the length of the transmission line.

10.20 Eddy Currents in Integrated Inductors

The performance degradation of planar spiral inductors is caused bythe following effects:

1) Trace resistance due to its finite conductivity ($\sigma < \infty$).
2) Eddy-current losses due to magnetic field penetration into the substrate and adjacent traces.
3) Substrate ohmic losses.
4) Capacitance formed by metal trace, silicon oxide, and substrate. s capacitance conducts displacement currents, which flow through the substrate and the metal trace.

When a time-varying voltage is applied between the terminals of the inductor, a time-varying current flows along the inductor conductor and induces a magnetic field, which penetrates the metal trace in the direction normal to its surface. As a result, eddy currents are produced within the metal trace. These currents are concentrated near the edges of the trace. Eddy currents subtract from the inductor external current on the outside trace edge and add to it on the inside trace edge, near the center of the spiral. Therefore, the current density is not uniform. It is high on the inside edge and it is low on the outside edge, increasing the effective resistance of the metal trace. The magnetic field also induces eddy currents circulating below the spiral metal trace in the semiconductor substrate. In CMOS technologies, the substrate has a low resistivity, typically $\rho_{sub} = 0.015\,\Omega\,\text{cm}$. Therefore, eddy current losses are dominant components of the overall loss. The inductor quality factor Q is usually in the range of 3–4. In BiCMOS technologies, the substrate resistivity is high, usually in the range of 10–30 Ω cm. Therefore, eddy current losses are reduced to negligible values. The quality factor Q is typically from 5 to 10. The turn-to-substrate capacitance conducts displacement currents, causing power losses. These losses can be reduced by placing patterned ground shields, etching away the semiconductor material below the spiral trace, separating the spiral trace from the substrate by placing a thick oxide layer, using a high-resistivity substrate bulk, or using an insulating substrate such as sapphire. At high frequencies, the effective resistance of the metal traces increases due to skin effect and proximity effects, resulting in current crowding.

A broken guard ring made up of n^+ or p^+ diffusion regions surrounding the coil and connected to ground is often used. The purpose of the guard ring is to provide a low-impedance path to ground for currents induced in the substrate, reducing the substrate loss and increasing the inductor quality factor.

Another method for reducing the loss in the substrate is to insert a high-conductivity ground shield between the spiral and the substrate. The ground shield is patterned with slots to reduce the paths for eddy currents.

If n-well is fabricated beneath the oxide, the p-substrate and the n-well form a pn junction. If this junction is reverse biased, it represents a capacitance C_J. The oxide capacitance C_{ox} and the junction capacitance C_J are connected in series, reducing the total capacitance. Therefore, the current flowing from the metal trace to the substrate is reduced, reducing substrate loss and improving the quality factor.

Various structures may be used to block eddy currents. Eddy currents flow in paths around the axis of the spiral. This can be accomplished by inserting narrow strips of n^+ regions perpendicular to the eddy current flow to create blocking pn junctions.

The magnetic field reaches the maximum value near the axis of the spiral inductor. The current density is higher in the outer turns than that in the inner turns. Therefore, the series resistance and the power loss can be reduced if the outer turns are wider than the inner turns, as illustrated in Fig. 10.29. Also, the spacing between the turns can be increased from the center of the spiral to the outer turns.

Figure 10.29 RF IC inductor with a variable trace width.

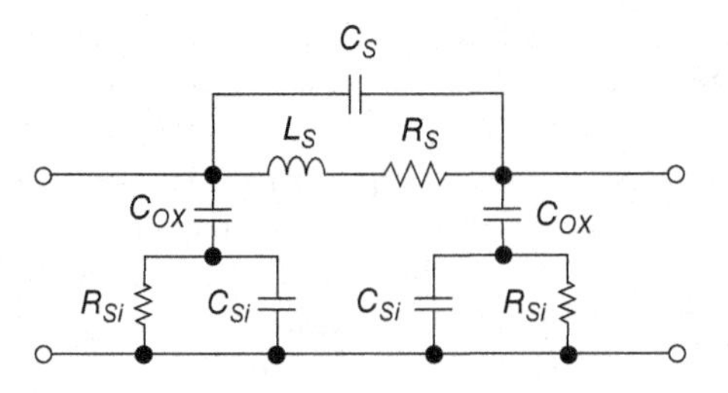

Figure 10.30 Lumped-parameter physical model of an RF IC spiral inductor.

10.21 Model of RF Integrated Inductors

A two-port π-model of RF IC planar spiral inductors is shown in Fig. 10.30 [26]. The components L_s, R_s, and C_s represent the inductance, resistance, and capacitance of the metal trace. The capacitance C_{ox} represents the metal-oxide-substrate capacitance. The components R_{Si} and C_{Si} represent the resistance and capacitance of the substrate. The inductance L_s is given by one of the equations given in Section 10.15. The ac resistance is given by (10.17).

The shunt stray capacitance is formed by the overlap of the underpass and the spiral inductor turns. It is given by

$$C_S = \epsilon_{ox} \frac{Nw^2}{t_{ox}} \tag{10.166}$$

The turn-to-turn capacitances C_{tt} are small even for very small spacing s between the turns because the thickness of the metal trace h is small, resulting in a small vertical area. The lateral turn-to-turn capacitances are connected in series, and therefore the equivalent interturn capacitance C_{tt}/N is small. The adjacent turns sustain a very small voltage difference. The electric energy stored in turn-to-turn capacitances is proportional to the square of voltage. Also, the displacement current through the turn-to-turn capacitance $i_C = C_{tt} dv/dt \approx C_{tt} \Delta v/\delta t$ is very low. Therefore, the effect of the turn-to-turn capacitances is negligible. The effect of the overlap capacitance is more significant because the potential difference between the underpass and the spiral turns is higher. The stray capacitance allows the current to flow directly from the input port to the output port without passing through the inductance. The current can also flow between one terminal and the substrate through capacitances C_{ox} and C_{Si}.

The capacitance between the metal trace and the silicon substrate is

$$C_{ox} = \epsilon_{ox} \frac{A}{2t_{ox}} = wl \frac{\epsilon_{ox}}{2t_{ox}} \tag{10.167}$$

where w is the trace width, l is the trace length, $A = wl$ is the area of the metal trace, t_{ox} is the thickness of the silicon dioxide (SiO_2), and $\epsilon_{ox} = 3.9\epsilon_0$ is the permittivity of the silicon dioxide. The typical value of t_{ox} is 1.8×10^{-8} m.

The capacitance of the substrate is

$$C_{Si} \approx \frac{wlC_{sub}}{2} \tag{10.168}$$

where $C_{sub} = 10^{-3}$ to 10^{-2} fF/μm^2.

The resistance representing the substrate dielectric loss is given by

$$R_{Si} = \frac{2}{wlG_{sub}} \tag{10.169}$$

where G_{sub} is the substrate conductance per unit area, typically $G_{sub} = 10^{-7}$ S/μm^2.

10.22 PCB Inductors

A printed circuit board (PCB) transformer is shown in Fig. 10.31. Planar inductors and transformers consist of a flat copper trace (foil) etched on a printed-circuit board and two pieces of flat ferrite core, one below and one above the coil. The windings are etched usually on both sides of a PCB. Typically, E cores, PQ cores, and RM cores with short legs are used in PCB inductor and transformers. The core are often wide. Planar magnetic component technology is suitable for a low number of turns. As the operating frequency increases, the number of turns decreases. At high frequencies, only several turns are often required. Eddy current losses are lower for thin copper foils than those for circular copper wires because it is easier to satisfy the condition $h < 2\delta_w$. Most planar inductors and transformers are gapped ferrite devices. The minimum total power losses are achieved by reducing the winding loss by lowering the current density and increasing the core loss by allowing for the flux density to be higher than that in traditional wire-wound magnetic components. The typical leakage inductance of planar transformers ranges from 0.1% to 1% of the primary winding inductance. Planar cores have a low winding window area and cover a large square area. Planar inductors have a higher level of electromagnetic interference/radio-frequency interference (EMI/RFI) than inductors with pot cores. Figure 10.32 shows a multilayer PCB inductor, where windings are etched in each layer of a printed-circuit board. Planar inductors and transformers are excellent devices for mass production.

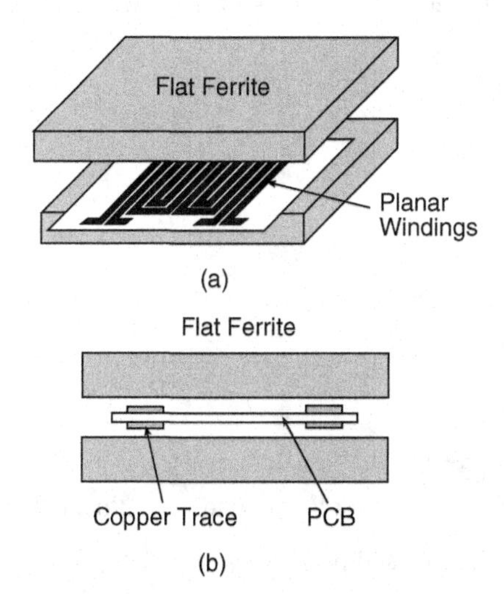

Figure 10.31 PCB transformer. (a) Top view of PCB transformer. (b) Side view of PCB transformer.

Figure 10.32 Multilayer PCB inductor.

The types of inductors that can be used for PCBs technology are as follows:

- Coreless inductors.
- Inductors with planar windings and magnetic plates.
- Closed-core structures.

The PCB inductors can be single-sided, double-sided, and multilayer board inductors. Photolithographic manufacturing processes are used, which have been developed for making PCBs. Photomasks are required to produce PCB inductors.

The simplest structure of a PCB inductor is a coreless planar winding. The maximum inductance is achieved by fitting the largest number of winding turns in each of the PCB layers. Typically, there are six layers in a PCB. The maximum power capability of the coreless PCB inductor is limited by the maximum temperature rise.

The planar inductors also use magnetic cores mounted on the top and bottom of the board. The magnetic plates on either side of the planar winding structure increase the inductance. The inductance also increases when the plate thickness increases. PCB inductors with magnetic plates offer better EMI performance than coreless inductors. PCB inductors have immense power density because they avoid the space consuming bobbins.

The closed-core structure of a PCB inductor consists of a core area provided by magnetic holes. The inductance increases with the inner core radius and the number of winding turns.

The advantages of planar inductors and transformers are as follows:

- Low profile
- Low volume
- High power density
- Small packages
- Excellent repeatability and consistency of performance
- High magnetic coupling
- Low leakage inductance
- Excellent mechanical integrity
- Low skin and proximity effect power loss due to a low thickness of metal traces
- Reduced winding ac resistance and power losses
- Low cost
- Good heat transfer because of a large surface area per unit volume.

10.23 Summary

- Inductors are key components of radio transmitters and other wireless communication circuits.
- Realization of integrated on-chip inductors with a high quality factor (low losses) is tough challenging problem.
- The family of integrated inductors includes planar spiral inductors, planar meander inductors, bondwire inductors, and MEMS inductors.

- Planar inductors are spiral inductors and meander inductors.
- Planar inductors are compatible with IC technology.
- The range of integrated inductances is from 0.1 to 20 nH.
- Integrated inductors require a large amount of chip area.
- Integrated inductors have a very low quality factor.
- Planar spiral RF IC inductors are the most widely used inductors.
- Meander inductors require only one metal layer.
- Meander inductors have a low inductance to surface area ratio.
- Planar spiral inductors usually require two metal layers.
- Planar spiral inductors have a high inductance per unit area.
- The direction of magnetic flux is perpendicular to the substrate and, Therefore, can interfere with the circuit.
- In square planar inductors, the total self-inductance is equal to the sum of self-inductances of all straight segments.
- In square planar inductors, the mutual inductance is only present between parallel segments.
- The mutual inductance between parallel segments is positive if the currents in parallel conductors flow in the same direction.
- The mutual inductance between parallel segments is negative if the currents in parallel conductors flow in the opposite directions.
- The inductance of planar inductors is approximately proportional to the trace length l.
- An increase in the conductor width w and spacing s reduces the inductance and the resistance.
- The conductor thickness does not affect the inductance value and significantly reduces the resistance.
- The magnetic flux of RF IC inductors penetrates the substrate, where it induces high-loss eddy currents.
- Patterned ground shield can be used to reduce eddy current loss.
- MEMS solenoid inductors are more complex to fabricate than the planar spiral inductors.
- MEMS inductors have a higher quality factor than the planar integrated inductors.
- The predictability of bondwire inductors is low.
- The parasitic capacitance of integrated planar inductors can be minimized by implementing the spiral with the topmost metal layer, increasing the SRF.

10.24 Review Questions

10.1 List the types of integrated inductors.

10.2 What is the range of integrated inductances?

10.3 Is it easy to achieve good performance of integrated inductors?

10.4 What kinds of integrated inductors are the most widely used?

10.5 What is the main disadvantage of planar integrated inductors?

10.6 How many metal layers are usually used in RF IC planar inductors?

10.7 What is an underpass in spiral planar inductors?

10.8 What is an air bridge in integrated inductors?

10.9 When the mutual inductance between two conductors is zero?

10.10 When the mutual inductance between two conductors is positive and when it is negative?

10.11 Is the quality factor high for integrated inductors?

10.12 What is the self-resonant frequency of an inductor?

10.13 What is the main disadvantage of bondwire inductors?

10.14 What are the advantages and disadvantages of planar inductors?

10.15 What are the advantages and disadvantages of MEMS inductors?

10.16 How eddy current losses can be reduced?

10.17 What are the components of the model of planar spiral inductors?

10.25 Problems

10.1 Calculate the skin depth at $f = 1$ GHz for (a) copper, (b) aluminum, (c) silver, and (d) gold.

10.2 Calculate the dc resistance of an aluminum trace with $l = 50\,\mu m$, $w = 1\,\mu m$, and $h = 1\,\mu m$. Find the ac resistance of the trace at 10 GHz. Calculate F_R.

10.3 Calculate the inductance of a straight trace with $l = 100\,\mu m$, $w = 1\,\mu m$, $h = 1\,\mu m$, and $\mu_r = 1$.

10.4 Calculate the inductance of a meander inductor with $N = 10$, $w = 40\,\mu m$, $s = 40\,\mu m$, $a = 40\,\mu m$, $h = 100\,\mu m$, and $\mu_r = 1$.

10.5 Calculate the inductance of a round straight inductor with $l = 2$ mm and $\mu_r = 1$ for $\delta \gg h$. Find the inductance of the above copper round straight conductor for $a = 20\,\mu m$ at $f = 2.4$ GHz.

10.6 Calculate the inductance of a bondwire with $l = 1$ mm and $a = 1\,\mu m$.

10.7 Find the inductance of the spiral planar inductor with $N = 10$ and $r = 100\,\mu m$.

10.8 Calculate an inductance of a square planar spiral inductor with $N = 10$, $s = 20\,\mu m$, $w = 30\,\mu m$, and $d = 40\,\mu m$. Use Bryan's formula.

10.9 Calculate a square planar spiral inductance with $N = 10$, $s = 20\,\mu m$, $w = 30\,\mu m$, and $d = 40\,\mu m$. Use Wheeler's formula.

10.10 Calculate a square planar spiral inductance with $N = 15$, $s = 20\,\mu m$, $w = 30\,\mu m$, $h = 20\,\mu m$, and $d = 40\,\mu m$. Use Greenhouse's formula.

10.11 Calculate a square planar spiral inductance with $N = 10$, $s = 20\,\mu m$, $w = 30\,\mu m$, $h = 20\,\mu m$, and $d = 40\,\mu m$. Use Rosa's formula.

10.12 Calculate a square planar spiral inductance with $N = 10$, $s = 20\,\mu\text{m}$, $w = 30\,\mu\text{m}$, $h = 20\,\mu\text{m}$, and $d = 40\,\mu\text{m}$. Use Cranin's formula.

10.13 Calculate a square planar spiral inductance with $N = 10$, $s = 20\,\mu\text{m}$, $w = 30\,\mu\text{m}$, $h = 20\,\mu\text{m}$, and $d = 40\,\mu\text{m}$. Use monomial formula.

10.14 Calculate a square planar spiral inductance with $N = N_i = 10$, $s = 20\,\mu\text{m}$, $w = 30\,\mu\text{m}$, $h = 20\,\mu\text{m}$, and $d = 40\,\mu\text{m}$. Use Jenei's formula.

10.15 Calculate a square planar spiral inductance with $N = 10$, $w = 30\,\mu\text{m}$, $h = 20\,\mu\text{m}$, and $D = 1000\,\mu\text{m}$. Use Terman's formula.

10.16 Calculate a hexagonal planar spiral inductance with $N = 10$, $s = 20\,\mu\text{m}$, $w = 30\,\mu\text{m}$, $h = 20\,\mu\text{m}$, and $d = 40\,\mu\text{m}$. Use Rosa's formula.

10.17 Calculate a octagonal planar spiral inductance with $N = 5$, $s = 20\,\mu\text{m}$, $w = 30\,\mu\text{m}$, $h = 20\,\mu\text{m}$, and $d = 60\,\mu\text{m}$. Use Rosa's formula.

10.18 Calculate a circular planar spiral inductance with $N = 10$, $d = 40\,\mu\text{m}$, $w = 30\,\mu\text{m}$, and $s = 20\,\mu\text{m}$. Use Rosa's formula.

10.19 Calculate the inductance of an MEMS solenoid with $N = 20$. The thickness of a square wire is $h = 10\,\mu\text{m}$ and the separation between the turns is $s = 10\,\mu\text{m}$. The solenoid has a square shape with $w = 100\,\mu\text{m}$.

References

[1] E. B. Rosa, "Calculation of the self-inductance of single-layer coils," *Bulletin of the Bureau of Standards*, vol. 2, no. 2, pp. 161–187, 1906.

[2] E. B. Rosa, "The the self and mutual inductances of linear conductors," *Bulletin of the Bureau of Standards*, vol. 4, no. 2, pp. 302–344, 1907.

[3] H. A. Wheeler, "Simple inductance formulas for radio coils," *Proceedings of the IRE*, vol. 16, no. 10, pp. 1398–1400, 1928.

[4] H. A. Wheeler, "Formulas for the skin effect," *Proceedings of the IRE*, vol. 30, pp. 412–424, 1942.

[5] F. E. Terman, *Radio Engineers' Handbook*. New York, NY: McGraw-Hill, 1943.

[6] R. G. Medhurst, "HF resistance and self-capacitance of single-layer solenoids," *Wireless Engineers*, pp. 35–43, 1947, and pp. 80–92, 1947.

[7] H. E. Bryan, "Printed inductors and capacitors," *Tele-Tech and Electronic Industries*, vol. 14, no. 12, p. 68–, 1955.

[8] F. W. Grover, *Inductance Calculations: Working Formulas and Tables*, Princeton, NJ: Van Nostrand, 1946; reprinted by Dover Publications, New York, NY, 1962.

[9] J. C. Maxwell, *A Treatise of Electricity and Magnetism, Parts III and IV*, 1st Ed., 1873, 3rd Ed, 1891; reprinted by Dover Publishing, New York, NY, 1954 and 1997.

[10] H. Dill, "Designing inductors for thin-film applications," *Electronic Design*, pp. 52–59, 1964.

[11] D. Daly, S. Knight, M. Caulton, and R. Ekholdt, "Lumped elements in microwave integrated circuits," *IEEE Transactions on Microwave Theory and Techniques*, vol. 15, no. 12, pp. 713–721, 1967.

[12] J. Ebert, "Four terminal parameters of HF inductors," *Bulletin de l'Academie Polonaise des. Sciences, Serie des Sciences Techniques, Bulletin de l'Academie Polonaise de Science*, No. 5, 1968.

[13] R. A. Pucel, D. J. Massé, and C. P. Hartwig, "Losses in microstrops," *IEEE Transactions on Microwave Theory and Techniques*, vol. 16, no. 6, pp. 342–250, 1968.

[14] H. M. Greenhouse, "Design of planar rectangular microelectronic inductors," *IEEE Transactions on Parts, Hybrids, and Packaging*, vol. PHP-10, no. 2, pp. 101–109, 1974.

[15] N. Saleh, "Variable microelectronic inductors," *IEEE Transactions on Components, Hybrids, and Manufacturing Technology*, vol. 1, no. 1, pp. 118–124, 1978.
[16] D. Schieber, "On the inductance of printed spiral coils," *Archiv fur Elektrotechnik*, vol. 68, pp. 155–159, 1985.
[17] B. Brooks, "Design of standards on inductance, and the proposed use of model reactors in the design of air-core and iron-core reactors," *Bureau Standard Journal Research*, vol. 7, pp. 289–328, 1931.
[18] P. Murgatroyd, "The Brooks inductor: a study of optimal solenoid cross-sections," *IEE Proceedings, Part B, Electric Power Applications*, vol. 133, no. 5, pp. 309–314, 1986.
[19] L. Weimer and R. H. Jansen, "Determination of coupling capacitance of underpasses, air bridges and crossings in MICs and MMICS," *Electronic Letters*, vol. 23, no. 7, pp. 344–346, 1987.
[20] N. M. Nguyen and R. G. Mayer, "Si IC-compatible inductors and LC passive filter," *IEEE Journal of Solid-State Circuits*, vol. 27, no. 10, pp. 1028–1031, 1990.
[21] P. R. Gray and R. G. Mayer, "Future directions in silicon IC's for RF personal communications," *Proceedings IEEE 1995 Custom Integrated Circuits Conference*, May 1995, pp. 83–90.
[22] J. Craninckx and M. S. J. Steyeart, "A 1.8 GHz CMOS low noise voltage-controlled oscillator with prescalar," *IEEE Journal of Solid-State Circuits*, vol. 30, pp. 1474–1482, 1995.
[23] J. R. Long and M. A. Copeland, "The modeling, characterization, and design of monolithic inductors for silicon RF IC's," *IEEE Journal of Solid-State Circuits*, vol. 32, no. 3, pp. 357–369, 1997.
[24] J. N. Burghartz, M. Soyuer, and K. Jenkins, "Microwave inductors and capacitors in standard multilevel interconnect silicon technology," *IEEE Transactions on Microwave Theory and Technique*, vol. 44, no. 1, pp. 100–103, 1996.
[25] K. B. Ashby, I. A. Koullias, W. C. Finley, J. J. Bastek, and S. Moinian, "High Q inductors for wireless applications in a complementary silicon bipolar process," *IEEE Journal of Solid-State Circuits*, vol. 31, no. 1, pp. 4–9, 1996.
[26] C. P. Yue, C. Ryu, J. Lau, T. H. Lee, and S. S. Wong, "A physical model for planar spiral inductors in silicon," *International Electron Devices Meeting Technical Digest*, December 1996, pp. 155–158.
[27] C. P. Yue and S. S. Wang, "On-chip spiral inductors with patterned ground shields for Si-bases RF ICs," *IEEE Journal of Solid-State Circuits*, vol. 33, no. 5, pp. 743–752, 1998.
[28] F. Mernyei, F. Darrer, M. Pardeon, and A. Sibrai, "Reducing the substrate losses of RF integrated inductors," *IEEE Microwave and Guided Wave Letters*, vol. 8. no. 9, pp. 300–3001, 1998.
[29] A. M. Niknejad and R. G. Mayer, "Analysis, design, and optimization of spiral inductors and transformers for Si RF IC's," *IEEE Journal of Solid-State Circuits*, vol. 33. no. 10, pp. 1470–1481, 1998.
[30] Y.-J. Kim and M. G. Allen, "Integrated solenoid-type inductors for high frequency applications and their characteristics," *1998 Electronic Components and Technology Conference*, 1998, pp. 1249–1252.
[31] C. P. Yue and S. S. Wong, "Design strategy of on-chip inductors highly integrated RF systems," *Proceedings of the 36th Design Automation Conference*, 1999, pp. 982–987.
[32] M. T. Thomson, "Inductance calculation techniques – Part II: approximations and handbook methods," *Power Control and Intelligent Motion*, pp. 1–11, 1999.
[33] S. S. Mohan, M. Hershenson. S. P. Boyd, and T. H. Lee, " Simple accurate expressions for planar spiral inductors," *IEEE Journal of Solid-State Circuits*, vol. 34, no. 10, pp. 1419–1424, 1999.
[34] Y. K. Koutsoyannopoulos and Y. Papananos, "Systematic analysis and modeling of integrated inductors and transformers in RF IC design" *IEEE Transactions on Circuits and Systems-II, Analog and Digital Signal Processing*, vol. 47, no. 8, pp. 699–713, 2000.
[35] W. B. Kuhn and N. M. Ibrahim, "Analysis of current crowding effects in multiturn spiral inductors," *IEEE Transactions on Microwave Theory and Techniques*, vol. 49, no. 1, pp. 31–38, 2001.
[36] A. Zolfaghati, A. Chan, and B. Razavi, "Stacked inductors and transformers in CMOS technology", *IEEE Journal of Solid-State Circuits*, vol. 36, no. 4, pp. 620–628, 2001.
[37] S. Jenei, B. K. J. Nauwelaers, and S. Decoutere, "Physics-based closed-form inductance expressions for compact modeling of integrated spiral inductors," *IEEE Journal of Solid-State Circuits*, vol. 37, no. 1, pp. 77–80, 2002.

[38] T.-S. Horng. K.-C. Peng, J.-K. Jau, and Y.-S. Tsai, "*S*-parameters formulation of quality factor for a spiral inductor in generalized tow-port configuration," *IEEE Transactions on Microwave Theory and Technique*, vol. 51, no. 11, pp. 2197–2202, 2002.

[39] Yu. Cao, R. A. Groves, X. Huang, N. D. Zamder, J.-O. Plouchart, R. A. Wachnik, T.-J. King, and C. Hu, "Frequency-independent equivalent-circuit model for on-chip spira; inductors, " *IEEE Journal of Solid-State Circuits*, vol. 38, no. 3, pp. 419–426, 2003.

[40] J. N. Burghartz and B. Rejaei, "On the design of RF spiral inductors on silicon," *IEEE Transactions on Electron Devices*, vol. 50, no. 3, pp. 718–729, 2003.

[41] J. Aguilera and R. Berenguer, *Design and Test of Integrated Inductors for RF Applications*. Boston, MA: Kluwer Academic Publishers, 2003.

[42] W. Y. Lin, J. Suryanarayan, J. Nath, S. Mohamed, L. P. B. Katehi, and M. B. Steer, "Toroidal inductors for radio-frequency integrated circuits," *IEEE Transactions on MIcrowave Circuits and Techniques*, vol. 52, no. 2, pp. 646–651, 2004.

[43] N. Wong, H. Hauser, T. O'Donnel, M. Brunet, P. McCloskey, and S. C. O'Mathuna, "Modeling of high-frequency micro-transformers," *IEEE Transactions on Magnetics*, vol. 40, pp. 2014–2016, 2004.

[44] M. Yamagouci, K. Yamada, and K. H. Kim, "Slit design consideration on the ferromagnetic RF integrated inductors," *IEEE Transactions on Magnetics*, vol. 42, pp. 3341–3343, 2006.

[45] S. Muroga, Y. Endo, W. Kodale, Y. Sasaki, K. Yoshikawa, Y. Sasaki, M. Nagata, and N. Masahiro, "Evaluation of thin film noise suupressor applied to noise emulator chip implemented in 65 nm CMOS technology," *IEEE Transactions on Magnetics*, vol. 48, pp. 44485–4488, 2011.

[46] T. Suetsugu and M. K. Kazimierczuk, "Integration of Class DE inverter for dc-dc converter on-chip power supplies," *IEEE International Symposium on Circuits and Systems*, Kos, Greece, May 21-24, 2006, pp. 3133–3136.

[47] T. Suetsugu and M. K. Kazimierczuk, "Integration of Class DE synchronized dc-dc converter on-chip power supplies," *IEEE Power Electronics Specialists Conference*, Jeju, South Korea, June 21-24, 2006.

[48] W.-Z. Chen, W.-H. Chen, and K.-C. Hsu, "Three-dimensional fully symmetrical inductors, transformers, and balun in CMOS technology," *IEEE Transactions on Circuits and Systems I*, vol. 54, no. 7, pp. 1413–1423, 2007.

[49] J. Wibben and R. Harjani, "A high-efficiency DC-DC converter using 2 nH integrated inductors," *IEEE Journal of Solid-State Circuits*, vol 43, no. 4, pp. 844–854, 2008.

[50] A. Massarini and M. K. Kazimierczuk, "Self-capacitance of inductors," *IEEE Transactions on Power Electronics*, vol. 12, no. 4, pp. 671–676, 1997.

[51] G. Grandi, M. K. Kazimierczuk, A. Massarini, and U. Reggiani, "Stray capacitance of single-layer solenoid air-core inductors," *IEEE Transactions on Industry Applications*, vol. 35, no. 5, pp. 1162–1168, 1999.

[52] G. Stojanovic, L. Zivanov, and M. Damjanovic, "Compact form of expressions for inductance calculation of meander inductors," *Serbian Journal of Electrical Engineering*, vol. 1, no. 3, pp. 57–68, 2004.

[53] J.-T. Kuo, K.-Y. Su, T.-Y. Liu, H.-H. Chen and S.-J. Chung, "Analytical calculations for dc inductances of rectangular spiral inductors with finite metal thickness in the PEEC formulation," *IEEE Microwave and Wireless Components Letters*, vol. 16, no. 2, pp. 69–71, February, 2006.

[54] A. Estrov, "Planar magnetics for power converters," *IEEE Transactions on Power Electronics*, vol. 4, pp. 46–53, 1989.

[55] D. van der Linde, C. A. M. Boon, and J. B. Klaasens, "Design of a high-frequency planar power transformer in the multilayer technology," *IEEE Transactions on Power Electronics*, vol. 38, no. 2, pp. 135–141, August 1991.

[56] M. T. Quire, J. J Barrett, and M. Hayes, "Planar magnetic component technology – A review," *IEEE Transactions on Components, Hybrides, and Manufacturing Technology*, vol. 15, no. 5, pp. 884–892, August 1992.

11

RF Power Amplifiers with Dynamic Power Supply

11.1 Introduction

This chapter studies the pulse-width modulated (PWM) buck switching-mode converter for applications in radio transmitters as a dynamic power supply or an amplitude modulator. In these applications, the converter output voltage is variable, and the dc input voltage is usually constant [1–61]. This circuit is also called Class S amplifier. Analysis is given for the buck converter in the continuous conduction mode (CCM) and for the synchronous buck converter. Current and voltage waveforms for all the components of the converter are derived. The dc voltage function is derived for CCM. Voltage and current stresses of the components are found. The boundary between CCM and discontinuous conduction mode (DCM) is determined. An expression for the output voltage ripple is derived. The power losses in all the components and the transistor gate drive power are estimated. The overall efficiency of the converter is determined. A design example is also given.

11.2 Dynamic Power Supply

In RF power amplifiers, the transistors are operated either as a dependent current source or as a switch. The drain efficiency of RF power amplifiers in which transistors are operated as dependent current sources is given by

$$\eta_D = k\left(\frac{V_m}{V_I}\right)^n \tag{11.1}$$

where k is a constant dependent on the type of amplifier $n = 2$ for the Class A RF power amplifier and $n = 1$ for Class AB, B, and C RF power amplifiers. For the Class A RF power amplifier, $k = \frac{1}{2}$, and for the Class B RF power amplifier, $k = \pi/4$. The lower the ratio V_m/V_I is, the higher

RF Power Amplifiers, Second Edition. Marian K. Kazimierczuk.

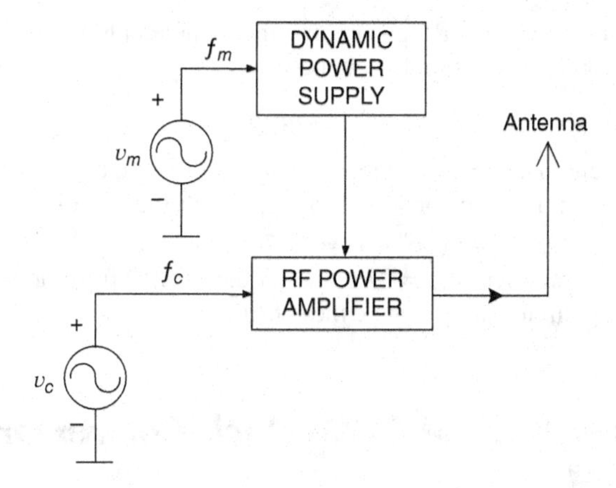

Figure 11.1 Block diagram of RF power amplifier with a high-efficiency dynamic power supply.

is the drain efficiency η_D. At a fixed dc supply voltage V_I, the drain efficiency η_D is low at low values of V_m and is relatively high as V_m becomes close to V_I. In order to improve the drain efficiency η_D in RF power amplifiers with a variable amplitude of the output voltage V_m (such as AM), the supply voltage can be made dependent on the value of the amplitude of the output voltage V_m so that V_m/V_I remains close to 1. In this case, the drain efficiency becomes

$$\eta_D = k\left[\frac{V_m(t)}{v_I(t)}\right]^n \tag{11.2}$$

A dynamic power supply may track the amplitude V_m. The method is called the envelope tracking (ET) method of improving the RF amplifier drain efficiency. The waveform of the output voltage $v_o(t)$ should be directly proportional to the ac component of the gate-to-source voltage $v_{gs}(t)$

$$v_o(t) = A_v v_{gs}(t) \tag{11.3}$$

where A_v is the voltage gain of the power amplifier. The voltage gain should be constant independent of amplitude of the output voltage. When the supply voltage V_I is too close to v_o, the distortion of the output voltage may occur because the amplifier may begin to saturate. The output voltage should be independent of v_I, that is, the power supply rejection ratio (PSRR) should be zero. The shape of the output voltage $v_o(t)$ should be controlled only by the gate-to-source voltage $v_{gs}(t)$. The output voltage of the dynamic power supply $v_I(t)$ should be proportional to the ac component of the gate-to-source voltage $v_{gs}(t)$

$$v_I(t) = a v_{gs}(t) \tag{11.4}$$

where a is such that

$$v_I(t) = v_o(t) + \Delta V \tag{11.5}$$

Typically, ΔV should be approximately 2 V. A block diagram of an RF power amplifier with a high-efficiency dynamic power supply is shown in Fig. 11.1.

11.3 Amplitude Modulator

In RF power amplifiers in which the transistors are operated as switches such as Class D and Class E amplifiers, amplitude modulation (AM) may be accomplished by varying the amplifier supply voltage V_I. In this type of RF power amplifiers, the amplitude of the output voltage V_m is directly proportional to the dc supply voltage V_I

$$V_m = M_{vo} V_I \tag{11.6}$$

where M_{vo} is a constant that depends on the type of amplifier. To obtain AM, the voltage source of the modulating signal may be connected in series with the dc voltage source V_I. For Class E RF power amplifier, $M_{vo} = (4/\sqrt{\pi^2 + 4}) \approx 1.074$ at the duty cycle $D = 0.5$. For the Class D half-bridge RF power amplifier, $M_{vo} = 2/\pi$, and for the Class D full-bridge RF power amplifier, $M_{vo} = 4/\pi$. Matching circuits may change these factors.

11.4 DC Analysis of PWM Buck Converter Operating in CCM

11.4.1 Circuit Description

A switching-mode power supply (SMPS) may be used as dynamic power supply. A circuit of the PWM buck dc–dc converter is depicted in Fig. 11.2(a). It consists of four components: a power MOSFET used as a controllable switch S, a rectifying diode D_1, an inductor L, and a filter capacitor C. Resistor R_L represents a dc load. Power MOSFETs are the most commonly used controllable switches in dc–dc converters because of their high speeds. In 1979, International Rectifier patented the first commercially viable power MOSFET, the HEXFET. Other power switches such as bipolar junction transistors (BJTs), isolated gate bipolar transistors (IGBTs), or MOS-controlled thyristors (MCTs) may also be used. The diode D_1 is called a *freewheeling diode*, a *flywheel diode*, or a *catch diode*.

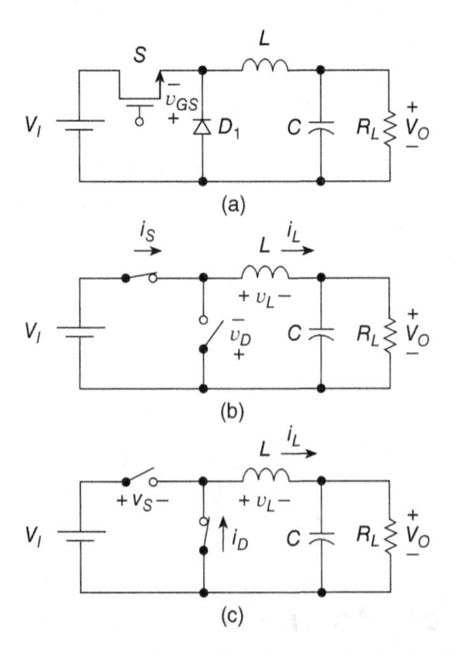

Figure 11.2 PWM buck converter and its ideal equivalent circuits for CCM. (a) Circuit. (b) Equivalent circuit when the switch is ON and the diode is OFF. (c) Equivalent circuit when the switch is OFF and the diode is ON.

The transistor and the diode form a single-pole, double-throw switch, which controls the energy flow from the source to the load. The task for the capacitor and the inductor is the energy storage and transfer. The switching network, composed of the transistor and the diode, "chops" the dc input voltage V_I and therefore the converter is often called a "chopper," which produces a reduced average voltage. The switch S is controlled by a *pulse-width modulator* and is turned on and off at the switching frequency $f_s = 1/T$. The duty cycle D is defined as

$$D = \frac{t_{on}}{T} = \frac{t_{on}}{t_{on} + t_{off}} = f_s t_{on} \tag{11.7}$$

where t_{on} is the time interval when the switch S is closed and t_{off} is the time interval when the switch S is open. Since the duty cycle D of the drive voltage v_{GS} is varied, so does the duty ratio of other waveforms. This permits the variation of the output voltage at a fixed dc supply voltage V_I and the load resistance R_L (or the load current I_O). The circuit L–C–R_L acts as a second-order low-pass filter whose corner frequency is $f_o = 1/(2\pi\sqrt{LC})$. The output voltage V_O of the buck converter is always lower than the input voltage V_I. Therefore, it is a *step-down* converter. The buck converter "bucks" the voltage to a lower level. Because the gate of the MOSFET is not referenced to ground, it is difficult to drive the transistor. The converter requires a floating gate drive.

The buck converter can operate in a CCM or in a DCM, depending on the waveform of the inductor current. In CCM, the inductor current flows for the entire cycle, whereas in DCM, the inductor current flows only for a part of the cycle. In DCM, it falls to zero, remains at zero for some time interval, and then starts to increase. Operation at the boundary between CCM and DCM is called the critical mode (CRM).

Let us consider the buck converter operation in the CCM. Figure 11.2(b) and (c) shows the equivalent circuits of the buck converter for CCM when the switch S is ON and the diode D_1 is OFF, and when the switch is OFF and the diode is ON, respectively. The principle of the converter operation is explained by the idealized current and voltage waveforms depicted in Fig. 11.3. At time $t = 0$, the switch is turned on by the driver. Consequently, the voltage across the diode is $v_D = -V_I$, causing the diode to be reverse biased. The voltage across the inductor L is $v_L = V_I - V_O$, and therefore the inductor current increases linearly with a slope of $(V_I - V_O)/L$. For CCM, $i_L(0) > 0$. The inductor current i_L flows through the switch, resulting in $i_S = i_L$ when the switch is ON. During this time interval, the energy is transferred from the dc input voltage source V_I to the inductor, capacitor, and the load. At time $t = DT$, the switch is turned off by the driver.

The inductor has a nonzero current when the switch is turned off. Because the inductor current waveform is a continuous function of time, the inductor current continues to flow in the same direction after the switch turns off. Therefore, the inductor L acts as a current source, which forces the diode to turn on. The voltage across the switch is V_I and the voltage across the inductor is $-V_O$. Hence, the inductor current decreases linearly with a slope of $-V_O/L$. During this time interval, the input source V_I is disconnected from the circuit and does not deliver energy to the load and the LC circuit. The inductor L and capacitor C form an energy reservoir that maintains the load voltage and current when the switch is OFF. At time $t = T$, the switch is turned on again, the inductor current increases and hence energy increases. PWM converters are operated at *hard switching* because the switch voltage waveform is rectangular and the transistor is turned on at a high voltage.

The power switch S and the diode D_1 convert the dc input voltage V_I into a square wave at the input of the L–C–R_L circuit. In other words, the dc input voltage V_I is chopped by the transistor–diode switching network. The L–C–R_L circuit acts as a second-order low-pass filter and converts the square wave into a low-ripple dc output voltage. Since the average voltage across the inductor L is zero for steady state, the average output voltage V_O is equal to the average voltage of the square wave. The width of the square wave is equal to the on-time of the switch S and can be controlled by varying the duty cycle D of the MOSFET gate drive voltage.

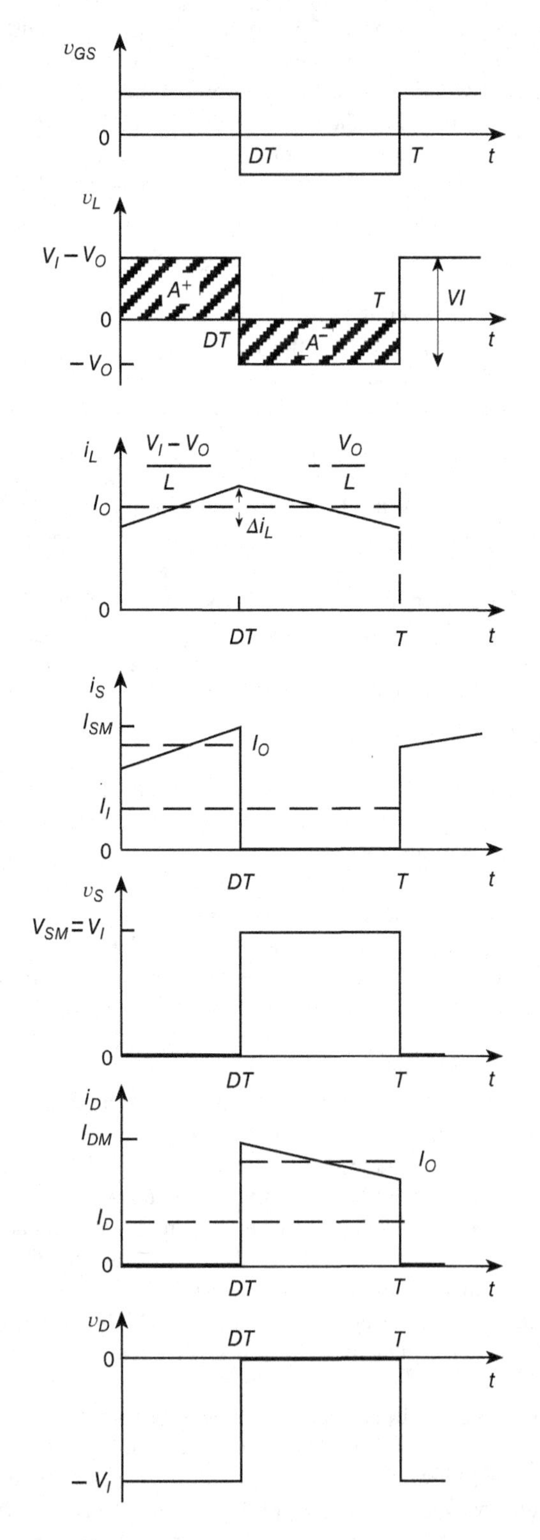

Figure 11.3 Idealized current and voltage waveforms in the PWM buck converter for CCM.

Thus, the square wave is a PWM voltage waveform. The average value of the PWM voltage waveform is $V_O = DV_I$, which depends on the duty cycle D and is almost independent of the load for CCM operation. Theoretically, the duty cycle D may be varied from 0% to 100%. This means that the output V_O ranges from 0 to V_I. Thus, the buck circuit is a *step-down* converter. Therefore, the amount of energy delivered from the input voltage source V_I to the load can be controlled by varying the switch on-duty cycle D. If the output voltage V_O and the load resistance R_L (or the load current I_O) are constant, the output power is also constant. The practical range of D is usually from 5% to 95% due to resolution. The duty cycle D is controlled by a PWM signal.

The inductor current contains an ac component that is independent of the dc load current in CCM and a dc component that is equal to the dc load current I_O. As the dc output current I_O flows through the inductor L, only one-half of the B–H curve of the inductor ferrite core is exploited. Therefore, the inductor L should be designed such that the core will not saturate. To avoid core saturation, a core with an air gap and sufficiently large volume may be required.

11.4.2 Assumptions

The analysis of the buck PWM converter of Fig. 11.2(a) begins with the following assumptions:

1) The power MOSFET and the diode are ideal switches.
2) The transistor output capacitance, the diode capacitance, and the lead inductances are zero, and thereby switching losses are neglected.
3) Passive components are linear, time invariant, and frequency independent.
4) The output impedance of the input voltage source V_I is zero for both dc and ac components.
5) The converter operates in a steady state.
6) The switching period $T = 1/f_s$ is much shorter than the time constants of reactive components.
7) The dc input voltage V_I and the load resistance R_L are constant, but the dc output voltage V_O is variable.
8) The converter is lossless.

11.4.3 Time Interval: $0 < t \leq DT$

During the time interval $0 < t \leq DT$, the switch S is ON and the diode D_1 is OFF. An ideal equivalent circuit for this time interval is shown in Fig. 11.2(b). When the switch is ON, the voltage across the diode v_D is approximately equal to $-V_I$, causing the diode to be reverse biased. The voltage across the switch v_S and the diode current are zero. The voltage across the inductor L is given by

$$v_L = V_I - V_O = L\frac{di_L}{dt} \tag{11.8}$$

Hence, the current through the inductor L and the switch S is

$$i_S = i_L = \frac{1}{L}\int_0^t v_L\, dt + i_L(0) = \frac{V_I - V_O}{L}\int_0^t dt + i_L(0) = \frac{V_I - V_O}{L}t + i_L(0) \tag{11.9}$$

where $i_L(0)$ is the initial current in the inductor L at time $t = 0$. The peak inductor current becomes

$$i_L(DT) = \frac{(V_I - V_O)DT}{L} + i_L(0) = \frac{(V_I - V_O)D}{f_s L} + i_L(0) \quad (11.10)$$

For a lossless buck converter, $M_{VDC} = V_O/V_I = D$. Hence, the peak-to-peak ripple current of the inductor L is

$$\Delta i_L = i_L(DT) - i_L(0) = \frac{(V_I - V_O)DT}{L} = \frac{(V_I - V_O)D}{f_s L} = \frac{(V_I - V_O)D}{f_s L} = \frac{V_I D(1-D)}{f_s L} \quad (11.11)$$

The diode voltage is

$$v_D = -V_I \quad (11.12)$$

Thus, the peak value of the diode reverse voltage is

$$V_{DM} = V_I \quad (11.13)$$

The average value of the inductor current is equal to the dc output current I_O. Hence, the peak value of the switch current is

$$I_{SM} = I_O + \frac{\Delta i_L}{2} \quad (11.14)$$

The increase in the magnetic energy stored in the inductor L during the time interval 0 to DT is given by

$$\Delta W_{L(in)} = \frac{1}{2}L\left[i_L^2(DT) - i_L^2(0)\right] = \frac{1}{2}L\left\{\frac{V_I D(1-D)}{f_s L}\left[\frac{V_I D(1-D)}{f_s L} + 2i_L(0)\right]\right\} \quad (11.15)$$

The time interval 0 to DT is terminated when the switch is turned off by the driver.

11.4.4 Time Interval: $DT < t \leq T$

During the time interval $DT < t \leq T$, the switch S is OFF and the diode D_1 is ON. Figure 11.2(c) shows an ideal equivalent circuit for this time interval. Since $i_L(DT)$ is nonzero at that instant, the inductor acts as a current source and turns the diode on. The switch current i_S and the diode voltage v_D are zero and the voltage across the inductor L is

$$v_L = -V_O = L\frac{di_L}{dt} \quad (11.16)$$

The current through the inductor L and the diode can be found as

$$i_D = i_L = \frac{1}{L}\int_{DT}^{t} v_L\, dt + i_L(DT) = \frac{1}{L}\int_{DT}^{t}(-V_O)dt + i_L(DT) = -\frac{V_O}{L}\int_{DT}^{t} dt + i_L(DT)$$
$$= -\frac{V_O}{L}(t - DT) + i_L(DT) \quad (11.17)$$

where $i_L(DT)$ is the initial condition of the inductor L at $t = DT$. The peak-to-peak ripple current of the inductor L is

$$\Delta i_L = i_L(DT) - i_L(T) = \frac{V_O T(1-D)}{L} = \frac{V_O(1-D)}{f_s L} = \frac{V_I D(1-D)}{f_s L} \quad (11.18)$$

Note that the peak-to-peak value of the inductor current ripple Δi_L is independent of the load current I_O in CCM and depends only on the dc input voltage V_I and thereby on the duty cycle D. The maximum value of the inductor ripple current at fixed V_I occurs for $D = 0.5$. Hence,

$$\Delta i_{Lmax} = \frac{V_I D(1-D)}{f_s L} = \frac{V_I \times 0.5 \times (1-0.5)}{f_s L} = \frac{V_I}{4f_s L} \quad (11.19)$$

The switch voltage v_S and the peak switch voltage V_{SM} are given by

$$v_S = V_{SM} = V_I \tag{11.20}$$

The diode and switch peak currents are given by

$$I_{DM} = I_{SM} = I_O + \frac{\Delta i_L}{2}. \tag{11.21}$$

This time interval ends at $t = T$ when the switch is turned on by the driver.

The decrease in the magnetic energy stored in the inductor L during time interval $DT < t \leq T$ is given by

$$\Delta W_{L(out)} = \frac{1}{2}L\left[i_L^2(DT) - i_L^2(T)\right] = \frac{1}{2}L\left[i_L^2(DT) - i_L^2(0)\right] = \Delta W_{L(in)} \tag{11.22}$$

For steady-state operation, the increase in the magnetic energy $\Delta W_{L(in)}$ is equal to the decrease in the magnetic energy $\Delta W_{L(out)}$.

The transient and steady-state waveforms in converters with commercial components can be obtained from computer simulations using SPICE, described in Appendix A.

11.4.5 Device Stresses for CCM

The maximum voltage and current stresses of the switch and the diode in CCM for steady-state operation are

$$V_{SMmax} = V_{DMmax} = V_{Imax} \tag{11.23}$$

and

$$I_{SMmax} = I_{DMmax} = I_{Omax} + \frac{\Delta i_{Lmax}}{2} = I_{Omax} + \frac{(V_{Imax} - V_O)D_{min}}{2f_sL}$$
$$= I_{Omax} + \frac{V_O(1 - D_{min})}{2f_sL} \tag{11.24}$$

11.4.6 DC Voltage Transfer Function for CCM

Consider a voltage transfer function of a lossless buck converter. The voltage and current across a linear inductor are related by Faraday's law in its differential form

$$v_L = L\frac{di_L}{dt} \tag{11.25}$$

For steady-state operation, the following boundary condition is satisfied

$$i_L(0) = i_L(T) \tag{11.26}$$

Rearranging (11.25),

$$\frac{1}{L}v_L\, dt = di_L \tag{11.27}$$

and integrating both sides yields

$$\frac{1}{L}\int_0^T v_L\, dt = \int_0^T di_L = i_L(T) - i_L(0) = 0 \tag{11.28}$$

The integral form of Faraday's law for an inductor under steady-state condition is

$$\int_0^T v_L\, dt = 0 \tag{11.29}$$

The average value of the voltage across an inductor for steady state is zero. Thus,

$$V_{L(AV)} = \frac{1}{T}\int_0^T v_L\,dt = 0 \tag{11.30}$$

This equation is also called a *volt-second balance* for an inductor, which means that "volt-second" stored is equal to "volt-second" released.

For PWM converters operating in CCM,

$$\int_0^{DT} v_L\,dt + \int_{DT}^T v_L\,dt = 0 \tag{11.31}$$

from which

$$\int_0^{DT} v_L\,dt = -\int_{DT}^T v_L\,dt \tag{11.32}$$

This means that the area encircled by the positive part of the inductor voltage waveform A^+ is equal to the area encircled by the negative part of the inductor voltage waveform A^-, that is,

$$A^+ = A^- \tag{11.33}$$

where

$$A^+ = \int_0^{DT} v_L\,dt \tag{11.34}$$

and

$$A^- = -\int_{DT}^T v_L\,dt \tag{11.35}$$

Referring to Fig. 11.3,

$$(V_I - V_O)DT = V_O(1-D)T \tag{11.36}$$

which simplifies to the form

$$V_O = DV_I \tag{11.37}$$

For a lossless converter, $V_I I_I = V_O I_O$. Hence, from (11.37), the dc voltage transfer function (or the voltage conversion ratio) of the lossless buck converter is given by

$$M_{VDC} \equiv \frac{V_O}{V_I} = \frac{I_I}{I_O} = D \tag{11.38}$$

The range of M_{VDC} is

$$0 \le M_{VDC} \le 1 \tag{11.39}$$

Note that the output voltage V_O is independent of the load resistance R_L. It depends only on the dc input voltage V_I and the duty cycle D. The sensitivity of the output voltage with respect to the duty cycle is

$$S \equiv \frac{dV_O}{dD} = V_I \tag{11.40}$$

In most practical situations, $V_O = DV_I$ is a constant, which means that if V_I is increased, D should be decreased by a control circuit to keep V_O constant, and *vice versa*.

The dc current transfer function is given by

$$M_{IDC} \equiv \frac{I_O}{I_I} = \frac{1}{D} \tag{11.41}$$

and its value decreases from ∞ to 1 as D is increased from 0 to 1.

From (11.14), (11.20), and (11.38), the switch and the diode utilization in the buck converter is characterized by the output-power capability

$$c_p \equiv \frac{P_O}{V_{SM}I_{SM}} = \frac{V_O I_O}{V_{SM}I_{SM}} \approx \frac{V_O}{V_{SM}} = \frac{V_O}{V_I} = D \tag{11.42}$$

As D is increased from 0 to 1, so does c_p.

11.4.7 Boundary Between CCM and DCM for Lossless Buck Converter

Figure 11.4 shows the inductor current waveforms at a constant dc input voltage V_I for three values of the dc output voltage V_O, which corresponds to three values of the duty cycle D. At $D = D_{min}$, the converter is at the boundary between the CCM and DCM. As the duty cycle D increases, the inductor ripple current Δi_L first increases, reaches its maximum value at $D = 0.5$, and then decreases. The average value of the inductor current also increases with D. Figure 11.5 depicts the inductor current waveform at the boundary between CCM and DCM, where $i_L(0) = 0$. This waveform is given by

$$i_L = \frac{V_I - V_O}{L}t \quad \text{for} \quad 0 < t \leq DT \tag{11.43}$$

For lossless buck converter, the efficiency $\eta = 1$, $M_{VDC} = V_O/V_I = D$, and $V_O = M_{VDC}V_I = DV_I$. Hence, the peak inductor current is

$$\Delta i_L = i_L(DT) = \frac{(V_I - V_O)DT}{L} = \frac{(V_I - V_O)D}{f_s L} = \frac{V_I D(1 - D)}{f_s L} \tag{11.44}$$

Figure 11.6 shows the normalized inductor ripple current $\Delta i_L/(V_I/f_s L)$ at the boundary between CCM and DCM as a function of the duty cycle D for buck converter with fixed V_I and variable V_O.

The load current at the boundary between the CCM and DCM is

$$I_{OB} = \frac{\Delta i_L}{2} = \frac{(V_I - V_O)D}{2f_s L} = \frac{V_I D(1 - D)}{2f_s L} \tag{11.45}$$

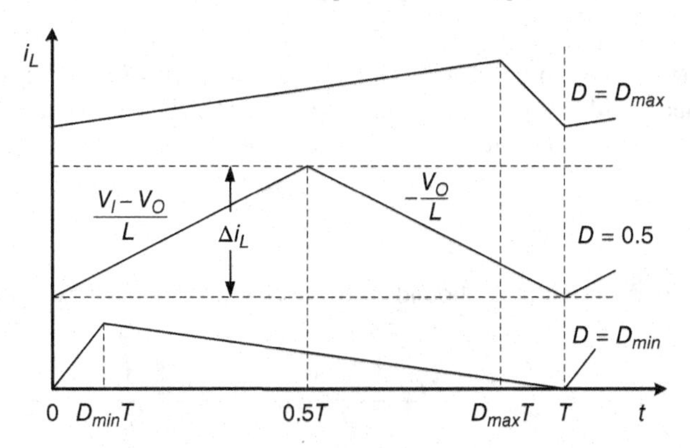

Figure 11.4 Waveforms of the inductor current at various values of the duty cycle, when the dc input voltage V_I is constant and the dc output voltage V_O varies.

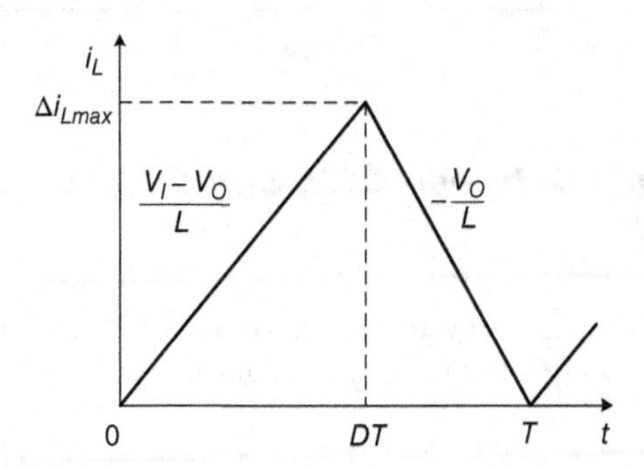

Figure 11.5 Waveforms of the inductor current at the boundary between CCM and DCM.

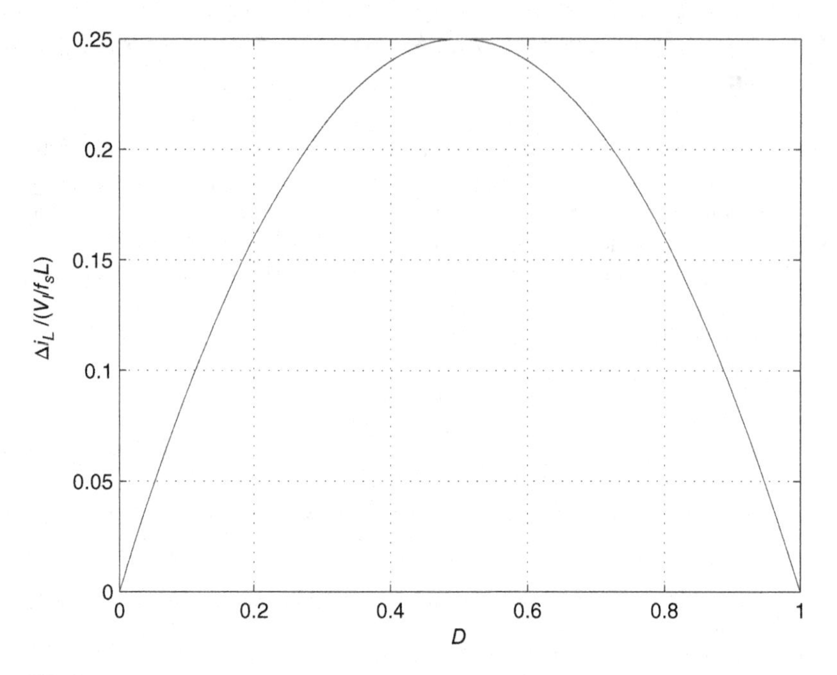

Figure 11.6 Normalized inductor ripple current $\Delta i_L/(V_I/f_sL)$ at the boundary between CCM and DCM as a function of the duty cycle D for buck converter at a fixed V_I and variable V_O.

Figure 11.7 shows the normalized load current $I_{OB}/(V_I/f_sL) = D(1-D)/2$ at the boundary between CCM and DCM as a function of the duty cycle D at fixed V_I, f_s, and L when V_O varies. The inductance required to operate the buck converter at the boundary between CCM and DCM is

$$L = \frac{(V_I - V_O)D}{2f_sI_{OB}} \tag{11.46}$$

The minimum load current at the boundary between CCM and DCM occurs at the minimum duty cycle D_{min}

$$I_{Omin} = I_{OBmin} = \frac{(V_I - V_{Omin})D_{min}}{2f_sL} = \frac{V_ID_{min}(1-D_{min})}{2f_sL} \tag{11.47}$$

Hence, the minimum inductance for achieving the buck converter operating in CCM is

$$L_{min} = \frac{V_ID_{min}(1-D_{min})}{2f_sI_{OBmin}} = \frac{V_ID_{min}(1-D_{min})}{2f_sI_{Omin}} = \frac{V_ID_{min}(1-D_{min})R_L}{2f_sV_{Omin}} \tag{11.48}$$

11.4.8 Boundary Between CCM and DCM for Lossy Buck Converter

The efficiency of the lossy buck converter η is lower than 1. Therefore, $M_{VDC} = V_O/V_I = \eta D$, yielding $V_O = M_{VDC}V_I = \eta DV_I$. The peak-to-peak inductor current is

$$\Delta i_L = i_L(DT) = \frac{(V_I - V_O)DT}{L} = \frac{(V_I - V_O)D}{f_sL} = \frac{V_ID\left(\frac{1}{\eta} - D\right)}{f_sL} \tag{11.49}$$

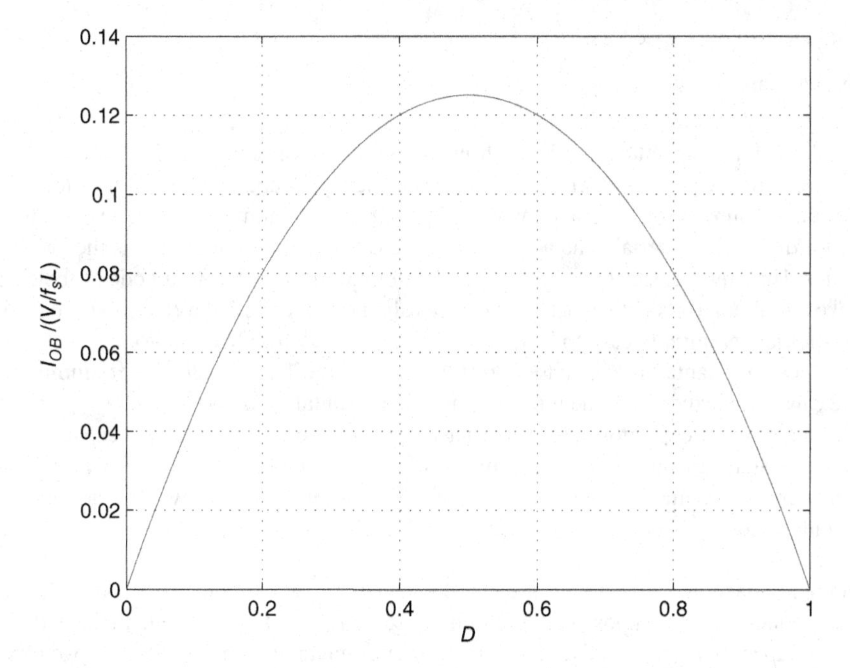

Figure 11.7 Normalized load current $I_{OB}/(V_I/f_sL)$ at the boundary between CCM and DCM as a function of the duty cycle D for buck converter for a fixed V_I, f_s, and L and variable V_O.

The load current of the lossy buck converter at the boundary between the CCM and DCM is

$$I_{OB} = \frac{\Delta i_L}{2} = \frac{(V_I - V_O)D}{2f_sL} = \frac{V_ID\left(\frac{1}{\eta} - D\right)}{2f_sL} \tag{11.50}$$

The converter efficiency η changes with D. The maximum efficiency η_{max} occurs at the maximum duty cycle D_{max} and the minimum efficiency η_{min} occurs at the minimum duty cycle D_{min}. The minimum load current of the lossy buck converter at the boundary between CCM and DCM occurs at the minimum duty cycle D_{min}

$$I_{Omin} = I_{OBmin} = \frac{(V_I - V_{Omin})D_{min}}{2f_sL} = \frac{V_ID_{min}\left(\frac{1}{\eta_{min}} - D_{min}\right)}{2f_sL} \tag{11.51}$$

Hence, the minimum inductance for achieving the lossy buck converter operating in CCM is

$$L_{min} = \frac{V_ID_{min}\left(\frac{1}{\eta_{min}} - D_{min}\right)}{2f_sI_{OBmin}} = \frac{V_ID_{min}\left(\frac{1}{\eta_{min}} - D_{min}\right)}{2f_sI_{Omin}} = \frac{V_ID_{min}\left(\frac{1}{\eta_{min}} - D_{min}\right)R_L}{2f_sV_{Omin}} \tag{11.52}$$

11.4.9 Capacitors

Capacitors are classified according to dielectric material used between the conductors. The following types of capacitors are used in switching-mode power supplies:

- Wet aluminum electrolytic capacitors.
- Wet tantalum electrolytic capacitors.

- Solid electrolytic capacitors.
- Ceramic capacitors.

Wet electrolytic capacitors can be built using aluminum or tantalum. They are made of two aluminum foils. A paper spacer soaked in wet electrolyte separates the two aluminum foils. One of the aluminum foils is coated with an insulating aluminum oxide layer, which forms the capacitor dielectric material. The aluminum foil coated in aluminum oxide is the anode of the capacitor. The liquid electrolyte and the second aluminum foil act as a cathode of the capacitor. The two aluminum foils with attached leads are rolled together with the electrolyte soaked paper in a cylindrical aluminum case to form a wet aluminum electrolyte capacitor.

Wet electrolyte tantalum capacitors are formed in a similar manner as wet aluminum electrolyte capacitors except that the dielectric material is tantalum oxide.

Solid electrolytic capacitors are constructed similarly to wet electrolytic capacitors except that a solid dielectric material is used in place of a wet dielectric material. These capacitors have moderate capacitances and a higher ripple current rating. Electrolytic capacitors are most commonly used in power electronics because of a high ratio of capacitance per unit volume and low cost.

Ceramic capacitors use ceramic dielectric to separate two conductive plates. The ceramic dielectric material is composed of titanium dioxide (Class I) or barium titanate (Class II). Ceramic capacitors can be disc capacitors or multilayer ceramic (MLC) capacitors. Disc capacitors have low capacitance per unit volume. Conductive material is placed on the ceramic dielectric material forming interlace fingers. Ceramic capacitors have lower capacitances than electrolytic capacitors. The capacitances of ceramic capacitors are usually below 1 μF. Ceramic capacitors have very low values of ESR. This property reduces voltage ripple and power loss.

Important parameters of capacitors are the capacitance C, the equivalent series resistance (ESR) r_C, and the series equivalent inductance (ESL) L_s, the self-resonant frequency f_r, and the breakdown voltage V_{BD}. The capacitance is

$$C = \frac{\epsilon_r \epsilon A}{d} \tag{11.53}$$

where A is the area of each conductor, d is the thickness of the dielectric, ϵ_r is the relative permittivity of the dielectric, and $\epsilon_0 = 10^{-9}/36\pi = 8.85 \times 10^{-12}$ F/m is the permittivity of free space. The ESR is the sum of the resistances of leads, the resistances of the contacts, and the resistance of the plate conductors. The ESL is the inductance of the leads. The self-resonant frequency is

$$f_r = \frac{1}{2\pi\sqrt{CL_s}} \tag{11.54}$$

The dissipation factor of a capacitor is

$$DF = \omega C r_C \tag{11.55}$$

The quality factor of a capacitor at a frequency $f = \omega/(2\pi)$ is

$$Q_C = \frac{1}{\omega C r_C} = \frac{1}{DF} \tag{11.56}$$

Capacitors are rated for the breakdown voltage and the maximum rms value of the ripple current. The maximum rms ripple current is the limit of ac current and is dependent on the temperature and frequency of the current conducted by a capacitor. The ripple current flowing through the ESR causes power loss $P_C = r_C I^2_{ac(rms)}$, which generates heat within the capacitor. Electrolytic tantalum capacitors have the highest values of ESR, whereas ceramic capacitors have the lowest ESR.

The performance of electrolytic capacitors is mainly affected by operating conditions, such as frequency, ac current, dc voltage, and temperature. ESR is frequency dependent. As the frequency increases, the ESR first decreases, usually reaches a minimum value at the self-resonant frequency, and then increases. For electrolytic capacitors, the ESR decreases as the dc voltage increases. It also decreases as the peak-to-peak ac ripple voltage increases. ESR is often measured by manufacturers at the capacitor self-resonant frequency. The ESR of capacitors controls the peak-to-peak value of the output ripple voltage. Also, the higher the ESR of the capacitor, the greater the heat generated due to the continuous flow of current through the ESR. This reduces the converter efficiency and life expectancy of the power supply. During aging process, the electrolytic liquid inside the capacitor gradually evaporates, causing an increase in ESR.

When a voltage is applied between the conductors and across the dielectric of a capacitor, an electric field is induced in the dielectric. The electric energy is stored in the electric field. The dielectric has a maximum value of the electric field strength $E_{BD} = V_{BD}/d$, resulting in a capacitor breakdown voltage V_{BD}.

11.4.10 Ripple Voltage in Buck Converter for CCM

A model of the filter capacitor consists of capacitance C, ESR r_C, and equivalent series inductance L_{ESL} The impedance of the capacitor model is

$$Z_C = r_C + j\left(\omega L_{ESL} - \frac{1}{\omega C}\right) = r_C\left[1 + jQ_{Co}\left(\frac{\omega}{\omega_r} - \frac{\omega_r}{\omega}\right)\right] = |Z_C|e^{\phi_C} \tag{11.57}$$

where the self-resonant frequency of the filter capacitor is

$$f_r = \frac{1}{2\pi\sqrt{CL_{ESL}}} \tag{11.58}$$

and the quality factor of the capacitor at its self-resonant frequency is

$$Q_{Co} = \frac{1}{\omega_r C r_C} \tag{11.59}$$

Figures 11.8 and 11.9 show plots of the magnitude $|Z_C|$ and phase ϕ_C of the capacitor for $C = 1\ \mu F$, $r_C = 50\ m\Omega$, and $L_{ESL} = 15\ nH$, respectively. The filter capacitor impedance is capacitive below the self-resonant frequency and inductive above the self-resonant frequency.

The input voltage of the second-order low-pass LCR output filter is rectangular with a maximum value V_I and a duty cycle D. This voltage can be expanded into a Fourier series

$$v = DV_I\left[1 + 2\sum_{n=1}^{\infty}\frac{\sin(n\pi D)}{n\pi D}\cos n\omega_s t\right]$$

$$= DV_I + 2DV_I\left[\frac{\sin \pi D}{\pi D}\cos \omega_s t + \frac{\sin 2\pi D}{2\pi D}\cos 2\omega_s t + \frac{\sin 3\pi D}{3\pi D}\cos 3\omega_s t + \cdots\right] \tag{11.60}$$

The components of this series are transmitted through the output filter to the load. It is difficult to determine the peak-to-peak output voltage ripple V_r using the Fourier series of the output voltage. Therefore, a different approach will be taken for deriving an expression for V_r.

A simpler derivation [46] is given later. A model of the output part of the buck converter for frequencies lower than the capacitor self-resonant frequency (i.e., $f \leq f_r$) is shown in Fig. 11.10. The filter capacitor in this figure is modeled by its capacitance C and its ESR designated by r_C. Figure 11.11 depicts current and voltage waveforms in the converter output circuit. The dc component of the inductor current flows through the load resistor R_L, while the ac component is divided between the capacitor C and the load resistor R_L. Consequently, the load ripple current is

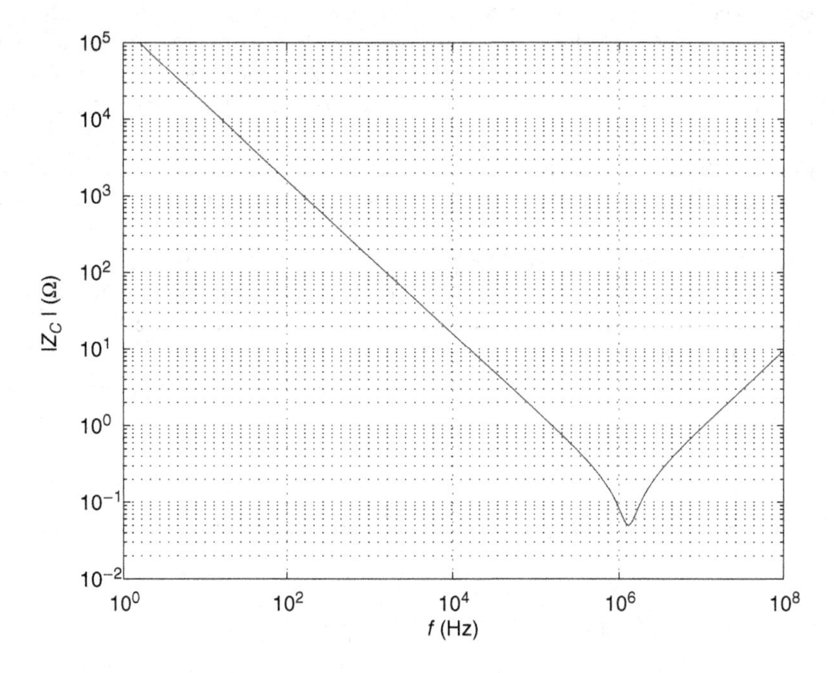

Figure 11.8 Magnitude of the capacitor impedance $|Z_C|$.

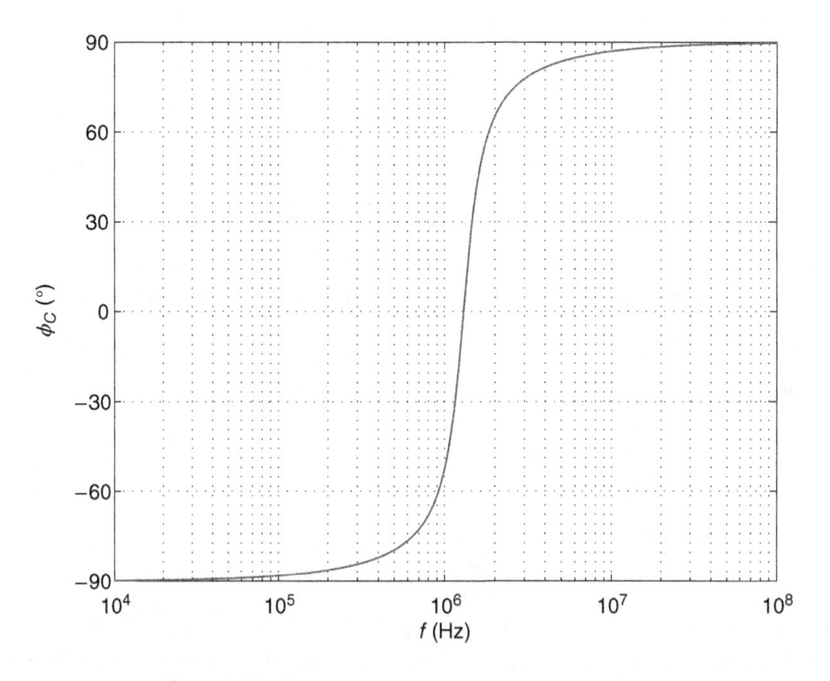

Figure 11.9 Phase of the capacitor impedance ϕ_C.

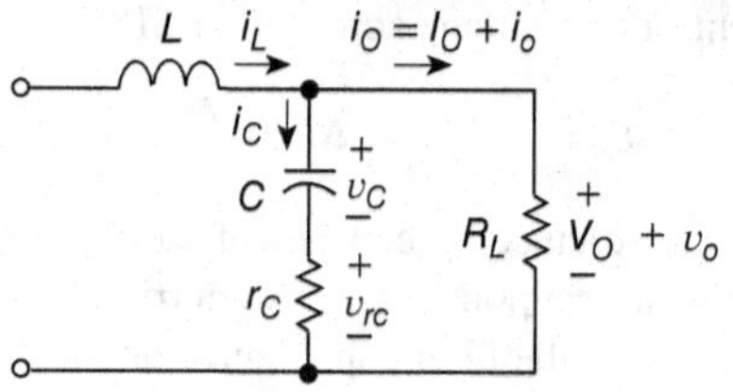

Figure 11.10 Model of the output circuit of the buck converter for frequencies lower than the self-resonant frequency of the filter capacitor.

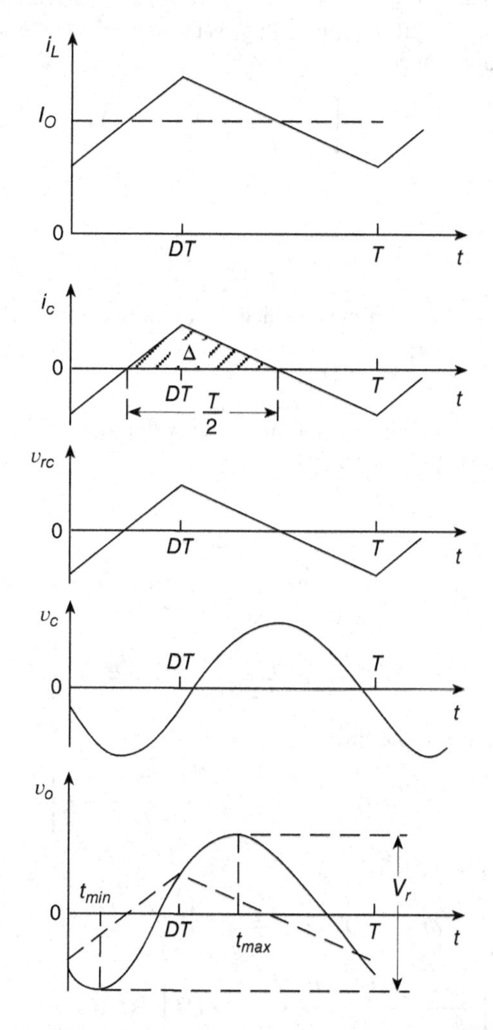

Figure 11.11 Waveforms illustrating the ripple voltage in the PWM buck converter.

very small and can be neglected. Thus, the current through the capacitor is approximately equal to the ac component of the inductor current, that is, $i_C \approx i_L - I_O$.

For the interval $0 < t \le DT$, when the switch is ON and the diode is OFF, the capacitor current is given by

$$i_C = \frac{\Delta i_L t}{DT} - \frac{\Delta i_L}{2} \tag{11.61}$$

resulting in the ac component of the voltage across the ESR

$$v_{rc} = r_C i_C = r_C \Delta i_L \left(\frac{t}{DT} - \frac{1}{2} \right) \tag{11.62}$$

The voltage across the filter capacitance v_C consists of the dc voltage V_C and the ac voltage v_c, that is, $v_C = V_C + v_c$. Only the ac component v_c may contribute to the output ripple voltage. The ac component of the voltage across the filter capacitance is found as

$$v_c = \frac{1}{C}\int_0^t i_C\,dt + v_c(0) = \frac{\Delta i_L}{C}\int_0^t \left(\frac{t}{DT} - \frac{1}{2} \right) dt + v_c(0) = \frac{\Delta i_L}{2C}\left(\frac{t^2}{DT} - t \right) + v_c(0) \tag{11.63}$$

For steady state, $v_c(DT) = v_c(0)$. The waveform of the voltage across capacitance C is a parabolic function. The ac component of the output voltage is the sum of voltage across the filter capacitor ESR r_C and the filter capacitance C

$$v_o = v_{rc} + v_c = \Delta i_L \left[\frac{t^2}{2CDT} + \left(\frac{r_C}{DT} - \frac{1}{2C} \right) t - \frac{r_C}{2} \right] + v_c(0) \tag{11.64}$$

Let us consider the minimum value of the voltage v_o. The derivative of the voltage v_o with respect to time is

$$\frac{dv_o}{dt} = \Delta i_L \left(\frac{t}{CDT} + \frac{r_C}{DT} - \frac{1}{2C} \right) \tag{11.65}$$

Setting this derivative to zero, the time at which the minimum value of v_o occurs is given by

$$t_{min} = \frac{DT}{2} - r_C C \tag{11.66}$$

The minimum value of v_o is equal to the minimum value of v_{rc} if $t_{min} = 0$. This occurs at a minimum capacitance, which is given by

$$C_{min(on)} = \frac{D_{max}}{2 f_s r_{Cmax}} \tag{11.67}$$

Consider the time interval $DT < t \le T$ when the switch S is OFF and the diode D_1 is ON. Referring to Fig. 11.11, the current through the capacitor is

$$i_C = -\frac{\Delta i_L (t - DT)}{(1-D)T} + \frac{\Delta i_L}{2} \tag{11.68}$$

resulting in the voltage across the ESR

$$v_{rc} = r_C i_C = r_C \Delta i_L \left[-\frac{t - DT}{(1-D)T} + \frac{1}{2} \right] \tag{11.69}$$

and the voltage across the capacitor

$$\begin{aligned} v_c &= \frac{1}{C}\int_{DT}^t i_C\,dt + v_c(DT) = \frac{\Delta i_L}{C}\int_{DT}^t \left[-\frac{t - DT}{(1-D)T} + \frac{1}{2} \right] dt + v_c(DT) \\ &= \frac{\Delta i_L}{2C}\left[-\frac{t^2 - 2DTt + (DT)^2}{(1-D)T} + t - DT \right] + v_c(DT) \end{aligned} \tag{11.70}$$

Adding (11.69) and (11.70) yields the ac component of the output voltage

$$v_o = r_c \Delta i_L \left[-\frac{t - DT}{(1-D)T} + \frac{1}{2} \right] + \frac{\Delta i_L}{2C}\left[-\frac{t^2 - 2DTt + (DT)^2}{T(1-D)} + t - DT \right] + v_c(DT) \tag{11.71}$$

The derivative of v_o with respect to time is

$$\frac{dv_o}{dt} = -\frac{r_C \Delta i_L}{(1-D)T} + \frac{\Delta i_L}{C}\left[-\frac{t - DT}{(1-D)T} + \frac{1}{2} \right] \tag{11.72}$$

Setting the derivative to zero, the time at which the maximum value of v_o occurs is expressed by

$$t_{max} = \frac{(1 + D)T}{2} - r_C C \tag{11.73}$$

The maximum value of v_o is equal to the maximum value of v_{rc} if $t_{max} = DT$. This occurs at a minimum capacitance, which is given by

$$C_{min(off)} = \frac{1 - D_{min}}{2f_s r_{Cmax}} \tag{11.74}$$

The peak-to-peak ripple voltage is independent of the voltage across the filter capacitance C and is determined only by the ripple voltage across the ESR if

$$C \geq C_{min} = max\{C_{min(on)}, C_{min(off)}\} = \frac{max\{D_{max}, 1 - D_{min}\}}{2f_s r_C} \tag{11.75}$$

Hence,

$$C_{min} = \frac{D_{max}}{2f_s r_C} \quad \text{for} \quad D_{min} + D_{max} > 1 \tag{11.76}$$

and

$$C_{min} = \frac{1 - D_{min}}{2f_s r_C} \quad \text{for} \quad D_{min} + D_{max} < 1 \tag{11.77}$$

For the worst case, $D_{min} = 0$ or $D_{max} = 1$. Thus, the earlier condition is satisfied at any value of D if

$$C \geq C_{min} = \frac{1}{2r_C f_s} \tag{11.78}$$

For the buck converter with a variable output voltage, the maximum inductor ripple current occurs at $D = 0.5$. Therefore, the minimum capacitance of the filter capacitor is

$$C_{min} = \frac{1}{4r_C f_s} \tag{11.79}$$

If condition (11.75) is satisfied, the maximum peak-to-peak ripple voltage of the buck converter occurs at $D = 0.5$ and is given by

$$V_r = r_C \Delta i_{Lmax} = \frac{r_C V_I D(1 - D)}{f_s L} = \frac{r_C V_I}{4f_s L} \tag{11.80}$$

For steady-state operation, the average value of the ac component of the capacitor voltage v_c is zero, that is,

$$\frac{1}{T}\int_0^T v_c \, dt = 0 \tag{11.81}$$

resulting in

$$v_c(0) = \frac{\Delta i_L (2D - 1)}{12 f_s C} \tag{11.82}$$

Waveforms of v_{rc}, v_c, and v_o are depicted in Fig. 11.12 for three values of the filter capacitance C. In Fig. 11.12(a), the peak-to-peak value of v_o is higher than the peak-to-peak value of v_{rc} because $C < C_{min}$. Figure 11.12(b) and (c) shows the waveforms for $C = C_{min}$ and $C > C_{min}$, respectively. For both these cases, the peak-to-peak voltages of v_o and v_{rc} are the same. For aluminum electrolytic capacitors, $Cr_C \approx 65 \times 10^{-6}$ s.

If condition (11.75) is not satisfied, both the voltage drop across the filter capacitor C and the voltage drop across the ESR contribute to the ripple output voltage. The ac component of the voltage across the filter capacitor increases when the ac component of the charge stored in capacitor is positive. The positive charge is equal to the area under the capacitor current

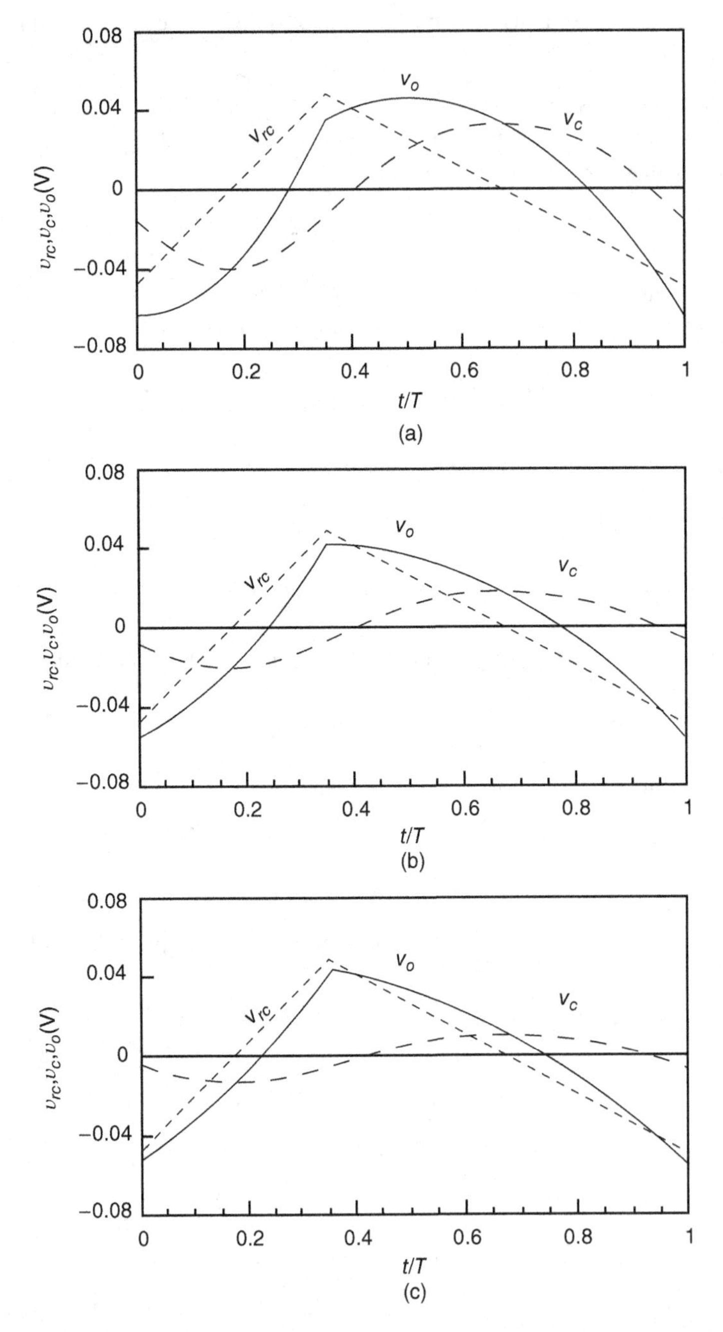

Figure 11.12 Waveforms of v_c, v_{rc}, and v_o at three values of the filter capacitor for CCM. (a) $C < C_{min}$. (b) $C = C_{min}$. (c) $C > C_{min}$.

waveform for $i_C > 0$. The capacitor current is positive during time interval $T/2$. The maximum increase of the charge stored in the filter capacitor in every cycle T is

$$\Delta Q = \frac{\frac{T}{2}\,\frac{\Delta i_{Lmax}}{2}}{2} = \frac{T\Delta i_{Lmax}}{8} = \frac{\Delta i_{Lmax}}{8f_s} \tag{11.83}$$

Hence, using (11.44), the voltage ripple across the capacitance C at $D = 0.5$ is

$$V_{Cpp} = \frac{\Delta Q}{C} = \frac{\Delta i_L}{8f_s C} = \frac{V_O(1-D)}{8f_s^2 LC} = \frac{V_I D(1-D)}{8f_s^2 LC} = \frac{V_I}{32f_s^2 LC} \tag{11.84}$$

where $f_o = 1/(2\pi\sqrt{LC})$ is the corner frequency of the output filter. The minimum filter capacitance required to reduce its peak-to-peak ripple voltage below a specified level V_{Cpp} is

$$C_{min} = \frac{\Delta i_L}{8f_s V_{Cpp}} = \frac{D(1-D)V_I}{8f_s^2 LV_{Cpp}} = \frac{V_I}{32f_s^2 LV_{Cpp}} \tag{11.85}$$

Thus, C_{min} is inversely proportional to f_s^2. Therefore, high switching frequencies are desirable to reduce the size of the filter capacitor.

Using (11.44), the peak-to-peak voltage ripple across the ESR is

$$V_{rcpp} = r_C \Delta i_L = \frac{r_C V_I D(1-D)}{f_s L} = \frac{r_C V_I(1-D)}{4f_s L} \tag{11.86}$$

Hence, the conservative estimation of the total voltage ripple is

$$V_r \approx V_{Cpp} + V_{rcpp} = \frac{V_O(1-D)}{8f_s^2 LC} + \frac{r_C V_I D(1-D)}{f_s L} \tag{11.87}$$

For $D = 0.5$,

$$V_r = \frac{V_I}{32f_s^2 LC} + \frac{r_C V_I}{32f_s L} \tag{11.88}$$

11.4.11 Switching Losses with Linear MOSFET Output Capacitance

Let us assume that the MOSFET output capacitance C_o is linear. First, we shall consider the transistor *turn-off transition*. During this time interval, the transistor is OFF, the drain-to-source voltage v_{DS} increases from nearly zero to V_I, and the transistor output capacitance is charged. Because $dQ = C_o dv_{DS}$, the charge transferred from the input voltage source V_I to the transistor output capacitance C_o during the turn-off transition is

$$Q = \int_0^T i_I\,dt = \int_0^{V_I} dQ = C_o \int_0^{V_I} dv_{DS} = C_o V_I \tag{11.89}$$

yielding the energy transferred from the input voltage source V_I to the converter during the turn-off transition as

$$W_{V_I} = \int_0^T p(t)dt = \int_0^T v_I i_I\,dt = V_I \int_0^T i_I\,dt = V_I Q = C_o V_I^2 \tag{11.90}$$

An alternative method for deriving an expression for the energy delivered from a dc source V_I to a series R–C_o circuit after turning on V_I is as follows. The input current is

$$i_I = \frac{V_I}{R} e^{-\frac{t}{\tau}} \tag{11.91}$$

where $\tau = RC_o$ is the time constant. Hence,

$$W_{V_I} = \int_0^\infty v_I i_I \, dt = V_I \int_0^\infty i_I \, dt = \frac{V_I^2}{R} \int_0^\infty e^{-\frac{t}{\tau}} dt = \frac{V_I^2 \tau}{R} = C_o V_I^2 \tag{11.92}$$

Using $dW_s = Q dv_{DS}/2$, the energy stored in the transistor output capacitance C_o at the end of the transistor turn-off transition, when $v_{DS} = V_I$, is given by

$$W_s = \int_0^{V_I} dW_s = \frac{1}{2} Q \int_0^{V_I} dv_{DS} = \frac{1}{2} Q V_I = \frac{1}{2} C_o V_I^2 \tag{11.93}$$

Thus, the energy lost in the parasitic resistance of the capacitor charging path is the turn-off switching energy loss described by

$$W_{turn-off} = W_{V_I} - W_s = C_o V_I^2 - \frac{1}{2} C_o V_I^2 = \frac{1}{2} C_o V_I^2 \tag{11.94}$$

which results in the *turn-off switching power loss* in the resistance of the charging path

$$P_{turn-off} = \frac{W_{turn-off}}{T} = f_s W_{turn-off} = \frac{1}{2} f_s C_o V_I^2 \tag{11.95}$$

After turn-off, the transistor remains in the off-state for some time interval and the charge W_s is stored in the output capacitance C_o. The efficiency of charging a linear capacitance from a dc voltage source is 50%.

Now consider the transistor *turn-on transition*. When the transistor is turned on, its output capacitance C_o is shorted out through the transistor on-resistance r_{DS}, the charge stored in C_o decreases, and the drain-to-source voltage decreases from V_I to nearly zero. As a result, all the energy stored in the transistor output capacitance is dissipated as heat in the transistor on-resistance r_{DS}. Therefore, the turn-on switching energy loss is

$$W_{turn-on} = W_s = \frac{1}{2} C_o V_I^2 \tag{11.96}$$

resulting in the turn-on switching power loss in the MOSFET

$$P_{turn-on} = P_{sw(FET)} = \frac{W_{turn-on}}{T} = f_s W_{turn-on} = \frac{1}{2} f_s C_o V_I^2 \tag{11.97}$$

The turn-on loss is independent of the transistor on-resistance r_{DS} as long as the transistor output capacitance is fully discharged before the turn-off transition begins.

The total switching energy loss in every cycle of the switching frequency during the process of first charging and then discharging of the output capacitance is given by

$$W_{sw} = W_{turn-off} + W_{turn-on} = W_{V_I} = C_o V_I^2 \tag{11.98}$$

and the total switching loss in the converter is

$$P_{sw} = \frac{W_{sw}}{T} = f_s W_{sw} = f_s C_o V_I^2 \tag{11.99}$$

For a linear capacitance, one-half of the switching power is lost in the MOSFET and the other half is lost in the resistance of the charging path of the transistor output capacitance, that is, $P_{turn-on} = P_{turn-off} = P_{sw}/2$.

The behavior of a diode is different from that of a transistor because a diode cannot discharge its parallel capacitance through its forward resistance. This is because a diode does not turn on until its voltage drops to the threshold voltage. However, the junction diodes suffer from the reverse recovery at turn-off.

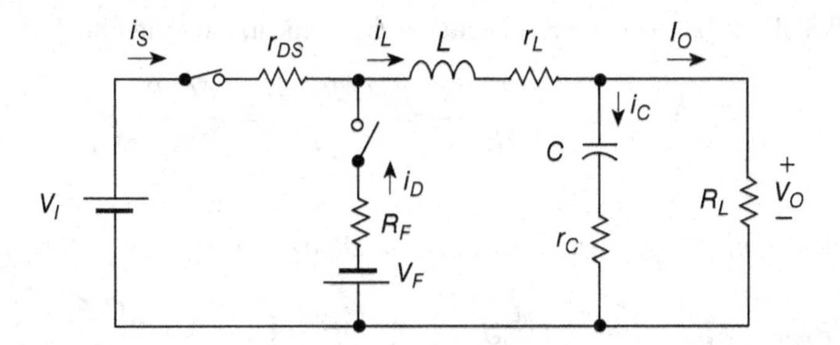

Figure 11.13 Equivalent circuit of the buck converter with parasitic resistances and the diode offset voltage.

11.4.12 Power Losses and Efficiency of Buck Converter for CCM

An equivalent circuit of the buck converter with parasitic resistances is shown in Fig. 11.13. In this figure, r_{DS} is the MOSFET on-resistance, R_F is the diode forward resistance, V_F is the diode threshold voltage, r_L is the ESR of the inductor L, and r_C is the ESR of the filter capacitor C. The slope of the I_D-V_{DS} curves in the ohmic region is equal to the inverse of the MOSFET on-resistance $1/r_{DS}$. The MOSFET on-resistance r_{DS} increases with temperature because the mobility of electrons $\mu_n \approx K_1/T^{2.5}$ decreases with temperature T in the range from 100 to 400 K, where K_1 is a constant. Typically, r_{DS} doubles as the temperature rises by 100 °C.

The large-signal model of a diode consists of a battery V_F in series with a forward resistance R_F. The voltage across the conducting diode is $V_D = V_F + R_F I_D$. If a line is drawn along the linear high-current portion of the I_D-V_D curve (or log (I_D)-V_D) extending to the V_D-axis, the intercept on the V_D-axis is V_F and the slope is $1/R_F$. The threshold voltage V_F is typically 0.7 V for silicon (Si) *pn* junction diodes, and $V_F = 2.8$ V for silicon carbide (SiC) *pn* junction diodes. The threshold voltage $V_F = 0.3$–0.4 V for silicon Schottky diodes and $V_F = 2$ V for silicon carbide Schottky diodes. The threshold voltage V_F of silicon diodes decreases with temperature at the rate of 2 mV/°C. The series resistance R_F of *pn* junction diodes decreases with temperature, while resistance R_F of Schottky diodes increases with temperature.

The conduction losses will be evaluated assuming that the inductor current i_L is ripple-free and is equal to the dc output current I_O. Hence, the switch current can be approximated by

$$i_S = \begin{cases} I_O & \text{for} \quad 0 < t \le DT \\ 0 & \text{for} \quad DT < t \le T \end{cases} \tag{11.100}$$

which results in its rms value

$$I_{Srms} = \sqrt{\frac{1}{T}\int_0^T i_S^2\, dt} = \sqrt{\frac{1}{T}\int_0^{DT} I_O^2\, dt} = I_O\sqrt{D} \tag{11.101}$$

and the MOSFET conduction loss

$$P_{rDS} = r_{DS} I_{Srms}^2 = D r_{DS} I_O^2 = \frac{D r_{DS}}{R_L} P_O = \frac{r_{DS}}{R_L}\frac{V_O}{V_I} P_O \tag{11.102}$$

The transistor conduction loss P_{rDS} is proportional to the duty cycle D at a fixed load current I_O. At $D = 0$, the switch is OFF for the entire cycle, and therefore the conduction loss is zero. At $D = 1$, the switch is ON for the entire cycle, resulting in a maximum conduction loss.

Assuming that the transistor output capacitance C_o is linear, the switching loss is expressed by

$$P_{sw} = f_s C_o V_I^2 = \frac{f_s C_o V_O^2}{M_{VDC}^2} = \frac{f_s C_o R_L}{M_{VDC}^2} P_O = \frac{f_s C_o R_L}{\left(\frac{V_O}{V_I}\right)^2} P_O \tag{11.103}$$

Excluding the MOSFET gate drive power, the total power dissipation in the MOSFET is

$$P_{FET} = P_{rDS} + \frac{P_{sw}}{2} = D r_{DS} I_O^2 + \frac{1}{2} f_s C_o V_I^2 = \left(\frac{D r_{DS}}{R_L} + \frac{f_s C_o R_L}{2M_{VDC}^2}\right) P_O$$

$$= \left[\frac{r_{DS}}{R_L}\frac{V_O}{V_I} + \frac{f_s C_o R_L}{2\left(\frac{V_O}{V_I}\right)^2}\right] P_O \tag{11.104}$$

Similarly, the diode current can be approximated by

$$i_D = \begin{cases} 0 & \text{for} \quad 0 < t \leq DT \\ I_O & \text{for} \quad DT < t \leq T \end{cases} \tag{11.105}$$

yielding its rms value

$$I_{Drms} = \sqrt{\frac{1}{T}\int_0^T i_D^2 \, dt} = \sqrt{\frac{1}{T}\int_{DT}^T I_O^2 \, dt} = I_O\sqrt{1 - D} \tag{11.106}$$

and the power loss in R_F

$$P_{RF} = R_F I_{Drms}^2 = (1 - D) R_F I_O^2 = \frac{(1 - D) R_F}{R_L} P_O = \frac{\left(1 - \frac{V_O}{V_I}\right) R_F}{R_L} P_O \tag{11.107}$$

The average value of the diode current is

$$I_D = \frac{1}{T}\int_0^T i_D \, dt = \frac{1}{T}\int_{DT}^T I_O \, dt = (1 - D) I_O \tag{11.108}$$

which gives the power loss associated with the voltage V_F

$$P_{VF} = V_F I_D = (1 - D) V_F I_O = \frac{(1 - D) V_F}{V_O} P_O = \left(1 - \frac{V_O}{V_I}\right)\frac{V_F}{V_O} P_O \tag{11.109}$$

Thus, the overall diode conduction loss is

$$P_D = P_{VF} + P_{RF} = (1 - D) V_F I_O + (1 - D) R_F I_O^2 = \left(1 - \frac{V_O}{V_I}\right)\left(\frac{V_F}{V_O} + \frac{R_F}{R_L}\right) P_O \tag{11.110}$$

The diode conduction loss P_D decreases when the duty cycle D increases at a fixed load current I_O. At $D = 0$, the diode is ON for the entire cycle, resulting in a maximum conduction loss. At $D = 1$, the diode is OFF for the entire cycle, and therefore the conduction loss is zero.

Typically, the power loss in the inductor core can be ignored and only the copper loss in the inductor winding should be considered. The inductor current can be approximated by

$$i_L \approx I_O \tag{11.111}$$

leading to its rms value

$$I_{Lrms} = I_O \tag{11.112}$$

Therefore, the inductor conduction loss is

$$P_{rL} = r_L I_{Lrms}^2 = r_L I_O^2 = \frac{r_L}{R_L} P_O \tag{11.113}$$

and the maximum power loss in the inductor is

$$P_{rLmax} = r_L I_{Omax}^2 = \frac{r_L}{R_{Lmin}} P_{Omax} \tag{11.114}$$

Using (11.18), (11.61), and (11.68), the rms current through the filter capacitor is found to be

$$I_{Crms} = \sqrt{\frac{1}{T}\int_0^T i_C^2\, dt} = \frac{\Delta i_L}{\sqrt{12}} = \frac{V_O(1-D)}{\sqrt{12} f_s L} \tag{11.115}$$

and the power loss in the filter capacitor

$$P_{rC} = r_C I_{Crms}^2 = \frac{r_C \Delta i_L^2}{12} = \frac{r_C V_O^2 (1-D)^2}{12 f_s^2 L^2} = \frac{r_C R_L \left(1 - \frac{V_O}{V_I}\right)^2}{12 f_s^2 L^2} P_O \tag{11.116}$$

The maximum power loss in the capacitor is

$$P_{rCmax} = \frac{r_C \Delta i_{Lmax}^2}{12} = \frac{r_C V_O^2 (1-D_{min})^2}{12 f_s^2 L^2} \approx \frac{r_C R_L \left(1 - \frac{V_O}{V_{Imax}}\right)^2}{12 f_s^2 L^2} P_{Omax} \tag{11.117}$$

The overall power loss is given by

$$\begin{aligned} P_{LS} &= P_{rDS} + P_{sw} + P_D + P_{rL} + P_{rC} \\ &= D r_{DS} I_O^2 + f_s C_o V_I^2 + (1-D)(V_F I_O + R_F I_O^2) + r_L I_O^2 + \frac{r_C \Delta i_L^2}{12} \\ &= \left[\frac{r_{DS}}{R_L}\frac{V_O}{V_I} + \frac{f_s C_o R_L}{\left(\frac{V_O}{V_I}\right)^2} + \left(1 - \frac{V_O}{V_I}\right)\left(\frac{V_F}{V_O} + \frac{R_F}{R_L}\right) + \frac{r_L}{R_L} + \frac{r_C R_L \left(1 - \frac{V_O}{V_I}\right)^2}{12 f_s^2 L^2} \right] P_O \end{aligned} \tag{11.118}$$

Thus, the buck converter efficiency in CCM is

$$\begin{aligned} \eta &= \frac{P_O}{P_I} = \frac{P_O}{P_O + P_{LS}} = \frac{1}{1 + \frac{P_{LS}}{P_O}} \\ &= \frac{1}{1 + \frac{r_{DS}}{R_L}\frac{V_O}{V_I} + \frac{f_s C_o R_L}{\left(\frac{V_O}{V_I}\right)^2} + \left(1 - \frac{V_O}{V_I}\right)\left(\frac{V_F}{V_O} + \frac{R_F}{R_L}\right) + \frac{r_L}{R_L} + \frac{r_C R_L}{12 f_s^2 L^2}\left(1 - \frac{V_O}{V_I}\right)^2} \end{aligned} \tag{11.119}$$

For $D = 0$, the switch is OFF and the diode is ON, yielding the converter efficiency

$$\eta = \frac{1}{1 + \frac{R_F + r_L}{R_L} + \frac{V_F}{V_O} + \frac{f_s C_o R_L}{M_{VDC}^2} + \frac{r_C R_L}{12 f_s^2 L^2}} \tag{11.120}$$

For $D = 1$, the switch is ON and the diode is OFF, resulting in the converter efficiency

$$\eta = \frac{1}{1 + \frac{r_{DS} + r_L}{R_L} + f_s C_o R_L} \tag{11.121}$$

If the inductor peak-to-peak current ripple $\Delta i_L = V_O(1 - D)/(f_s L) = D(1 - D)V_I/(f_s L)$ is taken into account, the rms value of the switch current is given by

$$I_{Srms} = \sqrt{\frac{D}{3}(I_{Smin}^2 + I_{Smin}I_{Smax} + I_{Smax}^2)} = I_O\sqrt{D}\sqrt{1 + \frac{1}{12}\left(\frac{\Delta i_L}{I_O}\right)^2} \tag{11.122}$$

where $I_{Smin} = I_O - \Delta i_L/2$ and $I_{Smax} = I_O + \Delta i_L/2$. Similarly, the rms value of the diode current is

$$I_{Drms} = \sqrt{\frac{1-D}{3}(I_{Dmin}^2 + I_{Dmin}I_{Dmax} + I_{Dmax}^2)} = I_O\sqrt{1-D}\sqrt{1 + \frac{1}{12}\left(\frac{\Delta i_L}{I_O}\right)^2} \tag{11.123}$$

where $I_{Dmin} = I_O - \Delta i_L/2$ and $I_{Dmax} = I_O + \Delta i_L/2$. The rms value of the inductor current is

$$I_{Lrms} = \sqrt{\frac{1}{3}(I_{Lmin}^2 + I_{Lmin}I_{Lmax} + I_{Lmax}^2)} = I_O\sqrt{1 + \frac{1}{12}\left(\frac{\Delta i_L}{I_O}\right)^2} \tag{11.124}$$

For example, for $\Delta i_L/I_O = 0.1$, $I_{Lrms} = 1.0017 I_O$, and for $\Delta i_L/I_O = 0.5$, $I_{Lrms} = 1.0408 I_O$.

Assuming that the resistances r_L, r_{DS}, and R_F are constant and frequency independent, the conduction power loss in the MOSFET is given by

$$P_{rDS} = r_{DS}I_{Srms}^2 = r_{DS}DI_O^2\left[1 + \frac{1}{12}\left(\frac{\Delta i_L}{I_O}\right)^2\right] = \frac{r_{DS}D}{R_L}\left[1 + \frac{1}{12}\left(\frac{\Delta i_L}{I_O}\right)^2\right]P_O \tag{11.125}$$

The conduction power loss in the diode-forward resistance is

$$P_{RF} = R_F I_{Drms}^2 = R_F(1 - D)I_O^2\left[1 + \frac{1}{12}\left(\frac{\Delta i_L}{I_O}\right)^2\right] = \frac{R_F(1 - D)}{R_L}\left[1 + \frac{1}{12}\left(\frac{\Delta i_L}{I_O}\right)^2\right]P_O \tag{11.126}$$

Assuming that the inductor resistance r_L is independent of frequency, the power loss in the inductor winding is given by

$$P_{rL} = r_L I_{Lrms}^2 = r_L I_O^2\left[1 + \frac{1}{12}\left(\frac{\Delta i_L}{I_O}\right)^2\right] = \frac{r_L}{R_L}\left[1 + \frac{1}{12}\left(\frac{\Delta i_L}{I_O}\right)^2\right]P_O \tag{11.127}$$

The overall power loss is

$$P_{LS} = \left\{\frac{Dr_{DS} + (1 - D)R_F + r_L}{R_L}\left[1 + \frac{1}{12}\left(\frac{\Delta i_L}{I_O}\right)^2\right] + \frac{f_s C_o R_L}{M_{VDC}^2} + \frac{(1 - D)V_F}{V_O} + \frac{r_C R_L (1 - D)^2}{12 f_s^2 L^2}P_O\right\} \tag{11.128}$$

Hence, the converter efficiency is

$$\eta = \frac{1}{1 + \frac{Dr_{DS} + (1-D)R_F + r_L}{R_L}\left[1 + \frac{1}{12}\left(\frac{\Delta i_L}{I_O}\right)^2\right] + \frac{(1-D)V_F}{V_O} + \frac{f_s C_o R_L}{M_{VDC}^2} + \frac{r_C R_L (1-D)^2}{12 f_s^2 L^2}}$$

$$= \frac{1}{1 + \frac{Dr_{DS} + (1-D)R_F + r_L}{R_L}\left[1 + \frac{1}{12}\left(\frac{\Delta i_L R_L}{DV_I}\right)^2\right] + \frac{(1-D)V_F}{DV_I} + \frac{f_s C_o R_L}{D^2} + \frac{r_C R_L (1-D)^2}{12 f_s^2 L^2}}. \quad (11.129)$$

For example, for $\Delta i_L / I_O = 0.1$,

$$P_{rL} = r_L I_{Lrms}^2 = r_L I_O^2 \left[1 + \frac{1}{12}\left(\frac{1}{10}\right)^2\right] = r_L I_O^2 \left(1 + \frac{1}{1200}\right) = 1.0008333 r_L I_O^2 \quad (11.130)$$

For $\Delta i_L / I_O = 0.2$,

$$P_{rL} = r_L I_{Lrms}^2 = r_L I_O^2 \left[1 + \frac{1}{12}\left(\frac{1}{5}\right)^2\right] = r_L I_O^2 \left(1 + \frac{1}{300}\right) = 1.00333 r_L I_O^2 \quad (11.131)$$

In the buck converter, part of the dc input power is transferred directly to the output and is converted to ac power, which is then converted back to dc power. It can be shown that the amount of power which is converted to ac power is

$$P_{AC} = (1-D)P_O \quad (11.132)$$

and the amount of the dc power that directly flows to the output is

$$P_{DC} = DP_O \quad (11.133)$$

11.4.13 DC Voltage Transfer Function of Lossy Converter for CCM

The dc component of the input current is

$$I_I = \frac{1}{T}\int_0^T i_S \, dt = \frac{1}{T}\int_0^{DT} I_O \, dt = DI_O \quad (11.134)$$

leading to the dc current transfer function of the buck converter

$$M_{IDC} \equiv \frac{I_O}{I_I} = \frac{1}{D} \quad (11.135)$$

This equation holds true for both lossless and lossy converters. The converter efficiency can be expressed as

$$\eta = \frac{P_O}{P_I} = \frac{V_O I_O}{V_I I_I} = M_{VDC} M_{IDC} = \frac{M_{VDC}}{D} \quad (11.136)$$

from which the voltage transfer function of the lossy buck converter is

$$M_{VDC} = \frac{\eta}{M_{IDC}} = \eta D$$

$$= \frac{D}{1 + \frac{Dr_{DS}}{R_L} + \frac{f_s C_o R_L}{M_{VDC}^2} + (1-D)\left(\frac{V_F}{V_O} + \frac{R_F}{R_L}\right) + \frac{r_L}{R_L} + \frac{r_C R_L (1-D)^2}{12 f_s^2 L^2}}$$

$$= \frac{D}{1 + \frac{r_{DS}}{R_L}\frac{V_O}{V_I} + \frac{f_s C_o R_L}{\left(\frac{V_O}{V_I}\right)^2} + \left(1 - \frac{V_O}{V_I}\right)\left(\frac{V_F}{V_O} + \frac{R_F}{R_L}\right) + \frac{r_L}{R_L} + \frac{r_C R_L}{12 f_s^2 L^2}\left(1 - \frac{V_O}{V_I}\right)^2} \tag{11.137}$$

For $D = 1$, $M_{VDC} = \eta < 1$.

From (11.137), the on-duty cycle is

$$D = \frac{M_{VDC}}{\eta} = \frac{V_O}{\eta V_I} \tag{11.138}$$

The duty cycle D at a given dc voltage transfer function is higher for the lossy converter than that of a lossless converter. This is because the switch S must be closed for a longer period of time for the lossy converter to transfer enough energy to supply both the required output energy and the converter losses.

Substitution of (11.138) into (11.119) gives the converter efficiency

$$\eta = \frac{N_\eta}{D_\eta} \tag{11.139}$$

where

$$N_\eta = 1 + M_{VDC}\left(\frac{V_F}{V_O} + \frac{r_C R_L}{6 f_s^2 L^2} - \frac{r_{DS} - R_F}{R_L}\right) + \left\{\left[1 + M_{VDC}\left(\frac{V_F}{V_O} + \frac{r_C R_L}{6 f_s^2 L^2} - \frac{r_{DS} - R_F}{R_L}\right)\right]^2 - \frac{M_{VDC}^2 r_C R_L}{3 f_s^2 L^2}\left(1 + \frac{R_F + r_L}{R_L} + \frac{V_F}{V_O} + \frac{f_s C_o R_L}{M_{VDC}^2} + \frac{r_C R_L}{12 f_s^2 L^2}\right)\right\}^{\frac{1}{2}} \tag{11.140}$$

and

$$D_\eta = 2\left(1 + \frac{R_F + r_L}{R_L} + \frac{V_F}{V_O} + \frac{f_s C_o R_L}{M_{VDC}^2} + \frac{r_C R_L}{12 f_s^2 L^2}\right) \tag{11.141}$$

11.4.14 MOSFET Gate Drive Power

For the transistor driven by a square-wave voltage source, the MOSFET gate drive power is associated with charging the transistor input capacitance when the gate-to-source voltage increases, and discharging this capacitance when the gate-to-source voltage decreases. Unfortunately, the input capacitance of power MOSFETs is highly Nonlinear, and therefore it is difficult to determine the gate drive power using the transistor input capacitance. In data sheets, a total gate charge Q_g stored in the gate-to-source capacitance and the gate-to-drain capacitance is given at a specified gate-to-source voltage V_{GS} (usually, $V_{GS} = 10$ V) and a specified drain-to-source voltage V_{DS} (usually, $V_{DS} = 0.8$ of the maximum rating). Using a square-wave voltage source to drive the MOSFET gate, the energy transferred from the gate drive source to the transistor is

$$W_G = Q_g V_{GSpp} \tag{11.142}$$

This energy is lost during one cycle T of the switching frequency $f_s = 1/T$ for charging and discharging the MOSFET input capacitance. Thus, the MOSFET gate drive power is

$$P_G = \frac{W_G}{T} = f_s W_G = f_s Q_g V_{GSpp} \tag{11.143}$$

The gate drive power P_G is proportional to the switching frequency f_s.

The *power gain* is defined by

$$k_p = \frac{P_O}{P_G} \tag{11.144}$$

The *power-added efficiency* (PAE) incorporates the gate drive power P_G by subtracting it from the output power P_O and is defined by

$$\eta_{PAE} = \frac{P_O - P_G}{P_I} \tag{11.145}$$

If the power gain k_p is high, $\eta_{PAE} \approx \eta$. If the power gain $k_p < 1$, $\eta_{PAE} < 0$.

The *total efficiency* is defined by

$$\eta_t = \frac{P_O}{P_I + P_G} \tag{11.146}$$

The *average efficiency* is defined by

$$\eta_{AVG} = \frac{P_{OAVG}}{P_{IAVG}} \tag{11.147}$$

In order to determine this efficiency, the probability-density functions of the average input and output powers are required.

11.4.15 Design of Buck Converter Operating as Amplitude Modulator Operating in CCM

Design a PWM buck converter operating in CCM as an AM to meet the following specifications: $V_I = 25$ V, $V_O = 3$ to 23 V, $P_{Omax} = 7$ W, $f_s = 20$ MHz, and $V_r/V_O \leq 1\%$ at $D = 0.5$.

Solution. The minimum, nominal, and maximum values of the output voltage $V_{Omin} = 3$ V, $V_{Onom} = 13$ V, and $V_{Omax} = 23$ V. The load resistance is

$$R_L = \frac{V_{Omax}^2}{P_{Omax}} = \frac{23^2}{7} = 75.571\ \Omega \tag{11.148}$$

Pick $R_L = 75\ \Omega$. The maximum load current is

$$I_{Omax} = \sqrt{\frac{P_{Omax}}{R_L}} = \sqrt{\frac{7}{75}} = 0.305\ \text{A} \tag{11.149}$$

Hence, the output power at $V_O = 13$ V is

$$P_{Onom} = \frac{V_{Onom}^2}{R_L} = \frac{13^2}{75} = 2.25\ \text{W} \tag{11.150}$$

The nominal load current is

$$I_{Onom} = \sqrt{\frac{P_{Onom}}{R_L}} = \sqrt{\frac{2.25}{75}} = 0.1732\ \text{A} \tag{11.151}$$

The output power at $V_O = 3$ V is

$$P_{Omin} = \frac{V_{Omin}^2}{R_L} = \frac{3^2}{75} = 0.12\ \text{W} \tag{11.152}$$

The minimum load current is

$$I_{Omin} = \sqrt{\frac{P_{Omin}}{R_L}} = \sqrt{\frac{0.12}{75}} = 0.04\ \text{A} \tag{11.153}$$

The minimum, nominal, and maximum values of the dc voltage transfer functions are

$$M_{VDCmin} = \frac{V_{Omin}}{V_I} = \frac{3}{25} = 0.12 \tag{11.154}$$

$$M_{VDCnom} = \frac{V_{Onom}}{V_I} = \frac{13}{25} = 0.52 \tag{11.155}$$

and

$$M_{VDCmax} = \frac{V_{Omax}}{V_I} = \frac{23}{25} = 0.92 \tag{11.156}$$

Assume the converter efficiency $\eta = 95\%$ at $V_O = 23$ V and $\eta = 0.85\%$ at $V_O = 13$ V. The maximum and nominal values of the duty cycle are

$$D_{max} = \frac{M_{VDCmax}}{\eta} = \frac{0.92}{0.95} = 0.9684 \tag{11.157}$$

$$D_{nom} = \frac{M_{VDCnom}}{\eta} = \frac{0.52}{0.85} = 0.6118 \tag{11.158}$$

Assume the converter efficiency $\eta = 30\%$ at $V_I = 3$ V. The minimum value of the duty cycle is

$$D_{min} = \frac{M_{VDCmin}}{\eta} = \frac{0.12}{0.3} = 0.4 \tag{11.159}$$

Assuming the switching frequency $f_s = 20$ MHz, the minimum inductance that is required to maintain the converter in CCM is

$$L_{min} = \frac{V_I D_{min}\left(\frac{1}{\eta} - D_{min}\right) R_L}{2 f_s V_{Omin}} = \frac{25 \times 0.3 \times \left(\frac{1}{0.3} - 0.4\right) \times 75}{2 \times 20 \times 10^6 \times 3} = 13.75\ \mu\text{H} \tag{11.160}$$

Let $L = 15$ μ and $H/r_L = 0.2\ \Omega$. The maximum inductor ripple current is

$$\Delta i_{Lmax} = \frac{V_I}{4 f_s L} = \frac{25}{4 \times 20 \times 10^6 \times 15 \times 10^{-6}} = 0.0208\ \text{mA} \tag{11.161}$$

The ripple voltage is

$$V_r = \frac{V_O}{100} = \frac{13}{100} = 130\ \text{mV} \tag{11.162}$$

If the filter capacitance is large enough, $V_r = r_{Cmax} \Delta i_{Lmax}$ and the maximum ESR of the filter capacitor is

$$r_{Cmax} = \frac{V_r}{\Delta i_{Lmax}} = \frac{130 \times 10^{-3}}{0.0208} = 6.31\ \Omega \tag{11.163}$$

Let $r_C = 500$ mΩ. The minimum value of the filter capacitance at which the ripple voltage is determined by the ripple voltage across the filter capacitor ESR is

$$C_{min} = \frac{1}{4 f_s r_C} = \frac{1}{4 \times 2 \times 10^7 \times 0.5} = 25\ \text{nF} \tag{11.164}$$

Pick $C = 27$ nF/25 V/500 mΩ.

The corner frequency of the output low-pass filter is

$$f_o = \frac{1}{2\pi\sqrt{LC}} = \frac{1}{2\pi\sqrt{15 \times 10^{-6} \times 27 \times 10^{-9}}} = 25\ \text{kHz} \tag{11.165}$$

Thus, $f_s/f_o = 20 \times 10^6/(25 \times 10^3) = 800$.

The voltage and current stresses of power MOSFET and diode are

$$V_{SM} = V_{DM} = V_I = 25\ \text{V} \tag{11.166}$$

and

$$I_{SM} = I_{DM} = I_{Omax} + \frac{\Delta i_{Lmax}}{2} = 0.305 + \frac{0.1736}{2} = 0.3918\ \text{A} \tag{11.167}$$

The drain current of the enhancement-mode MOSFET is given by

$$I_D = \frac{1}{2}\mu_{n0}C_{ox}\left(\frac{W}{L}\right)(v_{GS} - V_t)^2 = \frac{1}{2}K_n\left(\frac{W}{L}\right)(v_{GS} - V_t)^2 \tag{11.168}$$

where $K_n = \mu_{n0}C_{ox}$. Setting $V_{GS} = V_{GS(ON)}$ and $I_{Dmax} = aI_{SM}$, the MOSFET aspect ratio is

$$\frac{W}{L} = \frac{2I_{Dmax}}{\mu_{n0}C_{ox}(V_{GS(ON)} - V_t)^2} = \frac{2aI_{DM}K_n}{(V_{GS(ON)} - V_t)^2} \tag{11.169}$$

Assuming $V_t = 1$ V, $V_{GS(ON)} = 5$ V, $K_n = \mu_{n0}C_{ox} = 0.142 \times 10^{-3}$ A/V^2, and $a = 2$, we obtain

$$\frac{W}{L} = \frac{2aI_{DM}K_n}{(V_{GS(ON)} - V_t)^2} = \frac{2 \times 2 \times 0.3918}{0.142 \times 10^{-3}(5 - 1)^2} = 690 \tag{11.170}$$

Assuming the channel length $L = 0.18$ μm, the channel width is

$$W = \left(\frac{W}{L}\right)L = 690 \times 0.18 = 142.2\ \mu\text{m} \tag{11.171}$$

The selected switching components have the following parameters: MOSFET: $V_{DSS} = 40$ V, $I_{SM} = 0.4$ A, $r_{DS} = 5$ mΩ, $C_o = 25$ pF, and $Q_g = 11$ nC; Schottky barrier diode: $I_{DM} = 0.4$ A, $V_{DM} = 40$ V, $V_F = 0.4$ V, and $R_F = 25$ mΩ.

Assuming that the peak-to-peak gate-to-source voltage is $V_{GSpp} = 4$ V, the MOSFET gate drive power is

$$P_G = f_s Q_g V_{GSpp} = 20 \times 10^6 \times 11 \times 10^{-9} \times 4 = 0.88\ \text{W} \tag{11.172}$$

Efficiency at the maximum output voltage. The power losses and the efficiency will be calculated at the maximum dc output voltage $V_{Omax} = 23$ V, which corresponds to the minimum duty cycle $D_{max} = 0.9684$. The conduction power loss in the MOSFET is

$$P_{rDS} = D_{max} r_{DS} I_{Omax}^2 = 0.9684 \times 0.005 \times 0.305^2 = 0.45\ \text{mW} \tag{11.173}$$

and the switching loss is

$$P_{sw} = f_s C_o V_I^2 = 20 \times 10^6 \times 25 \times 10^{-12} \times 25^2 = 313\ \text{mW} \tag{11.174}$$

Hence, the maximum total power loss in the MOSFET is

$$P_{FETmax} = P_{rDS} + \frac{P_{sw}}{2} = 0.45 + \frac{313}{2} = 156.95\ \text{mW} \tag{11.175}$$

The diode loss due to V_F is

$$P_{VF} = (1 - D_{max})V_F I_{Omax} = (1 - 0.9684) \times 0.4 \times 0.305 = 3.855\ \text{mW} \tag{11.176}$$

the diode loss due to R_F is

$$P_{RF} = (1 - D_{max})R_F I_{Omax}^2 = (1 - 0.9684) \times 0.025 \times 0.305^2 = 0.0735\ \text{mW} \tag{11.177}$$

and the total diode conduction loss is

$$P_D = P_{VF} + P_{RF} = 3.855 + 0.0735 = 3.929\ \text{mW} \tag{11.178}$$

The power loss in the inductor with dc ESR $r_L = 0.2\ \Omega$ is

$$P_{rL} = r_L I_{Omax}^2 = 0.2 \times 0.305^2 = 18.605\ \text{mW} \tag{11.179}$$

The peak-to-peak inductor current ripple at $D_{max} = 0.9684$ is

$$\Delta i_L = \frac{V_{Omax}(1 - D_{max})}{f_s L} = \frac{23 \times (1 - 0.9684)}{20 \times 10^6 \times 1.8 \times 10^{-6}} = 0.0202 \text{ A} \qquad (11.180)$$

The power loss in the capacitor ESR is

$$P_{rC} = \frac{r_C (\Delta i_L)^2}{12} = \frac{0.5 \times 0.0202^2}{12} = 0.017 \text{ mW} \qquad (11.181)$$

The total power loss is

$$P_{LS} = P_{rDS} + P_{sw} + P_D + P_{rL} + P_{rC} = 0.45 + 313 + 3.929 + 18.605 + 0.017 = 336 \text{ mW} \qquad (11.182)$$

and the efficiency of the converter at full load is

$$\eta = \frac{P_O}{P_O + P_{LS}} = \frac{7}{7 + 0.336} = 95.4\% \qquad (11.183)$$

If the assumed efficiency is much different than the calculated efficiency in (11.183), another iteration is needed with a new assumed converter efficiency.

Figures 11.14–11.16 depict plots of the power losses in the parasitic components. Figure 11.17 shows a plot of efficiency η as a function of output voltage of the buck amplitude modulator at $R_L = 75\,\Omega$, $r_{DS} = 5\text{ m}\Omega$, $R_F = 25\text{ m}\Omega$, $V_F = 0.4$ V, $r_L = 0.2\,\Omega$, $r_C = 500\text{ m}\Omega$, $C_o = 25$ pF, $L = 1.8\,\mu$H, and $f_s = 20$ MHz.

Efficiency at the nominal output voltage. The power losses and the efficiency will be calculated at the nominal dc output voltage $V_{Onom} = 13$ V, which corresponds to the minimum duty cycle $D_{nom} = 0.6118$. The conduction power loss in the MOSFET is

$$P_{rDS} = D_{nom} r_{DS} I^2_{Onom} = 0.6118 \times 0.005 \times 0.1732^2 = 0.09176 \text{ mW} \qquad (11.184)$$

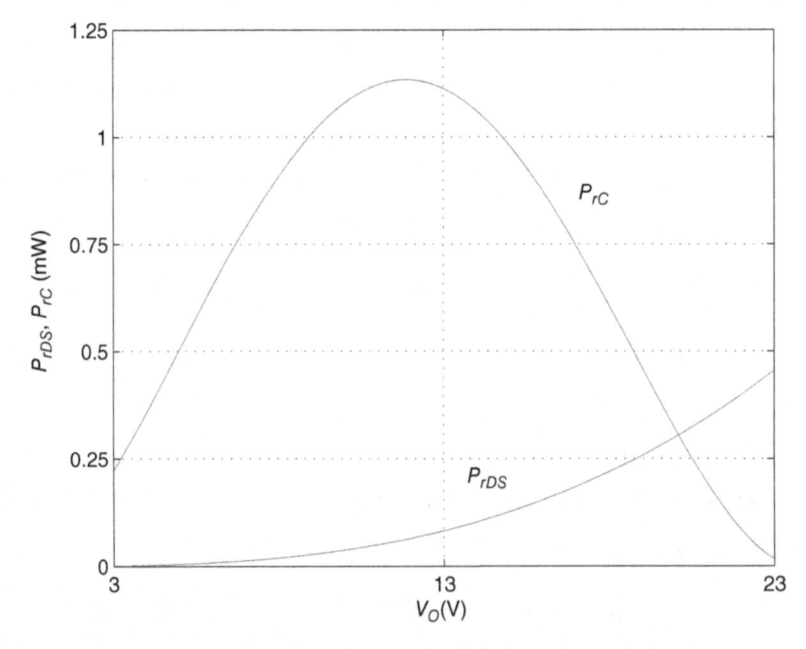

Figure 11.14 Conduction power loss in the MOSFET on-resistance P_{rDS} and in the capacitor resistance ESR P_{rC} as a function of V_O for the buck converter in CCM at $V_I = 25$ V, $r_{DS} = 5\text{ m}\Omega$, $r_C = 500\text{ m}\Omega$, and $f_s = 20$ MHz.

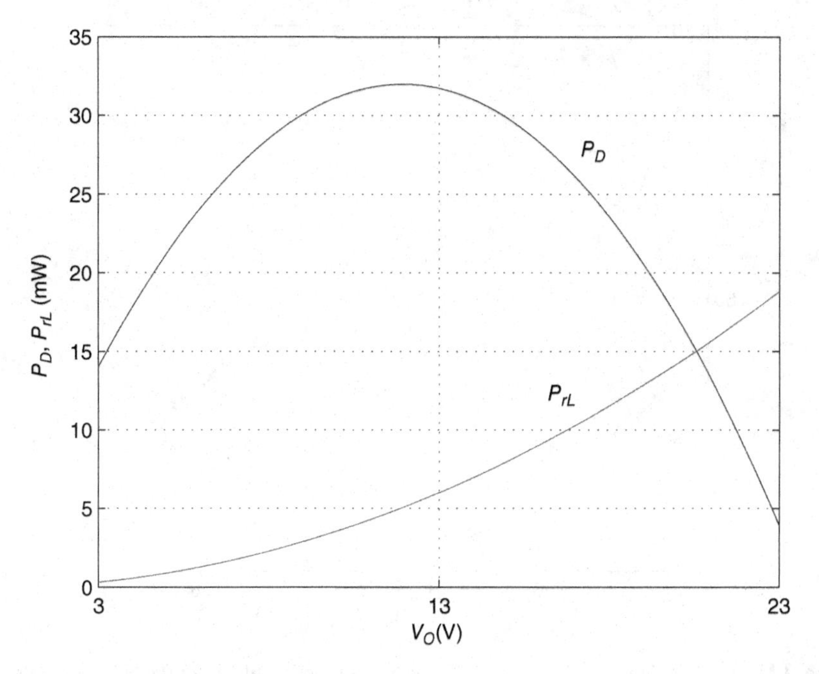

Figure 11.15 Conduction power loss in diode P_D and in the inductor resistance ESR P_{rL} as a function of V_O for the buck converter in CCM at $V_I = 25$ V, $R_F = 25$ mΩ, $V_F = 0.4$ V, $r_L = 0.2$ Ω, and $f_s = 20$ MHz.

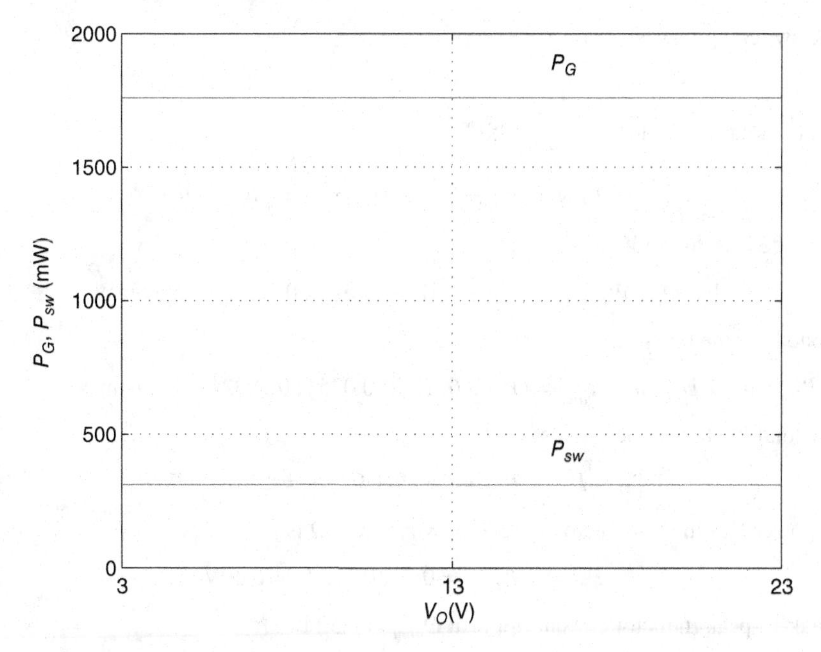

Figure 11.16 Switching power losses in the gate P_G and in the drain P_{sw} as a function of V_O for the buck converter in CCM at $Q_g = 11$ nC, $V_{GSpp} = 4$ V, $V_I = 25$ V, $C_o = 25$ pF, and $f_s = 20$ MHz.

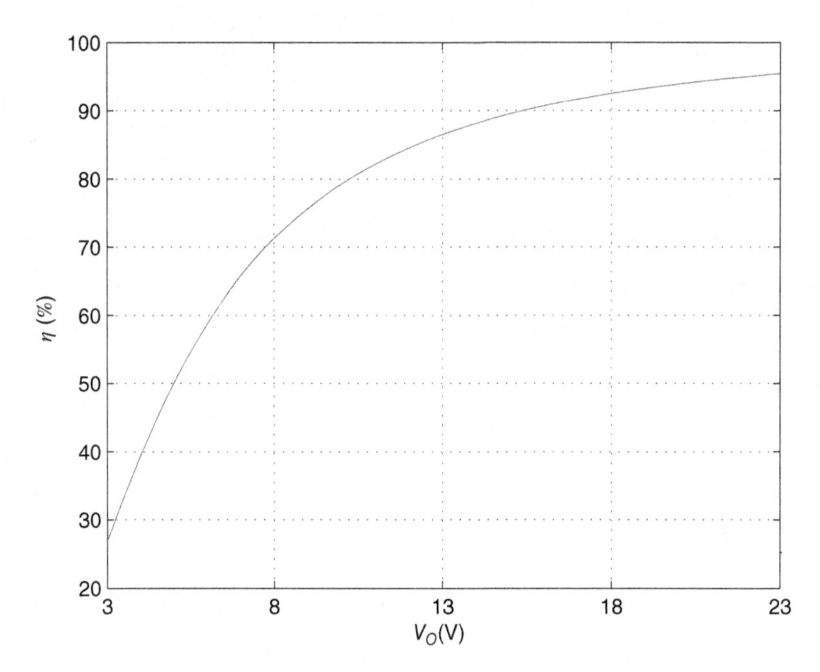

Figure 11.17 Efficiency of the buck converter η as a function of V_O for the buck converter in CCM at $V_I = 25$ V, $R_L = 75\ \Omega$, $r_{DS} = 5$ mΩ, $R_F = 25$ mΩ, $V_F = 0.4$ V, $r_L = 0.2\ \Omega$, $r_C = 500$ mΩ, $C_o = 25$ pF, $L = 1.8\ \mu$H, and $f_s = 20$ MHz.

and the switching loss is

$$P_{sw} = f_s C_o V_I^2 = 20 \times 10^6 \times 25 \times 10^{-12} \times 25^2 = 313 \text{ mW} \tag{11.185}$$

Hence, the total power loss in the MOSFET is

$$P_{FET} = P_{rDS} + \frac{P_{sw}}{2} = 0.09176 \times + \frac{313}{2} = 156.592 \text{ mW} \tag{11.186}$$

The diode loss due to V_F is

$$P_{VF} = (1 - D_{nom}) V_F I_{Onom} = (1 - 0.6118) \times 0.4 \times 0.1732 = 26.89 \text{ mW} \tag{11.187}$$

the diode loss due to R_F is

$$P_{RF} = (1 - D_{nom}) R_F I_{Onom}^2 = (1 - 0.6118) \times 0.025 \times 0.1732^2 = 0.291 \text{ mW} \tag{11.188}$$

and the total diode conduction loss is

$$P_D = P_{VF} + P_{RF} = 26.89 + 0.291 = 27.187 \text{ mW} \tag{11.189}$$

The power loss in the inductor with dc ESR $r_L = 0.2\ \Omega$ is

$$P_{rL} = r_L I_{Onom}^2 = 0.2 \times 0.1732^2 = 6 \text{ mW} \tag{11.190}$$

The peak-to-peak inductor current ripple at $D_{nom} = 0.52$ is

$$\Delta i_L = \frac{V_{Onom}(1 - D_{nom})}{f_s L} = \frac{13 \times (1 - 0.6118)}{20 \times 10^6 \times 1.8 \times 10^{-6}} = 0.1402 \text{ A} \tag{11.191}$$

The power loss in the capacitor ESR is

$$P_{rC} = \frac{r_C (\Delta i_L)^2}{12} = \frac{0.5 \times 0.1402^2}{12} = 0.819 \text{ mW} \tag{11.192}$$

The total power loss is

$$P_{LS} = P_{rDS} + P_{sw} + P_D + P_{rL} + P_{rC}$$
$$= 0.9176 + 313 + 27.187 + 6 + 0.819 = 347.294\ \text{mW} \quad (11.193)$$

and the efficiency of the converter at full load is

$$\eta = \frac{P_O}{P_O + P_{LS}} = \frac{2.25}{2.25 + 0.347924} = 86.61\% \quad (11.194)$$

Efficiency at the minimum output voltage. The power losses and the efficiency will be calculated at the minimum dc output voltage $V_{Omin} = 3$ V, which corresponds to the minimum duty cycle $D_{min} = 0.1263$. The conduction power loss in the MOSFET is

$$P_{rDS} = D_{min} r_{DS} I^2_{Omin} = 0.4 \times 0.005 \times 0.04^2 = 0.0032\ \text{mW} \quad (11.195)$$

and the switching loss is

$$P_{sw} = f_s C_o V_I^2 = 20 \times 10^6 \times 25 \times 10^{-12} \times 25^2 = 313\ \text{mW} \quad (11.196)$$

Hence, the total power loss in the MOSFET is

$$P_{FET} = P_{rDS} + \frac{P_{sw}}{2} = 0.0032 + \frac{313}{2} = 156.5032\ \text{mW} \quad (11.197)$$

The diode loss due to V_F is

$$P_{VF} = (1 - D_{min}) V_F I_{Omin} = (1 - 0.4) \times 0.4 \times 0.04 = 9.6\ \text{mW} \quad (11.198)$$

the diode loss due to R_F is

$$P_{RF} = (1 - D_{min}) R_F I^2_{Omin} = (1 - 0.4) \times 0.025 \times 0.04^2 = 0.024\ \text{mW} \quad (11.199)$$

and the total diode conduction loss is

$$P_D = P_{VF} + P_{RF} = 9.6 + 0.024 = 9.624\ \text{mW} \quad (11.200)$$

The power loss in the inductor with dc ESR $r_L = 0.2$ mΩ is

$$P_{rL} = r_L I^2_{Omin} = 0.2 \times 0.04^2 = 0.32\ \text{mW} \quad (11.201)$$

The peak-to-peak inductor current ripple at $D_{min} = 0.1263$ is

$$\Delta i_L = \frac{V_{Omin}(1 - D_{min})}{f_s L} = \frac{3 \times (1 - 0.4)}{20 \times 10^6 \times 1.8 \times 10^{-6}} = 0.006\ \text{A} \quad (11.202)$$

The power loss in the capacitor ESR is

$$P_{rC} = \frac{r_C (\Delta i_L)^2}{12} = \frac{0.5 \times 0.006^2}{12} = 0.0015\ \text{mW} \quad (11.203)$$

The total power loss is

$$P_{LS} = P_{rDS} + P_{sw} + P_D + P_{rL} + P_{rC}$$
$$= 0.0032 + 313 + 9.624 + 0.32 + 0.0015 = 322.556\ \text{mW} \quad (11.204)$$

and the efficiency of the converter at full load is

$$\eta = \frac{P_O}{P_O + P_{LS}} = \frac{0.12}{0.12 + 0.3273556} = 27.09\% \quad (11.205)$$

Figure 11.18 shows the plot of the dc voltage transfer function M_{VDC} as a function of the dc output voltage V_O.

Figures 11.19 and 11.20 show conduction losses as function of the duty cycle D. Figure 11.21 shows the efficiency of the designed buck converter η as a function of duty cycle D at a fixed dc input voltage V_I. Figure 11.22 shows the dc voltage transfer function $M_{VDC} = \eta D$ of the buck converter as a function of the duty cycle D.

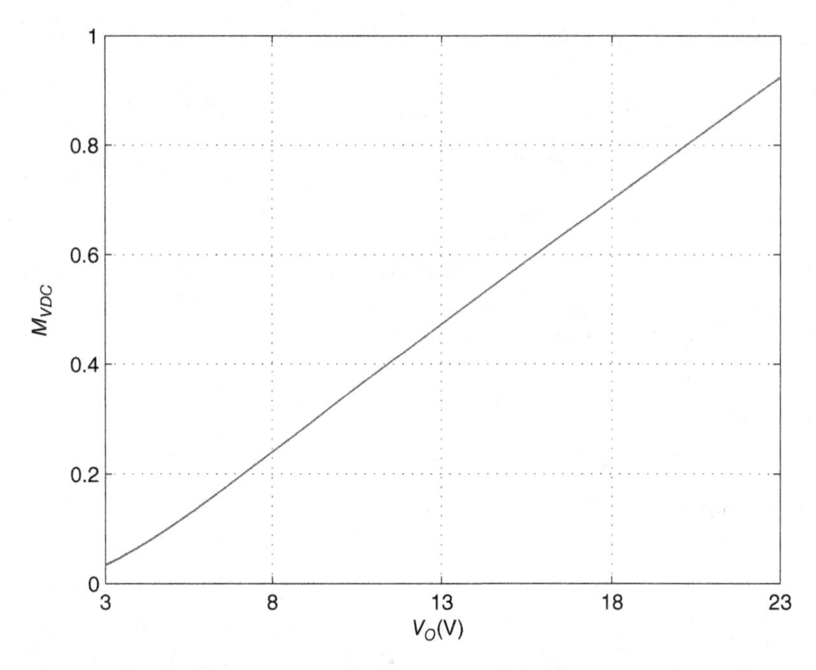

Figure 11.18 DC voltage transfer function $M_{VDC} = \eta D$ as a function of V_O for the buck converter CCM at $V_I = 25$ V, $R_L = 75\ \Omega$, $r_{DS} = 5$ mΩ, $R_F = 25$ mΩ, $V_F = 0.4$ V, $r_L = 0.2\ \Omega$, $r_C = 500$ mΩ, $C_o = 25$ pF, $L = 1.8\ \mu$H, and $f_s = 20$ MHz.

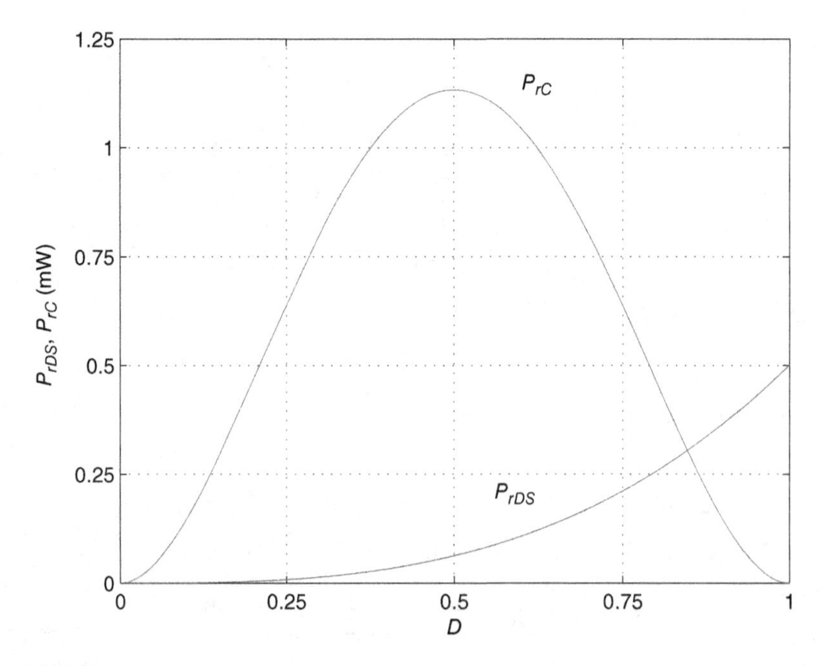

Figure 11.19 Conduction power loss in the MOSFET on-resistance P_{rDS} and in the filter capacitor resistance ESR P_{rC} as a function of the duty cycle D for the buck converter in CCM at $V_I = 25$ V, $r_{DS} = 5$ mΩ, $r_C = 500$ mΩ, and $f_s = 20$ MHz.

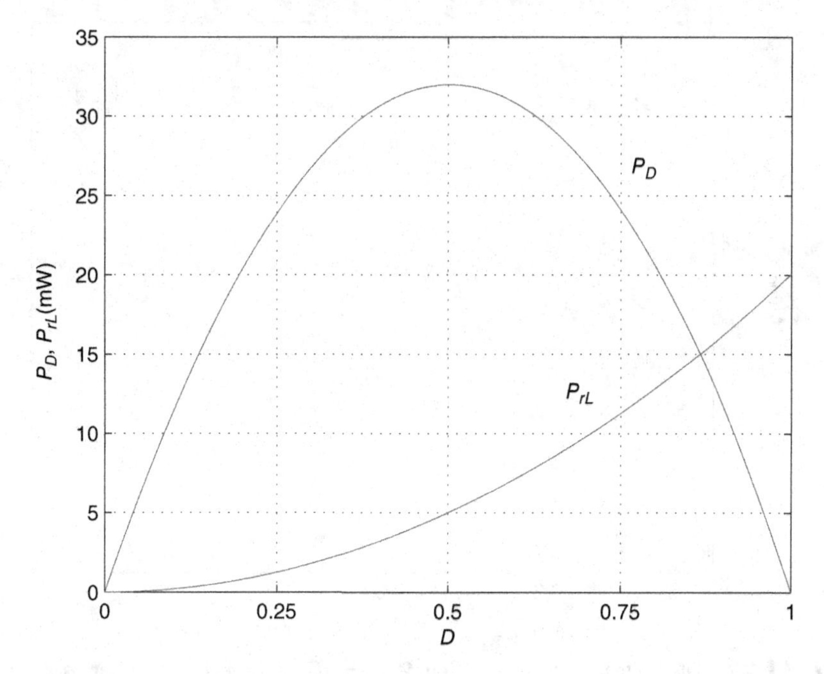

Figure 11.20 Conduction power loss in diode P_D and in the inductor resistance ESR P_{rL} as a function of the duty cycle D for the buck converter in CCM at $V_I = 25$ V, $R_F = 25$ mΩ, $V_F = 0.4$ V, $r_L = 0.2$ Ω, and $f_s = 20$ MHz.

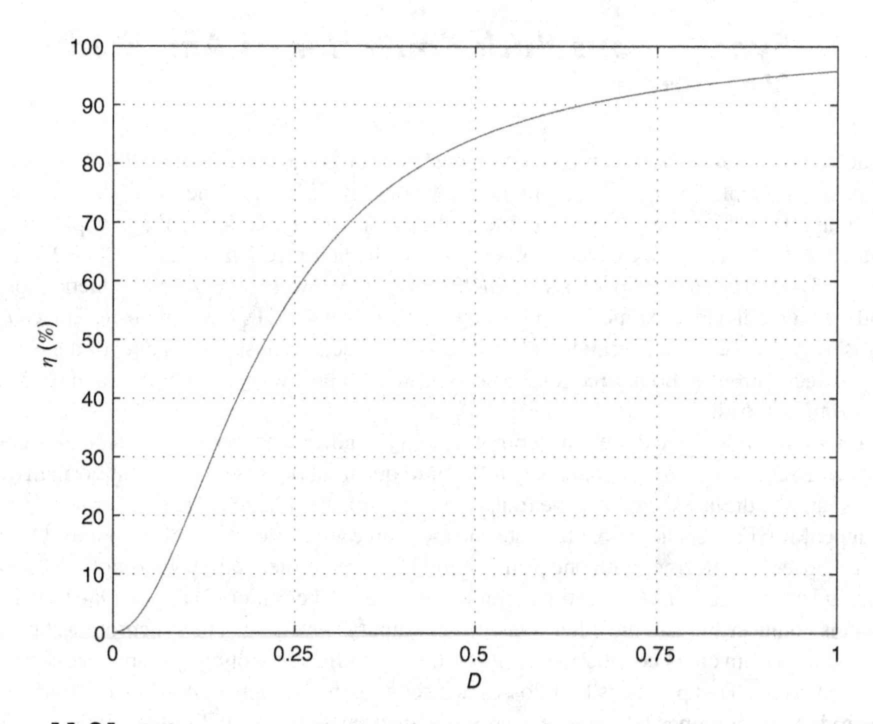

Figure 11.21 Efficiency η as a function of the duty cycle D for the buck converter in CCM at $V_I = 25$ V, $R_L = 75$ Ω, $r_{DS} = 5$ mΩ, $R_F = 25$ mΩ, $V_F = 0.4$ V, $r_L = 0.2$ Ω, $r_C = 500$ mΩ, $C_o = 25$ pF, $L = 1.8$ μH, and $f_s = 20$ MHz.

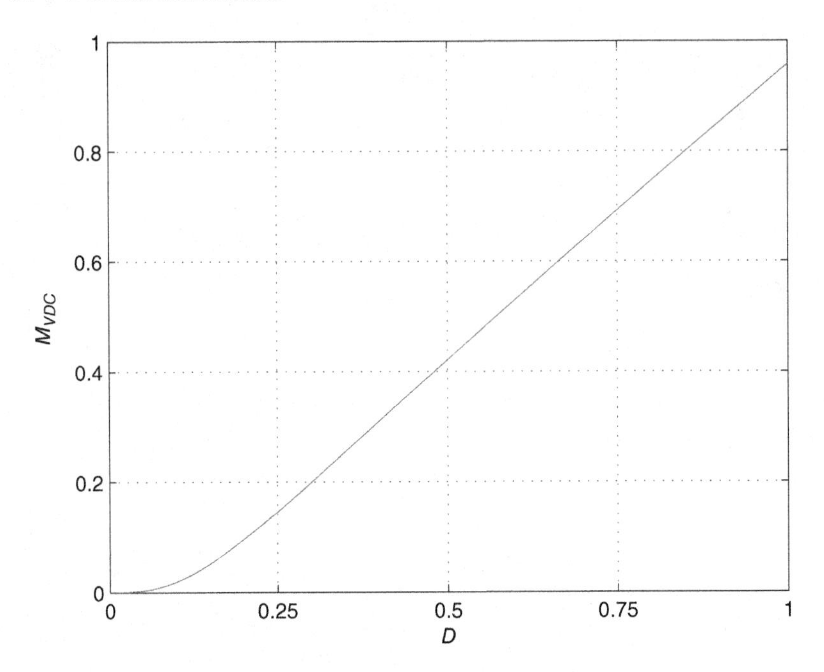

Figure 11.22 DC voltage transfer function $M_{VDC} = \eta D$ of the buck converter as a function of the duty cycle D for the buck converter CCM at $V_I = 25$ V, $R_L = 75\ \Omega$, $r_{DS} = 5$ mΩ, $R_F = 25$ mΩ, $V_F = 0.4$ V, $r_L = 0.2\ \Omega$, $r_C = 500$ mΩ, $C_o = 25$ pF, $L = 1.8\ \mu$H, and $f_s = 20$ MHz.

11.5 Synchronous Buck Converter as Amplitude Modulator

A buck converter topology with a synchronous rectifier is shown in Fig. 11.23(a). This circuit is obtained by replacing the diode with an n-channel MOSFET. In general, diodes have an offset voltage V_F, which may be comparable to the output voltage in low-voltage applications. In contrast, MOSFETs do not have an offset voltage. If the on-resistance of a MOSFET is low, then the forward voltage drop across the MOSFET is very low, reducing the conduction loss and yielding high efficiency. Some low-breakdown voltage MOSFETs have an on-resistance r_{DS} as low as 6 mΩ. In addition, operation in DCM can be avoided because the channel of the transistor can conduct current in both directions. The synchronous buck converter operates in CCM from no load to full load.

The two MOSFETs are driven in a complimentary manner. The low-side n-channel MOSFET replaces a Schottky diode and operates in the third quadrant because the current normally flows from source to drain. When both the transistors are n-channel MOSFETs, it is difficult to drive the upper MOSFET because both the gate and the source are connected to "hot" points. One solution is to use a transformer with one primary winding and two secondary windings. The primary winding is connected to a driver, for example, an integrated circuit (IC) driver. One transformer output is noninverting and the other transformer output is inverting. The synchronous buck converter suffers from cross-conduction (or shoot-through) effect, resulting in high-current spikes in both transistors. This produces high losses and reduces the efficiency. A nonoverlapping driver can produce a dead time and reduce the cross-conduction loss. During the dead time periods, the inductor current flows through the lower MOSFET body diode. This body diode has a very slow reverse recovery characteristic that can adversely affect the converter efficiency. An external Schottky diode can be connected in parallel with the low-side MOSFET to shunt the body diode

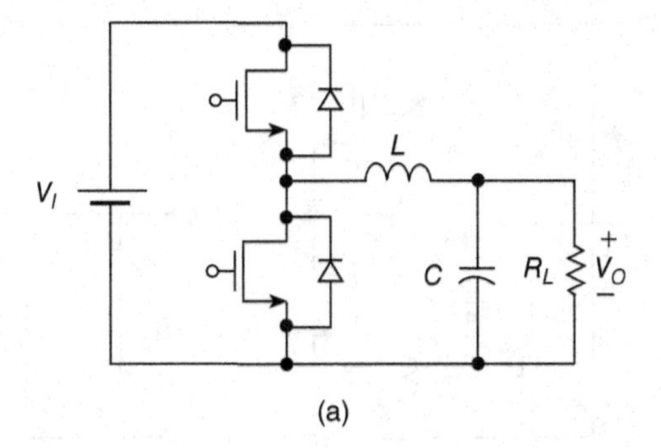

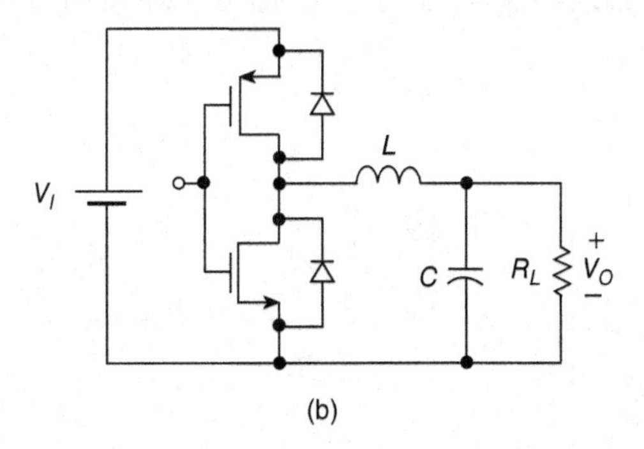

Figure 11.23 Buck converter with a synchronous rectifier. (a) With two *n*-channel MOSFETs. (b) CMOS buck converter.

and to prevent it from affecting the converter performance. The added Schottky diode can have a much lower current rating than the diode in the conventional nonsynchronous buck converter because it only conducts during the small dead time when both MOSFETs are OFF.

If the upper MOSFET is a PMOS and the lower MOSFET is an NMOS, then the circuit is similar to a digital CMOS inverter, as shown in Fig. 11.23(b). In this case, both transistors can be driven by the same gate-to-source voltage. The peak-to-peak gate-to-source voltage should be equal or close to the dc input voltage V_I. Therefore, the CMOS buck synchronous converter is a good topology for applications with a low dc voltage V_I. The whole converter can be integrated except for the filter capacitor C.

At a high voltage V_I, the peak-to-peak gate-to-source voltage is high and may break the MOSFET gate. The same gate-to-source voltage may cause cross-conduction of both transistors, generating high spikes and drastically reducing the converter efficiency. A dead time will reduce the current spikes, but this requires two nonoverlapping gate-to-source voltages to drive the MOSFETs.

The synchronous buck converter is especially attractive in power supplies with a very low output voltage (e.g., $V_O = 3.3$ V or $V_O = 1.8$ V) and/or a wide load range, including operation from no load to full load. Its main advantage is higher efficiency than that of the conventional buck converter. The synchronous buck converter may also be used as a bidirectional converter.

Figure 11.24 shows a synchronous buck converter with a transformer driver. If both MOSFETs are *n*-channel devices, then the upper output of the transformer should be noninverting and the other should be inverting. If the upper transistor is a PMOS and the bottom transistor is an NMOS, then both transformer outputs should be noninverting or inverting.

Figure 11.25 shows a synchronous buck converter with a voltage mirror driver. The voltage mirror driver acts as a voltage shifter for the ac voltage waveform so that the gate-to-source

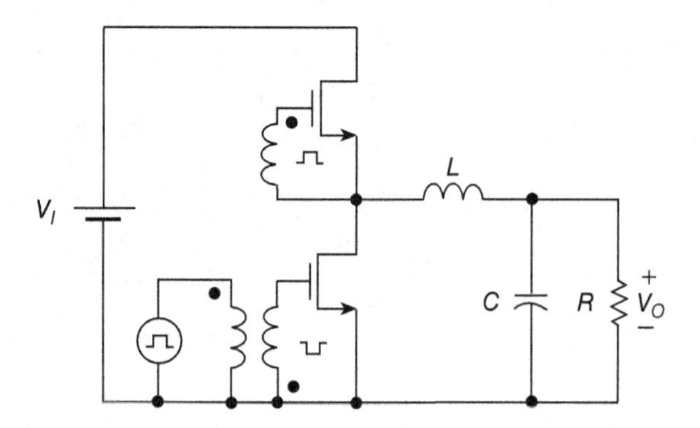

Figure 11.24 Synchronous buck converter with a transformer driver.

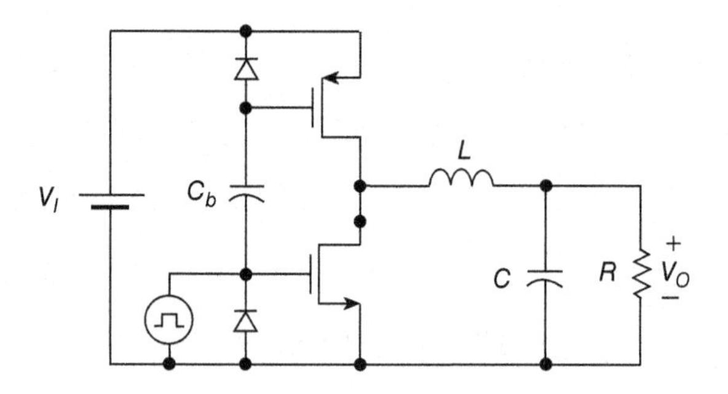

Figure 11.25 Synchronous buck converter with a voltage mirror driver.

voltage of the n-channel MOSFET is the same as the source-to-gate voltage of the p-channel MOSFET. Contrary to the CMOS synchronous buck converter, the peak-to-peak voltage of the gate-to-source voltage can be lower than the dc input voltage V_I. Therefore, this driver is good for applications with high values of V_I.

The minimum inductance for the synchronous buck converter is limited only by the inductor current. For CCM, the maximum inductor current ripple is

$$\Delta i_{Lmax} = \frac{V_O(1 - D_{0.5})}{f_s L_{min}} = \frac{V_I D_{0.5}(1 - D_{0.5})}{f_s L_{min}} = \frac{V_I}{4 f_s L_{min}} \tag{11.206}$$

Hence, the minimum inductance is given by

$$L_{min} = \frac{V_I}{4 f_s L_{min}} \tag{11.207}$$

Setting $V_F = 0$ and $R_F = r_{DS2}$ in (11.119), one obtains the efficiency of the synchronous buck converter is

$$\eta = \frac{P_O}{P_I} = \frac{1}{1 + \frac{D r_{DS1} + (1 - D) r_{DS2} + r_L}{R_L} \left[1 + \frac{1}{12}\left(\frac{\Delta i_L}{I_O}\right)^2\right] + \frac{f_s C_o R_L}{M_{VDC}^2} + \frac{r_C R_L (1 - D)^2}{12 f_s^2 L^2}} \tag{11.208}$$

where $\Delta i_L = V_O(1-D)/(f_s L) = D(1-D)V_I/(f_s L)$. If $r_{DS1} = r_{DS2}$, the converter efficiency becomes

$$\eta = \frac{1}{1 + \frac{r_{DS}+r_L}{R_L}\left[1 + \frac{1}{12}\left(\frac{\Delta i_L}{I_O}\right)^2\right] + \frac{f_s C_o R_L}{M_{VDC}^2} + \frac{r_C R_L (1-D)^2}{12 f_s^2 L^2}}$$

$$= \frac{1}{1 + \frac{r_{DS}+r_L}{R_L}\left[1 + \frac{1}{12}\left(\frac{\Delta i_L R_L}{V_O}\right)^2\right] + \frac{f_s C_o R_L}{(V_O/V_I)^2} + \frac{r_C R_L}{12 f_s^2 L^2}\left(1 - \frac{V_O}{V_I}\right)^2} \quad (11.209)$$

Figure 11.26 shows the efficiency η as a function of the dc output voltage V_O of the buck converter with synchronous rectifier.

Letting $V_F = 0$ and $R_F = r_{DS2}$ in (11.137), the dc voltage transfer function of the synchronous buck converter is obtained as

$$M_{VDC} = \frac{V_O}{V_I} = \eta D = \frac{D}{1 + \frac{r_{DS}+r_L}{R_L}\left[1 + \frac{1}{12}\left(\frac{\Delta i_L}{I_O}\right)^2\right] + \frac{f_s C_o R_L}{M_{VDC}^2} + \frac{r_C R_L (1-D)^2}{12 f_s^2 L^2}}$$

$$= \frac{D}{1 + \frac{r_{DS}+r_L}{R_L}\left[1 + \frac{1}{12}\left(\frac{\Delta i_L R_L}{V_O}\right)^2\right] + \frac{f_s C_o R_L}{(V_O/V_I)^2} + \frac{r_C R_L}{12 f_s^2 L^2}\left(1 - \frac{V_O}{V_I}\right)^2} \quad (11.210)$$

Figure 11.27 shows the dc voltage transfer function M_{VDC} as a function of the dc output voltage V_O at fixed supply voltage $V_I = 25$ V for the buck converter with synchronous rectifier. Figure 11.28 shows the efficiency η as a function of the duty cycle D at fixed $V_I = 25$ V for

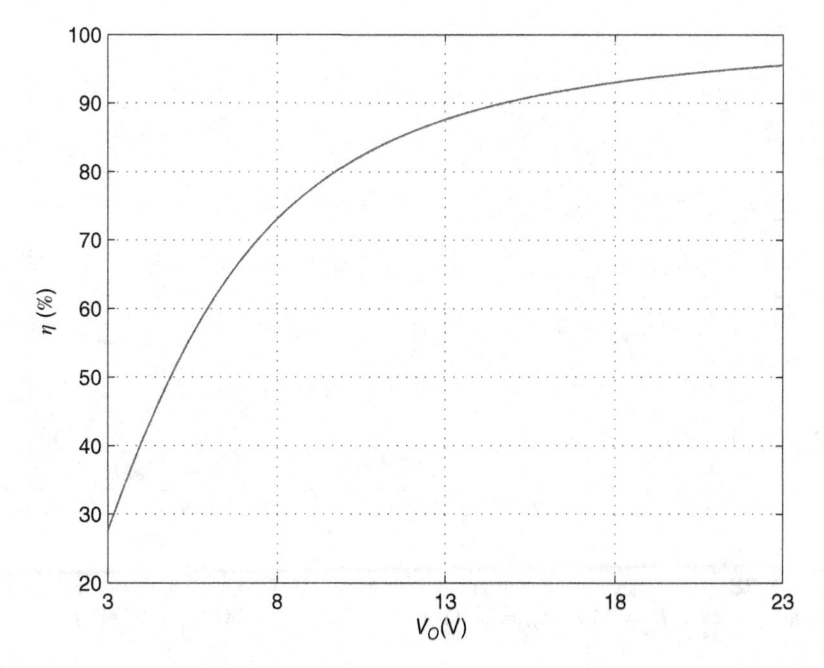

Figure 11.26 Efficiency η as a function of the dc output voltage V_O of the buck converter with synchronous rectifier at $V_I = 25$ V, $R_L = 75\ \Omega$, $r_{DS} = 5\ \text{m}\Omega$, $r_L = 0.2\ \Omega$, $r_C = 0.5\ \Omega$, $C_o = 25$ pF, $L = 1.8\ \mu$H, and $f_s = 20$ MHz.

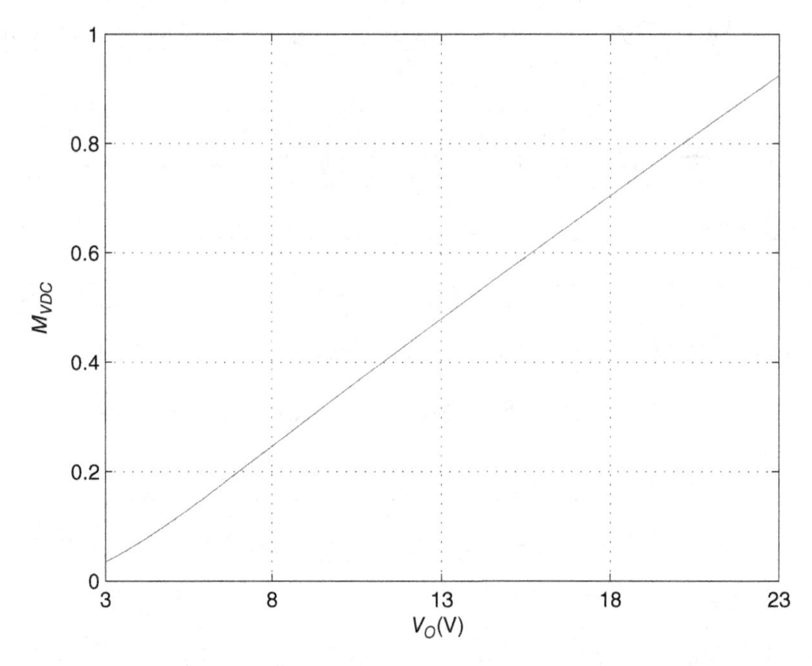

Figure 11.27 DC voltage transfer function M_{VDC} as a function of the dc output voltage V_O of the buck converter with synchronous rectifier at $V_I = 25$ V, $R_L = 75\ \Omega$, $r_{DS} = 5\ \text{m}\Omega$, $r_L = 0.2\ \Omega$, $r_C = 0.5\ \Omega$, $C_o = 25$ pF, $L = 1.8\ \mu$H, and $f_s = 20$ MHz.

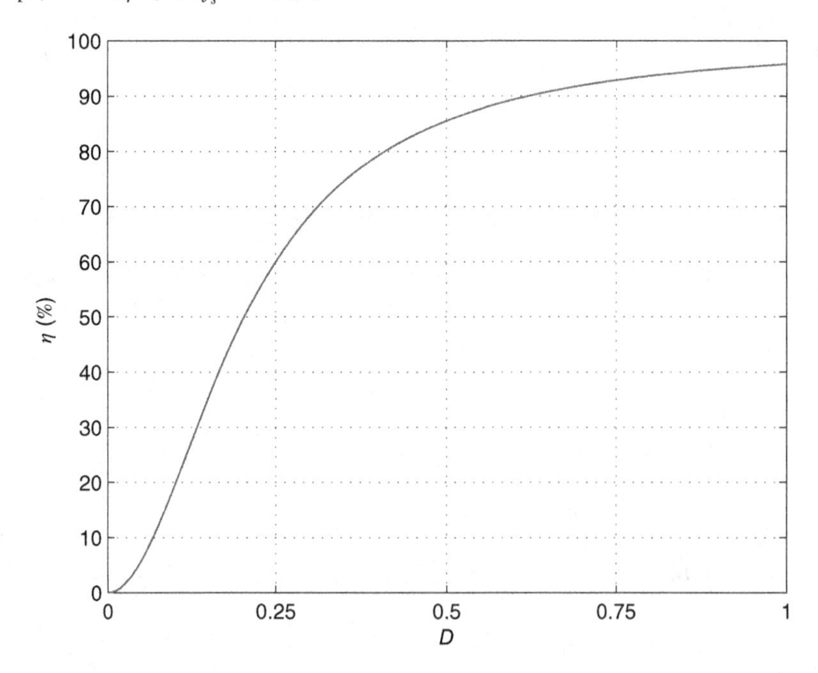

Figure 11.28 Efficiency η as a function of the duty cycle D of the buck converter with synchronous rectifier at $V_I = 25$ V, $R_L = 75\ \Omega$, $r_{DS} = 5\ \text{m}\Omega$, $r_L = 0.2\ \Omega$, $r_C = 0.5\ \Omega$, $C_o = 25$ pF, $L = 1.8\ \mu$H, and $f_s = 20$ MHz.

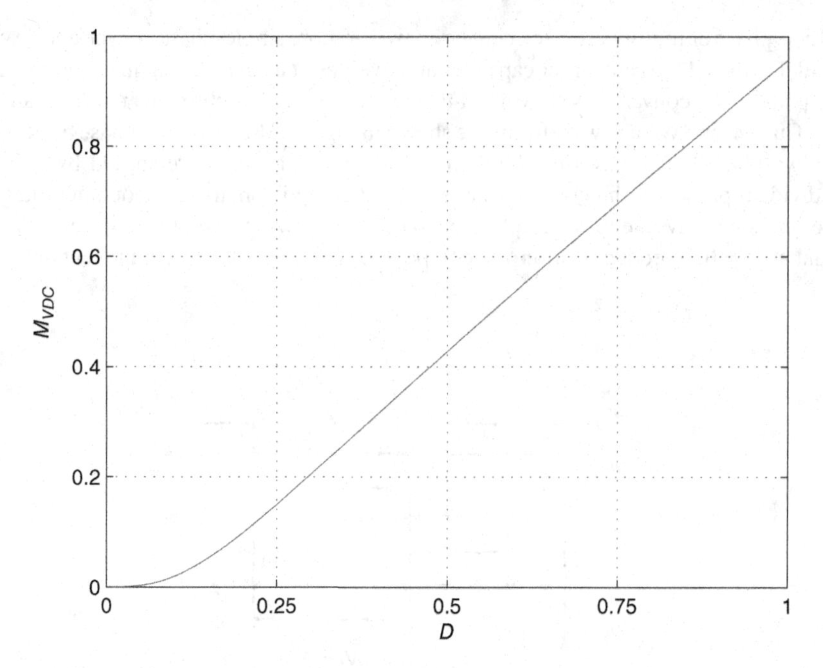

Figure 11.29 DC voltage transfer function M_{VDC} as a function of the duty cycle D of the buck converter with synchronous rectifier at $V_I = 25$ V, $R_L = 75\ \Omega$, $r_{DS} = 5$ mΩ, $r_L = 0.2\ \Omega$, $r_C = 0.5\ \Omega$, $C_o = 25$ pF, $L = 1.8\ \mu$H, and $f_s = 20$ MHz.

the buck converter with synchronous rectifier. Figure 11.29 shows the dc voltage transfer function M_{VDC} as a function of the dc output voltage V_O at $V_I = 25$ V of the buck converter with synchronous rectifier.

11.6 Multiphase Buck Converter

So far, we have studied a single-phase buck converter. This circuit requires a relatively large filter capacitor to reduce the output voltage ripple. A very wide bandwidth is required in AM used in RF transmitters. Multiphase buck converter has a smaller filter capacitor and therefore a wider bandwidth.

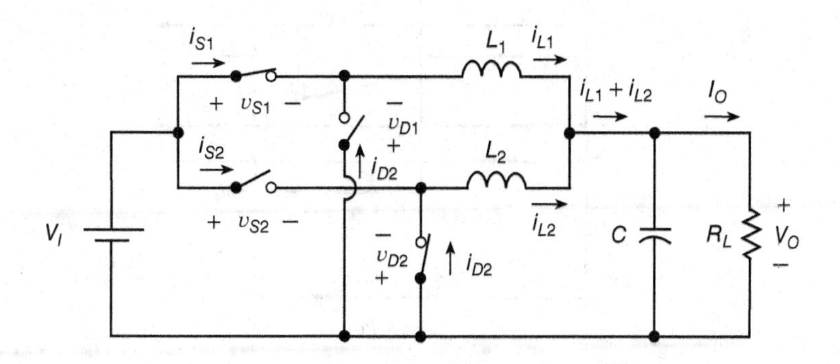

Figure 11.30 Two-phase buck converter.

In a polyphase or multiphase buck converter, two or more single-phase converters are operated in parallel and feed the same filter capacitor and load resistance, resulting in ripple cancellation. A two-phase buck converter is shown in Fig. 11.30. Usually, synchronous rectifiers are used as diodes. Current and voltage waveforms are shown in Fig. 11.31 for the two-phase buck converter. In the two-phase buck converter, the drive signals v_{GS1} and v_{GS2} are shifted by 180°. When the individual phases of the converter are switched complimentarily, the output voltage ripple reduces considerably due to the ripple cancellation. In a two-phase buck converter, $i_{L1} + i_{L2}$ is constant at $D = 0.5$, producing a zero ac component. Therefore, the ac component of the current

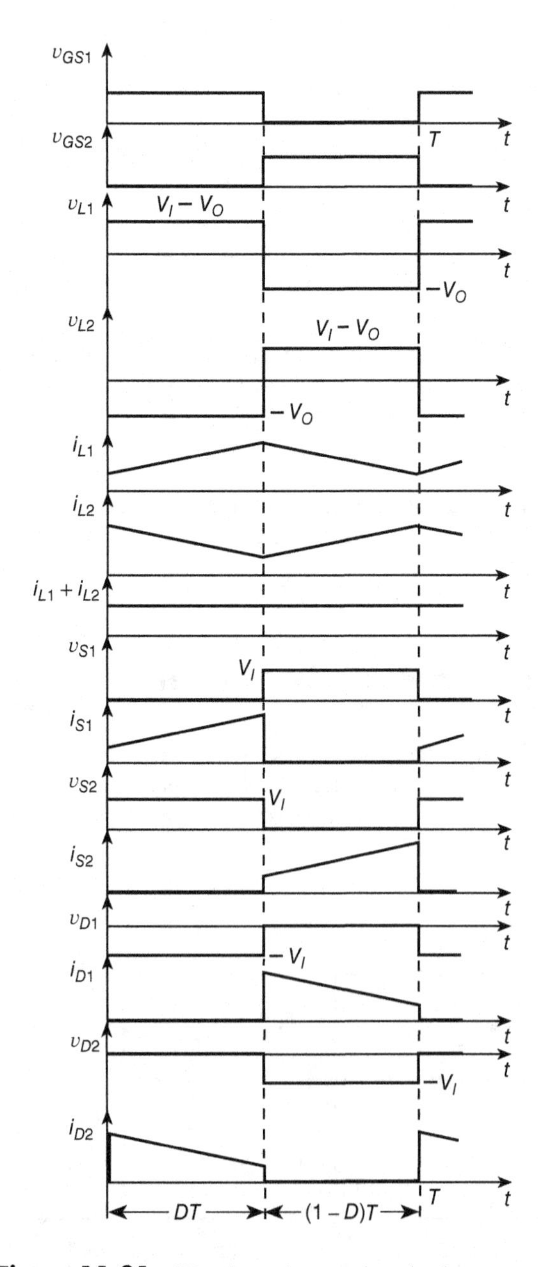

Figure 11.31 Waveforms in two-phase buck converter.

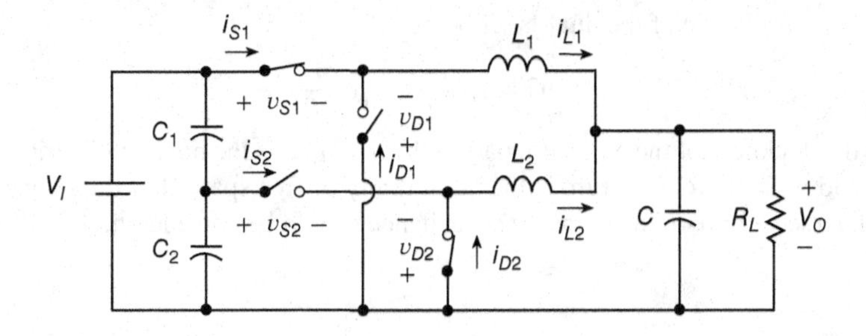

Figure 11.32 Two-phase buck converter with two input capacitors.

through the filter capacitor is also zero, resulting in zero ripple voltage. However, only partial ripple cancellation occurs at $D \neq 0.5$.

If n individual phases are operated in parallel, the frequency of the ripple in the output voltage is n times the switching frequency of each single-phase converter, that is, $f_r = nf_s$. The ripple cancellation occurs at $D = 1/n$. Due to the reduced magnitude and increased frequency of the output voltage ripple, the required filter capacitance is reduced significantly. This improves the transient response of the power supply.

Figure 11.32 shows a two-phase buck converter, with two large capacitors C_1 and C_2 at the input. The voltage stresses of the switches in this converter are reduced. The dc voltage transfer function is

$$M_{VDC} = \frac{V_O}{V_I} = \frac{D}{2} \tag{11.211}$$

11.7 Layout

The layout of the converter components is very important from EMI and power loss point of view. The dc current in each loop distributes to minimize the dc voltage drop around the loop, thus minimizing the dc conduction loss. The ac currents distribute to minimize energy stored in magnetic field for each current harmonic. Thus, the ac current flows through the path of the lowest inductance. Interconnections of components represent impedances. The dc and low-frequency resistance of a cooper trace of resistivity ρ, length l, width w, and thickness h is

$$R_{dc} = \frac{\rho l}{A} = \frac{\rho l}{wh} \tag{11.212}$$

The high-frequency resistance of the trace is

$$R_{HF} = \frac{\rho l}{2(w+h)\delta} = \frac{l}{2(w+l)}\sqrt{\pi \mu_r \mu_0 \rho f} \tag{11.213}$$

where $\delta_w = \sqrt{\rho/\pi\mu_r\mu_0 f}$ is the skin depth. Therefore, the traces between components should be as wide and short as possible. For $w \leq 2l$, the inductance of the rectangular trace is

$$L = \frac{\mu l}{2\pi}\left[\ln\left(\frac{2l}{w+h}\right) + \frac{w+h}{3l} + 0.50049\right] \quad \text{(H)} \tag{11.214}$$

The trace impedance is

$$Z = R + j\omega L \tag{11.215}$$

which for high frequencies is

$$Z_{HF} = R_{HF} + j\omega L = \frac{l}{2(w+l)}\sqrt{\pi \mu_r \mu_0 \rho f} + jf\mu l\left[\ln\left(\frac{2l}{w+h}\right) + \frac{w+h}{3l} + 0.50049\right] \quad (\Omega) \tag{11.216}$$

The parasitic capacitance of the dual-layer board is

$$C = \epsilon_r \epsilon_0 \frac{A}{t} = \epsilon_r \epsilon_0 \frac{wl}{t} \tag{11.217}$$

where t is the thickness of the dielectric part of the board, ϵ_r is the board dielectric constant, and $\epsilon_0 = 10^{-9}/36\pi = 8.85 \times 10^{-12}$ F/m is the permittivity of free space. For dual-layer boards, the effect of the trace width on the parasitic capacitance should be considered.

11.8 Summary

- The PWM buck converter is a step-down converter ($V_O < V_I$).
- The buck converter is a nonisolated converter. It does not provide dc isolation.
- It can operate in two modes: CCM or DCM.
- If the losses are neglected, the dc voltage transfer function of the buck converter is $M_{VDC} = V_O/V_I = D$ for CCM. It is independent of the load resistance R_L (or the load current I_O) and depends only on the switch on-duty cycle D. Therefore, the output voltage $V_O = DV_I$ is independent of the load resistance R_L and depends only on the dc input voltage V_I
- The converter has conduction losses and switching losses.
- For the lossy converter, the dc voltage transfer function of the buck converter is $M_{VDC} = V_O/V_I = \eta D$ for CCM.
- The duty cycle D of the lossy converter is greater than that of the lossless converter at the same dc voltage transfer function M_{VDC}.
- As the duty cycle is increased from 0 to 1, the efficiency increases from 0 to a maximum value, usually above 90%, in dynamic buck power converter with a fixed dc input voltage V_I.
- The buck converter operates most of the time at a low duty cycle, for example, $D = 0.3$, where the power efficiency is relatively low. Therefore, the long-term efficiency is also relatively low.
- The switching loss P_{sw} and gate drive power P_G remain constant for any output voltage. At high frequencies, these two components dominate the overall power loss.
- The peak-to-peak value of the current through the filter capacitor C is equal to the peak-to-peak inductor ripple current Δi_L.
- If the capacitance of the filter capacitor is sufficiently high, the output ripple voltage is determined only by the ESR of the filter capacitor and is independent of the capacitance of the filter capacitor. In order to reduce the output ripple voltage, it is necessary to choose a filter capacitor with a low ESR.
- The minimum value of the inductor is determined by the boundary between CCM and DCM, ripple voltage, or ac losses in the inductor and the filter capacitor.
- The inductance L required for the converter operation in CCM is larger for lossy converter than that for lossless converter.
- A disadvantage of the buck converter is that the input current is pulsating. However, an LC filter can be placed at the converter input to obtain a nonpulsating input current waveform.

- The corner frequency of the output filter $f_o = 1/(2\pi\sqrt{LC})$ is independent of the load resistance.
- It is relatively difficult to drive the transistor because neither the source nor the gate is referenced to ground. Therefore, a transformer, an optical coupler, or charge pump is required in the driver circuit.
- A filter capacitor with a very low ESR is required for the buck converter to achieve a low-ripple voltage.
- Only one-half of the *B*-*H* curve of the inductor core is utilized in the buck converter because the dc current flows through the inductor *L*.

11.9 Review Questions

11.1 Define the converter operation in the CCM and the DCM.

11.2 Is the back converter an isolating converter?

11.3 Is the input current of the basic buck converter pulsating?

11.4 How can the buck circuit be modified to obtain a nonpulsating input current?

11.5 Is the transistor driven with respect to ground in the buck converter?

11.6 How is the dc voltage transfer function M_{VDC} related to the duty cycle D for the lossless buck converter operated in CCM?

11.7 Is the duty cycle D of the lossy buck converter lower or greater than that of the lossless converter at a given value of M_{VDC} for CCM?

11.8 Does the dc voltage transfer function of the buck converter depend on the load resistance?

11.9 What determines the ripple voltage in the buck converter in CCM?

11.10 Is the corner frequency of the output filter dependent on the load resistance in the buck converter?

11.11 Is the efficiency high at heavy or light loads for the buck converter operated in CCM?

11.12 Are both halves of the *B*-*H* curve of the inductor core utilized in the buck converter?

11.13 Is the expression for the minimum inductance for CCM operation the same for the buck converter with a fixed dc input voltage and a variable dc output voltage as that in the buck converter with a fixed dc output voltage and a variable dc output voltage?

11.10 Problems

11.1 Derive an expression for the dc voltage transfer function of the lossless buck converter operating in CCM using the diode voltage waveform.

11.2 Design a buck converter with a variable output voltage to meet the following specifications: $V_I = 12$ V, $V_O = 2$ to 10 V, $P_{Omax} = 4$ W, $f = 20$ MHz, and $V_r/V_O \leq 1\%$.

References

[1] C. Buoli, A. Abbiati, and D. Riccardi, "Microwave power amplifier with "envelope controlled" drain power supply," *European Microwave Conference*, 1995, pp. 31–35.

[2] M. K. Kazimierczuk, *Pulse-Width Modulated DC-DC Power Converters*, 2nd Ed. New York, NY: John Wiley & Sons, 2012.

[3] G. Hanington, P. M. Asbeck, and L. E. Larsen, "High-efficiency power amplifier using dynamic power-supply voltage for CDMA applications," *IEEE Transactions on Microwave Theory and Techniques*, vol. 48, no. 8, pp. 1471–1476, 1999.

[4] B. Sahu and G. A. Rincon-Mora, "A high-efficiency linear RF power amplifier with a power-tracking dynamically adoptive buck-boost supply," *IEEE Transactions on Microwave Theory and Techniques*, vol. 52, no. 1, pp. 112–120, 2004.

[5] N. Wang. V. Yousefzadehi, D. Maksimović, S. Paijć, and Z. B. Popović, "60% efficient 10-MHz power amplifier with dynamic drain bias control," *IEEE Transactions on Microwave Theory and Techniques*, vol. 52, no. 3, pp. 1077–1081, 2004.

[6] Y. Eo and K. Lee, "High-efficiency 5 GHz CMOS power amplifier with adoptive bias control circuit," *IEEE RFIC Symposium Digest*, June 2004, pp. 575–578.

[7] I.-H. Chen, K. U. Yen, and J. S. Kenney, "An envelope elimination and restoration power amplifier using a CMOS dynamic power supply circuit," *IEEE MTT-S International Microwave Symposium Digest*, vol. 3, June 2004, pp. 1519–1522.

[8] N. Schlumpf, M. Declercq, and C. Dehollain, "A fast modulator for dynamic supply linear RF power amplifier," *IEEE Journal of Solid-State Circuits*, vol. 39, no. 8, pp. 1015–1025, 2004.

[9] P. Hazucha, G. Schrom, J. Hahn, B. A. Bloechel, P. Hack, G. E. Dermer, S. Narendra, D. Gardner, T. Karnik, V, De, and S. Borker, "A 233-MHz 80%-87% efficiency four-phase dc-dc converter utilizing air-core inductors on package," *IEEE Journal of Solid-State Circuits*, vol. 40, no. 4, pp. 838–845, 2005.

[10] W.-Y. Chu, P. M. Bakkaloglu, and S. Kiaei, "A 10 MHz bandwidth, 2 mV ripple PA for CDMA transmitter," *IEEE Journal of Solid-State Circuits*, vol. 43, no 12, pp. 2809–2819, 2008.

[11] V. Pinon, F. Hasbani, A. Giry. D. Pache, and C. Garnier, "A single-chip WCDMA envelope reconstruction LDMOS PA with 130 MHz switched-mode power supply," *International Solid-State Circuits Conference*, 2008, pp. 564–636.

[12] J. Wibben and R. Harjani, "A high-efficiency dc/dc converter using 2 nH integrated inductor," *IEEE Journal of Solid-State Circuits*, vol. 43, no. 4, pp. 844–854, 2008.

[13] J. Sun, D. Giuliano, S. Deverajan, J.-Q. Lu, T. P. Chow, and R. J. Gutmann, "Fully monolithic cellular buck converter design for 3-G power delivery," *IEEE Transactions on Very Large Scale Integration (VLSI) Systems*, vol. 17, no. 3, pp. 447–451, 2009.

[14] R. Shrestham, R. van der Zee, A. de Graauw, and B. Nauta, "A wideband supply modulator for 20 MHz RF bandwidth polar PAs in 65 nm CMOS," *IEEE Journal of Solid-State Circuits*, vol. 44, no. 4, pp. 1272–1280, 2009.

[15] J. Jeong, D. F. Kimball, P. Draxler, and P. M. Asbeck, "Wideband envelope tracking power amplifier with reduced bandwidth power supply waveforms and adoptive digital predistortion technique," *IEEE Transactions on Microwave Theory and Techniques*, vol. 57, no. 12, pp. 3307–3314, 2009.

[16] R. D. Middlebrook and S. Ćuk, *Advances in Switched-Mode Power Conversion*, vols. I, II, and III. Pasadena, CA: TESLAco, 1981.

[17] O. A. Kossov, "Comparative analysis of chopper voltage regulators with *LC* filter," *IEEE Transactions on Magnetics*, vol. MAG-4, no. 4, pp. 712–715, 1968.

[18] *The Power Transistor and Its Environment*. Thomson-CSF, SESCOSEM Semiconductor Division, 1978.

[19] E. R. Hnatek, *Design of Solid-State Power Supplies*, 2nd Ed. New York, NY: Van Nostrand, 1981.

[20] K. K. Sum, *Switching Power Conversion*. New York, NY: Marcel Dekker, 1984.

[21] G. Chryssis, *High-Frequency Power Supplies: Theory and Design*. New York, NY: McGraw-Hill, 1984.

[22] R. P. Severns and G. Bloom, *Modern DC-to-DC Switchmode Power Converter Circuits*. New York, NY: Van Nostrand, 1985.

[23] O. Kilgenstein, *Switching-Mode Power Supplies in Practice*. New York, NY: John Wiley & Sons, 1986.

[24] D. M. Mitchell, *Switching Regulator Analysis*. New York, NY: McGraw-Hill, 1988.

[25] K. Billings, *Switchmode Power Supply Handbook*. New York, NY: McGraw-Hill, 1989.

[26] M. H. Rashid, *Power Electronics, Circuits, Devices, and Applications*, 3rd Ed. Upper Saddle River, NJ: Prentice Hall, 2004.

[27] N. Mohan, T. M. Undeland, and W. P. Robbins, *Power Electronics: Converters, Applications and Design*, 3rd Ed. New York, NY: John Wiley & Sons, 2004.

[28] J. G. Kassakian, M. F. Schlecht, and G. C. Verghese, *Principles of Power Electronics*. Reading, MA: Addison-Wesley, 1991.

[29] A. Kislovski, R. Redl, and N. O. Sokal, *Analysis of Switching-Mode DC/DC Converters*. New York, NY: Van Nostrand, 1991.

[30] A. I. Pressman, *Switching Power Supply Design*. New York, NY: McGraw-Hill, 1991.

[31] M. J. Fisher, *Power Electronics*. Boston, MA: PWS-Kent, 1991.

[32] B. M. Bird, K. G. King, and D. A. G. Pedder, *An Introduction to Power Electronics*. New York, NY: John Wiley & Sons, 1993.

[33] D. W. Hart, *Introduction to Power Electronics*. Upper Saddle River, NJ: Prentice Hall, 1997.

[34] R. W. Erickson and D. Maksimović, *Fundamentals of Power Electronics*. Norwell, MA: Kluwer Academic Publisher, 2001.

[35] I. Batarseh, *Power Electronic Circuits*. New York, NY: John Wiley & Sons, 2004.

[36] A. Aminian and M. K. Kazimierczuk, *Electronic Devices: A Design Approach*. Upper Saddle River, NJ: Prentice Hall, 2004.

[37] M. K. Kazimierczuk and D. Czarkowski, *Resonant Power Converters*, 2nd Ed. New York, NY: John Wiley & Sons, 2011,

[38] A. Reatti, "Steady-state analysis including parasitic components and switching losses of buck and boost dc-dc converter," *International Journal of Electronics*, vol. 77, no. 5, pp. 679–702, 1994.

[39] M. K. Kazimierczuk, "Reverse recovery of power *pn* junction diodes," *International Journal of Circuits, Systems, and Computers*, vol. 5, no. 4, pp. 747–755, 1995.

[40] D. Maksimović and S. Ćuk, "Switching converters with wide dc conversion range," *IEEE Transactions on Power Electronics*, vol. 6, no. 1, pp. 151–157, 1991.

[41] D. A. Grant and Y. Darraman, "Watkins-Johnson converter completes tapped inductor converter matrix," *Electronic Letters*, vol. 39, no. 3, pp. 271–272, 2003.

[42] T. H. Kim, J. H. Park, and B. H. Cho, "Small-signal modeling of the tapped-inductor converter under variable frequency control," *IEEE Power Electronics Specialists Conference*, pp. 1648–1652, 2004.

[43] K. Yao, M. Ye, M. Xu, and F. C. Lee, "Tapped-inductor buck converter for high-step-down dc-dc conversion," *IEEE Transactions on Power Electronics*, vol. 20, no. 4, pp. 775–780, 2005.

[44] B. Axelord, Y. Berbovich, and A. Ioinovici, "Switched-capacitor/switched-inductor structure for getting transformerless hybrid dc-dc PWM converters," *IEEE Transactions on Circuits and Systems*, vol. 55, no. 2, pp. 687–696, 2008.

[45] Y. Darroman and A. Ferré, "42-V/3-V Watkins-Johnson converter for automotive use," *IEEE Transactions on Power Electronics*, vol. 21, no. 3, pp. 592–602, 2006.

[46] D. Czarkowski and M. K. Kazimierczuk, "Static- and dynamic-circuit models of PWM buck-derived converters," *IEE Proceedings, Part G, Devices, Circuits and Systems*, vol. 139, no. 6, pp. 669–679, 1992.

[47] P. A. Dal Fabbro, C. Meinen, M. Kayal, K. Kobayashi, and Y. Watanabe, "A dynamic supply CMOS RF power amplifier for 2.4 GHz and 5.2 GHz frequency bands," *Proceedings of IEEE Radio Frequency Integrated Circuits (RFIC) Symposium*, 2006, pp. 144–148.

[48] T.-W. Kwak, M.-C. Lee, and G.-H. Cho, "A 2 WCMOS hybrid switching amplitude modulator for edge polar transmitters," *IEEE Journal of Solid-State Circuits*, vol. 42, no 12, pp. 2666–2676, 2007.

[49] M. C. W. Hoyerb and M. A. E. Anderesn, "Ultrafast tracking power supply with fourth-order output filter and fixed-frequency hysteretic control," *IEEE Transactions on Power Electronics*, vol. 23, no. 5, pp. 2387–2398, 2008.

[50] M. Vlasić, O. Garcia, J. A. Oliver, P. Alou, D. Diaz, and J. A. Cobos, "Multilevel power supply for high-efficiency RF amplifiers," *IEEE Transactions on Power Electronics*, vol. 25, no. 4, pp. 1078–1089, 2010.

[51] P.Y. Wu and P. K. T. Mok, "A two-phase switching hybrid supply modulator for RF power amplifier with 9% efficiency improvement," *IEEE Journal of Solid-State Circuits*, vol. 45. no. 12, pp. 2543–2556, 2010.

[52] B. Kim, J. Moon, and I. Kim, "Efficiently amplified," *IEEE Microwave Magazine*, pp. 87–100, August 2010.

[53] C. Hsia, A. Zhu, J. J. Yan, P. Draxel, D. Kimball, S. Lanfranco, and P. A. Asbeck, "Digitally assisted dual-switch high-efficiency envelope amplifier," *IEEE Transactions on Microwave Theory and Technique*, vol. 59, no. 11, pp. 2943–2952, 2011.

[54] M. Bathily, B. Allard, F. Hasbani, V. Pinon, and J. Vedier, "Design flow for high switching frequency and large-bandwidth analog dc/dc step-down converters for a polar transmitter," *IEEE Transactions on Power Electronics*, vol. 27, no. 2, pp. 838–847, 2012.

[55] E. McCune, "Envelope tracking or polar - Which is it?," *IEEE Microwave Magazine*, pp. 54–56, 2012.

[56] S. Shinjo, Y.-P. Hong, H. Gheidi, D. F. Kimball, and P. A. Asbeck, "High speed, high analog bandwidth buck converter using GaN HEMTs for envelope tracking power amplifier applications," *IEEE Topical Conference on Wireless Sensors and Sensor Networks (WiSNet)*, 2013, pp. 13–15.

[57] T. Salvatierra and M. K. Kazimierczuk, "Inductor design for PWM buck converter operated as dynamic supply or amplitude modulator for RF transmitters," *56th IEEE Midwest Symposium on Circuits and Systems*, Columbus, OH, August 3-7, 2013, pp. 37–40.

[58] A. Ayachit and M. K. Kazimierczuk, "Two-phase buck converter as a dynamic power supply for RF power amplifiers applications," *56th IEEE Midwest Symposium on Circuits and Systems*, Columbus, OH, August 3-7, 2013, pp. 493–480.

[59] O. Garcia, M. Vlasić, P. Alou, J. A. Oliver, and J. A. Cobos, "An overview of fast dc-dc converters for envelope amplifiers in RF amplifiers," *IEEE Transactions on Power Electronics*, vol. 28, no. 10, pp. 4712–4720, 2013.

[60] M. Vasić, O. Garcia, J. A. Oliver, P. Alau, and J. A. Cobos, "Theoretical efficiency limits of a serial and parallel linear-assisted switching converter as an envelope amplifier," *IEEE Transactions on Power Electronics*, vol. 29, no. 2, pp. 719–728, 2014.

[61] P. F. Miaja, J. Sebastián, R. Marante, and J. A. Garcia, "A linear assisted switching envelope amplifier for a UHF polar transmitters," *IEEE Transactions on Power Electronics*, vol. 29, no. 4, pp. 1850–1858, 2014.

12

Oscillators

12.1 Introduction

An oscillator is a circuit that produces a periodic output signal without any ac input signal. Sinusoidal-tuned *LC* resonant oscillators [1–58] are widely used in instrumentation and test equipment as function generators and local signal oscillators in amplitude modulation (AM), frequency modulation (FM), and wireless receivers and transmitters operated in a wide range of frequencies. In radio systems, sinusoidal oscillators establish the transmitter carrier frequency and drive the mixer stages that convert signals from one frequency to another. *LC* oscillators also form an integral part of voltage-controlled oscillators (VCOs) used in communication and digital systems. Power *LC* oscillators also find a wide range of applications, such as radio transmitters, radars, and electronic ballasts. When the frequency of oscillations is required to be accurate and stable, crystal oscillators are used. In these oscillators, the oscillation frequency is precisely controlled by the vibrations of a quartz crystal. The crystal is a mechanical resonator. Electronic watches and other critical timing applications use crystal oscillators because they provide an accurate clock frequency. *LC* tuned oscillators are also used as power oscillators because power losses in reactive components are low, yielding high efficiency. An electronic oscillator is a nonlinear circuit with at least two energy storage components. A resonant circuit is a fundamental building block of *LC* oscillators because it acts a bandpass filter and determines the frequency of oscillations. In many circuits, it also forms a part of the feedback network or the amplifier or both. Operation of oscillators is based on the following principles:

- Positive feedback.
- Negative resistance.

Oscillators with positive feedback consist of a forward two-port network and a positive feedback two-port network. An oscillator family may be divided into three groups:

- Energy storage components are only in the positive feedback two-port network.
- Energy storage components are only in the forward two-port network.
- Energy storage components are in both the forward two-port network and the positive feedback two-port network.

RF Power Amplifiers, Second Edition. Marian K. Kazimierczuk.

The major parameters of oscillators are as follows:

- the oscillation frequency f_o,
- the frequency stability,
- the spectral purity of the output voltage (the content of harmonics),
- the output voltage, current, or power amplitude,
- the output voltage stability, and
- the range of frequency in VCOs.

In this chapter, we will consider the principle of operation of *LC* tuned oscillators with positive feedback and with a negative resistance. We will present various topologies of oscillators, using op-amps and single-transistor amplifiers.

12.2 Classification of Oscillators

Oscillators may be classified broadly into the following categories:

- sinusoidal oscillators,
- rectangular oscillators,
- triangular oscillators,
- fixed frequency oscillators,
- variable frequency oscillators,
- voltage-controlled (VC) oscillators,
- *LC* oscillators,
- *RC* oscillators,
- monostable and astable multivibrators,
- crystal oscillators,
- ring oscillators,
- signal oscillators,
- power oscillators.

12.3 General Conditions for Oscillations

12.3.1 Transfer Functions of Oscillators

A block diagram of an oscillator with positive feedback is shown in Fig. 12.1. The oscillator consists of an amplifier and a feedback network. The feedback network β in the oscillator produces an input voltage v_f to the amplifier A, which in turn produces an input voltage v_o to the feedback network. The input voltage of the amplifier (the error voltage) is

$$v_e = v_s + v_f \tag{12.1}$$

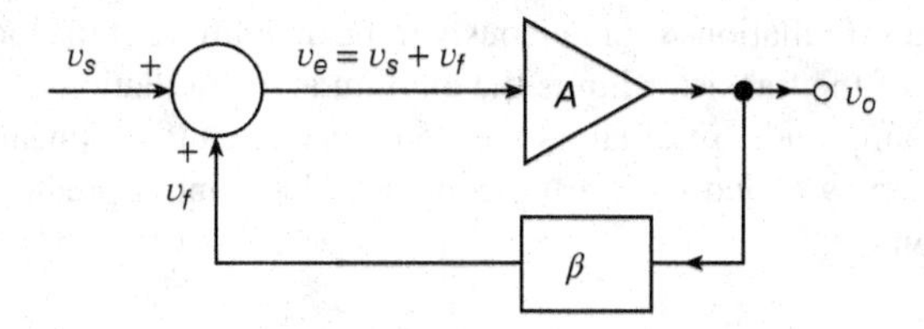

Figure 12.1 Block diagram of an oscillator with positive feedback.

The *amplifier voltage gain*, or the forward gain, is defined as

$$A = \frac{v_o}{v_e} = |A|e^{j\phi_A} \tag{12.2}$$

and the voltage transfer function of the feedback network, called the *feedback factor* is defined as

$$\beta = \frac{v_f}{v_o} = |\beta|e^{j\phi_\beta} \tag{12.3}$$

The output voltage of the oscillator is

$$v_o = Av_e = A(v_s + v_f) = Av_s + \beta A v_o \tag{12.4}$$

resulting in

$$v_o(1 - \beta A) = Av_s \tag{12.5}$$

Hence, the closed-loop voltage gain of the oscillator is

$$A_f = \frac{v_o}{v_s} = \frac{A}{1 - \beta A} = \frac{A}{1 - T} \tag{12.6}$$

where $1 - T$ is called the *amount of feedback.*

The *loop gain* of an oscillator is defined as

$$T = \frac{v_o}{v_s} = \frac{v_o}{v_f}\frac{v_f}{v_s} = \beta A = |T|e^{j\phi_T} \tag{12.7}$$

In an oscillator, a finite and nonzero output voltage v_o exists with no input voltage v_s applied, that is, $v_o \neq 0$ at $v_s = 0$. Hence, the amount of feedback required for steady-state oscillations is

$$1 - T(s) = 1 - \beta(s)A(s) = 0 \tag{12.8}$$

yielding

$$T(s) = \beta(s)A(s) = 1 \tag{12.9}$$

For steady-state oscillations with a constant amplitude of the output voltage, $s = j\omega_o$. Therefore,

$$1 - T(j\omega_o) = 1 - \beta(j\omega_o)A(j\omega_o) = 0 \tag{12.10}$$

The loop gain T must be unity at the oscillation frequency f_o for steady-state oscillations with a constant amplitude of the output voltage

$$T(j\omega_o) = \beta(j\omega_o)A(j\omega_o) = 1 \tag{12.11}$$

This relation is the Barkhausen criterion for sustained oscillation. The startup condition for oscillations is

$$T(\omega_o) = \beta(j\omega_o)A(j\omega_o) > 1 \tag{12.12}$$

Free electrons in resistors and transistors experience frequent collisions and generate noise within the oscillator circuit. The noise voltage is amplified by the amplifier in an oscillator. When a power supply of the oscillator is turned on, the transient signal is generated and the thermal noise voltage initiates oscillations spontaneously if the magnitude of the loop gain is greater than unity. Nonlinearity of the transistor limits the amplitude of oscillations, and the output voltage reaches a constant amplitude. Oscillators are nonlinear circuits. A linear analysis can predict the oscillation frequency f_o and the startup condition, but cannot predict the amplitude of the oscillator output voltage.

12.3.2 Polar Form of Conditions for Oscillation

The condition for steady-state oscillation at the oscillation frequency f_o is given by

$$T(f_o) = \beta(f_o)A(f_o) = |T(f_o)|e^{\phi_T(f_o)} = 1 = 1e^{j0} \tag{12.13}$$

The Barkhausen magnitude criterion for steady-state oscillation at the oscillation frequency f_o is given by

$$|T(f_o)| = 1 \tag{12.14}$$

and the Barkhausen phase criterion for sustained steady-state oscillation

$$\phi_T(f_o) = \phi_A + \phi_\beta = 0^\circ \pm 360^\circ n \quad \text{at} \quad f \neq 0 \quad \text{for} \quad n = 1, 2, 3, \ldots \tag{12.15}$$

At the oscillation frequency f_o, the magnitude of the loop gain must be unity and the phase shift (delay) around the loop must be zero or a multiple of 360°. A single-frequency solution is desired to achieve spectral purity. The Barkhausen criterion is only a necessary condition for oscillation, but not a sufficient one.

12.3.3 Rectangular Form of Conditions for Oscillation

The loop gain for steady-state oscillation can be expressed as

$$T(f_o) = \beta(f_o)A(f_o) = |T(f_o)|e^{\phi_T(f_o)} = Re\{T(f_o)\} + jIm\{T(f_o)\} = 1 + j0 \tag{12.16}$$

The criterion of the real part of loop gain for oscillation is given by

$$Re\{T(f_o)\} = 1 \tag{12.17}$$

and the criterion of the imaginary part of the loop gain for oscillation is given by

$$Im\{T(f_o)\} = 0 \tag{12.18}$$

At the oscillation frequency f_o, the real part of the loop gain must be unity and the imaginary part of the loop gain must be zero. One of these conditions determines the oscillation frequency the other the amplitude of the oscillator output voltage. The necessary condition for the startup of oscillation is the loop gain T greater than unity. Thermal noise causes oscillations to start.

Referring to Fig. 12.1,

$$v_f = \beta v_o \tag{12.19}$$

and

$$v_o = Av_f = \beta A v_o \tag{12.20}$$

Hence, the condition for sustained oscillation is

$$T = \beta A = 1 \tag{12.21}$$

12.3.4 Closed-Loop Gain of Oscillators

The gain of the amplifier can be expressed as

$$A(s) = \frac{N_A(s)}{D_A(s)} \tag{12.22}$$

The transfer function of the feedback network can be expressed as

$$\beta(s) = \frac{N_\beta(s)}{D_\beta(s)} \tag{12.23}$$

The loop gain is

$$T(s) = \beta(s)A(s) = \frac{N_\beta(s)N_A(s)}{D_\beta(s)D_A(s)} = \frac{N_T(s)}{D_T(s)} \tag{12.24}$$

The closed-loop transfer function of an oscillator is

$$A_f(s) = \frac{A(s)}{1 - T(s)} = \frac{A(s)}{1 - \beta(s)A(s)} = \frac{\frac{N_A(s)}{D_A(s)}}{1 - \left[\frac{N_\beta(s)}{D_\beta(s)}\right]\left[\frac{N_A(s)}{D_A(s)}\right]} = \frac{\frac{N_A(s)}{D_A(s)}}{\frac{D_\beta(s)D_A(s) - N_\beta(s)N_A(s)}{D_\beta(s)D_A(s)}}$$

$$= \frac{N_A(s)D_\beta(s)}{D_\beta(s)D_A(s) - N_\beta(s)N_A(s)} = \frac{D_T(s)}{D_T(s) - N_T(s)} = \frac{N_{Af}(s)}{D_{Af}(s)} \tag{12.25}$$

where $N_{Af}(s) = D_\beta(s)N_A(s)$ and $D_{Af}(s) = D_T(s) - N_T(s)$. The oscillator circuit is unstable and is able to start oscillations if the closed-loop gain $A_f(s)$ has poles in the right-half plane (RHP). The poles of the loop gain $T(s)$ are the same as the poles of $1 - T(s)$. If $T(s)$ has poles in the RHP, then $1 - T(s)$ also has poles in the RHP. The oscillator circuit is unstable (i.e., $A_f(s)$ has poles in the RHP) if $1 - T(s)$ has zeros in the RHP.

12.3.5 Characteristic Equation of Oscillators

The characteristic equation of an oscillator is defined as the equation obtained by setting the denominator polynomial of the closed-loop transfer function to zero, that is, $D_{Af}(s) = 0$. Thus, the roots of the characteristic equation are the poles of the closed-loop transfer function of an oscillator $A_f(s)$. The roots of the characteristic equation determine the character of the time response of the closed-loop system. The loop gain can be expressed in terms of the nominator $N_T(s)$ and the denominator $D_T(s)$

$$1 - T(s) = 1 - \frac{N_T(s)}{D_T(s)} = \frac{D_T(s) - N_T(s)}{D_T(s)} = \frac{D_\beta(s)D_A(s) - N_\beta(s)iN_A(s)}{D_\beta(s)D_A(s)} = 0 \tag{12.26}$$

The poles of $1 - T(s)$ are the same as the poles of the loop gain $T(s)$ because in both cases they are determined by $D_T(s) = D_\beta(s)D_A(s) = 0$. Therefore, if $T(s)$ has no poles in the RHP, then $1 - T(s)$ has no poles in the RHP.

The *characteristic equation* of an oscillator is

$$D_{Af}(s) = D_T(s) - N_T(s) = 0 \tag{12.27}$$

which can be expanded to the form

$$D_{Af}(s) = D_\beta(s)D_A(s) - N_\beta(s)N_A(s) = D_T(s) - N_T(s) = 0 \tag{12.28}$$

The roots of the nominator of the oscillator characteristic equation are equal to the poles of the closed-loop gain $A_f(s)$. The stability of the closed-loop gain $A_f(s)$ is determined by the roots (zeros) of $1 - T(s)$ if there is no cancellation of RHP poles and zeros.

As the amplitude of the oscillator output voltage v_o increases, σ decreases and the pair of complex conjugate poles moves from the RHP toward the imaginary axis. For steady-state oscillation, $\sigma = 0$ and $s = j\omega_o$. Hence,

$$D_{Af}(j\omega_o) = D_T(j\omega_o) - N_T(j\omega_o) = Re\{D_f(\omega_o)\} + jIm\{D_f(\omega_o)\} = 0 = 0 + j0 \quad (12.29)$$

yielding

$$Re\{D_f(\omega_o)\} = 0 \quad (12.30)$$

and

$$Im\{D_f(\omega_o)\} = 0 \quad (12.31)$$

Solution of (12.31) produces an expression for the steady-state oscillation frequency f_o and solution of (12.30) and substitution of $\omega = \omega_o$ into the result produces an expression for the magnitude condition for steady-state oscillation $|T(f_o)| = 1$.

The characteristic equation can be also derived from a small-signal model of an oscillator. The oscillator model can be described by a set of equations, using KCL or KLV. Since the oscillator circuit has no external excitation, the determinant of the matrix representing the set of equations is $\Delta(s) = D_f(s) = 0$.

12.3.6 Instability of Oscillators

An oscillator circuit must be unstable to start oscillations and marginally stable to sustain oscillations. For oscillations to start and build up to a desired level, it must initially have poles located in the RHP for very small amplitude of the output voltage v_o. The necessary and sufficient condition for steady-state sinusoidal oscillations is the location of a pair of complex conjugate poles on the imaginary axis with no poles present in the RHP. The following methods can be utilized for the investigation of unstability of oscillators:

- Bode plots of loop gain $T(s)$.
- Nyquist plots of loop gain $T(j\omega)$.
- Root locus of the characteristic equation of the closed-loop gain $A_f(s)$.
- Routh–Hurwitz criterion of stability using the characteristic equation $D_{Af}(s)$.

12.3.7 Root Locus of Closed-Loop Gain

The root locus of the closed-loop system is a set of trajectories of the roots of the characteristic equation in the complex-frequency s-plane as a system parameter is changed. It allows us to observe the motion of the poles of the closed-loop gain $A_f(s)$ in the complex s-plane as a function of the amplifier gain. The closed-loop transfer function of an oscillator is

$$A_f(s) = \frac{A(s)}{1 + T(s)} = \frac{A(s)}{1 + KP(s)} = \frac{N_{Af}(s)}{D_{Af}(s)} \quad (12.32)$$

where $P(s)$ is not a function of K and K is not a function of s. The denominator of the closed-loop transfer function, which is the characteristic equation, can be written as

$$D_{Af}(s) = 1 + T(s) = 1 + KP(s) = 1 + K\frac{N_P(s)}{D_P(s)} = \frac{D_P(s) + KN_P(s)}{D_P(s)} \quad (12.33)$$

Thus, the poles of the characteristic equation are the roots of the following equation:

$$D_P(s) + KP(s) = (s - p_1)(s - p_2)(s - p_3)\cdots = 0 \quad (12.34)$$

When K varies, the locations of the poles also vary. The trajectories of these poles form a root locus of the poles of the characteristic equation of the closed-loop transfer function $A_f(s)$.

To start and sustain oscillation, the circuit must be unstable. This means that the closed-loop transfer function $A_f(s)$ must have a pair of complex conjugate poles in the RHP, as shown in Fig. 12.2. The stability of the closed-loop gain of oscillators $A_f(s)$ is determined by the location of the poles of the denominator of $A_f(s)$, equal to $1 - \beta(s)A(s) = 1 - T(s)$. The poles of $1 - T(s)$ are the same as those of $T(s)$.

The dynamic behavior of an oscillator circuit can be described in terms of two parameters: the *damping ratio* ζ and the *natural undamped frequency* ω_o. These parameters lead to two other parameters: the *damping coefficient*

$$\sigma = -\zeta\omega_o \quad (12.35)$$

and the *natural damped frequency*

$$\omega_d = \omega_o\sqrt{1 - \zeta^2} \quad (12.36)$$

The oscillation process consists of three stages: startup, transient, and steady-state operation.

Startup of Oscillation. Figure 12.2 shows the trajectories of the poles for an oscillator during the startup when the oscillation amplitude is growing. In this case, the closed-loop transfer function $A_f(s)$ must contain a term

$$\frac{\omega_d}{(s-\sigma)^2 + \omega_d^2} = \frac{\omega_d}{(s+\zeta\omega_o)^2 + \omega_d^2} = \frac{\omega_d}{s^2 + 2\zeta\omega_o s + (\zeta\omega_o)^2 + \omega_o^2(1-\zeta^2)}$$
$$= \frac{\omega_d}{(s-p_1)(s-p_2)} = \frac{\omega_d}{[s - (\sigma + j\omega_d)][s - (\sigma - j\omega_d)]}$$
$$= \frac{\omega_d}{[s - (-\zeta\omega_o + j\omega_d)][s - (-\zeta\omega_o - j\omega_d)]}. \quad (12.37)$$

To start the oscillation, the damping ratio must be negative

$$\zeta < 0 \quad (12.38)$$

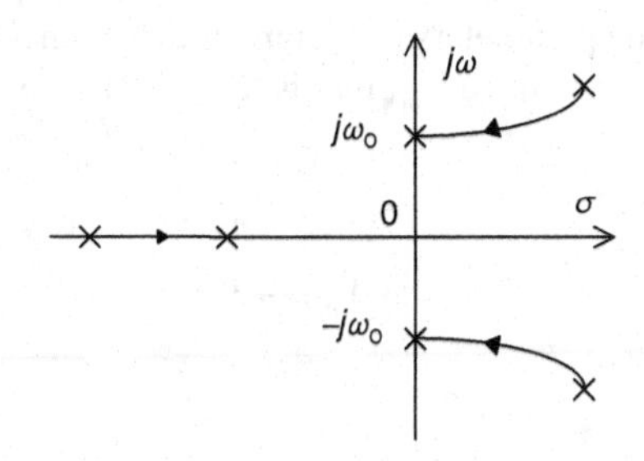

Figure 12.2 Trajectories of the poles for an oscillator when the oscillation amplitude is growing.

yielding a positive damping coefficient

$$\sigma = -\zeta\omega_o > 0 \tag{12.39}$$

Under these conditions, the denominator of the closed-loop transfer function $A_f(s)$ must contain at least a pair of complex conjugate poles located in the RHP

$$p_1, p_2 = \sigma \pm j\omega_o = -\zeta\omega_o \pm j\omega_o\sqrt{1-\zeta^2} \tag{12.40}$$

For this form of the closed-loop transfer function, noise inherent in the oscillator components (in transistors and resistors) is amplified around the loop, causing the amplitude of the output voltage to grow. In addition, when the dc power supply is turned on, the transient in the circuit may start oscillations. After oscillations will start growing, the output voltage contains the following component:

$$\mathcal{L}^{-1}\left\{\frac{\omega_d}{(s-\sigma)^2+\omega_d^2}\right\} = e^{\sigma t}\sin(\omega_d t) = e^{-\zeta\omega_o t}\sin(\omega_o\sqrt{1-\zeta^2 t}) \quad \text{for} \quad t \geq 0 \tag{12.41}$$

For the oscillator with a parallel LCR resonant circuit, the oscillation frequency is equal to the *natural damped frequency*

$$\omega_d = \sqrt{\frac{1}{LC} - \frac{1}{4R^2C^2}} = \sqrt{\frac{1}{LC}\left(1 - \frac{L}{4R^2C}\right)} = \sqrt{\frac{1}{LC}\left(1 - \frac{Z_o^2}{4R^2}\right)} = \sqrt{\frac{1}{LC}\left(1 - \frac{1}{4Q_L^2}\right)}$$

$$= \omega_o\sqrt{1 - \frac{1}{4Q_L^2}} = \omega_o\sqrt{1-\zeta^2} \tag{12.42}$$

where the *natural undamped frequency* is

$$\omega_o = \frac{1}{\sqrt{LC}} \tag{12.43}$$

and the loaded quality factor of the parallel resonant circuit is

$$Q_L = \frac{R}{\omega_o L} = \omega_o CR = \frac{R}{Z_o} \tag{12.44}$$

and the characteristic equation is

$$Z_o = \sqrt{\frac{L}{C}} \tag{12.45}$$

Transient Oscillation. After the oscillation startup, ζ and σ decrease to 0, the oscillation frequency ω_d approaches ω_o, and the pair of the complex conjugate poles is moving toward the imaginary value $j\omega$. As a result, the amplitude of the oscillator output voltage is growing. During the transient time interval, the parameters describing the dynamics of an oscillator are functions of the amplitude of the output voltage V_m, that is, $\zeta = f(V)m)$, $\sigma = f(V_m)$, and $\omega = f(V_m)$. All these parameters approach 0, that is,

$$\zeta(V_m) \to 0 \tag{12.46}$$

$$\sigma(V_m) \to 0 \tag{12.47}$$

and

$$\omega(V_m) \to 0 \tag{12.48}$$

Steady-State Oscillation. Figure 12.3 shows the location of poles for an oscillator in steady state, that is, with a constant amplitude of the output voltage v_o. For steady-state (sustained) oscillations,

$$\zeta = 0, \tag{12.49}$$

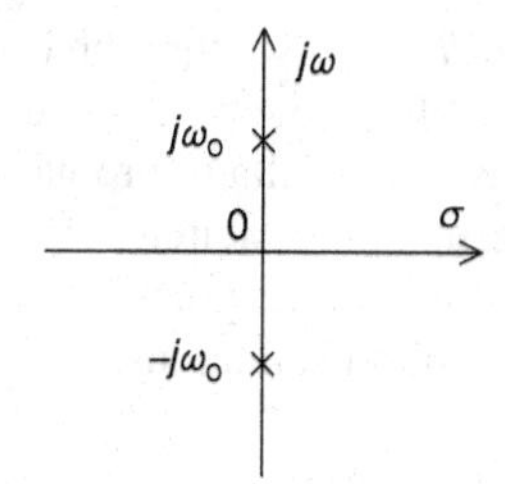

Figure 12.3 Locations of the poles for an oscillator for steady-state oscillations (i.e., when the circuit is marginally stable).

resulting in

$$\sigma = -\zeta\omega_o = 0 \tag{12.50}$$

and

$$\omega_d = \omega_o \tag{12.51}$$

Therefore, the amplitude of the output voltage is constant. In this case, the closed-loop transfer function $A_f(s)$ contains a term

$$\frac{\omega_o}{s^2 + \omega_o^2} = \frac{\omega_o}{(s - p_1)(s - p_2)} = \frac{\omega_o}{(s - j\omega_o)(s + j\omega_o)} \tag{12.52}$$

Hence, the characteristic equation of an oscillator has a pair of complex conjugate roots located on the imaginary axis

$$p_1, p_2 = \pm j\omega_o \tag{12.53}$$

The output voltage contains the term

$$\mathcal{L}^{-1}\left\{\frac{\omega_o}{s^2 + \omega_o^2}\right\} = \sin(\omega_o t) \quad \text{for} \quad t \geq 0 \tag{12.54}$$

12.3.8 Nyquist Plot of Oscillators

A Nyquist plot of an oscillator is a polar contour of the loop gain $T(j\omega)$ for $\sigma = 0$ and $s = j\omega$ as the frequency increases from $-\infty$ to ∞. It is the mapping or transformation of the s-plane into the $Re\{T(j\omega)\}$-$Im\{T(j\omega)\}$ plane. The loop gain can be expressed as

$$T(j\omega) = |T(\omega)|e^{\phi_T(\omega)} = |T(\omega)|\cos\phi_T(\omega) + j|T(\omega)|\sin\phi_T(\omega) = Re\{T(j\omega)\} + jIm\{T(j\omega)\} \tag{12.55}$$

The Nyquist plot is a plot of $Im\{T(j\omega)\}$ as a function of $Re\{T(j\omega)\}$. H. Nyquist developed the criterion of stability in 1932. The Nyquist plot for frequencies from $-\infty$ to 0 is a mirror about the real axis with that for frequencies from 0 to ∞. Therefore, it is often sufficient to draw the Nyquist plot for frequencies from 0 to ∞. The symmetry of the two parts of the Nyquist plot follows from the fact that $T(-j\omega) = T^*(j\omega)$. This is because the $T(s)$ has real coefficients, which are determined by the oscillator components. The Nyquist plot can be used as a test of stability of a circuit. The principle of the argument in the complex variable theory gives the relationship between the poles and zeros of the loop gain $T(j\omega)$ enclosed by the contour and the number of times the contour will encircle the origin. If the contour of the loop gain $T(j\omega)$ encircles the point $-1 + j0$ in a clockwise direction, then the oscillator circuit has poles in the RHP, is unstable, and is able to start oscillations. If this contour crosses the point $-1 + j0$, the circuit is marginally stable and produces steady-state oscillations.

If the Nyquist plot of the loop gain $T(s)$ passes through the $-1 + j0$ point, then the zeros of the characteristic equation $1 - T(s)$, which are the poles of the closed-loop gain $A_f(s)$, are located on the imaginary $j\omega$ axis. In this case, an oscillator produces steady-state oscillations.

If the Nyquist plot of the loop gain $T(j\omega)$ encircles the $-1 + j0$ point, then the zeros of the characteristic equation $1 - T(s)$, which are the poles of the closed-loop gain $A_f(s)$, are located in the RHP, and the circuit is able to start oscillations.

12.3.9 Stability of Oscillation Frequency

Variations in temperature, supply voltage, load impedance, transistor parameters, stray capacitances and inductances, and component values with time result in changes in the frequency of oscillation f_o. The frequency of oscillation is controlled by the phase of loop gain ϕ_T. A good measure of stability of the oscillation frequency or the precision of oscillation frequency is the slope

$$\left.\frac{\Delta\phi_T}{\Delta f_o}\right|_{\phi_T=0} = a \tag{12.56}$$

The frequency of oscillation f_o is determined solely by the phase ϕ_T of the loop gain T. The frequency of oscillations occurs at the frequency at which the loop-gain phase is zero, that is, $\phi(f_o) = 0$. The phase shift of the loop gain should shift appreciably whenever the oscillation frequency f_o changes by a small amount. Figure 12.4 shows the phase shift ϕ_T of the loop gain as the oscillation frequency changes by Δf. The oscillation frequency change is given by

$$\Delta f_o = \frac{\Delta\phi_T}{a} \tag{12.57}$$

As the slope $d\phi_T/df$ at $\phi_T = 0$ approaches ∞, Δf_o approaches 0. The larger the rate of the phase change $d\phi_T/df$ at $\phi_T = 0$, the more lower the oscillation frequency change df. In other words, the more vertical is the plot of the phase shift ϕ_T, the more stable is the frequency of oscillation f_o. As the slope a approaches infinity, the change in the frequency of oscillation Δf approaches zero. A high value of the loaded quality factor Q_L of the resonant circuit in LC oscillators makes the slope magnitude of the loop-gain phase ϕ_T at $f = f_o$ very high and, therefore, provides good stability of oscillation frequency. The larger the Q_L, the larger is the slope of the phase of the loop gain ϕ_T, and the more stable is the oscillation frequency f_o. It is important that the oscillation frequency is equal to the resonant frequency because the phase of loop gain has the largest slope at the resonant frequency. The motivation of using crystal oscillators is their high quality factor Q_L.

12.3.10 Stability of Oscillation Amplitude

The amplitude of the oscillator output signal is controlled by the magnitude of the loop gain $|T = \beta A|$. The small-signal model and linear theory do not predict the amplitude of the

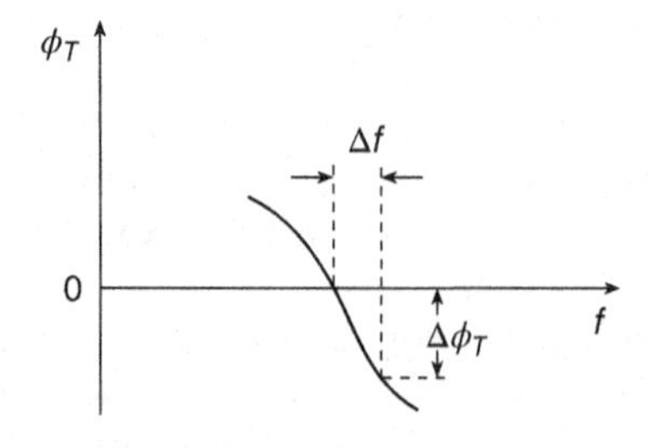

Figure 12.4 Phase shift $\Delta\phi_T$ of the loop gain as the frequency changes by Δf.

steady-state output voltage of an oscillator. In accordance with the linear theory, an oscillator may have any amplitude of the output voltage. In reality, an amplifier input–output characteristic $v_o = f(v_f)$ is nonlinear. The amplitude of the output voltage of an oscillator in steady state is determined by the solution of the equation $\beta A = 1$. When the amplitude of the oscillator output voltage increases, the nonlinearity of the amplifier sufficiently reduces the amplifier gain, which shifts the pair of complex conjugate poles from the right-hand plane to the left until they reside permanently on the imaginary axis. In this case, the loop gain $T(\omega_o) = |A(\omega_o||\beta(\omega_o)|$ becomes exactly 1 and the amplitude of the output voltage remains constant. A provision can be made in the external oscillator circuit, usually using diodes, to limit the amplifier gain and thereby to limit the amplitude of the output voltage. Unfortunately, the nonlinearity of the amplifier causes some distortion of the drain current, which contains odd harmonics. However, the resonant circuit acts as a bandpass filter, which reduces the amplitudes of harmonics. Therefore, the output voltage v_o is a sinusoid of high spectral purity if the loaded quality factor of the resonant circuit Q_L is high enough.

The voltage transfer function of the amplifier is

$$v_o = Av_f \tag{12.58}$$

The voltage transfer function of the feedback network is

$$v_f = \beta v_o \tag{12.59}$$

which can be presented in the form

$$v_o = \frac{1}{\beta} v_o \tag{12.60}$$

Equations (12.58) and (12.59) are illustrated in Fig. 12.5. The feedback network is a voltage divider, which consists of capacitors and/or inductors. Therefore, the transfer function of the feedback network β is linear. The amplifier contains a nonlinear active device (a MOSFET, a MESFET, or a BJT). Therefore, the voltage gain of the amplifier $A = A(v_f) = A(t)$ is nonlinear; it depends on the amplitude of the feedback voltage v_f. The small-signal gain of the amplifier at a given value of the feedback voltage v_f is equal to the slope of the plot $v_o = f(v_f)$ at the value of v_f

$$A = \frac{dv_o}{dv_f} \tag{12.61}$$

This gain can be lower or greater than $1/\beta$.

When v_f is in the range from 0 to $v_f(Q)$,

$$A > \frac{1}{\beta} \tag{12.62}$$

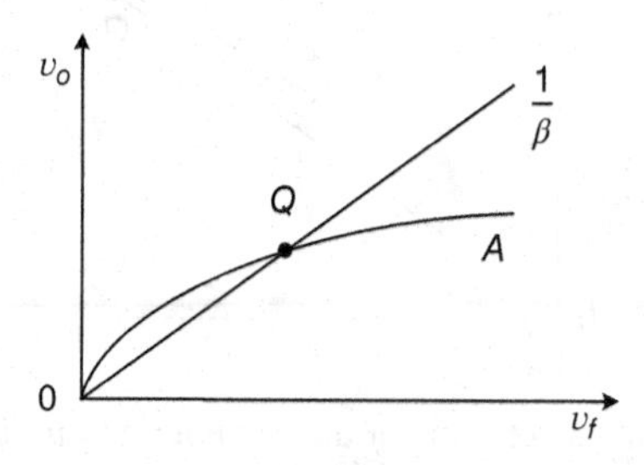

Figure 12.5 Input–output characteristic of the amplifier $v_o = f(v_f)$ and the inverted input–output characteristic of the feedback network $v_o = v_f/\beta$ for stable operating point Q, yielding the amplitude stability. The operating point at the origin is unstable, which satisfies the condition to start oscillations.

yielding

$$T = \beta A > 1 \tag{12.63}$$

In this case, the amplitude of the output voltage is growing until $v_o = v_o(Q)$. Then $\beta A = 1$, resulting in steady-state oscillation.

In contrast, for $v_f > v_f(Q)$,

$$A < \frac{1}{\beta} \tag{12.64}$$

producing

$$T = \beta A < 1 \tag{12.65}$$

In this case, the amplitude of the output voltage decreases until $v_o = v_o(Q)$. Then $\beta A = 1$, resulting in steady-state oscillation. Therefore, the operating point Q is a stable point, resulting in a constant (stable) amplitude of the output voltage of an oscillator.

The conditions for the stable amplitude of oscillation at the operating point Q are

$$A = \left.\frac{dv_o}{dv_f}\right|_{v_f \to v_f^-(Q)} > \frac{1}{\beta} \tag{12.66}$$

and

$$A = \left.\frac{dv_o}{dv_f}\right|_{v_f \to v_f^+(Q)} < \frac{1}{\beta} \tag{12.67}$$

To start the oscillation, the point at the origin must be unstable. The condition to start oscillation is

$$A = \left.\frac{dv_o}{dv_f}\right|_{v_f \to 0} > \frac{1}{\beta} \tag{12.68}$$

yielding

$$T = \beta A > 1 \tag{12.69}$$

This condition is satisfied for the plots depicted in Fig. 12.5.

Figure 12.6 shows plots of A and $1/\beta$, where the operating point at the origin is stable

$$A = \left.\frac{dv_o}{dv_f}\right|_{v_f \to 0} < \frac{1}{\beta} \tag{12.70}$$

resulting in

$$T = \beta A < 1 \tag{12.71}$$

Therefore, the condition to start the oscillation is not satisfied. For v_f in the range from 0 to the first intersection, the output voltage of the oscillator will decrease to zero. For v_f greater than its value at the first intersection, the amplitude of the oscillation will increase until the operating

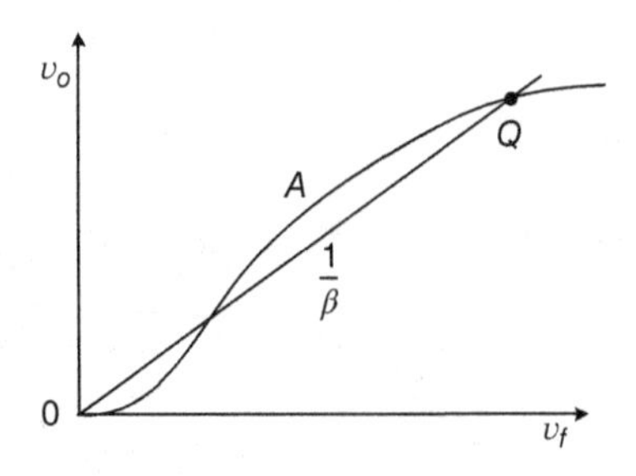

Figure 12.6 Input–output characteristic of the amplifier $A = v_o/v_f$ and the inverted input–output characteristic of the feedback network $1/\beta = v_o/v_f$. The first intersection point is unstable and the second intersection (operating) point Q is stable. The operating point at the origin is stable, which does not satisfy the condition to start oscillations.

point Q is reached, which is a stable point. To start the oscillation, the startup voltage must be higher than that at the first intersection.

12.4 Topologies of LC Oscillators with Inverting Amplifier

A block diagram of classic LC oscillators is shown in Fig. 12.7. It consists of an inverting amplifier A and a inverting feedback network β. The feedback network is inverting at the oscillation frequency f_o. The amplifier is loaded by a parallel resonant circuit, which consists of reactances X_1, X_2i, and X_3. The parallel resonant circuit behaves as a frequency-selective bandpass filter. Therefore, the oscillation frequency f_o is determined by the resonant frequency of the LC resonant circuit. The feedback network must be inverting at f_o because the amplifier is inverting. Reactances X_1 and X_2 form the feedback network reactances X_1 and X_2, which form a voltage

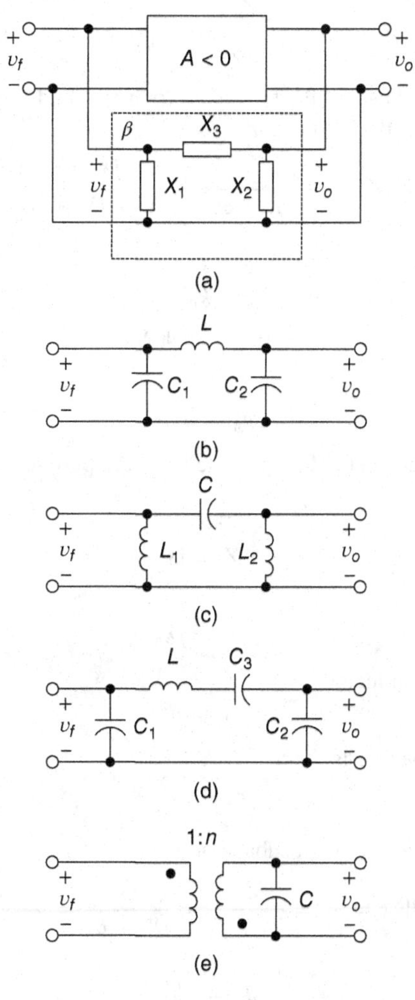

Figure 12.7 Block diagram of classic LC oscillators with inverting amplifier ($A < 0$) and inverting feedback networks ($\beta < 0$). (a) Block diagram of LC oscillators with $A < 0$. (b) Feedback network of Colpitts oscillator. (c) Feedback network of Hartley oscillator. (d) Feedback network of Clapp oscillator. (e) Feedback network of Armstrong oscillator (with inverting transformer).

divider. A feedback network β produces a feedback voltage v_f that is in phase with the amplifier input voltage. The magnitude of the feedback factor $|\beta|$ is determined by the ratio C_2/C_1 or L_1/L_2. A feedback voltage v_f is applied to the amplifier, which introduces a $-180°$ phase shift. The amplifier output voltage is applied to the input of the feedback network, which introduces another $-180°$ phase shift. As a result, the voltage is shifted in phase by $-360°$ as it travels around the loop. The $-360°$ phase shift is equivalent to the $0°$ phase shift. The family inverting amplifiers include inverting op-amp, common-source (CS) amplifier, and common-emitter amplifier.

The oscillator with the feedback network shown in Fig. 12.7(b) is called Colpitts oscillator. The oscillator with the feedback network shown in Fig. 12.7(c) is called Hartley oscillator. The oscillator with the feedback network shown in Fig. 12.7(d) is called Clapp oscillator. It is obtained from the Colpitts oscillator by adding a capacitor C_3 in series with the inductor L. The oscillator with the feedback network shown in Fig. 12.7(e) is called Armstrong oscillator. The feedback network contains an inverting transformer. A resonant capacitor is connected in parallel with the transformer. The magnetizing inductance L_m of the transformer and the capacitor form a parallel resonant circuit, which determines the oscillation frequency f_o.

The gain of the inverting amplifier is

$$A = |A|e^{j\phi A} = \frac{v_o}{v_f} = -g_m(r_o||R_L||Z) < 0 \tag{12.72}$$

where r_o is the output resistance of the transistor, R_L is the load resistance of the oscillator, and Z is the input impedance of the feedback network

$$Z = \frac{Z_3(Z_1 + Z_2)}{Z_1 + Z_2 + Z_3} \approx \frac{jX_3(jX_1 + jX_2)}{jX_1 + jX_2 + jX_3} \tag{12.73}$$

Ideally, $Z(f_o) = \infty$ and

$$A = -g_m(r_o||R_L) \tag{12.74}$$

$$|A| = g_m(r_o||R_L) \tag{12.75}$$

and

$$\phi_A = -180° \tag{12.76}$$

Assume that the impedances in the feedback network are pure reactances

$$Z_1 = jX_1 \tag{12.77}$$

$$Z_2 = jX_2 \tag{12.78}$$

and

$$Z_3 = jX_3 \tag{12.79}$$

where the reactance of an inductor is

$$X = \omega L \tag{12.80}$$

and the reactance of a capacitor is

$$X = -\frac{1}{\omega C} \tag{12.81}$$

The voltage transfer function of the feedback network is

$$\beta = |\beta|^{j\phi_\beta} = \frac{v_f}{v_o} = \frac{Z_1}{Z_1 + Z_3} = \frac{jX_1}{jX_1 + jX_3} = \frac{X_1}{X_1 + X_3} < 0 \tag{12.82}$$

where

$$\phi_\beta = -180° \tag{12.83}$$

The loop gain at the frequency of oscillation f_o is

$$T(f_o) = \beta(f_o)A(f_o) = 1 \tag{12.84}$$

Since $T(f_o) = \beta(f_o)A(f_o) = 1$ and $A(f_o) < 0$, $\beta(f_o) = 1/A(f_o) < 0$. At the resonant frequency,

$$X_1 + X_2 + X_3 = 0 \tag{12.85}$$

yielding

$$X_3 = -(X_1 + X_2) \tag{12.86}$$

Thus,

$$\beta = \frac{X_1}{X_1 + X_3} = \frac{X_1}{X_1 - X_1 - X_2} = -\frac{X_1}{X_2} = -\frac{X_1}{X_2} < 0 \tag{12.87}$$

yielding

$$\frac{X_1}{X_2} > 0 \tag{12.88}$$

Therefore, (1) X_1 and X_2 must have the same sign (i.e., either both positive or both negative) and (2) X_3 and $(X_1 + X_2)$ must have the opposite signs. These conditions lead to two topologies of *LC* oscillators:

$$X_3 > 0,\ X_1 < 0, \quad \text{and} \quad X_2 < 0 \tag{12.89}$$

or

$$X_3 < 0,\ X_1 > 0, \quad \text{and} \quad X_2 > 0 \tag{12.90}$$

If X_3 is an inductor, X_1 and X_2 are capacitors, resulting in the Colpitts oscillator. If X_3 is a capacitor, X_1 and X_2 are inductors, resulting in the Hartley oscillator.

12.5 Op-Amp Colpitts Oscillator

The feedback network satisfying the conditions in (12.89) is shown in Fig. 12.7(b). A tapped capacitor is used to form a feedback network. The forward gain amplifier *A* in the oscillator can be an op-amp or a single-transistor amplifier. In *LC* oscillators, op-amps are used at low RF frequencies, whereas single-transistor amplifiers are used as an active device at high frequencies. Figure 12.8 shows an op-amp Colpitts oscillator. Edwin Henry Colpitts invented this circuit in 1915 at Western Electric.

At the oscillation frequency f_o,

$$X_L = X_{C1} + X_{C2} \tag{12.91}$$

producing

$$\omega_o L = \frac{1}{\omega_o C_1} + \frac{1}{\omega_o C_2} = \frac{C_1 + C_2}{\omega_o C_1 C_2} \tag{12.92}$$

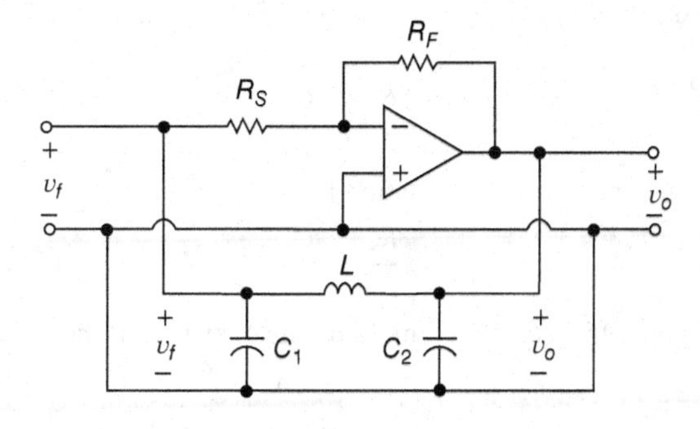

Figure 12.8 Colpitts oscillator.

Hence, the frequency of oscillation of the Colpitts oscillator is

$$f_o = \frac{1}{2\pi\sqrt{L\frac{C_1 C_2}{C_1 + C_2}}} \tag{12.93}$$

The transfer function of the feedback network is

$$\beta = \frac{v_f}{v_o} = |\beta|e^{j\phi_\beta} = \frac{j\omega L}{j\omega L - \frac{j}{\omega C_1}} = \frac{\omega L}{\omega L - \frac{1}{\omega C_1}} = \frac{1}{1 - \frac{1}{\omega^2 C_1 L}} \tag{12.94}$$

The transfer function of the feedback network at the oscillation frequency f_o is

$$\beta(f_o) = -\frac{X_{C1}}{X_{C2}} = -\frac{C_2}{C_1} = \frac{C_2}{C_1}e^{(-180^\circ)} \tag{12.95}$$

where

$$|\beta| = \frac{C_2}{C_1} \tag{12.96}$$

and

$$\phi_\beta = -180^\circ \tag{12.97}$$

The voltages across C_1 and C_2 are 180° out of phase. Therefore, the feedback voltage v_f is 180° out of phase with respect to the output voltage v_o.

Figure 12.8 shows a practical implementation of the Colpitts oscillator using an inverting op-amp. The voltage gain of this amplifier is

$$A = \frac{v_o}{v_f} = |A|e^{j\phi_A} = -\frac{R_F}{R_S} = \frac{R_F}{R_S}e^{j(-180^\circ)} \tag{12.98}$$

where

$$|A| = \frac{R_F}{R_S} \tag{12.99}$$

and

$$\phi_A = -180^\circ \tag{12.100}$$

The loop gain for steady-state oscillation at the oscillation frequency f_o is given by

$$T(f_o) = |T(f_o)|e^{j\phi_T(f_o)} = \beta(f_o)A(f_o) = \left(-\frac{C_2}{C_1}\right)\left(-\frac{R_F}{R_S}\right) = \left(\frac{C_2}{C_1}\right)\left(\frac{R_F}{R_S}\right) = 1 \tag{12.101}$$

where

$$|T(f_o)| = \left(\frac{C_2}{C_1}\right)\left(\frac{R_F}{R_S}\right) = 1 \tag{12.102}$$

and

$$\phi_T(f_o) = \phi_A(f_o) + \phi_\beta(f_o) = -180^\circ + (-180^\circ) = -360^\circ \tag{12.103}$$

Thus,

$$\frac{R_F}{R_S} = \frac{C_1}{C_2} \tag{12.104}$$

The startup oscillation condition for oscillation is

$$\frac{R_F}{R_S} > \frac{C_1}{C_2} \tag{12.105}$$

The maximum oscillation frequency is limited by the slow rate of the op-amp SR

$$f_{omax} = \frac{SR}{2\pi V_{omax}} \tag{12.106}$$

where V_{omax} is the maximum amplitude of the oscillator output voltage.

Example 12.1

Design an op-amp Colpitts oscillator for $f_o = 1$ MHz.

Solution. Let $C_1 = C_2 = 10$ nF. Hence, the inductance is

$$L = \frac{1}{4\pi^2 f_o^2 C_1 C_2/(C_1 + C_2)} = \frac{1}{4\pi^2 \times (10^6)^2 \times 5 \times 10^{-9}} = 5.066 \text{ mH} \tag{12.107}$$

Since $C_1/C_2 = 1$, we get $R_F/R_S = 1$. Pick $R_F = R_S = 10$ kΩ/0.25 W/1%.

12.6 Single-Transistor Colpitts Oscillator

General and various forms of ac circuits of classic oscillators with a single transistor in CS configuration are depicted in Fig. 12.9. The CS configuration is an inverting amplifier. Figure 12.10 shows a circuit LC oscillator with a single-transistor CS amplifier (or a common-emitter amplifier) for both the dc biasing and the ac component. Three representations of the ac circuit of the Colpitts oscillator are depicted in Fig. 12.11. For steady-state oscillation, the voltage gain

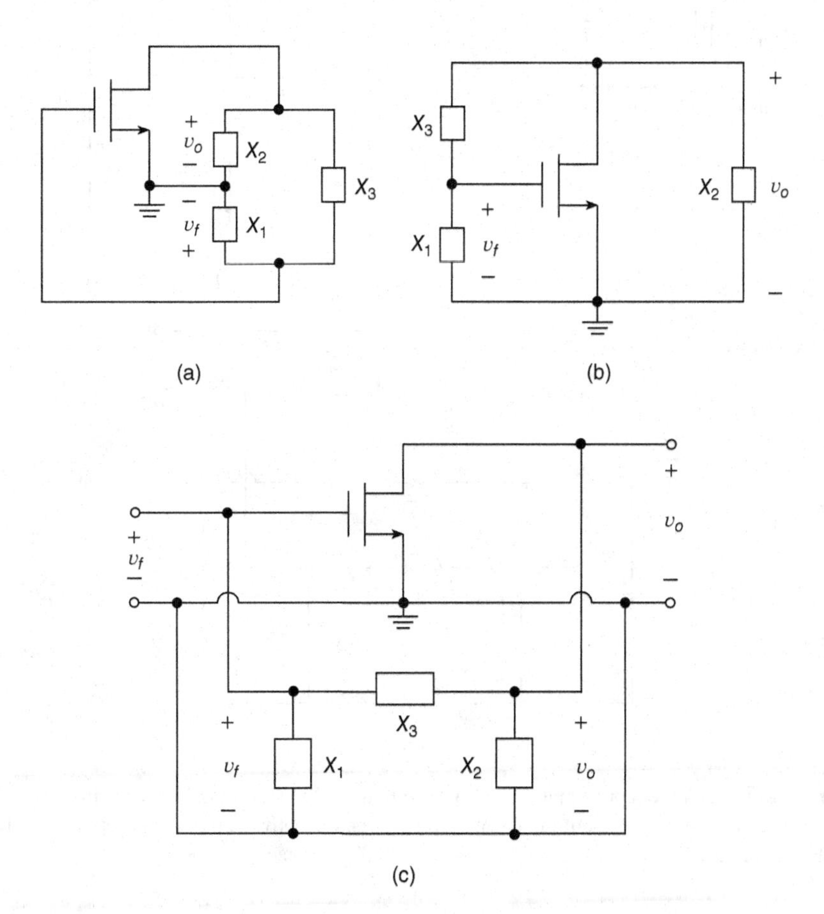

Figure 12.9 General forms of ac circuits of classic oscillators with an inverting amplifier implemented with a single transistor in the common-source (CS) configuration.

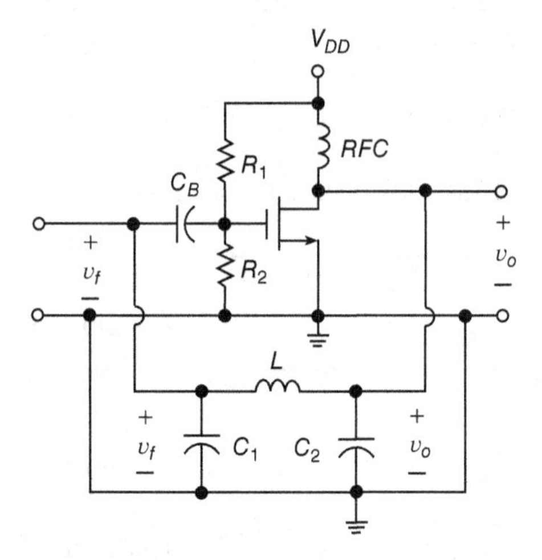

Figure 12.10 Colpitts oscillator with common-source amplifier.

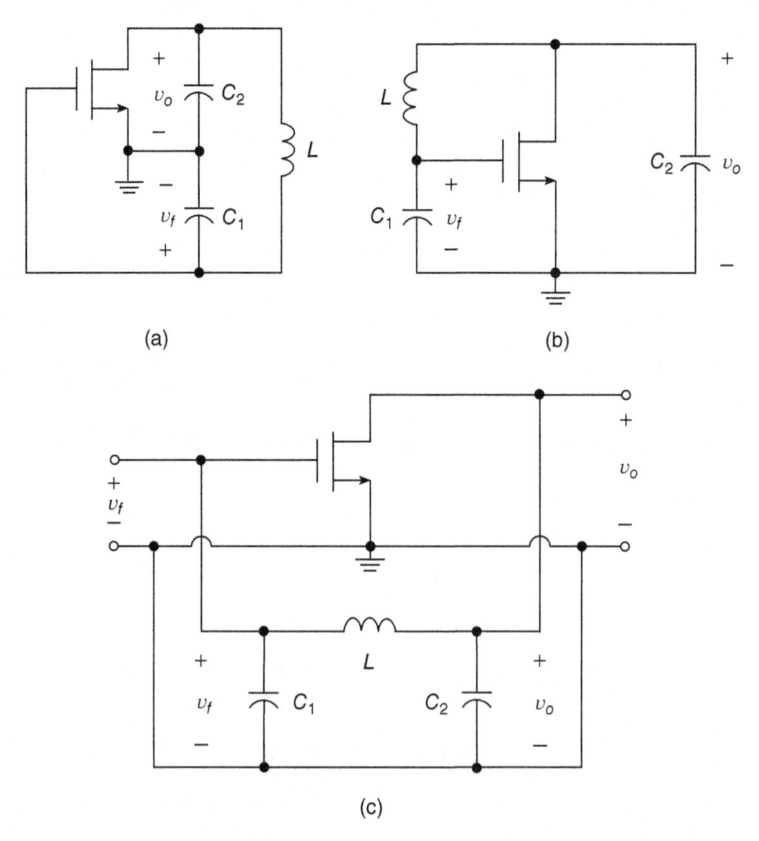

Figure 12.11 Three representations of the ac circuit of Colpitts oscillator with single transistor. (a) With the resonant circuit at the output. (b) With the resonant reactive components between the transistor terminals. (c) In the form of the forward amplifier and the feedback network.

of the amplifier in the Colpitts oscillator at the oscillation frequency f_o equal to the resonant frequency is

$$A_v(f_o) = \frac{v_o}{v_f} = -g_m(r_o||R_L) \tag{12.108}$$

where g_m is the small-signal transconductance, r_o is the small-signal output resistance of the MOSFET, and R_L is the load resistance. The voltage transfer function of the feedback network at the oscillation frequency f_o is

$$\beta(f_o) = \frac{v_f}{v_o} = -\frac{C_2}{C_1} \tag{12.109}$$

The loop gain for steady-state oscillation is

$$T(f_o) = \beta(f_o)A_v(f_o) = [-g_m(r||R_L)]\left(-\frac{C_2}{C_1}\right) = [g_m(r||R_L)]\left(\frac{C_2}{C_1}\right) = 1 \tag{12.110}$$

Hence, the ratio of capacitances required for steady-state oscillations is

$$g_m(r_o||R_L) = \frac{C_1}{C_2} \tag{12.111}$$

The startup conditions for oscillations is

$$g_m(r_o||R_L) > \frac{C_1}{C_2} \tag{12.112}$$

The ac circuits of the Colpitts oscillator for three configurations of the transistor are shown in Fig. 12.12. In Fig. 12.12(a), the MOSFET is operated in the CS configuration. In Fig. 12.12(b), the MOSFET is operated in the common-gate (CG) configuration. In Fig. 12.12(c), the MOSFET is operated in the common-drain (CD) configuration.

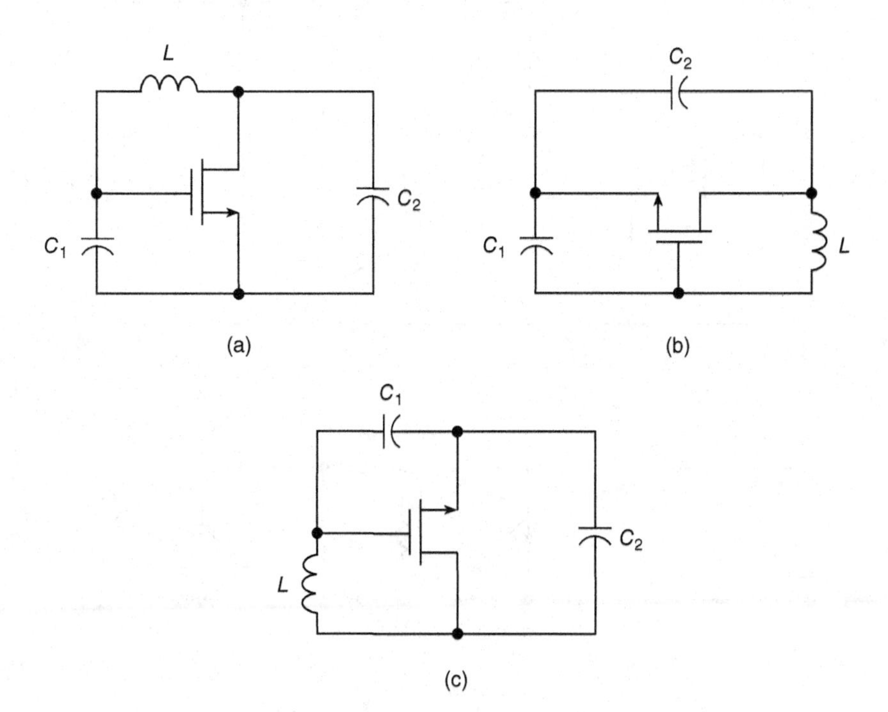

Figure 12.12 Colpitts oscillator ac circuits for three configurations of the transistor. (a) MOSFET in common-source (CS) configuration. (b) MOSFET in common-gate (CG) configuration. (c) MOSFET in common-drain (CD) configuration.

12.7 Common-Source Colpitts Oscillator

Figure 12.13(a) shows an ac circuit of the CS Colpitts oscillator. This circuit is referred to as the CS Colpitts oscillator. The load resistance R_L is connected between the drain and the source terminals and in parallel to the capacitor C_2. A small-signal model of the Colpitts oscillator in which the transistor is operated in the CS configuration is shown in Fig. 12.13(b).

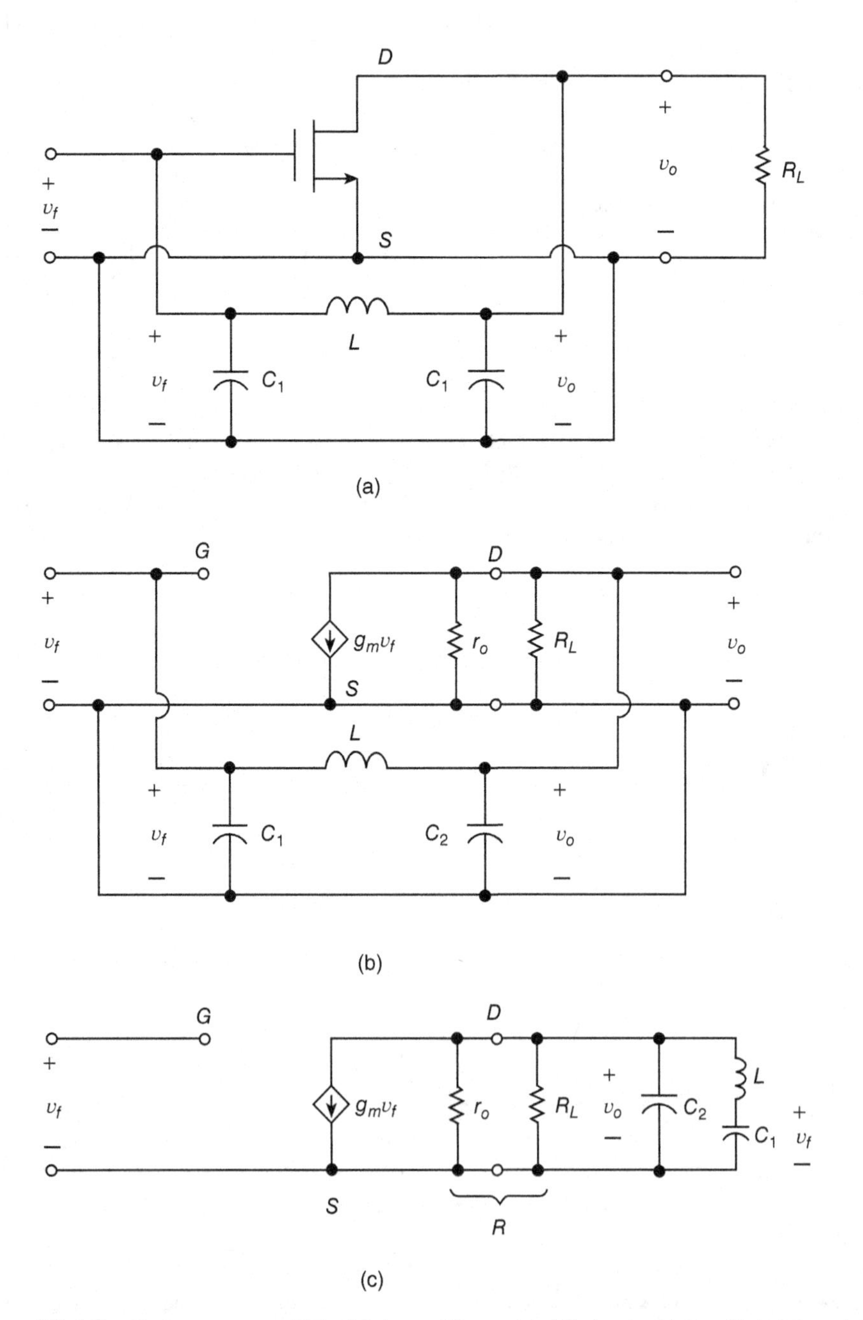

Figure 12.13 Common-source (CS) Colpitts oscillator. (a) AC circuit. (b) Small-signal model. (c) Small-signal model of the amplifier for determining the voltage gain A.

Figure 12.13(c) shows the small-signal model of the amplifier loaded by the feedback network. The gate-to-source capacitance C_{gs} is included in the capacitance C_1. The drain-to-source capacitance C_{ds} is included in the capacitance C_2. The gate-to-drain capacitance C_{gd} is neglected. The gate-to-source voltage v_{gs} is equal to the feedback voltage v_f, that is, $v_{gs} = v_f$.

12.7.1 Feedback Network Factor of Common-Source Colpitts Oscillator

Figure 12.14 shows a feedback network of the CS Colpitts oscillator. The voltage transfer function of the feedback network is

$$\beta(s) = \frac{v_f}{v_o} = \frac{\frac{1}{sC_1}}{sL + \frac{1}{sC_1}} = \frac{1}{1 + s^2LC_1} \tag{12.113}$$

The feedback network of the CS Colpitts oscillator is a second-order low-pass filter. Figures 12.15 and 12.16 show the magnitude $|\beta|$ and ϕ_β of the feedback factor as functions of frequency f at $L = 50$ μH and $C_1 = 10$ μF.

12.7.2 Amplifier Voltage Gain of Common-Source Colpitts Oscillator

Figure 12.13(c) shows a small-signal model of the amplifier of the CS Colpitts oscillator loaded by the feedback network. The total resistance is

$$R = r_o||R_L = \frac{r_oR_L}{r_o + R_L} \tag{12.114}$$

The load impedance seen by the MOSFET drain is

$$Z_L = \left(R||\frac{1}{sC_2}\right) || \left(sL + \frac{1}{sC_1}\right) = \frac{R(1 + s^2LC_1)}{s^3LC_1C_2R + s^2LC_1 + sR(C_1 + C_2) + 1} \tag{12.115}$$

The voltage gain of the amplifier loaded by the feedback network is

$$A(s) = \frac{v_o}{v_f} = -g_mZ_L = -\frac{g_mR(1 + s^2LC_1)}{s^3LC_1C_2R + s^2LC_1 + sR(C_1 + C_2) + 1} \tag{12.116}$$

Figures 12.17 and 12.18 show Bode plots of the amplifier voltage gain A at $L = 50$ μH, $C_1 = C_2 = 10$ μF, $R_L = 1$ kΩ, $r_o = 100$ kΩ, and $g_m = 1.01$ mA/V.

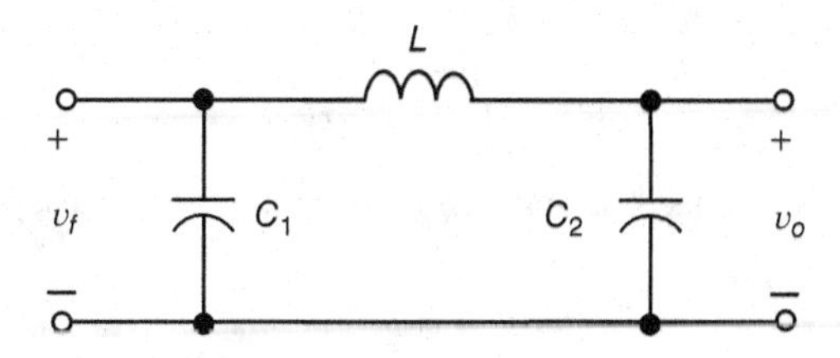

Figure 12.14 Feedback network of common-source (CS) Colpitts oscillator.

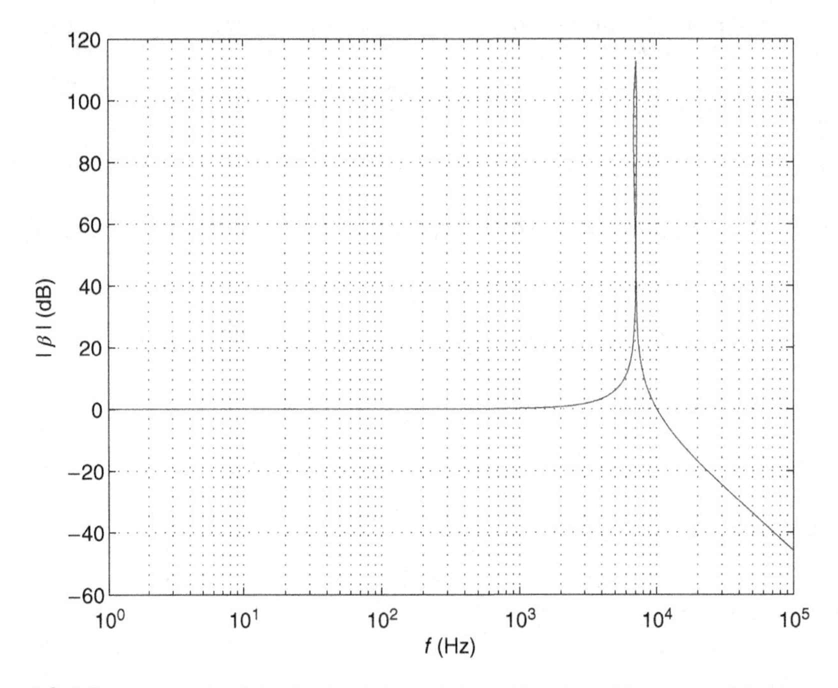

Figure 12.15 Magnitude of the feedback factor $|\beta|$ as a function of frequency f for the common–source Colpitts oscillator.

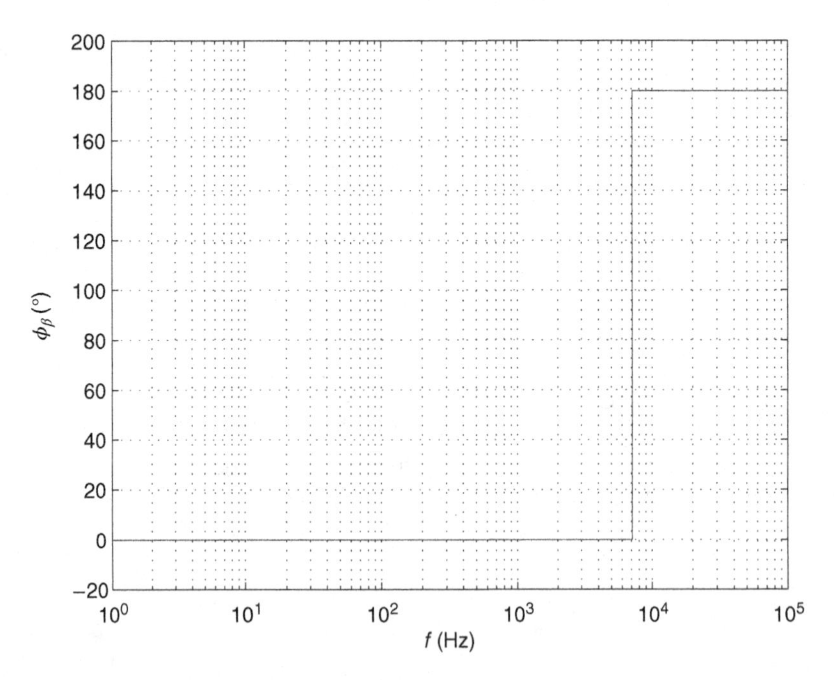

Figure 12.16 Phase of the feedback factor ϕ_β as a function of frequency f for the common-source Colpitts oscillator.

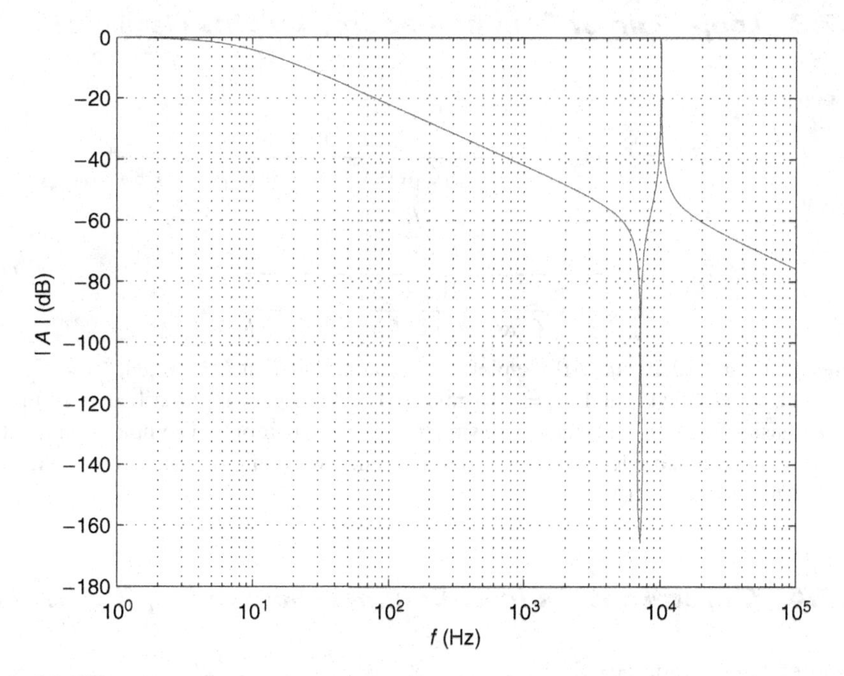

Figure 12.17 Magnitude of the amplifier voltage gain $|A|$ as a function of frequency f for the common-source Colpitts oscillator.

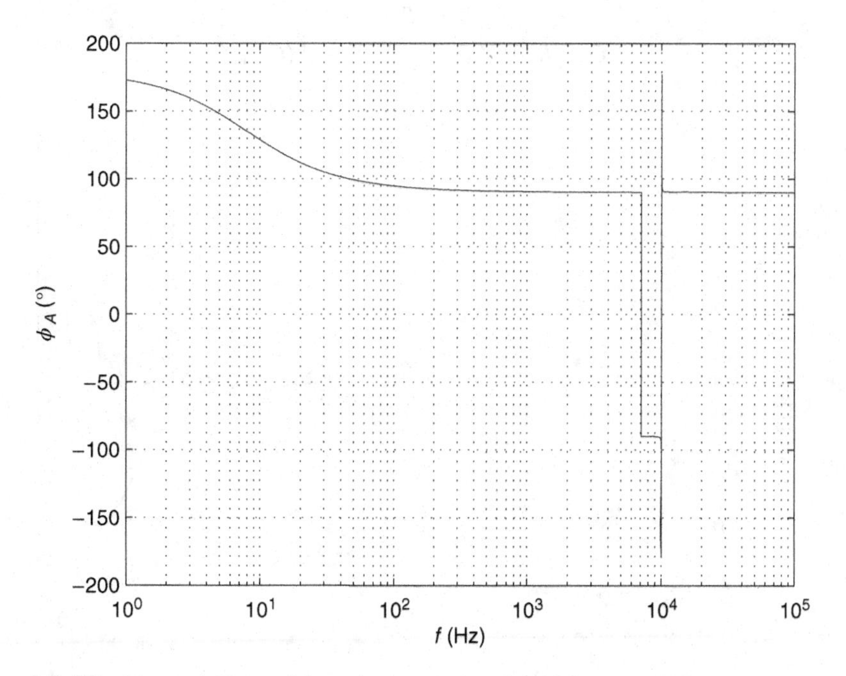

Figure 12.18 Phase of the amplifier gain ϕ_A as a function of frequency f for the common-source Colpitts oscillator.

12.7.3 Loop Gain of Common-Source Colpitts Oscillator

The loop gain is

$$T(s) = \beta(s)A(s) = -\frac{g_m R}{s^3 LC_1C_2R + s^2LC_1 + sR(C_1 + C_2) + 1}$$

$$= -\frac{\frac{g_m R}{LC_1C_2}}{s^3 + s^2\frac{1}{C_2R} + s\frac{(C_1 + C_2)}{LC_1C_2} + \frac{1}{LC_1C_2R}} \quad (12.117)$$

Figures 12.19 and 12.20 show Bode plots of the loop gain T at $L = 50$ μH, $C_1 = C_2 = 10$ μF, $R_L = 1$ kΩ, $r_o = 100$ kΩ, and $g_m = 1.01$ mA/V. The loop gain is described by a third-order voltage transfer function. It contains one simple pole and a pair of two complex conjugate poles. As g_m increases, the simple pole moves to left and the two complex poles move from the RHP to the left-half plane (LHP).

12.7.4 Closed-Loop Gain of Common-Source Colpitts Oscillator

The closed-loop gain of the CS Colpitts oscillator is

$$A_f(s) = \frac{A(s)}{1 - T(s)}$$

$$= -\frac{g_m R(1 + s^2LC_1)}{s^3LC_1C_2R + s^2LC_1(1 + g_mR) + sR(C_1 + C_2) + g_mR + 1} \quad (12.118)$$

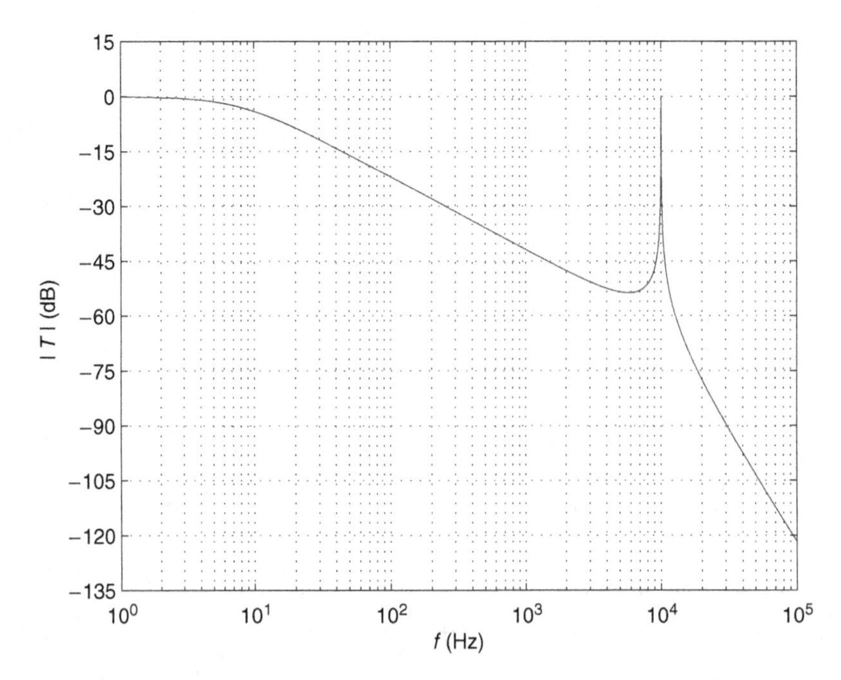

Figure 12.19 Magnitude of the loop gain $|T|$ as a function of frequency f for the common-source Colpitts oscillator.

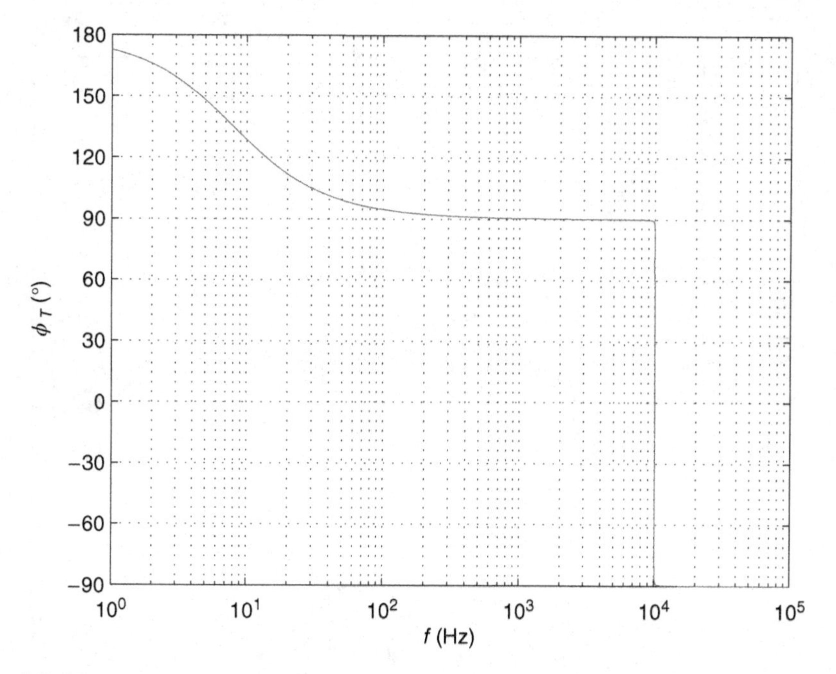

Figure 12.20 Phase of the loop gain ϕ_T as a function of frequency f for the common-source Colpitts oscillator.

Figures 12.21 and 12.22 show Bode plots of the closed-loop gain A_f at $L = 50$ μH, $C_1 = C_2 = 10$ μF, $R_L = 1$ kΩ, $r_o = 100$ kΩ, and $g_m = 1.01$ mA/V.

For $s = j\omega$, we obtain the feedback factor

$$\beta(\omega) = \frac{1}{1 - \omega^2 LC} \tag{12.119}$$

the amplifier voltage gain

$$A(j\omega) = \frac{g_m R(1 - \omega^2 LC_1)}{1 - \omega^2 LC_1 + j\omega R(C_1 + C_2 - \omega^2 LC_1 C_2)} \tag{12.120}$$

and the loop gain

$$T(j\omega) = \beta(j\omega)A(j\omega) = -\frac{g_m R}{(1 - \omega^2 LC_1) + j\omega R[(C_1 + C_2) - \omega^2 LC_1 C_2]} \tag{12.121}$$

Equating the imaginary part of the denominator of $T(j\omega)$ to zero, we get the oscillation frequency

$$\omega_o = \frac{1}{\sqrt{L\left(\frac{C_1 C_2}{C_1 + C_2}\right)}} \tag{12.122}$$

For $L = 50$ μH and $C_1 = C_2 = 1$ μF, $f_o = 10$ kHz. The feedback factor at the oscillation frequency f_o is

$$\beta(\omega_o) = \frac{1}{1 - \omega_o^2 LC_1} = -\frac{C_2}{C_1} \tag{12.123}$$

The amplifier voltage gain at the oscillation frequency is

$$A(\omega_o) = -g_m R \tag{12.124}$$

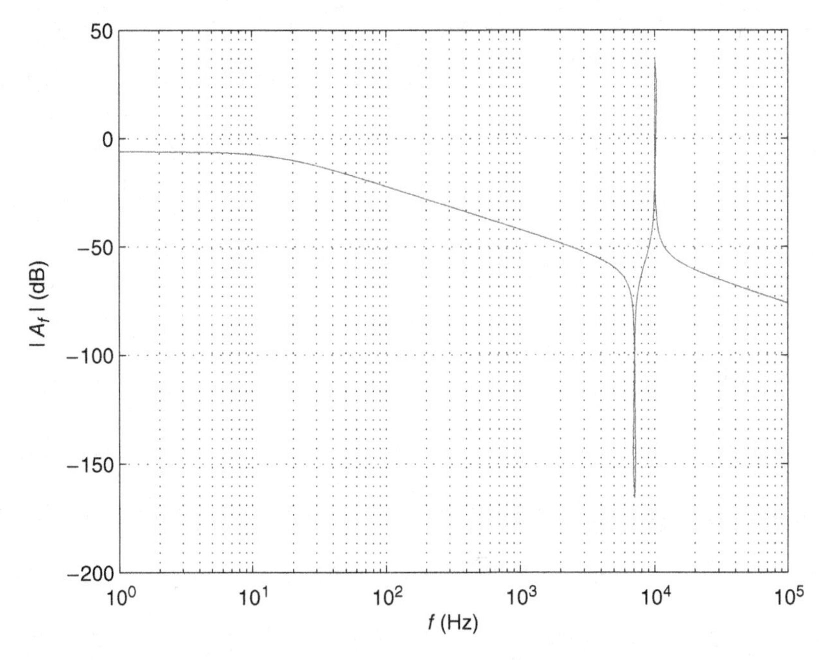

Figure 12.21 Magnitude of the closed-loop gain $|A_f|$ as a function of frequency f for the common–source Colpitts oscillator.

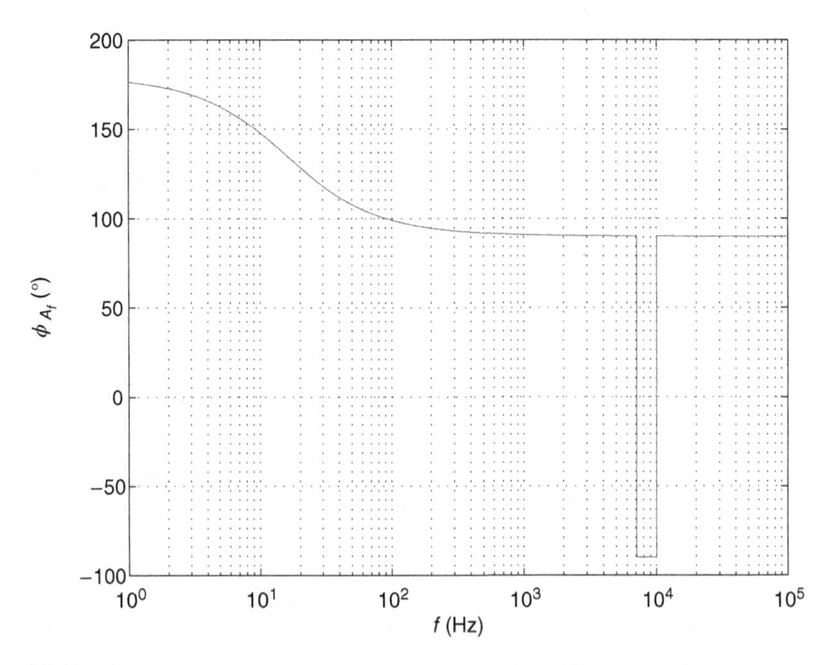

Figure 12.22 Phase of the closed-loop gain ϕ_{A_f} as a function of frequency f for the common-source Colpitts oscillator.

The loop gain at the oscillation frequency is

$$T(\omega_o) = g_m R \frac{C_2}{C_1} \tag{12.125}$$

For $C_1 = C_2 = 10\ \mu\text{F}$, $\beta(\omega_o) = -C_2/C_1 = -1/1 = -1$.

Setting $T(\omega_o) = 1$, we obtain the magnitude condition for steady-state oscillation

$$g_m R = \frac{C_1}{C_2} \tag{12.126}$$

The startup condition for oscillations is

$$g_m R > \frac{C_1}{C_2} \tag{12.127}$$

12.7.5 Nyquist Plot for Common-Source Colpitts Oscillator

Figure 12.23 shows a Nyquist plot of the loop gain $T(j\omega)$ for the CS Colpitts oscillator for steady-state oscillations at $g_m = 1.01$ mA/V. It was observed that the Nyquist plot of crosses the point $-1 + j0$. For $g_m > 1.01$ mA/V, the Nyquist plot of $T(j\omega)$ encircles the point $-1 + j0$.

12.7.6 Root Locus for Common-Source Colpitts Oscillator

Dividing the characteristic equation, which is the denominator of the closed-loop gain A_f, by

$$s^3 LC_1C_2R + s^2 LC_1 + sR(C_1 + C_2) + 1 \tag{12.128}$$

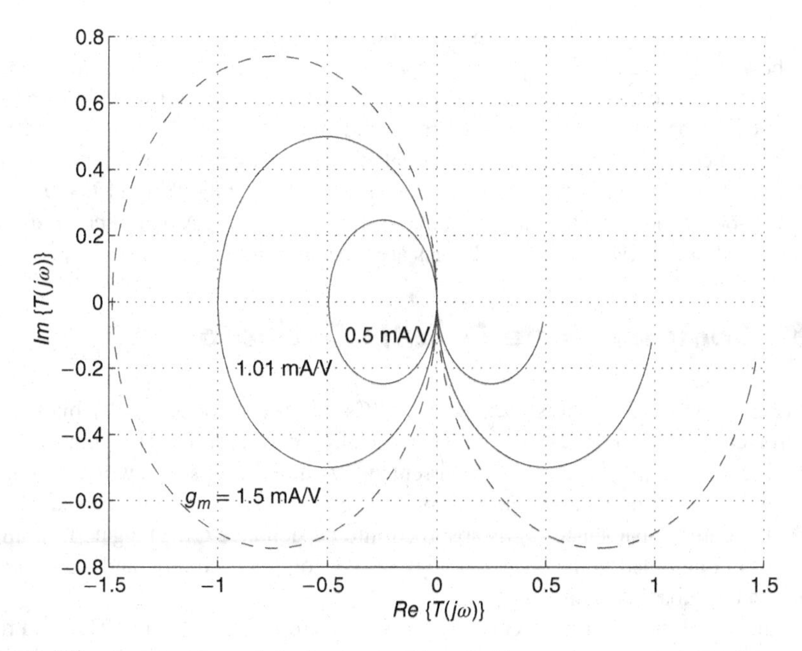

Figure 12.23 Nyquist plots of the loop gain $T(j\omega)$ for the common-source Colpitts oscillator at $g_m = 0.5$ mA/V (too low value to start oscillations), $g_m = 1.01$ mA/V (steady-state oscillations), and $g_m = 1.5$ mA/V (sufficiently large value to start oscillations).

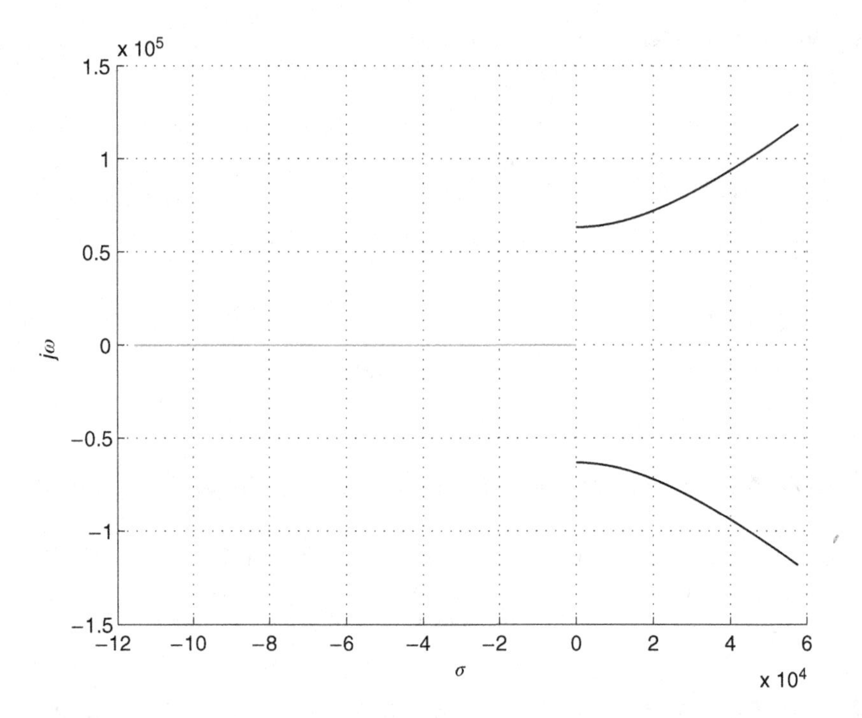

Figure 12.24 Root locus of the closed-loop gain $A_f(s)$ for the common-source Colpitts oscillator when g_m is changed from 0 to 10 A/V.

we obtain

$$1 + g_m \frac{R(1 + sLC_1)}{s^3LC_1C_2R + s^2LC_1 + sR(C_1 + C_2) + 1} \tag{12.129}$$

Figure 12.24 shows a root locus for the CS Colpitts oscillator. The same plot in an enlarged scale is shown in Fig. 12.25. For $g_m = 0$, the open-loop poles are $p_1, p_2 = -25.25 \pm j63245$ rad/s and $p_3 = -50.505$ rad/s. All poles are located in the LHP. For $g_m = 1.01$ mA/V, the poles are $p_1, p_2 = \pm j63245.55$ rad/s and $p_3 = -101$ rad/s and the circuit produces steady-state oscillations. A pair of complex conjugate poles is located on the imaginary axis and the real poles are located in the LHP. For $g_m = 10$ A/V, the poles are $p_1, p_2 = 57{,}683.98 \pm j118{,}300.37$ rad/s and $p_3 = -115{,}468.97$ rad/s and the circuit is able to start oscillations. A pair of complex conjugate poles is located in the RHP and the real poles are located in the LHP.

12.8 Common-Gate Colpitts Oscillator

Figure 12.26 shows CG Colpitts oscillators. In ICs, the transistor is usually biased by a current mirror connected in series with either the source or the drain. The current source I in Fig. 12.26 represents the current-mirror current sink connected in series with the source of the oscillator transistor. The gate-to-source capacitance C_{gs} is absorbed into capacitance C_1 and the drain-to-source capacitance C_{ds} is absorbed into capacitance C_2. The gate-to-drain capacitance C_{gd} is not included in the basic topology of the Colpitts oscillator, and for simplicity it is neglected in the subsequent analysis.

The small-signal model of the CG oscillator is depicted in Fig. 12.27(a). The load resistance R_L is connected between the drain and the gate terminals and in parallel with the resonant inductor L. The small-signal model shown in Fig. 12.27(a) is obtained by replacing the MOSFET between the drain and the source terminals by a voltage-dependent current source $g_m v_f$ and the

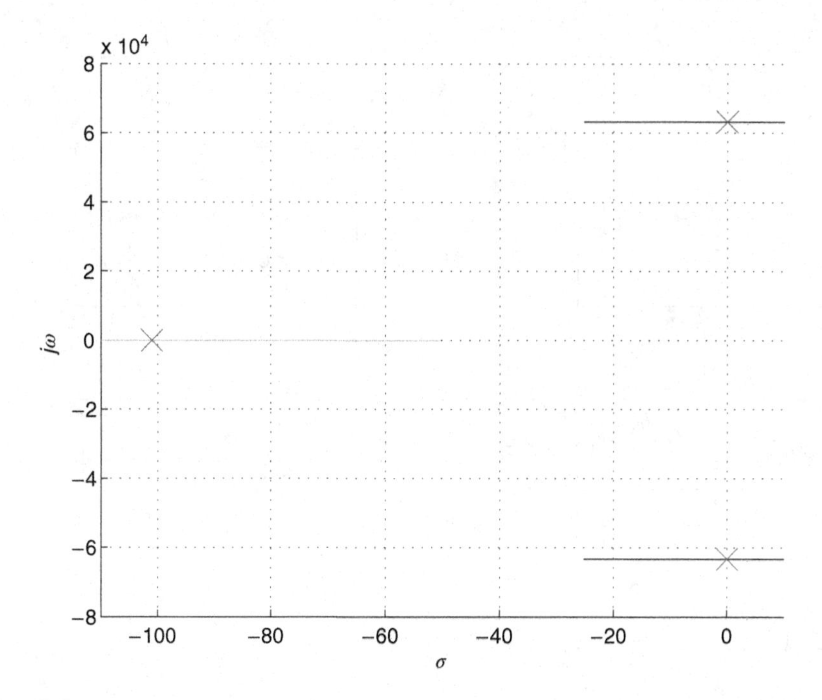

Figure 12.25 Root locus of the closed-loop gain A_f for the common-source Colpitts oscillator in enlarged scale.

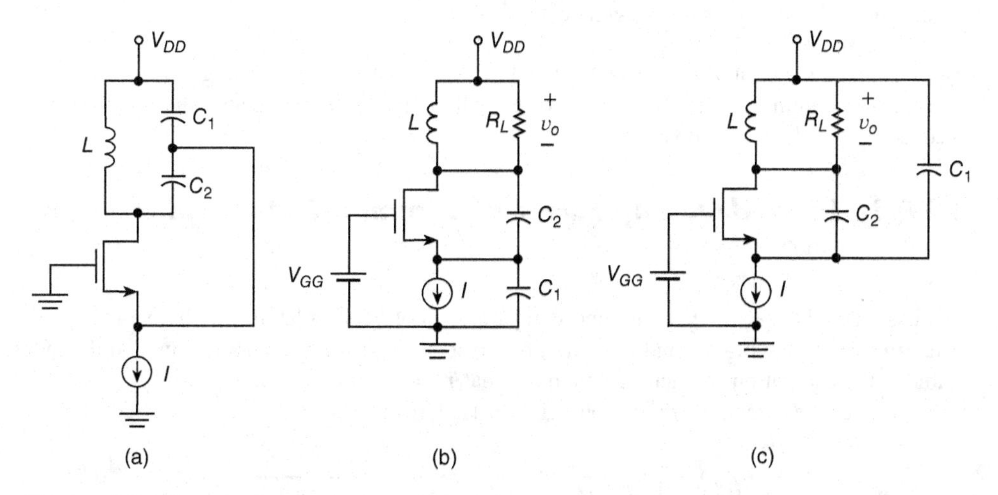

Figure 12.26 Colpitts oscillators with a transistor operating in the common-gate (CG) configuration and biased by a current source I suitable for ICs. (a) Colpitts oscillator with capacitor C_1 connected to power supply V_{DD}. (b) Colpitts oscillator with capacitor C_1 connected to ground. (c) Colpitts oscillator with capacitor C_1 connected to dc power supply.

MOSFET output resistance r_o, where $v_f = v_{gs}$ and g_m is the MOSFET transconductance. Using the current-splitting theorem, the voltage-dependent current source $g_m v_f$ connected between the drain and the source terminals is split into two voltage-dependent current sources $g_m v_f$, one dependent source is connected between the source and the gate terminal and the other source is connected between the drain and the gate terminals, as shown in Fig. 12.27(b). In Fig. 12.27(c), the voltage-dependent current source $g_m v_f$ between the source and the gate terminals is driven

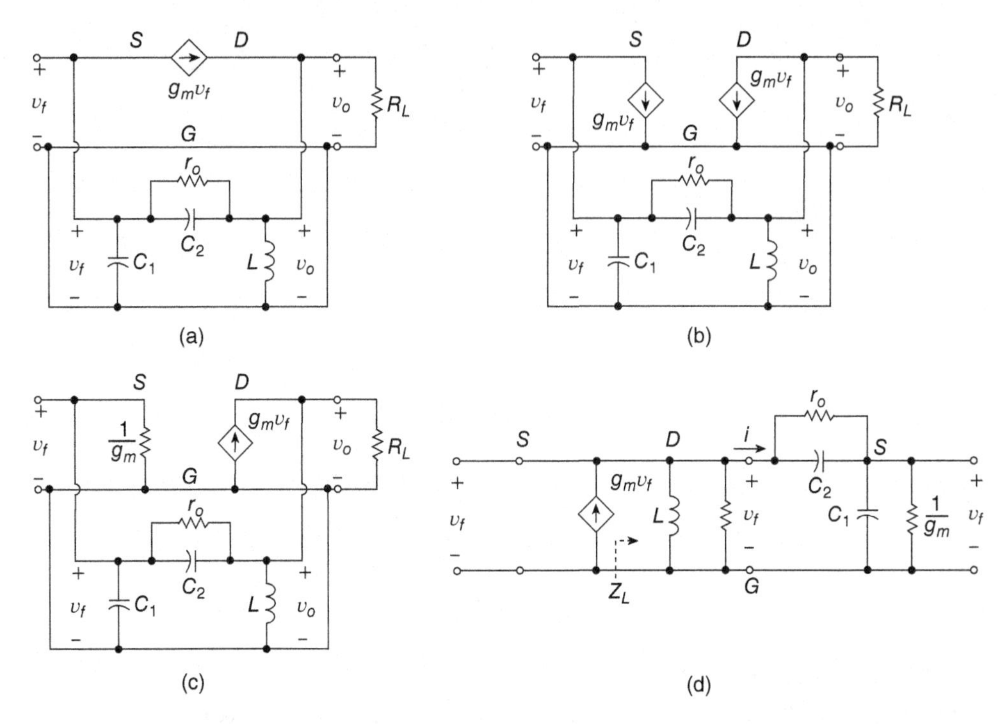

Figure 12.27 Small-signal model of common-gate Colpitts oscillator. (a) Small-signal model. (b) Current-splitting theorem. (c) Source-absorption theorem. (d) Small signal for determining the gain of the amplifier with the transistor configuration.

by its own voltage v_f and is replaced by an equivalent resistance $1/g_m$, using the current-source absorption theorem. Figure 12.27(d) shows a small-signal model of the amplifier loaded by the feedback network. The feedback network is loaded by resistance $1/g_m$.

12.8.1 Loaded Quality Factor of Common-Gate Colpitts Oscillator

The loaded quality factor Q_L is an important parameter of LC oscillators. The higher the Q_L, the better the selectivity of the bandpass filter, the steeper the phase of the loop gain, and the more stable is the oscillation frequency. Assuming that $r_o = \infty$ and denoting $R_{gm} = 1/g_m$, the total impedance of the resonant circuit shown in Fig. 12.27(d) is

$$Z(s) = \left(R_{gm} || \frac{1}{sC_1}\right) + \frac{1}{sC_2} = \frac{1 + sR_{gm}(C_1 + C_2)}{s^2 R_{gm} C_1 C_2 + sC_2} \tag{12.130}$$

Setting $s = j\omega$,

$$Z(j\omega) = R_{eq} + \frac{1}{j\omega C_{eq}} = \frac{1 + j\omega R_{gm}(C_1 + C_2)}{-\omega^2 R_{gm} C_1 C_2 + j\omega C_2} \tag{12.131}$$

Hence, the admittance of the resonant circuit is

$$Y = \frac{1}{Z} = \frac{\omega^2 R_{gm} C_2^2 + j\omega C_2 - j\omega^3 R_{gm}^2 C_1 C_2 (C_1 + C_2)}{1 + \omega^2 R_{gm}^2 (C_1 + C_2)^2}$$

$$= \frac{\omega^2 R_{gm} C_2^2}{1 + \omega^2 R_{gm}^2 (C_1 + C_2)^2} + j\frac{\omega C_2 - \omega^3 R_{gm}^2 C_1 C_2 (C_1 + C_2)}{1 + \omega^2 R_{gm}^2 (C_1 + C_2)^2} = G + jB \tag{12.132}$$

Thus, the equivalent resistance at the oscillation frequency $\omega_o = 1/\sqrt{LC_1C_2/(C_1+C_2)}$ is

$$R_{eq} = \frac{1}{G} = \frac{1+\omega^2 R_{gm}^2 (C_1+C_2)^2}{\omega^2 R_{gm} C_2^2} = R_{gm}\left[\frac{\dfrac{1}{\omega^2 R_{gm}^2} + (C_1+C_2)^2}{C_2}\right]$$

$$= \frac{1}{g_m}\left[\frac{\dfrac{C_1^2}{q_{C1}^2} + (C_1+C_2)^2}{C_2^2}\right] \tag{12.133}$$

where $q_{C1} = \omega R_{gm} C_1$ is the reactance factor of the parallel combination of capacitor C_1 and resistance $1/g_m$. For $C_1/q \ll C_1 + C_2$, the equivalent resistance can be approximated by

$$R_{eq} \approx \frac{1}{g_m}\left(1 + \frac{C_1}{C_2}\right)^2 \tag{12.134}$$

The total resistance connected in parallel with the inductor L is

$$R_p = R_L || R_{eq} = R_L || \frac{1}{g_m}\left(1 + \frac{C_1}{C_2}\right)^2 \tag{12.135}$$

The reactance of the equivalent capacitance is

$$X_{Ceq} = \frac{1}{B} = \frac{1+\omega^2 R_{gm}^2 (C_1+C_2)^2}{\omega[C_2 - \omega^2 R_{gm}^2 C_1 C_2 (C_1+C_2)]} = -\frac{1}{\omega C_{eq}} \tag{12.136}$$

resulting in

$$C_{eq} = \frac{\omega^2 R_{gm}^2 C_1 C_2 (C_1+C_2) - C_2}{1+\omega^2 R_{gm}^2 (C_1+C_2)^2} = \frac{C_1 C_2 (C_1+C_2) - \dfrac{C_1^2 C_2}{q_{C1}^2}}{(C_1+C_2) + \dfrac{C_1^2}{q_{C1}^2}} \tag{12.137}$$

For $(C_1/q)^2 \ll (C_1+C_2)^2$, the equivalent capacitance can be approximated by

$$C_{eq} \approx \frac{C_1 C_2}{C_1 + C_2} \tag{12.138}$$

The loaded quality factor of the resonant circuit at the oscillation frequency f_o is

$$Q_L = \frac{R_p}{\omega_o L} = \frac{R_L || \left[\dfrac{1}{g_m}\left(1+\dfrac{C_1}{C_2}\right)^2\right]}{\omega_o L} = \omega_o R_{eq} C_{eq}. \tag{12.139}$$

12.8.2 Feedback Factor of Common-Gate Colpitts Oscillator

Using KCL at the S terminal in Fig. 12.27(c), we obtain

$$i_{C1} + i_{C2} + i_{ro} + g_m v_f = 0 \tag{12.140}$$

producing

$$sC_1 v_f + sC_2(v_f - v_o) + \frac{v_f - v_o}{r_o} + g_m v_f = 0 \tag{12.141}$$

Rearrangement of this equation gives

$$v_o = v_f \frac{sr_o(C_1 + C_2) + g_m r_o + 1}{sr_o C_2 + 1} \quad (12.142)$$

Hence, the feedback factor loaded by resistance $1/g_m$ is given by

$$\beta(s) = \frac{v_f}{v_o} = \frac{sC_2 r_o + 1}{sr_o(C_1 + C_2) + g_m r_o + 1} = \frac{C_2}{C_1 + C_2} \frac{\left(s + \frac{1}{r_o C_2}\right)}{\left[s + \frac{g_m r_o + 1}{r_o(C_1 + C_2)}\right]} = \frac{C_2}{C_1 + C_2} \frac{s + \omega_{z\beta}}{s + \omega_{p\beta}} \quad (12.143)$$

where the frequency of the zero

$$\omega_{z\beta} = \frac{1}{r_o C_2} \quad (12.144)$$

and the frequency of the pole, which is the 3-dB corner frequency of the high-pass filter, is given by

$$\omega_{p\beta} = \omega_L = \frac{g_m r_o + 1}{r_o(C_1 + C_2)} \approx \frac{g_m}{C_1 + C_2} \quad (12.145)$$

For $s = 0$,

$$\beta(0) = \frac{1}{g_m r_o + 1} \quad (12.146)$$

At high frequencies,

$$\beta_o = \beta(\infty) \approx \frac{C_2}{C_1 + C_2} \quad (12.147)$$

The feedback network behaves as a first-order high-pass filter. Figures 12.28 and 12.29 show the Bode plots of the feedback factor β for the CG Colpitts oscillator at $L = 4.75$ mH, $C_1 = C_2 = 1$ μH, $R_L = 1$ kΩ, $r_o = 100$ kΩ, and $g_m = 4.09$ mA/V. For these data, we obtain $\beta(0) =$

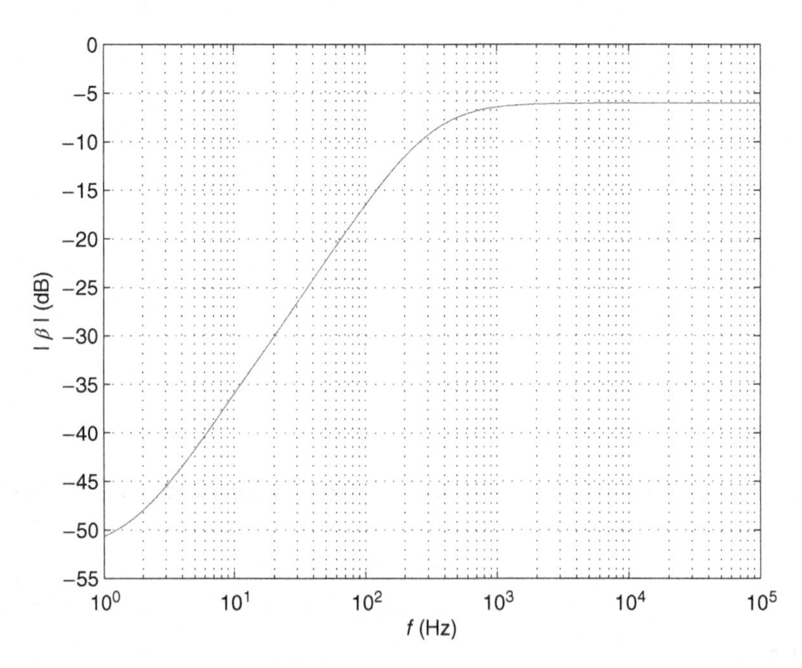

Figure 12.28 Magnitude of the feedback factor $|\beta|$ as a function of frequency f for the common–gate Colpitts oscillator.

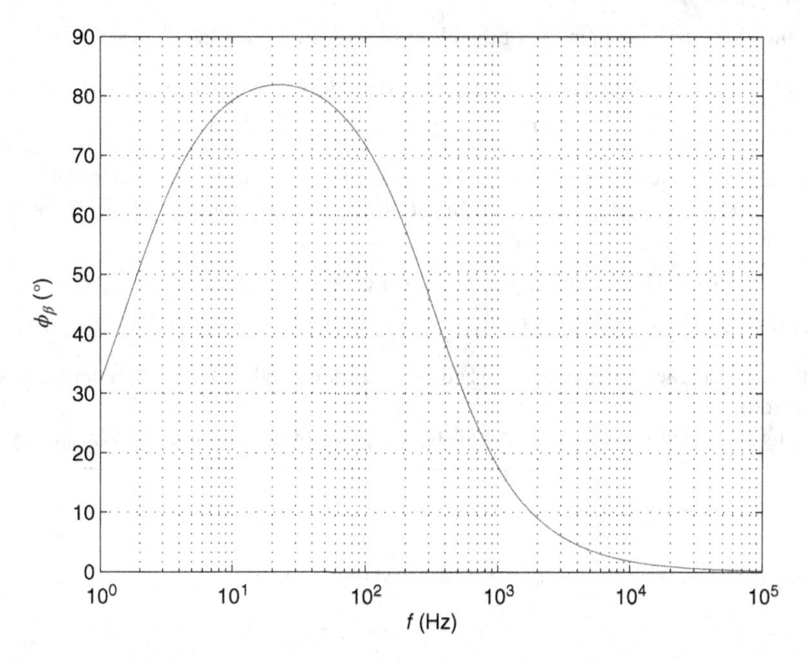

Figure 12.29 Phase of the feedback factor ϕ_β as a function of frequency f for the common-gate Colpitts oscillator.

$1/(g_m r_o + 1) = 1/(402 \times 100 + 1 = 1/403 = -52.08$ dB, $\beta_o = \beta(\infty) = C_2/(C_1 + C_2) = 1/(1+1) = 0.5 = -6$ dB, $f_{z\beta} = 1/(2\pi r_o C_1) = 1/(2\pi 100 \times 10^3 \times 10^{-6}) - 1.59$ Hz, and $f_{p\beta} = g_m/[2\pi(C_1 + C_2)] = 4.09 \times 10^{-3}/[2\pi(1+1) \times 10^{-6}] = 325.47$ Hz.

12.8.3 Characteristic Equation for Common-Gate Colpitts Oscillator

Applying KCL at the D terminal in Fig. 12.27(c), we get

$$i_{C2} + i_{ro} + g_m v_f = i_{RL} + i_L \tag{12.148}$$

which gives

$$sC_2(v_f - v_o) + \frac{v_f - v_o}{r_o} + g_m v_f = \frac{v_o}{R_L} + \frac{v_o}{sL} \tag{12.149}$$

Rearrangement of this equation yields

$$v_f\left(\frac{sC_2 r_o + g_m}{r_o} + g_m\right) = v_o\left(\frac{sL + R_L}{sLR_L} + \frac{sC_2 r_o + 1}{r_o}\right) \tag{12.150}$$

Substitution of (12.142) into (12.149) produces

$$v_f\left(\frac{sC_2 r_o + g_m}{r_o} + g_m\right) - v_f\left[\frac{sr_o(C_1 + C_2) + g_m r_o + 1}{sr_o C_2 + 1}\right]\left(\frac{sL + R_L}{sLR_L} + \frac{sC_2 r_o + 1}{r_o}\right) = 0 \tag{12.151}$$

When the oscillations have begun, v_f is not zero. Dividing both sides of (12.151) by v_f, we obtain

$$\left(\frac{sC_2 r_o + g_m}{r_o} + g_m\right) - \left[\frac{sr_o(C_1 + C_2) + g_m r_o + 1}{sr_o C_2 + 1}\right]\left(\frac{sL + R_L}{sLR_L} + \frac{sC_2 r_o + 1}{r_o}\right) = 0 \tag{12.152}$$

Hence, the characteristic equation of the closed-loop gain A_f for the CG oscillator is

$$s^3LC_1C_2R_Lr_o + s^2[Lr_o(C_1+C_2) + LC_1R_L] + s[R_Lr_o(C_1+C_2) + L(g_mr_o+1)] + R_L(g_mr_o+1) = 0 \tag{12.153}$$

The characteristic equation for the CG Colpitts oscillator is a third-order function.

For steady-state oscillations, $s = j\omega_o$. Substituting this equation into (12.153) and separating the real and imaginary terms, we obtain

$$R_L(g_mr_o+1) - \omega_o^2[Lr_o(C_1+C_2) + LC_1R_L] + j\{\omega_o[R_Lr_o(C_1+C_2) + L(g_mr_o+1)] - \omega_o^3(LC_1C_2R_Lr_o)\} = 0 = 0 + j0 \tag{12.154}$$

Since the total characteristic equation is equal to zero, its both real and imaginary parts are also equal to zero.

Equating the imaginary part of (12.154) to zero, we obtain the oscillation frequency

$$\omega_o = \sqrt{\frac{C_1+C_2}{LC_1C_2} + \frac{g_m}{C_1C_2R_L} + \frac{1}{C_1C_2R_Lr_o}} = \sqrt{\frac{C_1+C_2}{LC_1C_2} + \frac{1}{C_1C_2R_L}\left(g_m + \frac{1}{r_o}\right)} \tag{12.155}$$

For $C_1C_2R_L \gg g_m$ and $C_1C_2R_Lr_o \gg 1$, (12.155) simplifies to

$$\omega_o \approx \frac{1}{\sqrt{\frac{LC_1C_2}{C_1+C_2}}} \tag{12.156}$$

For $L = 4.75$ mH and $C_1 = C_2 = 1$ μH, $f_o = 3.26$ kHz.

Equating the real part of (12.154) to zero, we obtain the condition for steady-state oscillations

$$g_mR = \frac{C_1}{C_2} + \frac{r_o}{R_L+r_o}\left(\frac{C_2}{C_1} + 2\right) \tag{12.157}$$

where $R = R_Lr_o/(R_L+r_o)$. For $R_L \ll r_o$, (12.157) can be approximated by

$$g_mR \approx 2 + \frac{C_1}{C_2} + \frac{C_2}{C_1} \tag{12.158}$$

For example, for $C_1 = C_2$, we get $g_mR = 2 + 1 + 1 = 4$ V/V. Figure 12.30 illustrates the function $2 + C_1/C_2 + C_2/C_1$ as a function of C_1/C_2.

In order for the oscillations to start and build up, the magnitude of the loop gain at the oscillation frequency f_o must be greater than 1, that is, $|T(f_o)| > 1$. The startup condition for oscillations is

$$g_mR > \frac{C_1}{C_2} + \frac{r_o}{R_L+r_o}\left(\frac{C_2}{C_1} + 2\right) \approx 2 + \frac{C_1}{C_2} + \frac{C_2}{C_1} \tag{12.159}$$

12.8.4 Amplifier Voltage Gain of Common-Gate Colpitts Oscillator

Figure 12.27(d) shows the small-signal model of the amplifier A loaded by the feedback network. The total impedance seen by the MOSFET drain is

$$Z_L = (sL||R_L)||\left[\left(r_o||\frac{1}{sC_2}\right) + \left(\frac{1}{g_m}||\frac{1}{sC_1}\right)\right]$$
$$= \frac{sLR_L[sr_o(C_1+C_2) + g_mr_o + 1]}{s^3LC_1C_2R_Lr_o + s^2[Lr_o(C_1+C_2) + LR_L(C_1 + C_2g_mr_o)] + s[Lg_mR_L + R_Lr_o(C_1+C_2) + L(g_mr_o+1)] + R_L(g_mr_o+1)} \tag{12.160}$$

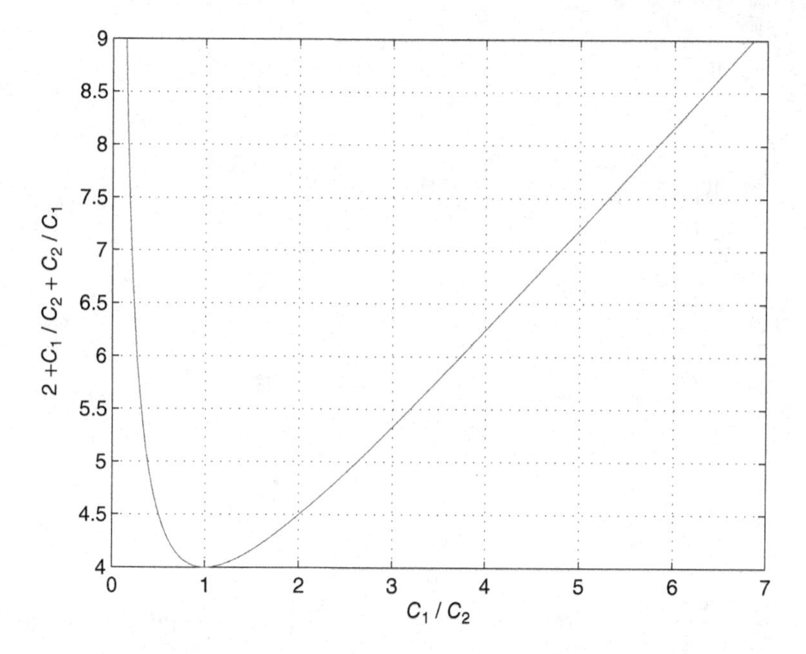

Figure 12.30 Function $2 + C_1/C_2 + C_2/C_1$ as a function of C_1/C_2.

The voltage gain of the amplifier loaded by the feedback network is

$$A(s) = \frac{v_o}{v_f} = g_m Z_L$$

$$= \frac{sLg_mR_L[sr_o(C_1 + C_2) + g_mr_o + 1]}{s^3LC_1C_2R_Lr_o + s^2[Lr_o(C_1 + C_2) + LR_L(C_1 + C_2g_mr_o)] + s[Lg_mR_L + R_Lr_o(C_1 + C_2) + L(g_mr_o + 1)] + R_L(g_mr_o + 1)} \quad (12.161)$$

Figures 12.31 and 12.32 show Bode plots of the amplifier voltage gain A for the CG Colpitts oscillator at $L = 4.75$ mH, $C_1 = C_2 = 1$ μH, $R_L = 1$ kΩ, $r_o = 100$ kΩ, and $g_m = 4.09$ mA/V. The amplifier gain at the oscillation frequency is $|A(\omega_o)| = 2 = 6$ dB.

12.8.5 Loop Gain of Common-Gate Colpitts Oscillator

Using (12.143) and (12.161), we obtain the loop gain of the CG Colpitts oscillator

$$T(s) = \beta A$$

$$= \frac{sLg_mR_L(sr_oC_2 + 1)}{s^3LC_1C_2R_Lr_o + s^2[Lr_o(C_1 + C_2) + LR_L(C_1 + C_2g_mr_o)] + s[Lg_mR_L + R_Lr_o(C_1 + C_2) + L(g_mr_o + 1)] + R_L(g_mr_o + 1)} \quad (12.162)$$

The loop gain is described as a third-order transfer function. It has two finite real zeros and three poles. One zero is at the origin, $z_1 = 0$, and the other zero is given by $z_2 = -1/(r_oC_2)$. One pole is real and two poles are either real or complex conjugate. When the loop gain T increases, both the real and imaginary parts of the complex poles also increase, and thereby their damping coefficient decreases and the frequency increases. Figures 12.33 and 12.34 show Bode plots of loop gain T for the CG Colpitts oscillator at $L = 4.75$ mH, $C_1 = C_2 = 1$ μH, $R_L = 1$ kΩ, $r_o = 100$ kΩ, and $g_m = 4.09$ mA/V. The loop gain at the oscillation frequency is $|T(\omega_o)| = 1 = 0$ dB.

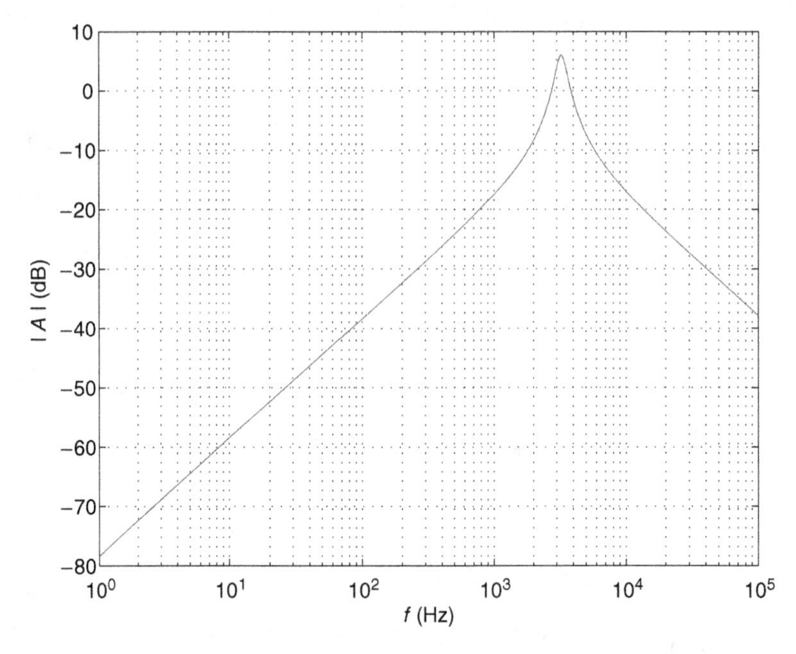

Figure 12.31 Magnitude of the amplifier voltage gain $|A|$ as a function of frequency f for the common-gate Colpitts oscillator.

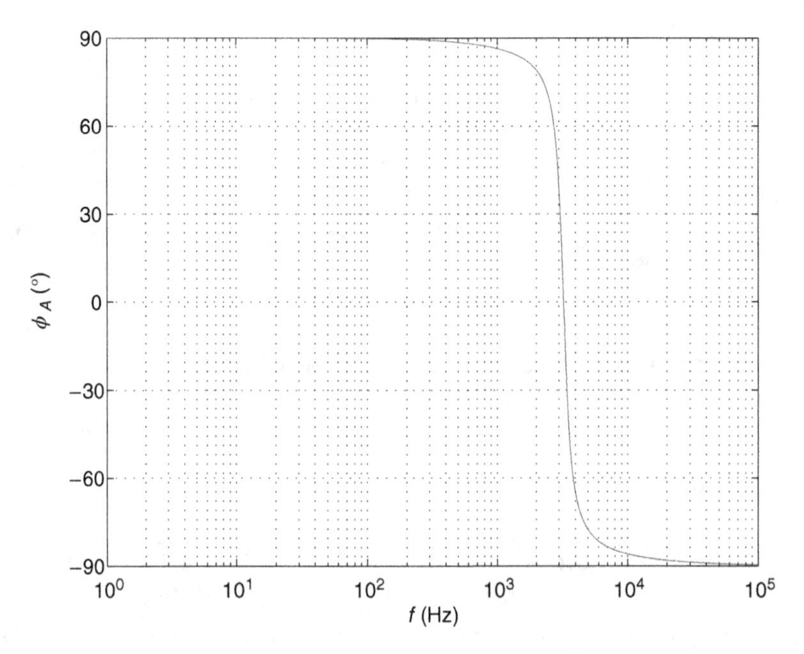

Figure 12.32 Phase of the amplifier gain ϕ_A as a function of frequency f for the common-gate Colpitts oscillator.

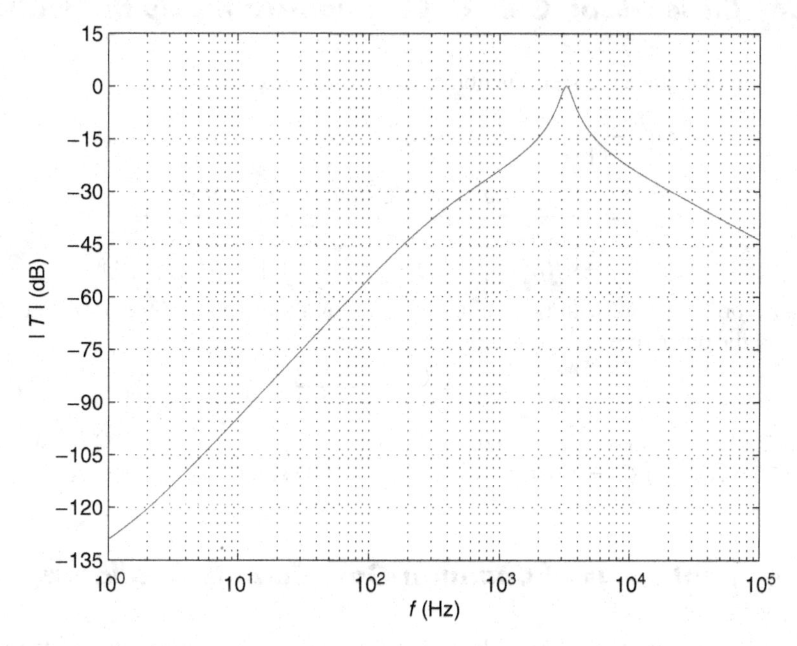

Figure 12.33 Magnitude of the loop gain $|T|$ as a function of frequency f for the common-gate Colpitts oscillator.

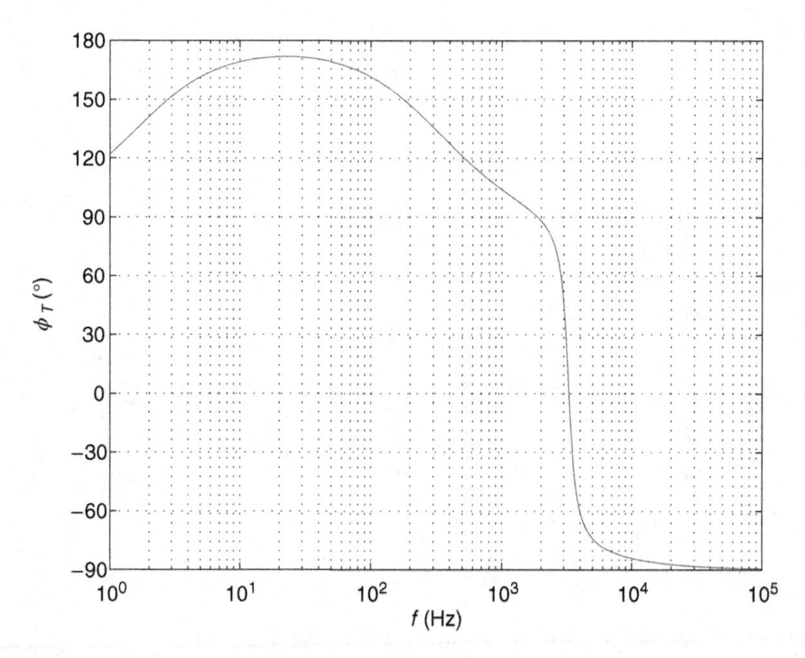

Figure 12.34 Phase of the loop gain ϕ_T as a function of frequency f for the common-gate Colpitts oscillator.

12.8.6 Closed-Loop Gain of Common-Gate Colpitts Oscillator

The closed-loop voltage transfer function of the CG oscillator is

$$A_f(s) = \frac{A(s)}{1 - T(s)}$$
$$= \frac{sLg_m R_L[sr_o(C_1 + C_2) + g_m r_o + 1]}{s^3 LC_1C_2R_L r_o + s^2 L[r_o(C_1 + C_2) + C_1R_L] + s[R_L r_o(C_1 + C_2) + L(g_m r_o + 1)] + R_L(g_m r_o + 1)} \quad (12.163)$$

It is a third-order system with two finite real zeros and three poles. One zero is at the origin, $z_1 = 0$, and the other zero is

$$z_3 = -\frac{1 + g_m r_o}{r_o(C_1 + C_2)} \quad (12.164)$$

Figures 12.35 and 12.36 show Bode plots of the closed-loop gain A_f for the CG Colpitts oscillator at $L = 4.75$ mH, $C_1 = C_2 = 1$ μH, $R_L = 1$ kΩ, $r_o = 100$ kΩ, and $g_m = 4.09$ mA/V.

12.8.7 Root Locus of Common-Gate Colpitts Oscillators

To plot the root locus of the closed-loop transfer function A_f as a function of the MOSFET transconductance g_m, we divide the characteristic equation by

$$s^3 LC_1C_2R_L r_o + s^2 L[r_o(C_1 + C_2) + C_1R_L] + s[R_L r_o(C_1 + C_2) + L] + R_L \quad (12.165)$$

to obtain

$$1 + g_m \frac{r_o(sL + R_L)}{s^3 LC_1C_2R_L r_o + s^2 L[r_o(C_1 + C_2) + C_1R_L] + s[R_L r_o(C_1 + C_2) + L] + R_L} = 0 \quad (12.166)$$

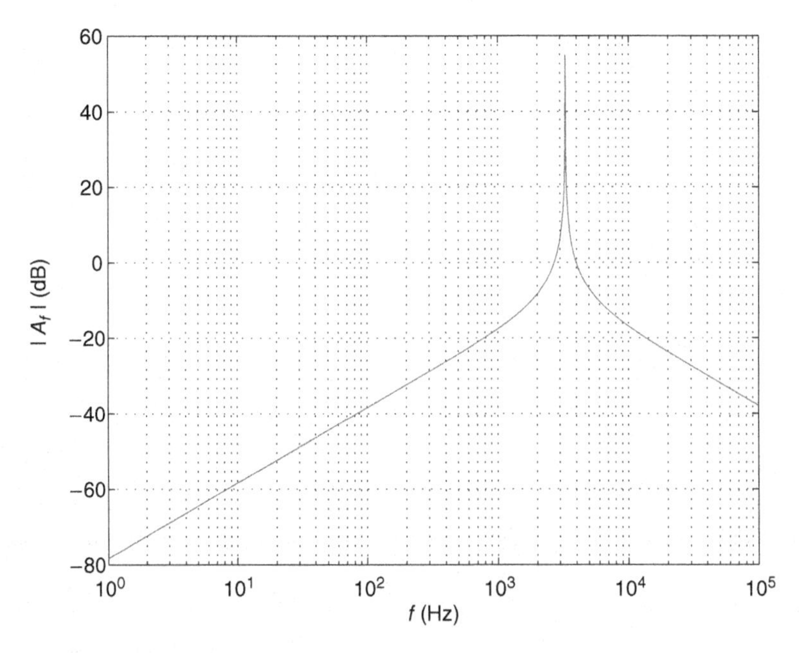

Figure 12.35 Magnitude of the closed-loop gain $|A_f|$ as a function of frequency f for the common–gate Colpitts oscillator.

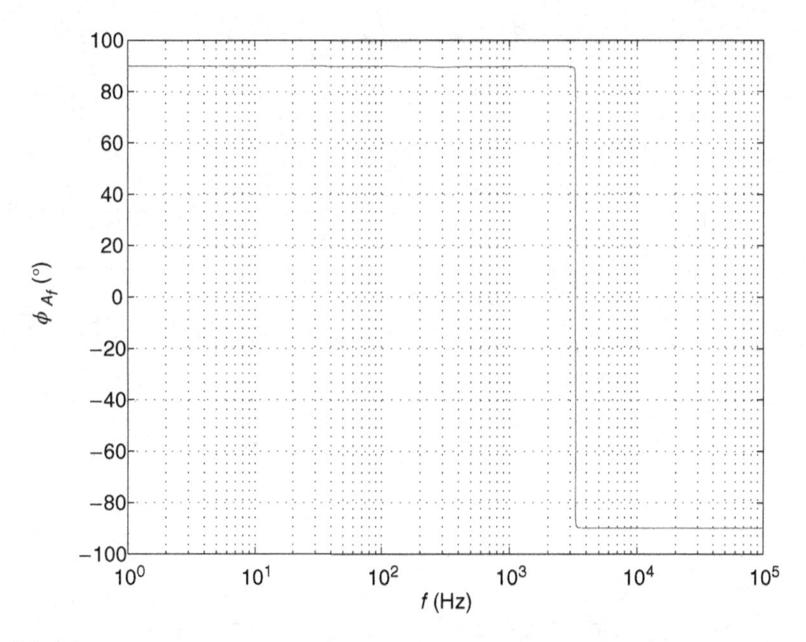

Figure 12.36 Phase of the closed-loop gain ϕ_{A_f} as a function of frequency f for the common-gate Colpitts oscillator.

Figure 12.37 shows a root locus for the CG Colpitts oscillator when g_m varies from 0 to 10 A/V. Figure 12.38 shows the root locus for steady-state oscillation, which occur at $g_m = 4.09$ mA/V. For $g_m = 0$, the open-loop poles are $p_1, p_2 = -1012.65 \pm j20,598.063$ rad/s and $p_3 = -2030.23$ rad/s. All poles are located in the LHP. Therefore, the loop gain is too low to start oscillations. For $g_m = 4.09$ mA/V, the poles are $p_1, p_2 = \pm j20,723$ rad/s and $p_3 = -5$ rad/s and the circuit produces steady-state oscillations. A pair of complex conjugate poles is located on the imaginary axis. For $g_m = 1$ A/V, the poles are $p_1, p_2 = 25,139.11 \pm j58,595.41$ rad/s and $p_3 = -52,308.58$ rad/s. In this case, a pair of complex conjugate poles is located in the RHP and the circuit is able to start oscillations.

12.8.8 Nyquist Plot of Common-Gate Colpitts Oscillator

Figure 12.39 shows the Nyquist plots for the CG Colpitts oscillator at $L = 4.75$ mH, $C_1 = C_2 = 1$ μH, $R_L = 1$ kΩ, and $r_o = 100$ kΩ for $g_m = 2$ mA/V, $g_m = 4.09$ mA/V, and $g_m = 10$ mA/V. The Nyquist plot encircles the point $-1 + j0$ at $g_m = 10$ mA/V and the oscillator can start oscillations. As the amplitude of the oscillator voltage increases after the startup, the nonlinearity of the transistor reduces the amplifier gain. Therefore, the Nyquist plot will change such that it crosses the critical point $-1 + j0$ at steady-state oscillations.

12.9 Common-Drain Colpitts Oscillator

Figure 12.40 shows a common-drain Colpitts oscillator, its dc circuit, and its ac circuit. An ac circuit of the Colpitts oscillator with the transistor operated in the CD configuration is depicted in Fig. 12.41(a). The load resistance R_L is connected between the source and the drain terminals and is in parallel with the capacitor C_2. A small-signal model of the CD Colpitts oscillator is shown

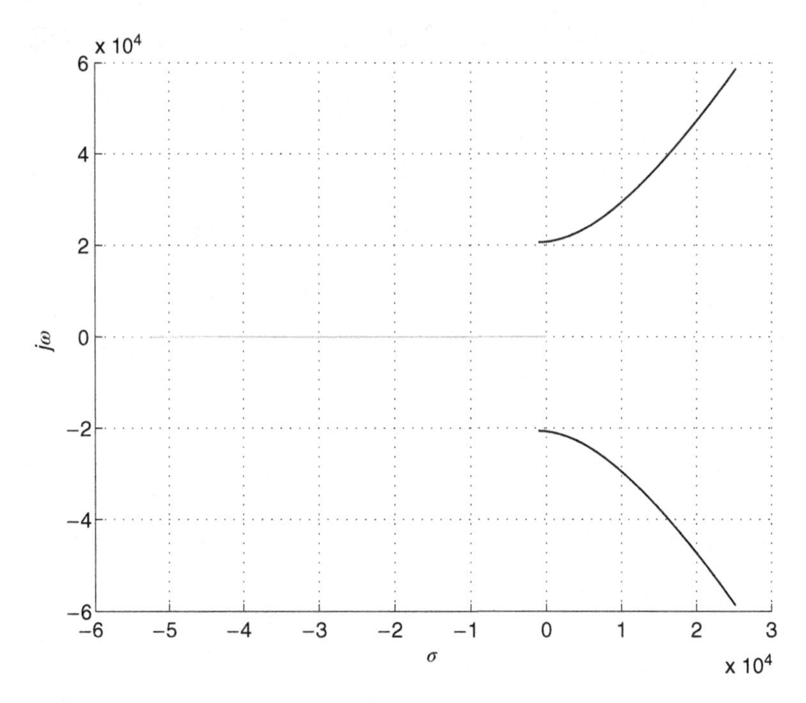

Figure 12.37 Root locus of the closed-loop gain $A_f(s)$ for the common-gate Colpitts oscillator for variations of g_m from 0 to 1 A/V.

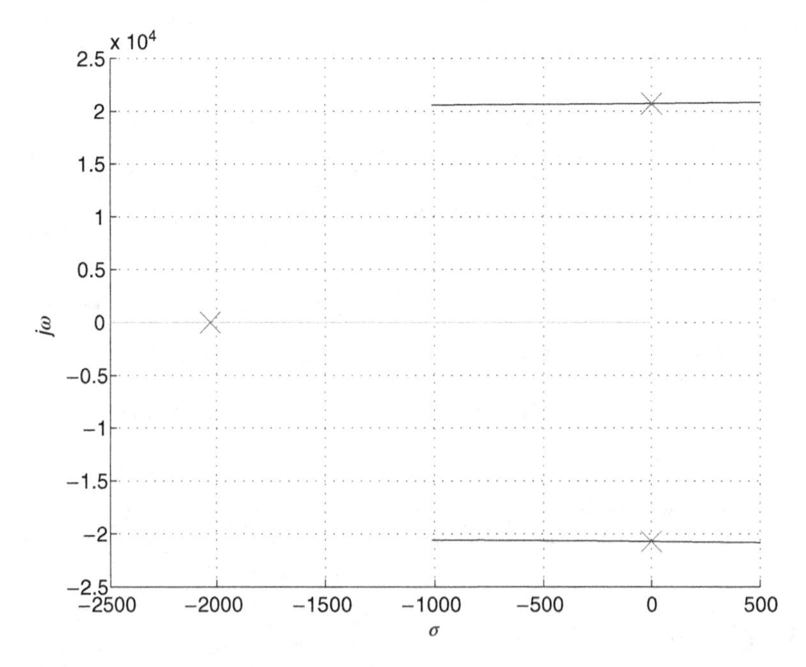

Figure 12.38 Root locus for the common-gate Colpitts oscillator for steady-state oscillations at $g_m = 4.09$ mA/V.

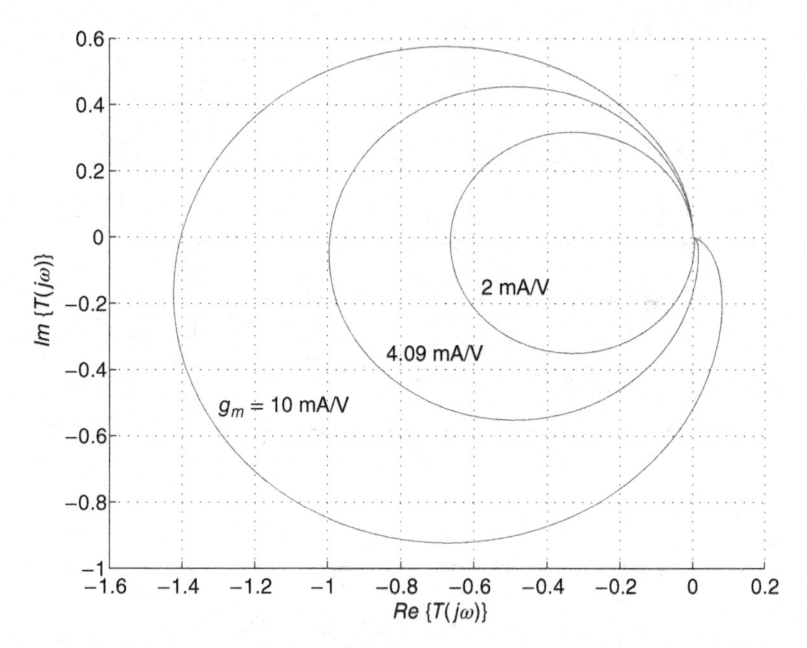

Figure 12.39 Nyquist plot of the loop gain $T(j\omega)$ for the common-gate Colpitts oscillator for $g_m = 2$ mA/V (too low value to start oscillations), $g_m = 4.09$ mA/V (steady-state oscillations), and $g_m = 10$ mA/V (sufficiently large value to start oscillations).

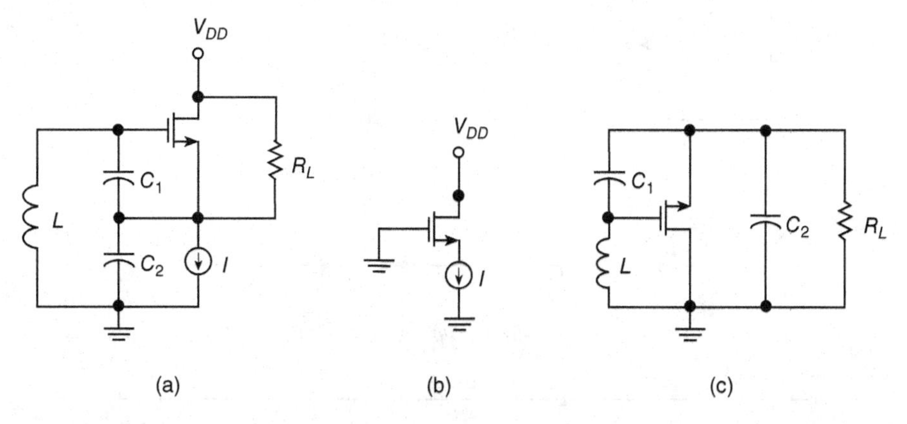

Figure 12.40 Common-drain (CD) Colpitts oscillator: (a) circuit, (b) DC circuit, and (c) small-signal model.

in Fig. 12.41(b). It should be noted that $v_{gs} = v_f - v_o$. Figure 12.41(c) shows a small-signal model of the amplifier loaded by the feedback network.

12.9.1 Feedback Factor of Common-Drain Colpitts Oscillator

Applying KCL to the small-signal model of the Colpitts oscillator shown in Fig. 12.41(b), we can describe the current through the inductor L

$$i = \frac{v_f}{sL} \tag{12.167}$$

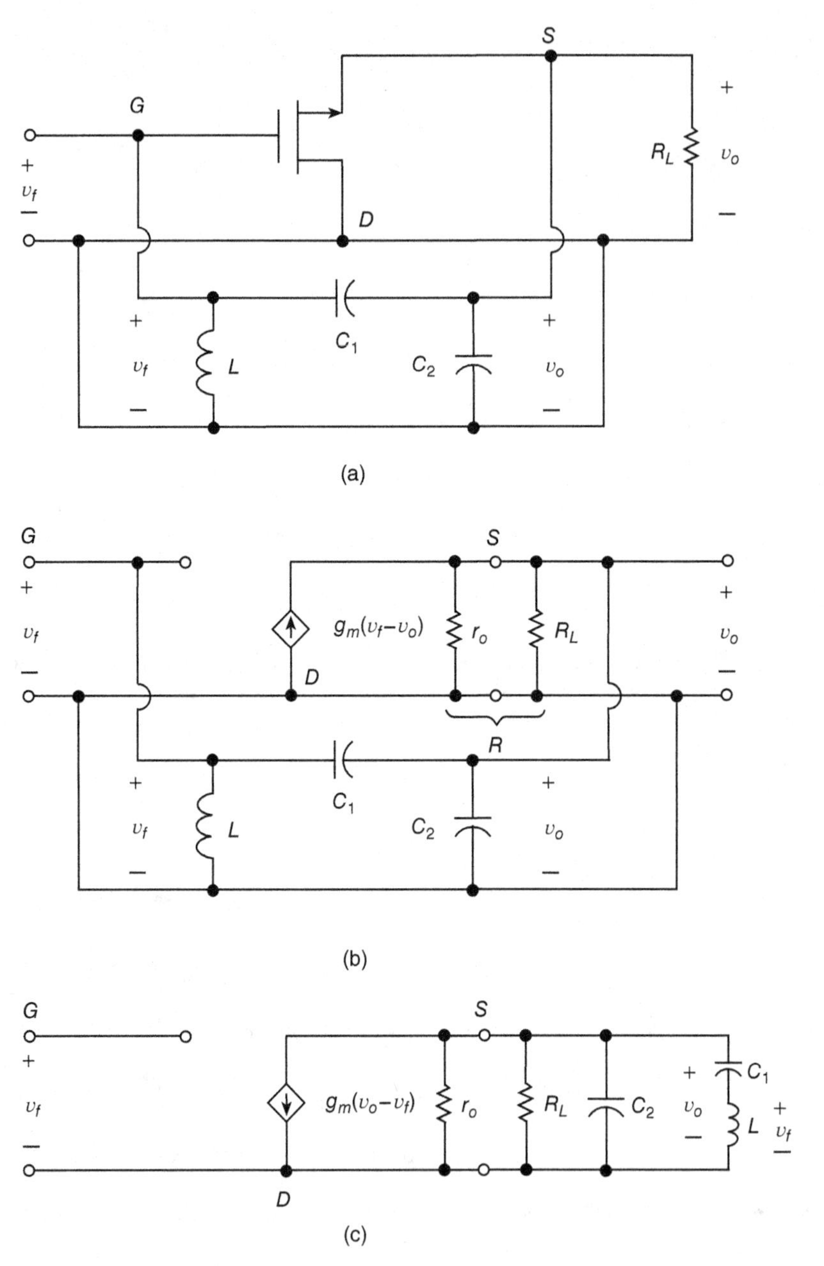

Figure 12.41 Common-drain (CD) Colpitts oscillator: (a) ac circuit, (b) small-signal model, and (c) small-signal model of the amplifier of the CD Colpitts oscillator loaded by the feedback network.

and the current through the series combination of inductor L and capacitor C_1

$$i = \frac{v_o}{sL + \frac{1}{sC_1}} = v_o \left(\frac{sC_1}{1 + s^2 LC_1} \right) \tag{12.168}$$

Equating the right-hand side of these expressions, we get

$$v_o = v_f \left(\frac{s^2 LC_1 + 1}{s^2 LC_1} \right) \tag{12.169}$$

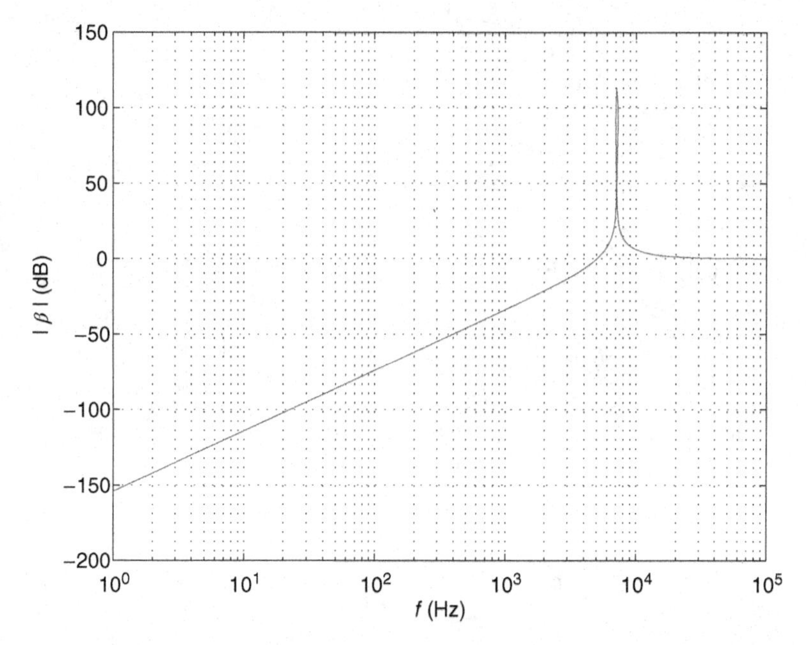

Figure 12.42 Magnitude of the feedback factor $|\beta|$ as a function of frequency f for the common-drain Colpitts oscillator.

resulting in the voltage transfer function of the feedback network

$$\beta(s) = \frac{v_f}{v_o} = \frac{s^2LC_1}{s^2LC_1 + 1} \tag{12.170}$$

Alternatively, the voltage transfer function of the feedback network can be derived as

$$\beta(s) = \frac{v_f}{v_o} = \frac{sL}{sL + \dfrac{1}{sC_1}} = \frac{s^2LC_1}{s^2LC_1 + 1} \tag{12.171}$$

The feedback network of the CD Colpitts oscillator is a second-order high-pass filter. Figures 12.42 and 12.43 show the magnitude $|\beta|$ and phase ϕ_β of the feedback factor as functions of frequency f.

12.9.2 Characteristic Equation of Common-Drain Colpitts Oscillator

Using KCL at the source terminal S in Fig. 12.41(b), we obtain

$$v_o\left(\frac{sC_1}{s^2LC_1 + 1}\right) + v_o\left(\frac{sC_2R + 1}{R}\right) + g_m(v_o - v_f) = 0 \tag{12.172}$$

producing

$$v_o\left(\frac{sC_1}{s^2LC_1 + 1} + \frac{sC_2R + 1}{R} + g_m\right) - g_m v_f = 0 \tag{12.173}$$

Substitution of (12.169) into (12.173) gives

$$v_f\left(\frac{s^2LC_1 + 1}{s^2LC_1}\right)\left(\frac{sC_1}{s^2LC_1 + 1} + \frac{sC_2R + 1}{R} + g_m\right) - g_m v_f = 0 \tag{12.174}$$

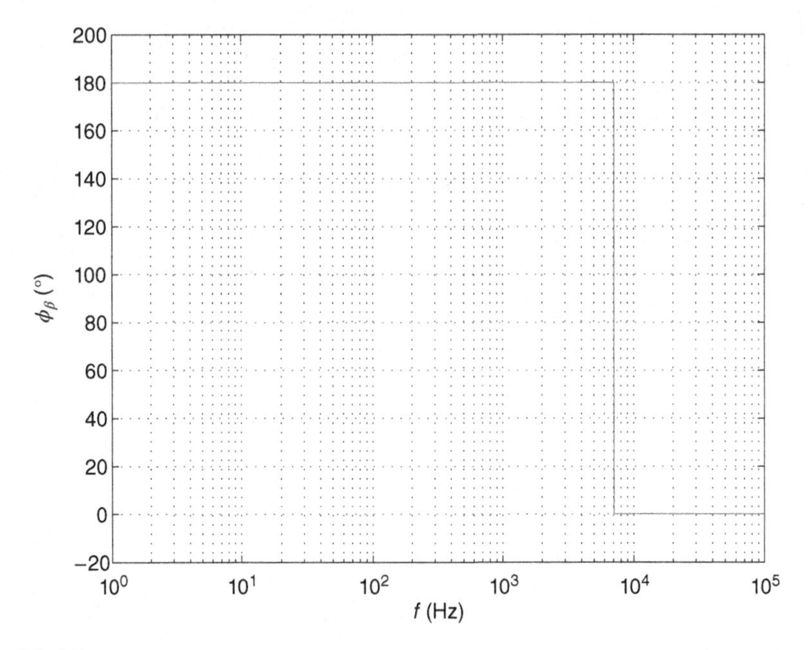

Figure 12.43 Phase of the feedback factor ϕ_β as a function of frequency f for the common-drain Colpitts oscillator.

When the oscillator oscillates, $v_f \neq 0$. Hence, dividing both sides of (12.174) by v_f, we obtain

$$\left(\frac{s^2LC_1+1}{s^2LC_1}\right)\left(\frac{sC_1}{s^2LC_1+1}+\frac{sC_2R+1}{R}+g_m\right)-g_m=0 \tag{12.175}$$

Rearrangement of this equation yields the characteristic equation of the CD oscillator

$$s^3LC_1C_2R+s^2LC_1+sR(C_1+C_2)+g_mR+1=0 \tag{12.176}$$

For steady-state oscillations, $s=j\omega_o$. Hence, collecting the like terms, we get

$$1+g_mR-\omega_o^2LC_1+j\omega_oR(C_1+C_2-\omega_o^2LC_1C_2)=0=0+j0 \tag{12.177}$$

Equating the imaginary part of (12.177) to zero, we can derive an expression for the frequency of oscillations

$$\omega_o=\frac{1}{\sqrt{\dfrac{LC_1C_2}{C_1+C_2}}} \tag{12.178}$$

Equating the real parts of (12.177) to zero and substituting ω_o, the condition for steady-state oscillations is obtained as

$$g_mR=\frac{C_1}{C_2} \tag{12.179}$$

The startup condition for oscillations is

$$g_mR>\frac{C_1}{C_2} \tag{12.180}$$

12.9.3 Amplifier Gain of Common-Drain Colpitts Oscillator

Figure 12.41(c) shows a small-signal model of the amplifier loaded by the feedback network. The output voltage of the amplifier in the CD Colpitts oscillator is

$$v_o = -(v_o - v_f)Z_L \tag{12.181}$$

resulting in

$$v_o(1 + g_m Z_L) = -v_f Z_L \tag{12.182}$$

Hence, the voltage gain of the amplifier is

$$A(s) = \frac{v_o}{v_e} = -\frac{g_m Z_L}{1 + g_m Z_L} = -\frac{1}{1 + \frac{1}{g_m Z_L}} \tag{12.183}$$

The load impedance of the amplifier is

$$Z_L = \left(R||\frac{1}{sC_2}\right) || \left(sL + \frac{1}{sC_1}\right) = \frac{R(1 + s^2LC_1)}{s^3LC_1C_2R + s^2LC_1 + sR(C_1 + C_2) + 1} \tag{12.184}$$

where $R = R_L r_o/(R_L + r_o)$. The voltage transfer function of the amplifier loaded by the feedback network is

$$A = \frac{v_o}{v_f} = -\frac{g_m R(1 + s^2LC_1)}{s^3LC_1C_2R + s^2LC_1(1 + g_m R) + sR(C_1 + C_2) + g_m R + 1}. \tag{12.185}$$

Figures 12.44 and 12.45 show Bode plots of the amplifier voltage gain.

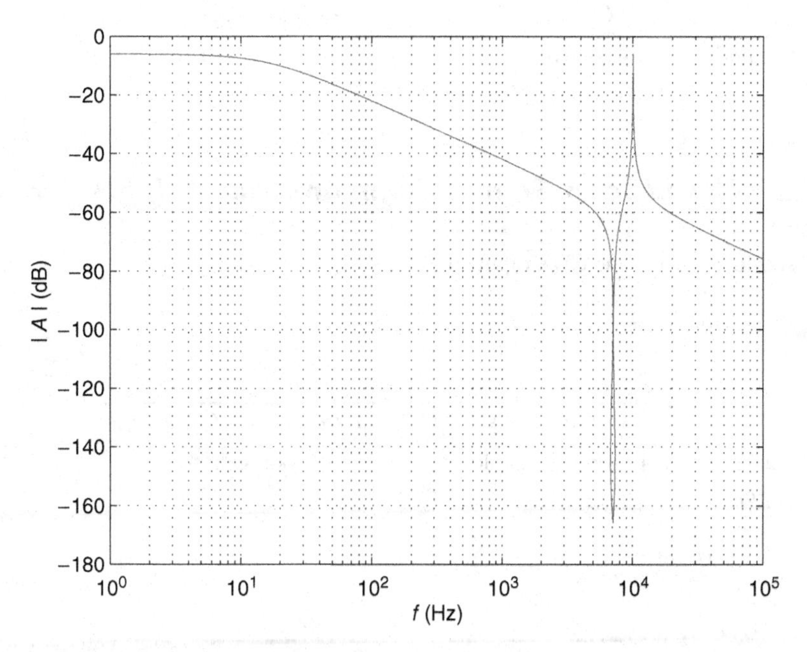

Figure 12.44 Magnitude of the amplifier voltage gain $|A|$ as a function of frequency f for the common-drain Colpitts oscillator.

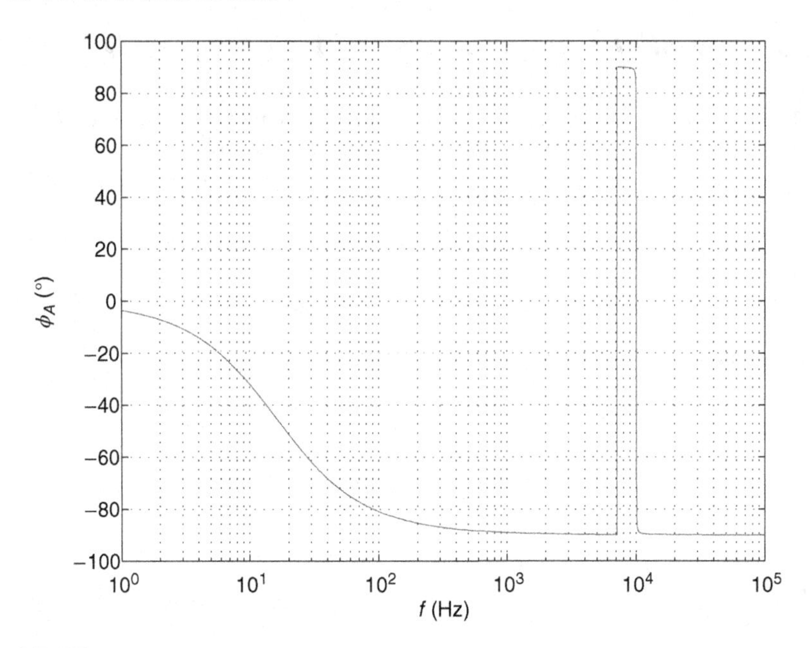

Figure 12.45 Phase of the amplifier gain ϕ_A as a function of frequency f for the common-drain Colpitts oscillator.

12.9.4 Loop Gain of Common-Drain Colpitts Oscillator

The loop gain of the CD Colpitts oscillator is

$$T(s) = \beta A = \frac{s^2 g_m RLC_1}{s^3 LC_1 C_2 R + s^2 LC_1(1 + g_m R) + sR(C_1 + C_2) + g_m R + 1} \tag{12.186}$$

The loop gain is a third-order function. Figures 12.46 and 12.47 show Bode plots of the loop gain T.

12.9.5 Closed-Loop Gain of Common-Drain Colpitts Oscillator

The closed-loop gain of the CD Colpitts oscillator is

$$A_f(s) = \frac{A(s)}{1 - T(s)}$$

$$= -\frac{g_m R(1 + s^2 LC_1)}{s^3 LC_1 C_2 R + s^2 LC_1(1 + g_m R) + sR(C_1 + C_2) + 2g_m R + 1} \tag{12.187}$$

Figures 12.48 and 12.49 show Bode plots for the closed-loop gain A_f.

For $s = j\omega$,

$$\beta(\omega) = \frac{\omega^2 LC_1}{\omega^2 LC_1 - 1} \tag{12.188}$$

$$A(j\omega) = \frac{g_m R(1 - \omega^2 LC_1)}{(1 - \omega^2 LC_1)(1 + g_m R) + j[\omega R(C_1 + C_2) - \omega^3 LC_1 C_2 R]} \tag{12.189}$$

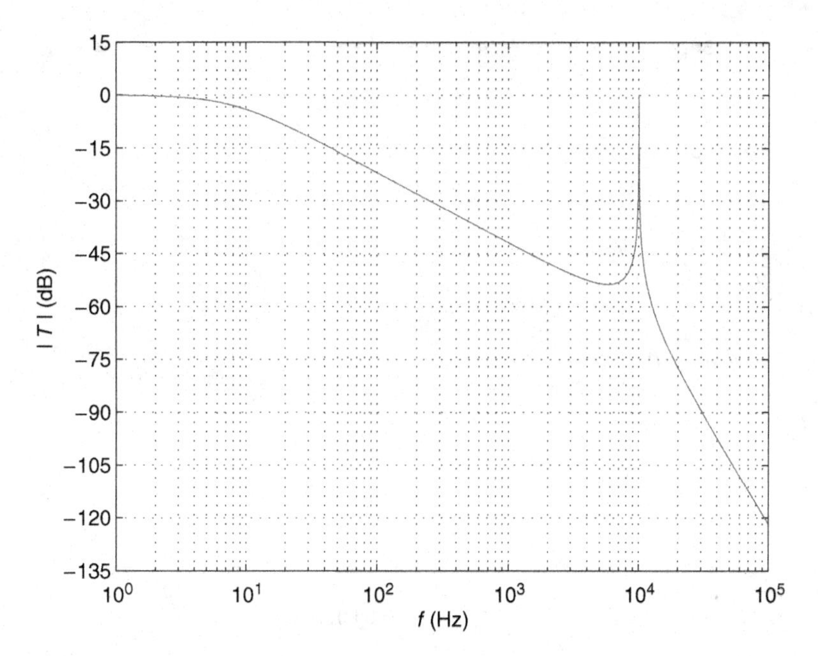

Figure 12.46 Magnitude of the loop gain $|T|$ as a function of frequency f for the common-drain Colpitts oscillator.

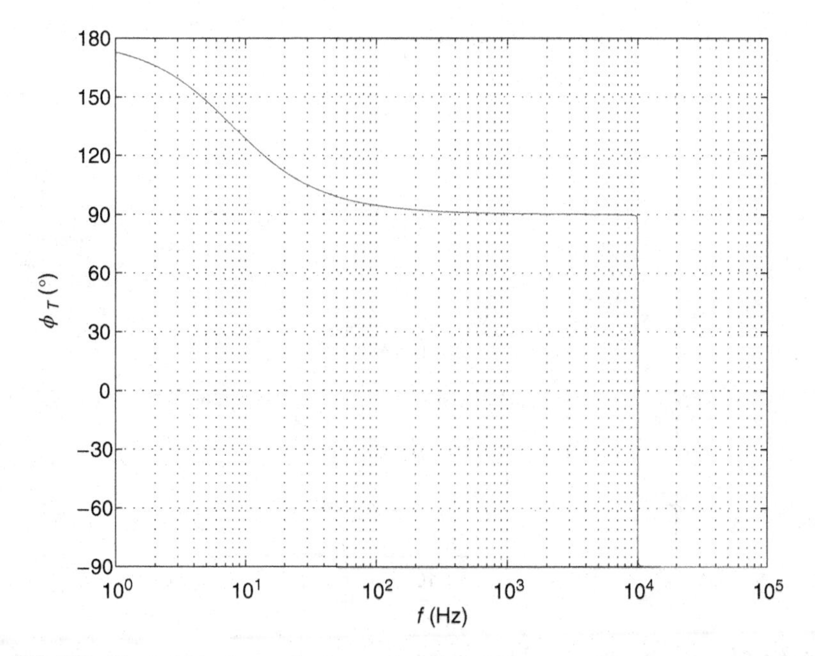

Figure 12.47 Phase of the loop gain ϕ_T as a function of frequency f for the common-drain Colpitts oscillator.

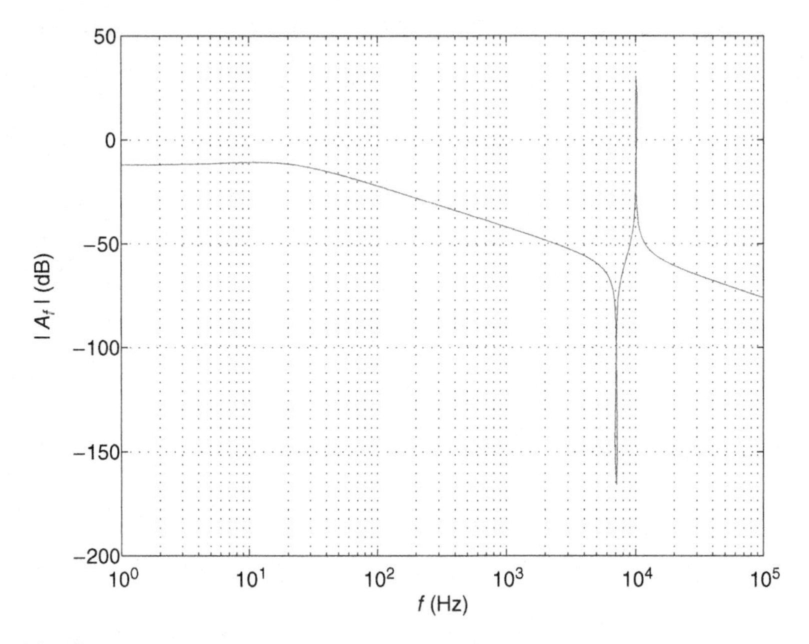

Figure 12.48 Magnitude of the closed-loop gain $|A_f|$ as a function of frequency f for the common-drain Colpitts oscillator.

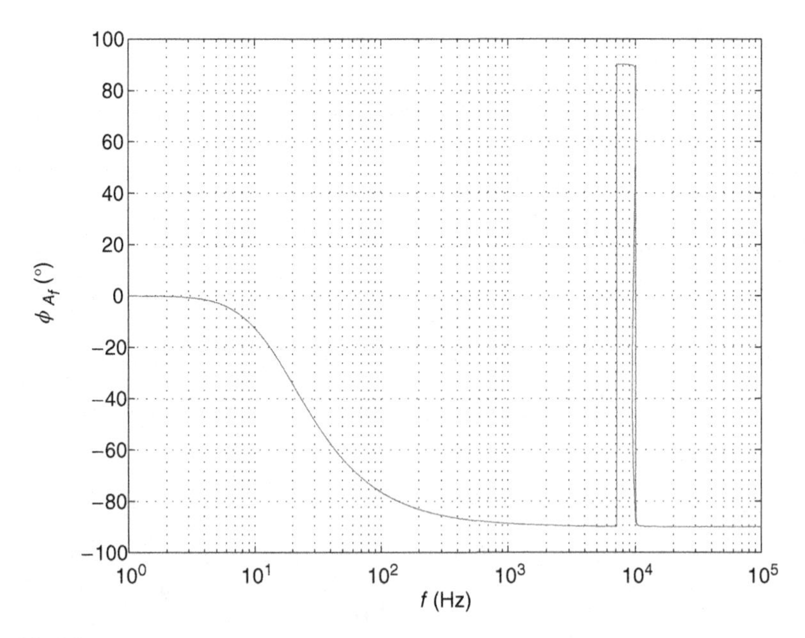

Figure 12.49 Phase of the closed-loop gain ϕ_{A_f} as a function of frequency f for the common-drain Colpitts oscillator.

and

$$T(j\omega) = \frac{-g_m R}{(1-\omega^2 LC_1)(1+g_m R) + j\omega R[(C_1+C_2) - \omega^2 LC_1 C_2]} \tag{12.190}$$

The imaginary part of the denominator of T is zero for steady-state oscillation. Hence, the oscillation frequency is

$$\omega_o = \frac{1}{\sqrt{L\frac{C_1 C_2}{C_1+C_2}}} \tag{12.191}$$

For $\omega = \omega_o$,

$$\beta(\omega_o) = \frac{\omega_o^2 LC_1}{\omega_o^2 LC_1 - 1} = 1 + \frac{C_2}{C_1} \tag{12.192}$$

$$A(\omega_o) = \frac{g_m R}{1+g_m R} \tag{12.193}$$

and

$$T(\omega_o) = \beta(\omega_o)A(\omega_o) = \frac{g_m R}{1+g_m R}\left(1+\frac{C_2}{C_1}\right) = 1 \tag{12.194}$$

Hence, the amplitude condition for steady-state oscillation is

$$g_m R = 1 + \frac{C_1}{C_2} \tag{12.195}$$

The startup condition for oscillations is

$$g_m R > 1 + \frac{C_1}{C_2} \tag{12.196}$$

12.9.6 Nyquist Plot of Common-Drain Colpitts Oscillator

Figure 12.50 depicts a Nyquist plot for the CD Colpitts oscillator at three values of g_m.

12.9.7 Root Locus of Common-Drain Colpitts Oscillator

Dividing the characteristic equation of the CD Colpitts oscillator given in (12.9.2) by

$$s^3 LC_1 C_2 R + s^2 LC_1 + sR(C_1+C_2) + g_m R + 1 = 0 \tag{12.197}$$

we obtain

$$1 + \frac{g_m R}{s^3 LC_1 C_2 R + s^2 LC_1 + sR(C_1+C_2) + 1} \tag{12.198}$$

Figure 12.51 shows the root locus for the CD Colpitts oscillator when g_m varies from 0 to 10 A/V. Figure 12.52 shows the root locus for steady-state oscillations when $g_m = 1.01$ mA/V. For $g_m = 0$, the poles are $p_1, p_2 = -25.25 \pm j199{,}999.99$ rad/s and $p_3 = -50.5$ rad/s. For $g_m = 1.01$, the poles are $p_1, p_2 = \pm j199{,}999.99$ rad/s and $p_3 = -101.005$ rad/s. For $g_m = 10$, the poles are $p_1, p_2 = 111{,}442.82 \pm j277{,}994$ rad/s and $p_3 = -222{,}986.65$ rad/s.

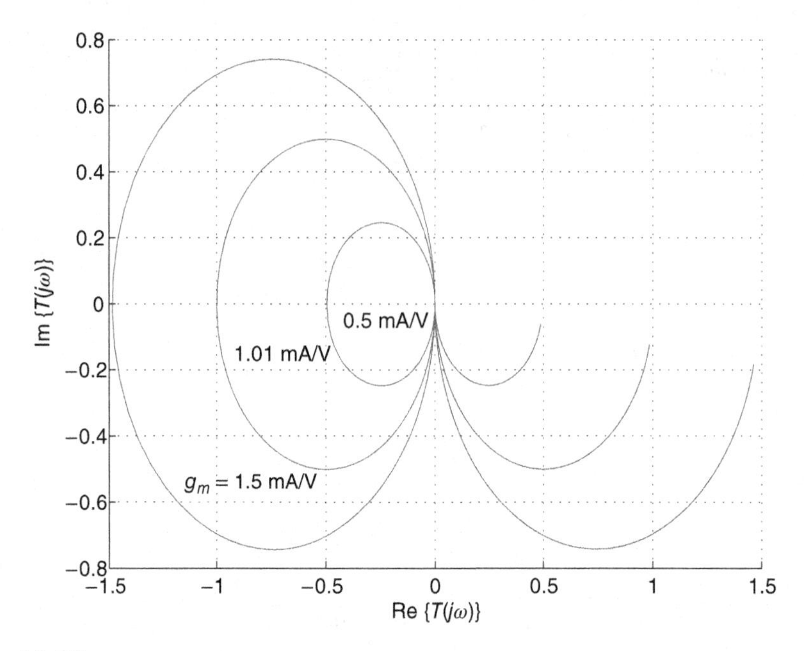

Figure 12.50 Nyquist plot of the loop gain $T(j\omega)$ for the common-drain Colpitts oscillator at $g_m = 0.5$ mA/V (too low value to start oscillations), $g_m = 1.01$ mA/V (steady-state oscillations), and $g_m = 1.5$ mA/V (sufficiently large value to start oscillations).

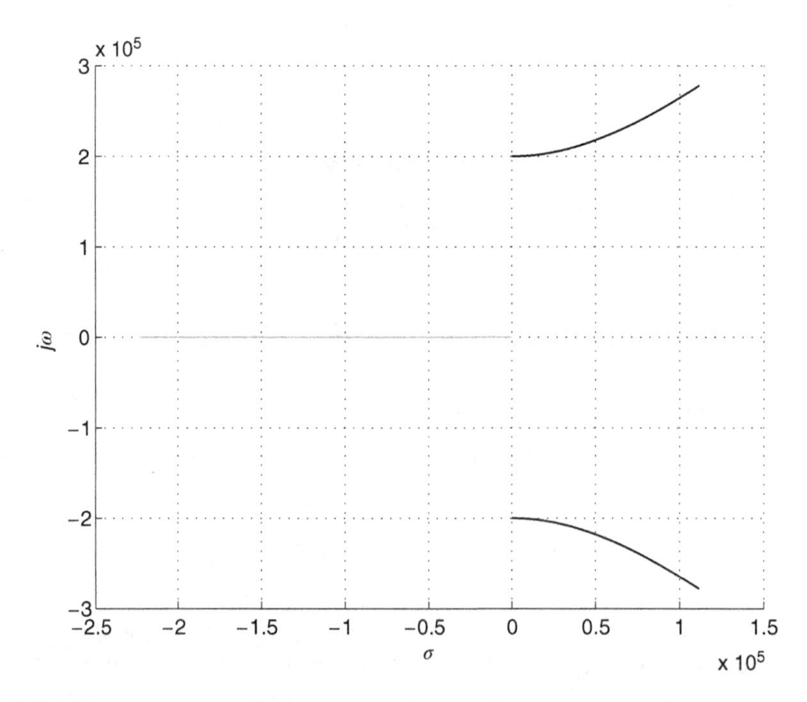

Figure 12.51 Root locus of the loop gain T for the common-drain Colpitts oscillator as g_m varies from 0 to 10 A/V.

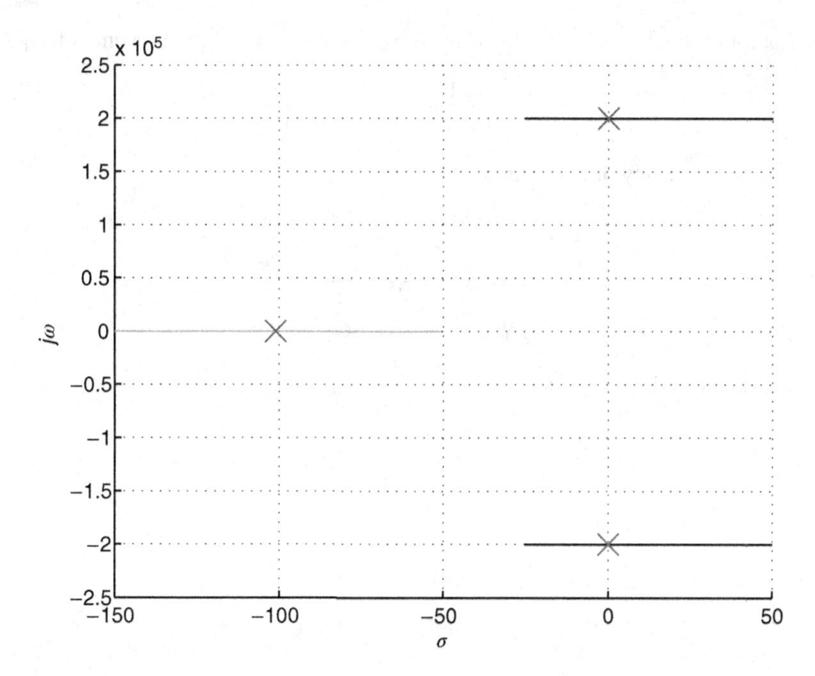

Figure 12.52 Root locus of the loop gain T for the common-drain Colpitts oscillator for steady-state oscillations at $g_m = 1.01$ mA/V.

12.10 Clapp Oscillator

Figure 12.53 shows the circuit of the Clapp oscillator. This oscillator is a modification of the Colpitts oscillator by adding capacitor C_3 in series with the inductor L. If the inductance L in the Colpitts oscillator becomes very small, then the quality factor Q is very low, causing a low stability of the oscillation frequency. A capacitor C_3 connected in series with the inductance L gives a larger inductance L and the quality factor Q, while maintaining the same effective inductive reactance. The amplifier input capacitance C_i is absorbed into capacitance C_1 and the

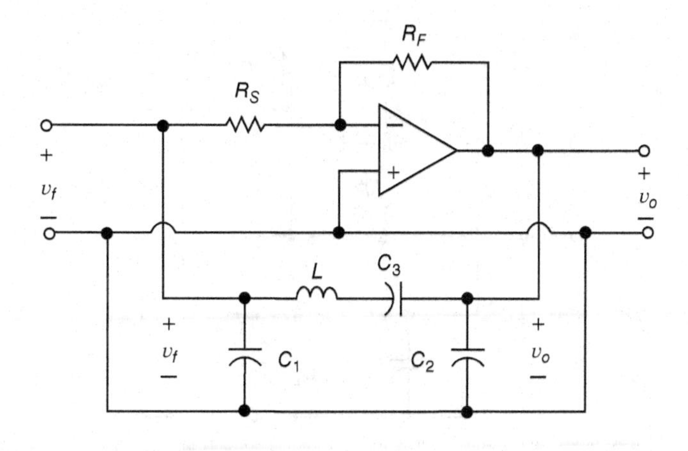

Figure 12.53 Clapp oscillator.

amplifier output capacitance C_o is absorbed into capacitance C_2. At the resonant frequency f_o,

$$\omega_o L = \frac{1}{\omega_o C_1} + \frac{1}{\omega_o C_2} + \frac{1}{\omega_o C_3} \tag{12.199}$$

resulting in the frequency of oscillation

$$f_o = \frac{1}{2\pi\sqrt{L\frac{1}{\frac{1}{C_1}+\frac{1}{C_2}+\frac{1}{C_3}}}} = \frac{1}{2\pi\sqrt{LC}} \tag{12.200}$$

The total capacitance is

$$C = \frac{1}{\frac{1}{C_1}+\frac{1}{C_2}+\frac{1}{C_3}} \tag{12.201}$$

For $C_3 \ll C_1$ and $C_3 \ll C_2$,

$$C \approx C_3 \tag{12.202}$$

and the frequency of oscillation is nearly independent of C_1 and C_2

$$f_o = \frac{1}{2\pi\sqrt{LC}} \approx \frac{1}{2\pi\sqrt{LC_3}} \tag{12.203}$$

Therefore, the frequency of oscillation is nearly independent of the input and output capacitances of the amplifier. The capacitance C_3 and the inductance L set the oscillation frequency f_o and the ratio of capacitors is

$$\frac{C_1}{C_2} = \frac{R_F}{R_S} \tag{12.204}$$

Figure 12.54 shows a circuit of the Clapp oscillator with a single transistor.

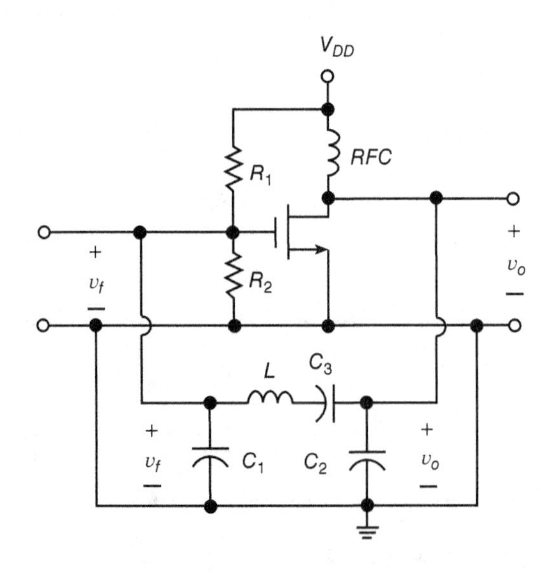

Figure 12.54 Clapp oscillator with single transistor.

12.11 Crystal Oscillators

Some crystals found in nature exhibit piezoelectric effect. Piezoelectricity is the mechanical-to-electrical or electrical-to-mechanical transduction discovered by Jacques and Pierre Curie. The quartz crystal is made of silicon dioxide (SiO_2). The same material is used to make an insulating layer for MOSFETs gates. A natural shape of a quartz crystal is a hexagonal prism with pyramids at the ends. Manufacturers slice rectangular slabs out of natural crystal. A thin slice of a quartz crystal is placed between two conducting plates, such as those of a capacitor. When a mechanical stress is applied across one set of faces, a mechanical deformation along one crystal axis will occur and difference of potential develops between the opposite crystal faces. Deforming a crystal will separate charges and produce a voltage. Conversely, a voltage applied across one set of faces of the crystal causes mechanical distortion in the crystal shape. When ac voltage is applied to a crystal, mechanical vibrations occur at the frequency of applied voltage. The crystal has electromechanical resonance. It can vibrate in a number of mechanical modes. The mode with the lowest resonant frequency is called the *fundamental mode*. Higher order modes are called *overtons*. When the crystal is vibrating, it acts as a tuned circuit. The crystal can be represented by an electrical equivalent circuit. The frequency range of vibrations is from a few kilohertzs to a few hundred megahertzs. The frequency of vibrations and the quality factor Q depend on the physical dimensions of a quartz crystal. These dimensions depend on temperature. The fundamental frequency of oscillation is given by

$$f_o \approx \frac{1670}{t} \tag{12.205}$$

where t is the quartz thickness. Since the fundamental frequency is inversely proportional to the crystal thickness, there is a limit to the highest fundamental frequency. The thinner the crystal, the more likely it is to break when vibrating. Quartz crystals work well up to 10 MHz at the fundamental frequency. To achieve higher frequencies, crystals can vibrate on overtons (harmonics of the fundamental frequency). Overtons can be used up to 100 MHz.

The frequency of oscillation is extremely stable with respect to time. The quartz crystal can be used to build crystal-controlled oscillators (CCO). These oscillators have a very stable frequency of oscillation. The electrical equivalent circuit of the quartz crystal is shown in Fig. 12.55, where C is the series capacitance of the crystal, L is the series inductance of the crystal, R is the series resistance of the crystal, and C_p represents the electrostatic capacitance between the electrodes with the crystal serving as a dielectric. The inductance L, the capacitance C, and the resistance

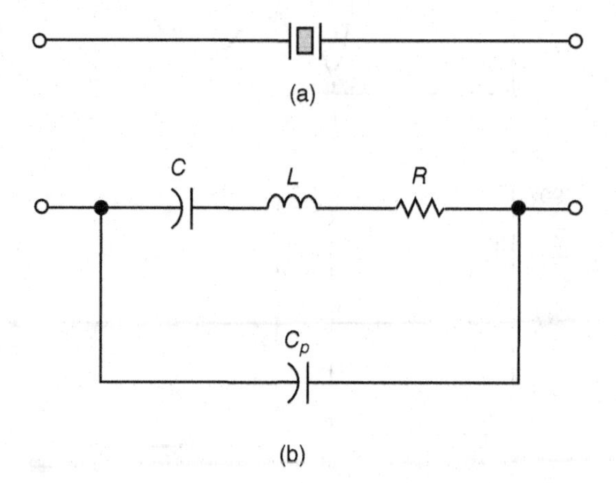

Figure 12.55 Equivalent circuit of a quartz crystal.

R represent electrical equivalents of mass, compliance, and internal friction, respectively. The parallel capacitance C_p represents a static capacitance created by the quartz metal plates that make electrical contacts with it. The impedance of the crystal equivalent circuit is given by

$$Z(s) = \frac{\frac{1}{sC_p}\left(s^2 + \frac{1}{sC} + R\right)}{\frac{1}{sC_p} + sL + \frac{1}{sC} + R} = \frac{1}{sC_p}\frac{s^2 + s\frac{R}{L} + \frac{1}{LC}}{s^2 + s\frac{R}{L} + \frac{1}{L\frac{CC_p}{C + C_p}}} \approx \frac{1}{sC_p}\frac{s^2 + \frac{1}{LC}}{s^2 + \frac{1}{L\frac{CC_p}{C + C_p}}} = \frac{1}{sC_p}\frac{s^2 + \omega_s^2}{s^2 + \omega_p^2} \quad (12.206)$$

Figure 12.56 shows the magnitude and phase of the crystal impedance. Neglecting the resistance R, the impedance of the crystal can be approximated by a reactance

$$Z \approx jX = \frac{j}{\omega C_p}\frac{\omega^2 - \omega_s^2}{\omega^2 - \omega_p^2} \quad (12.207)$$

Figure 12.57 shows the reactance of the quartz crystal as a function of frequency. At low frequencies, the impedance of the crystal is dominated by large reactances of C and C_p. The series-resonant frequency of the crystal impedance is given by

$$f_s = \frac{1}{2\pi\sqrt{LC}} \quad (12.208)$$

and the parallel-resonant frequency is

$$f_p = \frac{1}{2\pi\sqrt{L\frac{CC_p}{C + C_p}}} = f_s\sqrt{1 - \frac{C}{C_p}} \quad (12.209)$$

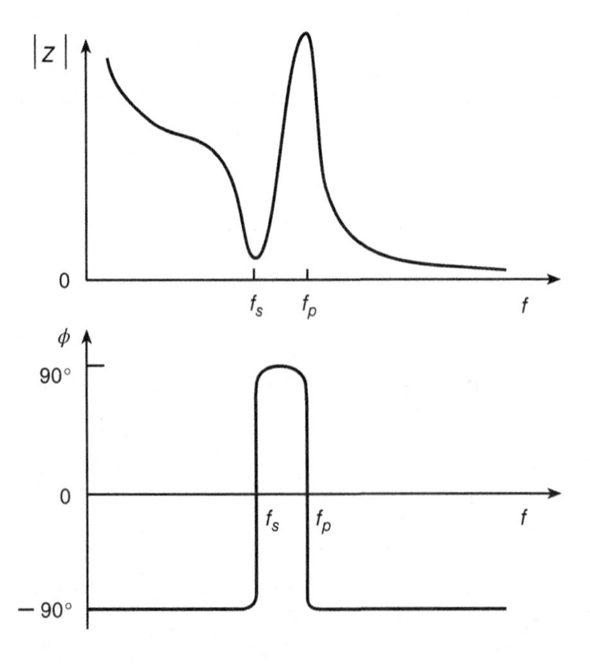

Figure 12.56 Impedance of the quartz crystal as a function of frequency.

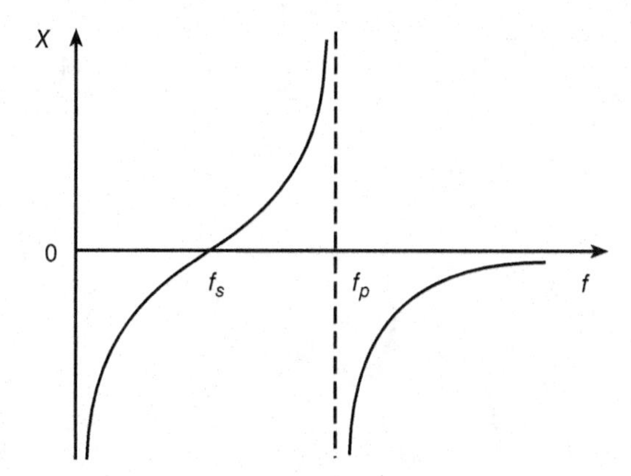

Figure 12.57 Reactance of the quartz crystal as a function of frequency.

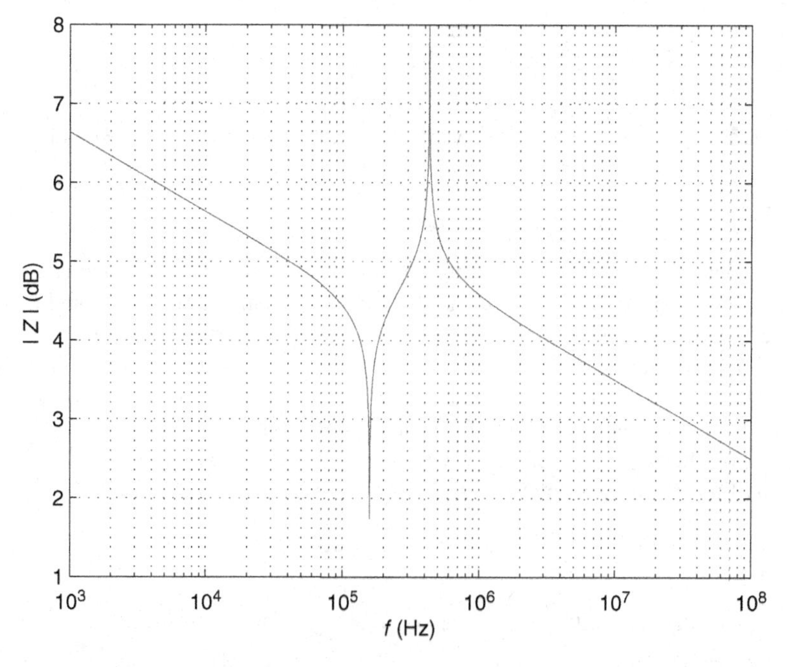

Figure 12.58 Magnitude $|Z|$ of the quartz crystal as a function of frequency at $L = 31.8$ mH, $C = 31.8$ fF, $R = 50\ \Omega$, and $C_p = 5$ pF.

Since $C_p \gg C, f_p \approx f_s$. Usually, f_p is from 0.1% to 1% higher than f_s. The difference between the resonant frequencies is

$$\Delta f = f_p - f_s = f_s \left(1 - \sqrt{1 + \frac{C}{C_p}}\right) \quad (12.210)$$

Example values of the quartz crystal are as follows: $L = 31.8$ mH, $C = 31.8$ fF, $R = 50\ \Omega$, $C_p = 5$ pF, $Q = 20,000$, $f_s = 5$ MHz, $f_p = 5.02$ MHz, and $f_p/f_s = 1.004$. This is 0.4% difference between the two resonant frequencies. Figures 12.58 and 12.59 show Bode plots of the impedance Z for the quartz crystal at $L = 31.8$ mH, $C = 31.8$ fF, $R = 50\ \Omega$, and $C_p = 5$ pF. For another example, $f_o = 1$ MHz, $L = 8$ H, and $C = 3.2$ fF. At a series-resonant frequency f_s, the

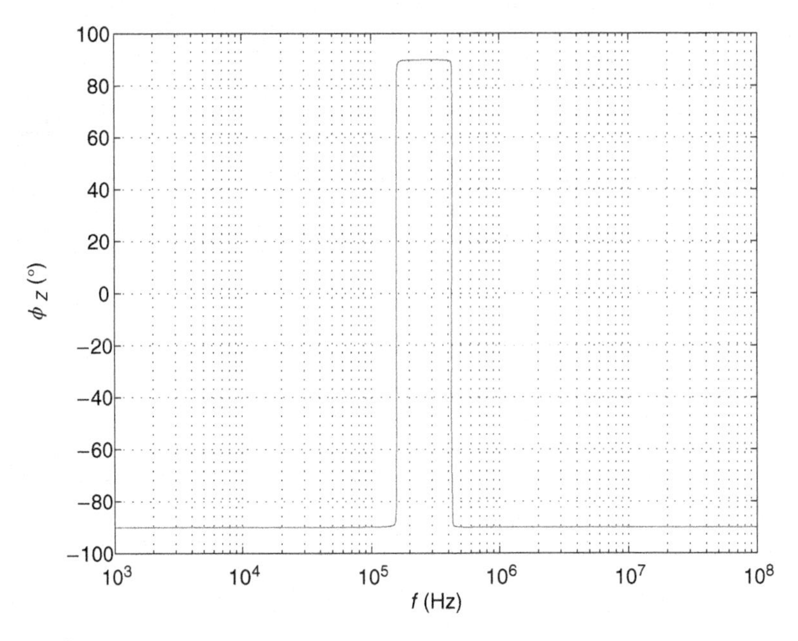

Figure 12.59 Phase ϕ_Z of the quartz crystal as a function of frequency at $L = 31.8$ mH, $C = 31.8$ fF, $R = 50\ \Omega$, and $C_p = 5$ pF.

crystal reactance drops to nearly zero; it is equal to R. As the frequency increases beyond f_s, the net reactance rapidly increases. At the same time, the combination of C_p and L approaches resonance. At the parallel-resonant frequency f_p, the impedance of the crystal approaches infinity. At frequencies above f_p, the reactance of C_p causes the crystal reactance to decrease. Typically, f_p is only 0.2–0.4% higher than f_s. For $f_s < f < f_p$, the reactance X is inductive, and it is capacitive outside this range. Therefore, the crystal can be used to replace the inductor L in the Colpitts oscillator for the very narrow frequency band between f_s and f_p. The frequency of oscillation of the crystal oscillator is

$$f_o = \frac{1}{\sqrt{LC_{eq}}} \tag{12.211}$$

where the equivalent capacitance of the resonant circuit of the oscillator is

$$C_{eq} = \frac{C\left(C_p + \dfrac{C_1C_2}{C_1 + C_2}\right)}{C + C_p + \dfrac{C_1C_2}{C_1 + C_2}} \tag{12.212}$$

Since

$$C \ll C_p + \frac{C_1C_2}{C_1 + C_2} \tag{12.213}$$

the frequency of oscillation of the crystal oscillator can be approximated by

$$f_o = \frac{1}{\sqrt{LC}} \tag{12.214}$$

The series inductance L is very high and the series capacitance C is very low. Therefore, the quality factor Q is very large, typically in the range from 10^4 to 10^6. The resonant frequency is given by

$$f_o = \frac{1}{2\pi\sqrt{LC}} \frac{1}{\sqrt{1 + \frac{1}{Q^2}}} \tag{12.215}$$

As Q approaches infinity, f_r approaches f_s.

The frequency of oscillations is related to the resistance R

$$f_o \approx \frac{5 \times 10^8}{R} \tag{12.216}$$

Frequency stability is a measure of the degree to which an oscillator maintains the same value of frequency over a specific time interval. A quartz crystal-controlled oscillator is shown in Fig. 12.60. It is a Colpitts or Clapp oscillator. The crystal acts as an inductor that resonates with C_1 and C_2. The oscillation frequency f_o lies between f_s and f_r. A crystal exhibits a rapid variation of reactance with frequency in the band from f_s to f_p. This high rate of change of impedance stabilizes the oscillation frequency f_o because any change in operating frequency causes a large change in the feedback loop phase shift and prevents changes of oscillating frequency. The intention is to improve the frequency stability by reducing the effect of stray capacitances. This is accomplished by reducing the effect of the amplifier input and output capacitances and stray capacitances as well as the effects of aging and temperature. If the transistor input and output capacitances are used as C_1 and C_2, the circuit is called a Pierce crystal oscillator. Crystal-controlled oscillators are fixed-frequency circuits. The stability of the frequency of oscillation of crystal oscillators is very high because the quality factor Q is incredibly high and the phase of the crystal impedance is very sensitive to frequency. Therefore, the frequency of oscillations is very stable. The derivative of the phase shift of the feedback network θ with respect to frequency is given by

$$\frac{d\theta}{d\omega} = -\frac{\omega_0}{2Q} \quad \text{for} \quad \omega = \omega_o \tag{12.217}$$

As Q approaches infinity, $d\theta/d\omega$ approaches zero. Time and temperature stability of crystal oscillators is $\pm 0.001\% = \pm 10/10^6 = \pm 10$ parts per million (ppm). Frequency stability for 1 year at constant temperature is 1 ppm. The temperature coefficient of f_o is in the range 1–2 ppm over 1 °C. The drift of frequency of oscillation f_o with time is very small, typically less than one part in 10^6 per day. A clock with this drift will take 0.76 year to lose or gain 1 s. Crystal resonators offer a significant advantage over lumped LC resonant circuits. The crystal oscillators are widely used in digital watches as a basic timing device and employ a torsional model of vibrations to allow resonance at a low frequency of 32.768 kHz in a small size. These oscillators are used in

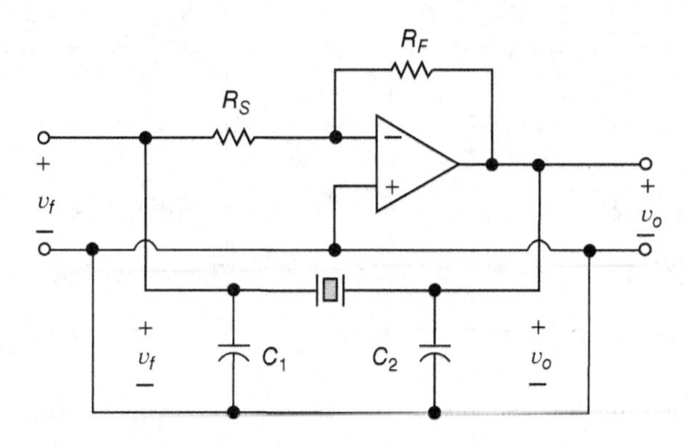

Figure 12.60 Crystal-controlled oscillator (CCO).

communication transmitters and receivers. For high frequencies, above 20 MHz, the crystals are expensive to make for the fundamental mode, and therefore overtone modes are used.

Piezoelectric may be also used for energy harvesting. As the piezoelectric is subject to oscillations, the mass causes the piezoelectric layer to be stretched. By doing so, it generates electric power than can supply a small load.

12.12 CMOS Oscillator

A CMOS crystal-controlled oscillator is shown in Fig. 12.61. A resistor R_F is connected between the gates and the inverter output to set the dc operating point Q in the high-gain region of the CMOS inverter so that the MOSFETs are in the pinched-off region, that is, they are neither fully on nor fully off. The inverter input and output capacitances are absorbed into capacitances C_1 and C_2, respectively. This is a Pierce oscillator, which can be derived from Colpitts oscillator. The inverting amplifier has a phase shift between the input and the output by $-180°$. Capacitances C_1 and C_2 provide a $-180°$ phase shift. Therefore, the total phase shift in the loop is $-360°$. The quartz crystal provides an inductive reactance. The quartz crystal and capacitances C_1 and C_2 form a bandpass filter. The oscillation frequency is determined by the crystal.

12.12.1 Deviation of Oscillation Frequency Caused by Harmonics

The frequency of oscillations is affected by harmonics of current flowing in the resonant circuit [4]. For Colpitts oscillator, the relative decrease in the oscillation frequency Δf from the resonant frequency f_o is given by

$$\frac{\Delta f}{f_o} = -\frac{1}{2Q_L^2}\sum_{n=2}^{\infty}\frac{1}{n^2-1}\left(\frac{I_n}{I_1}\right)^2 \tag{12.218}$$

where Q_L is the loaded quality factor at the resonant frequency f_o, n is the order of the harmonic, I_n is the amplitude of the nth harmonic flowing through the resonant circuit, and I_1 is the amplitude of the fundamental component flowing through the resonant circuit. For example, for $Q_L = 10$, $n = 2$, and $I_2/I_1 = 0.1$, we get $\Delta f/f_o = -1/6000$.

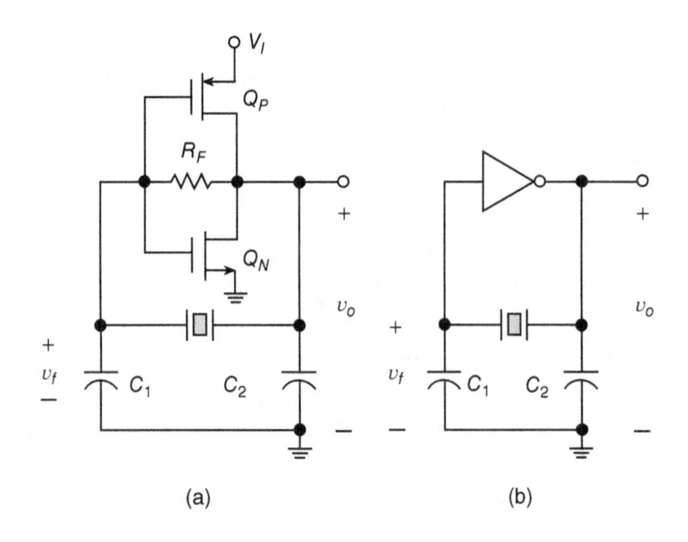

Figure 12.61 CMOS Pierce crystal-controlled oscillator.

For the differential LC oscillator, the relative shift of the oscillation frequency from the resonant frequency is

$$\frac{\Delta f}{f_o} = -\frac{1}{2Q_L^2}\sum_{n=2}^{\infty}\frac{n^2}{n^2-1}\left(\frac{I_n}{I_1}\right)^2 \tag{12.219}$$

12.13 Hartley Oscillator

12.13.1 Op-Amp Hartley Oscillator

The feedback network that satisfies the conditions in (12.90) is shown in Fig. 12.7(c). Figure 12.62 shows the Hartley oscillator with an inverting op-amp. A tapped inductor is used to form a feedback network. Ralph Vinton Lyon Hartley invented this oscillator on February 10, 1915.

At the oscillation frequency f_o,

$$X_C = X_{L1} + X_{L2} \tag{12.220}$$

resulting in

$$\frac{1}{\omega_o C} = \omega_o L_1 + \omega_o L_2 = \omega_o(L_1 + L_2) \tag{12.221}$$

Hence, the frequency of oscillation of the Hartley oscillator is

$$f_o = \frac{1}{2\pi\sqrt{C(L_1+L_2)}} \tag{12.222}$$

The transfer function of the feedback network is

$$\beta = \frac{j\omega L_1}{j\omega L_1 - \dfrac{j}{\omega C}} = \frac{\omega L_1}{\omega L_1 - \dfrac{1}{\omega C}} = \frac{1}{1 - \dfrac{1}{\omega C L_1}} \tag{12.223}$$

The transfer function of the feedback network at the oscillation frequency f_o is

$$\beta(f_o) = -\frac{X_{L1}}{X_{L2}} = -\frac{L_1}{L_2} \tag{12.224}$$

The voltages across L_1 and L_2 are 180° out of phase. Therefore, the feedback voltage v_f is 180° out of phase with respect to the output voltage v_o.

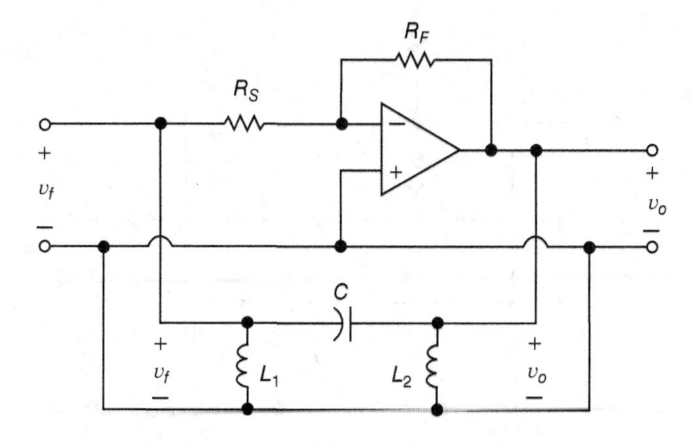

Figure 12.62 Hartley oscillator.

Figure 12.62 shows a practical implementation of the Hartley oscillator using an inverting op-amp. The loop gain for steady-state oscillation at the oscillation frequency f_o is given by

$$T(f_o) = \beta(f_o)A(f_o) = \left(-\frac{L_1}{L_2}\right)\left(-\frac{R_F}{R_S}\right) = \left(\frac{L_1}{L_2}\right)\left(\frac{R_F}{R_S}\right) = 1 \tag{12.225}$$

which gives

$$\frac{R_F}{R_S} = \frac{L_2}{L_1} \tag{12.226}$$

The startup condition for oscillation is

$$\frac{R_F}{R_S} = \frac{L_2}{L_1} \tag{12.227}$$

Example 12.2

Design an op-amp Hartley oscillator for $f_o = 100$ kHz.

Solution. Let $L_1 = L_2 = 120$ μH. Hence, the capacitance is

$$C = \frac{1}{4\pi^2 f_o^2 (L_1 + L_2)} = \frac{1}{4\pi^2 \times (10^5)^2 \times (120 + 120) \times 10^{-6}} = 10.55 \text{ nF} \tag{12.228}$$

Since $L_1/L_2 = 1$, we get $R_F/R_S = 1$. Pick $R_F = R_S = 100$ kΩ/0.25 W/1%.

12.13.2 Single-Transistor Hartley Oscillator

The circuit of a Hartley oscillator with a single transistor is shown in Fig. 12.63. Figure 12.64 shows ac circuits of the Hartley oscillator. Figure 12.65 depicts common-source, common-gate,

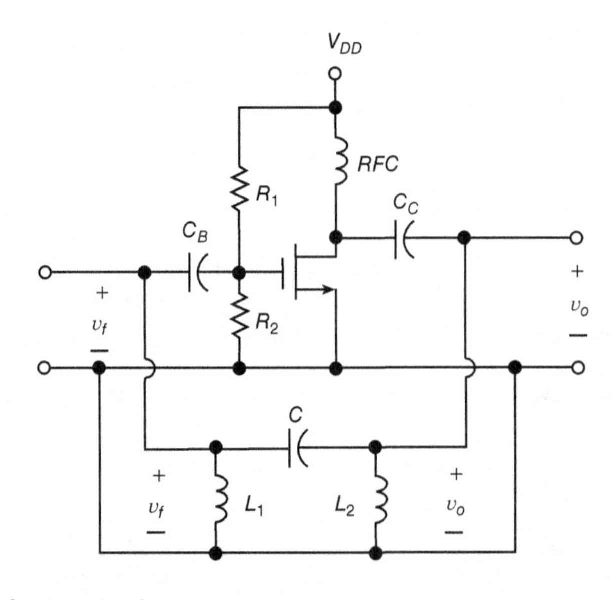

Figure 12.63 Hartley oscillator with common-source amplifier.

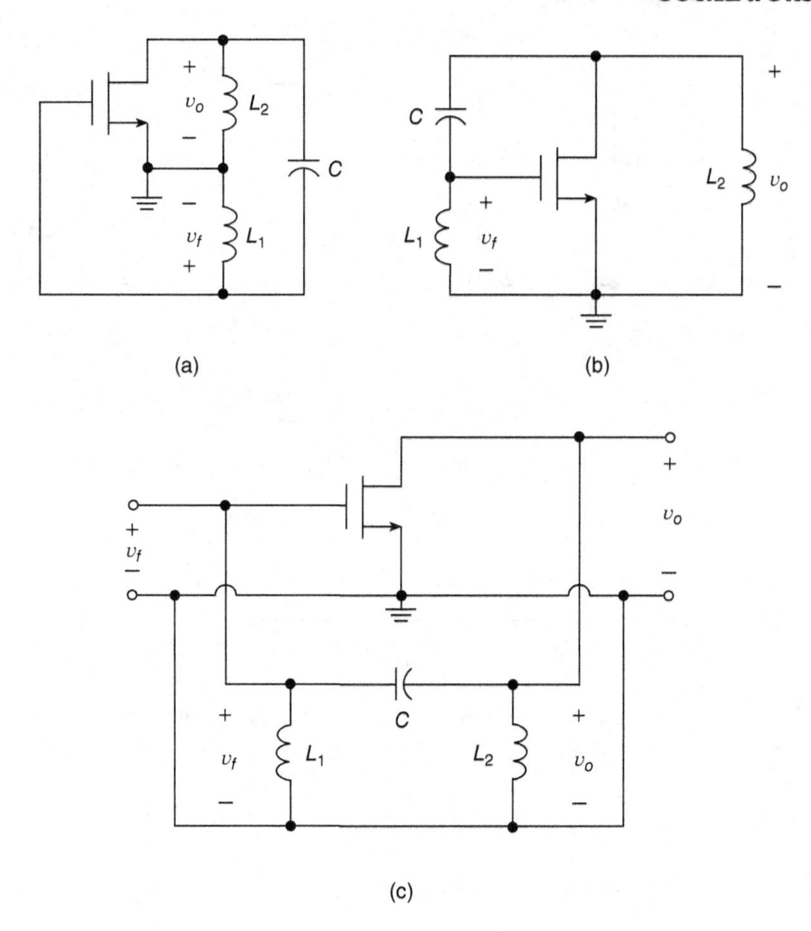

Figure 12.64 An ac circuit of the Hartley oscillator with common-source amplifier.

and common-drain Hartley oscillators. The drain-to-gate capacitance is absorbed into capacitance C. However, the gate-to-source capacitance C_{gs} and the drain-to-source capacitance C_{ds} are not included in the Hartley oscillator topology. The voltage gain of the amplifier in the Hartley oscillator at the oscillation frequency f_o is

$$A_v(f_o) = \frac{v_o}{v_f} = -g_m(r_o||R_L) \tag{12.229}$$

The voltage gain of the feedback network at the resonant frequency is

$$\beta(f_o) = \frac{v_f}{v_o} = -\frac{C_1}{C_2} \tag{12.230}$$

The loop gain at the oscillation frequency is

$$T(f_o) = \beta(f_o)A_v(f_o) = [-g_m(r||R_L)]\left(-\frac{L_1}{L_2}\right) = [g_m(r||R_L)]\left(\frac{L_1}{L_2}\right) = 1 \tag{12.231}$$

Hence, the ratio of inductances is

$$g_m(r_o||R_L) = \frac{L_2}{L_1} \tag{12.232}$$

The startup condition for oscillations is

$$g_m(r_o||R_L) > \frac{L_2}{L_1} \tag{12.233}$$

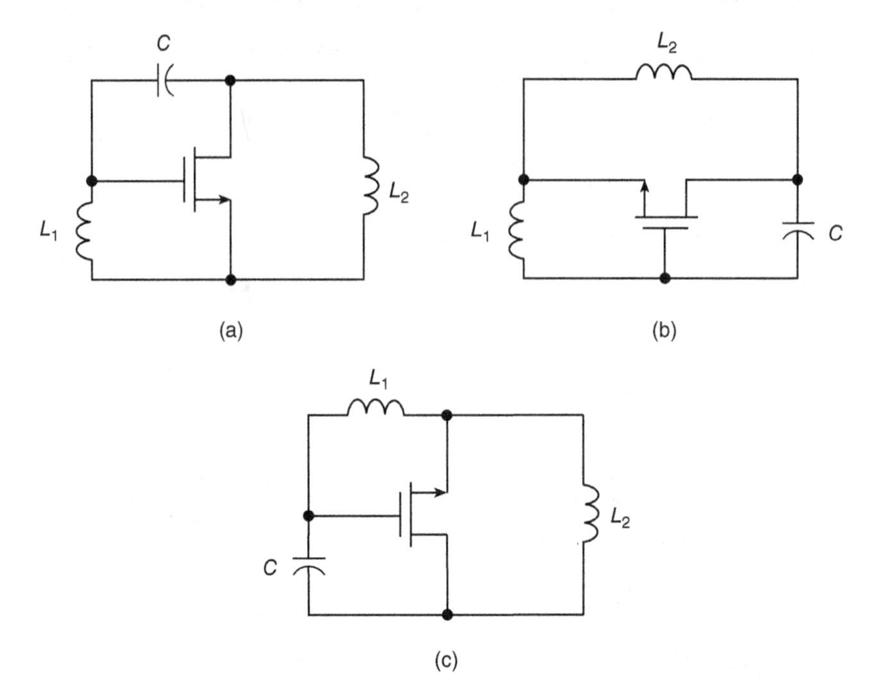

Figure 12.65 Hartley oscillator ac circuits for three configurations of the transistor. (a) In common-source (CS) configuration. (b) In common-gate (CG) configuration. (c) In common-drain (CD) configuration.

12.14 Armstrong Oscillator

12.14.1 Op-Amp Armstrong Oscillator

Figure 12.66 shows the Armstrong oscillator with an inverting op-amp. It consists of an inverting amplifier, an inverting transformer, and a capacitor C. The transformer magnetizing inductance L_m and the capacitor C form a parallel resonant circuit, which acts as a bandpass filter.

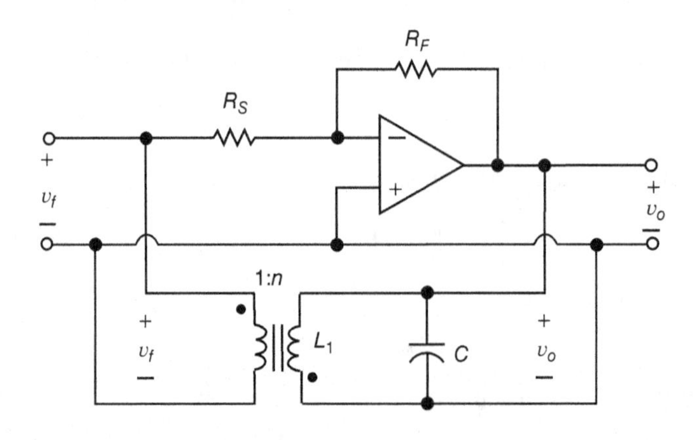

Figure 12.66 Armstrong oscillator.

The frequency of oscillation is given by

$$f_o = \frac{1}{\sqrt{L_m C}} \tag{12.234}$$

The voltage transfer function of the feedback network is

$$\beta = \frac{v_f}{v_o} = -n \tag{12.235}$$

where n is the transformer turn ratio. The loop gain at the resonant frequency is

$$T(f_o) = \beta(f_o)A(f_o) = (-n)\left(-\frac{R_F}{R_S}\right) = (n)\left(\frac{R_F}{R_S}\right) = 1 \tag{12.236}$$

Hence,

$$\frac{R_F}{R_S} = \frac{1}{n} \tag{12.237}$$

12.14.2 Single-Transistor Armstrong Oscillator

Figure 12.67 shows the Armstrong oscillator with a single transistor in the CS configuration. The voltage gain of the amplifier in the Armstrong oscillator at the oscillation frequency is

$$A_v(f_o) = \frac{v_o}{v_f} = -g_m(r_o \| R_L) \tag{12.238}$$

The voltage gain of the feedback network is

$$\beta = \frac{v_f}{v_o} = -n \tag{12.239}$$

The loop gain of the Armstrong oscillator for steady-state operation at the resonant frequency is

$$T(f_o) = \beta A_v(f_o) = [-g_m(r_o \| R_L)](-n) = [g_m(r_o \| R_L)](n) = 1 \tag{12.240}$$

Hence, the transformer turn ratio is

$$n = \frac{1}{g_m(r_o \| R_L)} \tag{12.241}$$

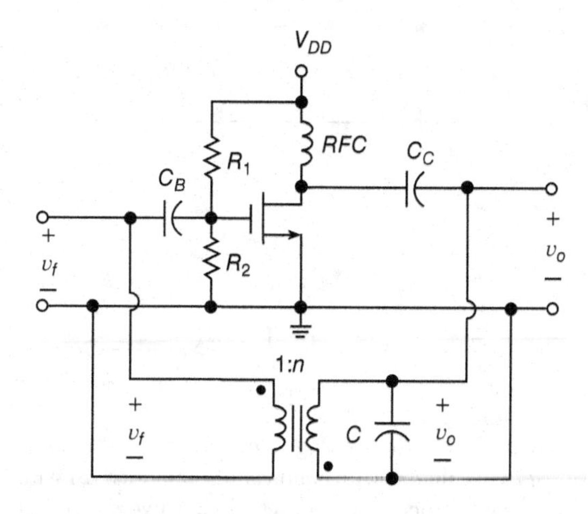

Figure 12.67 Armstrong oscillator with common-source amplifier.

12.15 LC Oscillators with Noninverting Amplifier

Figure 12.68 shows a block diagram of LC oscillators with noninverting amplifier ($A > 0$) and noninverting feedback networks ($\beta > 0$). The oscillator consists of a noninverting amplifier A and a noninverting feedback network β. The feedback network in the oscillator shown in Fig. 12.68(b) is a capacitive voltage divider C_1 and C_2. The feedback network in the oscillator

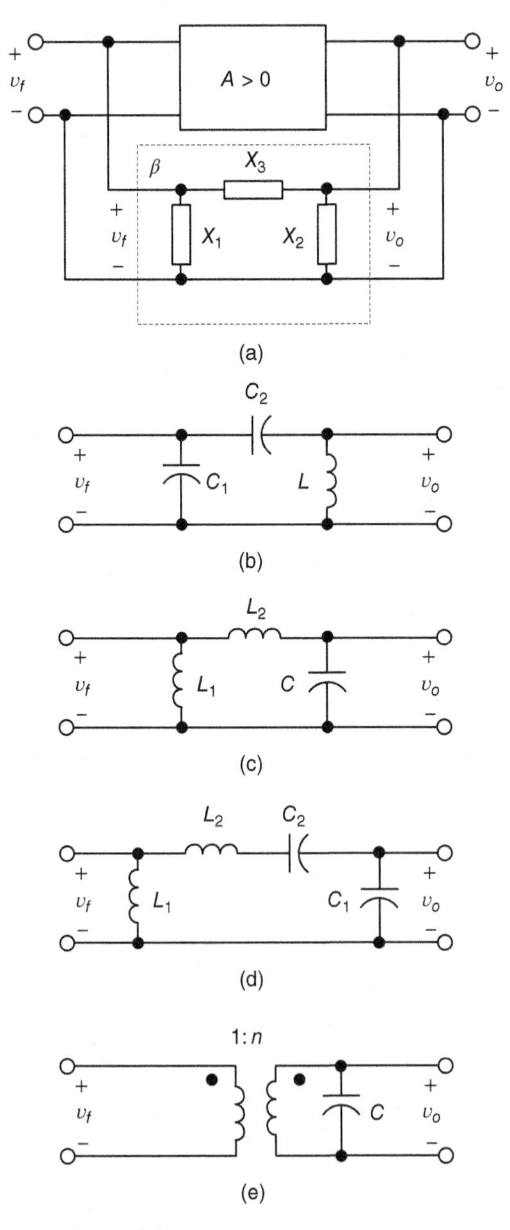

Figure 12.68 LC oscillators with noninverting amplifier $A > 0$ and noninverting feedback networks $\beta > 0$. (a) Block diagram of LC oscillators with noninverting amplifier. (b) With output tapped-capacitor feedback network. (c) With output tapped-inductor feedback network. (d) With output tapped-inductor tapped-capacitor feedback network. (e) With noninverting transformer.

shown in Fig. 12.68(c) is an inductive voltage divider L_1 and L_2. The feedback network in the oscillator shown in Fig. 12.68(d) is a modification of the feedback network of Fig. 12.68(b), which is obtain by adding a capacitor C_2 in series with the inductor L_2. The feedback network in the oscillator shown in Fig. 12.68(e) is made up of a noninverting transformer and a capacitor C.

The voltage gain of the amplifier is defined as

$$A = |A|e^{j\phi_A} = \frac{v_o}{v_f} > 0 \tag{12.242}$$

where

$$\phi_A = 0 \tag{12.243}$$

The feedback network consists of the reactances X_1, X_2, and X_3, where $X = j\omega L$ or $X = -j/(\omega C)$. The feedback network is a bandpass filter. The oscillation frequency f_o is approximately equal to the resonant frequency of the feedback network. Hence,

$$X_1 + X_2 + X_3 = 0 \tag{12.244}$$

Hence,

$$X_3 = -(X_1 + X_2) \tag{12.245}$$

Since $|T(f_o)| = 1$ and $A > 0$, then $\beta > 0$. The voltage transfer function of the feedback network at the frequency of oscillations f_o is

$$\beta = |\beta|e^{j\phi_\beta} = \frac{v_f}{v_o} = \frac{X_1}{X_1 + X_2} = -\frac{X_1}{X_3} > 0 \tag{12.246}$$

where

$$\phi_\beta = 0 \tag{12.247}$$

Thus,

$$\frac{X_1}{X_3} < 0 \tag{12.248}$$

From (12.244),

$$X_2 = -(X_1 + X_3) \tag{12.249}$$

This leads to the following sets of conditions for the components of the feedback network:

$$X_1 > 0,\ X_2 > 0, \quad \text{and} \quad X_3 < 0 \tag{12.250}$$

or

$$X_1 < 0,\ X_2 < 0, \quad \text{and} \quad X_3 > 0 \tag{12.251}$$

or

$$X_1 > 0,\ X_2 < 0,\ X_3 < 0, \quad \text{and} \quad X_1 > -X_2 \tag{12.252}$$

or

$$X_1 < 0,\ X_2 > 0,\ X_3 > 0, \quad \text{and} \quad X_1 < -X_2 \tag{12.253}$$

Conditions in (12.250) are satisfied by the noninverting feedback network shown in Fig. 12.68(b). Conditions in (12.251) are satisfied by the noninverting feedback network shown in Fig. 12.68(c). For the feedback network of Fig. 12.68(b) with tapped capacitor, the voltage transfer function is given by

$$\beta = \frac{v_f}{v_o} = \frac{X_1}{X_1 + X_2} = \frac{\dfrac{-j}{\omega C_1}}{\dfrac{-j}{\omega C_1} + \dfrac{-j}{\omega C_2}} = \frac{C_2}{C_1 + C_2} = \frac{1}{1 + \dfrac{C_1}{C_2}} \tag{12.254}$$

For the feedback network of Fig. 12.68(c) with tapped inductor, the voltage transfer function is given by

$$\beta = \frac{v_f}{v_o} = \frac{X_1}{X_1 + X_2} = \frac{\omega L_1}{\omega L_1 + \omega L_2} = \frac{L_1}{L_1 + L_2} = \frac{1}{1 + \frac{L_2}{L_1}} \tag{12.255}$$

12.15.1 LC Single-Transistor Oscillators with Noninverting Amplifier

Figure 12.69 shows examples of oscillators with CG amplifier. For the oscillator of Fig. 12.69(a), the loop gain is

$$T = \beta A = \frac{1}{1 + \frac{C_1}{C_2}} = 1 \tag{12.256}$$

yielding

$$\frac{C_1}{C_2} = \frac{1}{A} - 1 \tag{12.257}$$

For the oscillator of Fig. 12.69(b), the loop gain is

$$T = \beta A = \frac{1}{1 + \frac{L_2}{L_1}} A = 1 \tag{12.258}$$

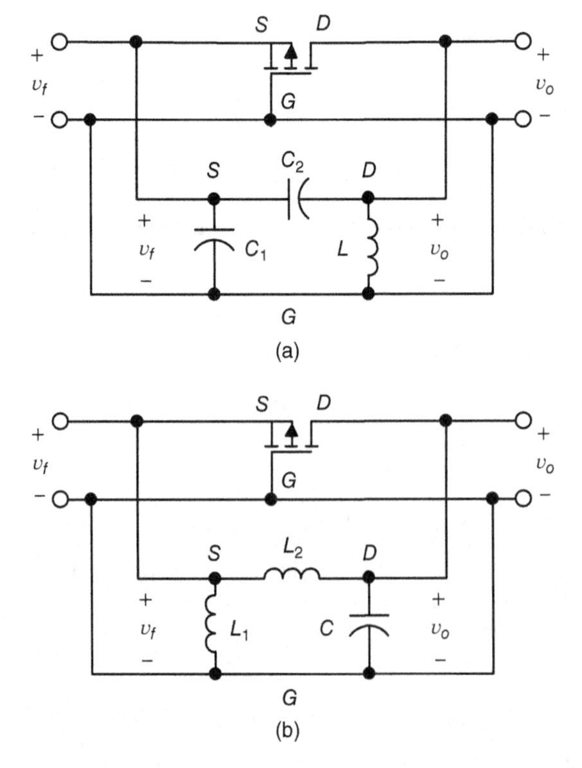

Figure 12.69 *LC* oscillators with the common-gate amplifier. (a) With output tapped-capacitor. (b) With output tapped-inductor.

producing

$$\frac{L_2}{L_1} = \frac{1}{A} - 1 \tag{12.259}$$

12.15.2 LC Oscillators with Noninverting Op-Amp

Figure 12.70 shows two examples of oscillators that use noninverting op-amp. The gain of the noninverting op-amp of Fig. 12.70(a) is given by

$$A = \frac{v_o}{v_f} = 1 + \frac{R_F}{R_S}. \tag{12.260}$$

The loop gain of the oscillator shown in Fig. 12.70(a) at f_o is given by

$$T = \beta A = \frac{1}{1 + \frac{C_1}{C_2}}\left(1 + \frac{R_F}{R_S}\right) = 1 \tag{12.261}$$

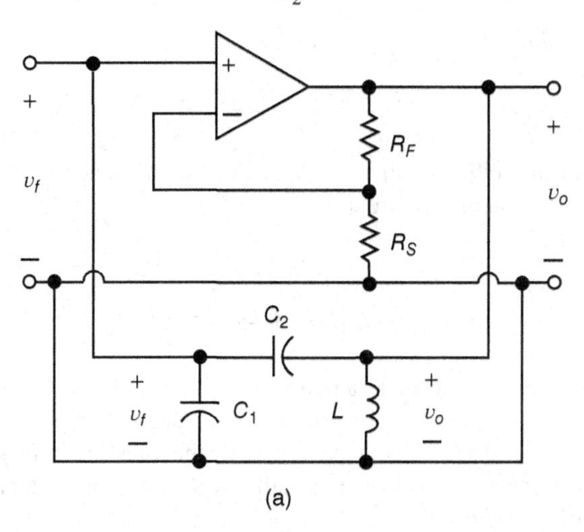

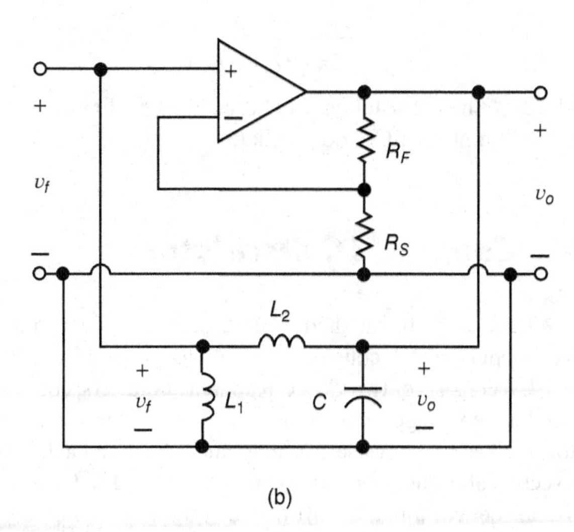

Figure 12.70 Oscillators with noninverting op-amp. (a) With output tapped-capacitor. (b) With output tapped-inductor.

Hence,

$$\frac{R_F}{R_S} = \frac{C_1}{C_2} \tag{12.262}$$

The frequency of oscillations is

$$f_o = \frac{1}{2\pi\sqrt{\frac{LC_1C_2}{C_1 + C_2}}} \tag{12.263}$$

Similarly, the loop gain of the oscillator shown in Fig. 12.70(b) at f_o is given by

$$T = \beta A = \frac{1}{1 + \frac{L_2}{L_1}}\left(1 + \frac{R_F}{R_S}\right) = 1 \tag{12.264}$$

Hence,

$$\frac{R_F}{R_S} = \frac{L_2}{L_1}. \tag{12.265}$$

The frequency of oscillations is

$$f_o = \frac{1}{2\pi\sqrt{C(L_1 + L_2)}}. \tag{12.266}$$

Figure 12.71 shows an oscillator with a noninverting A circuit and an $LLCC$ feedback network. The frequency of oscillation of this circuit is

$$f_o = \frac{1}{2\pi\sqrt{(L_1 + L_2)\frac{C_1C_2}{C_1 + C_2}}} \tag{12.267}$$

Figure 12.72 shows an oscillator with a noninverting A circuit and a transformer in feedback network. The inductance L_1 is formed by the transformer magnetizing inductance, and the inductance L_2 is formed by the leakage inductance formed by the transformer magnetizing inductance, and the inductance L_2 is formed entirely or partially by the transformer leakage inductance. The frequency of oscillation of this circuit is

$$f_o = \frac{1}{2\pi\sqrt{(L_m + L_2)C}} \tag{12.268}$$

Figures 12.73 and 12.74 show crystal-controlled oscillators. Figures 12.75 and 12.76 show implementations of single-transistor CG LC oscillators.

12.16 Cross-Coupled LC Oscillators

The conditions for oscillations can be satisfied in multistage amplifiers arranged in a closed loop. One example of this concept is a cross-coupled LC oscillator, also called a differential LC oscillator, shown in Fig. 12.77. It consists of two CS LC resonant amplifiers, connected in a closed loop, such that the drain-to-source voltage of one transistor v_{ds} is equal to the gate-to-source voltage of the other transistor v_{gs}. The two transistors arrangement forms a latch. Consequently, the first amplifier drives the second amplifier, and *vice versa.* transistor. The CS amplifier is an inverting circuit, which ideally introduces a phase shift between the drain and the gate equal to $-180°$. Therefore, two cascaded stages of the CS amplifiers introduce the phase shift (delay) equal to $-360°$. Therefore, the phase condition for oscillation is satisfied. The load of each transistor is a

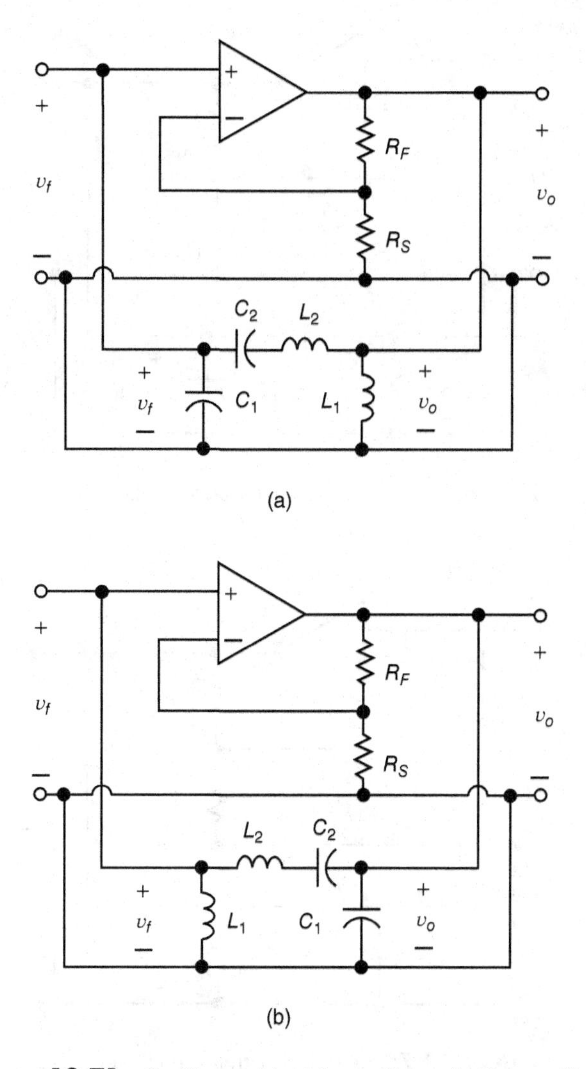

Figure 12.71 Oscillators. (a) Alpha oscillator. (b) Beta oscillator.

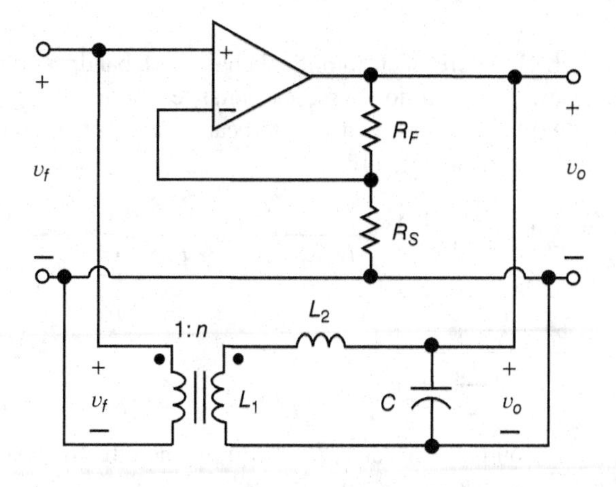

Figure 12.72 Oscillator with $A > 0$ and a transformer.

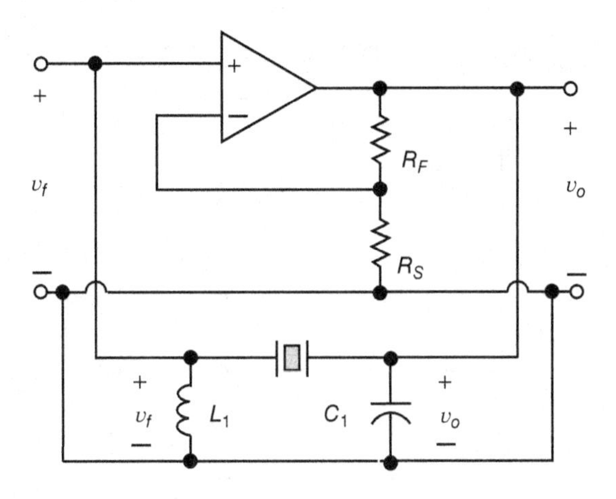

Figure 12.73 Crystal-controlled oscillator type 1.

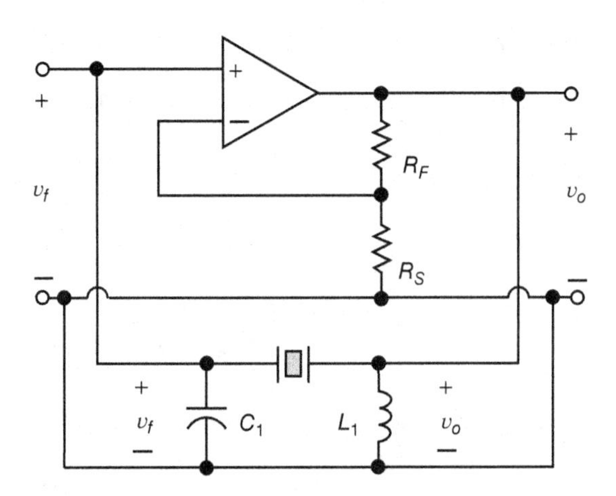

Figure 12.74 Crystal-controlled oscillator type 2.

parallel-resonant circuit. Therefore, each amplifier behaves as a bandpass filter, which can select the oscillation frequency. The drain-to-source capacitance C_{ds} and the gate-to-source capacitance C_{gs} of each MOSFET are absorbed into the capacitance of the resonant circuits. The impedance of the parallel-resonant circuit is

$$Z(s) = (R||r_o)||(sL)||\left(\frac{1}{sC}\right) = \frac{sLR_L}{LCR_L s^2 + sL + R_L} = \frac{1}{C}\frac{s}{s^2 + \frac{s}{CR_L} + \frac{1}{LC}}$$

$$= \frac{R_L\omega_o}{Q_L}\frac{s}{s^2 + \frac{\omega_o}{Q_L}s + \omega_o^2} \quad (12.269)$$

where r_o is the MOSFET output resistance, $R_L = R||r_o$ is the effective load resistance of each transistor, $\omega_o = 1/\sqrt{LC}$, and $Q_L = \omega_o CR_L = R_L/(\omega_o L)$ is the loaded quality factor of the parallel-resonant circuit. Setting $s = j\omega$, we obtain the impedance of the parallel-resonant

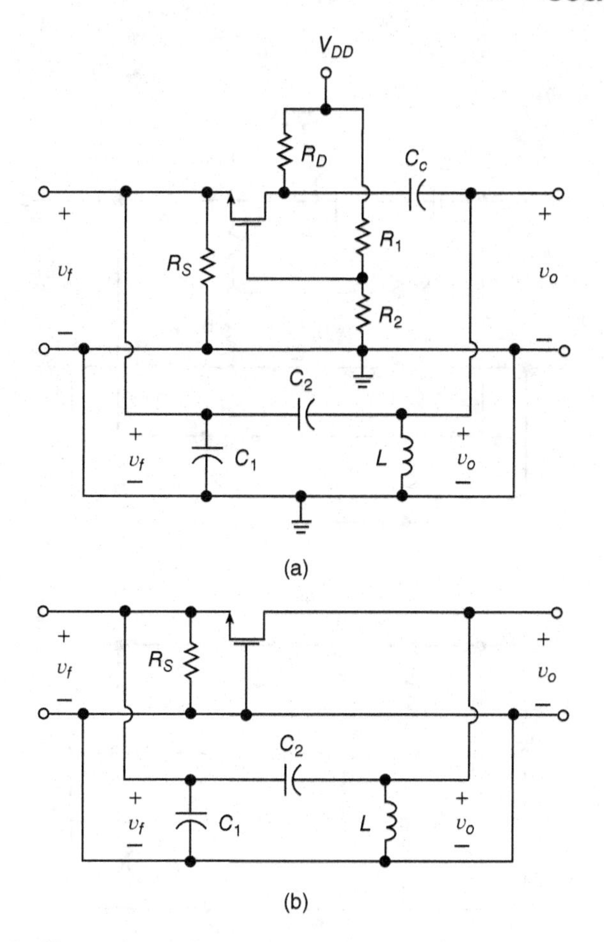

Figure 12.75 Oscillator with a common-gate amplifier, two capacitors, and single inductor. (a) Complete circuit. (b) Equivalent circuit for the ac component.

circuit in the frequency domain

$$Z(j\omega) = \frac{j\omega L R_L}{R_L(1-\omega^2 LC) + j\omega L} = \frac{j\omega L R_L}{R_L\left[1-\left(\frac{\omega}{\omega_o}\right)^2\right]^2 + j\omega L} \tag{12.270}$$

At the resonant frequency $\omega_o = 1/\sqrt{LC}$,

$$Z(j\omega_o) = R_L \tag{12.271}$$

The voltage gain of each amplifier in the s-domain is

$$A_{v1}(s) = \frac{v_{ds1}}{v_{gs1}} = -g_m Z(s) = -g_m \frac{sLR_L}{LCR_L s^2 + sL + R_L} = -g_m \frac{1}{C} \frac{s}{s^2 + \frac{s}{CR_L} + \frac{1}{LC}}$$

$$= -g_m R_L \frac{\omega_o}{Q_L} \frac{s}{s^2 + \frac{\omega_o}{Q_L} s + \omega_o^2} \tag{12.272}$$

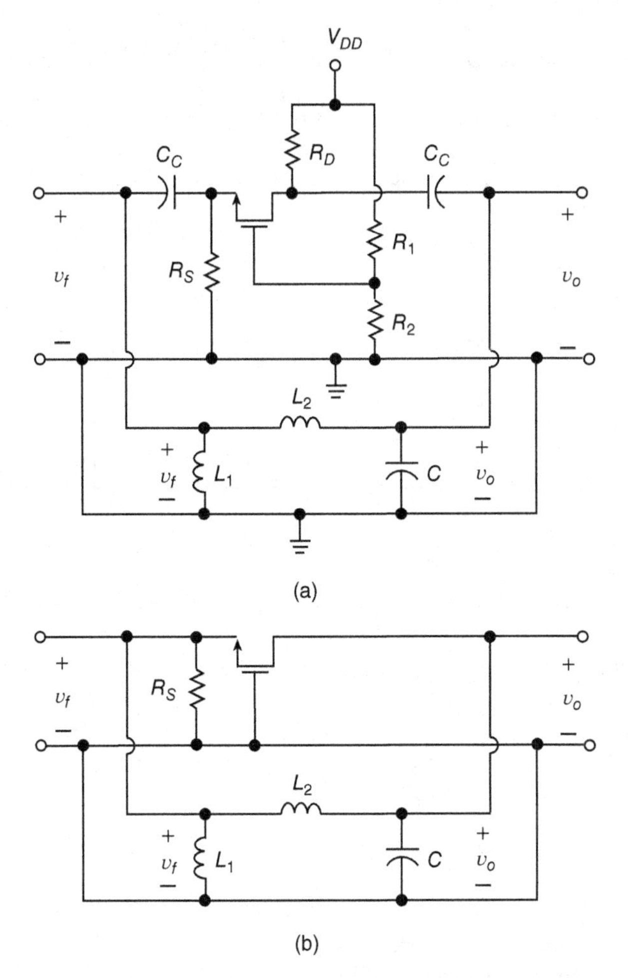

Figure 12.76 Oscillator with a common-gate amplifier, two inductors, and single capacitor. (a) Complete circuit. (b) Equivalent circuit for the ac component.

and in the frequency domain is

$$A_{v1}(j\omega) = \frac{v_{ds1}}{v_{gs1}} = -g_m Z(j\omega) = -g_m \frac{j\omega L R_L}{R_L(1 - \omega^2 LC) + j\omega L} \tag{12.273}$$

The voltage gain of both amplifiers in the s-domain is

$$A_v(s) = \frac{v_{ds2}}{v_{gs1}} = \frac{v_{ds1}}{v_{gs1}}\frac{v_{ds2}}{v_{gs2}} = [A_{v1}(s)]^2 = [-g_m Z(s)]^2 = \left[-g_m \frac{sLR_L}{LCR_L s^2 + sL + R_L}\right]^2$$

$$= \left[g_m R_L \frac{\omega_o}{Q_L} \frac{s}{s^2 + \frac{\omega_o}{Q_L}s + \omega_o^2}\right]^2 \tag{12.274}$$

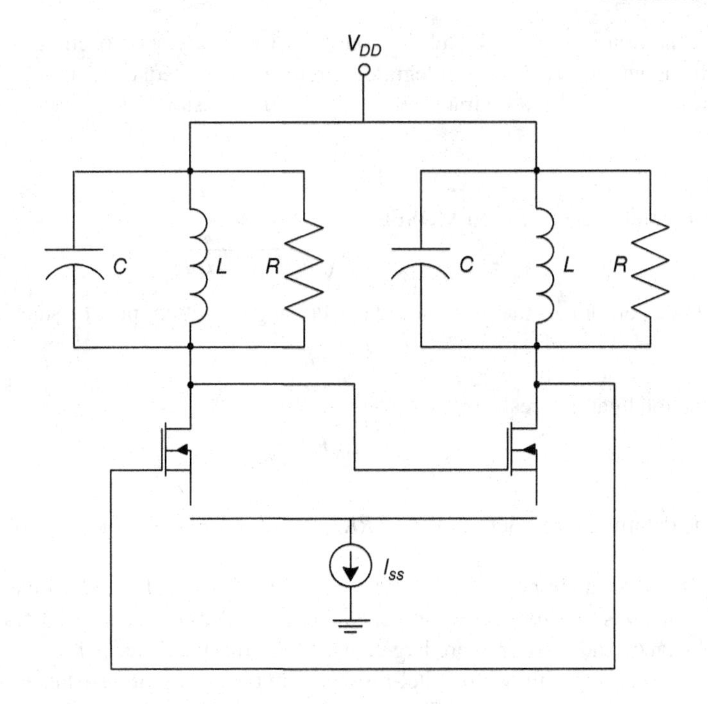

Figure 12.77 Cross-coupled *LC* oscillator.

and in the frequency domain is

$$A_v(j\omega) = \frac{v_{ds2}}{v_{gs1}} = [-g_m Z(j\omega)]^2 = \left[-g_m \frac{j\omega L R_L}{R_L(1-\omega^2 LC) + j\omega L}\right]^2$$

$$= \left[g_m \frac{\omega L R_L}{R_L(1-\omega^2 LC) + j\omega L}\right]^2 \qquad (12.275)$$

At the resonant frequency $\omega_o = 1/\sqrt{LC}$,

$$A_v(f_o) = (g_m R_L)^2 \qquad (12.276)$$

The voltage gain of the feedback network is

$$\beta = 1 \qquad (12.277)$$

The loop gain is

$$T = \beta A_v = A_v = A_{v1}^2 = \left[g_m R_L \frac{\omega_o}{Q_L} \frac{s}{s^2 + \frac{\omega_o}{Q_L} s + \omega_o^2}\right]^2 \qquad (12.278)$$

At the resonant frequency $\omega_o = 1/LC$, the loop gain is

$$T(f_o) = \beta A_v(f_o) = [A_v(f_o)]^2 = (g_m R_L)^2 \qquad (12.279)$$

The startup condition is $g_m R_L > 1$ and the condition for steady-state oscillation is $g_m R_L = 1$. This oscillator is widely used as an integrated circuit in high-frequency applications, such as wireless communications in radio transceivers. The two transistors form a negative resistance

$$R_N = -\frac{2}{g_m} \tag{12.280}$$

where the transconductance of each MOSFET is

$$g_m = g_{m1} = g_{m2} = \sqrt{\mu_n C_{ox}(W/L)I_{SS}} \tag{12.281}$$

For steady-state operation of the oscillator, the following condition must be satisfied

$$-R_N = R \tag{12.282}$$

resulting is the total parallel resistance

$$R_T = \frac{-R_N R}{-R_N + R} = \infty \tag{12.283}$$

Therefore, the damping coefficient of the $LCRR_N$ circuit is $\zeta = 0$ and the two poles are located on the imaginary axis.

Figure 12.78 shows a simple version of the cross-coupled oscillator. Two parallel-resonant circuits connected in series for the ac component are equivalent to one parallel-resonant circuit. The total inductance and resistance are larger, and the total capacitance is lower by a factor of 2 than that of the oscillator with two parallel-resonant circuits. The load resistance consists of the parasitic resistance of the inductor r_L. The equivalent parallel resistance is $R = r_L(1 + Q_{Lo})^2$.

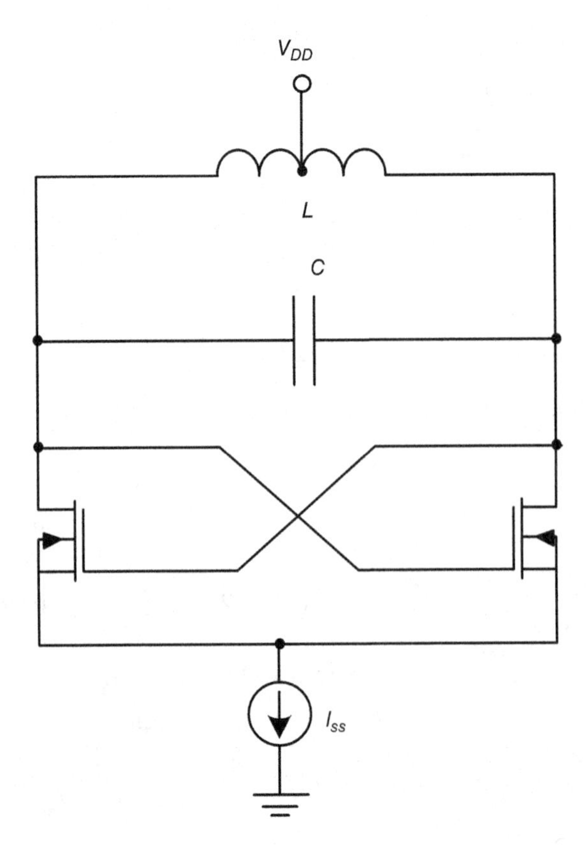

Figure 12.78 Cross-coupled *LC* oscillator with a single parallel-resonant circuit.

The differential oscillator has several advantages:

1. It has a simple topology.
2. Even harmonics are zero for ideal circuit symmetry.
3. The circuit is capable of very high-frequency operation and, therefore, is widely used in RF analog integrated circuits.

The disadvantages of the oscillator are as follows:

1. It requires two matched inductors.
2. Integrated inductors have a very low quality factor Q_{Lo} and are very lossy.
3. It is difficult to achieve a high loaded quality factor Q_L and, therefore, to achieve a low level of phase noise.

12.17 Wien-Bridge RC Oscillator

12.17.1 Loop Gain

Figure 12.79 shows a circuit of the Wien-bridge oscillator. The low-frequency voltage gain of the op-amp is

$$A_o = \frac{v_o}{v_f} = 1 + \frac{R_F}{R_S} \tag{12.284}$$

The impedance of the series combination of the resistor and the capacitor Z_s and the parallel combination of the resistor and the capacitor Z_p are

$$Z_s = R + \frac{1}{sC} = \frac{sRC + 1}{sRC} \tag{12.285}$$

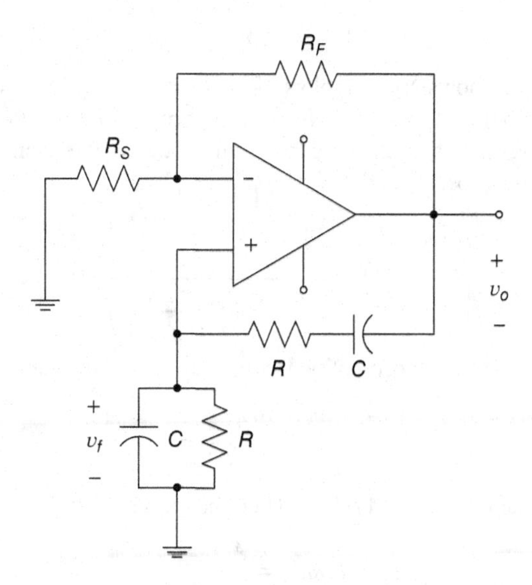

Figure 12.79 Wien-bridge oscillator.

and

$$Z_p = \frac{R\frac{1}{sC}}{R + \frac{1}{sC}} = \frac{R}{sRC + 1} \tag{12.286}$$

Hence, the voltage transfer function of the feedback network is

$$\beta(s) = \frac{v_f(s)}{v_o(s)} = \frac{Z_p}{Z_s + Z_p} = \frac{sRC}{(RC)^2 s^2 + 3RCs + 1} = \frac{s}{RC\left[s^2 + \frac{3s}{RC} + \frac{1}{(RC)^2}\right]}$$

$$= \frac{\omega_o s}{s^2 + 3\omega_o s + \omega_o^2} = \frac{1}{3 + RCs + \frac{1}{sRC}} = \frac{1}{3 + \frac{s}{\omega_o} + \frac{\omega_o}{s}} \tag{12.287}$$

where the frequency of oscillation is

$$\omega_o = \frac{1}{RC} \tag{12.288}$$

The loop gain is

$$T(s) = A_o \beta(s) = \frac{s}{(RC)\left[s^2 + \frac{3s}{RC} + \frac{1}{(RC)^2}\right]} = A_o \frac{\omega_o s}{s^2 + 3\omega_o s + \omega_o^2}$$

$$= A_o \frac{\omega_o s}{s^2 + 2\varsigma_T \omega_o s + \omega_o^2} = \frac{A_o}{3 + RCs + \frac{1}{sRC}} = \frac{A_o}{3 + \frac{s}{\omega_o} + \frac{\omega_o}{s}} \tag{12.289}$$

where $\varsigma_T = 1.5 = RC/2$ and $Q = 1/(2\varsigma) = 1/3$. The poles of $\beta(s)$ and $T(s)$ are

$$p_1, p_2 = \frac{-3 \pm \sqrt{5}}{2}\omega_o = -2.618\omega_o, -0.382\omega_o. \tag{12.290}$$

For $s = j\omega$,

$$T(j\omega) = \frac{A_o}{3 + j\left(\frac{\omega}{\omega_o} - \frac{\omega_o}{\omega}\right)} \tag{12.291}$$

Figures 12.79 and 12.80 shows Nyquist plots of the loop gain $T(j\omega)$ for the Wien-bridge oscillator at $A_o = 2$, 3, and 5. For $A_o = 3$, the Nyquist plot crosses the point $(-1, 0)$ and will sustain steady-state oscillation. For $A_o = 2$, the circuit will not start the oscillation. For $A_o = 5$, the circuit will start a growing oscillation.

For $s = j\omega$, the loop gain is

$$T(j\omega) = \frac{A_o}{3 + j\left(\omega RC - \frac{1}{\omega RC}\right)} \tag{12.292}$$

Since $T(j\omega_o)$ must be real at the oscillation frequency f_o, the imaginary part of the denominator of $T(j\omega)$ must be zero

$$\omega_o RC - \frac{1}{\omega_o RC} = 0 \tag{12.293}$$

the frequency of oscillation is $\omega_o = 1/(RC)$. Then the loop gain at the oscillation frequency is

$$T(j\omega_o) = \frac{A_o}{3} = 1 \tag{12.294}$$

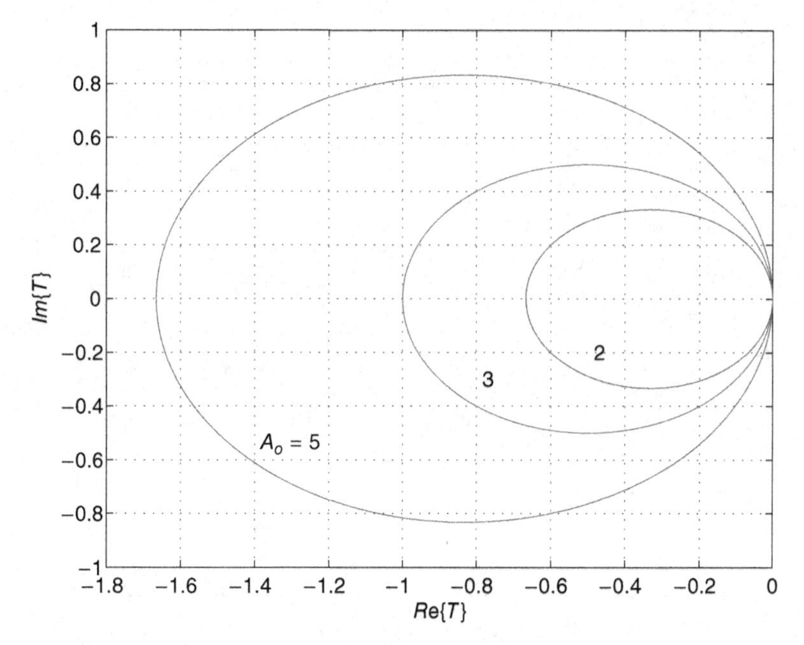

Figure 12.80 Nyquist plots of the loop gain $T(j\omega)$ for Wien-bridge oscillator at selected values of A_o.

producing the amplifier gain for steady-state oscillations

$$A_o = 3 \tag{12.295}$$

To initiate oscillation spontaneously, it is required to have $A_o > 3$.

The feedback network transfer function is

$$\beta(j\omega) = \frac{1}{3 + j\left(\dfrac{\omega}{\omega_o} - \dfrac{\omega_o}{\omega}\right)} \tag{12.296}$$

and the loop gain is

$$T(j\omega) = \frac{3}{3 + j\left(\dfrac{\omega}{\omega_o} - \dfrac{\omega_o}{\omega}\right)} = \frac{1}{1 + j\dfrac{1}{3}\left(\dfrac{\omega}{\omega_o} - \dfrac{\omega_o}{\omega}\right)} \tag{12.297}$$

Figures 12.81 and 12.82 show Bode plots of the loop gain T for the Wien-bridge oscillator at $A_o = 3$ and $f_o = 1$ kHz. Since $2\varsigma = 3$, $\varsigma = 1.5$ and $Q = 1/(2\varsigma) = 1/3$.

12.17.2 Closed-Loop Gain

The closed-loop gain is

$$A_f(s) = \frac{A_o}{1 - T(s)} = \frac{A_o(s^2 + 3\omega_o s + \omega_o^2)}{s^2 + (3 - A_o)\omega_o s + \omega_o^2}. \tag{12.298}$$

Since $2\varsigma = 3 - A_o$, $\varsigma = (3 - A_o)/2$ and $Q = 1/(2\varsigma) = 1/(3 - A_o)$. As A_o increases, ς decreases and Q increases. For $A_o < 3$, $\varsigma > 0$ and $Q > 0$, in which case, the oscillations will not start. At $A_o = 3$, $\varsigma = 0$ and $Q = \infty$, in which case, the circuit produces steady-state oscillations with a constant amplitude. For $A_o > 3$, $\varsigma < 0$ and $Q < 0$, the amplitude of oscillations increases.

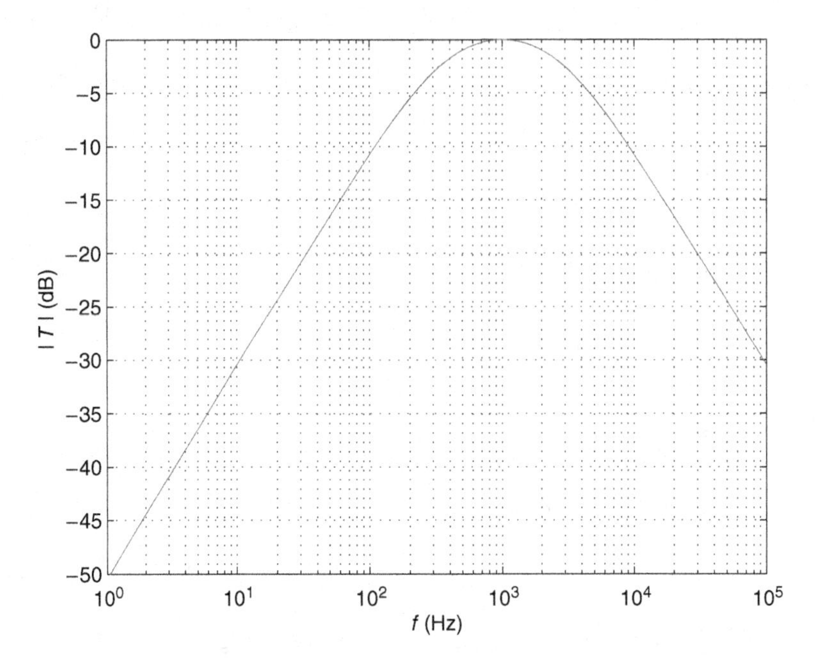

Figure 12.81 Magnitude of the loop gain T for Wien-bridge oscillator at $A_o = 3$ and $f_o = 1$ kHz.

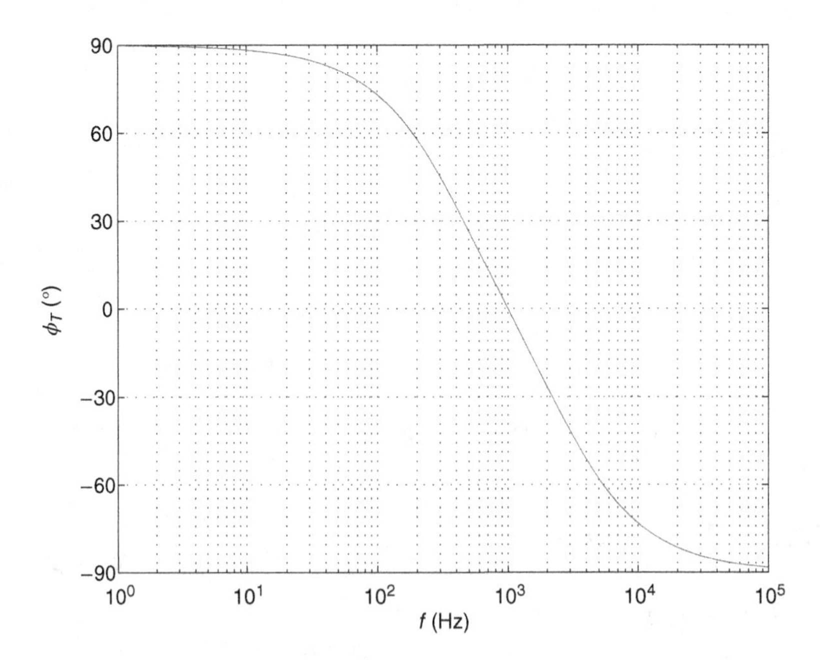

Figure 12.82 Phase of the loop gain T for Wien-bridge oscillator at $A_o = 3$ and $f_o = 1$ kHz.

The poles of the closed-loop gain $A_f(s)$ are

$$p_1, p_2 = -\frac{(3 - A_o)\omega_o}{2} \pm \frac{1}{2}\sqrt{(3 - A_o)^2\omega_o^2 - 4\omega_o^2}$$
$$= -\frac{(3 - A_o)\omega_o}{2} \pm \frac{\omega_o}{2}\sqrt{(3 - A_o)^2 - 4} \quad \text{for} \quad 0 \le A_o \le 1 \quad \text{and} \quad A_o \ge 5 \tag{12.299}$$

or

$$p_1, p_2 = -\frac{(3 - A_o)\omega_o}{2} \pm j\frac{1}{2}\sqrt{4\omega_o^2 - (3 - A_o)^2\omega_o^2}$$
$$= -\frac{(3 - A_o)\omega_o}{2} \pm j\frac{\omega_o}{2}\sqrt{4 - (3 - A_o)^2} \quad \text{for} \quad 1 < A_o < 5 \tag{12.300}$$

For $A_o = 0$,

$$p_1, p_2 = \frac{-3 \pm \sqrt{5}}{2}\omega_o \tag{12.301}$$

As A_o is increased from 0 to 1, the poles are real and move to each other. The two poles are equal and real when the imaginary part is zero,

$$(3 - A_o)^2 = 4 \tag{12.302}$$

yielding

$$A_o = 1 \tag{12.303}$$

or

$$A_o = 5 \tag{12.304}$$

and the poles are

$$p_1 = p_2 = -\omega_o \tag{12.305}$$

As A_o is increased from $A_o = 1$ to 3, the poles are complex conjugate and are located in the LHP. For $A_o = 3$, the poles are

$$p_1, p_2 = \pm j\omega_o \tag{12.306}$$

In this case, the oscillator circuit is marginally stable and produces steady-state oscillation at ω_o. As A_o is increased from $A_o = 3$ to 5, the poles are complex conjugate and are located in the RHP. In this case, the oscillations will start and increase until they reach steady state. For $A_o = 5$, the poles are real, equal, and located in the RHP. For $A_o \ge 5$, the poles are real and located in the RHP.

Figures 12.83 and 12.84 show Bode plots of the closed-loop gain $A_f(s)$ for $A_o = 3$ and f_o = 1 kHz. Figure 12.85 shows the root locus of the closed-loop gain $A_f(s)$ for the Wien-bridge oscillator as A_o varies from 0 to ∞. The less negative real pole p_1 moves from $(-3 + \sqrt{5})\omega_o/2$ for $A_o = 0$ to the left, becomes equal to the more positive pole p_2 at $-\omega_o$ for $A_o = 1$, then becomes complex with a negative imaginary part, reaches $-j\omega_o$ for $A_o = 3$, becomes real again and equal to p_2 at ω_o for $A_o = 5$, and moves to infinity. The more negative real pole p_2 moves from $(-3 - \sqrt{5})\omega_o/2$ for $A_o = 0$ to the right, becomes equal to the less negative pole p_1 at $-\omega_o$ for $A_o = 1$, then becomes complex with a positive imaginary part, crosses $j\omega_o$ for $A_o = 3$, becomes real again and equal to p_1 at ω_o for $A_o = 5$, and moves to the origin.

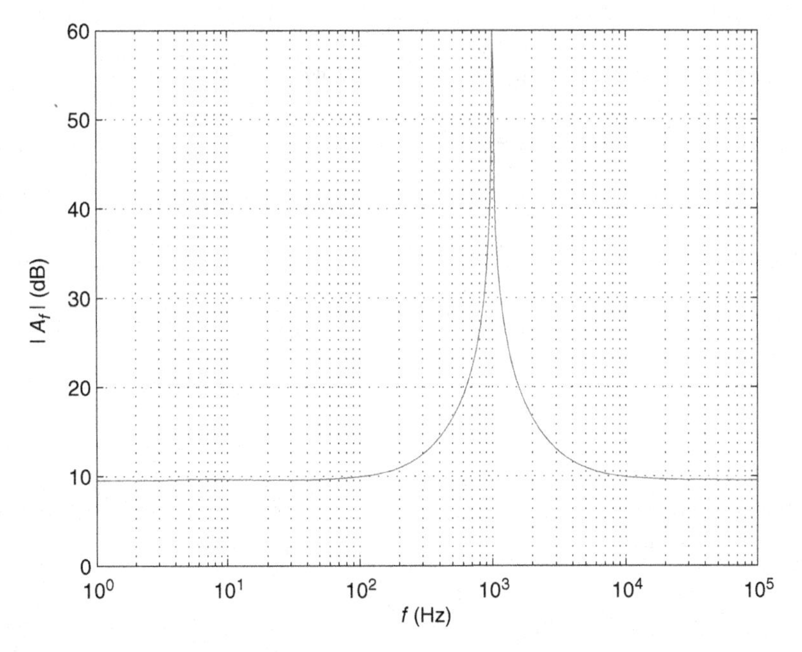

Figure 12.83 Magnitude of the closed-loop gain A_f for Wien-bridge oscillator at $A_o = 3$ and $f_o = 1$ kHz.

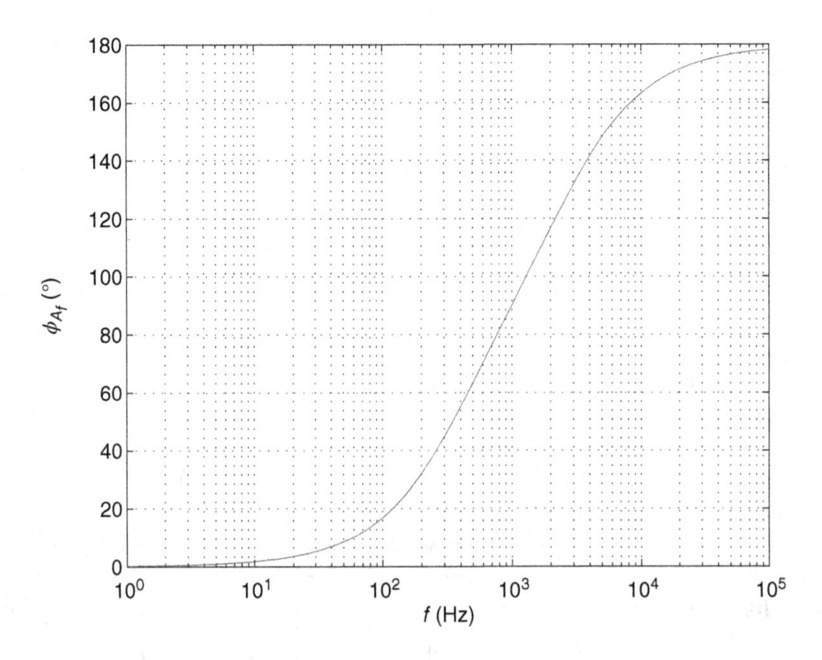

Figure 12.84 Phase of the closed-loop gain for Wien-bridge oscillator at $A_o = 3$ and $f_o = 1$ kHz.

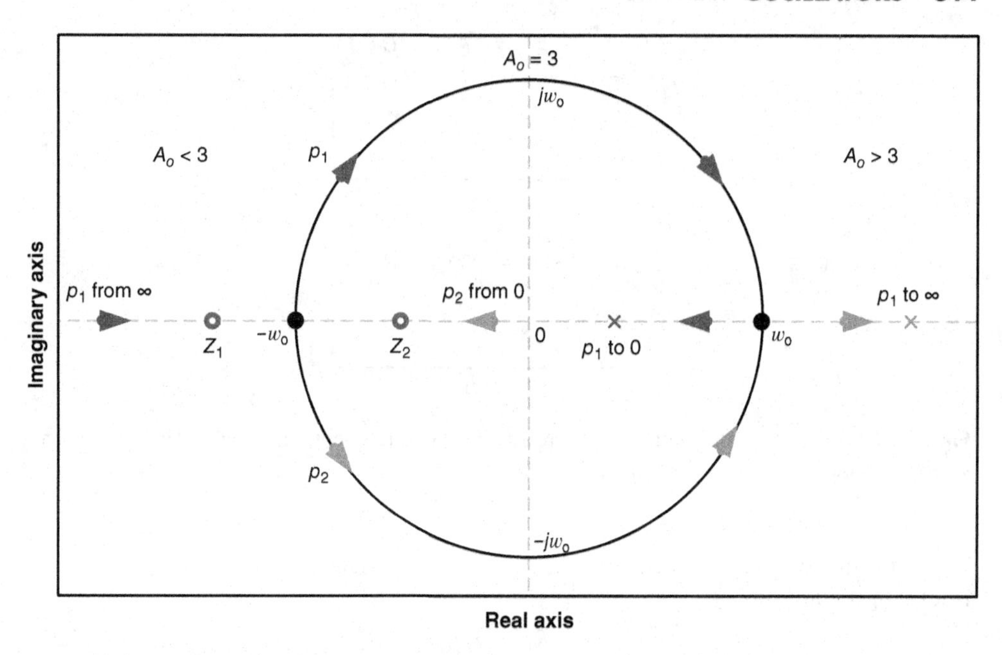

Figure 12.85 Root locus of the closed-loop gain $A_f(s)$ for Wien-bridge oscillator.

Example 12.3

Design a Wien-bridge oscillator for $f_o = 100$ kHz.

Solution. Let $C = 100$ pF. Hence, the resistance is

$$R = \frac{1}{2f_o\pi C} = \frac{1}{2\pi 100 \times 10^3 \times 100 \times 10^{-12}} = 15.9\ \text{k}\Omega \tag{12.307}$$

Pick $R = 16$ kΩ/0.25 W/1%. Let $R_s = 5.1$ kΩ/0.25 W/1%. The feedback resistance required for steady-state oscillation is

$$R_F = 2R_S = 2 \times 5.1 = 10.2\ \text{k}\Omega \tag{12.308}$$

Pick $R_F = 11$ kΩ/0.25 W/1%. To start oscillation, we may select Pick $R_F = 16$ kΩ/0.25 W/1% and two Zener diodes 1N5231 connected back-to-back in parallel with R_F. Select a 741 op-amp. Use $\pm V = \pm 12$ V.

12.18 Oscillators with Negative Resistance

A circuit of an oscillator with a negative resistance is shown in Fig. 12.86. It consists of a series-resonant circuit and a nonlinear current-controlled dependent-voltage source. The differential equation describing this circuit is as follows:

$$L\frac{di^2}{dt^2} + R\frac{di}{dt} + \frac{i}{C} = -\frac{dv}{dt} \tag{12.309}$$

The nonlinear voltage source is described by

$$v = v(i) = v_o + R_1 i + R_2 i^2 + R_3 i^3 + \cdots \tag{12.310}$$

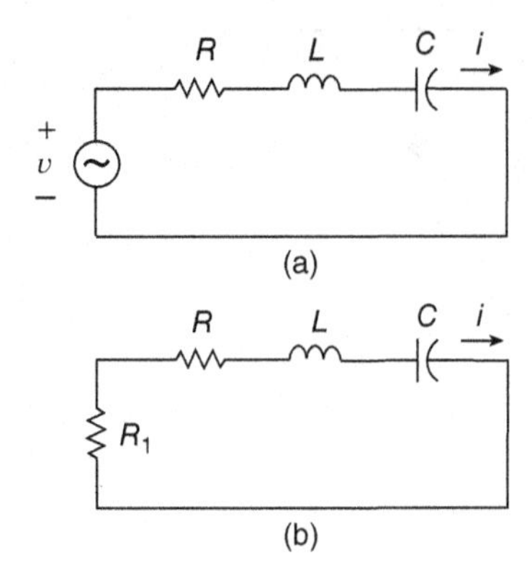

Figure 12.86 Oscillator with negative resistance. (a) Equivalent circuit with dependent-voltage source. (b) Equivalent circuit with negative resistance.

Taking the first two terms, we can find the small-signal resistance (i.e., slope of this characteristic) at operating point Q

$$R_1 = \left.\frac{dv}{di}\right|_Q \tag{12.311}$$

The small-signal resistance may become negative.

The static resistance of the nonlinear device is always positive and, therefore, the device always absorbs dc power. In contrast, an ac signal applied to the diode at the operating point Q will see a negative small-signal resistance R_1 and oscillations could develop. If the rms current through the series resonant circuit is I, the device with the negative incremental resistance R_1 delivers power $I^2|R_1|$, and the load absorbs the power I^2R. The oscillator is in steady state. If the power delivered is greater than the power absorbed, the amplitude of the oscillations will increase. This will increase the swing of the ac signal on either side of the operating point, and the instantaneous incremental resistance of the nonlinear device may become positive over part of each cycle, when the amplitude of oscillation becomes sufficiently large. This will decrease $|R_1|$ and decrease the power delivered until equilibrium is established.

Thus,

$$L\frac{di^2}{dt^2} + R\frac{di}{dt} + \frac{i}{C} = -R_1\frac{di}{dt} \tag{12.312}$$

This equation simplifies to the form:

$$L\frac{di^2}{dt^2} + (R + R_1)\frac{di}{dt} + \frac{i}{C} = 0 \tag{12.313}$$

This equation can be converted into the s-domain

$$s^2I(s) + s\frac{R + R_1}{L}I(s) + \frac{I(s)}{LC} = 0 \tag{12.314}$$

yielding

$$s^2 + s\frac{R + R_1}{L} + \frac{1}{LC} = 0 \tag{12.315}$$

or

$$\left(s + \frac{R + R_1}{2L}\right)^2 + \frac{1}{LC} - \left(\frac{R + R_1}{2L}\right)^2 = 0 \tag{12.316}$$

which can be written as

$$(s + \alpha)^2 + \omega_d^2 = 0 \tag{12.317}$$

where

$$\alpha = \frac{R + R_1}{2L} \tag{12.318}$$

and

$$\omega_d = \sqrt{\frac{1}{LC} - \left(\frac{R + R_1}{2L}\right)^2} \tag{12.319}$$

The poles of this equation are

$$p_1, p_2 = \alpha \pm j\omega_o = -\frac{R + R_1}{2L} \pm j\sqrt{\frac{1}{LC} - \left(\frac{R + R_1}{2L}\right)^2} \tag{12.320}$$

For $\alpha < 0$, the poles are located in the LHP of s and the oscillations decay. For $\alpha = 0$, the poles are located in the imaginary axis, and the steady-state oscillations with a constant amplitude occur. For $\alpha > 0$, the poles are located in the RHP and the amplitude of oscillations will increase until steady-state operation is reached due to nonlinearity of the semiconductor device and the poles are moved to the imaginary axis.

In order to start and grow oscillations, the total resistance in the series-resonant circuit must be negative

$$R_T = R + R_1 < 0 \tag{12.321}$$

or

$$R_1 < -R \tag{12.322}$$

In this case, $\alpha > 0$ and the poles are located in the RHP, causing growing oscillation.

The condition for steady-state oscillations is

$$R + R_1 = 0 \tag{12.323}$$

or

$$R_1 = -R \tag{12.324}$$

In this case, the energy lost in resistance R in every cycle is delivered by the active device. Therefore, the circuit is lossless and is described by the equation

$$\frac{di^2}{dt^2} + \frac{i}{LC} = 0 \tag{12.325}$$

This equation in the s-domain is

$$s^2 I(s) + \frac{I(s)}{LC} = 0 \tag{12.326}$$

which simplifies to the form

$$s^2 + \frac{1}{LC} = 0 \tag{12.327}$$

which can be written as

$$(s + \omega_o)^2 = 0 \tag{12.328}$$

Thus, the circuit produces a steady-state oscillation. This situation is equivalent to the transfer function with two complex conjugate poles

$$p_1, p_2 = \pm j\omega_o = \pm j\frac{1}{\sqrt{LC}} \tag{12.329}$$

located on the imaginary axis.

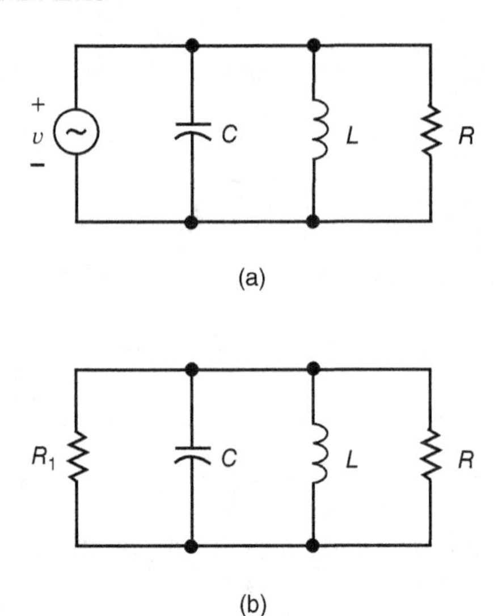

Figure 12.87 Negative resistance oscillator with a parallel-resonant circuit: (a) circuit and (b) equivalent circuit.

Negative-resistance oscillators can be also represented by a parallel-resonant circuit *LCR* connected across a nonlinear device with a negative resistance R_1, as shown in Fig. 12.87.

The startup condition for oscillations is $|-R_1| < R$.

$$R_T = \frac{R_1 R}{R_1 + R} < 0 \tag{12.330}$$

For R greater than $|-R_1|$, $R_T < 0$ and $\alpha < 0$, and oscillations will start and grow. For R less than $|-R_1|$, $R_T > 0$ and $\alpha > 0$, and oscillations will decay. The condition for steady-state oscillations is $|-R_1| = R$, yielding a lossless resonant circuit, where

$$R_T = \frac{R_1 R}{R_1 + R} = \infty \tag{12.331}$$

If R is equal to $|R_1|$, R_T is infinity, and the circuits will produce steady-state sinusoidal voltage.

Semiconductor diodes can be used as devices with a negative dynamic resistance. Examples of such devices are Gunn diodes, tunnel diodes, impact ionization avalanche transit time (IMPATT) diodes, and TRAPATT diodes. Figures 12.88 and 12.89 show the *I–V* characteristics of the Gunn

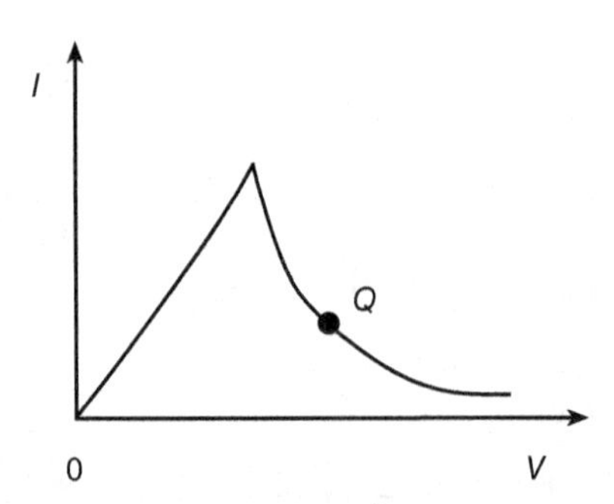

Figure 12.88 *I–V* characteristic of the Gunn diode.

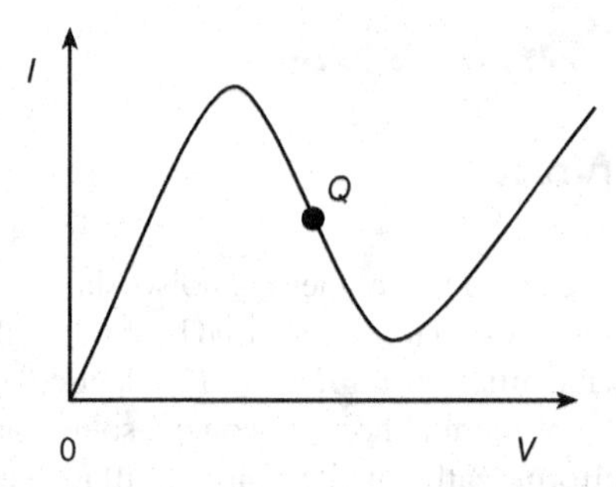

Figure 12.89 *I–V* characteristic of the tunnel diode.

diode and tunnel diode, respectively. It can be seen that there is a range of voltage V, where the slope of the characteristic is negative. If the operating point Q is chosen in this range, the dynamic resistance R_1 is also negative.

The Gunn diode is also called *transfer electron device* (TED) or the *limited space-charge accumulation device* (LSA). It is built using the n-type gallium arsenide crystal (GaAs) for frequencies from 1 to 35 GHz and output power up to 4 W for continuous operation and 1000 W for pulsed operation. An indium phosphide (InP) crystal is used for frequencies from 35 to 140 GHz and 0.5 W of output power. Varactor diodes may be used to obtain FM. The maximum frequency deviation is $\Delta f = 0.04 f_c$.

The IMPATT pn junction diode is operated in the avalanche breakdown region. It can produce 10 W at 10 GHz and is capable of operating up to 300 GHz.

12.19 Voltage-Controlled Oscillators

A VCO or a current-controlled oscillator (ICO) is a circuit whose instantaneous oscillation frequency f_o is controlled electronically by a voltage or a current, respectively. VCOs are widely used in various applications, including wireless transceivers, such as FM, frequency-shift keying (FSK), phase modulation (PM), and phase-locked loops (PLLs). In LC sinusoidal VCOs, the resonant frequency f_o is usually controlled by a voltage. This is accomplished by controlling electronically a voltage-dependent capacitance of the resonant circuit $RLC(v_C)$. The oscillation frequency is

$$f_o(t) = K_o v_C(t) \tag{12.332}$$

where K_o is a constant in Hz/V. The junction capacitance of a reverse-biased pn junction diode is used as a variable capacitance

$$C_j = \frac{C_{j0}}{\left(1 - \frac{v_D}{V_{bi}}\right)^m} \tag{12.333}$$

These kinds of diodes are known as *varactors*. The frequency of oscillation is

$$f_o = \frac{1}{2\pi\sqrt{LC(v_D)}} = \frac{1}{2\pi}\sqrt{\frac{\left(1 - \frac{v_D}{V_{bi}}\right)^m}{LC_{j0}}} \tag{12.334}$$

Colpitts, Clapp, and cross-coupled oscillators are the most commonly used circuits as VCOs.

12.20 Noise in Oscillators

12.20.1 Thermal Noise

There are different kinds of electrical noise: thermal noise, shot noise, flicker noise ($1/f$ noise), burst noise, and avalanche noise. A shot noise is caused by random fluctuations of charge carriers in semiconductor devices. A metallic resistor $R = 1/G$ is a source of thermal noise. Figure 12.90 shows two equivalent sources of thermal noise: a voltage source and current source. The spectrum of thermal noise is uniform, called white noise, as illustrated in Fig. 12.91. At ambient temperature $T > 0$ K, resistors contain free electrons, which move randomly in different directions with different velocities and experience collisions. Thermal noise (also called Johnson noise [2], Nyquist noise [3], or Johnson–Nyquist noise) is caused by the random motion of charge carriers in a resistor due to thermal agitation, with a kinetic energy that is proportional to the temperature T. These motions are called Brownian motions of free electrons. Random statistical fluctuations of electrons cause the existence of transient differences in electron densities at the two terminals of a resistor. This leads to the fluctuating voltage difference between the resistor terminals. As a result, there are small random voltage and current fluctuations. The thermal velocity of the carriers increases, when the absolute temperature T and resistance R increase. Noise corrupts (contaminates) the desired signal. Noise is a random or stochastic signal. Electrical noise can be observed as a voltage waveform v_n or a current waveform i_n. It is defined by a probability density function.

From quantum mechanics, Planck's blackbody radiation law describes the thermal noise power density as

$$P(f) = \frac{4hf}{e^{\left(\frac{hf}{kT}\right)} - 1} df \tag{12.335}$$

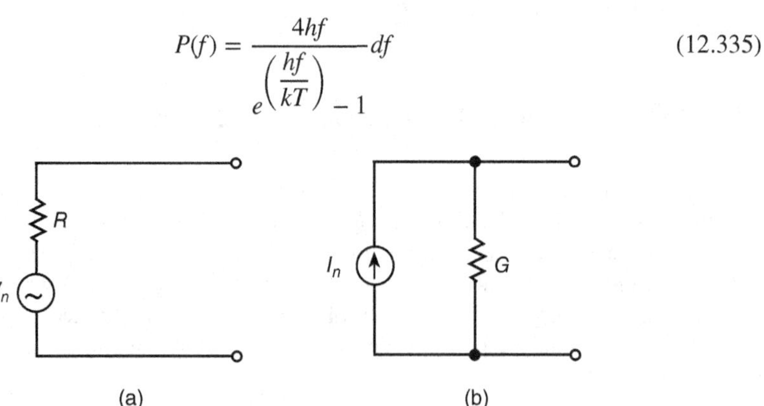

Figure 12.90 Equivalent circuits representing thermal (or white) noise in a resistor. (a) Noise voltage source. (b) Noise current source.

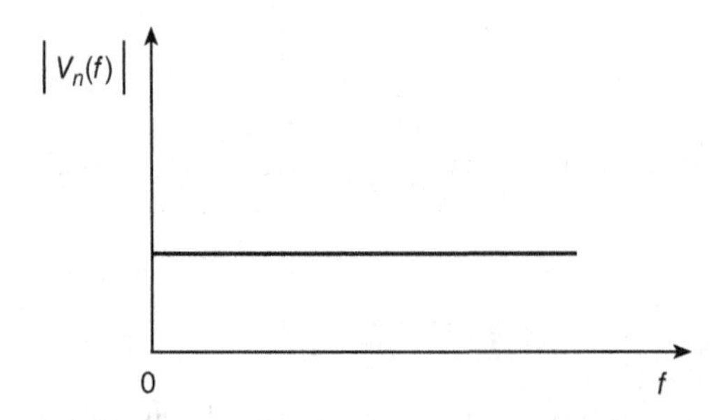

Figure 12.91 Spectrum of white noise.

where $h = 6.62617 \times 10^{-34}$ J·s is Planck's constant, $k = 1.38 \times 10^{-23}$ J/K is Boltzmann's constant, and T is the absolute temperature. Using the approximation $e^x \approx 1 + x$ for $x \ll 1$, one obtains the Rayleigh–Joans approximation

$$P(f) = \frac{4hf\,df}{\left(\frac{hf}{kT}\right) + \frac{1}{2!}\left(\frac{hf}{kT}\right)^2 + \frac{1}{3!}\left(\frac{hf}{kT}\right)^3 + \cdot} \approx kT\,df \quad \text{for} \quad hf/kT \ll 1 \tag{12.336}$$

The normalized thermal noise power density can be expressed as

$$\frac{P(f)}{4kTdf} = \frac{hf}{kT\,df\left[e^{\left(\frac{hf}{kT}\right)} - 1\right]} \tag{12.337}$$

Figure 12.92 illustrates the normalized thermal noise power density as a function of frequency at $T = 300$ K. It can be observed that the normalized thermal noise power density is flat and equal to kT for frequencies from 0 to 10^{11} Hz. The thermal noise power decreases above the frequency 100 GHz.

The *thermal noise* is also called *white noise* because it has a uniform power spectral density for frequencies from zero up to 10^{11} Hz = 100 GHz, that is, there is equal power in each incremental frequency interval df. The spectrum of white noise is shown in Fig. 12.91. Figure 12.90(a) and (b) shows equivalent circuits representing thermal noise in a resistor in the forms of voltage and current sources with internal resistance R, respectively. Each practical resistor behaves as a small voltage source V_n with a series noiseless resistor R or a small current source I_n with a parallel noiseless resistor R, which generate voltage and current at all frequencies, up to 10^{11} Hz. The circuit with the current source is the Norton equivalent of the circuit with the voltage source. The power spectral density of thermal noise is the noise power in 1-Ω resistance and in 1-Hz bandwidth; it is given by $N_0 = kT$. The rms values of the noise voltage and noise current are, respectively,

$$V_n = \sqrt{4kTRB} \tag{12.338}$$

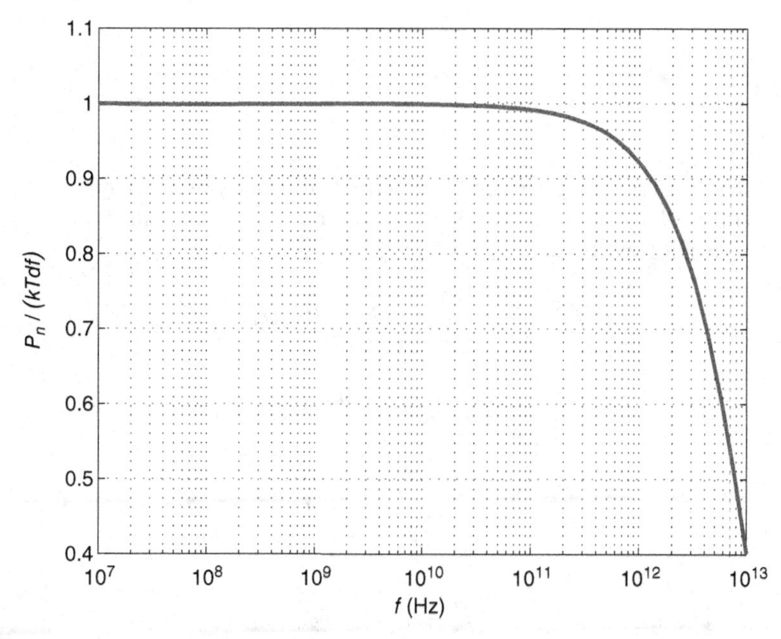

Figure 12.92 Normalized thermal noise power density $P(f)/(4kTdf)$ as a function of frequency f at $T = 300$ K.

and

$$I_n = \frac{V_n}{R} = \sqrt{4kTGB} = \sqrt{\frac{4kTB}{R}} \tag{12.339}$$

where $k = 1.38 \times 10^{-23}$ J/K is Boltzmann's constant, T is the absolute temperature of the resistor R in K, and B is the bandwidth in which the noise is observed. The higher the temperature T, the higher the thermal agitation and thereby the thermal noise level. The rms noise voltage V_n increases with temperature T, bandwidth B, and resistance R. The noise voltage source polarity and the noise current source direction are meaningless. For $R = 1$ kΩ at $T = 300$ K, $V_n = 4$ nA/$\sqrt{\text{Hz}}$.

A short-circuited resistor R dissipates a noise power given by

$$P_{nR} = \frac{V_n^2}{R} = 4kTB \tag{12.340}$$

The noise power in a resistor P_n in dBm (decibels relative to 1 mW) is

$$P_n = 10 \log \left(\frac{kTB}{10^{-3}}\right) \text{ (dBm)} \tag{12.341}$$

At room temperature $T = 300$ K,

$$P_n = 10 \log \left(\frac{V_n^2}{10^{-3}R}\right) = 10 \log (1000 \times 4kTB) = -174 + \log (B) \text{ (dBm)} \tag{12.342}$$

The normalized time-average noise power produced by a 1 Ω resistor is given by

$$P_n = \frac{V_n^2}{R} = \frac{V_n^2}{1\ \Omega} = V_n^2 = \frac{1}{T}\int_0^T v_n^2(t)dt = 4kTB \tag{12.343}$$

and

$$P_n = I_n^2 R = I_n^2 \times 1\ \Omega = I_n^2 = \frac{1}{T}\int_0^T i_n^2(t)dt = 4kTB \tag{12.344}$$

Thus,

$$V_n^2 = RP_n \tag{12.345}$$

and

$$I_n^2 = \frac{P_n}{R} \tag{12.346}$$

Noise voltage and current have power distributed continuously over a frequency range. Since the time-average voltage of noise is zero, power spectrum $V_n^2(f)$ is specified for each frequency interval df. The noise spectral power density is given by

$$S_v = \frac{V_n^2}{B} = 4kTR \text{ (V}^2\text{/Hz)} \tag{12.347}$$

and

$$S_i = \frac{P_n}{B} = 4kTG = \frac{4kT}{R} \text{ (A}^2\text{/ Hz)} \tag{12.348}$$

Thus,

$$V_n^2 = \int_0^\infty S_v(f)df \tag{12.349}$$

and

$$I_n^2 = \int_0^\infty S_i(f)df \tag{12.350}$$

Both V_n^2 and I_n^2 are independent of frequency f.

The *mean-square noise voltage* is equal to the rms value of the thermal noise at the output of a network containing both resistive and reactive elements

$$V_n^2 = 4kT \int_B R(f)df = \frac{4kT}{2\pi} \int_B R(\omega)d\omega = \frac{2kT}{\pi} \int_B R(\omega)d\omega \tag{12.351}$$

where $R(f) = Re\{Z(f)\}$ is the real (resistive) part of the input impedance at frequency f, and the integration is performed over the bandwidth B.

The signal-to-noise ratio (SNR) is defined as

$$SNR = \frac{S}{N} = \frac{P_s}{P_n} = \frac{V_{s(rms)}^2}{V_n^2} = 10 \log \left(\frac{P_s}{P_n}\right) \text{ (dB)} \tag{12.352}$$

where P_s is the time-average power of the signal and $V_{s(rms)}$ is the rms source voltage. The SNR is a measure of the detectability of the signal in the presence of noise. The larger the SNR, the less signal is corrupted by the noise.

From thermodynamics and statistical mechanics, the average energy associated with one degree of freedom of a system in thermal equilibrium is

$$E_{av} = \frac{1}{2}kT \tag{12.353}$$

The average electric energy stored in the capacitor C connected in parallel with a resistor R is

$$\frac{1}{2}CV_{n(rms)}^2 = \frac{1}{2}kT \tag{12.354}$$

yielding

$$V_{n(rms)}^2 = \frac{kT}{C} \tag{12.355}$$

Similarly, the average magnetic energy stored in the inductor connected in series with a resistor R at temperature T is

$$\frac{1}{2}LI_{n(rms)}^2 = \frac{1}{2}kT \tag{12.356}$$

resulting in

$$I_{n(rms)}^2 = \frac{kT}{L} \tag{12.357}$$

When white noise is transmitted through a network (e.g., a filter) described by a transfer function $A_v = |A_v|e^{\phi_{Av}}$, the output noise is not white anymore because it is attenuated. The power-spectral density of the output noise is determined by the network transfer function and the power-spectral density of the noise at the input. In the case of the network with a voltage transfer function, the input-to-output spectral relationship is

$$S_{v(out)}(f) = |A_v(f)|^2 S_{v(in)}(f) \tag{12.358}$$

The power of the output noise is

$$P_{n(out)} = \int_0^\infty S_{v(out)}df = S_{v(in)} \int_0^\infty |H(f)|^2 \, df \tag{12.359}$$

The *noise-effective bandwidth* of a network described by frequency response A_v is defined as

$$B_n = \frac{\int_0^\infty |A_v(f)|^2 \, df}{|A_m|^2} \tag{12.360}$$

where A_m is the maximum value of $|A_v(f)|$. The integral describes the total area under the curve of $|A_v|^2$. The noise-effective bandwidth B_n for many filters is related to the filter 3-dB bandwidth BW by

$$B_n = \frac{\pi}{2} \times BW \tag{12.361}$$

In general, the noise-equivalent bandwidth B_n is wider than the 3-dB signal bandwidth BW.

Example 12.4

Determine the rms value of the noise voltage V_n, noise current I_n, and noise power P_{nR} produced by a 1 MΩ resistor in a bandwidth of 1 MHz at $T = 27$ °C.

Solution. The rms value of the noise voltage is

$$V_n = \sqrt{4kTRB} = \sqrt{4 \times 1.38 \times 10^{-23} \times 300 \times 10^6 \times 10^6} = 128.7\ \mu\text{V rms} \quad (12.362)$$

The rms value of the noise current is

$$I_n = \sqrt{\frac{4kTB}{R}} = \sqrt{\frac{4 \times 1.38 \times 10^{-23} \times 300 \times 10^6}{10^6}} = 12.87\ \mu\text{A rms} \quad (12.363)$$

The noise power produced by resistor R is

$$P_{nR} = 4kTB = 4 \times 1.38 \times 10^{-23} \times 300 \times 10^6 = 0.01658\ \text{pW} \quad (12.364)$$

The noise power of the oscillator in dBm is

$$P_{nR} = 10 \log (1000 \times 4kTB) = 10 \log (1.656 \times 10^{-14} \times 10^3) = -114\ \text{dBm} \quad (12.365)$$

12.20.2 Phase Noise

The effect of noise is critical to the performance of oscillators. In communications systems, noise determines the threshold for the minimum signal that can reliably be detected by a receiver. A noisy local oscillator leads to downconversion or upconversion of undesired signals. From noise point of view, there are two parameters of oscillators:

1. Phase noise.
2. SNR at the oscillator output.

For an ideal noiseless oscillator, the output voltage is a pure single-frequency sinusoid given by

$$v_o(t) = V_m \cos(\omega_o t + \phi) \quad (12.366)$$

The spectrum of the pure sinusoid is a single spectral line $|V_o| = V_c \delta(f_o)$. In any practical oscillator circuit, unwanted random and extraneous signals are present in addition to the desired signal. The existence of noise in oscillators may affect the amplitude and the frequency of the oscillator output signal $v_o(t)$. The influence of the noise on amplitude of the output signal is usually negligible because of the active device saturation due to nonlinearity. However, the influence of noise on the oscillator frequency of the output signal f_o is important. In a practical oscillator, the phase $\phi_n(t)$ is time dependent. It is a small random excess phase, representing variation in the oscillation period. The phase noise causes random angle modulation. It is a short-term frequency fluctuations of the oscillator output signal. A low phase noise is required for any receiver and transmitter. The output voltage of a practical oscillator is

$$v_o(t) = V_m(t) \cos[\omega_o t + \phi_n(t)] = [V_m + \epsilon(t)][\cos \omega_o t \cos \phi_n(t) - \sin \omega_o t \sin \phi_n(t)] \quad (12.367)$$

where the phase $\phi_n(t)$ is time dependent and $\omega_o t + \phi_n(t)$ is the total phase. An instantaneous phase variation is indistinguishable from variation in oscillation frequency. The AM noise is reduced by the saturation of the active device in the oscillator because of its nonlinearity. Only the fluctuations of the phase broaden the spectrum of the oscillator output voltage. For $\phi_n(t) \ll 1$, $\cos\phi_n(t) \approx 1$ and $\sin\phi_n(t) \approx \phi_n(t)$.

Hence,

$$v_o(t) \approx V_m(t)[\cos\omega_o t - \phi(t)\sin\omega_o t] \tag{12.368}$$

If the phase is described by

$$\phi_n(t) = m\cos\omega_m t \tag{12.369}$$

the output voltage of the practical oscillator becomes

$$\begin{aligned} v_o(t) &= V_m(t)\cos(\omega_o t + m\cos\omega_m t) \\ &= V_m(t)[\cos(\omega_o t)\cos(m\cos\omega_m t) - \sin(\omega_o t)\sin(m\cos\omega_m t)] \end{aligned} \tag{12.370}$$

This is an analog angle-modulated signal. Its exact description involves Bessel functions. However, for $m \ll 1$ rad, $\cos(m\cos\omega_m t) \approx 1$ and $\sin(m\cos\omega_m t) \approx m\cos\omega_m t$. Thus,

$$\begin{aligned} v_o(t) &\approx V_m(t)[\cos\omega_o t - m\sin\omega_o t\cos\omega_m t] \\ &= V_m(t)\cos\omega_o t - \frac{mV_m(t)}{2}\sin[(\omega_o + \omega_m)t] - \frac{mV_m(t)}{2}\sin[(\omega_o - \omega_m)t] \end{aligned} \tag{12.371}$$

This equation represents a PM signal. For the single-frequency modulation noise, the phase noise of an oscillator is defined as the ratio of the noise power in a single sideband at frequency $f_o + f_m$ or $f_o - f_m$ to the carrier power

$$\mathcal{L}(f_m) = \frac{P_n}{P_c} = \frac{\frac{1}{2R_L}\left(\frac{mV_m}{2}\right)^2}{\frac{V_{om}^2}{2R_L}} = \frac{m^2}{4}\left(\frac{V_m}{V_{om}}\right)^2 \tag{12.372}$$

where $P_c = V_{om}^2/2R_L$ is the carrier power.

In general, the phase noise of an oscillator is defined as

$$\text{Phase Noise} = 10\log\left(\frac{\text{Power in 1 Hz bandwidth at } \Delta f \text{ offset}}{\text{Power of the carrier}}\right)\left(\frac{\text{dBc}}{\text{Hz}}\right) \tag{12.373}$$

or

$$\mathcal{L}(\Delta f) = 10\log\left[\frac{P_{sideband}(f_o + \Delta f) \text{ in 1 Hz bandwidth}}{P_{carrier}}\right]\left(\frac{\text{dBc}}{\text{Hz}}\right) \tag{12.374}$$

Figure 12.93 shows the spectra of the output voltage of ideal and practical oscillators. For example, the required phase noise in GSM is −138 dBc/Hz at the offset $\Delta f = 1$ MHz.

The use of oscillators with large phase noise in transmitters causes distortion in neighboring channels. An oscillator with finite phase noise in receivers causes a problem in downconversion. The spectrum of the local oscillator participates in downconversion. As a result, both wanted and unwanted signals are down-converted.

Figure 12.94 shows an equivalent circuit of the parallel-resonant circuit with the thermal noise produced by the resistor. The resistor is modeled by current source I_n and a noiseless resistor R. The impedance of an ideal parallel-resonant circuit RLC is

$$\begin{aligned} Z(j\omega) &= \frac{V_{on}}{I_{in}} = \frac{1}{\frac{1}{R} + j\omega C + \frac{1}{j\omega L}} = \frac{R}{1 + jQ_L\left(\frac{\omega}{\omega_o} - \frac{\omega_o}{\omega}\right)} = R\frac{1 - jQ_L\left(\frac{\omega}{\omega_o} - \frac{\omega_o}{\omega}\right)}{1 + Q_L^2\left(\frac{\omega}{\omega_o} - \frac{\omega_o}{\omega}\right)^2} \\ &= R_s(f) - jX_s(f) \end{aligned} \tag{12.375}$$

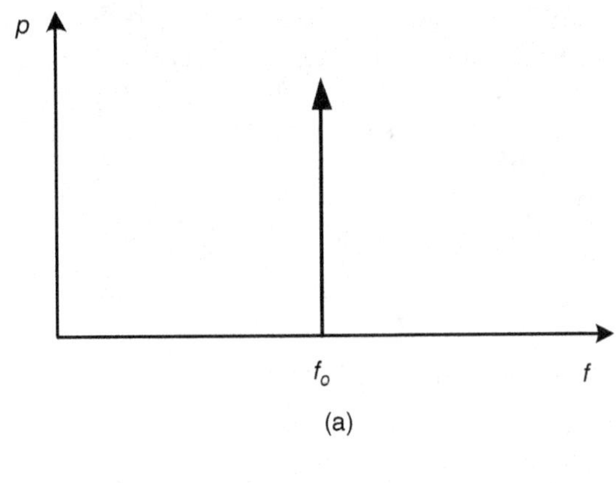

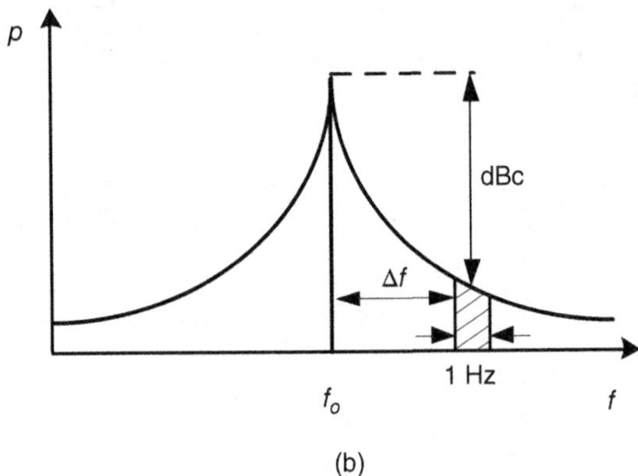

Figure 12.93 Phase noise. (a) The spectrum of the output voltage of an ideal oscillator. (a) The spectrum of the output voltage of a practical oscillator.

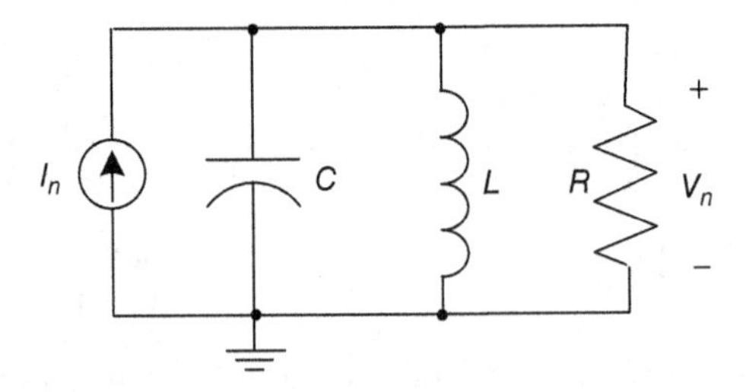

Figure 12.94 Equivalent circuit for determining the thermal (or white) noise at the output of an oscillator with a parallel-resonant circuit.

where $Q_L = R/\omega_o L = \omega_o CR = R\sqrt{C/L}$ is the loaded-quality factor. The modulus of the impedance of the parallel-resonant circuit is

$$|Z(\omega)| = \frac{R}{\sqrt{1 + Q_L^2\left(\frac{\omega}{\omega_o} - \frac{\omega_o}{\omega}\right)^2}} \tag{12.376}$$

The noise-equivalent bandwidth of the parallel-resonant circuit RLC is

$$B_n = \int_0^\infty \left|\frac{Z(\omega)}{R}\right|^2 df = \int_0^\infty \frac{1}{1 + Q_L^2\left(\frac{\omega}{\omega_o} - \frac{\omega_o}{\omega}\right)^2} df = \frac{1}{4RC} \tag{12.377}$$

Another method for finding B_n is as follows. The 3-dB bandwidth BW of a parallel-resonant circuit RLC is $BW = f_o/Q_L$. The noise-equivalent bandwidth of a parallel-resonant circuit RLC is

$$B_n = \frac{\pi}{2} \times BW = \frac{\pi f_o}{2Q_L} = \frac{\pi f_o}{4\pi f_o RC} = \frac{1}{4RC} \tag{12.378}$$

The noise current at the input of the parallel-resonant circuit is

$$I^2_{n(in)} = \frac{4kTB_n}{R} = \frac{kT}{R^2C} \tag{12.379}$$

The noise voltage at the output of the parallel-resonant circuit is

$$V^2_{n(out)} = I^2_{n(in)}|Z(f_o)|^2 = I^2_{n(in)}R^2 = \frac{4kTB_n}{R} \times R^2 = \frac{kT}{R^2C} \times R^2 = \frac{kT}{C} \tag{12.380}$$

The power of the desired signal at the oscillator output is

$$P_s = \frac{V^2_{o(rms)}}{R} = \frac{V^2_m}{2R} \tag{12.381}$$

Hence, the SNR at the oscillator output is

$$SNR = \frac{S}{N} = \frac{V^2_{o(rms)}}{V^2_{n(out)}} = \frac{V^2_m C}{2RkT} = \frac{P_s Q_L}{kT\omega_o} = \frac{P_s}{2\pi kT(BW)} \tag{12.382}$$

The SNR increases when both the signal power P_s and the loaded quality factor Q_L increase, or when the signal power P_s increases and the oscillator signal bandwidth BW decreases.

An expression for the phase noise can be also derived as follows [32]. The voltage waveform across the parallel-resonant circuit RLC is

$$v_C = V_m \cos\omega t = \sqrt{2}V_{rms}\cos\omega t \tag{12.383}$$

The maximum energy of the signal (as a carrier) stored in the parallel-resonant circuit RLC at the resonant frequency is given by

$$W = \frac{1}{2}CV^2_m = CV^2_{rms} \tag{12.384}$$

yielding

$$V^2_{rms} = \frac{W}{C} \tag{12.385}$$

The impedance of the parallel-resonant circuit RLC at offset frequency $\omega_o + \Delta\omega$ is

$$Z[j(\omega_o + \Delta\omega)] = \frac{R}{1 + jQ_L\left(\dfrac{\omega_o + \Delta\omega}{\omega_o} - \dfrac{1}{\dfrac{\omega_o + \Delta\omega}{\omega_o}}\right)} = \frac{R}{1 + jQ_L\left(1 + \dfrac{\Delta\omega}{\omega_o} - \dfrac{1}{1 + \dfrac{\Delta\omega}{\omega_o}}\right)} \tag{12.386}$$

Since $1/(1 + x) \approx 1 - x$ for $x \ll 1$, the impedance of the parallel-resonant circuit can be approximated by

$$Z[j(\omega_o + \Delta\omega)] \approx \frac{R}{1 + jQ_L\left(1 + \dfrac{\Delta\omega}{\omega_o} - 1 + \dfrac{\Delta\omega}{\omega_o}\right)} \approx \frac{R}{1 + j2Q_L\dfrac{\Delta\omega}{\omega_o}} \approx \frac{R}{j2Q_L\dfrac{\Delta\omega}{\omega_o}} \tag{12.387}$$

The modulus of the impedance of the parallel-resonant circuit RLC at offset frequency $\omega_o + \Delta\omega$ is

$$|Z(\omega_o + \Delta\omega)| = \frac{R}{\sqrt{1 + Q_L^2\left(1 + \frac{\Delta\omega}{\omega_o} - \frac{1}{1 + \frac{\Delta\omega}{\omega_o}}\right)^2}} \approx \frac{R}{\sqrt{1 + 4Q_L^2\left(\frac{\Delta\omega}{\omega_o}\right)^2}} \approx \frac{R}{2Q_L\Delta\omega/\omega_o} \tag{12.388}$$

Hence,

$$V_{n(out)}^2 = I_{n(in)}^2|Z(\omega_o + \Delta\omega)|^2 = \frac{4kT}{R} \times \frac{R^2\omega_o^2}{4Q_L^2\Delta\omega^2} = \frac{kTRf_o^2}{Q_L^2\Delta f^2} \tag{12.389}$$

The spectral power density of the output noise voltage is

$$S_v = \frac{V_{n(out)}^2}{B_n} = \frac{4kT}{R}|Z(\omega_o + \Delta\omega_o)|^2 = \frac{kT}{R}\left(\frac{Rf_o}{Q_L\Delta f}\right)^2 = kTR\left(\frac{f_o}{Q_L\Delta f}\right)^2 \tag{12.390}$$

The spectral power density of the oscillator output voltage noise normalized to the mean square signal is

$$\frac{S_v}{V_{o(rms)}^2} = \frac{V_{n(out)}^2/B_n}{V_{o(rms)}^2} = \frac{2kT}{V_m^2 R}\left(\frac{Rf_o}{Q_L\Delta f}\right)^2 = \frac{kT}{P_s}\left(\frac{f_o}{Q_L\Delta f}\right)^2 \tag{12.391}$$

The phase noise of an oscillator is

$$\mathcal{L}(\Delta f) = \log\left[\frac{V_{n(out)}^2/B_n}{V_{o(rms)}^2}\right] = \log\left[\frac{kT}{V_m^2 R}\left(\frac{Rf_o}{Q_L\Delta f}\right)^2\right] = \log\left[\frac{kT}{2P_s}\left(\frac{f_o}{Q_L\Delta f}\right)^2\right] \left(\frac{\text{dBc}}{\text{Hz}}\right) \tag{12.392}$$

The total noise at the oscillator output can be divided into the phase noise and the amplitude noise, for example, 70% and 30%, respectively.

Using Leeson's model [18], the phase noise of an oscillator is

$$\mathcal{L}(\Delta f) = 10\log\left\{\frac{2FkT}{P_s}\left[1 + \left(\frac{f_o}{2Q_L\Delta f}\right)^2\right]\left(1 + \frac{f_1}{|\Delta f|}\right)\right\} \left(\frac{\text{dBc}}{\text{Hz}}\right) \tag{12.393}$$

where F is the noise factor that takes into account both the noise generated by the active device and resistance R, f_1 is the flicker corner frequency of the active device, and P_s is the signal power of the oscillator. Usually, $1 \le F \le 2$.

12.21 Summary

- There are two categories of LC tuned oscillators: positive feedback oscillators and negative resistance oscillators.
- Thermal noise or transients caused by turning on a power supply may start oscillations.
- For oscillations to start and build up, it must have poles located in the RHP at very small amplitude of the output voltage.
- A necessary and sufficient condition for steady-state sinusoidal oscillations is the presence of a pair of complex conjugate poles on the imaginary axis with no poles in the RHP.
- The startup of oscillations requires that the Nyquist plot of the loop gain $T(j\omega)$ encloses the point $-1 + j0$.

- The steady-state oscillations requires that the Nyquist plot of the loop gain $T(j\omega)$ crosses the point $-1 + j0$ at the oscillation frequency.
- For steady-state oscillations, a pair of complex conjugate poles must be located on the imaginary axis of the complex-frequency s-plane.
- *LC* oscillators with positive feedback may have an inverting or noninverting amplifier.
- The oscillator loop network acts as a bandpass filter to determine the oscillation frequency. Either the feedback network or the amplifier loaded by the feedback network acts as a bandpass filter.
- Classic *LC* oscillators include Colpitts, Hartley, Clapp, and Armstrong topologies.
- The Colpitts oscillator is described by a third-order characteristic equation.
- The Clapp oscillator is a modified Colpitts oscillator, which is obtained by adding a capacitor in series with the inductor.
- The frequency of the Clapp oscillator is less sensitive to the input and output capacitances of the amplifier than that of the Colpitts oscillator. This is because the oscillation frequency is determined by an inductor and its series capacitor.
- In the Clapp oscillator, the quality factor Q of the resonant circuit is higher than that in the Colpitts oscillator, resulting in a greater stability of the oscillation frequency.
- Crystal-controlled oscillators offer a very accurate and stable frequency of oscillation because of a very high quality factor Q.
- The Pierce oscillator is a modified Colpitts oscillator.
- The frequency of oscillation f_0 is determined solely by the phase characteristic of the feedback loop. The loop oscillates at the frequency for which the total phase shift around the loop is zero.
- At the oscillation frequency f_o, the magnitude of the loop gain must be equal to or greater than unity.
- In crystal oscillators, the oscillation frequency is precisely controlled by the vibrations of a quartz crystal.
- The crystal behaves as an inductor above the series-resonant frequency and below the parallel-resonant frequency of the fundamental tone. In crystal oscillators, the oscillation frequency is precisely
- A nonlinear circuit is required in oscillators for gain control to limit the amplitude of oscillations.
- If the magnitude of the negative series resistance in the series-resonant circuit is greater than the positive series resistance, then the circuit will start to oscillate.
- If the magnitude of the negative parallel resistance in the parallel-resonant circuit is lower than the positive parallel resistance, then the circuit will start to oscillate.
- A resistor is a noise generator.
- The thermal noise has a flat spectrum.
- The power spectral density is the noise power per unit frequency. It is independent of frequency for white noise.

12.22 Review Questions

12.1 Give the conditions for steady-state oscillations.

12.2 How does a sinusoidal oscillator gives output voltage without an input voltage?

12.3 Give the conditions for the startup of oscillations.

12.4 How does an oscillator start if there is no input voltage?

12.5 What is the characteristic equation of an oscillator?

12.6 What is the root locus of an oscillator?

12.7 What is the Nyquist plot of an oscillator?

12.8 What is the location of poles of the closed-loop gain of oscillators to start oscillations?

12.9 What is the location of poles of the closed-loop gain of oscillators for steady-state oscillations?

12.10 How does the Nyquist plot look like to start oscillations?

12.11 How does the Nyquist plot look like for steady-state oscillations?

12.12 Give a classification of *LC* oscillators.

12.13 List applications of *LC* oscillators.

12.14 List the topologies of *LC* oscillators with inverting amplifier.

12.15 What is the advantage of the Clapp oscillator?

12.16 What kind of oscillator is used to achieve a precise oscillation frequency?

12.17 What is the advantage of the crystal-controlled oscillator?

12.18 How much is the total phase shift in an oscillator loop?

12.19 What is the principle of operation of oscillators with a negative resistance?

12.23 Problems

12.1 Design a Colpitts oscillator for $f_o = 1$ MHz.

12.2 A quartz crystal is specified to have $L = 130$ mH, $C = 3$ fF, $C_p = 5$ pF, and $R = 120\ \Omega$. Find f_s, f_p, Q, and f_o.

12.3 A quartz crystal has $L = 58.48$ mH, $C = 54$ fF, $R = 15\ \Omega$. Calculate f_s, f_p, and $\Delta f = f_p - f_s$.

12.4 Design a Hartley oscillator for $f_o = 800$ kHz.

12.5 Design a Clapp oscillator for $f_o = 1.5$ MHz.

12.6 Design a Wien-bridge oscillator for $f_o = 500$ kHz.

References

[1] H. Barkhausen, *Lehrbuch der Elektronen-Rohre*, 3 Band. Rückkopplung: Verlag S. Hirzwl, 1935.
[2] J. B. Johnson, "Thermal agitation of electricity in conductors," *American Physics Society*, vol. 32, no. 1, pp. 97–109, 1928.
[3] H. Nyquist, "Thermal agitation of electric charge in conductors," *Physical Review*, vol. 32, no. 1, pp. 110–113, 1928.
[4] J. Groszkowski, "The interdependence of frequency variation and harmonic content, and the problem of the constant-frequency oscillators," *Proceedings of the IRE*, vol. 27, no. 7, pp. 958–981, 1933.
[5] J. Groszkowski, *Frequency of Self-Oscillations*. New York, NY: Pergamon Press, Macmillan, 1964.
[6] J. K. Clapp, "An inductance capacitance oscillator of unusual frequency stability," *Proceedings of the IRE*, vol. 36, p. 356–, 1948.
[7] D. T. Hess and K. K. Clark, *Communications Circuits: Analysis and Design*, Reading, MA: Addison-Wesley, 1971.
[8] R. W. Rhea, *Oscillator Design and Computer Simulations*, 2nd Ed. New York, NY: McGraw-Hill, 1997.
[9] J. R. Westra, C. J. M. Verhoeven, and A. H. M. van Roermund, *Oscillations and Oscillator Systems – Classification, Analysis and Synthesis*. Boston, MA: Kluwer Academic Publishers, 1999.
[10] A. S. Sedra and K. C. Smith, *Microelectronic Circuits*, 6th Ed. New York, NY: Oxford University Press, 2010.
[11] R C. Jaeger and T. N. Blalock, *Microelectronic Circuits Design*, 3rd Ed. New York, NY: McGraw-Hill, 2006.
[12] A. Aminian and M. K. Kazimierczuk, *Electronic Devices: A Design Approach*. Upper Saddle River, NJ: Prentice-Hall, 2004.
[13] T. H. Lee, *The Design of Radio Frequency Integrated Circuits*, 2nd Ed. New York, NY: Cambridge University Press, 2004.
[14] E. Rubiola, *Phase Noise and Frequency Stability in Oscillators*, New York, NY: Cambridge University Press, 2008.
[15] J. R. Pierce, "Physical source of noise," *Proceedings of the IEEE*, vol. 44, pp. 601–608, 1956.
[16] W. R. Bennett, "Methods of solving noise problems," *Proceedings of the IRE*, vol. 44, pp. 609–637, 1956.
[17] W. A. Edson, "Noise in oscillators," *Proceedings of the IRE*, vol. 48, pp. 1454–1466, 1960.
[18] D. B. Leeson, "A simple method for feedback oscillator noise spectrum," *Proceedings of the IEEE*, vol. 54, no. 2, pp. 329–330, 1966.
[19] E. Hafner, "The effects of noise in oscillators," *Proceedings of the IEEE*, vol. 54, no. 2, pp. 179–198, 1966.
[20] M. Lux, "Classical noise. Noise in self-sustained oscillators," *Physical Review*, vol. 160, pp. 290–307, 1967.
[21] J. Ebert and M. Kazimierczuk, "Class E high-efficiency tuned power oscillator," *IEEE Journal of Solid-State Circuits*, vol. 16, no. 1, pp. 62–66, 1981.
[22] M. Kazimierczuk, "A new approach to the design of tuned power oscillators," *IEEE Transactions on Circuits and Systems*, vol. 29, no. 4, pp. 261–267, 1982.
[23] F. R. Karner, "Analysis of white and $f^{-\infty}$ noise in oscillators," *International Journal of Circuits Theory and Applications*, vol. 18, pp. 485–519, 1990.
[24] N. M. Nguyen and R. G. Mayer, "A 1.8 GHz monolithic *LC* voltage-controlled oscillator," *IEEE Journal of Solid-State Circuits*, vol. 27, no. 3, pp. 444–450, 1992.

[25] E. Bryetorn, W. Shiroma, and Z. B. Popović, "A 5-GHz high-efficiency Class-E oscillator," *IEEE Microwave and Guided Wave Letters*, vol. 6, no. 12, pp. 441–443, 1996.

[26] A. Hajimiri and T. H. Lee, "A general theory of phase noise in electrical oscillators," *IEEE Journal of Solid-State Circuits*, vol. 33, no. 2, pp. 179–194, 1998.

[27] A. Hajimiri, S. Limotyrakis, and T. H. Lee, "Jitter and phase noise in ring oscillators," *IEEE Journal of Solid-State Circuits*, vol. 34, no. 6, pp. 790–804, 1999.

[28] G. M. Magio, O. D. Feo, and M. P. Kennedy, "Nonlinear analysis of the Colpitts oscillator and applications to design," *IEEE Transactions on Circuits and Systems, Part I, Fundamental Theory and Applications*, vol. 46, no. 9, pp. 1118–1129, 1982.

[29] S. Pasupathy, "Equivalence of *LC* and *RC* oscillators," *International Journal of Electronics*, vol. 34, no. 6, pp. 855–857, 1973.

[30] B. Linares-Barranco, A. Rodrigez-Vázquez, R. Sánchez-Sinencio, and J. L. Huertas, "CMOS OTA-C high-frequency sinusoidal oscillator," *IEEE Journal of Solid-State Circuits*, vol. 26, no. 2, pp. 160–165, 1991.

[31] A. M. Niknejad and R. Mayer, *Design, Simulations and Applications of Inductors and Transformers for Si RF ICs*. New York, NY: Kluwer, 2000.

[32] T. H. Lee and A. Hajimiri, "Oscillator phase noise, a tutorial," *IEEE Journal of Solid-State Circuits*, vol. 35, no. 3, pp. 326–336, 2000.

[33] A. Hajimiri and T. H. Lee, *The Design of Low Noise Oscillators*. New York, NY: Kluwer, 2003.

[34] K. Mayaram, "Output voltage analysis for MOS Colpitts oscillator," *IEEE Transactions on Circuits and Systems, Part I, Fundamental Theory and Applications*, vol. 47, no. 2, pp. 260–262, 2000.

[35] A. Demir, A. Mehrotra, and J. Roychowdhury, "Phase noise in oscillators: a unifying theory and numerical methods for characterization," *IEEE Transactions on Circuits and Systems, Part I, Regular Papers*, vol. 47 no. 5, pp. 655–6674, 2000.

[36] A. Buonomo, M. Pennisi, and S. Pennisi, "Analyzing the dynamic behavior of RF oscillators," *IEEE Transactions on Circuits and Systems I: Regular Papers*, vol. 49, no. 11, pp. 1525–1533, 2002.

[37] M. K. Kazimierczuk, V. G. Krizhanovski, J. V. Rossokhina, and D. V. Chernov, "Class-E MOSFET tuned power oscillator design procedure," *IEEE Transactions on Circuits and Systems I: Regular Papers*, vol. 52, no. 6, pp. 1138–1147, 2005.

[38] M. K. Kazimierczuk, V. G. Krizhanovski, J. V. Rossokhina, and D. V. Chernov, "Injected-locked Class-E oscillator," *IEEE Transactions on Circuits and Systems I: Regular Papers*, vol. 53, no. 6, pp. 1214–1222, 2006.

[39] X. Li, S. Shekhar, and D. J. Allstot, "Gm-boosted common-gate LNA and differential Colpitts VCO/QVCO in 0.18 μm CMOS," *IEEE Journal of Solid-State Circuits*, vol. 40, no. 12, pp. 2609–2619, 2005.

[40] I. M. Filanovsky and C. J. M. Verhoeven, "On stability of synchronized van der Pol oscillators," *Proceedings ICECS'06*, Nice, France, 2006, pp. 1252–1255.

[41] N. T. Tchamov, S. S. Broussev, I. S. Uzunov, and K. K. Rantala, "Dual-band *LC* VCO architecture with a fourth-order resonator," *IEEE Transactions on Circuits and Systems II: Express Briefs*, vol. 54, no. 3, pp. 277–281, 2007.

[42] F. Tzang, D. Pi, A. Safarian, and P. Heydari, "Theoretical analysis of novel multi-order *LC* oscillators," *IEEE Transactions on Circuits and Systems II: Express Briefs*, vol. 54, no. 3, pp. 287–291, 2007.

[43] I. M. Filanovsky, C. J. M. Verhoeven, and M. Reja, "Remarks on analysis, design and amplitude stability of MOS Colpitts oscillator," *IEEE Transactions on Circuits and Systems II: Express Briefs*, vol. 54, no. 9, pp. 800–804, 2006.

[44] V. G. Krizhanovski, D. V. Chernov, and M. K. Kazimierczuk, "Low-voltage electronic ballast based on Class E oscillator," *IEEE Transactions on Power Electronics*, vol. 22, no. 3, pp. 863–870, 2007.

[45] R. Devine and M.-R. Tofichi, "Class E Colpitts oscillator for low power wireless applications," *IET Electronic Letters*, vol. 44, no. 21, pp. 549–551, 2008.

[46] W. Tangsrirat, D. Prasertsom, T. Piyatat, and W. Surakompontorn, "Single-resistance-controlled quadrature oscillators using current differencing buffered amplifiers," *International Journal of Electronics*, vol. 95, no. 11, pp. 1119–1126, 2008.

[47] G. Palumbo, M. Pennesi, and S. Pennesi, "Approach to analysis and design of nearly sinusoidal oscillators," *IET Circuits, Devices and Systems*, vol. 3, no. 4, pp. 204–221, 2009.

[48] J. Roger and C. Plett, *Radio Frequency Integrated Circuit Design*, 2nd Ed. Norwood, MA: Artech House, 2010.

[49] G. Weiss, "Network theorem for transistor circuits," *IEEE Transactions of Education*, vol. 37, no. 1, pp. 36–41, 1994.

[50] M. K. Kazimierczuk and D. Murthy-Bellur, "Loop gain of the common-drain Colpitts oscillator," *International Journal of Electronics and Telecommunications*, vol. 56, no. 4, pp. 423–426, 2010.

[51] M. K. Kazimierczuk and D. Murthy-Bellur, "Loop gain of the common-gate Colpitts oscillator," *IET Circuits, Devices and Systems*, vol. 5, no. 4, pp. 275–284, 2011.

[52] P. Andreani and X. Wang, "On the phase-noise and phase-error performances of multiphase *LC* CMOS VCOs," *IEEE Journal of Solid-State Circuits*, vol. 39, no. 11, pp. 1883–1893, 2004.

[53] P. Andreani, X. Wang, L. Vandi, and A. Fard, "A study of phase noise in Colpitts and *LC*-tank CMOS oscillators," *IEEE Journal of Solid-State Circuits*, vol. 40, no. 5, pp. 1107–1118, 2005.

[54] P. Andreani and A Fard, "More on the $1/f^2$ phase noise performance of CMOS differential-pair *LC*-tank oscillators," *IEEE Journal of Solid-State Circuits*, vol. 41, no. 12, pp. 2703–2712, 2006.

[55] P. Andreani and A Fard, "An analysis of $1/f^2$ phase noise in bipolar Colpitts oscillators (with degression on bipolar differential-pair *LC* oscillators)," *IEEE Journal of Solid-State Circuits*, vol. 42, no. 2, pp. 374–384, 2007.

[56] A. Mazzanti and P. Andreani, "Class-C harmonic CMOS VCOs with a general result on phase noise," *IEEE Journal of Solid-State Circuits*, vol. 43, no. 12, pp. 2716–2729, 2008.

[57] A. Bevilaqua and P. Andreani, "On the bias noise to phase noise conversion in harmonic oscillators using Groszkowski theory," *IEEE International Symposium on Circuits and Systems*, 2011, pp. 217–220.

[58] M. K. Kazimierczuk and D. Murthy-Bellur, "Synthesis of *LC* oscillators," *International Journal of Engineering Education*, pp. 26–41, 2012.

Appendices

SPICE Model of Power MOSFETs

Figure A.1 shows a SPICE large-signal model for *n*-channel enhancement MOSFETs. It is a model of integrated MOSFETs, which can be adopted to power MOSFETs. SPICE parameters of the large-signal model of enhancement-type *n*-channel MOSFETs are given in Table A.1.

The diode currents are

$$i_{BD} = IS\left(e^{\frac{v_{BD}}{V_T}} - 1\right) \tag{A.1}$$

and

$$i_{BS} = IS\left(e^{\frac{v_{BS}}{V_T}} - 1\right) \tag{A.2}$$

The junction capacitances in the voltage range close to zero are

$$C_{BD} = \frac{(CJ)(AD)}{\left(1 - \frac{v_{BD}}{PB}\right)^{MJ}} \quad \text{for} \quad v_{BD} \leq (FC)(PB) \tag{A.3}$$

and

$$C_{BS} = \frac{(CJ)(AS)}{\left(1 - \frac{v_{BS}}{PB}\right)^{MJ}} \quad \text{for} \quad v_{BS} \leq (FC)(PB) \tag{A.4}$$

where CJ is the zero-bias junction capacitance per unit area, AD is the drain area, AS is the source area, PB is the built-in potential, and MJ is the grading coefficient.

The junction capacitances in the voltage range far from zero are

$$C_{BD} = \frac{(CJ)(AD)}{(1 - FC)^{1+MJ}}\left[1 - (1 + MJ)FC + MJ\frac{v_{BD}}{PB}\right] \quad \text{for} \quad v_{BD} \geq (FC)(PB) \tag{A.5}$$

and

$$C_{BS} = \frac{(CJ)(AS)}{(1 - FC)^{1+MJ}}\left[1 - (1 + MJ)FC + MJ\frac{v_{BD}}{PB}\right] \quad \text{for} \quad v_{BS} \geq (FC)(PB) \tag{A.6}$$

RF Power Amplifiers, Second Edition. Marian K. Kazimierczuk.

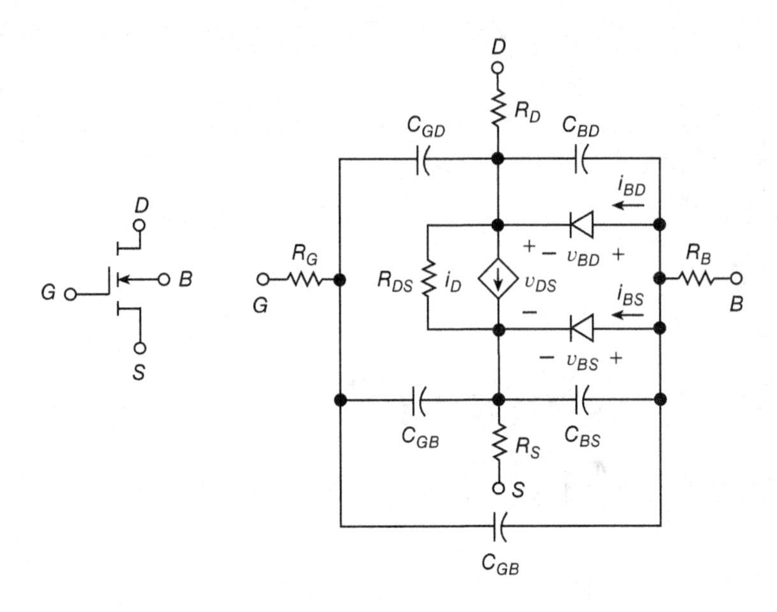

Figure A.1 SPICE large-signal model for n-channel MOSFET.

Table A.1 Selected SPICE level 1 NMOS large-signal model parameters.

Symbol	SPICE S.	Model parameter	Default value	Typical value
V_{to}	VTO	Zero-bias threshold voltage	0 V	0.3–3 V
μC_{ox}	KP	Process constant	2×10^{-5} A/V^2	20–346 μA/V^2
λ	Lambda	Channel-length modulation	0 V^{-1}	0.5–10^{-5} V^{-1}
γ	Gamma	Body-effect V_t parameter	0 V$^{\frac{1}{2}}$	0.35 V$^{\frac{1}{2}}$
$2\phi_F$	PHI	Surface potential	0.6 V	0.7 V
R_D	RD	Drain series resistance	0 Ω	0.2 Ω
R_S	RS	Source series resistance	0 Ω	0.1 Ω
R_G	RG	Gate series resistance	0 Ω	1 Ω
R_B	RB	Body series resistance	0 Ω	1 Ω
R_{DS}	RDS	Drain-source shunt R	∞	1 MΩ
R_{SH}	RSH	Drain-source diffusion sheet R	0	20 Ω/Sq.
I_S	IS	Saturation current	10^{-14} A	10^{-9} A
M_j	MJ	Grading coefficient	0.5	0.36
C_{j0}	CJ	Zero-bias bulk junction C/m^2	0 F/m^2	1 nF/m^2
V_{bi}	PB	Junction potential	1 V	0.72 V
M_{jsw}	MJSW	Grading coefficient	0.333	0.12
C_{j0sw}	CJSW	Zero-bias junction perimeter C/m	0 F/m	380 pF/m
V_{BSW}	PBSW	Junction sidewell potential	1 V	0.42 V
C_{GDO}	CGDO	Gate-drain overlap C/m	0 F/m	220 pF/m
C_{GSO}	CGSO	Gate-source overlap C/m	0 F/m	220 pF/m
C_{GBO}	CGBO	Gate-bulk overlap C/m	0 F/m	700 pF/m
F_C	FC	Forward-biased C_J coefficient	0.5	0.5
t_{ox}	TOX	Oxide thickness	∞	4.1–100 nm
μ_{ns}	UO	Surface mobility	600 cm^2/Vs	600 cm^2/Vs
n_{sub}	NSUB	Substrate doping	0 cm^{-3}/Vs	0 cm^{-3}/Vs

The typical values:
$C_{ox} = 3.45 \times 10^{-5}$ pF/μm
$t_{ox} = 4.1 \times 10^{-3}$ μm
$\varepsilon_{ox(SiO_2)} = 3.9\varepsilon_0$
$C_{j0} = 2 \times 10^{-4}$ F/m^2
$C_{jsw} = 10^{-9}$ F/m
$C_{GBO} = 2 \times 10^{-10}$ F/m
$C_{GDO} = C_{GSO} = 4 \times 10^{-11}$ F/m

SPICE NMOS Syntax:

Mxxxx D G S B MOS-model-name L = xxx W = yyy

Example:

M1 2 1 0 0 M1-FET L = 0.18um W = 1800um

SPICE NMOS Model Syntax:

.model model-name NMOS (parameter=value ...)

Example:

.model M1-FET NMOS (Vto = 1V Kp = E-4)

SPICE PMOS Model Syntax:

.model model-name PMOS (parameter=value ...)

SPICE Subcircuit Model Syntax:

xname N1 N2 N3 model-name

Example:

x1 2 1 0 IRF840

Copy and paste the obtained device model.

.SUBCKT IRF840 1 2 3

and the content of the model.

Introduction to SPICE

SPICE is an abbreviation for *Simulation Program for Integrated Circuit Emphasis*. PSPICE is the PC version of SPICE. Analog and digital electronic circuit designs are verified widely by both industries and the academia using PSPICE. It is used to predict the circuit behavior.

Passive Components: Resistors, Capacitors, and Inductors

Rname N+ N- Value [IC = TC1]

Lname N+ N- Value [IC = Initial Voltage Condition]

Cname N+ N- Value [IC = Initial Current Condition]

Examples:

R1 1 2 10K

L2 2 32M

C3 3 4 100P

Transformer

Lp Np+ Np- Lpvalue

Ls Ns+ Ns- Lsvalue

Kname Lp Ls Kvalue

Example:

Lp 1 0 1 mH

Ls 2 4 100 uH

Kt Lp Ls 0.999

RF Power Amplifiers, Second Edition. Marian K. Kazimierczuk.

Temperature

.TEMP list of temperatures

Example:

.TEMP 27 100 150

Independent DC Sources

Vname N+ N- DC Value

Iname N+ N- DC Value

Examples:

Vin 1 0 DC 10

Is 1 0 DC 2

DC Sweep Analysis

.DC Vsource-name Vstart Vstop Vstep

Example:

.DC VD 0 0.75 1m

Independent Pulse Source for Transient Analysis

Vname N+ N- PULSE (VL VH td tr tf PW T)

Example:

VGS 1 0 PULSE(0 10 0 0 0 10E-6 100e-6)

Transient Analysis

.TRAN time-step time-stop

Example:

.TRAN 0.1ms 100ms 0ms 0.2ms

Independent AC Sources for Frequency Response

Vname N+ N- AC Vm Phase

Iname N+ N- AC Im Phase

Example:

Vs 2 3 AC 2 30

Is 2 3 AC 0.5 30

Independent Sinusoidal AC Sources for Transient Analysis

Vname N+ N- SIN (Voffset Vm f T-delay Damping-Factor Phase-delay)

Iname N+ N- SIN (Ioffset Im f T-delay Damping-Factor Phase-delay)

Examples:

Vin 1 0 SIN (0 170 60 0 -120)

Is 1 0 SIN (0 2 120 0 45)

AC Frequency Analysis

.AC DEC points-per-decade fstart fstop

Example:

.AC DEC 100 20 20k

Operating Point

.OP

Getting Started the SPICE Program

1) Open the PSpice A/D Lite window (**Start** > **Programs** > **Orcad9.2 Lite Edition** > **PSpice AD Lite**).
2) Create a new text file (**File** > **New** > **Text File**).
3) Type the example code.
4) Save the file as fn.cir (for example, Lab1.cir), file type: all files, and simulate by pressing the appropriate icon.
5) To include the Spice code of a commercial device model, visit the web site, e.g., http://www.irf.com, http:www.onsemi.com, or http://www.cree.com. For example, for IRF devices, click on (**Design** > **Support** > **Models** > **Spice Library**).

Example Program

```
Diode I-V Characteristics
*Joe Smith
VD 1 0 DC 0.75
D1N4001 1 0 Power-Diode
.model Power-Diode D (Is=195p n=1.5)
.DC VD 0 0.75 1m
.TEMP 27 50 100 150
.probe
.end
```

Introduction to MATLAB®

MATLAB® is an abbreviation for MATrix LABoratory. It is a very powerful mathematical tool used to perform numerical computation using matrices and vectors to obtain two- and three-dimensional graphs. MATLAB can also be used to perform complex mathematical analysis.

Getting Started

1. Open MATLAB by clicking **Start** > **Programs** > **MATLAB** > **R2006a** > **MATLAB R2006a**.
2. Open a new M-file by clicking **File** > **New** > **M-File**.
3. Type the code in the M-File.
4. Save the file as fn.m (e.g., Lab1.m).
5. Simulate the code by doing one of the following:

(a) Click on **Debug** > **Run**.

(b) Press F5

(c) On the tool bar, click the icon **Run**.

Use **HELP** by pressing F1.

Use % at the beginning of a line for comments.

Generating a *x*-axis Data

x=Initial-Value: Increment:Final-Value;

Example:

x=1:0.001:5;

or

x=[list of all the values];

RF Power Amplifiers, Second Edition. Marian K. Kazimierczuk.

Example:

x = [1, 2, 3, 5, 7, 10];

or

x = linspace(start-value, stop-value, number-of-points);

Example:

x = linspace(0, 2*pi, 90);

or x = logspace(start-power, stop-power, number-of-points);

Example:

x = logspace(1, 5, 1000);

Semilogarithmic Scale

semilogx(x-variable, y-variable);

semilogy(x-variable, y-variable);

grid on

Log-log Scale

loglog(x, y);

grid on

Generate a *y*-axis Data

y = f(x);

Example:

y = cos(x);

z = sin(x);

Multiplication and Division

A dot should be used in front of the operator for matrix multiplication and division.

c = a.*b;

or

c = a./b;

Symbols and Units

Math symbols should be in italic. Math signs (like ( ), =, and +) and units should not be in italic. Leave one space between a symbol and a unit.

***x*-axis and *y*-axis Labels**

xlabel('{\it x} (unit) ')

ylabel('{\it y} (unit) ')

Example:

xlabel('{\it v_{GS}} (V)')

ylabel('{\it i_{DS}} (A)')

***x*-axis and *y*-axis Limits**

set(gca, 'xlim', [xmin, xmax])

set(gca, 'xtick', [xmin, step, xmax])

set(gca, 'xtick', [x1, x2, x3, x4, x5])

Example:

```
set(gca, 'xlim', [1, 10])
set(gca, 'xtick', [0:2:10])
set(gca, 'xtick', [-90 -60 -30 0 30 60 90])
```

Greek Symbols

Type: \alpha , \beta , \Omega , \omega , \pi , \phi , \psi , \gamma , \theta , and \circ to obtain: α, β, Ω, ω, π, ϕ, ψ, γ, θ, and $\circ$.

Plot Commands

```
plot (x, y, '.-', x, z, '--')
set(gca, 'xlim', [x1, x2]);
set(gca, 'ylim', [y1, y2]);
set(gca, 'xtick', [x1:scale-increment:x2]);
text(x, y, '{\it symbol} = 25 V');
plot (x, y), axis equal
```

Examples:

```
set(gca, 'xlim', [4, 10]);
set(gca, 'ylim', [1, 8]);
set(gca, 'xtick', [4:1:10]);
text(x, y, '{\it V} = 25 V');
```

3-D Plot Commands

```
[X1, Y1] = meshgrid(x1,x2);
mesh(X1, Y1,z1);
```

Example:

```
t = linspace(0, 9*pi);
xlabel('sin(t)')
ylabel('cos(t)')
zlabel('t')
plot(sin(t), cos(t), t)
```

Bode Plots

```
f = logspace(start-power, stop-power, number-of-points)
NumF = [a1 a2 a3]; %Define the numerator of polynomial in s-domain.
DenF = [a1 a2 a3]; %Define the denominator of polynomial in s-domain.
[MagF, PhaseF] = bode(NumF, DenF, (2*pi*f));
figure(1)
semilogx(f, 20*log10(MagF))
F = tf(NumF, DenF) %Converts the polynomial into transfer function.
[NumF, DenF] = tfdata(F) %Converter transfer function into polynomial.
```

Step Response

```
NumFS = D*NumF;
t = [0:0.000001:0.05];
[x, y] = step(NumFS, DenF, t);
figure(2)
plot(t, Initial-Value + y);
```

To Save Figure

Go to File, click Save as, go to EPS file option, type the file name, and click Save.

Example Program

```
clear all
clc
x = linspace(0, 2*pi, 90);
y = sin(x);
z = cos(x);
grid on
xlabel('{\it x}')
ylabel(' {\it y }, {\it z }')
plot(x , y, '-.', x, z, '--')
```

Polynomial Curve Fitting

```
x = [0 0.5 1.0 1.5 2.0 2.5 3.0];
y = [10 12 16 24 30 37 51];
p = polyfit(x, y, 2)
yc = polyval(p, x);
plot(x, y, 'x', x, yc)
xlabel('x')
ylabel('y'), grid
legend('Actual data', 'Fitted polynomial')
```

Bessel Functions

```
J0 = besselj(0, x);
```

Modified Bessel Functions

```
I0 = besseli(0, x):
Example:
model = [1 2 3]:
rro = -1:0.00001:1;
kr = (1 + j)*(rodel)'*(rro);
JrJ0 = besseli(0,kr); figure, plot(rro, abs(JrJ0)) figure, plot(rro,angle(JrJ0)*180/pi)
```

Trigonometric Fourier Series

A periodic function $f(t)$ satisfies the condition

$$f(t) = f(t \pm nT) \tag{D.1}$$

or

$$f(\omega t) = f(\omega t \pm 2\pi n) \tag{D.2}$$

where $f = 1/T$ is the *fundamental frequency* of function $f(t)$, $T = 1/f$ is the period of function $f(t)$, $n = 1, 2, 3, \ldots$ is an integer, and $\omega = 2\pi f = 2\pi/T$.

Any nonsinusoidal periodic function can be expressed as an infinite sum of sinusoidal and cosinusoidal functions. The *trigonometric Fourier series* of a periodic function $f(t)$ is given by

$$f(t) = a_0 + \sum_{n=1}^{\infty}(a_n \cos n\omega t + b_n \sin n\omega t)$$
$$= a_0 + a_1 \cos \omega t + b_1 \sin \omega t + a_2 \cos 2\omega t + b_2 \sin 2\omega t + a_3 \cos 3\omega t + b_3 \sin 3\omega t + \cdots \tag{D.3}$$

where the Fourier coefficients are

$$a_0 = \frac{1}{T}\int_0^T f(t)dt = \frac{1}{2\pi}\int_0^{2\pi} f(\omega t)d(\omega t) \tag{D.4}$$

$$a_n = \frac{2}{T}\int_0^T f(t)\cos n\omega t\, dt = \frac{1}{\pi}\int_0^{2\pi} f(\omega t)\cos n\omega t\, d(\omega t) \tag{D.5}$$

and

$$b_n = \frac{2}{T}\int_0^T f(t)\sin n\omega t\, dt = \frac{1}{\pi}\int_0^{2\pi} f(\omega t)\sin n\omega t\, d(\omega t) \tag{D.6}$$

The term a_0 is the dc component of function $f(t)$ or the time-average value of function $f(t)$. The terms a_n and b_n are the amplitudes of cosinusoids and sinusoids, respectively.

RF Power Amplifiers, Second Edition. Marian K. Kazimierczuk.

The *amplitude-phase form* of the trigonometric Fourier series is

$$f(t) = a_0 + \sum_{n=1}^{\infty}(c_n \cos n\omega t \cos \phi_n - \sin n\omega t \sin \phi_n)$$

$$= a_0 + \sum_{n=1}^{\infty}(c_n \cos \phi_n) \cos n\omega t - (c_n \sin \phi_n) \sin n\omega t$$

$$= a_0 + \sum_{n=1}^{\infty} c_n \cos(n\omega t + \phi_n) \tag{D.7}$$

where the amplitude of the fundamental component or the nth harmonic is

$$c_n = \sqrt{a_n^2 + b_n^2} \tag{D.8}$$

the phase of the fundamental component or the n-th harmonic is

$$\phi_n = -\arctan\left(\frac{b_n}{a_n}\right) \tag{D.9}$$

$$a_n = c_n \cos \phi_n \tag{D.10}$$

and

$$b_n = -c_n \sin \phi_n \tag{D.11}$$

The complex form of the amplitude is

$$c_n e^{\phi_n} = a_n - jb_n \tag{D.12}$$

An alternative form of the *amplitude-phase form* of the trigonometric Fourier series is

$$f(t) = a_0 + \sum_{n=1}^{\infty} c_n(\sin n\omega t \cos \phi_n + \cos n\omega t \sin \phi_n)$$

$$= a_0 + \sum_{n=1}^{\infty}(c_n \cos \phi_n) \cos n\omega t + (c_n \sin \phi_n) \sin n\omega t$$

$$= a_0 + \sum_{n=1}^{\infty} c_n \sin(n\omega t + \phi_n) \tag{D.13}$$

where

$$c_n = \sqrt{a_n^2 + b_n^2} \tag{D.14}$$

$$\phi_n = \arctan\left(\frac{a_n}{b_n}\right) \tag{D.15}$$

$$a_n = c_n \sin \phi_n \tag{D.16}$$

and

$$b_n = c_n \cos \phi_n \tag{D.17}$$

The plot of the amplitude c_n versus frequencies nf is the *amplitude spectrum* of function $f(t)$. The plot of the phase ϕ_n versus frequencies nf is the *phase spectrum* of function $f(t)$. The plots of both the amplitude c_n and the phase ϕ_n versus frequencies nf are the *frequency spectra* of function $f(t)$.

D.1 Even Symmetry

The function $f(t)$ is even if its plot is symmetrical about the vertical axis. For the *even symmetry* of function $f(t)$, the following condition is satisfied:

$$f(t) = f(-t) \tag{D.18}$$

and the *Fourier cosine series* of function $f(t)$ is

$$f(t) = a_0 + \sum_{n=1}^{\infty} a_n \cos n\omega t \tag{D.19}$$

where

$$a_0 = \frac{2}{T}\int_0^{T/2} f(t)dt = \frac{1}{\pi}\int_0^{2\pi} f(\omega)d(\omega t) \tag{D.20}$$

$$a_n = \frac{4}{T}\int_0^{T/2} f(t)\cos n\omega t\, dt = \frac{1}{\pi}\int_0^{\pi} f(\omega t)\cos n\omega t\, d(\omega t) \tag{D.21}$$

and

$$b_n = 0 \tag{D.22}$$

D.2 Odd Symmetry

The function $f(t)$ is odd if its plot is antisymmetrical about the vertical axis. For the *odd symmetry* of function $f(t)$, the following condition is satisfied:

$$f(-t) = -f(t) \tag{D.23}$$

and the *Fourier sine series* of function $f(t)$ is

$$f(t) = \sum_{n=1}^{\infty} b_n \sin n\omega t \tag{D.24}$$

where

$$a_0 = a_n = 0 \tag{D.25}$$

and

$$b_n = \frac{4}{T}\int_0^{T/2} f(t)\sin n\omega t\, dt = \frac{2}{\pi}\int_0^{\pi} f(\omega t)\sin n\omega t\, d(\omega t) \tag{D.26}$$

A function often becomes odd, when its dc component a_0 is made equal to zero. Then

$$f(t) - a_0 = \sum_{n=1}^{\infty} b_n \sin n\omega t \tag{D.27}$$

D.3 Generalized Trigonometric Fourier Series

The classical trigonometric Fourier series of a periodic function gives sinusoidal and cosinusoidal components for every harmonics, where each component has zero phase. The *generalized trigonometric Fourier series* of a periodic function $f(t)$ for every harmonics, where each component has phase ϕ, is given by

$$f(t) = a_0 + \sum_{n=1}^{\infty}[a_n \cos(n\omega t + \phi) + b_n \sin(n\omega t + \phi)] \tag{D.28}$$

where the Fourier coefficients are

$$a_0 = \frac{1}{T}\int_0^T f(t)dt = \frac{1}{2\pi}\int_0^{2\pi} f(\omega t)d(\omega t) \tag{D.29}$$

$$a_n = \frac{2}{T}\int_0^T f(t)\cos(n\omega t + \phi)dt = \frac{1}{\pi}\int_0^{2\pi} f(\omega t)\cos(n\omega t + \phi)d(\omega t) \tag{D.30}$$

and

$$b_n = \frac{2}{T}\int_0^T f(t)\sin(n\omega t + \phi)dt = \frac{1}{\pi}\int_0^{2\pi} f(\omega t)\sin(n\omega t + \phi)d(\omega t) \tag{D.31}$$

Circuit Theorems

E.1 Generalized Ohm's Law Theorem

Figure E.1 illustrates a generalized Ohm's law. A traditional representation of the Ohm's law in the form of a resistor $R = V/I$ is depicted in Fig. E.1(a). The resistor R can be replaced by a *voltage-controlled current source* (VCCS) controlled by its own voltage V and, therefore, its current is $I = GV = V/R$, as shown in Fig. E.1(b). The resistor R can also be replaced by a *current-controlled voltage source* (CCVS) controlled by its own current I and the voltage across the voltage source is $V = RI$, as depicted in Fig. E.1(c).

E.2 Current-Source Absorption Theorem

The source absorption theorems permit the replacement of a controlled current or voltage source by a resistance or impedance. The current-source absorption theorem is illustrated in Fig. E.2. Since the voltage-controlled current source is controlled by its own voltage V, it can be replaced by a resistor

$$R = \frac{V}{I} = \frac{V}{g_m V} = \frac{1}{g_m} \tag{E.1}$$

E.3 Voltage-Source Absorption Theorem

The voltage-source absorption theorem is illustrated in Fig. E.3. Because this source is controlled by its own current I, it can be replaced by an equivalent resistance r_m

$$R = \frac{V}{I} = \frac{r_m I}{I} = r_m \tag{E.2}$$

RF Power Amplifiers, Second Edition. Marian K. Kazimierczuk.

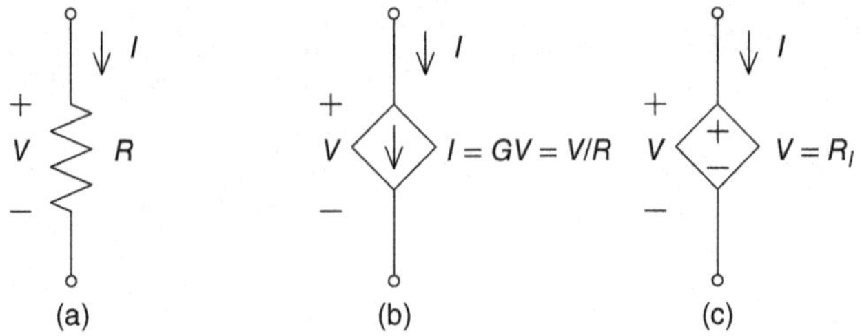

Figure E.1 Generalized Ohm's law or resistor absorption theorem.

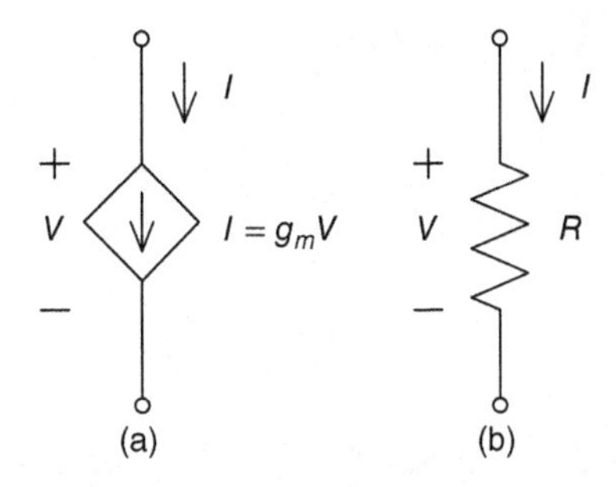

Figure E.2 Current-source absorption theorem.

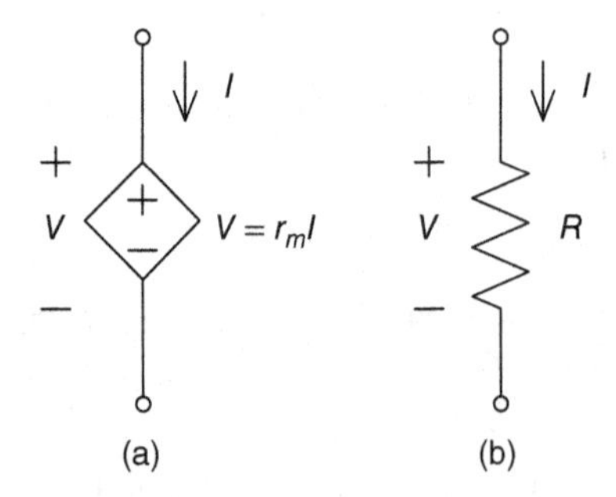

Figure E.3 Voltage-source absorption theorem.

E.4 Current-Source Splitting Theorem

Figure E.4(a) shows an ideal current source I connected between two networks N_1 and N_2. This source can be either independent or dependent. The branch with a single ideal current source can be replaced with a branch with any number of identical ideal current sources I connected in series. Figure E.4(b) shows two identical current sources. In the simplest case, an ideal current source can be replaced by two ideal current sources connected in series. The point between two current sources can be connected to any other point in the circuit. The current through the conductor between these two points is zero, as shown in Fig. E.4(c). The single conductor can be replaced by two conductors, as shown in Fig. E.4(d). The conductor at the bottom can be cut and removed from the circuit, as shown in Fig. E.4(f).

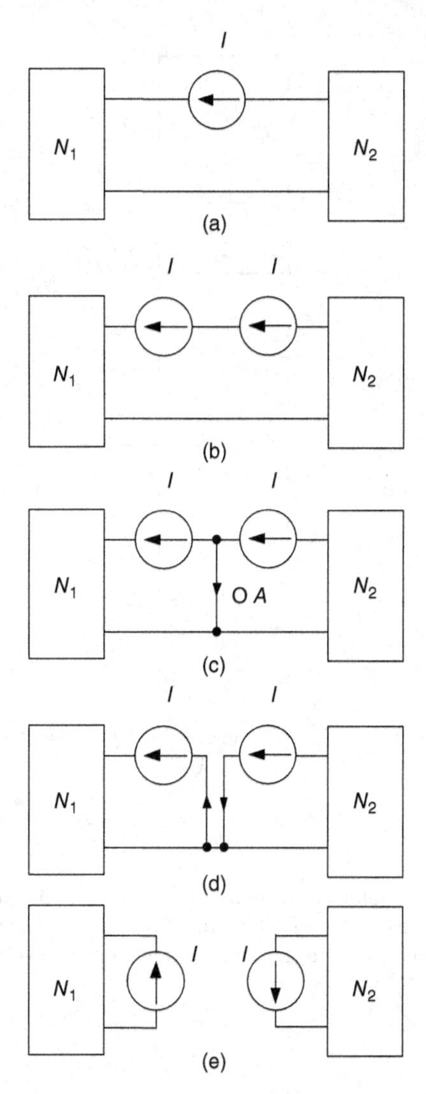

Figure E.4 Current-source splitting theorem. (a) Circuit with a horizontal current source between two networks N_1 and N_2. (b) Circuit with two identical current sources connected in series. (c) Circuit with two identical current sources and a conductor that conducts no current and is connected to the bottom conductor. (d) Circuit with two identical current sources connected to bottom terminal. (e) Equivalent circuit with two identical current sources.

E.5 Voltage Source Splitting Theorem

Figure E.5(a) shows an ideal voltage source V connected to two different points in the network N. A single conductor can be replaced by two conductors, as shown in Fig. E.5(b). The single ideal voltage source V can be replaced by any number of identical ideal voltage sources V connected in parallel, as shown in Fig. E.5(c). The current flowing in the conductor between two identical voltage sources is zero. Therefore, the conductor connecting the two voltage sources can be removed from the circuit, as shown in Fig. E.5(d).

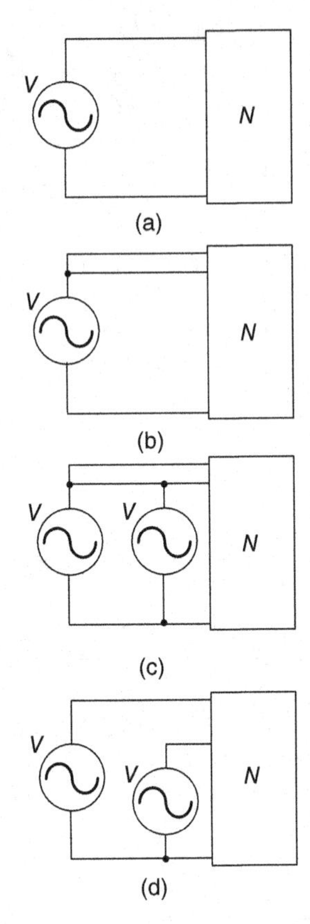

Figure E.5 Voltage-source splitting theorem. (a) Circuit with one voltage source. (b) Circuit with one voltage source and two parallel conductors. (c) One voltage source is replaced by two identical voltage sources connected in parallel. (d) Equivalent circuit with two voltage sources.

SABER Circuit Simulator

SABER is a circuit simulator. It enables circuit analysis, schematic design, mixed-mode circuit simulation, waveform analysis, and report generation.

Starting SABER

1) In the Start menu, locate and click "XWin32 2011." In the same way, find "Secure Shell" and click "Secure Shell Client." A terminal console will appear. Hit Enter. Enter "thor.cs.wright.edu" as host name, the UNIX log-on name as username and click OK. Enter your Password when prompted. Now, you are connected remotely.
2) Type "sketch" on the terminal console. SABER will start automatically.

Setting up a circuit on SABER

1) SABER created many files during netlisting and simulations. In order to keep everything in order, create a new folder in your home directory called "Saber" (or "Lab1" and so on). Use the SSH Secure File Transfer Client, or type "mkdir Saber" at the home directory command prompt.
2) On the SABER circuit design window, construct a given circuit/schematic using the parts in the component window. The parts can be found in the component library at the bottom icon tray on your screen under "Select and place parts," and you can search for parts individually. Left clock at the terminal to connect in your schematic and you can search for parts individually. Left click at the terminals to connect various parts in the schematic. It is recommended to save your work regularly with a suitable file name. Make sure to save all your work in a designated folder in the Secure Shell Transfer Client because SABER generated netlists that are required for circuit simulations.
3) Provide appropriate values to all the components on the schematic and name the wire that are used to connect these components, such as Vout, Vin, iL, and so on. This can be done by right clicking on the selected wire and then select "attribute."

RF Power Amplifiers, Second Edition. Marian K. Kazimierczuk.

Transient Analysis of a Circuit with SABER

1) Select the show/hide Saber Guide icon bar to view the simulation tool bar.
2) Select an Operating Point/Transient Analysis to run transient analysis versus time.
   (a) Click the Basic Bab.
   (b) Set a simulation end time. This time is different for each experiment. Assign appropriate end times based on the tame taken for the system to reach steady-state operation. Set a time step of 0.01 μs. If the end time is long, it is recommended to increase the time step so that the simulation runs faster.
   (c) Set Monitor Progress = **100**. Plot after Analysis = **Yes Open Only**.
   (d) Click **Input/Output** Tab. Select **Input/Output** Tab. Select **All Signals for the Signal List**.
   (e) Select **Across and through variables** in order to save both voltage and current information.
   (f) Select OK to run the analysis. It will take 1 or 2 min to run. If you want to view the status of the simulation, click the Simulation Transcript and Command Line button in the top bar.
   (g) Once the analysis is completed, it should open Cosmos Scope to view the simulation results.

Plotting with SABER

1) The Cosmos Scope pops up with two more dialog boxes. In the Signal Manager window, under Plot files, you will see the file you have been working on, which contains the results of the transient analysis.
2) All types of waveform can be plotted and various mathematical operations can be performed on them. You should be able to find the required waveforms under the drop-down option corresponding to the component under consideration.
3) Each lab will have different sets of waveforms to be plotted.

Printing with SABER

1) Make sure all your prints appear on white background, both for your schematic and plots.
2) Always choose russ3 as the designed or default printer.
3) If possible create a world document, paste all the required plots and schematic, and save as a single file.
4) Your plots and schematics can be exported (saved) in the format you desire. Export option is available under the File tab. Your saved file will be present in the Secured Shell File Transfer Window.

Answers to Problems

CHAPTER 1

1.1 $THD = 1\%$.

1.2 $h_a = 1.5625$ cm.

1.3 $P_{AM} = 11.25$ kW.

1.4 (a) $V_m = 12.5$ V.
(b) $P_{AM} = 7.031125$ W.

1.5 (a) $f_{max} = 555$ kHz.
(b) $f_{min} = 545$ kHz.
(c) $BW = 10$ kHz.

1.6 (a) $k_{IMP} = 5$.
(b) $k_{IMP} = 5$.

1.7 $h_a = 3.125$ cm.

1.8 $P_{AV} = 100$ W.

1.9 $P_{AM} = 15$ W.

1.11 $P_{LSB} = P_{USB} = 22.5$ W.

1.12 $BW = 10$ kHz.

1.13 $P_O = 10$ kW.

1.14 $m_f = 2$.

1.15 $BW = 90$ kHz.

1.16 $BW = 20$ kHz.

RF Power Amplifiers, Second Edition. Marian K. Kazimierczuk.

CHAPTER 2

2.1 $L_{max} = 33$ nm.

2.2 (a) $\eta_D = 12.5\%$. (b) $\eta_D = 40.5\%$.

2.3 $P_I = 10$ W.

2.4 $L = 0.0224$ nH, $C = 196.2$ pF, $L_f = 2.24$ nH, $I_m = 0.3846$ A, $I_{DM} = 0.7692$ A, $V_{DSM} = 2.8$ V, $P_{LS} = 1.2048$ W, $\eta = 15.05\%$, $\eta = 18.7\%$, $W/L = 13061$, $W = 4.5713$ mm, $W_1 = 45.713$ mm, and $P_{Q1} = 5.769$ mW.

2.5 $C_1 = 6.631$ pF, $C_2 = 4.872$ pF, $L = 1.4117$ nH.

2.6 (a) $\eta_D = 32\%$ and $P_D = 6.8$ W.
(b) $\eta_D = 8\%$ and $P_D = 18.4$ W.

2.7 (a) $\eta_D = 40.5\%$.
(b) $\eta_D = 12.5\%$.
(c) $\eta_D = 0.5\%$.

CHAPTER 3

3.1 $R = 4.805\ \Omega$, $V_m = 3.1$ V, $I_m = 0.66$ A, $I_I = 0.42$ A, $I_{DM} = 1.32$ A, $P_D = 0.386$ W, $P_I = 1.386$ W, $\eta_D = 72.15\%$, $L = 0.03117$ nH, and $C = 141$ pF.

3.2 $R = 50\ \Omega$, $V_m = 47$ V, $I_m = 0.94$ A, $I_I = 0.7116$ A, $I_{DM} = 1.7527$ A, $P_I = 34.1568$ W, $P_D = 12.1568$ W, $\eta_D = 64.4\%$, $L = 0.8841$ nH, and $C = 35.36$ pF.

3.3 $R = 19.22\ \Omega$, $V_m = 3.1$ V, $I_m = 0.161$ A, $I_I = 0.08976$ A, $I_{DM} = 0.4117$ A, $P_D = 0.0462$ W, $P_I = 0.2962$ W, $\eta_D = 84.4\%$, $L = 0.1274$ nH, and $C = 34.5$ pF.

3.4 $R = 10\ \Omega$, $V_m = 11$ V, $I_m = 1.1$ A, $I_I = 0.5849$ A, $I_{DM} = 3.546$ A, $P_D = 1.0188$ W, $P_I = 7.0188$ W, $\eta_D = 85.48\%$, $Q_L = 10$, $L = 0.0663$ nH, and $C = 66.31$ pF.

CHAPTER 4

4.1 $L = 2.215$ nH, $C = 11.437$ pF, $I_m = 1.057$ A, $V_{Cm} = V_{Lm} = 14.71$ V, $\eta = 90.7\%$.

4.2 $f_0 = 1$ MHz, $Z_o = 529.2\ \Omega$, $Q_L = 2.627$, $Q_o = 365$, $Q_{Lo} = 378$, $Q_{Co} = 10583$.

4.3 $Q = 65.2$ VA, $P_O = 24.8$ W.

4.4 $V_{Cm} = V_{Lm} = 262.7$ V, $I_m = 0.4964$ A, $Q = 65$ VA, and $f_o = 1$ MHz.

4.5 $\eta_r = 99.28\%$.

4.7 $V_{SM} = 400$ V.

4.8 $f_{Cm} = 230.44$ kHz, $f_{Lm} = 233.87$ kHz.

4.9 $R = 162.9\ \Omega$, $L = 675\ \mu$H, $C = 938$ pF, $I_I = 0.174$ A, $I_m = 0.607$ A, $f_o = 200$ kHz, $V_{Cm} = V_{Lm} = 573$ V.

4.10 $P_I = 88.89$ W, $R = 82.07\ \Omega$, $L = 145.13\ \mu$H, $C = 698.14$ pF.

CHAPTER 5

5.1 $R = 6.281\ \Omega$, $C_1 = 4.65$ pF, $L = 5$ nH, $C = 6.586$ pF, $L_f = 43.56$ nH, $V_{SM} = 11.755$ V, $I_{SM} = 0.867$ A, $V_{Cm} = 13.629$ V, $V_{Lm} = 17.719$ V, $V_{Lfm} = 8.455$ V, $\eta = 97.09\%$.

5.2 $R = 6.281\ \Omega$, $L = 5$ nH, $C_1 = 4.65$ pF/12 V, $C_2 = 20.99$ pF/10 V, $C_3 = 8.4$ pF/10 V.

5.3 $R = 10.63\ \Omega$, $I_{SM} = 7.44$ A, $V_{SM} = 170.976$ V, $L = 4.23\ \mu$H, $C_1 = 1.375$ nF, $L_f = 36.9\ \mu$H.

5.4 $V_{SM} = 665$ V.

5.5 $V_{SM} = 1274.5$V.

5.7 $f_{max} = 0.95$ MHz.

5.8 $R = 8.306\ \Omega$, $C_1 = 1.466$ pF, $L = 5.51$ nH, $V_{SM} = 42.744$ V, $I_{SM} = 2.384$ A, $L_f = 24$ nH, $C = 0.9024$ pF, $V_{Cm} = 114$ V, $V_{Lm} = 128.79$ V, and $\eta = 96.39\%$.

5.9 $R = 8.306\ \Omega$, $C_1 = 1.466$ pF, $L = 5.51$ nH, $C_2 = 1.21$ pF, $C_3 = 2.97$ pF, and $V_{C2m} = 85.07$ V.

CHAPTER 6

6.1 $R = 1.3149\ \Omega$, $L_1 = 1.2665$ nH, $C = 16.8$ pF, $L = 0.862$ nH, $V_{SM} = 42.931$ V, $I_I = 0.66$ A, $I_{SM} = 2.351$ A, $I_m = 3.9$ A, $R_{DC} = 22.72\ \Omega$, $f_{o1} = 1.32$ GHz. $f_{o2} = 841.64$ MHz.

6.2 $V_{SM} = 973.1$ V.

6.3 $R = 7.57\ \Omega$, $L = 4.25\ \mu$H, $L_1 = 32.8\ \mu$H, $C = 21$ nF, $I_{SM} = 4.95$ A. $V_{SM} = 515.2$ V.

6.4 $R = 9.23\ \Omega$, $V_I = 125.7$ V, $V_{SM} = 359.7$ V.

6.5 $R = 11.7\ \Omega$, $L_1 = 10.1\ \mu$H, $C = 3.02$ nF, $L = 0.383\ \mu$H, $V_{SM} = 286.2$ V, $I_I = 0.5$ A, $I_{SM} = 1.78$ A, $I_m = 2.92$ A, $R_{DC} = 200\ \Omega$.

CHAPTER 7

7.1 $R = 1.266\ \Omega$, $C_1 = 5$ pF, $C = 3.73$ pF, $L = 0.504$ nH, $f_o = 3.67$ GHz, $P_I = 1.0638$ W, $I_I = 0.212$ A, $V_m = 1.591$ V, $I_{SMmax} = 1.257$ A, $V_{SM} = 5$ V, and $V_{Cmax} = 13.4$ V.

7.2 $C_s = 10$ pF. $C_{s(ext)} = 8$ pF.

CHAPTER 8

8.1 $V_m = 133.875$ V, $V_{m3} = 14.875$ V, $R = 89.613\ \Omega$, $V_{DSmax} = 240$ V, $I_m = 1.494$ A, $I_{DM} = 2.988$ A, $I_I = 0.951$ A, $P_I = 114.12$ W, $P_D = 14.12$ W, $\eta_D = 87.63\%$, and $R_{DC} = 125.12\ \Omega$.

8.2 $V_m = 137.409$ V, $V_{m3} = 22.902$ V, $R = 94.406\ \Omega$, $V_{DSmax} = 240$ V, $I_m = 1.4555$ A, $I_{DM} = 2.911$ A, $I_I = 0.9266$ A, $P_I = 111.192$ W, $P_D = 11.192$ W, $\eta_D = 89.93\%$, and $R_{DC} = 128.3921\ \Omega$.

8.3 $V_m = 6.26$ V, $V_{m2} = 1.565$ V, $R = 19.59\ \Omega$, $V_{DSmax} = 13.33$ V, $I_m = 0.319$ A, $I_{DM} = 0.501$ A, $I_I = 0.25$ A, $P_I = 1.25$ W, $P_D = 0.25$ W, $\eta_D = 80\%$, and $R_{DC} = 18.7\ \Omega$.

CHAPTER 9

9.1 $A_{p1} = 1$ for $0 \le \upsilon_s \le 1$ V, $A_{p2} = 3$ for $1\text{ V} < \upsilon_s \le 2$ V,

9.2 $A_{f1} = 9.89$. $A_{f2} = 9.836$. $A_{f3} = 9.6774$.

9.3 $\upsilon_o = \dfrac{\upsilon_s}{\beta} - \dfrac{1}{\beta}\sqrt{\dfrac{\upsilon_o}{A}}$.

9.4 $\upsilon_o = 100\left[\upsilon_s - \beta\upsilon_o + \dfrac{(\upsilon_s - \beta\upsilon_O)^2}{2}\right]$.

CHAPTER 10

10.1 $\delta_{Cu} = 2.0897\ \mu$m, $\delta_{Al} = 2.59\ \mu$m, $\delta_{Ag} = 2.0069\ \mu$m, $\delta_{Au} = 2.486\ \mu$m.

10.2 $R_{dc} = 1.325\ \Omega$, $R_{ac} = 2.29\ \Omega$, $\delta_{Al} = 0.819\ \mu$m, $F_R = 1.728$.

10.3 $L = 0.1022$ nH.

10.4 $L = 1.135$ nH.

10.5 $L = 0.1$ nH, $\delta = 1.348\ \mu$m, $L_{HF} = 0.0134$ nH.

10.6 $L = 1.37$ nH.

10.7 $L = 39.47$ nH.

10.8 $D = 1000\ \mu$m, $L = 42.64$ nH.

10.9 $D = 1000\ \mu$m, $L = 43.21$ nH.

10.10 $D = 1500\ \mu$m, $L = 354.8$ nH.

10.11 $D = 1000\ \mu$m, $L = 45$ nH.

10.12 $D = 1000\ \mu$m, $l = 21060\ \mu$m, $L = 59.13$ nH.

10.13 $D = 1000\ \mu$m, $L = 41.856$ nH.

10.14 $l = 22140\ \mu$m, $b = 183.33\ \mu$m, $L = 39.015$ nH.

10.15 $L = 496.066$ nH.

10.16 $D = 1000\ \mu$m, $L = 36.571$ nH.

10.17 $D = 520\ \mu$m, $L = 5.75$ nH.

10.18 $D = 1000\ \mu$m, $L = 122.38$ nH.

10.19 $l_c = 380\ \mu$m, $L = 13.228$ nH.

Index